Progress in Thermochemical Biomass Conversion

Volume 2

Edited by

A. V. Bridgwater

Director of the Bio-Energy Research Group in
Chemical Engineering and Applied Chemistry
Aston University, Aston Triangle
Birmingham, UK

b

Blackwell
Science

Editorial Offices:
Osney Mead, Oxford OX2 0EL
25 John Street, London WC1N 2BS
23 Ainslie Place, Edinburgh EH3 6AJ
350 Main Street, Malden
MA 02148 5018, USA
54 University Street, Carlton
Victoria 3053, Australia
10, rue Casimir Delavigne
75006 Paris, France

Other Editorial Offices:

Blackwell Wissenschafts-Verlag GmbH
Kurfürstendamm 57
10707 Berlin, Germany

Blackwell Science KK
MG Kodenmacho Building
7-10 Kodenmacho Nihombashi
Chuo-ku, Tokyo 104, Japan

Iowa State University Press
A Blackwell Science Company
2121 S. State Avenue
Ames, Iowa 50014-8300, USA

First published 2001

Printed and bound in Great Britain by
MPG Books Ltd, Bodmin, Cornwall

DISTRIBUTORS [SMALL CAPS]

Marston Book Services Ltd
PO Box 269
Abingdon
Oxon OX14 4YN
(*Orders:* Tel: 01235 465500
Fax: 01235 465555)

USA
Blackwell Science, Inc.
Commerce Place
350 Main Street
Malden, MA 02148 5018
(*Orders:* Tel: 800 759 6102
781 388 8250
Fax: 781 388 8255)

Canada
Login Brothers Book Company
324 Saulteaux Crescent
Winnipeg, Manitoba R3J 3T2
(*Orders:* Tel: 204 837-2987
Fax: 204 837-3116)

Australia
Blackwell Science Pty Ltd
54 University Street
Carlton, Victoria 3053
(*Orders:* Tel: 03 9347 0300
Fax: 03 9347 5001)

A catalogue record for this title is available from the British Library

ISBN 0-632-05533-2

For further information on Blackwell Science, visit our website:
www.blackwell-science.com

Contents

Volume 1

Gasification

Combustion

Volume 2

Pyrolysis

Preface

There has been considerable progress in the science and technology of thermo-chemical biomass conversion since the previous conference on *Developments in Thermochemical Biomass Conversion* in Banff, Canada, in 1996. This fifth conference again covers all aspects of thermal biomass conversion systems from fundamental research through applied research and development to commercial applications to reflect the progress made in the last four years.

The programme was divided into three major thermal conversion technologies: gasification, combustion and pyrolysis, and this division is reflected in the structure of these proceedings. Each main area of the conference was preceded by a state-of-the-art review to provide a focus for the ensuing presentations and an authoritative reference. One of the major features of this conference series is the high quality of the presentations and papers, which is achieved by subjecting all contributions to a full peer review process.

An important aim was to provide the widest opportunities to exchange ideas, discuss problems with fellow researchers, and to hear about the latest research and development. This was achieved in two ways: all posters were formally presented in short presentations to provide all authors and all delegates with the opportunity to meet each other face to face and benefit from the ensuing intimate interaction of a small and interested group; and the workshop programme further encouraged this interaction in those areas of interest selected by participants. The resultant report at the end of the book provides a summary of the outcomes from the workshops.

The research community will continue to provide the lead in developing new science and technology and in stimulating the development of new ideas. The benefits from the interactions at this conference will eventually translate into new and/or better products and processes, which will not only provide the justification for continued research and development, but will also more rapidly translate into commercial processes and products and help to deliver the promised benefits of the bio-energy sector.

Tony Bridgwater
March 2001

Acknowledgements

The following organisations provided considerable financial support to the conference, which was much appreciated by the organiser and delegates:

- IEA Bioenergy through the Executive Committee, and the Combustion, Pyrolysis and Techno-Economic Assessment Tasks;
- Austrian Federal Ministry of Transport, Innovation and Technology;
- Austrian Federal Ministry of Agriculture, Forestry, Environment and Water Management;
- Department of Trade and Industry, UK;
- Natural Resources Canada;
- VTT – Technical Research Centre of Finland

The scientific committee listed below provided whole-hearted support throughout the preparation and running of the conference. Their encouragement and efforts in publicising the meeting, refereeing papers, providing constructive feedback on the programme, organising workshops and chairing sessions was invaluable:

Michael Antal, USA	Larry Baxter, USA
Ton Beenackers, Netherlands	Dave Boocock. Canada
Robert Brown, USA	Esteban Chornet, Canada
Mike Connor, Australia	Jose Corella, Spain
Columba Di Blasi, Italy	Doug Elliott, USA
Hermann Hofbauer, Austria	Johan Hustad, Norway
Walter Kaminsky, Germany	Barbara Krieger-Brockett, USA
Kyriakos Maniatis, Belgium	Robert Manurung, Indonesia
Rainer Marutzky, Germany	Dietrich Meier, Germany
Tom Milne, USA	Thomas Nussbaumer, Switzerland
Ralph Overend, USA	Jan Piskorz, Canada
Christian Roy, Canada	Ann Segerborg-Fick, EC, Belgium
Kai Sipila, Finland	Krister Sjostrom, Finland
Yrjo Solantausta, Finland	Josef Spitzer, Austria
Erik Suuberg, USA	Chuangzhi Wu, China
Shin-ya Yokoyama, Japan	

Particular thanks are due to Nina Ahrendt and Claire Humphreys who provided all the Conference Administration throughout the preparation and running of the conference. Their contribution was invaluable and much appreciated.

An Overview of Fast Pyrolysis

AV Bridgwater [1], S Czernik [2], J Piskorz [3]

[1] *Bio-Energy Research Group, Aston University, Birmingham B4 7ET, UK*

[2] *NREL, 1617 Cole Boulevard, Golden, Colorado 80401, USA*

[3] *RTI Ltd, 110 Baffin Place, Unit 5, Waterloo, Ontario, N2V 1Z7, Canada*

ABSTRACT The process of fast pyrolysis is one of the most recent renewable energy processes to have been introduced and offers the advantages of a liquid product, bio-oil, that can be readily stored and transported and that can also be used for production of chemicals as well as being a fuel. Thermal biomass conversion has been investigated for many years as a source of renewable solid, gaseous and liquid fuels. Compared to combustion, which is widely practised commercially and gasification, which is being extensively demonstrated around the world, fast pyrolysis is at a relatively early stage of development. The technology has now achieved some commercial success for production of chemicals and is being actively developed for producing liquid fuels. Bio-oils have been successfully tested in engines, turbines and boilers, and have been upgraded to high quality hydrocarbon fuels although at an unacceptable energetic and financial cost. The paper critically reviews scientific and technical developments and applications to date paying particular attention to the research and developments reported in this book. It concludes with some suggestions for strategic developments.

INTRODUCTION

Renewable energy is of growing importance in satisfying environmental concerns over fossil fuel usage. Wood and other forms of biomass are some of the main renewable energy resources available and provide the only source of renewable liquid, gaseous and solid fuels. Wood and biomass can be used in a variety of ways to provide energy:

- by direct combustion to provide heat. This technology is commercially available and presents minimum risk to investors. The product is heat, which must be used immediately for heat and/or power generation. Overall efficiencies to power tend to be rather low, although in many California power plants high efficiencies have been reported but there have been problems with ash build-up in many installations. The status is reviewed in these proceedings (1).
- by gasification to provide a fuel gas for combustion for heat, or in an engine or turbine for electricity generation. The fuel gas quality requirements, for turbines in

particular, are very high, although there is now extensive experience available from the Varnamo plant (2). The gas is very costly to store or transport so it has to be used immediately. Hot gas efficiencies (total energy in raw product gas divided by energy in feed) can be as high as 95-97% for close coupled turbine and boiler applications, and up to 85% for cold gas efficiencies. The status is reviewed in (3). There is renewed interest in synthesis of liquid fuels from the product gas, but costs are still very high.

- by fast pyrolysis to provide a liquid fuel that can substitute for fuel oil in any static heating or electricity generation application. The liquid can also be used to produce a range of speciality and commodity chemicals. The key advantage is that a liquid can be readily stored and/or transported.

This review explains how the technology is developing and how it depends on an improved understanding of the underlying science as reviewed four years ago (4).

FAST PYROLYSIS

Pyrolysis is by definition thermal decomposition occurring in the absence of oxygen. It is always also the first step in combustion and gasification processes where it is followed by total or partial oxidation of the primary products.

Fast pyrolysis occurs in time of few seconds or less. Therefore, not only chemical reaction kinetics but also heat and mass transfer processes, as well as phase transition phenomena, play important roles. The critical issue is to bring the reacting biomass particle to the optimum process temperature and minimise its exposure to the intermediate (lower) temperatures that favour formation of charcoal. This objective can be achieved by using small particles, thus reducing the time necessary for heat up. This option is used in fluidised bed processes that are described later. Another possibility is to transfer heat very fast only to the particle surface that contacts the heat source. Because of the low thermal conductivity the deeper parts of the particles will be maintained at temperatures lower than necessary for char production. The products that form on the surface are immediately removed exposing that way consecutive biomass layers to the contact with the heat source. This second method is applied in ablative processes that are described later.

Fast pyrolysis is not an equilibrium process. During fast pyrolysis dramatic changes occur in specific volume between the reactants (biopolymers) and the products (by a factor of x 500) causing the volatile products to leave the pyrolysis zone at considerable velocities. This results in the entrainment of solid particles and aerosols, which normally would not volatilise at the process temperature. All these phenomena have important implications on pyrolysis technologies and are discussed in this paper.

MECHANISMS

Biomass is a complex polymeric material and its thermal decomposition is a multistage complicated process. Many pathways and mechanisms have been proposed to illustrate/explain the fundamental steps in pyrolysis (e.g. 5, 6, 7, 8, 9, 10, 11, 12, 13). Broido-Shafizadeh type kinetic models are perhaps the most widely used for cellulose pyrolysis but they can be also applied, at least qualitatively, to the whole biomass (see Figure 1).

Correct estimation of the kinetic constants is essential in optimising process parameters for maximizing liquids production. Nevertheless, kinetic data does not contribute any predictive powers aiming to optimise yields of specific chemicals or their classes. As fast pyrolysis is a non-steady state process, the derivation of isothermal kinetic data for modelling is extremely difficult, except in very small samples, as mass transfer processes, phase transition and heat transfer phenomena play important roles. The so-called "black box" engineering approach dominates our understanding in this area at present.

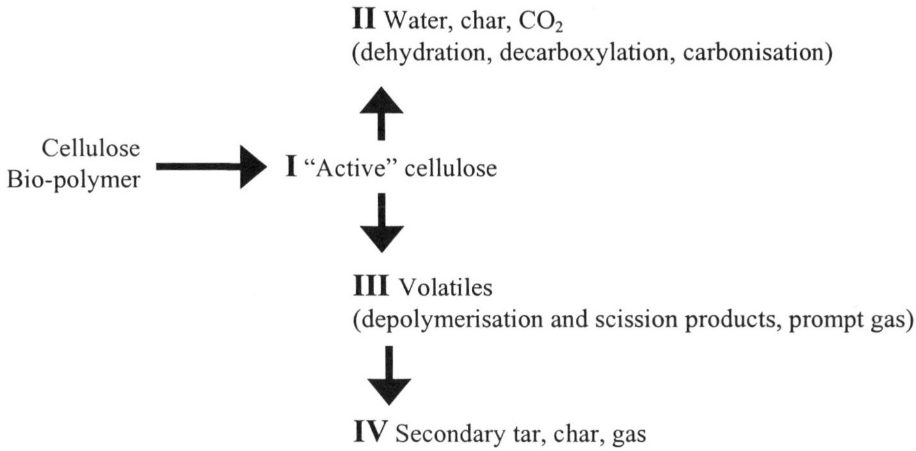

Figure 1 Typical cellulose decomposition model

As shown in the model, pyrolysis of cellulose results in solid, liquid and gaseous products. However, the proportions of the product yields can change depending on the process conditions. The knowledge of thermodynamics and kinetics of the reaction pathways allows us to adjust the conditions to maximize the yield of the desired products.

Dehydration of cellulose is exothermic while depolymerisation and secondary vapour cracking are endothermic and have higher activation energy than dehydration. Therefore, lower process temperature and longer vapour residence times will favour the production of charcoal. High temperature and longer residence time will increase the biomass conversion to gas and moderate temperature and short vapour residence time, necessary to minimize secondary cracking, are optimum for producing liquids. Table 1 provides data on the product distribution obtained from different modes of pyrolysis process.

Table 1 Typical product yields obtained by different modes of pyrolysis of wood

	Liquid	Char	Gas
FAST PYROLYSIS	75%	12%	13%
moderate temperature and short residence time, particularly vapour			
CARBONISATION	30%	35%	35%
low temperature and long residence time			
GASIFICATION	5%	10%	85%
high temperature and long residence times			

PRINCIPLES

Fast pyrolysis is a high temperature process in which biomass is rapidly heated in the absence of oxygen. As a result it decomposes to generate mostly vapours and aerosols and some charcoal. After cooling and condensation, a dark brown mobile liquid is formed which has a heating value about half that of conventional fuel oil. While it is related to the traditional pyrolysis processes for making charcoal, fast pyrolysis is an advanced process, with carefully controlled parameters to give high yields of liquid.

The essential features of a fast pyrolysis process for producing liquids are:

- very high heating and heat transfer rates at the reaction interface, which usually requires a finely ground biomass feed,
- carefully controlled pyrolysis reaction temperature of around 500°C and vapour phase temperature of 400-450°C,
- short vapour residence times of typically less than 2 seconds,
- rapid cooling of the pyrolysis vapours to give the bio-oil product.

The main product, bio-oil, is obtained in yields of up to 75% wt on dry feed basis, together with by-product char and gas which are used within the process so there are no waste streams other than flue gas and ash.

Figure 2 shows a conceptual schematic of fast pyrolysis process that includes the necessary steps of drying the feed to typically less than 10% water to minimise the water in the product liquid oil (although up to 15% can be acceptable), grinding the feed (to around 2 mm in the case of fluid bed reactors) to give sufficiently small particles to ensure rapid reaction, pyrolysis reaction, separation of solids (char), and collection of liquid product (bio-oil).

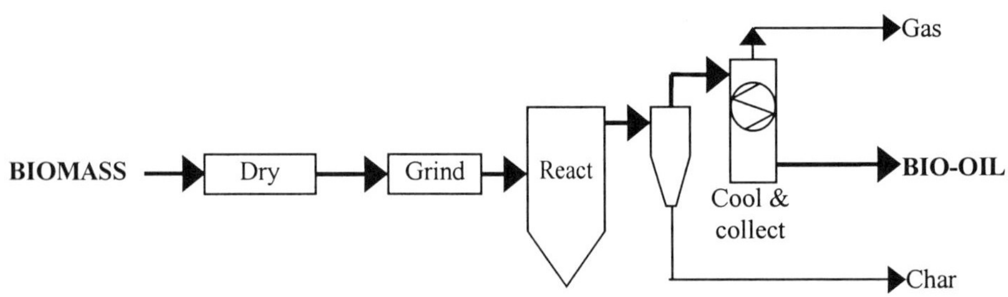

Figure 2. Conceptual Fast Pyrolysis Process

REACTORS

The heart of a fast pyrolysis process is the reactor. Although it probably represents at most only about 20% of the total capital cost, almost all research and development has focussed on the reactor. The rest of the process consists of biomass reception, storage and handling, biomass drying and comminution, and product collection, storage and, when relevant, upgrading. The key aspects of these peripheral steps are described later.

The critical features of successful pyrolysis reactors have been defined above as very high heating rates, carefully controlled temperatures and rapid cooling or quenching of the gaseous product which includes vapours, aerosols and gases.

A comprehensive survey of fast pyrolysis processes has recently been published that lists and describes all the pyrolysis processes for liquids production that have been built and tested in the last 10-15 years (14). Table 2 summarises the current significant operational fast pyrolysis plants for production of liquids.

Table 2 Operational pyrolysis units for liquids (see (14) for further details). Throughputs are on a dry wood basis.

Fluid bed	250 kg/h at Wellman
	75 kg/h at Dynamotive
	20 kg/h at RTI
	Many research units including in these proceedings (15)
Transported bed	about 1500 kg/h at Red Arrow (Ensyn)
	650 kg/h at ENEL (Ensyn)
	20 kg/h at VTT (Ensyn)
CFB	10 kg/h at CRES (22)
Rotating cone	150 kg/h at BTG (28)
Ablative	20 kg/h at NREL
	20 kg/h at Aston
Vacuum	3500 kg/h at Pyrovac (29)

Bubbling fluid beds

Bubbling fluid beds have been selected for further development by several companies including Union Fenosa (16) who have built and operated a 200 kg/h pilot unit in Spain, Dynamotive who have a 75 kg/h unit in Canada based on a RTI design (17) with a 10 t/d plant under construction (18) and Wellman who have built a 250 kg/h unit in the UK. Many research units have been built as they are relatively easy to construct and operate and give good results (14, and in these proceedings).

Bubbling fluid beds have many attractive features, particularly for research and development, which include:

- Simple construction and operation,
- Good temperature control,
- Very efficient heat transfer to biomass particles due to high solids density,
- Easy scaling,
- Well understood technology,
- Good and consistent performance with high liquid yields: of typically 70-75%wt. from wood on a dry feed basis,

Particular features that require consideration in design and operation include:

- Heating can be achieved in a variety of ways as shown in Figure 3,
- Residence time of solids and vapours is controlled by the fluidising gas flow rate and is higher for char than for vapours,
- Char acts as an effective vapour cracking catalyst at fast pyrolysis reaction temperatures so rapid and effective char separation/elutriation is important,
- Small biomass particle sizes are needed to achieve high biomass heating rates, and particle sizes generally need to be less than 2 to 3 mm,
- Good char separation is important and this is usually achieved by ejection and entrainment followed by separation in one or more cyclones,
- Heat transfer to bed at large scale has to be considered carefully due to scale-up limitations (see also Figure 3).

A typical plant configuration is shown in Figure 4, in this case based on a bubbling fluid bed, but similar principles apply to most reactors.

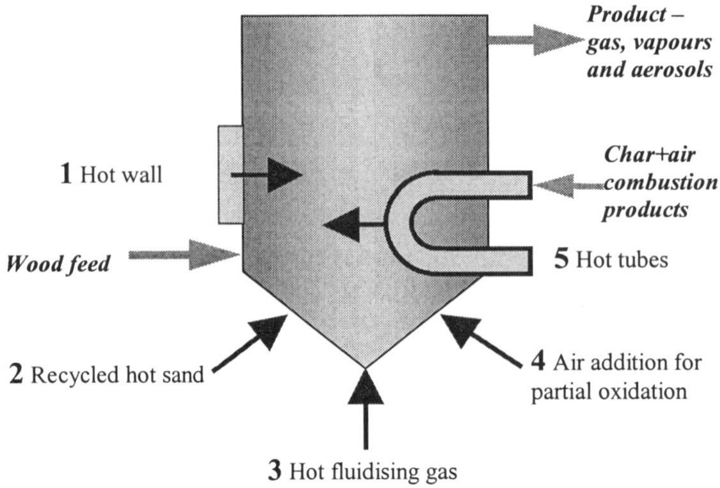

Figure 3. Means of providing heat to a pyrolysis reactor

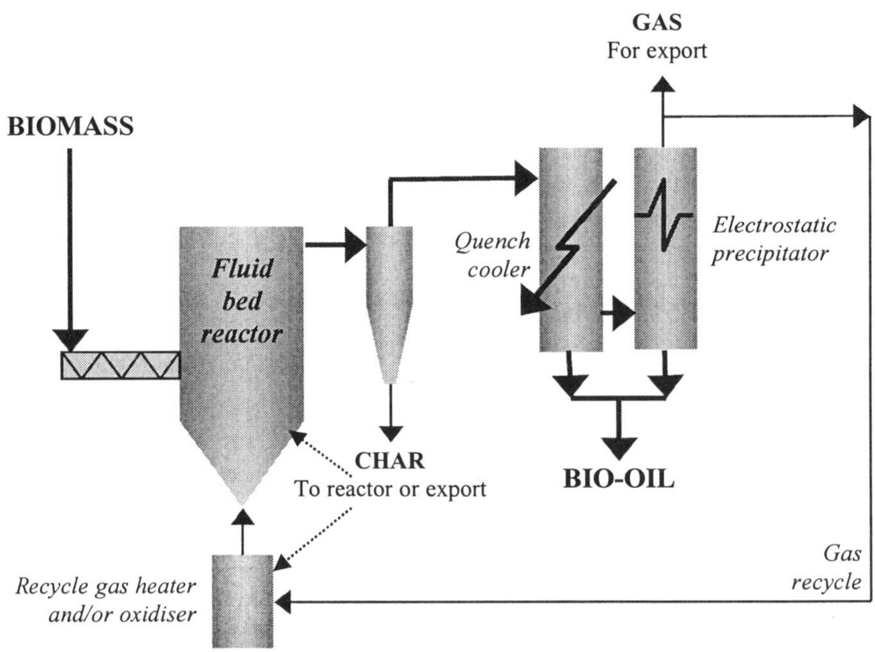

Figure 4. Bubbling fluid bed reactor

Circulating fluid beds and transported bed

Circulating fluid beds and transported bed reactors have been developed to commercial status by Ensyn and their process is used commercially by Red Arrow in the USA for food flavourings in several plants of 1 to 1.5 t/h (19). Ensyn have also supplied a 650 kg/h unit to ENEL in Italy (20) and a 20 kg/h system to VTT in Finland (21). In some of these plants it is believed that only sand is recycled while char is separated and collected, in others, operating in a dual reactor configuration, char is also recycled and burned in the second reactor to provide process energy by reheating and recycling the sand.

CRES is operating a 10 kg/h circulating fluid bed unit with the char combustor integrated into the base of the CFB riser as a bubbling fluid bed (22) as indicated in Figure 5. Liquid yields of 60-70% wt on a dry feed basis have been achieved.

The principles of a circulating fluid bed system are shown in Figure 5. As discussed above, a second rector can be required for char combustion and sand reheating (Ensyn) or the char combustion can occur in the lower part of the pyrolyser (CRES) thus eliminating the need for the second vessel.

Particular features of circulating fluid bed and transported bed reactors include:

- Good temperature control can be achieved in reactor,
- Residence time for the char is almost the same as for vapours and gas,
- CFBs are suitable for very large throughputs,
- Well understood technology,
- Hydrodynamics more complex,
- Char is more attrited due to higher gas velocities; char separation is by cyclone,
- Closely integrated char combustion in a second reactor requires careful control,
- Heat transfer at large scale has to be proven.

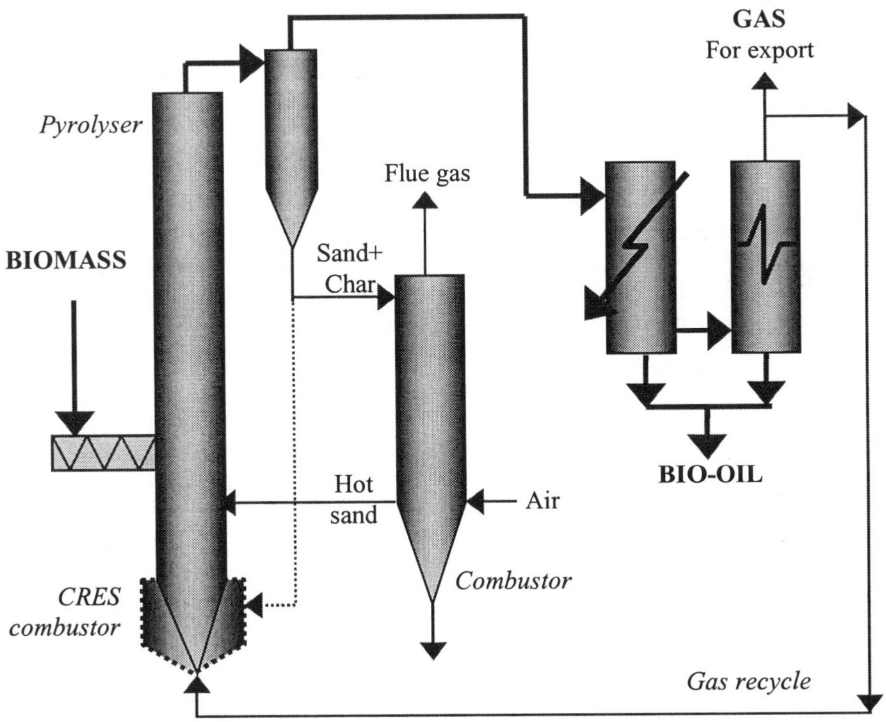

Figure 5. Circulating fluid bed reactor

Ablative pyrolysis

Ablative pyrolysis is substantially different in concept compared to the other methods of fast pyrolysis. In all these other methods, the rate of reaction is limited by the rate of heat transfer through a biomass particle, which is why small particles are required. The mode of reaction in ablative pyrolysis is analogous to melting butter in a frying pan, when the rate of melting can be significantly enhanced by pressing down and moving the butter over the heated pan surface. In ablative pyrolysis heat is transferred from the hot reactor wall to "melt" wood that is in contact with it under pressure. The pyrolysis front thus moves unidirectionally through the biomass particle. As the wood is mechanically moved away, the residual oil film both provides lubrication for successive biomass particles and also rapidly evaporates to give pyrolysis vapours for collection in the same way as other processes. The rate of reaction is strongly influenced by pressure, the relative velocity of wood on the heat exchange surface and the reactor surface temperature.

The key features of ablative pyrolysis are therefore as follows:

- High pressure of particle on hot reactor wall, achieved due to centrifugal force (NREL) or mechanically (Aston)
- High relative motion between particle and reactor wall,
- Reactor wall temperature less than 600°C.

Particular features of ablative reactor systems include:

- Use of large feed sizes,
- Inert gas is not required, so the processing equipment is smaller, (in case of mechanically applied pressure)
- The reaction system is more intensive,
- The process is limited by the rate of heat supply to the reactor rather than the rate of heat absorption by the pyrolysing biomass as in other reactors.
- Reaction rates are limited by heat transfer to the reactor, not to the biomass,
- The process is surface area controlled so scaling is more costly,
- The process is mechanically driven so the reactor is more complex.

Ablative pyrolysis is interesting as much larger particle sizes can be employed than in other systems and there is no requirement for inert gas. Both lead to a potentially lower cost system.

Much of the pioneering work on ablative pyrolysis reactors was performed by CNRS at Nancy where extensive basic research has been carried out onto the relationships between pressure, motion and temperature (23). NREL developed an ablative vortex reactor (see Figure 6 and (14)), in which the biomass is accelerated to supersonic velocities to derive high tangential pressures inside a heated cylinder. Unreacted particles are recycled and the vapours and char fines leave the reactor axially for collection. Liquid yields of 60-65%wt. on dry feed basis are typically obtained.

More recent developments have been carried out at Aston University with a prototype rotating blade reactor in which pressure and motion is derived mechanically thus obviating the need for a carrier gas (see Figure 7 and (14)). Liquid yields of 70-75%wt. on dry feed are typically obtained. A second-generation reactor has recently been built and commissioned. Other configurations include the coiled tube at Castle Capital (14) (now owned by Enervision) and cyclonic reactors (24).

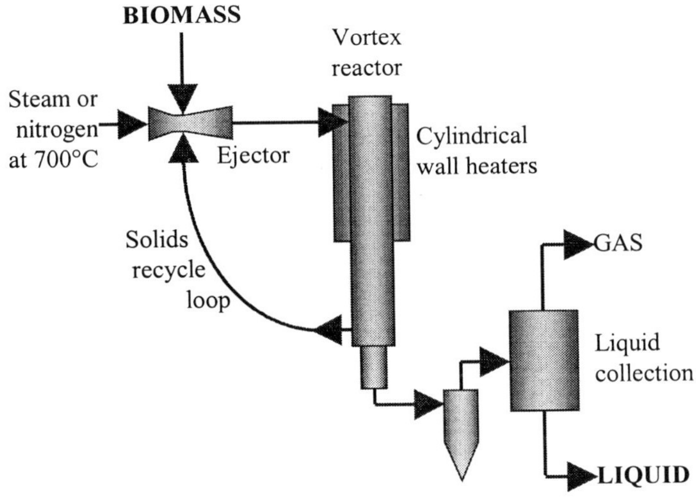

Figure 6. NREL Vortex ablative reactor

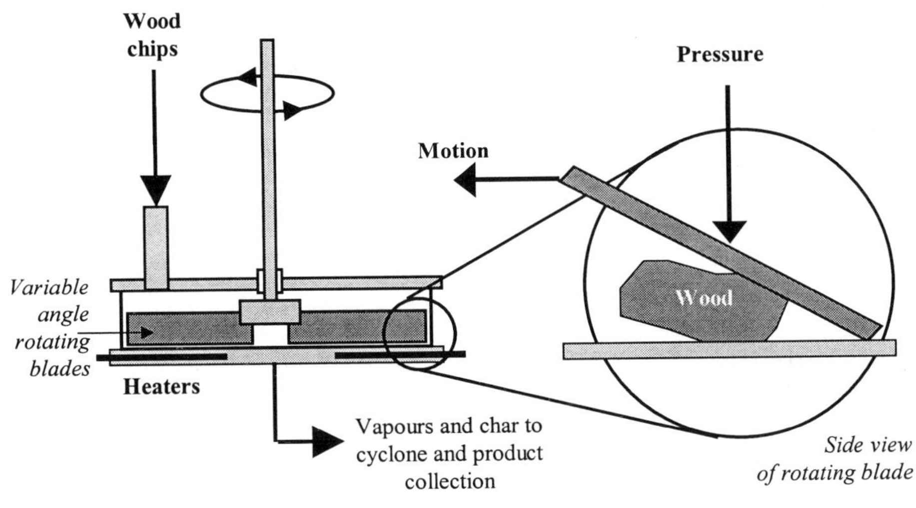

Figure 7. Aston University rotating blade ablative reactor

Entrained flow

Entrained flow fast pyrolysis was developed at Georgia Tech Research Institute (25) and scaled up by Egemin (26). Further details are available in (14). Neither the GTRI nor Egemin process is now operational and there are no known plans for further development, probably because of the difficulties that have been encountered in achieving good heat transfer from a gaseous heat carrier to solid biomass. Liquid yields of 50-60%wt. on dry feed have been obtained.

Rotating cone

The rotating cone reactor, invented at the University of Twente (27) and being developed by BTG, is a recent development and effectively operates as a transported bed reactor, but with transport effected by centrifugal forces rather than gas (28). A 250 kg/h unit is now operational, and plans for scale-up to 10 t/d have recently been announced. The basic principle is shown in Figure 8 and further details are available in (14) and (28). The key features of this technology are:

- Centrifugation at 600 rpm drives hot sand and biomass up a rotating heated cone,
- Vapours are collected and processed conventionally,
- Char is burned in a secondary bubbling fluid bed combustor. The hot sand is recirculated to the pyrolyser,
- Carrier gas requirements in the pyrolysis reactor are much less than for fluid bed and transported bed systems (however, more gases are needed for char burn off and for sand transport)
- Complex integrated operation of three subsystems: rotating cone pyrolyser, bubbling bed char combustor, and riser for sand recycling.
- Liquid yields of 60-70% on dry feed are typically obtained.

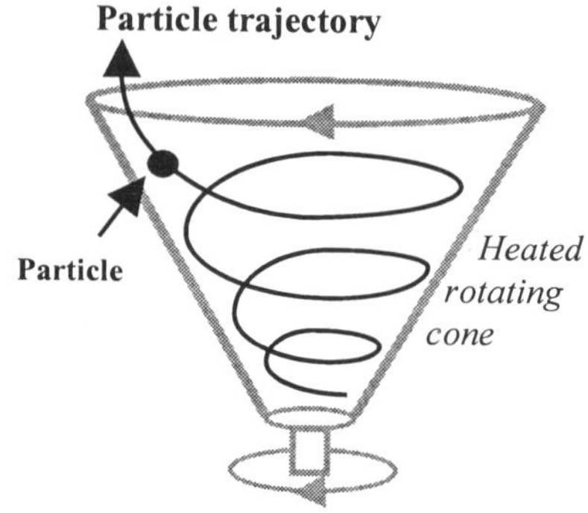

Figure 8. Principle of rotating cone pyrolysis reactor

Vacuum pyrolysis

Vacuum pyrolysis is arguably not a true fast pyrolysis as the heat transfer rate to and through the solid biomass is much slower than in the previously described reactors, but the vapour residence time is comparable.

The basic technology was developed at the University of Laval using a multiple hearth furnace but is now based on a purpose designed horizontal moving bed. A 50 kg/h unit is available for research and the technology has been scaled up by Pyrovac to a 3.5 t/h unit, which is currently operating at Jonquiere in Canada (29). The technology was selected for some of the Non Fossil Fuel Obligation contracts in the UK where six plants totalling about 70MWe are at various stages of permitting and design (30).

Key features of the process include:

- It can process larger particles than most fast pyrolysis reactors,
- There is less char in the liquid product due to lower gas velocities,
- There is no requirement for a carrier gas,
- Liquid yields of 35-50% on dry feed are typically obtained with higher char yields than fast pyrolysis systems. Conversely, the liquid yields are higher than in slow pyrolysis technologies because of fast removal of vapours from the reaction zone,
- The process is relatively complicated mechanically,

CHAR REMOVAL

Char acts as a vapour cracking catalyst so rapid and effective separation from the pyrolysis product vapours is essential. Cyclones are the usual method of char removal and two are usually provided – the first to remove the bulk of the material and the second to remove as much of the residual fines as possible. However, some fines always pass through the cyclones and collect in the liquid product where they accelerate aging and exacerbate the instability problem, which is described below.

There are several solutions to the problem of char entrainment. NREL has shown that hot vapour filtration, analogous to hot gas filtration in gasification processes, gives a high quality char free product (31), however the liquid yield is reduced by about 10-20% due to the char accumulating on the filter surface that cracks the vapours. VTT has also been developing hot vapour filtration (32).

Pressure filtration of the liquid is very difficult due to the complex interaction of the char and pyrolytic lignin, which appears to form a gel-like phase that rapidly blocks the filter. Modification of the liquid micro-structure by addition of solvents such as methanol or ethanol that solubilise the less soluble constituents will improve this problem and also contribute to improvements in liquid stability as described below.

LIQUID COLLECTION

The product of fast pyrolysis is vapours, aerosols and gases from decomposition of holocellulose and lignin with any carrier gases from fluidisation or transport. Aerosols consist of sub-micron liquid droplets and they present a severe problem in the successful recovery of the pyrolysis oils. These aerosols appear visually as smoke. The aerosols are probably formed directly from pyrolysing biomass, especially from submicron biomass particles that are rapidly depolymerised. The liquid product can then be entrained out of the reactor before it is vaporised. Another mechanism proposed for the formation of aerosols in the pyrolysis reactor involves the ejection of liquid droplets from internally pressurised cell capillaries of a pyrolysing particle (33).

The vapours require rapid cooling to minimise secondary reactions and to condense the true vapours, while the aerosols require coalescence or agglomeration. Simple heat exchange can cause preferential deposition of lignin derived components leading to liquid fractionation and eventually blockage.

Quenching in product oil or in an immiscible hydrocarbon solvent is widely practiced. Orthodox aerosol capture devices such as demisters and other commonly used impingement devices are not very effective and electrostatic precipitation is currently the preferred method.

The vapour product from fluid bed and transported bed reactors has a low partial pressure of collectible products due to the large volumes of fluidising gas, and this is an important design consideration in liquid collection.

The exit gas temperature from the liquid condensation train has a definite effect on the properties of the recovered oil product. As the exit gas temperature is lowered, progressively more of the volatile organic vapours and water will be condensed to become the pyrolysis oil product. These volatiles can be considered solvents as they have very low viscosities and thus have a very beneficial impact on lowering the viscosity of the oil, even at low concentrations. The recovery of these volatile, oxygenated organic solvents is thought to decrease the tendency of the oil to phase separate. Use of higher collection temperatures to minimise water condensation will result in lower yields and in higher viscosity of less stable bio-oil.

PYROLYSIS LIQUID - BIO-OIL

Pyrolysis liquid is referred to by many names including pyrolysis oil, bio-oil, bio-crude-oil, bio-fuel-oil, wood liquids, wood oil, liquid smoke, wood distillates, pyroligneous tar, pyroligneous acid, and liquid wood. The crude pyrolysis liquid is dark brown and approximates to biomass in elemental composition. It is composed of a very complex mixture of oxygenated hydrocarbons with an appreciable proportion of water from both the original moisture and reaction product. Solid char and dissolved alkali metals from ash (34) may also be present.

The complexity arises from the degradation of lignin, cellulose, hemicellulose and any other organics in the feed material, giving a broad spectrum of phenolic and many other classes of compounds that result from uncontrolled degradation as described below. The liquid from fast or flash pyrolysis has significantly different physical and chemical properties compared to the liquid from slow pyrolysis processes, which is more like a tar.

LIQUID PRODUCT CHARACTERISTICS

Composition

The liquid is formed by rapidly quenching and thus "freezing" the intermediate products of flash degradation of hemicellulose, cellulose and lignin as shown in Table 3 and Table 4. The liquid thus contains many reactive species, which contribute to its unusual attributes. Bio-oil can be considered a micro-emulsion in which the continuous phase is an aqueous solution of holocellulose decomposition products, that stabilizes the discontinuous phase of pyrolytic lignin macro-molecules through mechanisms such as hydrogen bonding. Aging or instability is believed to result from a breakdown in this emulsion. In some ways it is analogous to asphaltenes found in petroleum.

Table 3 Degradation products of fast pyrolysis of biomass constituents (12)

Hemicellulose	produces acetic acid, furfural, furan
Cellulose	produces levoglucosan, 5-hydroxymethylfurfural, hydroxyacetaldehyde, acetol, formaldehyde
Lignin	produces small amount of monomeric phenols (including phenols, cresols, guaiacols, syringols) but mostly oligomeric product of molecular mass ranging from few hundred to several thousand Da.
Extractives	produces molecules of waxy components such as fatty acids and rosin acids that are difficult to identify and are mostly immiscible with bio-oil. They are relatively thermally stable and volatile. The content depends on the biomass feed and can reach 5-15wt% levels
<u>Compounds classes in bio-oil</u>	
C1	formic acid, methanol, formaldehyde
C2-C4	linear hydroxyl and oxo substituted aldehydes and ketones
C5-C6	hydroxyl, hydroxymethyl and/or oxo substituted furans, furanones and pyranones
Anhydrosugars	e.g. levoglucosan, anhydro-oligosaccharides
Substituted phenols	Monomeric and dimeric methoxyl substituted phenols
Pyrolytic lignin	
Waxes, resins, fatty acids, terpenoid derivatives	

Table 4 Representative chemical composition of fast pyrolysis liquid

Major Components	Mass %
Water	20-30
Lignin fragments: insoluble pyrolytic lignin	15-30
Aldehydes: formaldehyde, acetaldehyde, hydroxyacetaldehyde, glyoxal, methylglyoxal	10-20
Carboxylic acids: formic, acetic, propionic, butyric, pentanoic, hexanoic, glycolic, (hydroxy acetic))	10-15
Carbohydrates: cellobiosan, α- D- levoglucosan, oligosaccharides, 1.6 anhydroglucofuranose	5-10
Phenols: phenol, cresols, guaiacols, syringols	2-5
Furfurals	1-4
Alcohols: methanol, ethanol	2-5
Ketones acetol (1-hydroxy-2-propanone), cyclo pentanone	1-5

Fast pyrolysis liquid has a higher heating value of about 17 MJ/kg as produced with about 25% wt. water that cannot readily be separated. The liquid is often referred to as "oil" or "bio-oil" or "bio-crude" although it will not mix with any hydrocarbon liquids. It is composed of a complex mixture of oxygenated compounds that provide both the potential and challenge for utilisation. There are some important characteristics of this liquid that are summarised in Table 5 and discussed briefly below.

Table 5 Typical properties and characteristics of wood derived crude bio-oil

Physical property		Typical value
Moisture content		15-30%
pH		2.5
Specific gravity		1.20
Elemental analysis	C	55-58%
	H	5.5-7.0%
	O	35-40%
	N	0-0.2%
	Ash	0-0.2%
HHV as produced (depends on moisture)		16-19 MJ/kg
Viscosity (at 40°C and 25% water)		40-100 cp
Solids (char)		1%
Vacuum distillation residue		up to 50%

Characteristics

- Liquid fuel,
- Ready substitution for conventional fuels in many static applications such as boilers, engines, turbines,
- Heating value of 17 MJ/kg at 25% wt. water, is about 40% that of fuel oil / diesel
- Does not mix with hydrocarbon fuels,
- Not as stable as fossil fuels,
- Quality needs definition for each application,

Appearance

Pyrolysis oil typically is a dark brown free flowing liquid. Depending upon the initial feedstock and the mode of fast pyrolysis, the colour can be almost black through dark red-brown to dark green, being influenced by the presence of micro-carbon in the liquid and by the chemical composition. Hot vapour filtration gives a more translucent red-brown appearance due to the absence of char. High nitrogen contents in the liquid can give it a dark green tinge.

Odour

The liquid has a distinctive odour - an acrid smoky smell, which can irritate the eyes if exposed for a prolonged period to the liquids. The cause of this smell is due to the low2 molecular weight aldehydes and acids. The liquid contains several hundred different chemicals in widely varying proportions, ranging from formaldehyde and acetic acid to complex high molecular weight phenols, anhydrosugars and other oligosaccharides.

Miscibility

The liquid contains varying quantities of water which forms a stable single phase mixture, ranging from about 15 wt% to an upper limit of about 30-50wt% water, depending on how it was produced and subsequently collected. Pyrolysis liquids can tolerate the addition of some water, but there is a limit to the amount of water, which can be added to the liquid before phase separation occurs, in other words the liquid cannot be dissolved in water. It is miscible with polar solvents such as methanol, acetone, etc. but totally immiscible with petroleum-derived fuels.

Density

The density of the liquid is very high at around 1.2 kg/litre compared to light fuel oil at around 0.85 kg/litre. This means that the liquid has about 42% of the energy content of fuel oil on a weight basis, but 61% on a volumetric basis. This has implications on the design and specification of equipment such as pumps.

Viscosity

The viscosity of the bio-oil as produced can vary from as low as 25 cSt to as high as 1000 cSt (measured at 40°C) or more depending on the feedstock, the water content of the oil, the amount of light ends that have been collected and the extent to which the oil has aged. Viscosity is important in many fuel applications (35).

Distillation

Pyrolysis liquids cannot be completely vaporised once they have been recovered from the vapour phase. If the liquid is heated to 100°C or more to try to remove water or distil off lighter fractions, it rapidly reacts and eventually produces a solid residue of around 50wt% of the original liquid and some distillate containing volatile organic compounds and water. The liquid is, therefore, chemically unstable, and the instability increases with heating, so it is preferable to store the liquid at room temperature. These changes do also occur at room temperature, but much more slowly and can be accommodated in a commercial application.

Unusual Properties

The complexity and nature of bio-oil causes some unusual behavior, specifically that the following properties tend to change with time:

- Viscosity increases,
- Volatility decreases,
- Phase separation and deposition of gums can occur.

This is due to a complex interaction of physical and chemical processes such as:

- Polymerization/condensation
- Esterification and etherification
- Agglomeration of oligomeric molecules

UPGRADING OF PYROLYSIS LIQUID

The properties that negatively affect bio-oil fuel quality are foremost low heating value, incompatibility with conventional fuels, solids content, high viscosity, and chemical instability. The heating value can be significantly increased, but it requires extensive changes to the chemical structure of bio-oils, which is technically feasible but not economic. The other undesired characteristics can be improved using simpler, physical methods. Both options are reviewed below.

Physical Methods

Hot-gas filtration can reduce the ash content of the oil to less than 0.01% and the alkali content to less than 10 ppm - much lower than reported for biomass oils produced in systems using only cyclones. Diesel engine tests performed on crude and on hot-filtered oil showed a substantial increase in burning rate and a lower ignition delay for the latter, due to the lower average molecular weight for the filtered oil (36). Hot gas filtration has not yet been demonstrated over a long-term process operation. Pyrolysis oils are not miscible with hydrocarbon fuels but with the aid of surfactants they can be emulsified with diesel oil.

A process for producing stable micro-emulsions with 5-30% of bio-oil in diesel has been developed at CANMET (37). The resultant emulsions showed promising ignition characteristics. A drawback of this approach is the cost of surfactants and the high energy required for emulsification. Other work on bio-emulsions is being carried out by the University of Florence, Italy, in an EC sponsored project, using proportions of bio-oil ranging from 5% to 75% (38).

Polar solvents have been used for many years to homogenize and to reduce viscosity of biomass oils. The addition of solvents, especially methanol, also showed a significant effect on the oil stability. It was observed (39) that the rate of viscosity increase ("aging") for the oil with 10 wt. % of methanol was almost 20 times less than for the oil without additives.

Chemical Methods

The chemical reactions that can occur between the bio-oil and methanol or ethanol are esterification and acetalization. Although not favoured thermodynamically, they can proceed to a significant extent if appropriate conditions are applied. For example, in the presence of an acid catalyst and molecular sieves (to adsorb water and to shift the reaction equilibria), bio-oil reacted with ethanol forming ethyl acetate, ethyl formate, and diethoxyacetal of hydroxyacetaldehyde at the expense of formic acid, acetic acid, and hydroxyacetaldehyde (40). Eventually, in addition to the decrease in viscosity and in the aging rate, other desirable changes such as reduced acidity, improved volatility, heating value, and miscibility with diesel fuels were also achieved. This field has been thoroughly reviewed (41).

The chemical/catalytic upgrading processes aim at the removal of oxygen, which is the main cause of instability and other unwanted characteristics of bio-oils. They are more complex and expensive than physical methods, but offer significant improvements ranging from simple stabilization to high-quality fuel products (42). Full deoxygenation to high-grade products such as transportation can be accomplished by two main routes: hydrotreating and catalytic vapour cracking. Chemical methods for upgrading bio-oil by hydro-treating and zeolite cracking have been reviewed (43, 44).

Hydrotreating of bio-oil carried out at high temperature, high hydrogen pressure, and in the presence of catalysts results in elimination of oxygen as water and in hydrogenation-hydrocracking of large molecules. The catalysts (sulphided CoMo or NiMo supported on alumina) and the process conditions are similar to those used in the refining of petroleum cuts (45).

Catalytic vapour cracking makes deoxygenation possible through simultaneous dehydration-decarboxylation over acidic zeolite catalysts. At 450°C and the atmospheric pressure oxygen is rejected as H_2O, CO_2, and CO producing mostly aromatics (46). The low H/C ratio in the bio-oils imposes a relatively low limit on the hydrocarbon yield and, in addition, the technical feasibility is not yet completely proven. The catalyst deactivation still raises many concerns for both routes. The processing costs are high and the products are not competitive with fossil fuels (47).

APPLICATIONS FOR BIO-OIL

Bio-oil can substitute for fuel oil or diesel in many static applications including boilers, furnaces, engines and turbines for electricity generation. The possibilities are summarised in Figure 8. There is also a range of chemicals that can be extracted or derived including food flavourings, specialities, resins, agri-chemicals, fertilisers, and emissions control agents. Upgrading bio-oil to transportation fuels is feasible but currently not economic.

Table 6 Demonstrated applications (with references to papers in these proceedings. See also Table 7)

Electricity	diesel, turbine, Stirling (48);
Heat	CHP and boiler (49);
Transport fuels	upgrading, emulsions (38);
Bulk chemicals	e.g. resins (50, 51), fertilizers, hydrogen (52)
Fine chemicals	e.g. levoglucosenone (53)
Emissions control	Calcium enriched bio-oil (54)

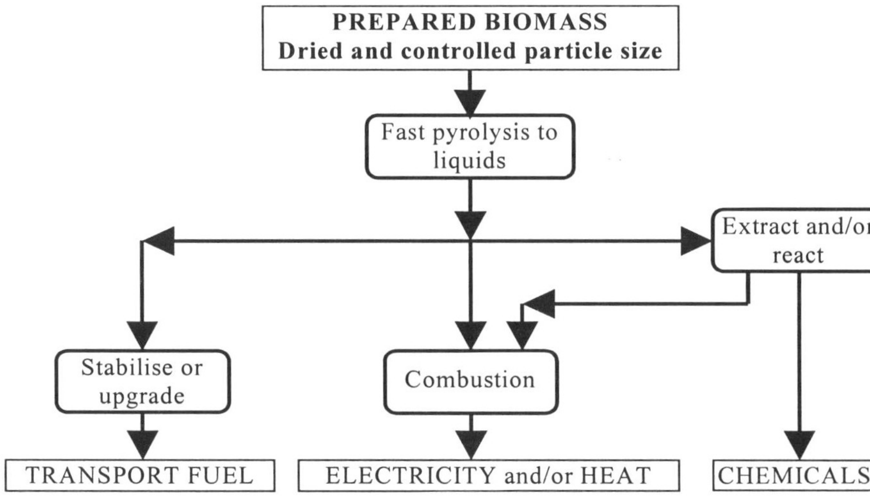

Figure 8. Applications for Bio-oil.

ELECTRICITY PRODUCTION

At least 500 hours operation has been achieved in the last few years on various engines from laboratory test units to 1.4 MWe modified dual fuel diesel engines. One such engine is a 250 kWe dual fuel engine on which nearly 400 hours have been logged in total, including several runs of over 9 hours, and with electricity being generated for 320 hours (55). A 2.5 MWe gas turbine has been modified and successfully run on bio-oil (56), and work is being carried out with bio-oil in gas turbine combustors (57).

CHEMICALS

A range of chemicals can also be produced from specialities such as levoglucosan to commodities such as resins and fertilisers as summarised in Table 7. Food flavourings are commercially produced from wood pyrolysis products in many countries. All chemicals are attractive possibilities due to their much higher added value compared to fuels and energy products, and lead to the possibility of a bio-refinery concept in which the optimum combinations of fuels and chemicals are produced.

Table 7 Chemicals from fast pyrolysis (with examples of references)

Acetic acid (58)	Adhesives (50, 51)	Calcium enriched bio-oil (54)
Food flavourings (19)	Hydrogen (52)	Hydroxyaceladehyde (59)
Levoglucosan (60)	Levoglucosenone (53)	Preservatives (61)
Resins (50, 51)	Slow release fertilisers (62)	Sugars (7)

SUMMARY AND CONCLUSIONS

The liquid bio-oil product from fast pyrolysis has the considerable advantage of being storable and transportable as well as the potential to supply a number of valuable chemicals, but there are many challenges facing fast pyrolysis that relate to technology, product and applications. The problems facing the sector include the following:

- Cost of bio-oil, which is 10% to 100% more than fossil fuel,
- Availability: there are limited supplies for testing,
- There is a lack of standards for use and distribution of bio-oil and inconsistent quality inhibits wider usage. Considerable work is required to characterise and standardise these liquids and develop a wider range of energy applications.
- Bio-oil is incompatible with conventional fuels,
- Users are unfamiliar with this material,
- Dedicated fuel handling systems are needed,
- Pyrolysis as a technology does not enjoy a good image.

Therefore, more research is needed in the field of fast pyrolysis. The most important issues that need to be addressed seem to be:

- Scale-up,
- Cost reduction,
- Improving product quality including setting norms and standards for producers and users,
- Environment health and safety issues in handling, transport and usage,
- Encouragement for developers to implement processes; and users to implement applications.
- Information dissemination.

There is much potential for further development and optimisation. Chemicals offer more interesting commercial opportunities and are likely to be a major focus of continuing research and development effort.

REFERENCES

1 Baxter, L, "Biomass combustion: Status quo and challenges", these proceedings.
2 Stahl K, Neergaard M, Nieminen J, "Final report: Varnamo demonstration programme", these proceedings.
3 Maniatis K, "Progress in biomass gasification: An overview", these proceedings.
4 Diebold JP and Bridgwater AV, "Overview of Fast Pyrolysis of Biomass for the production of Liquid Fuels", pp 5-26, Developments in Thermochemical Biomass Conversion, Bridgwater, AV and Boocock, DGB (Eds.) (Blackie Academic & Professional, London 1997)
5 Bradbury AGW, Sakai Y, Shafizadeh F, "Kinetic Model for Pyrolysis of Cellulose", J. Appl. Polym. Sci., Vol. 23, pp.3271, 1979.
6 Boroson ML, Howard JB, Longwell JP, Peters WA, "Heterogeneous Cracking of Wood Pyrolysis Tars over Fresh Wood Char Surfaces", Energy & Fuels, 3, 735-740, 1989.
7 Radlein D, Piskorz J, Scott DS, " Fast Pyrolysis of Natural Polysaccharides as a Potential Industrial Process", J. Anal. Appl. Pyrol., 19, 41-63, 1991.
8 Vladars-Usas A, "Thermal Decomposition of Cellulose", M.A.Sc. Thesis, University of Waterloo, Ont. Canada, 1993.
9 Diebold J, "A Unified Global Model for the Pyrolysis of Cellulose", Biomass and Bioenergy Vol. 7, Nos. 1-6, pp.75-85, 1994.

10 Piskorz J, Scott D, Radlein D, "Mechanisms of the Fast Pyrolysis of Biomass: Comments on Some Sources of Confusion", presented at Frontiers of Pyrolysis: Biomass Conversion and Polymer Recycling Conference, Breckenridge, 1995, included in written form in Minutes of the 2nd PYRA Meeting, La Coruna, 1995.
11 Narayan R, Antal MJ Jr, "Thermal Lag, Fusion and the Compensation Effect during Biomass Pyrolysis", Ind. Eng. Chem. Res. 35, pp.1711-1721. 1996.
12 Radlein D, "The Production of Chemicals from Fast Pyrolysis Bio-oils", Fast Pyrolysis of Biomass: A Handbook, A. Bridgwater et al. pp164, CPL 1999.
13 Boutin O, Lede J, "Use of a Concentrated Radiation for the Determination of the Elementary Mechanisms of Cellulose Thermal Decomposition", these proceedings.
14 Bridgwater AV and Peacocke GVC, "Fast pyrolysis processes for biomass", Sustainable and Renewable Energy Reviews, 4 (1) 1-73 (Elsevier, 1999)
15 Gerdes C, Meier D, Kaminsky W, "Flash pyrolysis of industrial biomass wastes",
Meier D, Ollesch T, Faix O, "Fast pyrolysis of impregnated waste wood - The fate of hazardous components",
Hata T, Meier D, Kajimoto T, Kikuchi H, Imamura Y, "Fate of arsenic after fast pyrolysis of chromium-copper-arsenate (CCA) treated wood",
All, these proceedings
16 Cuevas A, Reinoso C and Scott DS, "Pyrolysis oil production and its perspectives", in Proc. Power production from biomass II, Espoo, March 1995 (VTT)
17 PyNe Newsletter 4, September 1997, Aston University, UK
18 Dynamotive Press release, 3 October 2000.
19 Underwood G, "Commercialisation of fast pyrolysis products", in 'Biomass thermal processing', Eds. Hogan E, Robert, J, Grassi G and Bridgwater AV, pp 226-228, (CPL Scientific Press, 1992)
20 Rossi C and Graham RG. "Fast pyrolysis at ENEL". In: Kaltschmitt MK and Bridgwater AV, editors, Biomass gasification and pyrolysis: State of the art and future prospects, CPL Press, 1997, p 300-306.
21 PyNE Newsletter No. 4, September 1997, Aston University, UK
22 Boukis I, Gyftopoulou ME, Papamichael I, "Biomass fast pyrolysis in an air-blown circulating fluidized bed reactor", these proceedings
23 Lédé J, Panagopoulos J, Li HZ and Villermaux J, "Fast Pyrolysis of Wood: direct measurement and study of ablation rate", in Fuel, 1985, vol. 64, pp. 1514-1520.
24 Lédé J, "The Cyclone: A Multifunctional Reactor for the Fast Pyrolysis of Biomass", Ind. Eng. Chem. Res., 2000, 39, 893-903.
25 Kovac R.J. and O'Neil DJ, "The Georgia Tech Entrained Flow Pyrolysis Process," Pyrolysis and Gasification, G.L. Ferrero, K. Maniatis, A. Buekens, and A.V. Bridgwater, eds., Elsevier Applied Science 1989, pp. 169-179.
26 Maniatis K, Baeyens J, Peeters H and Roggeman G, "The Egemin flash pyrolysis process: commissioning and results", pp 1257-1264 in Advances in thermochemical biomass conversion, Ed. AV Bridgwater (Blackie 1993)
27 Prins W and Wagenaar BM, In Biomass gasification and pyrolysis, Eds. Kaltschmitt MK and Bridgwater AV, pp 316-326 (CPL 1997)
28 Wagenaar BM, Venderbosch RH, Carrasco J, Strenziok R, van der Aa BJ, "Rotating cone bio-oil production and applications", these proceedings
29 Yang J, Blanchette D, de Caumia B, Roy C "Modelling, scale-up and demonstration of a vacuum pyrolysis reactor", These proceedings.
30 OFFER announcement, February 1997, (Office of Electricity Regulation, UK)

31 Diebold JP, Czernik S, Scahill JW, Philips SD and Feik CJ. "Hot-gas filtration to remove char from pyrolysis vapours produced in the vortex reactor at NREL", in: Milne, TA, editor, Biomass Pyrolysis Oil Properties and Combustion Meeting, NREL, 1994, p. 90-108.
32 PyNe Newsletter 3, March 1997, Aston University, UK pp21
33 Lédé J, Diebold JP, Peacocke GVC, Piskorz J, (1996) "The Nature and Properties of Intermediate and Unvaporized Biomass Pyrolysis Materials," in Developments In Thermochemical Biomass Conversion, Bridgwater AV and Boocock DGB, (Eds.) (Blackie Academic & Professional, London 1997).
34 Huffman, D.R., Vogiatzis, A.J. and Bridgwater, A.V., "The characterisation of RTP bio-oils", Advances in Thermochemical Biomass Conversion, Ed. A V Bridgwater, (Elsevier, 1993)
35 Diebold JP, Milne TA, Czernik S, Oasmaa A, Bridgwater AV, Cuevas A, Gust S, Huffman D, Piskorz J, "Proposed Specifications for Various Grades of Pyrolysis Oils", In Developments in Thermochemical Biomass Conversion; Bridgwater AV and Boocock DGB, Eds. (Blackie Academic & Professional, London 1997), pp. 433-447.
36 Shihadeh AL, Rural Electrification from Local Resources: Biomass Pyrolysis Oil Combustion in a Direct Injection Diesel Engine, *Ph.D. Thesis*, (Massachussetts Institute of Technology, 1998)
37 Ikura M, Slamak M, Sawatzky H, Pyrolysis Liquid-in-Diesel Oil Microemulsions, US Patent 5,820,640, (1998)
38 Baglioni P, Chiaramonti D, Bonini M, Soldaini I, Tondi G, "BCO/Diesel oil emulsification: Main achievements of the emulsification process and preliminary results of tests on diesel engine", These proceedings
39 Diebold, J.P. and Czernik, S., Additives to Lower and Stabilize the Viscosity of Pyrolysis Oils during Storage, *Energy & Fuels*, (1997), 11, 1081-1091.
40 Radlein D, Piskorz J, Majerski P, Method of upgrading biomass pyrolysis liquids for use as fuels and as a source of chemicals by reaction with alcohols, European Patent # 0718392, (1999.
41 Diebold JP, "A review of the chemical and physical mechanisms of the storage stability of fast pyrolysis bio-oils", Report for PyNe, 1999. (To be published in 2001)
42 Maggi R and Elliott D, In Developments in Thermochemical Biomass Conversion, Bridgwater AV and Boocock DGB Eds. (Blackie Academic & Professional, London 1997), pp. 575-588.
43 Bridgwater AV, "Production of high grade fuels and chemicals from catalytic pyrolysis of biomass", Catalysis Today, 29, 285-295 (1996)
44 Bridgwater AV, "Catalysis in thermal biomass conversion", Applied Catalysis A, 116, (1-2), pp5-47, (1994)
45 Elliott DC and Baker E, In Energy from Biomass and Wastes X, Klass D, Ed., (IGT, 1983), pp. 765-782.
46 Chang C and Silvestri AJ, Catalysis, (1977), 47, p. 249.
47 Bridgwater AV and Cottam ML, "Costs and Opportunities for Biomass Pyrolysis Liquids Production and Upgrading", Proc 6th conference on Biomass for Energy, Industry and the Environment, Athens, (April 1991).
48 Bandi A, Baumgart F, "Stirling engine with flox burner fuelled with fast pyrolysis with fast pyrolysis liquid", These proceedings
49 Oasmaa A, Kyto M, Sipila K, "Pyrolysis oil combustion tests in an industrial boiler", These proceedings

50 Pakdel H, Murwanashyaka J N, Roy C, "Fractional vacuum pyrolysis of biomass for high yields of phenolic compounds", These proceedings
51 Himmelblau A "Combined chemicals and energy production from biomass pyrolysis", These proceedings
52 Czernik S, French R, Feik C, Chornet E, "Production of hydrogen from biomass-derived liquids", These proceedings
53 Dobele G, Rossinskaja G, Telysheva G, Meier D, Radtke S, Faix O, "Levoglucosenone - A product of catalytic fast pyrolysis of cellulose", These proceedings
54 Venderbosch RH, Wagenaar BM, Gansekoele E, Sotirchos S, Moss HDT, "Co-firing of bio-oil with simultaneous SOx and NOx reduction", These proceedings
55 Leech J, "Running a dual fuel engine on pyrolysis oil", pp 495-497, In: Kaltschmitt MK and Bridgwater AV, editors, Biomass gasification and pyrolysis: State of the art and future prospects, (CPL Press, 1997)
55 PyNE Newsletter No. 4; (Aston University, UK)
56 Andrews R., Patnaik PC, Liu Q. and Thamburaj, "Firing Fast Pyrolysis Oils in Turbines", Proceedings of the Biomass Pyrolysis Oil Properties and Combustion Meeting, Eds. Milne, T., (NREL/CP-430-7215, 1994), pp. 383-391.
57 Strenziok R, Hansen U, Kunstner H, "Combustion of bio-oil in a gas turbine", These proceedings
58 PyNe Newsletter 4, pp 21, (September 1997), Aston University, UK
59 Radlein, D. and Piskorz, J., "Production of chemicals from bio-oil". In: Kaltschmitt M, Bridgwater AV, editors, Biomass Gasification and Pyrolysis, (CPL Press, 1997), p 471-481.
60 Pernikis P, Zandersons J, and Lazdina B, "Obtaining Levoglucosan by Fast Thermolysis of Cellolignin-Pathways of Levoglucosan Use," in Developments In Thermochemical Biomass Conversion, Bridgwater AV and Boocock DGB, (Eds.) (Blackie Academic & Professional, London 1997)
61 Meier D, Andersons B, Irbe I, Tshirkova J, Faix O, "Preliminary study of fungicide and sorption effects of fast pyrolysis liquids used as wood preservative", These proceedings
62 Radlein, D, Piskorz J and Majerski, P, "Method of Producing Slow-Release Nitrogenous Organic Fertilizer from Biomass" US Patent 5,676,727, 1997 and European Patent Application 0716056.

Test Bed to Turnkey – The Introduction of New Thermal Renewable Energy Technologies

I Burdon
PB Power Ltd (Merz and McLellan), Amber Court, William Armstrong Drive, Newcastle Business Park, Newcastle upon Tyne, NE4 7YQ, United Kingdom

ABSTRACT. The step between a laboratory-scale research project and a full-scale, commercially-viable, scheme which will attract financial support from investors is large. Considerable effort needs to be expended by the developer to demonstrate that the risks involved in moving from test-bed into a competitive, environmentally-sensitive, real world are mitigated to the maximum extent and borne by the project participants most able to control them. The paper considers a generic fuel-energy conversion process and describes the commercial climate in which such a project has to develop, grow and survive. The trends in the world's electricity supply industry, from state-controlled utilities to liberalised, commercial, regimes in which competition plays an important part are explained. Key ingredients for a successful project are identified and the way in which independently-owned power (ie electricity) projects are structured is described. The requirements of the financiers of such projects are highlighted. The author presents his views from the standpoint of an "independent engineer"; his organization often being employed in such a role by project financiers to appraise energy projects employing novel technologies. It is anticipated that the information given in the paper will be of use to individuals or organisations who are seeking financial support to take an advanced-technology process from a proven design concept to commercial operational service.

INTRODUCTION

Commercial viability is a fundamental requirement for any process or project which is intended to supply useful energy into the competitive market places which, nowadays, characterise the electricity supply industries in many parts of the world.

Most independent power projects which compete in these markets are financed "off-balance sheet" ie they rely on debt provided by banks and similar financial institutions to provide 70-85 per cent of the project's funding. The banks will require all risks associated with the project to be identified and mitigated or accepted by others; they will have a thorough appraisal of the project's technical and commercial aspects conducted by independent engineers and lawyers with all "due diligence".

This paper describes the project development process and attempts to highlight the commercial climate in which an energy project, based on a thermal conversion process, has to become established and survive. The detail of the thermal conversion process itself, ie the "technical" process, is not relevant to the conception or the understanding of the project development process, ie the "financial" process, although the former process will have a major influence on the project's risk profile which will come under intense scrutiny in the latter process.

A GENERIC PROCESS MODEL

Almost all thermochemical biomass conversion processes can be considered and thus analysed as a "black box" with a feedstock input and an energy output (either thermal or electrical). Useful co-products will arise in the operation of the process (eg sulphur, acetic acid, methanol even food flavouring!) from which revenue may be earned. Relatively useless by-products (eg ash, frit and other emissions to air, land and water) which arise may incur a cost in their disposal if a market for their use cannot be found.

Certain defined design and operational parameters will be intimately associated with the black box and these are shown on the diagrammatic representation of the overall process shown in Fig 1. The parameters can be considered of two types:

(1) "Hard" - ie those which are amenable to scientific analysis and, therefore, fairly precise calculation or prediction.
(2) "Soft" - which may be largely qualitative and, accordingly, subject to opinion and be affected by judgement or simple prejudice.

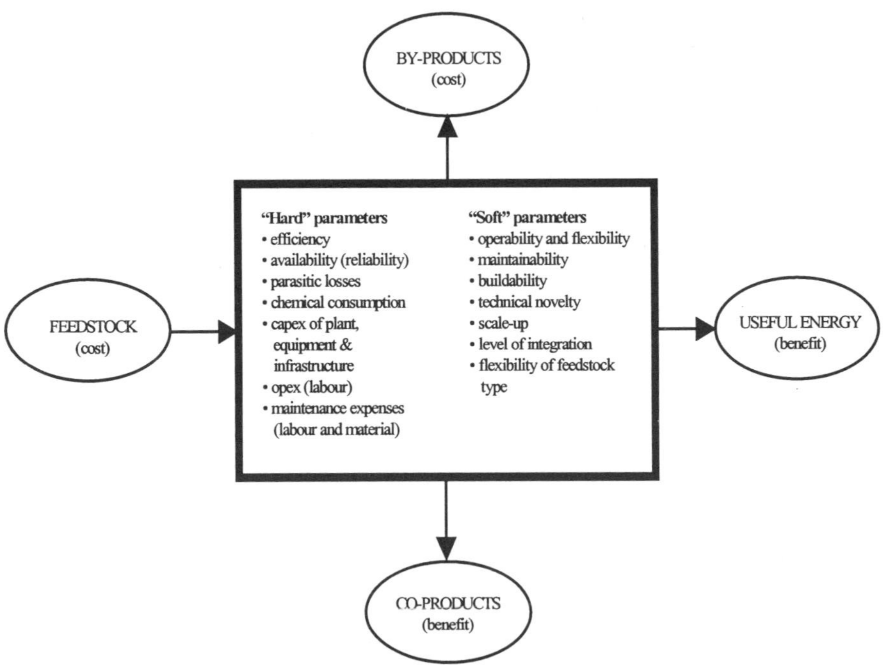

Fig. 1. Black Box Approach to project operational evaluation and risk assessment

All of the hard parameters will need to be taken into account in the financial model of the scheme which will evaluate the cashflows, in terms of costs and revenues to calculate the financial return on the investment in the project.

Two vitally important hard parameters associated with any conversion process, which have a critical effect on project viability, are efficiency and availability. These are defined as follows:

$$\textbf{efficiency} = \frac{\textbf{useful electricity exported}}{\textbf{heat value of feedstock}}$$

$$\textbf{availability} = \frac{\textbf{period time - maintenance time - forced outage time}}{\textbf{period time}}$$

NB efficiency should be measured at defined ISO-type conditions and forced outage time includes time at reduced capacity and the attainment of high availability is critical in maximizing the energy produced from the process and, therefore, the revenue.

The financial model is characterized, in summary, as follows:

(1) Based on selected technology
(2) Includes all project development costs
(3) Includes costs of the site, insurances, working capital and contingency allowance
(4) Allows for inflation and taxes
(5) Gives IRR/NPV, cashflow and cover ratios (ie the ratio of net project revenue to debt service costs) on project investment.

The model invariably involves many iterations before a satisfactory combination of project parameters is assembled; it will set the scene for the business case which the investor or the financier will consider prior to committing funds to the project. Further discussion on the financial model is given in the section on risks and critical issues later.

THE REAL WORLD

Once upon a time, electricity supply was a monopoly activity, the industry's assets were owned by the state in most countries or, where they were not state-owned, such as in the US or Germany or Japan, they were heavily regulated.

In the 1980s, fundamental changes took place in the state-owned sector with the so-called "privatization" of the industry. First in Chile, then in the UK, the assets of the publicly-owned electricity authorities were vested in companies which issued equity shares to the general public. This change in ownership was coupled with a fundamental shift in the way electricity was produced and sold. New, cheaper, technologies such as gas-fired, combined cycle, gas turbine plant, which could be built more quickly than the traditional coal-fired steam turbine plant, and which is less labour intensive in its operation, was introduced extensively in the UK. The separate producers of electricity competed with each other on price (per kWh) to win market share. It is alleged, with some justification in the UK, that such competition has reduced the price to the end-user.

In such a highly competitive market place, expensive technology, and that which is unproven or of a small-scale, will find it very difficult to compete with tried and tested CCGT plant fired on plentiful and (currently) cheap natural gas.

In the developing world, the requirement is for large increases in reliable capacity to provide power for industrial development, new townships and agriculture. Governments in such countries, in most parts of the world, are encouraging private sector developers to invest in the national electricity system and sell power to the distribution companies for onward supply to consumers. Again, conventional technology, utilising lowest cost, often indigenous, fuel such as coal, is a powerful imperative in the thinking of private sector investors.

Thus, it is suggested, expensive complex technology will be hard pushed to find a secure place in a market place where the emphasis is on low capital cost and technical efficiency for large-scale power production.

One area which, however, does offer some chance of success for advanced conversion technology lies in the field of renewable energy. The ability of a process to accept a feedstock which is sustainable, such as biomass, or can be considered, to all intents and purposes, as a renewable such as waste arisings eg sewage sludge, municipal solid waste or food waste, from large centres of population, can be beneficial for technological development in the context of a national renewable energy policy.

Such a policy manifests itself, for example, in the UK at the present time where Government has set targets for 5 per cent of all electricity production to be from renewable energy sources by 2003, rising to 10 per cent in 2010.

Thermal conversion of biomass will play a part in meeting these commitments, along with wind energy, hydro, etc. Premiums paid for renewable energy from such sources can compensate for the inevitable high capital cost of the technology and the disadvantages arising from dis-economies of scale.

A further important issue to be considered in a "real world" environment is not only competition in the market place for the output of the conversion process (ie electricity or heat) but competition between rival technologies for the attention and sponsorship of project developers. Factors which are likely to be considered by developers in selecting a particular technology will include:

(1) Previous operating experience - where the "technical" process has been developed from laboratory prototypes, data will need to be gathered on the performance of the process in terms of its thermal efficiency in recovering useful energy from the feedstock and the production of by-products (and co-products if relevant), particularly discharges to air, water and land which will be of keen interest to the environmental regulatory agency. Inevitably, such operating experience will, generally, relate to a small scale plant and predictions on performance and operating experience will have to be made where scaling-up is envisaged (see below).

(2) Scale of process – those investing in a project where significant scaling-up is needed to attain critical "commercial" mass will require evidence that such a scale-up is a practical proposition and that the relevant factors have been taken into account, both from a mechanical engineering aspect and from the thermochemical conversion standpoint. Sometimes, a multiple parallel-flow process may be adopted to avoid having to stretch the design too far. The risks inherent in the scaling-up will be of special interest to those involved in the independent project appraisal.

(3) Current status of technology – the development status of the technology is interrelated with the feedstock type to a large extent. For example, whilst the gasification of a coal feedstock is a relatively well understood and well tried process, the gasification of many biomass-type wastes is not, certainly outside of a research laboratory. The question generally asked under this heading in the appraisal process is "how long has the technology been around and where can we see it working?"

(4) Cost/MW of capacity – given that the process will be competing for developers funds against a number of other technologies, there will have to be significant advantages, in terms of efficiency or environmental emissions, to be gained from relatively expensive processes if the attention of a developer is to be caught.

(5) Potential application – it should be remembered that a conversion process can, fundamentally have one of three potential uses:-
   - electricity generation
   - feedstock utilization or disposal
   - production of co-products.

Each of these aspects needs to be quantified in the project evaluation process.

FINANCING A PROJECT

The majority of independent energy projects which have been promoted in countries such as the UK, which have a non-parastatal electricity supply industry, have been funded on the so-called "project finance" basis. Under such an arrangement (which was developed originally in connection with North Sea oil projects in the 1960s) the majority (in the region of 80 %) of the finance required to implement the project comprises debt borrowed over a period of, typically, 12-15 years. Bond financing can extend this period to 25 years in certain cases. The debt is repaid from the net revenues of the project. The remainder of the finance required for the project comprises equity provided by the project sponsors.

SPECIAL PURPOSE COMPANY

In order to provide a vehicle to take the project forward on a project finance basis, a special purpose company is established by the project sponsors. This company (the "energy company") becomes the counter-party to the various agreements for goods and services which, essentially, comprise the project (see Figure 2). The primary agreements in a typical renewable energy project are:

(1) Fuel or feedstock supply
(2) heat and electricity off-take
(3) design and construction of the facility
(4) operation and maintenance.

There are other secondary agreements covering lease or sale of land, finance etc as well as statutory consents such as planning, integrated pollution control, consents to build a power station, licence to generate electricity etc.

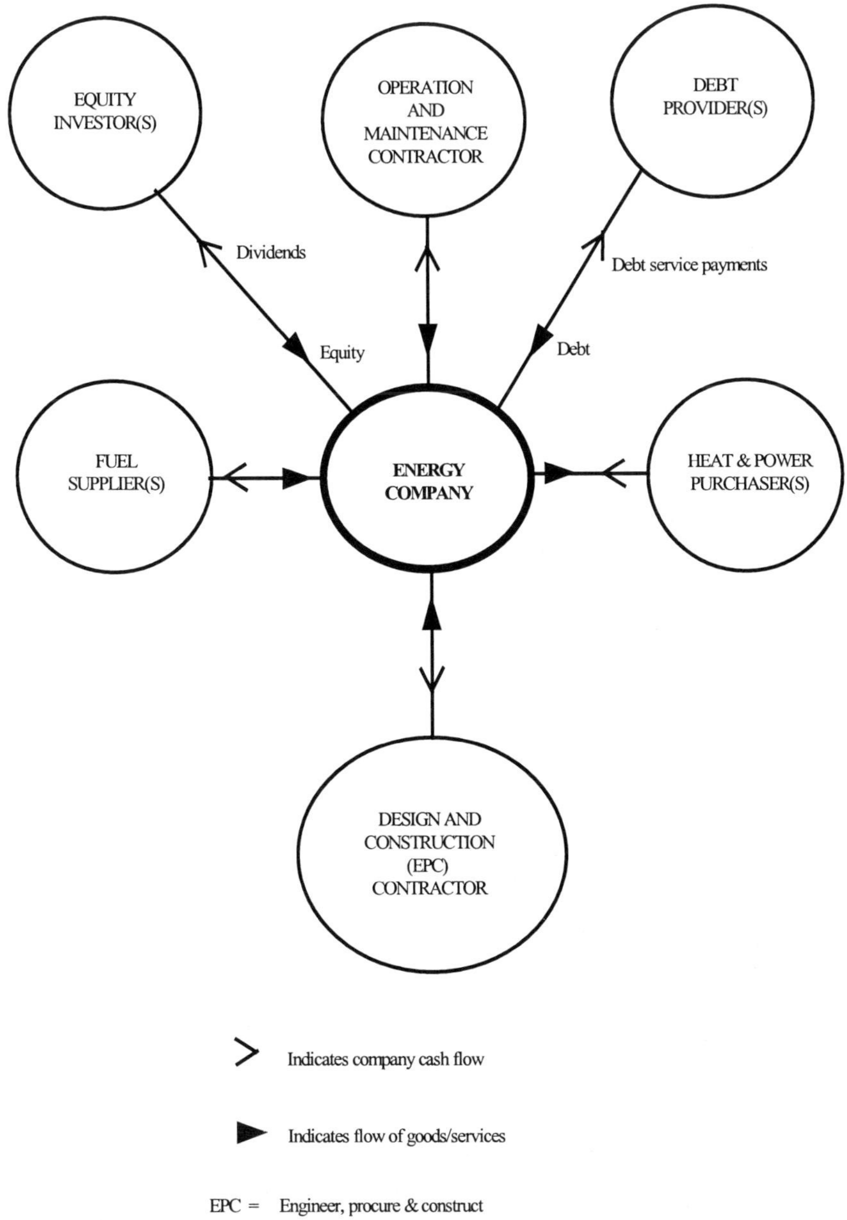

Fig 2 Contractual arrangements for independently financed energy conversion projects

The primary agreements are written in such a way that each complements one another eg the specification and quantity of feedstock supplied corresponds with that required by the plant to be built under the design and construction (D&C) contract to deliver the electrical energy output specified in the electricity off-take contract. At the same time, the obligations contained within the operation and maintenance (O&M) contract require specific levels of fuel conversion efficiency and plant availability to be attained by the operator in line with the terms and conditions of the electricity off-take contract.

The essential elements of each of these primary contracts are summarized below:

FUEL SUPPLY

Specific quantities of feedstock, of a defined quality, are to be made available under specified conditions at a particular price. A "gate fee" ie a payment from the feedstock supplier to the project company is generally paid in the case of a project in which the disposal of a waste (eg a by-product from a food processing industry) is a fundamental purpose of the project. Some form of "put or pay" arrangement is generally included in the contract in such circumstances.

Where the feedstock has an intrinsic value e.g. it can be used as a fuel in other processes (and, thus, there is some competition between procurers), the independent assessors of the project will require to see evidence of its long term, reliable, delivery in sufficient quantity to satisfy the needs of the project. Some measure of price stability will be necessary, for example by linking fuel purchase price to recognized and appropriate cost indices.

Diversity of supplier is generally desirable, whether there is a gate fee payable by the supplier or the feedstock has to be paid for by the project company, so as to mitigate the risk of a single supplier going out of business. The duration of the fuel supply agreement usually exceeds the term of the finance.

ELECTRICITY OFF-TAKE

An electricity off-take contract in a renewable energy project in many countries which offer support to renewable energy is usually a pro-forma agreement with one or other of the generating companies (or perhaps electricity supply companies) for a fixed term sometimes at a premium price. In the UK, for example, the Non-Fossil Fuel Obligation under the 1989 Electricity Act obliged the electricity distribution companies to procure a certain amount of the electricity that they supplied to their customers from renewable sources. Similar obligations exist under the recently introduced Utilities Act.

DESIGN AND CONSTRUCTION

The D&C contract in any project financed on a project finance basis is, invariably, based on a turnkey arrangement. This ensures that a single contractor assumes an undivided responsibility for completion date, final cost and performance of the facility. Liquidated damages are payable by the contractor for delay in completion or shortfall in performance. Such a contract is based on a carefully-written performance specification issued by the project company against which the contractor submits a fixed price tender in competition with other contractors. The contract is generally arranged so that the turnkey contractor accepts all the major construction risks.

OPERATION AND MAINTENANCE

Given that the payments made to the O&M contractor are likely to be a small fraction of the revenues earned by the project as a whole from electricity sales, it is not usually possible to secure damages from the O&M contractor for failure to achieve set performance targets for output (ie efficiency and availability) which are commensurate with the value of the lost electricity output. The essence of an O&M contract, therefore, is the establishment of, and commitment by the O&M contractor to, an agreed quality assurance system covering the activities of the contractor such as organization, documentation, safety, training, reporting, planning etc.

Regular audits of the contractor's activities and procedures give confidence to the owner that the contractor is likely to achieve the required performance targets. There may also be financial incentives included in the contract to encourage the O&M contractor to meet the desired level of performance and, conversely, penalties for non-achievement.

CONSENTS

Certain consents will be required by the project owner or operator prior to putting the facility into use. These, typically, include in the UK for example Integrated Pollution and Prevention Control Authorization (IPPCA) for thermal conversion processes, water abstraction licences for hydro electricity schemes etc. Assuming the use of proven technology and competent D&C and O&M contractors, such consents may not be difficult to obtain, unless the project is sited in a particularly environmentally sensitive area. In the case of pyrolysis processes, it will be necessary to convince the environmental regulators that the organic constituents of a pyrolysis oil-fuel (some of which might be carcinogenic) are destroyed in the overall energy production process.

An environmental statement is required under planning legislation in certain countries for all electricity generation projects and consultation with appropriate organizations may be necessary as part of the procedure for obtaining planning permission.

RISKS AND CRITICAL ISSUES

The main categories of risk associated with a renewable energy project are listed in the Appendix. The majority of risks associated with a given project can be covered in the contractual arrangements described in the preceding section. In this way, risks are allocated to the party who is best able to deal with them. The construction contractor, for instance, will typically be called upon to provide a guarantee of the performance of the plant, with appropriate damages for non-performance which protects other participants from financial loss. Project financiers will wish to see a suitable spreading of risks and mitigation of them (perhaps by insurance) over all the project participants. It is highly unlikely that the banks responsible for the provision of debt to the project will agree to bear any of the risks themselves.

RISK ANALYSIS

The interests of lenders, insurers and contractors require a very careful analysis of risk in the context of the various contractual obligations that go to make up the structure of a project.

Matters which would typically be regarded by lenders as representing a potential risk include:

(1) Technology - lenders will expect an independent engineer to report on the technical specification for the plant and will require evaluation of the

appropriateness and track record of the relevant technology in the proposed application. The developer may be enthusiastic about the potential benefits of a new technology whereas the lender would regard untried features as an unacceptable risk;

(2) Contractors - lenders' interests extend beyond just the identity of the principal contractors. They will consider their reputation and financial strength in assessing whether there is any risk of default in relation to any function which is material to the project's success;

(3) Security of cash flow - lenders will analyse all factors which may have a bearing on the ability of the project to maintain the project cash flow comfortably beyond the prospective term of the borrowing including, for instance, as mentioned in the previous section, the security of the fuel supply and contingency arrangement for alternative fuel;

(4) statutory authorizations - lenders will check on all factors affecting the consents and permits necessary for the commencement and continuation of the project.

The success of the developer in structuring and obtaining finance for projects is closely linked with their ability to identify areas of risk and to design technical solutions and allocate contractual responsibility so as to create the greatest comfort in the light of perceived risk. Much of this can be achieved by seeking initial expert advice in each relevant field so that there is a coherent strategy from an early stage. Some of the relevant issues which relate to risk analysis include:

(1) Financial model - a crucial test is to see how the economics of a project stand up to the perceived areas of risk. The development of any project will involve the production of a financial model which shows how the sensitivities of the various parameters at risk affect the economics. This is a process which will be followed by any prospective lender in relation to a funding proposal.

(2) Contractual framework - the contractual framework for any project will need to take account of the risk factors which are relevant to the specific features of the project. It involves careful planning on the identity of the contracting parties and assessment of their capacity to discharge their obligations, the scope of the positive obligations under each contract and the effect of limitations of liability.

(3) Financial structure - the funding structure will require a detailed evaluation of the capital requirement and the balance between equity or "risk" capital and loans. The proposed "gearing" will indicate the potential reward for the providers of equity and will also be an important indication in any proposal for project finance of the way in which the developers have considered the allocation of risk. Lenders like to see proposals which show that these principles have been understood and will gain confidence in the developers if the proposed financial structure of the project takes them into account.

(4) Insurance - insurance cover will be available for many of the areas of perceived risk and it is important that the scope of the cover is planned with reference to the specific detail of the project. This will, in any case, be a requirement of the lender. Specialist insurance brokers can assist in creating an effective insurance programme covering relevant interests with an indication of the sort of scope of cover that lenders will wish to see.

As a final note on this subject, the difference between risk and uncertainty is frequently misunderstood and the terminology, accordingly, mis-applied. The following note gives a clear enunciation of the difference:

> *Risk has its place in a calculus of probabilities. It applies to a specific course of action. The risk of an action is the likelihood that it will produce an* unwanted *result. Risk lends itself to quantitative expression, as when we say that the chances of failing to strike oil in a field are better than fifty-fifty or that the chances of finding a defective part in a batch are two out of a hundred. In the framework of benefit-cost analysis, the risk of an innovation is how much we stand to lose if we fail, multiplied by the probability of failure.*
>
> *Uncertainty is quite another matter. A situation is uncertain when it requires action but resists analysis of risks. A gambler takes a risk in an honest game of blackjack when, knowing the odds, he calls for another card. The same gambler, unsure of the odds and of the honesty of the game, is in a situation of uncertainty. He can act, but he cannot estimate the risks or rewards of his action. Even so, he operates in a risklike situation because, at any rate, he has two well-defined alternatives - to call for another card or not to call for it. He must act even though relevant alternatives are undefined. He must invent what to do. He has no way of calculating with any precision the risks of action. He has only rough guidelines of skill and experience to help him. (1)*

CONCLUSIONS

Translating thermal conversion research into a full-scale, practical, process plant in which all technical risks are understood and capable of quantification is bound to be a difficult task, but every new development has to start somewhere.

The commercial world of electricity supply, in which the thermal conversion process hopes to find an application, is changing rapidly in the developed world and the support and sustenance that prototype processes might have received in years gone by from central authorities is now vanishing.

In the developing world, the requirement is generally for tried and tested, reliable and easily maintained technology capable of providing a supply of electricity utilizing indigenous fuels. Advanced technologies, not long out of the research stage, are unlikely to satisfy their demands.

The drive to utilize more non-fossil fuels in electricity production, in order to meet climate change accords, is likely to provide the main stimulus to technological development of new thermal conversion processes which can utilize biomass or waste products as a feedstock in the developed world.

The financing of such projects, unless heavily underwritten as a stimulus to their commercial application from grant support of their capital cost from public funds, will require intensive analysis of the intrinsic technical risks of the process. The satisfaction of such a risk analysis will require the production of comprehensive data (and its intensive scrutiny) on operational performance and design, particularly in relation to scaling-up from the test-bed to the turnkey, commercial, installation.

ACKNOWLEDGEMENTS

Thanks are due to the Directors of PB Power Ltd (which now includes the power businesses of both Merz and McLellan and Kennedy & Donkin) for permission to publish this paper. It is emphasized that the opinions expressed are entirely those of the author and in no way reflect the corporate attitudes or views of PB Power.

BIBLIOGRAPHY

1. *A guide to finance for the UK renewable energy industry.* ETSU for the DTI, London, 1997.

REFERENCES

1. Schon DA (1967) *Technology and Change: The New Heraclitus*, Delacorte Press, New York

APPENDIX

Categories of risk in a typical renewable energy project

Major category	Specific aspect
Sponsor	• track record and reputation as a developer • degree of financial commitment • competing interests • insurability
Construction	• time, cost, quality • sponsor's ability to arrange construction • experience and standing of contractor • terms of main construction contract - fixed price/turnkey - penalties/bonuses - performance warranties and damages - insurances - incentives - force majeure - commissioning - subcontracting • competency of contractor's organization
Completion	• site conditions • infrastructure • competence of project manager • labour availability and suitability
Technology	• previous applications • competitive options • robustness • flexibility • adequacy of guarantees and warranties • achievability of performance targets • maintainability
Fuel supply	• availability • specification • price and exposure to cost increases • delivery and deliverability • terms and duration of fuel supply agreement
Operation and maintenance	• experience of operator • adequacy of infrastructure • maintenance of performance levels • access to replacement supplies and parts • cost constraints • exposure to cost increases • quality management system • appropriate maintenance schedules • terms of O&M agreement

Major category	**Specific aspect**
	• emission monitoring
Electricity and heat off-take	• quantity • product reliability • capacity charge/energy charge • market price levels and stability • competition • indexation • term of contract • predictability of heat and electricity loads • take or pay • penalties for non-supply • status of off-taker • potential for export off-site
Political	• energy policy • legislation • taxation • environmental controls • consents and authorizations

Woody and Herbaceous Biomass Feeds - how can we study their composition and their pyrolysis products?

B. Krieger-Brockett* and I. Rodriguez,
University of Washington, Dept of Chemical Engineering, Box 351750, Seattle, WA, 98195-1750 USA

ABSTRACT: Laboratories have renewed the quest for robust measures of alternative feedstock composition that are predictive of the pyrolysis products, particularly oil (sometimes called biocrude). A level beyond that predicted by pyrolysis conditions is sought. By alternative feedstocks we mean agricultural residues and other herbaceous or woody species that are infrequently commercially exploited. We have obtained a unique sample set of over 1100 feedstocks -- collected, pretreated, and characterized by the US Department of Agriculture principally for their ash and extractives (protein, polyphenol, oil, hydrocarbon) content and composition. We subsequently studied the sample set's pyrolysis behavior -- temperatures, gross product yields and compositions within the product fractions -- using an instrumented single particle reactor according to an experimental design. The composition attributes and pyrolysis conditions were systematically varied over a range of industrial interest. The experiments were designed using Principal Components Analysis to find feedstock compositions that were unambiguously attributable to one of the extractives. The resulting pyrolysis product slate was then similarly (statistically) analyzed for patterns that related to feed composition. Findings indicate that the native protein content, ash, and oil are the most important extractables in determining the pyrolysis products (oil, char, and gases) from these under-utilized herbaceous and agricultural residue species.

INTRODUCTION AND BACKGROUND

Expanding renewable and biomass feedstocks to non-traditional agricultural residues requires answering the question posed in the title, restated:

> What are unambiguous ways to study biomass composition and its relationship to pyrolysis rate and products? Are traditional, operational definitions of composition such as cellulose, Klason or Kraft lignin, and extractives weight fraction useful? What other composition metrics are useful especially as under-utilized or transgenic biomass species are examined?

The traditional pulp/paper composition metrics (e.g. specific lignin and cellulose contents) are based on chemical reaction (depolymerization) of the native biomass into fractions. Each separated fraction is then independently pyrolysed, and the pyrolysis yields are fit to models, and the composite used to predict other woody species' pyrolysis product slates. For example, powders from the *same* species (sweet gum hardwood) were used in the study of cellulose (Hajaligol, et al. 1982), lignin and whole wood (Nunn, et al. 1985a, 1985b) pyrolysis product slates. Each sample's decomposition products were measured using a heated grid apparatus. Gross product yields from the wood were correlated with parallel independent reaction kinetic parameters for each of the traditional components, lignin and cellulose. This construct was used even though lignin is not a component, but highly dependent in both molecular weight and structure upon the process used for its separation (pulping). This approach makes use of the extensive knowledge base in the forest products industries and has the merit that it is a familiar approach. However, even moderate pulping alters the polymeric constituents sufficiently that separate pyrolysis of wood fractions does not reproduce the native biomass pyrolysis chemistry and nor reproduce the physical microstructure changes that occur within large biomass wood chips during pyrolysis. Thus, a narrow range of applicability limits the traditional approaches.

To answer the question posed in the paper title, we offer a way to study composition effects when non-traditional biomass feedstocks are used in pyrolysis or energy production. Our method builds on traditional approaches and the whole or native biomass compositions as studied by scientists (e.g., Bagby et al, 1980) at the US Department of Agriculture (USDA) in a 14-year screening program for "botanochemicals". Bagby and co-workers' objectives included identifying and exploiting *minor* chemicals, such as extractives, (we will use this term for collectively referring to protein, ash, polyphenol, and oil fractions) from agricultural residues with the remainder, mostly cellulose and lignin, available for pulp/paper or fuels/energy. We ask a related question, i.e., is extraction, or alternatively leaving the components within the biomass, an advantage to pyrolysis and what pyrolysis product fractions are altered?

Our methods and experiments (UW) previously addressed composition effects in pyrolysis of RDF (Lai, et al 1993) and wood (Krieger-Brockett, et al 1997). In those papers and this one, even minor components are shown to alter pyrolysis slate when appropriate statistical methods are used. This paper briefly summarizes our work on pyrolysis product slates resulting from large- or macro-particle devolatilization (in which heat transfer is a slow process) of *native* biomass compositions in *under-utilized* species. The method has general applicability and owing to the limited scope of this article, the reader is referred to Somasundaram (1990), Lai (1991) and Rodriguez (1996) for details and extensive literature reviews with only a few relevant articles mentioned here.

SELECTED LITERATURE AND PREVIOUS WORK

Studies have focused on pyrolysis of wood's *separated* polymeric constituents', i.e., powdered wood, cellulose, hemicellulose and lignin (e.g. Antal, et al 1985; Nunn, et al 1985). The effects that minor components such as extractives' have on whole biomass pyrolysis products have not often been studied, except for ash (e.g., Richards and Zheng, 1991).

Using regression analysis to quantify the *intrinsic* cellulose, lignin, and ash fractions' effects on a few products, MacKay and Roberts (1982) studied the pyrolytic behavior of 20 unmodified or "native" powdered lignocellulosic materials. Linear regression models for total mass and carbon yields of the products were developed, assuming each constituent fraction of the biomass was pyrolyzed independently. They concluded that char yield increased with substrate carbon content; char yield from lignin content was found to be three times that of cellulose content. They did not study the gas nor tar composition in their experiments. In addition their particle sizes were inapplicable (powder) to industrial biomass processes wherein large particles and large temperature gradients ("thermally thick" particles) predominate.

Davis et al (1985) screened 57 different woody and herbaceous biomass feedstocks for their production of liquid fuel constituents from indirect liquefaction of powdered materials in a fluidized bed. Composition metrics in Davis' studies were ash, proteins, polyphenols, oil, and hydrocarbon content as defined and measured by the USDA (for example, in Adams, et al, 1986; Buchanan, et al, 1980a; Roth et al, 1984; Swanson, et al 1979). We will call the USDA-measured compositions *"traditional"* for purposes of this article. These feed compositions were correlated with pyrolysis gas production of hydrogen, CO, C_2H_4, total olefins, paraffins, and H_2/CO ratio using direct regression equations of the quadratic type, but tar/char production was not addressed.

To characterize agricultural species with chemical compositions suitable for conversion to bio-based chemicals and liquid fuels, a project lasting more than 14 years was conducted at the USDA. Over 1100 promising biomass species were systematically harvested and characterized for their content of ash, polyphenol, oil, protein, rubber, etc (and other "traditional" composition measures such as botanical family) (Bagby, 1989). In our current research, we make use of the USDA homoscedastic (based on standard methods performed by the same lab) biomass composition database to select species of contrasting and suitable chemical composition. Then, using statistical analysis to treat co-varying composition variables, we relate the resulting pyrolysis product yields to the traditional compositions of biomass, based on the USDA measurements. Because of the taxonomic information, as well as the extractives' fractions provided by USDA, the composition metrics are likely to have wide applicability.

Other related research was conducted at the National Renewable Energy Laboratory (NREL) to characterize and study the performance (after harvest and storage) of certain herbaceous biomass species with respect to the production of chemicals and hydrocarbons (e.g., Agblevor, et al, 1995,1992; Evans and Milne, 1987a,b). Of course, interest has increased regarding pyrolysis of genetically-engineered (transgenic) biomass species with special traits such as high isoprenoid content or easily-separated lignin.

Biomass pyrolysis oils have received attention owing to their potential value once upgraded (Bridgwater and Cottam, 1992) and their similarity to conventional fuels. Biomass oils are a very complex mixture of organic acids, oxygenated hydrocarbons, phenolic compounds and aromatics with precise composition being dependent on pyrolysis process, the rate of reaction and the feedstock chemical composition. Rodriguez (1996) reviews the literature on biomass pyrolysis oil analysis, with only a few studies mentioned here. Workers have used solid phase extraction, gel permeation chromatography (GPC), GC-MS, infrared spectroscopy, elemental analysis and solubility classification for characterization. Solid phase extraction on silica gel is useful for the separation of pyrolysis oils into functional fractions; GPC provides molecular weight distributions. Elliott et al (1988) have observed as many as 360 peaks in a pyrolysis oil by using GC-MS. Williams and Horne (1995) identify and quantified 161 compounds in biomass pyrolysis oils by using solid phase extraction and GC-MS. Although we too measured oil composition, space limits our discussion to results pertaining to *gross* product fractions (char, tar or oil, and gas yields) as affected by the native composition as discussed below.

MATERIALS AND METHODS

SAMPLE PREPARATION

The USDA characterized a large number of biomass species for their traditional composition fractions of ash, crude protein, polyphenols, oils and hydrocarbons among others. These fractions were defined *operationally* by the USDA "botanochemical screening" project; their sample analysis and fraction partitioning scheme is summarized in Fig. 1. Of interest were extractives, components that can be separated/partitioned from the plant by solvents. The major extractives included various oils, terpenes, fatty acids, unsaponifiables, aromatic compounds, tannins, and quinones. In exceptional cases extractives composed over 15% of the biomass (especially ash) but generally they did not exceed 5-10%.

The samples collected by the USDA (Fig. 1) consisted of whole plants, and for trees and shrubs, the latest 2 to 3 years' growth. All portions were included except the roots. Each sample was dried in still air at ambient conditions. The entire sample (about 500-1000 g, moisture free basis) was milled (Wiley) and screened using 1-mm-diameter holes (Carr *et al.* 1986). A subsample of each milled species was analyzed for moisture, ash, and crude protein (% Kjeldahl nitrogen x 6.25). Another portion of the sample (approx. 100 g) was extracted in a Soxhlet apparatus, first with acetone for 48 h and then with hexane for another 48 h. The acetone from the first extraction was allowed to evaporate before proceeding with the second extraction. The acetone-extracted fraction was partitioned between hexane and water/ethanol (1:7) to obtain the oil and polyphenol fractions. The residue after acetone extraction was extracted with hexane to yield the hydrocarbon fraction. The yields in all cases were determined gravimetrically after solvent removal and oven drying, and are reported as percents of the moisture-free and ash-free plant sample weight (Carr *et al.* 1986). We adhere to these operational definitions of the fractions and use the USDA data and samples directly.

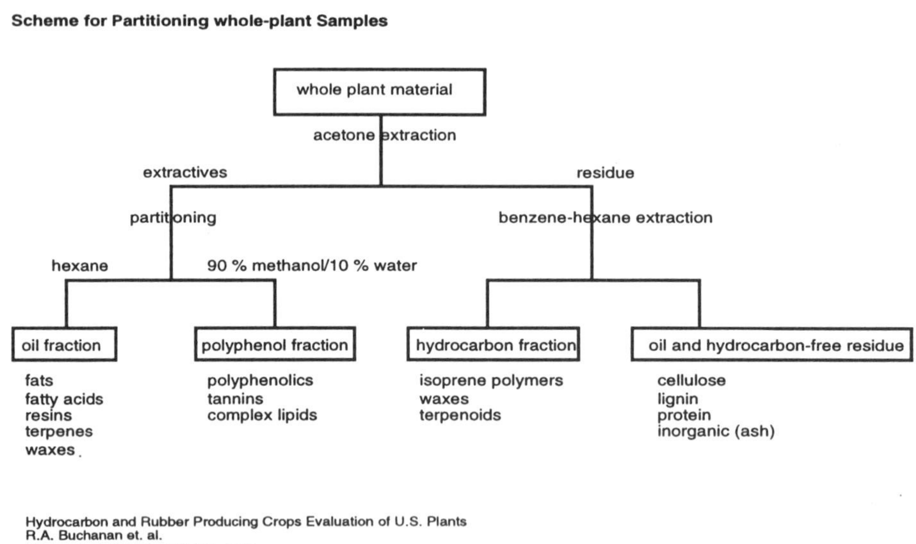

Figure 1 USDA scheme for extraction and partitioning of whole-plant material

SINGLE PARTICLE REACTOR

A brief description of the University of Washington single particle reactor, analysis system, and methodology is here, but the reader is referred to the more complete descriptions in Lai et al (1993), Lai (1991) and Krieger-Brockett et al (1997), and references cited therein. The glass portion consists of a reactor head with inlet ports for carrier gas, and an exit port for volatiles from the reaction zone. The fused silica window of the reactor allows for maximum transmission of radiant heat. Helium is fed though the inlet ports to the pellet front to sweep the volatiles away from the reacting zone and to the back face of the pellet to force volatiles "forward" for analysis. A glass sleeve holds the biomass pellet tightly while it is heated axially by a high intensity Xenon arc lamp. The arc lamp provides a uniform, constant radiant heat flux on the front of the biomass pellet and the lamp is calibrated absolutely, spatially, and frequently. During the heating, we visually observe the volatiles swept out of the reactor through the top port of the reactor head. The fluid residence time is short (measured by Chan, et al 1988; Lai, 1991). The volatiles pass through an accumulating tar trap at ~ - 40 ^{o}C (dry ice-acetone bath) which also traps water. The permanent gases pass through the trap and go to a semi-automatic GC analysis system.

Cylindrical pellets are made from the USDA samples expressly to remove consideration of native biomass morphology. We have shown biomass morphology alters pyrolysis slate (Chan et al 1988). If necessary, the sample is ground further in a Wiley mill and reformed in a cylindrical die to the native density, approximately 1.0 g/cm^{3}. Three holes are made radially in the pellet with a small drill and thermocouples are inserted at distances of 2, 4 and 6 mm from the heated face of the pellet. The pellet is then fitted tightly in a glass sleeve joined to the reactor via an O-ring seal. Real time intraparticle temperatures are displayed on the computer monitor and digitized at a rate of 1 Hz. The moisture content of the pellet is constant at a 5.0 wt %, equilibrium moisture except for some data in Chan et al (1988).

We operationally define pyrolysis products as gases, liquids (pyrolysis oil or biocrude) and solids (char+unreacted sample). Gases, 13-26 samples taken at pre-programmed times, are sampled on-line then later injected automatically into the GC for analysis. The pyrolysis liquids that condense in the reactor, tar-trap and lines are each determined quantitatively by weighing before and after the experiment. Thereafter, the pyrolysis oils are recovered by washing with a known amount of THF and analyzing each fraction. The char is determined quantitatively by separating the char and the unreacted part of the pellet with a microtome. The mass balance is then closed by knowing the initial weight of the pellet, the unreacted pellet, the quantitative oil, char, and gas produced.

Experimental conditions for the pyrolyses are chosen to be similar to industrial conditions in various gasification, liquefaction, and combustion reactors used with large particles or wood "chips". These process conditions are kept constant for all pyrolysis experiments reported here. The pellet length (thermal thickness) is 1.0 cm, the applied heat flux is 4.0 cal/cm^{2}-sec, chosen to be in the range of both pyrolysis as well as combustion and fire levels (2.8-27.4 cal/cm^{2}-sec). The reaction time of 12 min. is chosen because we wish to study the devolatilization process only. After 12 min., the temperature inside the reacting pellet is high enough to induce secondary tar reactions within the char layer that we wish to keep to a minimum. The particle surface experiences rapid pyrolysis, but the particle interior experiences slow pyrolysis, in keeping with most industrial-scale biomass thermal conversion processes.

Several new techniques for pyrolysis oil analysis were tried and results are presented in Rodriguez (1996). Low resolution, or low recovery, or both owing to the oil's complexity (more that 200 compounds are probably present) plagued all analyses.

Capillary GC results are pertinent and this technique provided the most convenient method, though recovery was low (a mean of 7% of the oil was analyzable). The oils were injected as a whole sample (0.5 µl-100:1 split) into the capillary column without derivatization. A DB-5 column (J & W), temperature gradient, and automatic data analysis were used (Rodriguez, 1996). Additional methods we use appear in Lai (1991).

RESULTS

SAMPLE SELECTION FOR CONTRASTING COMPOSITIONS

Traditional composition data from the USDA partitioning scheme are plotted (directly) in Fig. 2a top. Most species cluster at the origin consistent with low extractives content. Species names at some extreme compositions are bold (Latin names).

We have quantified in previous articles (Lai, et al, 1993) that some traditional extractives are correlated; for example ash and protein are positively correlated (r=0.26), whereas ash and polyphenol are negatively correlated (r= -0.32). These correlation coefficients imply that extractives' compositions clearly *co-vary*, or vary in concert, not independently. This could be seen in Fig. 2a if we rotated it to prominently show the axis along which compositions show the maximum covariance. However, we prefer to identify native species' compositions in which we can *independently contrast* the effect of high- versus low- ash samples, or high- versus low-oil contents, on the pyrolysis product distribution. That is, we prefer *orthogonal contrasts*. Thus, ideal sets of species are represented in Fig 2b (with an expanded scale) by the clusters (bold dots) that show clear contrasts of composition. For example on the two shaded surfaces marked A and B, we have 4 clusters of species whose comparisons along the axis marked "variable 2, Z" represent 4 independent observations of the "effect of variable 2 or Z" where Z or variable 2 may not be a traditional measure of composition. Cluster analysis and other multivariate statistics methods can provide such clusters and axes. Results of our statistical analysis of the composition database are described later and in detail in Lai (1991) and Rodriguez (1996).

Statistical manipulations on the USDA database (cluster analysis, principal component analysis with varimax rotation e.g., Everitt, 1980) revealed "subsets" of representative species, as idealized in Fig 2b, but with different variables (orthogonal principal components) than traditional fractions as measured by USDA. A set of species from each orthogonal subset appears in Table 1. The Latin names, and where available, the common names of the biomass species are given. The extractives' ranges are: ash content, 4 to 17%; protein content, 5 to 14%; polyphenol, 3 to 11%; and oil content, 1 to 4%. However no species contains extremes of all 4 variables. Nor can species be found, retaining native compositions, at extremes of just one extractive composition, while the other fractions are present at constant levels. Thus we use orthogonal but non-intuitive compositions in this work, then rank pyrolysis effects in terms of traditional extractives content to get an understanding of their impact on biomass pyrolysis.

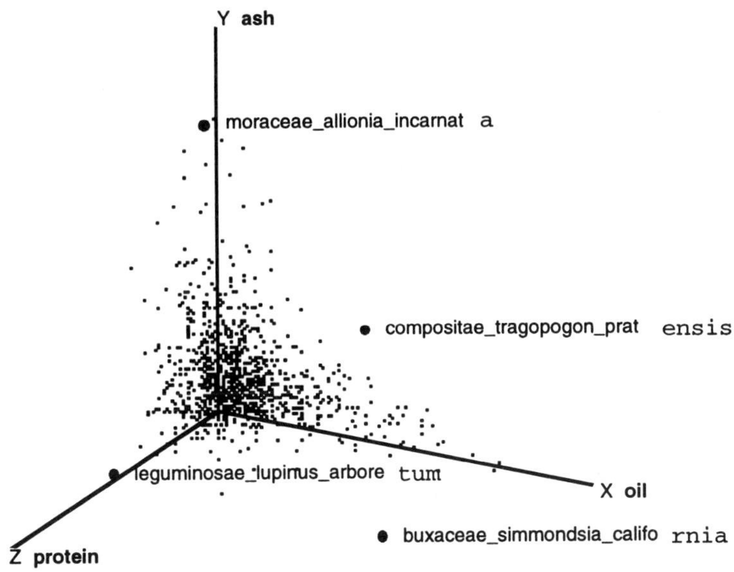

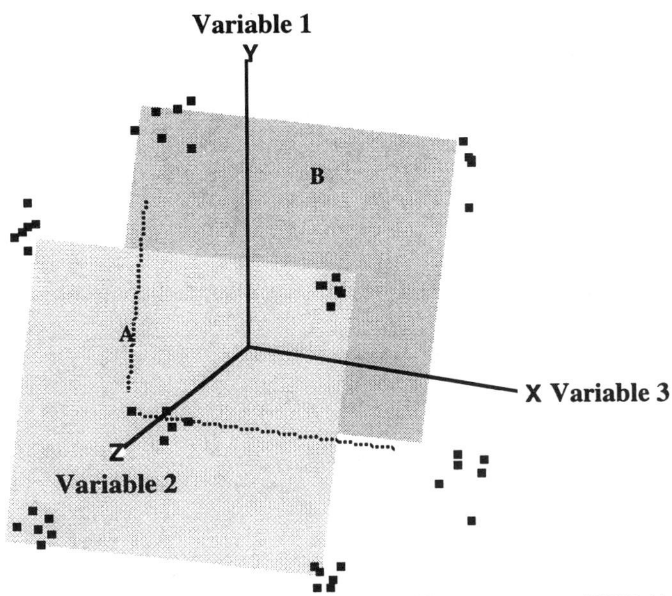

Fig. 2a and b - Three of 4 "traditional" composition measures (USDA) crossplotted (above); Idealized, contrasting clusters versus "orthogonal" compositions (below)

PYROLYSIS YIELDS FROM CONTRASTING COMPOSITIONS

We performed repeated single particle pyrolyses on one species to estimate the reproducibility of the entire experiment, i.e., the sampling, preparation, pyrolysis, product and data analysis. That species was a choice whose composition represented nearly the grand average of all the orthogonal composition measures -- *Ongracae Oenothera Biennis*, the so-called "centerpoint" [CP] of the composition fractions. It is listed at the bottom of Table 1. The replicates, and some comparison runs, are included in Table 2 wherein only the % reacted and overall product fractions are presented. Note that the char, oil or biocrude, and gas sum to 100% within 0.6% (round off+experimental error).

The replicate results, Table 2, show product yields with good reproducibility and low experimental error. The average conversion for the 12-minute one-dimensional heating is 84.6%, with standard deviation of 1.8 and a relative standard deviation of 2.1%. Note that experiment D* was carried out without extractives (the extractives were removed by using the procedure shown in Figure 1). Even though the extractives were removed, the overall yields are comparable with the results of experiments with extractives retained in the ground biomass (experiments A, B and C). The char yield average is 34.5%; the liquids or oil yield average is 47.1% with a relative standard deviation of 4.3%. These results show that on average about 47% of the total biomass converted goes into bio-oil (biocrude) which is of interest to this work. The average gas yield is 18.3% and shows the highest experimental error. Experiment E** was carried out with holocellulose (cellulose + hemicellulose). That is, the Klason lignin was removed by treating the extracted biomass with sulfuric acid (TAPPI Standard Method T 222 om-82). These data were not included in the statistics calculations. The pyrolysis experiments using biomass with extractives, without extractives, and with holocellulose only demonstrate that the pyrolysis product slates from these starting materials are different to extents greater than the experimental error.

Table 1 Biomass species chosen from the 16 orthogonal compositional subsets

[Subset #] *Biomass Species* (common name-USDA)	%ash	%pro-tein	%poly-phenol	%oil
[S1]*Compositae Haplopappus* (Greene Blake)	4.10	5.56	4.89	1.05
[S2]*Leguminosae. Desmodium Cuspidatum* (Tick Trefoil)	5.90	12.80	4.42	1.50
[S3]*Compositae Liatris Spicata* (Marsh Blazing Star)	6.10	6.57	3.75	2.25
[S4]*Leguminosae Galactia Wrighttii* (Wright Milk Pea)	6.90	14.00	5.95	2.23
[S5]*Polygonaceae Polystichum Munitum* (W. Sword Fern)	4.30	5.65	10.24	1.24
[S6]*Magnoliaceae Magnolia Acuminata* (Cucumber Tree)	4.20	9.01	10.44	1.53
[S7]*Compositae Solidago Graminifolia* (G L Goldenrod)	4.90	6.66	10.37	2.28
[S8]*Thymelaeaceae Largerstromia Indica* (Crepe Myrtle)	5.80	8.76	10.17	2.63
[S9]*Compositae Dyssodia Acerosa* (N/A)	13.4	5.29	3.69	1.00
[S10]R*anunculaceae Rananculus Acris* (Tall Buttercup)	17.1	12.4	3.89	1.16
[S11]*Balsaminace Impatients Pallida* (Yellow Jewelweed)	11.6	6.81	3.71	2.21
[S12]*Papaveraceae Argemone Platycera* (N/A)	11.2	10.7	4.65	2.68
[S13]*Valerianaceae Valerianella Intermedia* (Corn Salad)	9.50	6.79	10.05	1.53
[S14]*Araliaceae Hereda Canariensis* (Algerian Ivy)	10.5	9.51	10.19	1.42
[S15]*Carnaceae Cornus Drummondi* (N/A)	9.99	6.93	11.43	2.43
[S16]L*abiatae Nepeta Cataria* (Catnip)	11.4	9.66	8.77	2.21
[CP]*Ongraceae Oenothera Biennis* (Evening Primrose)	7.41	6.94	5.37	4.44

Table 2 Overall Product yields from species *Ongraceae Oenothera Biennis* [CP]

Experiment	%, Reacted	% Char	% Biocrude	% Gas
A	83.8	33.3	50.2	16.3
B	87.0	36.7	45.9	17.3
C	82.7	31.8	46.6	20.9
D*	84.8	36.1	45.8	18.6
Mean	84.6	34.5	47.1	18.3
St. Deviation	1.8	2.3	2.0	1.9
Variance	3.2	5.3	4.0	3.6
St. Error	0.9	1.1	1.0	1.0
%Rel. St. Dev.	2.1	6.7	4.3	10.8
E**	80.5	33.3	27.96	38.7

Note: Experiment D* extractives-free and E** holocellulose only

Given in Table 3 are the total yields of pyrolysis products from the 16 "contrasting" (orthogonal composition) samples of Table 1. The experimental pyrolysis conditions were kept constant to estimate only composition effects. The average % reacted or conversion was 71.76%. The species from three subsets, S7 (approximately: low ash, high protein, high polyphenol, and low oil), S9 (~low ash, low protein, low polyphenol, and high oil), and S10 (~high ash, low protein, low polyphenol, and high oil) show conversions of over 80%, while the species from two subsets, S6 (~high ash, low protein, high polyphenol, and low oil), and S16 (~high ash, high protein, high polyphenol, and high oil) show low conversions of 49 and 56% respectively. Pyrolysis oils were the major product in all cases, with an average of ~48% (Table 3). The biomass species from subset S7 (~low ash, high protein, high polyphenol, and low oil) shows a char yield of 48.5% and a biocrude oil yield of 45.2%. This was the only species that produced more char than biocrude. Char yield was the second highest product with an average of 38.4%. Gas was the lowest yield in all cases, with an average of 13.8%, although several species produced gas yields of almost 20%.

Table 3 Overall yields from pyrolysis experiments on 16 "orthogonal" compositions

(Subset) Species	% Reacted	% Biocrude	% Char	% Gas
(S1)*Compositae Haplopappus*	79.61	43.95	36.66	19.38
(S2)*Leguminosae DesmodiumCuspidatum*	69.90	50.20	40.30	9.40
(S3)*Compositae Liatris Spicata*	74.53	47.90	32.22	19.76
(S4)*Leguminosae Galactia Wrighttii*	77.27	51.58	32.44	15.96
(S5)*Polygonaceae Polystichum Munitum*	71.91	44.80	41.17	14.01
(S6)*Magnoliaceae Magnolia Acuminata*	49.15	56.49	31.14	12.35
(S7)*Compositae Solidago Graminifolia*	81.73	45.21	48.53	6.24
(S8)*Thymelaeaceae Largerstromia Indica*	64.69	49.91	40.57	9.50
(S9)*Compositae Dyssodia Acerosa*	85.13	45.25	39.88	14.86
(S10)*Ranunculaceae Rananculus Acris*	87.40	43.30	41.60	14.90
(S11)*Balsaminaceae Impatients Pallida*	73.46	43.92	39.55	16.71
(S12)*Papaveraceae Argemone Platycera*	60.95	52.40	38.27	10.95
(S13)*Valerianaceae Valerianella Interme.*	75.50	46.19	34.67	19.12
(S14)*Araliaceae Hereda Canariensis*	79.32	44.20	38.73	17.05
(S15)*Carnaceae Cornus Drummondi*	61.04	47.45	43.48	9.05
(S16)*Labiatae Nepeta Cataria*	56.57	52.24	35.83	11.92
Minimum	49.15	43.29	31.13	6.23
Maximum	87.40	56.49	48.52	19.76
Mean	71.76	47.81	38.42	13.82

EMPIRICAL MODELS FOR PYROLYSIS YIELDS

Pyrolysis product slates were analyzed using as regressors the traditional extractives as well as orthogonal principal components, i.e., quantitative variables linearly related to ash, protein, polyphenol and oil. The methods are well presented in Lai (1991) as a model for the analysis by Rodriguez (1996). Empirical models were calculated from the product slates by using linear biased regression and other appropriate methods (Ridge Regression, Principal Components Regression) to summarize the composition effects as coefficients or "contrasts" (Box, et al, 1978). As judged from Fig. 3 (representative), the agreement between the measured % reacted (circles) and the model prediction (line) is quite good. The replicates' spread joined by a horizontal line shows the small experimental uncertainty. The coefficients of determination (R^2) for all models are over 0.8, with F values over 20. Because all the experimental data was used in developing the models, cross-validation of the models was not carried out. The residual plots showed an even distribution of residuals without outliers. Therefore, the models were considered adequate. The traditional variables (regressors) ash, protein, polyphenol and oil, are *correlated.* The practical implication of correlated regressors is that the model coefficients are inflated by the extent of correlation among the variables, and thus they are ambiguous and not reported here. However, nearly the same trends were found for the fits and model coefficients based on *true orthogonal composition contrasts* (principal components) as regressors - but these are not intuitive variables to the unfamiliar user. Because the principal components were not intuitive, we present qualitative rankings of effects of the traditional compositions or extractives on pyrolysis in Table 4 and the reader is referred to forthcoming publications.

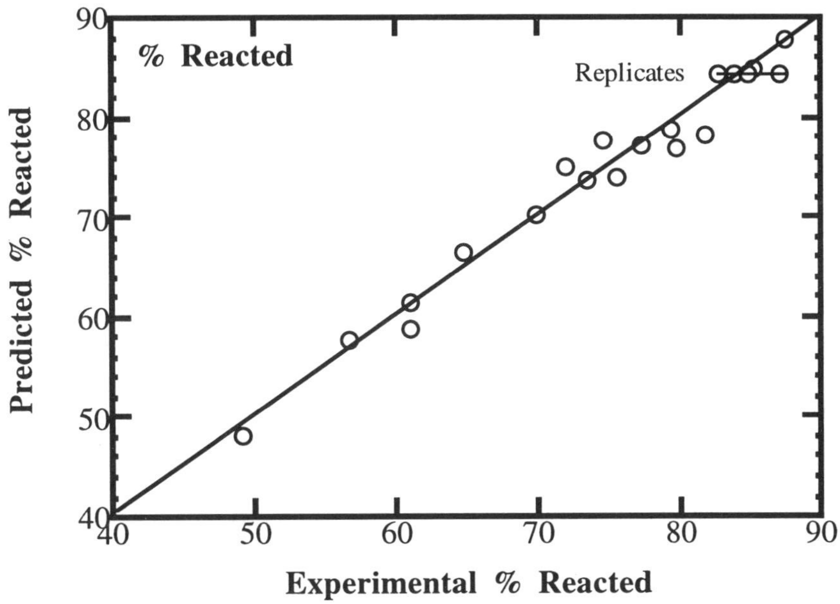

Figure 3 Experimental *vs*. Predicted biomass reacted

Table 4 Ranking of the traditional compositions on pyrolysis product distributions

Product Fraction	First Rank*	Second*	Third*	Significant Interactions*?
% Reacted	↓ Protein	↑ Ash	↓ Polyphenol	Yes
Biocrude	↑ Protein same as	↓ Ash		Yes
Char	↓ Oil	↑ Ash	↓ Protein	Yes
Gases	↑ Oil	↓ Protein	↓ Ash	No

*Greater than 90% confidence.

The qualitative influence of extractive composition on the percent of the particle that reacted, as well as on the fractions of biocrude, char, and gas produced, are given in relative order of their importance, i.e., as rankings. When the extractives are increased over the contrasts studied, the indicated gross pyrolysis product fractions also increase (↑) or decrease (↓). The ranked effects are shown in Table 4 using a 90% confidence level for the significance test. For example, even increasing the minor protein content from 4 to 17% reduced the % reacted for these experiments. Increasing the ash content over the range given in Table 1 *increased* the % reacted. The oil content of the reactant biomass is the most influential extractive on the char and gas produced from pyrolyis of the species in Table 1. However, as seen from the arrows, high oil content reduces char but increases gas produced in the macroparticle pyrolysis. The fourth column indicates that there are significant *interactions* (non-additive effects) among the extractives that alter all pyrolysis outcomes except gas production, which appears to be increased by oil in the reactant, but decreased by the presence of high protein or high ash in the biomass reactant.

Recall that our experimental design has “hidden replication”. Box, et al (1978) discuss the advantages of hidden replication and it is schematically shown in Fig. 2b, by the shaded surfaces A and B. That is, by choosing orthogonal subsets and clusters, we contrast 4 independent (orthogonal) experiments (surface A) with the four experiments represented on surface B. These 4 contrasts give 4 replicates to estimate the effect of Variable 2. For the sixteen species in Table 1, we have eight replications of the effect of a high or low extractable component of biomass (in principal component space). Thus we can detect small but statistically significant changes in pyrolysis slate due to native biomass composition differences.

Intraparticle temperature plays a key role in the pyrolysis process. In the single particle reactor the "thermally thick" biomass pellet is heated on one face by the arc lamp at the applied heat flux. Although not shown here, the measured temperature inside the pellet rises at distances of 2, 4 and 6 mm from the heated face. The temperature profiles observed in this work are variable from species to species, and even within the same species (*Oenothera Biennis*), owing to natural sample variability even though we homogenize the sample and the experimental conditions are identical. The temperature rises during the first 6-8 minutes of the experiment until it eventually “plateaus” as long as devolatilization is still occurring deeper in the pellet. At 2 mm inside, the temperature plateau is about 600° C, with the deeper temperature plateaus being about 380° at 4 mm and 150° at 6 mm. During the first 2-4 minutes at temperatures around 200-300° C, the pyrolysis process is most intense. This is confirmed by visual observation of the evolution of volatiles from the heated face of the pellet. After 4 minutes, the pyrolysis reactions are still taking place but the evolution of volatiles from the pellet interior diminishes in intensity. Thus, our results, while representative of conditions experienced by large biomass particles in industrial reactors, cannot be categorized as either slow or fast pyrolysis, but rather as a pyrolysis dominated by large gradients indicative of important heat and mass transfer rate limitations.

CONCLUSIONS AND SIGNIFICANCE

We have answered the question posed in the title of this paper, at least for a well-chosen subset of underutilized biomass residues. Our method of studying herbaceous and woody biomass composition effects on pyrolysis product slates is generally applicable and we again stress the importance of using statistically designed experiments, especially for correlated variables. We gratefully recognize that for the feedstock composition data and samples, we are fortunate to have the cooperative agreement with the USDA and our conclusions rest on their measurements and sample collections, and complement their findings in the cited papers. Other biomass sampling protocols and laboratory composition analyses might uncover different members of the subsets, or different pyrolysis product slate results for the same conditions. However, this paper is intended to focus on the question raised in the title. We stress the power of the statistics method and its ability to detect small differences within the range of natural sample variability. We also recognize that *other* contributions to the product slates may be significant, such as the 60-85% of the sample that is cellulose and lignin. As found by MacKay and Roberts, these two fractions strongly influence char yield. However, by the use of orthogonal variables and subset choices, our findings on the extractives' influence on pyrolysis yields should be valid in spite of this. The importance of orthogonal or uncorrelated independent variables is further illustrated in Lai (1991) and forthcoming publications.

ACKNOWLEDGMENTS

The authors wish to thank NREL and the USDA for their support and cooperation. Additionally, BKB acknowledges the inspiration and advice of MKB regarding the analysis and reviewers for their helpful suggestions.

CITED LITERATURE

Adams, R.P., M.F. Balandrin, K.J. Brown, G.A. Stone, S.M. Gruel (1986) Extraction of Liquid Fuels and Chemicals from Terrestrial Higher Plants, I. Yields from a Survey of 614 Western U.S. Plant Taxa, *Biomass 9* 255-292.

Buchanan, R.A., F.H. Otey, G.E. Hamerstrand (1980a) Multi-Use Botanochemical Crops, an Economic Analysis and Feasibility Study, *I&EC Prod. Res. and Devel., 19*(4) 480-496.

Agblevor, F.A., B. Rejai, D. Wang, A. Wiselogel, and H.L.Chum (1992) Thermochemical Conversion of Biomass to Fuels and Chemicals, the Effect of Storage Conditions on Pyrolysis Products, *Proceedings Alternative Energy-Liquid Fuels from Renewable Resources*, Dec. 13-15, 1992, Nashville, TN.

Agblevor, F.A., S. Besler, and A.E. Wiselogel (1995) Fast Pyrolysis of Stored Biomass Feedstocks, *Energy and Fuels*, **9**, 635-640.

Antal, Michael J., Jr. (1985) Biomass Pyrolysis: A Review of the Literature: Part 2 - Lignocellulose Pyrolysis, In: *Advances in Solar Energy,* Karl W. Boer and John A. Duffie Eds, **2** 175-255.: American Solar Energy Society, Inc., Plenum Press,. Boulder, CO.

Bagby, M.O., R.A. Buchanan, and F.H. Otey (1980) Multi-Use Crops and Botanochemical Production, *ACS Symp. Ser. 144, Biomass as a NonFossil Fuel Source*, D.L. Klass, Ed., 125-136.

Bagby, M.O.(1989) personal communication and cooperative agreement with UW & Northern Regional Research Center of the USDA.

Box, GEP, J.S. Hunter, W.G. Hunter, *Statistics for Experimenters*, Wiley, 1978.

Bridgwater, A.V., and M.L. Cottam (1992) Opportunities for Biomass Pyrolysis Liquids Production and Upgrading, *Energy and Fuels* **6**(2), 113-120.

Carr, M.E., M.O. Bagby, and W.B. Roth (1986) High Oil- and Polyphenol-Producing Species of the Northwest, *JAOCS* **63**(11), 1460-1464.

Chan, W.C.R, M. Kelbon, and B. Krieger-Brockett (1988) Single Particle Biomass Pyrolysis: Correlations of Reaction Products with Process Conditions, *Ind. Eng. Chem. Res.* **27** 2261-2275.

Davis, E.A., J.L. Kuester, and M.O. Bagby (1984). Biomass Conversion to Liquid Fuels: Potential of Some Arizona Chaparral Brush and Tree Species, *Nature* **307**(5953), p 726-728.

Elliot, DC, LJ Sealock, and RS Butner (1988) Product Analysis from Direct Liquefaction of Several High-Moisture Biomass Feedstocks *in Pyrolysis Oils from Biomass: Producing, Analyzing, and Upgrading*, Eds: Soltes, EJ, and Milne, T.A., *ACS Symp. Seri 376*, p 55-65,. 179-187.

Evans, R.J., TA Milne (1987) Molecular Characterization of the Pyrolysis of Biomass. 1., Fundamentals, *Energy and Fuels* **1** (3), 123-137.

Evans, R.J., TA Milne (1987) "Molecular Characterization of the Pyrolysis of Biomass. 2., Applications, *Energy and Fuels* **1** (4), 311-319.

Everitt, B.S. (1980) *Cluster Analysis* 2nd Ed, Heineman Ed. Books, Ltd.,. London.
Hajaligol, MR J.B. Howard, JP Longwell, WA Peters (1982) Product Compositions and Kinetics for Rapid Pyrolysis of Cellulose, *Ind.Eng Chem. Res.*, **21**, 457-465.
B. Krieger-Brockett, W.C. Lai and W.C. Chan (1997) Comparisons in Biomass and Refuse Derived Fuel Pyrolysis.Multivariate Analysis applied to an Experimental Design. *Developments in Biomass Thermochemical Biomass Research*, Elsevier, p. 43-61.
Lai, WC (1991) Ph.D. Thesis, Reaction Engineering of Heterogeneous Feeds: Municipal Solid Waste as a Model, Univ. of Washington, Dept of Chemical Engineering
Lai, WC, I. Rodriguez, and B. Krieger-Brockett (1993) Composition Effects on the Devolatilization Behavior of Biomass and Municipal Solid Waste, *Advances in Thermochemical Biomass Conversion,* Blackie Academic Publishers, London, pp. 818-832.
MacKay, DM and PV Roberts (1982) The Dependence of Char and Carbon Yield on Lignocellulose Precurors Composition, *Carbon,* **20** 95-104.
Nunn, TR., J B. Howard, JP. Longwell, and WA. Peters (1985a) Product Compositions and Kinetics in the Rapid Pyrolysis of Sweet Gum Hardwood, *Ind. Eng. Chem. Process Des. Dev.,* **24**, 836-844.
Nunn, TR., J B. Howard, JP. Longwell, and WA. Peters(1985b) Product Compositions and Kinetics in the Rapid Pyrolysis of Milled Wood Lignin, *Ind. Eng. Chem. Process Des. Dev.*, **24**, 844-852.
Richards, GN and G. Zheng (1991) Influence of Metal Ions and Salts on Products from Pyrolysis of Woods: Applications to Thermochemical processing of newsprint and Biomass, *J. Anal. and Appl. Pyrolysis* **21** (1-2) 133-146.
Rodriguez, I. (1996) Ph.D. Thesis Composition Related Effects of Thermal Reactivity of Organic Feedstocks, Univ. of Washington, Dept of Chemical Engineering.
Roth, WB, ME Carr, EA Davis, and MO Bagby (1984) Evaluation of 107 Legumes for Renewables Sources of Energy,. *Economic Botany* **38**(3) 358-364.
Somasundaram, S, (1990) MS Thesis, Biomass Pyrolysis: Temperature Distribution and Composition Effects , Univ. of Washington, Dept of Chemical Engineering.
Swanson, C.L., R.A. Buchanan, and F.H. Otey (1979) Molecular Weights of Natural Rubbers from Selected Temperature Zone Plants, *J. Appl. Poly. Sci.* **23** 743-748.
Williams, PT and PA Horne (1995) Analysis of Aromatic Hydrocarbons on Pyrolytic Oil Derived from Biomass, *J. Anal. and Appl. Pyrolysis* **31** 15-37.

Biomass Selection Criteria for Pyrolytic Conversion Processes

Anuradda Ganesh and Raveendran K.
Energy Systems Engineering, Indian Institute of Technology, Bombay - 400 076, India

ABSTRACT: Composition of biomass is proved to play a definitive role in determining the pyrolysis product distribution and their properties. In this paper, an attempt is made to suggest criterion for selecting biomass to obtain optimum products such as char, liquids, gases and char adsorbent through pyrolysis process. In order to facilitate the implementation of such a guideline, Rating Indices (RI) are defined based on the product yield and their properties which are specific to each of the end use, namely, pyrolysis-, carbonization, liquefaction and gasification. Based on the RI obtained for a particular application, different biomass are ranked for their suitability for each of the conversion process. This ranking is specific to end-use and is different for different application. Further, a stepwise procedure is presented to obtain the RI of any biomass from their initial composition.

INTRODUCTION

Biomass pyrolysis though is one of the first process man developed, is being studied extensively since the last two decades to obtain liquid, gaseous and solid fuels and chemicals. It is well recognised that pyrolysis plays a key role in any of the thermochemical conversion process be it combustion, gasification, liquefaction, production of char or active carbon. In this context, selection of feedstock and optimal utilisation of the products can play a vital role [1]. This paper suggests a criteria to select an appropriate biomass or find its relative suitability to the conversion processes.

Biomass is composed of various components such as cellulose, hemicellulose, lignin, extractives and mineral water. The composition of biomass plays a definitive role in altering the product distribution and their properties [2-3]. As is shown in earlier publications [4-8] different biomass, on pyrolysis, give different product yield with different product properties. In order to choose a biomass for a particular process (carbonisation, liquefaction, gasification or adsorbent char) knowledge on the product distribution and properties for various biomass are essential.

Literature does not provide information on (i) suitability of a particular biomass for a particular process or vice-versa, or (ii) a methodology to evaluate the suitability of a biomass or a process.

To address the above mentioned "suitability criteria" this paper introduces the concept of "Rating Indices", which is based on the desirable properties of the products for a particular application.

Further, it is also possible to obtain the suitability of biomass knowing only the initial composition of biomass. For this purpose various correlations developed and reported in previous publications [4-8] are employed in order to obtain the required product yield and their properties so as to enable rating of a biomass for a process.

RATING INDEX (RI)

DEFINITION

Rating Index (RI) has been defined separately for each of the four processes i.e. carbonisation, liquefaction and gasification and char adsorption (activation), based on the product yield and their properties as follows:

Carbonisation

Rating index for carbonisation is defined as the product of char yield, char heating value and total fraction of carbon present in the char, where all values are on as-measured basis which is shown in equation (1).

$$RIc = Yc \times Hc \times Cc \qquad (1)$$

where RIc is rating index for carbonisation, Yc is char yield (as fraction), Hc is higher heating value of char (in MJ/kg) and Cc is carbon content (percentage weight) of char.

Liquefaction

Rating index for liquefaction is given in equation (2) which is defined as product of liquid yield and higher heating value of pyrolysis liquids divided by mole of hydrogen required per mole of carbon in biomass liquids for upgrading to premium fuel.

$$RI\ell = Y\ell \times H\ell \times 1/Hr \qquad (2)$$

where $RI\ell$ is rating index Rating Index for liquefaction, $Y\ell$ is pyrolysis liquids yields (in fraction), $H\ell$ is higher heating value of pyrolysis liquids (in MJ/kg) and Hr is hydrogen requirement for upgradation of premium fuel. More details of Hr are given in section 2.2.

Gasification

Rating Index for gasification is defined as product of gas yield, heating value of the gases and the H/O ratio of the gases and the same is represented in equation (3).

$$RIg = Yg \times Hg \times H/Og \quad (3)$$

where RIg is rating index for gas, Yg is gas yield (in fraction), Hg is higher heating value of pyrolysis gases (in MJ/nm^3) and H/Og is ratio of hydrogen to oxygen of gases (by percentage weight).

In addition to the above three, rating index is also defined for the char to be used for adsorption purpose i.e. as activated charcoal.

Char Adsorption

Rating index for making char adsorbents is defined as product of char yield and char iodine adsorption value (taken in percentage weight on as-measured basis).

$$RIca = Yc \times Ic \quad (4)$$

where RIca is rating index for utilisation of char as carbon adsorbent and Ic is the percentage of char iodine adsorption on as-measured basis.

WHAT IS Hr?

Figure 1 shows the Van Krevelan diagram for liquefaction presented by Chronet and Overend [8]. The atomic H/C and O/C ratios of petroleum and also that of biomass pyrolysis liquids are indicated in the same figure. It is observed from the figure that the biomass pyrolysis liquids are characterised by very high O/C ratios (more than that of even the original biomass itself). However, biomass pyrolysis liquids have H/C ratios comparable to those of petroleum products.

It is important to note from Van Krevelan diagram that for the biomass liquids, upgradation possibility rests as a function of the methods of oxygen removal so as to achieve products with low O/C ratios which are compatible with the existing highly capital - intensive oil based transportation. For a carbon-rich, and oxygen-, hydrogen-poor fuel such as coal or coal liquids the problem is to add hydrogen on, rather than to remove oxygen.

The stoichiometric hydrogen requirement to upgrade biomass pyrolysis liquids to premium fuels has been calculated through the following relation presented by Chronet and Overend [8]:

$$Hr = \frac{\text{mole of H required}}{\text{mole of C in feed}} = 2\Delta R \left[R_{\text{product}} - R_{\text{reaction}} \right] \quad (5)$$

where R is the number of oxygen molecules required to combust completely one molecule of biomass pyrolysis liquids to water and carbon dioxide.

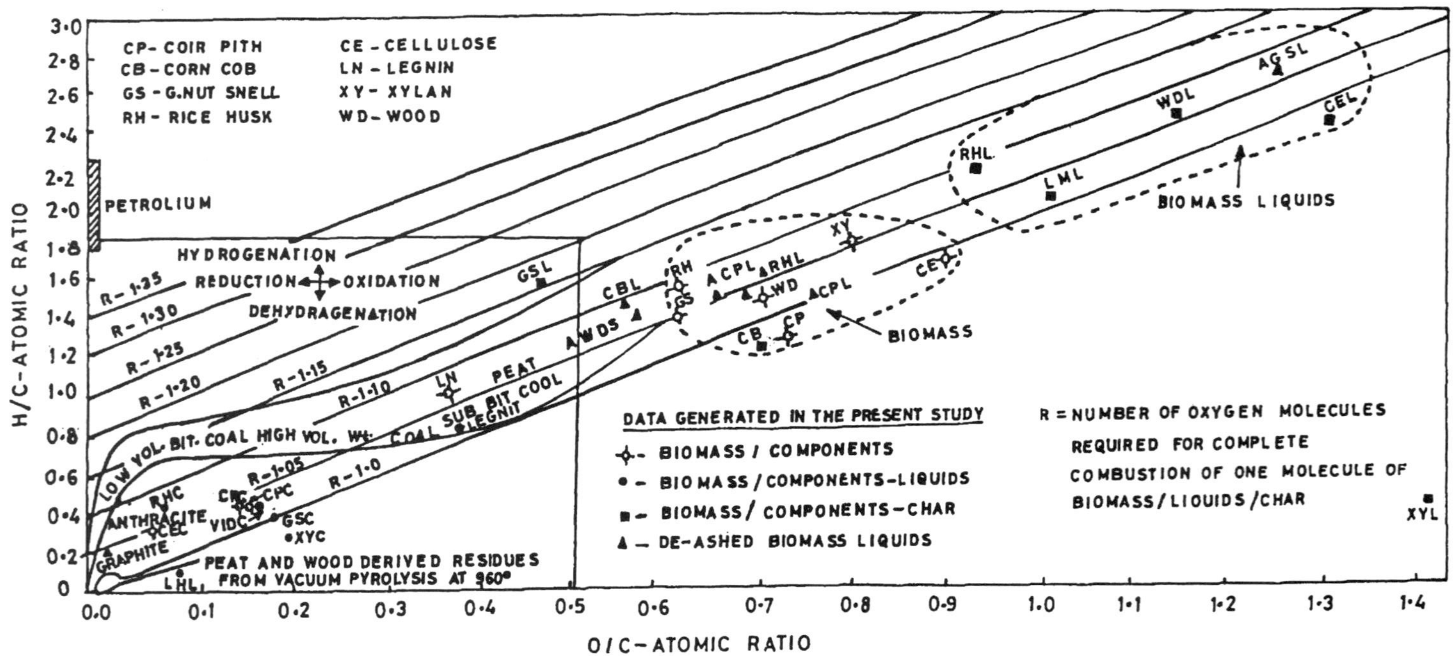

FIG.1 VAN KREVELEN DIAGRAM FOR LIQUEFACTION

NORMALISED RATING INDEX (NRI)

The Rating Indices obtained from all the biomass studied are normalised to obtain values (NRI) ranging from 0-1 for the purpose of comparison using the following relation.

$$NRI = [RIb-RImin] / [RImax-RImin] \quad (6)$$

where RIb is Rating Index for a given biomass, RImin is the lowest value of rating index obtained, and RImax is the highest value of rating index obtained among the biomass studied.

DISCUSSIONS

The calculated Rating Indices are tabulated in Table 1. It can be seen from Table 1 that lignin has the highest RI for carbonisation process as against xylan which has low carbonisation index. For liquefaction both cellulose and lignin show similar index. As expected, xylan has the highest RI for gasification process and for making char adsorbents. Among the five biomass studied in Packed Bed Pyrolysis (3,4,5), coir pith and groundnut shell have the highest rating index for carbonisation which is about 4.5 whereas groundnut shell and rice husk have the highest rating index for liquefaction having RI values of 2.63 and 2.34, respectively. On the other hand, wood has the highest rating index for gasification (0.70) and rice husk has the highest rating index (16.4) for making char adsorbents. De-ashed biomass are suitable for liquefaction and for making char adsorbents.

Considering liquefaction, groundnut shell shows the highest rating (2.63) followed by corn cob (1.65). Both cellulose and lignin shows higher rating for liquefaction (about 0.94). Under de-ashed condition also, groundnut shell and coir pith have higher RI for carbonisation. Wood has a higher RI for gasification, however on de-ashing shows a higher RI for carbonisation.

As may be seen from Table 1, groundnut shell, coir pith and corn cob are very attractive feedstock for making carbon adsorbents.

STEP WISE PROCEDURE TO CALCULATE RATING INDICES

In order to calculate Rating Indices, the data required are those of char yield, liquid yield, gas yield and that of respective heating values and elemental composition of each hydrogen requirement for upgradation of liquids is necessary. Also, to calculate RI for char adsorbent making, char iodine adsorption value is required.
Correlations have been developed to obtain each one of the above candidate values, as functions of the biomass compositions and are detailed in previous publications [4-7]. The correlations used for this purpose are presented here for the sake of completeness. Based on these, the stepwise procedure to calculate the different rating indices can be laid down as follows:

Table 1 Rating Indices of biomass for char, liquid and gas production.

Biomass/ Components	**Ratings**			
	RIc	**RIℓ**	**RIg**	**RIca**
Components				
Cellulose	3.22	0.96	0.03	3.8
Lignin	11.95	0.94	0.18	3.7
Xylan	5.13	0.17	1.04	10.2
Biomass				
Coir Pith	4.62	0.71	0.09	9.9
Corn Cob	4.21	1.65	0.03	5.3
G'nut Shell	4.39	2.63	0.08	5.7
Rice Husk	1.70	2.34	0.22	16.4
Wood	3.61	0.85	0.70	2.3
De-ashed Biomass				
Coir Pith	5.63	1.52	0.02	13.4
Corn Cob	2.49	2.66	0.10	8.6
G'nut Shell	6.34	2.23	0.07	11.6
Rice Husk	1.77	2.32	0.22	14.9
Wood	2.77	2.42	0.20	2.3

(i) determine chemical composition of biomass, particularly, cellulose, lignin, hemicellulose, ash and silica content,
(ii) calculate the product yield using the equation (7),

$$Y_{b,i} = \left[C_3 \left(\frac{X_c}{X_l} \right)^{nn_1} X_a^{nn_2} X_{si}^{nn_3} \right] \qquad (7)$$

where Y_b is the product yield and suffix i takes the values for char, liquids and gases; X is a percentage of each component suffixes c, l, a, si represents cellulose, lignin, ash and silica respectively values of constants c and nn are given below:

Product	C_3	nn_1	nn_2	nn_3
char	49.90	-0.0258	0.258	-0.044
liquids	102.86	0.7028	0.335	0.033
gases	13.93	-0.4168	-0.337	-0.002

(iii) calculate the heating values of products using the equations (8), (9) and (10).

$$H_c = 34.0138 X_1^{0.19275} X_a^{-0.18165} X_{Si}^{-0.134073}; \mathrm{SE} = 0.063 \text{ and } R^2 = 0.984 \qquad (8)$$

$$H_l = 91.6128 X_c^{1.18} X_l^{0.2529} X_a^{-0.0256}; \ \mathrm{SE} = 0.04 \text{ and } R^2 = 0.96 \qquad (9)$$

where H_c, H_l and H_g are higher heating values of biomass pyrolysis char, liquids and gases respectively and SE is standard error value and R^2 is R^2 value.

(iv) calculate the elemental composition of the products using the equations (11) & (12)

$$H_g = 0.963 X_c^{-2.533} X_l^{1.749} X_a^{-0.6143}; \ SE = 0.108 \text{ and } R^2 = 0.988 \quad (10)$$

$$(H/C)_g = 0.2335 X_l^{-3.07} X_a^{1.37} X_{si}^{-0.2476} \quad (11)$$

where H/C and O/C are percent weight ratios in percentages and suffixes c, l and g

$$(O/C)_g = X_c^{-0.486} X_l^{-4.4284} X_a^{4.962} X_{si}^{0.388} \quad (12)$$

represents char, liquids and gases respectively.

(v) calculate the carbon content of char using equation (13)

$$C_c = 0.4433 \times X_c^{-0.3350} X_l^{-0.1612} X_a^{-1.0225} \quad (13)$$

where C_c is the carbon content in the char.

(vi) calculate the hydrogen required for upgrading the liquids to premium fuels from van Krevelen's diagram (Fiqure 1 and equation (5))

(vii) calculate the char iodine adsorption values using the equation (14), and

$$I = 49.93 \times (Y_v \times R_m \times X_{si})^{0.2981} \quad (14)$$

where I is iodine adsorption, Y_v is volatile yield in fraction, R_m is maximum rate of volatilisation, X_{si} is silica content,

(viii) use equations (15) and (16) to generate the weight loss curve biomass to obtain volatile yield and maximum rate as de-volatilisation.

$$\left[\Delta W_b/T\right]_E = \left[\Delta W_b/T\right]_C \times \left(C_1 - \left[\left(\frac{X_a}{X_l} \right)^{n_1} X_{si}^{n_2} \right] \times \exp\left(\frac{-1}{T} \right) \right) \quad (15)$$

$$\left[\left(\frac{dw}{dt} \right)_{b/T} \right]_E = \left[\left(\frac{dw}{dt} \right)_{b/T} \right]_C \times \left(C_1 - \left[\left(\frac{X_a}{X_t} \right) X_{si}^{n_2} \right] \times \exp\left(\frac{-1}{T} \right) \right) \quad (16)$$

where $\left[\Delta W_{b/T}\right]_E$ & $\left[\left(\frac{dw}{dt} \right)_{b/T} \right]_E$ are the experimental weight loss data and derivative weight loss data for any given biomass and X_a , X_{si} and X_l are initial silica free ash, silica and lignin content, respectively, in fraction, present in dry biomass on weight basis and $C_1 = 0.5$, $n_1 = 8.5$ and $n_2 = 7.0$

Substituting the values of product yield and their corresponding product properties in equation (1) through (4) rating indices can be calculated.

BIOMASS CHARACTERISATION BASED ON NRI

Using the aforesaid procedure, the RIs for the other biomass (whose product properties are not experimentally studied) have been calculated. The RI values obtained are normalised to obtain NRI as per equation 5 and are presented in Table 2. The values of the maximum rate of devolatilisation is obtained from the actual TGA curves. (These may also be calculated with a high level of accuracy using the chemical analysis of biomass as described in steps (ii) and (iii) in Section 3). The last column indicates the remarks regarding their utilisation or end-product based on the ranking systems suggested.

It may be seen from Table 2 that for the purpose of charcoal production, coir pith, groundnut shell, corncob and cashewnut shell are better suited candidates than the other biomass considered for the present study.

Similarly, in the case of liquid fuel production, coconut shell, cotton gin waste, groundnut shell, coconut coir and rice husk show better properties.

Coconut shell, groundnut shell, wood, coconut coir and rice husk are better suited candidates compared to the other biomass studied for pyrolytic gasification applications.

For active carbon manufacture, rice husk, corn stalks, wheat straw, coir pith and cotton gin waste are better rated candidates.

The method, presented here is the initial step in choosing or selecting a biomass from various available sources. However, further studies have to be conducted for optimising process parameters and also to obtain the suitability in terms of physical properties (e.g. solid flow properties of biomass).

CONCLUSIONS

Criteria for evaluating the suitability of biomass for a pyrolysis conversion process to obtain solid, liquid and gaseous fuels are developed based on the properties of the pyrolysis products.

Based on this method, suitability of various biomass for pyrolytic conversion is studied and is found that based on the composition of biomass, different biomass are suited for different applications such as carbonisation, liquefaction, gasification and making of char adsorbent.

Table 2 Ranking Biomass for Pyrolysis Conversion Processes.

Biomass	Char		Liquids		Gases		Adsorbent		Remarks
	A	B	A	B	A	B	A	B	good feed for
Bagasse	.21	7	.32	11	.13	6	.44	6	-
Cashewnut Shell	.22	4	.50	6	.80	2	.14	12	C & G
Coconut Coir	.22	5	.62	4	.38	4	.19	10	L & G
Cococnut shell	.20	8	1.00	1	1.00	1	.04	13	L & G
Coir Pith	.31	1	.15	14	.09	7	.54	4	C
Corn cob	.27	3	.46	7	.02	9	.22	9	C
Corn Stalks	.13	9	.37	9	.01	10	.58	2	CA
Cotton gin waste	.03	11	.92	2	.04	14	.48	5	L & CA
Groundnut shell	.29	2	.79	3	.07	8	.24	8	C & L
Millet husk	.03	12	.33	10	.01	11	.30	7	-
Rice Husk	.04	10	.69	5	.20	5	1.00	1	CA
Rice Straw	.00	14	.39	8	.01	12	.16	11	-
Wood	.22	6	.19	12	.67	3	.00	14	G
Wheat Straw	.05	13	.17	13	.00	13	.55	3	GA

A - NRI for each of the product
B- Rank for biomass for each product
C- Carbonisation
G - Gasification
L - Liquefaction
CA - Carbon Adsorbent

REFERENCES

1. Antal Jr., M. J., "Biomass Pyrolysis: A Review of the Literature, Part I : Carbohydrate Pyrolysis (vol. 1, p175-239, 1982); Part II: Lignocellulose Pyrolysis (vol. 2, p 1-96, 1985), In Advances in Solar Energy, Eds. K. W. Boer and J. A. Duffie, American Solar Energy Society, New York.
2. Ganesh, A., "Studies on Characterisation of Biomass for gasification", Ph. D. Thesis, 1990, Department of Chemical Engineering, Indian Institute of Technology, New Delhi, India.
3. Raveendran, K. "Studies on Influence of Biomass Composition on Pyrolysis", Ph. D. Thesis, 1995, Interdisciplinary Program in Energy Systems Engineering, Indian Institute of Technology, Bombay, India.
4. Raveendran, K., Ganesh, A. and Khilar K. C., "Pyrolysis Characteristics of Biomass and Biomass Components", Fuel, v-75, n-8, pp.987-997, 1996.
5. Raveendran, K., Ganesh, A. and Khilar K. C., "Influence of Mineral Matter on Biomass Pyrolysis Characteristics", Fuel, v-74, n-12, pp.1812-1821, 1995.
6. Raveendran, K. and Ganesh, A. "Adsorption Characteristics and Pore Development of Biomass - Pyrolysis Char", Fuel, v-77, n-7, pp.769-781, 1998.
7. Raveendran, K. and Ganesh, A. "Heating Value of Biomass and Biomass Pyrolysis Products", Fuel, v-75, n-15, pp.1715-1720, 1996.
8. Chronet, E. and Overend, R. P., "Liquid Fuels from Biomass", In Biomass Regenerable Energy, (Eds. D. O. Hall and R. P. Overend), John Willey and Sons (1987).

Use of a Concentrated Radiation for the Determination of Cellulose Thermal Decomposition Mechanisms

O. Boutin and J. Lédé
Laboratoire des Sciences du Génie Chimique LSGC-CNRS-ENSIC-INPL, 1, rue Grandville - BP 451 - 54001 Nancy Cedex France

ABSTRACT The purpose of this paper is at first to summarize the main experimental results obtained with an original device (image furnace) in which pellets of pressed cellulose are subjected to a concentrated radiation. The time of pyrolysis and also the absorbed heat flux densities are perfectly controlled. The fractions and chemical compositions of all the products are determined as a function of the reaction time. It is hence possible to distinguish the primary products from those resulting from secondary reactions. The results bring new evidences for the validity of the Broido Shafizadeh model at atmospheric pressure as well as indications about the locations where the gases, liquids and solids are produced. It is then suggested to take advantage of these results for studying more usual pyrolysis reactors, on the basis of the heat flux densities that are locally exchanged. It is for example possible to predict the operating conditions in which no char is formed and hence almost only water soluble products are obtained.

INTRODUCTION

A great number of papers have been and continue to be published in the field of biomass thermochemical conversion. They are related to:

- fundamental studies (kinetic schemes, Arrhenius and thermodynamic constants [1 - 6])
- the optimization of reactors (cyclone, fluidized bed, rotating cone, ... [7-12])
- the modelling of the phenomena (chemical, transfer and hydrodynamic processes) occuring at the particle and at the reactor levels [13-18].

All the processes of biomass thermal conversion (pyrolysis, gasification and combustion) begin with elementary steps of decompositions of each of the components of the starting material (cellulose, hemicellulose and lignin). It is hence necessary to well understand the kinetics of the corresponding fast elementary reactions of pyrolysis. At the present time, no consensus has been reached, even for a model component such as cellulose. Most of the authors consider several kinds of competitive steps giving rise to gases, tar and char followed by subsequent consecutive reactions. The most often mentioned schemes rely on the Broido-Shafizadeh model [1]. However, according to some authors, a step giving rise to intermediate species (sometimes called "active cellulose") is not necessary for the prediction of the global kinetics of cellulose pyrolysis [19]. Other authors completely deny the existence of such species and just consider a single step giving rise to vapours from virgin cellulose [20-21]. Recently, Lédé et al. [4] gathered a great number of strong further evidences for the existence of these intermediate species. Several reasons may explain these controversies. Indeed, the studies are usually performed in systems (mainly thermogravimetric analysis) having several drawbacks :

- they are incompatible with the rapidity of the reactions.
- they are unable to evidence reactions occuring without mass loss.
- the heat and mass transfer efficiencies are often poor, and hence the pure chemical processes are not rate controlling.
- secondary reactions (crackings, recondensations ...) are possible and do not permit studying the initial short life time species.

The main purpose of the present paper is to report several new (qualitative and quantitative) results obtained in an original set up, in order to better understand the mechanisms of cellulose pyrolysis.

EXPERIMENTAL METHODS

A description of these methods has been given by Boutin [22] and we shall merely remind the basic principles of the experiments.

A small cylinder of cellulose (obtained from pressed cellulose powder) is placed in the middle of a quartz reactor fed at steady state by a flow of argon at room temperature. The end cross section of the pellet is positioned in the focal volume of an image furnace. Such a device relies on the use of a 5 kW xenon high pressure arc lamp associated with two elliptical mirors. The available flux density obtained in the focal volume and on the surface of the pellet can be quantitatively changed through the interposition of diaphragms [6]. The values of the absorbed heat flux densities used ($2.1 \times 10^5 \leq q\ (W\ m^{-2}) \leq 7.4 \times 10^6$) are known through calibration measurements. The cellulose sample is subjected to the concentrated radiation during perfectly controlled flash times ($5 \times 10^{-2} \leq t(s) \leq 5$). After the flash, the mass loss of the pellet is measured. The fractions of all the products (remaining on the pellet and escaping from the reactor) are determined as well as their chemical composition (HPLC, HPLC/MS, GC and elementary analysis). The reproducibility of the experiments and the mass balances is very good: around 98 %.

NATURE AND CONDITIONS OF FORMATION OF THE PRODUCTS

The products are recovered at the surface of the pellet and downstream of the reactor (condensed vapours and gases).

PRODUCTS RECOVERED AT THE SURFACE OF THE PELLET

After the reaction, the surface of the pellet is covered with a solid brown matter. Its microscopic observation reveals that the fibrillar structure of virgin cellulose (figure 1) has disappeared and that the products have the appearance of large networks of more or less agglomerated melted particles (figure 2). These observations clearly show that, in our conditions, the thermal decomposition of cellulose passes through an intermediate stage involving liquid compounds (ILC). However, although ILC appears to be liquid at the temperature of reaction it is solid at room temperature. It is soluble in water proving that it is not melted cellulose. From the temperature of glass transition of cellulose (510 K) it is possible to calculate a theoretical melting temperature of cellulose close to 720 K [23]. This temperature is very close to its pyrolysis temperature (around 740 K) and hence, even if cellulose melts, this possible liquid state would have a too short life time for being detectable in our device. ILC could be compared to the "active cellulose" postulated more than 20 years ago by Bradbury et al. [1] even if these authors had few chances to observe a liquid phase because their experiments were made at too low temperatures.

Direct and non destructive analysis (HPLC, and HPLC coupled with MS) shows that ILC contains anhydro-oligosaccharides with degrees of polymerization ranging from 7 to 2 and levoglucosan. These species result from the depolymerization of the cellulose molecules. It is important to note that the chemical composition of ILC does not depend on the operating conditions (heat flux densities and flash times).

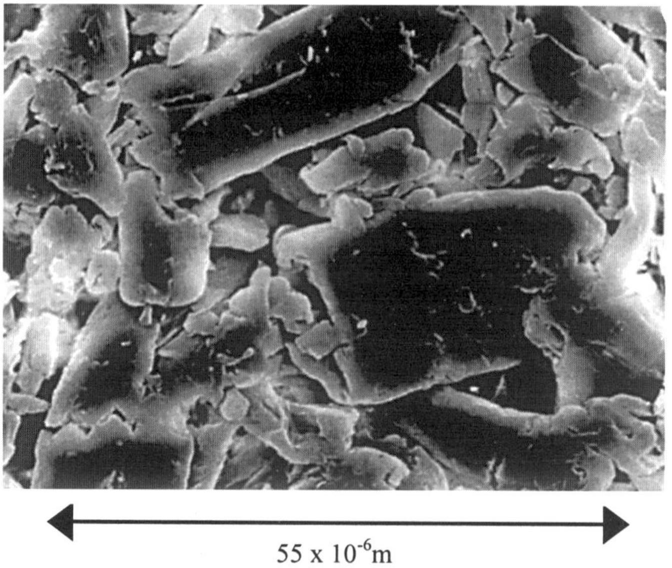

Fig. 1 SEM photograph of the surface of a cellulose pellet before the reaction (× 800)

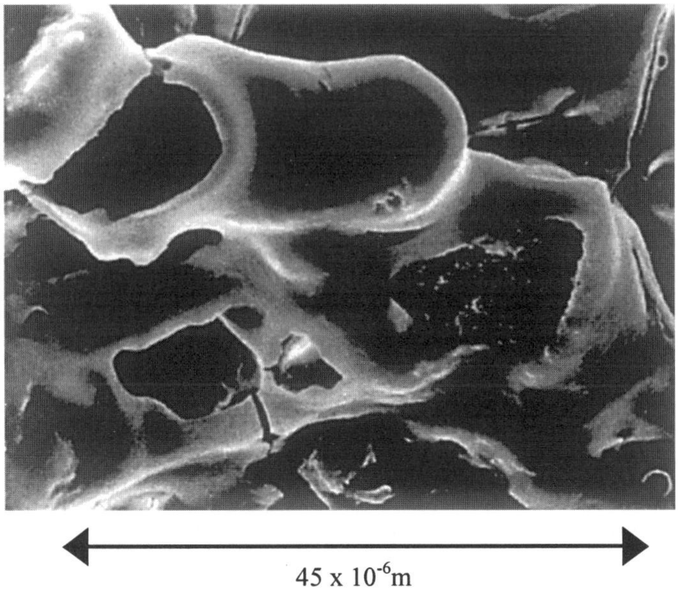

Fig.2 SEM photograph of the surface of a cellulose pellet after a flash of concentrated radiation (× 1000)

Studied as a function of the flash times of pyrolysis, the thickness of the layer of ILC first increases until it reaches a constant value ranging from 100 and 200 μm (steady state conditions of pyrolysis). The mass loss from the pellet that begins just after the appearance of ILC, linearly increases as soon as the extent of the layer of ILC has reached the steady state.

CHAR

For absorbed heat flux densities lower than approximately 9 x 10^5 W m^{-2}, the layer of ILC becomes progressively covered with amorphous black products (char) that are not soluble in water. The quantity of char increases with the flash time. Its mass fraction may range between 0 and 10 % according to the absorbed heat flux densities. The elementary analysis of char gives the following mean molar composition : C : 6 ; H : 7.6 ; O : 3.3, indicating that the char is far from being pure carbon. The image furnace provides a given flux density under the form of a concentrated radiation and not of a given temperature. Actually the effective temperature of the products formed inside the focal zone depends on a great number of chemical and physical (ex. optical) characteristics. It is hence very difficult, without the help of an accurate mathematical model, to estimate, a priori, the temperature reached by the char.

CONDENSED VAPOURS

The condensed matter (including water) trapped by the filters placed downstream of the reactor contains qualitatively the same species as those previously analysed in ILC. The presence of heavy molecules can be explained by mechanical ejections from the surface of the pellet (in the probable form of aerosols). We can expect that the highest fractions of levoglucosan and cellobiosan result from their vaporization as soon as they are formed from secondary crackings of the heaviest molecules inside the layer of ILC.

The quantity of the products trapped by the filters linearly increases with the flash time after the beginning of the steady state conditions of the pellet pyrolysis. The total mass fraction of these condensed vapours may reach 80 %.

GASES

A non negligible fraction of products (up to 10 %) is recovered in the form of gases. Their composition is given on a dry basis by the tables 1 and 2. The results do not depend on the flash time for a given heat flux density, but seems to change with this flux. No noticeable quantity of CO_2 is detected.

Table 1 Molar composition (%) of the gases for several absorbed heat flux densities (W m^{-2})

	No Char formation			Char formation	
Absorbed flux	4.0×10^6	3.4×10^6	2.7×10^6	4.6×10^5	2.2×10^5
CO	41.8	48.4	50.5	63.5	51.9
H_2	25.8	17.0	22.5	21.3	31.3
H_2 + CO	67.6	65.4	73.0	84.8	83.2
Methane	15.5	16.1	13.2	9.7	10.3
Propylene	1.5	0.5	1.4	1.7	1.3
Ethane	2.2	1.1	2.0	1.7	3.0
Ethylene	8.7	8.2	6.5	0.8	1.2
n pentane	1.1	7.3	1.1	0.3	0.2
Propane	1.7	0.4	1.1	0.1	0.4
Isobutane	1.4	0.7	0.9	0.5	0.3
n butane	0.3	0.3	0.8	0.4	0.1

Table 2 Mass composition (%) of the gases for several absorbed heat flux densities (W m^{-2})

	No Char formation			Char formation	
Absorbed flux	4.0×10^6	3.4×10^6	2.7×10^6	4.6×10^5	2.2×10^5
H_2 + CO	55.0	54.0	62.0	82.0	79.0
Hydrocarbons	45.0	46.0	38.0	18.0	21.0

When no char is formed, the mass fractions of the hydrocarbons may represent almost 50 % of all the gases. When char is formed, the molar fraction of H_2 + CO reaches 85 %. It is however difficult to determine if such behaviour is due to the differences in the values of the flux densities or to the presence or not of char, the two phenomena being connected each others.

Table 3 gives different results of the values of CO / H_2 and $H_2 + CO$ (on molar basis) reported by the literature. They are obtained in very different experimental conditions, and their comparison with those presented in this paper is difficult. They reveal that the ratio CO / H_2 may vary between 1.5 and 37 even if the total fraction $CO + H_2$ is relatively constant (roughly between 60 and 80 %). However we can observe that the results of the present work are similar to those obtained by Antal [24] and also in cyclone reactors. In each case, the vapour phase formed near the hot walls and / or near the reacting particles is rapidly mixed with a colder carrier gas (quench).

Another interesting observation is that in our image furnace experiments, the gases are detected only after the beginning of the steady state conditions of pyrolysis of the pellet.

Table 3 Molar composition of the gases as reported in the literature for several fast pyrolysis processes

References	Process	T (K)	CO / H_2	$(CO + H_2)$ (%)
Antal [24]	Solar furnace	973	2.9	70
Li [25]	Spinning disc	873	12.5	81
Lédé et al. [7]	Cyclone	893	1.5	76
Diebold et al. [26]	Vortex reactor	900	2.9	70
Graham et al. [27]	Vortactor	700	37	76
Peacocke [28]	Ablation	824	14.2	61
Boutin et al. [29]	Rotating cylinder	973	15.5	66
This study (no char)	Image furnace	-	2.1	69
This study (char)	Image furnace	-	2.2	84

OVERALL INTERPRETATIONS

All these observations show that after a period of simple heating, cellulose primarily decomposes into intermediate and short life time species (ILC) resulting from partial thermal depolymerization processes. After a short transient period these species remaining on the sample reach a constant thickness. This steady state regime indicates the existence of an equilibrium between the rates of ILC formation and disappearance (through probable ejection of aerosols and vaporizations). It is difficult to know the exact nature (i.e. degree of polymerization) of the primary molecules formed at the first stages of cellulose decomposition. For example, we cannot definitely conclude if the analysed species having DP ranging from 3 to 7 are of primary nature or not. However, it is likely that the lighter products such as levoglucosan and cellobiosan (analysed on the pellet and in the filters) are formed from secondary cracking, occurring inside the ILC layer. These products rapidly vaporize after their formation. Hence, we can expect that the lighter molecules, trapped by the filters, are not of primary natures as often stated in the literature.

Gases and char are also formed but only after the first appearance of ILC indicating that they also result from secondary reactions of the vapours and of ILC.

Above flux densities of about 9 x 10^5 W m^{-2}, no char is formed indicating that under these circumstances the pyrolysis of cellulose gives rise to about 90 % of water soluble products (ILC and vapours) and about 10 % of gases. The late formation of gases indicates that, under our conditions, no "prompt gas" would be formed as sometimes suggested in the literature [27].

The large hydrocarbons and H_2 contents of the gases, as well as the absence of CO_2, bring another proof of the secondary nature of the gases (formed by thermal cracking reactions). All these results provide evidence for the validity of the Broido Shafizadeh type model for representing the elementary processes of cellulose pyrolysis (Fig. 3).

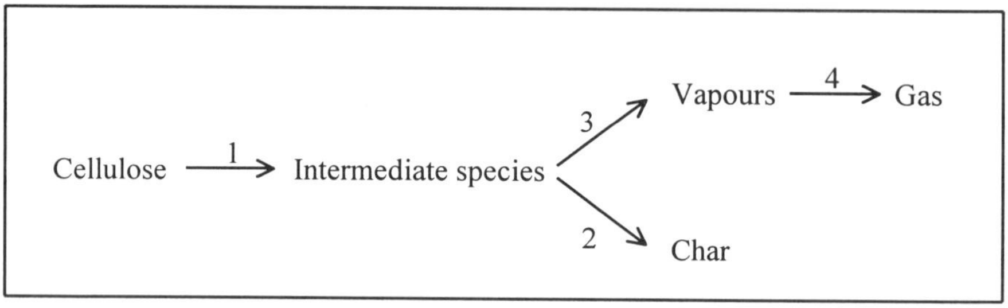

Fig. 3 Simplified cellulose thermal decomposition pathways at atmospheric pressure

The question of the mechanisms and locations of gas formation (reaction 4) deserves to be raised. In a recent bibliographical review, Lédé [8] gathers several data related to the kinetics of the thermal cracking of vapours formed during cellulose and biomass pyrolysis. These values can be associated to the short (around 0.5 s) residence time and low (around 500 K) theoretical mixing temperature of the cold carrier gas and hot vapours flowing through the quartz reactor in the present work. The calculations show that, in our conditions, these vapours have only few chances to thermally crack during their crossing of the reactor (quench effect).

The cracking of the vapours may, a priori, occur inside or downstream of the layer of ILC. If we expect that the coalescence of the small droplets forming ILC is not perfect, this layer has a porous structure (as proved by its low density) through which the vapours have chances to partially crack before they escape outside of this layer. After liberation from ILC, the vapours do not mix immediately with the cold carrier gas and probably form a more or less developed volume (boundary layer or free jet) inside which they can also partially crack. Such a situation in which pyrolysis products evolved from a high temperature reacting solid, must mix with a much colder medium is conceptually similar to that existing in a cyclone reactor [8] heated at its walls. In this last case, it has been shown that the vapours are partially cracked into gases inside the thin and hot boundary layer existing close to the hot inner surface of the cyclone.

We could hence assume that in the present work, the gases have chances to be formed partly inside the porous ILC but also during the brief crossing of the vapours through a hot volume separating the outer surface of the ILC from the colder gas flowing through the quartz reactor. These phenomena are still more complicated when char is formed because these vapours can also react during their transit through the thin porous layer of char. These additional processes could bring a possible explanation for the different gas compositions observed in the presence or absence of char.

These mechanisms of gas formation are probably valid for many other experimental devices where solid biomass particles react inside a colder medium. Hence, it seems difficult to completely avoid gas formation in most pyrolysis reactors because of the bad micro-mixing efficiencies existing in the vicinity of the particles.

APPLICATION OF THESE RESULTS TO THE COMPARISON OF PYROLYSIS REACTORS ON THE BASIS OF HEAT TRANSFER MECHANISMS

The performances of pyrolysis reactors are often compared on the basis of the fractions of the different recovered products (gases, oils, char). In addition to the maximum biomass throughputs, the comparison criteria usually rely on the values of the residence times and temperatures, two parameters that are often poorly known and defined, and that may be very different for each of the phases involved. The hydrodynamics are often very different from plug flows and generally the true reaction temperatures cannot be accurately measured.

It could be interesting to consider another comparison parameter relying on the exchanged heat flux densities. As already analysed by Reed et al. [30], among the main elementary processes of biomass heating, we can mention : radiation from a hot source, direct contact with a hot surface and exchange with a gas. It is possible to estimate the corresponding heat flux densities q in each of these elementary cases and to compare their values as a function of the heat source temperature T_W. The following rough calculations are made on the basis of a cellulose reacting temperature $T_R = 740$ K [4] and of 400 μm particle sizes.

- Heating by radiation (particles flowing through a reactor having a wall temperature T_W, but without contact with it) :

$$q_r = \sigma (T_W^4 - T_R^4) \quad (1)$$

The view factor and emissivity are expected close to 1, in order to calculate an upper limit for q_r.

- Ablative pyrolysis (a piece of biomass is pressed against a hot moving surface at T_W, [13]) :

$$q_a = 0.017\, P\, (T_W - T_R) \quad (2)$$

The calculations are made with $P = 10^6$ Pa.

A special case is that of a cyclone reactor in which the reacting particles flow against its heated walls. The heat transfer coefficient h_{cy} is estimated close to 2700 W m^{-2} K^{-1} [31] and hence :

$$q_{cy} = 2700\, (T_W - T_R) \quad (3)$$

- Heating by exchange with a hot gas :

The Nusselt number representing the exchange between a particle and a hot flowing gas is often calculated by [32] :

$$Nu_{Co} = 2 + 0.6\ Re_p^{\ 0.5}\ Pr^{\ 0.33} \qquad (4)$$

Assuming a low particle Reynolds number Re_p, Nu_{Co} is close to 2 and hence :

$$q_{Co} = 2\ \lambda / d_p\ (T_W - T_R) \qquad (5)$$

The values of the thermal conductivity λ are taken from Franck [33].

Fig. 4 reports the theoretical variations of q_a, q_{Co} , q_{cy} and q_r, as a function of T_W. The limit of char formation determined in the present work ($q_{lim} = 9 \times 10^5$ W m^{-2}) is also represented. It can be seen that ablation is the most efficient process of heating during which char would be formed only at low wall temperatures ($\leq$ 800 K). Gas exchange and non concentrated radiation lead to the lowest values of q and as a matter of fact, char is usually observed when each of these two situations occurs separately.

However, the available flux densities often result from several contributions as for example the sum of radiation from a hot wall and exchange with a hot gas. Having in mind that the equation (5) related to this last process represents a minimum (q_{Co} can be much higher in the case of high particle Reynolds numbers), the values of $q_r + q_{Co}$ can approach the limit flux density defined in this work. In these conditions, we can expect that low char fractions will be formed.

The results for the cyclone reactor are also interesting because the values of q_{cy} are not far from q_{lim}. In that case, the formation of char can be expected only for T_W lower than about 1000 K [7] while only condensible vapours and gases would be observed at higher temperatures. These results for the cyclone are only qualitative indications, the heat transfer coefficients h_{cy} depending on d_p and biomass flowrate.

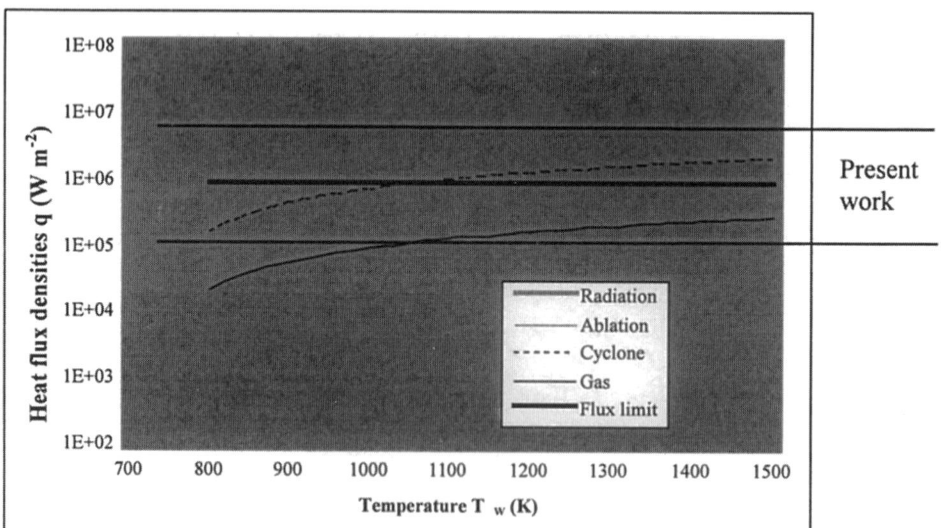

Fig. 4 Heat flux densities transferred to a particle for different elementary processes of heating

These very simple calculations give only first indications. A more sophisticated approach would need to take into account the fact that lignin gives rise to higher fractions of char than cellulose. Also, the existence of mass transfer resistances must not be ignored: if the primary products are not efficiently removed from the reacting zone, they have more chances to take part in subsequent reactions, with formation of char.

CONCLUSIONS

Cellulose can be pyrolysed under well characterised conditions of absorbed heat flux densities by using a concentrated radiation. Experiments have been performed with cellulose pellets submitted to controlled times of irradiation at the focal zone of an image furnace.

The composition and rate of formation of all the products have been studied as a function of flash times. The mass balances are excellent. The results clearly show that the reaction begins with depolymerization processes giving rise to a short life time intermediate compound (ILC) that is liquid at reaction temperature but solid at room temperature. This product partially decomposes into condensible vapours. A fraction of them undergoes a thermal cracking producing gases inside or in the vicinity of the sample. For heat flux densities higher than approximately 9×10^5 W m^{-2}, no char is observed.

With the help of these results, a comparison of the performances of pyrolysis reactors can be made on the basis of the types and extents of the elementary heat exchanges processes (radiation, ablation, convection). The calculations show that in some practical cases (ablative pyrolysis and cyclone reactor), it is possible to reach high flux densities and hence avoid the formation of char.

NOTATION

d_p	particle diameter (μm)
h_{cy}	heat transfer coefficient (particle / walls of a cyclone) (W m^{-2} K^{-1})
ILC	Intermediate Liquid Compound
Nu_{Co}	Nusselt number (exchange with a gas)
P	pressure (biomass against a wall) (Pa)
Pr	Prandtl number (close to 0.7 for a gas)
q	heat flux density (W m^{-2})
q_a, q_{Co}, q_{cy}, q_r	heat flux densities (ablation, gas exchange, cyclone, radiation) (W m^{-2})
q_{lim}	limit of the heat flux density for char formation (W m^{-2})
Re_p	particle Reynolds number
t	flash time (s)
T_R, T_W	temperature (cellulose pyrolysis, heat source) (K)
λ	gas thermal conductivity (W m^{-1} K^{-1})
σ	Boltzmann constant (5.67 10^{-8} W m^{-2} K^{-4})

ACKNOWLEDGMENTS

The authors want to thank ADEME (Agence de l'Environnement et de la Maîtrise de l'Energie) who has funded this research in the framework of AGRICE (AGRIculture pour la Chimie et l'Energie) under the contract n° 96 01 048.

REFERENCES

1. Bradbury A.G.W., Sakai Y. and Shafizadeh F. (1979) A Kinetic Model for Pyrolysis of Cellulose. *J. Appl. Polym. Sci.*, **23**, 3271-3280.
2. Milosavljevic I. and Suuberg E.M. (1995) Cellulose Thermal Decomposition Kinetics : Global Mass Loss Kinetics. *Ind. Eng. Chem. Res.*, **34** , 1081-1091.
3. Grönli M. (1996) *A Theoretical and Experimental Study of the Thermal Degradation of Biomass*, Thesis, The Norwegian University of Science and Technology, Faculty of Mechanical Engineering, Trondheim (N).
4. Lédé J., Diebold J.P., Peacocke G.V.C., Piskorz J. (1999) The Nature and Properties of Intermediate and Unvaporized Biomass Pyrolysis Materials. In *Fast Pyrolysis of Biomass : A Handbook*, (Ed. by A.V. Bridgwater ; S. Czernik ; J. Diebold ; D. Meier ; A. Oasmaa ; C. Peacocke ; J. Piskorz ; D. Radlein), pp. 51-65, CPL Press, Birmingham.
5. Antal M.J., Varhegyi G.and Jakab E. (1998) Cellulose Pyrolysis Kinetics : Revisited. *Ind. Eng. Chem. Res.*, **37**, 1267-1279.
6. Boutin O., Ferrer M. and Lédé J. (1998) Radiant Flash Pyrolysis of Cellulose - Evidence for the Formation of Short Life Time Intermediate Liquid Species. *J. Anal. Appl. Pyrolysis* , **47**, 13-31.
7. Lédé J.,Verzaro F., Antoine B., Villermaux J. (1986) Flash Pyrolysis of Wood in a Cyclone Reactor. *Chem. Eng. Process.*, **20**, 309-317.
8. Lédé J. (2000) The Cyclone : A Multifunctional Reactor for the Fast Pyrolysis of Biomass. *Ind. Eng. Chem. Res.*, **39**, 893 - 903.
9. Bilbao A., Millera A. and Arauzo J. (1988) Product Distribution in the Flash Pyrolysis of Lignocellulosic Materials in a Fluidised Bed, *Fuel*, **67**, 1586-1588.
10. Bridgwater A.V. (1999) Principles and Practice of Biomass Fast Pyrolysis Processes for Liquids. *J. Anal. Appl. Pyrolysis*, **51**, 3 -22.
11. Bridgwater A.V,. Peacocke G.V.C (2000) Fast Pyrolysis Processes for Biomass. *Renewable and Sustainable Energy Reviews*, **4**, 1-73.
12. Venderbosh R.H., Janse A.M.C., Radovanic M., Prins W., Van Swaaij W.P.M. (1997) Pyrolysis of Pine Wood in a Small Integrated Pilot Plant Rotating Cone Reactor. In *Biomass Gasification and Pyrolysis*, (Ed. by M. Kaltschmitt & A.V. Bridgwater), pp. 345-353, CPL Press, Newbury.
13. Lédé J., Panagopoulos J. , Li H.Z., Villermaux J. (1985) Fast Pyrolysis of Wood : Direct Measurement and Study of Ablation Rate. *Fuel*, **64**, 1514-1520.
14. Lédé J. (1994) Reaction Temperature of Solid Particules Undergoing an Endothermal Volatilization. Application to the Fast Pyrolysis of Biomass. *Biomass and Bioenergy*, **7**, 49-60.
15. Lédé J. (1996) Influence of the Heating Conditions on the Thermal and Chemical Behaviour of Solid Undergoing a Fast Endothermic Decomposition. *J. of Thermal Anal.*, **46**, 67-84.

16. Diebold J.P. (1994) A Unified, Global Model for the Pyrolysis of Cellulose. *Biomass and Bioenergy*, **7**, 75-85.
17. Di Blasi C. (1996) Heat, Momentum and Mass Transport through a Shrinking Biomass Particle Exposed to Thermal Radiation. *Chem. Eng. Sci.*, **51**, 1121-1132.
18. Miller R.S. and Bellan J. (1997) A Generalized Biomass Pyrolysis Model Based on Surimposed Cellulose, Hemicellulose and Lignin Kinetic. *Combust. Sci. Tech.*, **126**, 97-112.
19. Lanzetta M, Di Blasi C., Buonanno F. (1997) An Experimental Investigation of Heat Transfer Limitations in the Flash Pyrolysis of Cellulose *Ind. Eng. Chem. Res.*, **36**, 542-552.
20. Varhegyi G., Jakab M., Antal M.J. (1994) Is the Broido-Shafizadeh Model for Cellulose Pyrolysis True? *Energy and Fuels*, **8**, 1345-1352.
21. Antal M.J. and Varhegyi G. (1995) Cellulose Pyrolysis Kinetics : the Current State of Knowledge. *Ind. Eng. Chem. Res.*, **34,** 703-717.
22. Boutin O. (1999*) Analyse des processus primaires de dégradation thermochimique de la biomasse*. INPL thesis, Nancy, France.
23. Nordin S.B., Nyren J.O. and Back E.L. (1974) An Indication of Molten Cellulose Produced in a Laser Beam. *Textil Research J.,* 152-154
24. Antal M.J. (1983) Effects of Reactor Severity on the Gas Phase Pyrolysis of Cellulose, and Kraft Lignin, devired Volatile Matter. *Ind. Eng. Chem. Prod. Res. Dev.*, **22**, 366-375
25. Li H.Z. (1984) *Pyrolyse éclair de baguettes de bois - Modèle de fusion*. DEA INPL, Nancy, France.
26. Diebold J.P., Power A.J. (1988) Engineering Aspects of the Vortex Pyrolysis Reactor to Produce Primary Pyrolysis Oil Vapours for use in Resins and Adhesives. In : *Research in Thermochemical Biomass Conversion*, (Ed. by A.V. Bridgwater & J.L. Kuester), pp. 609-628, Elsevier Applied Science Publishers, London and New York.
27. Graham R.G., Bergougnou M.A., Freel B.A. (1994) The Kinetics of Vapour-Phase Cellulose Fast Pyrolysis Reactions. *Biomass and Bioenergy*, **7**, 33-47.
28. Peacocke G. (1994) *Ablative Pyrolysis of Biomass*. PhD Thesis, Aston University, Birmingham (England).
29. Boutin O., Lédé J. , Li H.Z. and Kiener P. (1997) Temperature of Ablative Pyrolysis of Wood. Comparison of Spinning disc and Rotating Cylinder Experiments. In : *Biomass Gasification and Pyrolysis*, (Ed. by M. Kaltschmit & A.V. Bridgwater), pp. 336 – 344, CPL Press, Newbury.
30. Reed T., Diebold J., Desrosiers R. (1980) Heat Transfer Perspectives. In *Specialists' Workshop on Fast Pyrolysis of Biomass*, SERI/CP-622-1096, pp. 7-20, Copper Mountain.
31. Lédé J., Li H.Z. , Soulignac F., Villermaux J. (1992) Le cyclone réacteur. IV: Mesure de l'efficacité des transferts de chaleur entre les parois et les phases gazeuse et solide. *Chem. Eng. J.*, **48**, 83-99.
32. Levenspiel O. (1972) *Chemical Reaction Engineering,* 2nd Edn, John Wiley & Sons, Canada.
33. Franck M.W. (1984) *Heat Transfer*. Adson-Wesley Publishing Company, London (England).

Modelling and measurements of drying and pyrolysis of large wood particles

Larfeldt J.[*], Leckner B.[*] and Melaaen M. Chr.[**]
[*]*Department of Energy Conversion, Chalmers University of Technology, S-412 96 Göteborg, Sweden, +46 31 7721430, +46 31 7723592 (fax),*
[**]*Telemark Institute of Technology, Porsgrunn, Norway, +47 35 575286, +47 35 575100 (fax).*

ABSTRACT: Internal temperature distribution and mass were measured during drying and pyrolysis of cylindrical samples of wood of 0, 14 and 44% moisture. A one-dimensional model of drying and pyrolysis is modified to reflect the anisotropy of the wood. Inclusion of an instant axial convective mass flow is shown to reduce the time of conversion compared to simulations with no axial flow. This mass flow, contrary to a convective mass flow through a porous structure, is not in thermal equilibrium with the solid phase. The gas leaves at a lower temperature compared to the temperature of the solid phase and is thus neglected in the energy equation.

Intrinsic gas permeabilities from literature were found to produce unrealistically high interior pressures. Simulations show that the permeability of liquid has a large influence on the time of conversion of the moist wood samples. Liquid permeabilities 10^4 times lower than the gas permeability were found. There was a discrepancy between measured and simulated temperature gradient found for moist samples probably explained by radial variations in material properties and structural dependence of the heat transfer properties.

BACKGROUND

This work concerns drying and pyrolysis of thermally thick wood particles. The work applies to, for instance, log firing in small scale boilers for house heating. In these boilers the rate of combustion is governed by each wood log. The boundary condition for a log is given by the design of the furnace and by the surrounding logs. In previous work structural changes during pyrolysis were included, which was shown to reduce the time of pyrolysis substantially [1]. Furthermore, the heat transfer properties of charcoal, produced from the same samples as those treated here, were investigated [2, 3]. These properties become important for a correct description of the rate of pyrolysis in the interior of large biomass particles. In earlier work the influence of anisotropy of wood on the convective flow of gas was neglected. In this work the drying process is modelled together with the pyrolysis and compared with measurements. The one-dimensional model is modified to reflect the anisotropy of the convective flow in wood, especially important during drying.

EXPERIMENTAL

WOOD SAMPLES AND EXPERIMENTAL PROCEDURE

Cylindrical pieces of birch wood with a radius of 25 mm, a length of 300 mm, moisture contents of 44%, 14% and 0% and an average dry density of 410 kg/m^3 were pyrolysed in an inert atmosphere using a furnace wall temperature of 700 °C. Initially the radiative heat flux to the sample was approximately 50 kW/m^2, decreasing as the sample surface temperature increase. The wood pieces were turned from stems of approximately 150 mm diameter, the centre of the stem coinciding with the centre of the resulting cylinders. The samples were heated in an electrical furnace with an inner diameter of 15 cm and a height of 110 cm. The experiments were separated into two types carried out in sequence, measurements of internal temperature distribution and mass, since it was found that the thermocouples mounted in the wood samples interfered with the mass measurement. The temperature was measured in five radial positions in the interior of the wood piece and at the surface by interior thermocouples mounted in axially drilled holes with a length of 150 mm. Due to difficulties to accomplish straight bore holes, the positions of the thermocouples were determined by X-ray analysis [4]. The normalised mass was given by an amplified voltage signal from a strain gauge.

RESULTS FROM EXPERIMENTS

The measured temperature distributions for 44, 14 and 0 % moisture are shown in Figure 1a-c. The profiles represent three or, in the case of dried wood, two measurements. In case of 44% moisture the total time of heating was approximately 2000 seconds, for 14 % moisture 1000 seconds and for dry wood 600 seconds. Drying plateaus develop in the wet samples at a temperature of around 100°C. The wood cylinders were converted into charcoal samples of the same length, approximately 300 mm, with a radius of 20±2 mm for all moisture contents. The samples cracked at relatively regular intervals. The cracking was more intense in the case of 44 % moisture, and the charcoal structure "fell apart" completely as the sample holder was removed. This was not the case for the dry sample and the sample of 14% moisture, where cracks were formed across the cylinder, distributed in the axial direction at about 20 mm distance. The structural changes during drying and pyrolysis are related to the initial moisture content of the sample and to the structure of the wood. The intense cracking in the case of 44 % moisture is also seen as irregularities in the interior temperature profiles at 21, 18 and 8.5 mm radii in Figure 1a. Also in the case of dry wood an irregularity is seen for the profile at 20 mm radius. All irregularities occur in the temperature range of 200 to 300 °C, where the pyrolysis is initiated. The irregularities are explained by sudden changes in the position of the thermocouples as cracks are formed during pyrolysis.

The instantaneous mass of a total of nine samples, normalised to their initial mass, are shown in Figure 1d. The time of pyrolysis is in agreement with the time of heating reported above; 500 seconds for dry wood, 1000 for 14 % moisture and 2000 seconds for samples of 44% moisture. In the case of dry samples and samples of 44 % moisture the measured masses are in complete agreement but at 14 % moisture there is a certain scatter. The dry samples have a char yield of approximately 20% of the initial dry mass. Assuming a char yield of 20%, as in the dry case, the total mass remaining in the case of 14 and 44 % moisture should be 11 and 17% respectively. This agrees with Figure 1d.

MATHEMATICAL MODEL

A model describing one-dimensional heat transfer in porous material [5] is used and modified to study drying and pyrolysis of large wood particles. The model solves the conservation equations of energy, solid phase, liquid water, bound water and the concentrations of species in the gas phase of the porous material. The phases are assumed to be in thermal equilibrium and the variables are calculated as volume averages [6].

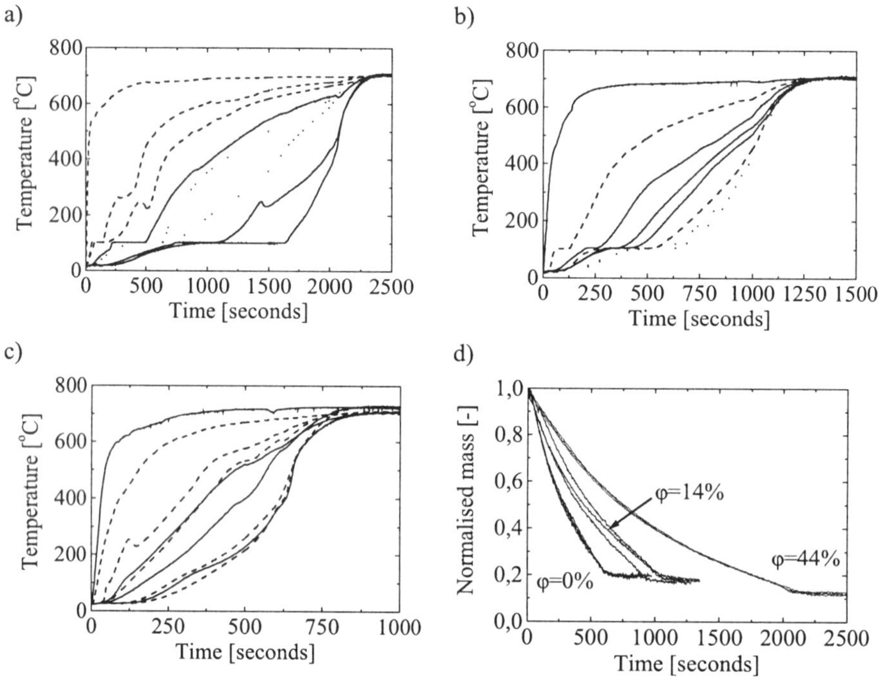

Fig. 1. Measured internal temperature in wood samples (sample no. I – solid, II - dashed and III – dotted lines) of;

a. 44% moisture at the radial positions of 25, 21, 18, 17, 14, 14, 8.5 and 6 mm.
b. 14 % moisture at the radial positions of 25, 21, 18, 15, 10, 9, 7 and 2 mm.
c. 0 % moisture at the radial positions of 25, 23, 20, 16, 15, 11, 5, 2 and 2 mm.
d. Measured normalised total mass of nine wood samples; three of 0%, three of 14% and three of 44% moisture

CONSERVATION EQUATIONS

The conservation of solid phase is written:

$$\frac{\partial\langle\rho_s\rangle}{\partial t}=\omega_p-\frac{\langle\rho_s\rangle}{V}\frac{\partial V}{\partial t} \tag{1}$$

where char contributes to the solid phase and is formed from the reacting wood during pyrolysis, ω_p, at a given char yield, χ. The third term on the right hand side accounts for structural changes during pyrolysis (described below). The concentration of species j (j=gas and tar from pyrolysis, water vapour from drying and air) in the gas phase is calculated from:

$$\frac{\partial}{\partial t}\left(\varepsilon_g\langle\rho_j\rangle^g\right)+\frac{1}{r}\frac{\partial}{\partial r}\left(ru_r^g\langle\rho_j\rangle^g\right)$$

$$=\frac{1}{r}\frac{\partial}{\partial r}\left(rD_{eff}\langle\rho_g\rangle^g\frac{\partial\left(\langle\rho_j\rangle^g/\langle\rho_g\rangle^g\right)}{\partial r}\right)+\omega_j \tag{2}$$

$$-\frac{\varepsilon_g\langle\rho_j\rangle^g}{V}\frac{\partial V}{\partial t}-\frac{2u_z^g\langle\rho_j\rangle^g}{H}$$

where the last two terms on the right hand side estimate the shrinking and the axial (i.e. 2D) flow of gases in the samples (discussed elsewhere [7]). The diffusion of gas species is neglected in the axial direction [7]. Water is described by convection of liquid water and diffusion of bound water in the wood structure as:

$$\frac{\partial}{\partial t}\left(\langle\rho_l\rangle+\langle\rho_b\rangle\right)+\frac{1}{r}\frac{\partial}{\partial r}\left(ru_r^l\langle\rho_l\rangle^l\right)=\frac{1}{r}\frac{\partial}{\partial r}\left(rD_b\langle\rho_w\rangle\frac{\partial\left(\langle\rho_b\rangle/\langle\rho_w\rangle\right)}{\partial r}\right)$$

$$+\omega_l+\omega_b-\frac{\left(\langle\rho_l\rangle+\langle\rho_b\rangle\right)}{V}\frac{\partial V}{\partial t}-\frac{2u_z^l\langle\rho_l\rangle^l}{H} \tag{3}$$

Also here terms are included to account for shrinking and axial flow of liquid water. The axial diffusion of bound water is neglected in the equations [7]. The temperature, T, is solved from the non-conservative form of the energy equation:

$$\left(c_{Pg}\varepsilon_g\langle\rho_g\rangle^g+c_{Pl}\langle\rho_l\rangle+c_{Pl}\langle\rho_b\rangle+c_{Ps}\langle\rho_s\rangle\right)\frac{\partial\langle T\rangle}{\partial t}$$

$$+\left(c_{Pg}\langle\rho_g\rangle^g u_r^g+c_{Pl}\langle\rho_l\rangle^l u_r^l-c_{Pl}\langle\rho_w\rangle D_b\frac{\partial\left(\langle\rho_b\rangle/\langle\rho_w\rangle\right)}{\partial r}\right)\frac{\partial\langle T\rangle}{\partial r} \tag{4}$$

$$=\frac{1}{r}\frac{\partial}{\partial r}\left(rk_{eff}\frac{\partial\langle T\rangle}{\partial r}\right)-(\dot{\omega}_l+\dot{\omega}_b)\Delta h_v+\dot{\omega}_p\Delta h_p-q_z$$

It has been shown that shrinking will not appear in the energy equation [1]. The axial flow of gas is included in the energy equation as the last term on the right hand side.

CONVECTIVE FLOWS

The velocities of liquid water and gas are calculated from Darcy's law, and the radial velocities are expressed as:

$$u_r^g = -\frac{\kappa_r^g \kappa_r^{g,rel}}{\mu_g} \frac{\partial \langle P_g \rangle^g}{\partial r} \tag{5}$$

$$u_r^l = -\frac{\kappa_r^l \kappa_r^{l,rel}}{\mu_l} \frac{\partial \left(\langle P_g \rangle^g - \langle P_{cap} \rangle \right)}{\partial r} \tag{6}$$

The axial velocities are given by:

$$u_z^g = -\frac{\kappa_z^g \kappa_z^{g,rel}}{\mu_g} \frac{\left(\langle P_g \rangle^g - \langle P_g^{surf} \rangle^g \right)}{H/2} \tag{7}$$

$$u_z^l = -\frac{\kappa_z^l \kappa_z^{l,rel}}{\mu_l} \frac{\left(\left(\langle P_g \rangle^g - \langle P_{cap} \rangle \right) - \left(\langle P_g^{surf} \rangle^g - \langle P_{cap}^{surf} \rangle \right) \right)}{H/2} \tag{8}$$

DRYING

Liquid water, sometimes referred as free liquid water, is present in the porous structure at moisture contents above the fibre saturation point, here assumed constant at all temperatures: 0.2 kg water per kg dry wood. The rate of evaporation in Equation (3), $\omega_b+\omega_l$, is indirectly solved from Equation (2) for water vapour, j=H_2O. In the presence of liquid water equilibrium vapour pressure is assumed in the gas phase described as [8]:

$$\langle p_{H_2O} \rangle^{g,sat} = 101325 \exp\left(17.58 - 5769/T - 5.686 \cdot 10^{-3} T\right) \tag{9}$$

At moisture contents below the fibre saturation point the vapour pressure sinks as [9]:

$$\langle p_{H_2O} \rangle^{g} = \langle p_{H_2O} \rangle^{g,sat} \left(1 - \left(1 - X_b / X_{fsp}\right)^{6.453 \cdot 10^{-3} T}\right) \tag{10}$$

The capillary pressure in Equation (6 and 8) is calculated as [10]:

$$P_{cap} = 1.364 \cdot 10^5 \left(0.128 - 1.85 \cdot 10^{-4} T\right)\left(X_l + 1.2 \cdot 10^{-4}\right)^{0.63} \tag{11}$$

PYROLYSIS

Four independent, parallel reactions with kinetic data, K_1-K_4, empirically derived from measurements of the mass loss of small samples of birch wood [9] have previously been shown to describe the pyrolysis reaction [1].

$$Wood \xrightarrow{K1+K2+K3+K4} f(1-\chi)Gas + (1-f)(1-\chi)Tar + \chi Char \quad (12)$$

The four reactions reflect the constituents of birch; two types of hemicellulose (13% and 36%), cellulose (39%) and lignin (11%) [9]. The rate of pyrolysis, ω_p, is given as the sum of these four reactions. The pyrolysis reactions result in gas, tar (f=0.68) and char according to the right hand side in Equation (12) and the production terms for Equation (2) are found as $\omega_{gas} = f(1-\chi)\omega_p$, and $\omega_{tar} = (1-f)(1-\chi)\omega_p$.

STRUCTURAL CHANGES

A shrinking model based on three empirical shrinking parameters describes the evolution of the solid and gas volumes during pyrolysis [11] as,

$$V_s = V_s^i \frac{(\langle \rho_w \rangle + \alpha \langle \rho_c \rangle)V}{\langle \rho_s^i \rangle V^i} \quad (13)$$

$$V_g = V_g^i(\eta + (1-\eta)\gamma) + \beta(V_s^i - V_s) \quad (14)$$

The shrinking parameters; α, β and γ (1.0, 0.5 and 0.66 in all cases treated) are related to the final sample radius, r_c, and char yield found in the experiments [1]. The volumes of gas and solid are related to the initial volumes of the wood sample. The shrinking during drying is neglected, motivated by results from [12]. The degree of pyrolysis, η, is calculated as the mass of virgin wood minus the char fraction divided by the initial mass of volatiles.

ADDITIONAL EQUATIONS

The gas pressure is calculated for an ideal gas and the gas density from the sum of gas species:

$$\langle P_g \rangle^g = \langle \rho_g \rangle^g R_0 \langle T \rangle / M_g \quad (15)$$

$$\langle \rho_g \rangle^g = \sum_{j=1}^{N_g} \langle \rho_j \rangle^g \quad (16)$$

The sum of the volume fractions of all phases is equal to 1:

$$\varepsilon_g + \varepsilon_l + \varepsilon_b + \varepsilon_s = 1 \tag{17}$$

The liquid and bound water volume fractions are calculated from the water content divided by the intrinsic density of water (1000 kg/m^3). The volume fraction of gas, i.e. the porosity, is given both by the gas law (i.e the gas density in Equation 14) and the volume of gas (i.e. structural changes arising from Equation 13). The solid volume fraction is calculated from Equation (12).

SOLUTION PROCEDURE

From Equations (1) to (16) the solid density, liquid water density, bound water density, the concentrations of species in the gas phase, temperature, gas density, pressure, solid, liquid and gas phase volumes in a cylinder of wood are calculated versus drying and pyrolysis time. The convective terms in Equations (2) and (6) are discretised by first-order upwinding, while central differentiation is used for the diffusive terms. Grid independence was tested. The time integration is managed by the code DASSL [13] that implicitly solves a set of differential and algebraic equations. The effect of the shrinking sample volume on the gas velocity was found to be negligible and is therefore not included in the equations. The temperature of the sample's surface is measured during the experiments and is given as the boundary condition in the simulations. The pressure at the surface equals the surrounding atmospheric pressure, also given as a boundary condition.

RESULTS AND DISCUSSION

AXIAL FLOW OF GAS

The permeability of hardwoods (such as birch treated here) is 10^4 to 10^8 times higher in the axial direction than in the radial and tangential directions [14]. Therefore the axial flow was included in the conservation equations (Equation 2, 3 and 4). A discussion on the anisotropy of wood, the inclusion of axial convective terms and the omission of axial diffusive terms are presented in [7]. In the conservation of gas species and of liquid phase the axial flow was estimated by Equation 7 and 8 assuming, for simplicity, a linear pressure gradient in the axial direction of the sample. For the energy equation (Equation 4) two cases were studied, q_z=0 and an axial heat loss according to:

$$q_z = 4c_{Pg}\langle\rho_g\rangle^g u_z^g \frac{\langle T^{surf}\rangle - \langle T\rangle}{H} \tag{18}$$

where T^{surf} is the temperature of the gas leaving the sample. Figure 2a shows the measured case of dry wood and three simulated interior temperatures for Case 1, 2 and 3 in Table 1. Case 1 corresponds to a case without axial flow. Including axial flows in Equation 2 and 3 and using Equation 18 in Equation 4 the time of conversion is seen to increase, Case 2 in Figure 2a. If, instead, the axial flow in the energy equation is neglected the time of conversion is decreased and the simulated case shows a better fit to experimental data. Figure 2b shows the measured case of 14% moisture and two simulated interior temperatures for Case 4 and Case 5 in Table 1. Neglection of the axial flow in the energy equation also here is shown to agree with measured data, while Equation 18 overestimates the conversion time of the sample.

In theory, all convective flows will increase the conversion time since energy is released from the sample. A convective flow of instantaneously released gas through cracks formed during pyrolysis is illustrated in Figure 3. In these cracks the gas is no longer in thermal equilibrium with the solid structure. The gas temperature is thus lower at instant convection compared to convective flow through the porous structure, which explains the negligible influence on the energy equation. Another explanation may be that the energy loss due to axial flow of gas is compensated for by an axial conduction of heat in the samples not accounted for in the model. From comparison of Case 1 and 3 in Figure 2a it is shown that including an instantaneous convective mass flow will reduce the time of conversion compared to the case of no axial flow.

Table 1. Simulated cases

Case	Moisture	Axial flow in Eq. 2 and 3	Axial flow in Eq. 4; q_z	Charcoal radius, r_c [mm]	κ_z^l [m^2]
1	0%	No	-	18	-
2	0%	Yes	Eq. 18	18	-
3	0%	Yes	0	18	-
4	14%	Yes	Eq. 18	20	-
5	14%	Yes	0	20	-
6	44%	Yes	0	20	10^{-13}
7	44%	Yes	0	20	10^{-14}

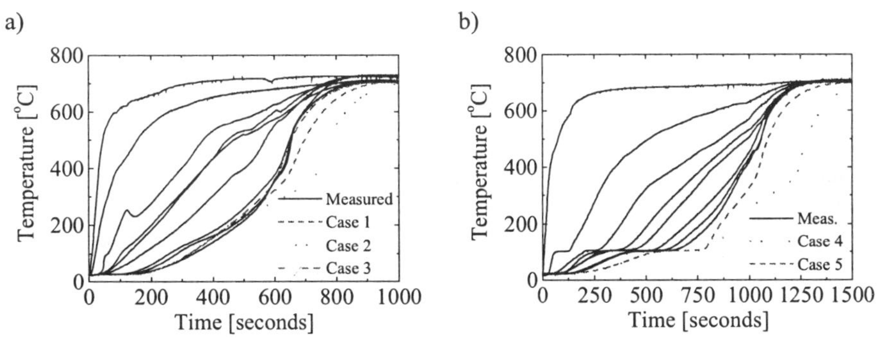

Fig. 2. Influence from axial flow. Comparison of

a. measured temperatures (solid lines) from Figure 1c and simulated centre temperatures of the wood sample for Case 1,2 and 3, Table 1, 0% moisture.
b. measured temperatures (solid lines) from Figure 1b and simulated centre temperatures of the wood sample for Case 4 and 5, Table 1, 14% moisture.

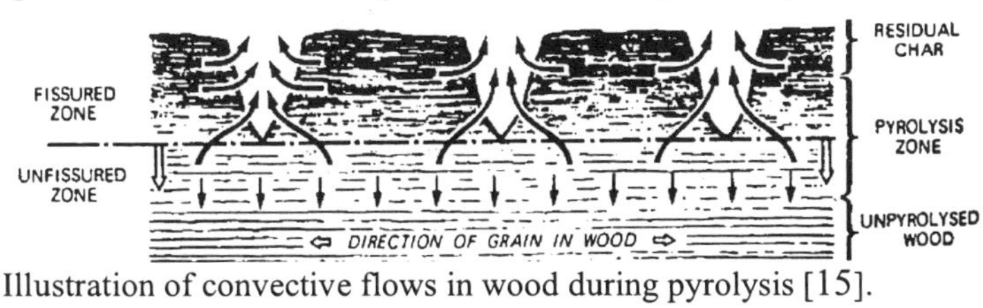

Fig. 3. Illustration of convective flows in wood during pyrolysis [15].

INFLUENCE FROM GAS PERMEABLILTY ON VAPOUR PRESSURE

Figure 4a shows simulated temperatures, Case 5, compared to measurements. The model fails to predict the drying plateaus at the sample's surface – but in the interior of the sample the drying plateaus agree with the measured ones. Figure 5a shows the drying plateaus in a larger scale. All measured plateaus fall in the range of 100 to 105°C, which corresponds to vapour pressures of 1 to 1.2 times the atmospheric pressure in the structure. The intrinsic permeability of gas in Case 5 was 10^{-10} in the axial direction and 10^{4} times lower in the radial direction. This resulted in interior pressures in the same range as seen in measurements. Also for the case of 44% moisture the measured drying plateaus, shown in Figure 5b, fall in the 100-105°C range. The intrinsic gas permeability is thus similar in the case of 44 % moisture. The intrinsic gas permeability of 10^{-10} found in the simulations here is, however, higher than reported values in literature of 10^{-11} to 10^{-14} in the axial direction and 10^{-15} to 10^{-17} in the radial direction [9]. The lower intrinsic permeability resulted in unrealistic high vapour pressures in the structure (for instance 6 times the atmospheric pressure).

It is noted that the temperature of the drying plateaus tend to increase towards the centre of the 14% moisture samples which is not the case for the 44% moisture samples. The simulated mass is in agreement with measured data as shown in Figure 4b.

INFLUENCE FROM LIQUID PERMEABILITY ON TIME OF CONVERSION

In the case of 44 % moisture there is liquid water present in the structure and also the permeability of liquid influences the conversion time of the sample. Figure 6 shows the measured temperature profiles of Figure 1 and the two simulated Cases 6 and 7 in Table 1, simulated for liquid axial permeabilities of 10^{-13} and 10^{-14} respectively. In both cases the radial permeability is assumed to be 10^{4} lower. It is seen from Figure 6 that the intrinsic permeability of liquid has a large influence on the pyrolysis time. A higher permeability leads to a larger transport of water through the wood and less water evaporates inside the sample, which reduces the time of conversion.

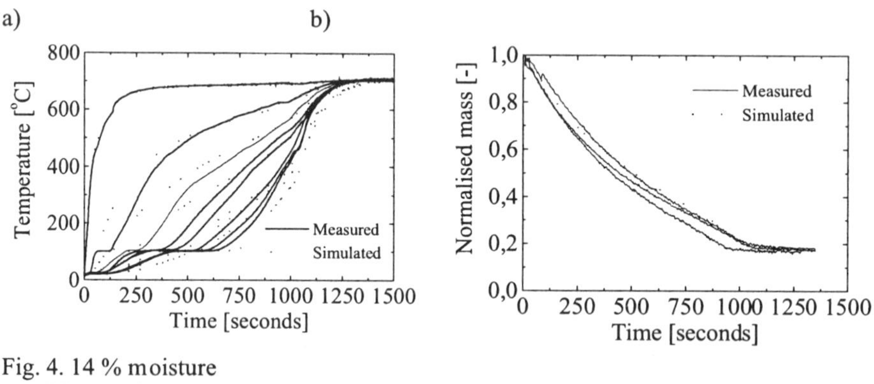

Fig. 4. 14 % moisture

a. Measured temperatures (solid lines) previously shown in Figure 1b and 2b and simulated temperatures (dotted lines) Case 5, Table 1, 14% moisture.
b. Measured normalised total mass, previously shown in Figure 1d, compared to simulated mass Case 5, Table 1, 14% moisture.

Comparison with measured data in Figure 6 shows that the liquid permeability is in the range of 10^{-13} and 10^{-14}. Case 7 for the lower liquid permeability of 10^{-14} is shown together with measured data in Figure 7. Also here the drying plateaus close to the surface of the sample were not well predicted, similar to the case of 14 % moisture. The simulated mass has a larger loss in the beginning and a slower at the end compared to the measured ones. Possibly the drying model included will overestimate the axial flow in the simulation. It is concluded that the transport of water in the wood structure has a large influence on the drying and pyrolysis processes. Experimental data on transport properties of birch wood are needed for more accurate simulations of wet fuels. Also the permeability may vary with the radius of the sample.

INFLUENCE FROM VARIOUS FACTORS

The model uses material properties and models of these properties, such as; thermal conductivity, permeability, diffusivity, specific heats, heat of pyrolysis, final sample radius and so on. The appendix gives an overview of the data used in the simulations.

a) b)

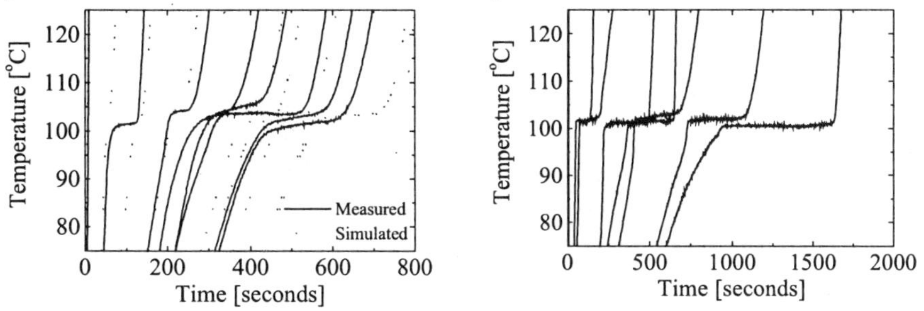

Fig. 5. Measured drying plateaus
a. and simulated (dotted lines) at 14 % moisture from Figure 4a
b. at 44 % moisture from Figure 1a.

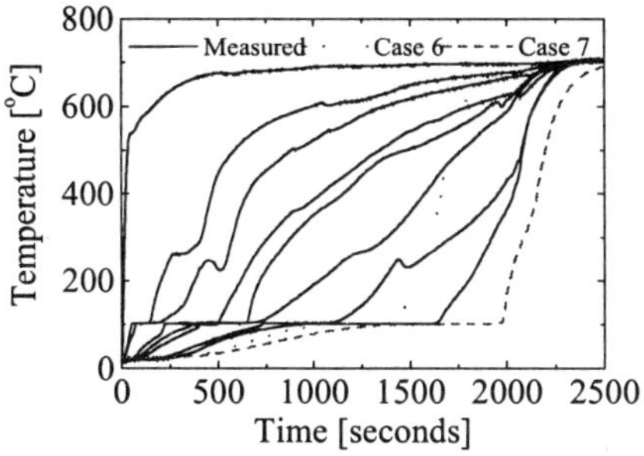

Fig. 6. Influence from liquid permeability. Measured temperatures (solid lines) from Figure 1a and 5b, and simulated centre temperatures (dotted lines) of the wood sample for Case 6 and 7, Table 1, 44% moisture.

From Figure 4a and 7a it is seen that after drying, above the drying plateau, the measured temperature gradients are steeper than the simulated ones. The material properties used in simulations are based on previous work where dry wood and charcoal were investigated. Using the data according to the Appendix, a case of dry wood is simulated, Case 3 in Table 1, and compared to measured data in Figure 8. Here the simulated and measured temperature profiles show better agreement. Also the simulated mass of the sample agrees with the measured mass of the samples.

There are several explanations for the discrepancy in measured and simulated temperatures in the initially moist samples. For instance the structural dependence of the heat transfer properties. The charcoal porosity can be divided in pores, originating from the virgin wood structure, and cracks formed during pyrolysis [3]. The formation of these pores and cracks, i.e. the structural changes, are known to vary with the heating rate during pyrolysis and the initial moisture content. The heating rate is known to vary with the sample radius. No radial variations in the structural changes or any effects of variations in the distribution between large cracks and small pores were accounted for. Large cracks may increase the radiation heat transfer in the porous structure while a high yield of pores may reduce the specific heat of char. Also the finite sample length leads to axial heating of the ends that in turn cause initial rapid drying of the ends which is not accounted for in the model.

MODEL SENSITIVITY

In earlier work a sensitivity analysis of simulations of dry wood shows an 8% reduction in time of pyrolysis at 1 mm reduction of the final charcoal radius [1]. This was shown to be in the same range as the influence from an exothermic heat of pyrolysis (150 kJ/kg). In this paper the inclusion of an axial convective flow is shown to influence the time of pyrolysis in the same range, a reduction of 6% (evaluated from Figure 2a where the times of conversion of Case 1 and Case 3 are estimated to 750 and 708 seconds respectively). For wood with a moisture content above the fibre saturation point the axial liquid permeability is shown to dominate the influence on the conversion time.

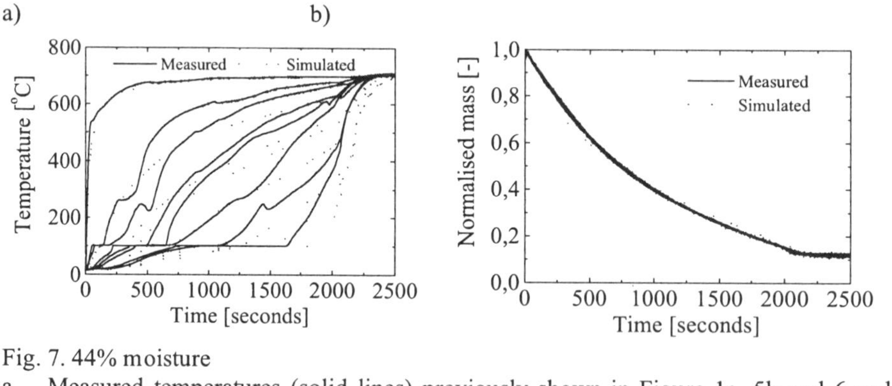

Fig. 7. 44% moisture

a. Measured temperatures (solid lines) previously shown in Figure 1a, 5b and 6 and simulated temperatures (dotted lines) for Case 7, Table 1.
b. Measured normalised total mass, previously shown in Figure 1d, compared to simulated mass for Case 7, Table 1.

a) b)

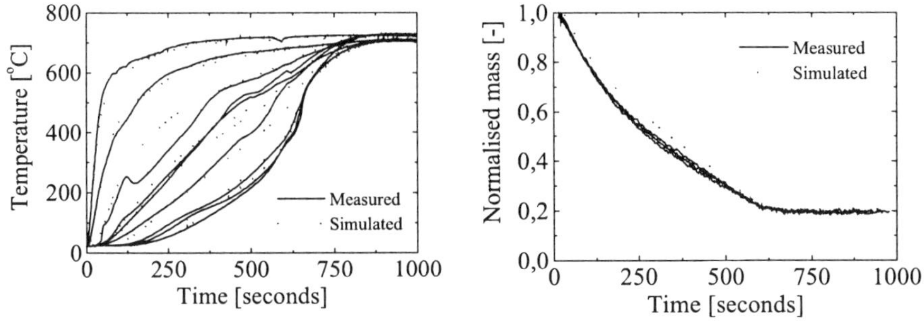

Fig. 8. Dry wood.

a. Measured temperatures (solid lines) previously shown in Figure 1c and 2a, and simulated temperatures (dotted lines) for Case 3, Table 1.
b. Measured normalised mass, previously shown in Figure 1d, compared to simulated mass for Case 3, Table 1.

CONCLUSION

The internal temperature distribution and the mass were measured during drying and pyrolysis of cylindrical samples of wood of 0, 14 and 44% moisture. A one-dimensional model of drying and pyrolysis is modified to reflect the anisotropy of wood. Inclusion of an instant axial convective mass flow is shown to reduce the time of conversion compared to simulations with no axial flow. This flow, contrary to a convective mass flow through a porous structure, is not in thermal equilibrium with the solid phase and is thus neglected in the energy equation.

Measured drying plateaus occurred in a temperature range of 100 - 105°C corresponding to vapour pressures of 1 –1.2 times atmospheric pressure. Intrinsic gas permeabilities from literature were found to produce unrealistically high interior pressures. Simulations show that the permeability of liquid has a large influence on the time of conversion of the moist wood samples. The simulations agreed with measured data for a liquid permeability 10^4 times lower than the gas permeability of 10^{-10} found.

For the moist samples, the drying plateau close to the surface of the sample and the temperature gradient above this plateau were not well predicted in the simulations. On the other hand a comparison of simulation and measurements on dry wood shows good agreement. The discrepancy for moist samples could be explained by structural dependence of the heat transfer properties.

ACKNOWLEDGEMENT

The Swedish contribution to this work was financed by the Small Scale Combustion Programme, Swedish Energy Administration.

NOMENCLATURE

c_p	Specific heat (J/kg K)	ε	Porosity (volume fraction)
D	Gas diffusivity (m^2/s)	ϕ	Phase
Δh	Enthalpy of formation (J/kg)	γ	Shrinking parameter
F	Fraction (-)	η	Degree of pyrolysis
H	Length of samples (m)	κ	Permeability (m^2)
K	Thermal conductivity (W/mK)	μ	Dynamic viscosity (kg/ms)
K	Reaction rate (1/s)	ω	Production rate (kg/m^3s)
M	Molecular mass (kg/mole)		
N	Number of species	***Index***	
P	Pressure (N/m^2)	b	Bound
Q	Heat loss (W/m^3)	c	Char
R	Radius (m)	eff	Effective
R_0	Universal gas const. (J/mole K)	fsp	Fibre saturation point
T	Temperature (K)	g	Gas
T	Time (s)	i	Initial
U	Velocity (m/s)	j	Gas specie
V	Volume (m^3)	l	Liquid
X	Moisture (kg water/kg wood)	p	Pyrolysis
		r	Radial
Greek		s	Solid
α	Shrinking parameter	sat	Saturated
β	Shrinking parameter	surf	Surface
χ	Char yield (kg char/kg wood)	w	Dry wood
ρ	Density (kg/m^3)	z	Axial

APPENDIX. DATA USED IN SIMULATIONS

Diffusivity of gas species in gas phase,

$$D^g_r = 0.15*10^{-4}(T/298)^{1.75}(P_0/P)/1000, \qquad D^g_z=20D_r$$

Diffusivity of bound water,

$$D_{b,r} = \exp(-9.9-4300/T+9.8(<\rho_b>/<\rho_w>)), \qquad D_{b,z}= 2.5D_{b,r}$$

The dynamic viscosity of gas, $\mu_g = 3.0*10^{-5}$ kg/ms, and water, $\mu_l = 5.0*10^{-4}$ kg/ms The initial sample porosity, ε_g^i, is 0.68 and the initial wood density, ρ_w^i, is 410 kg/m^3.

Table A1. Heat transfer properties used in simulations [1 and 3].

	Wood	Char	Water	Gas
k	0.20	0.09	0.658	0.026
c_p	2439	1000	4200	1000

Table A2. Data of gas and liquid permeability used in simulations.

	Intrinsic		Relative [16]		
	Wood	Char	$X^*=0$	$0<X^*<0.48$	$0.48<X^*<X_{sat}$
κ_r^g	10E-14	10E-11	1	$0.95(1-X/0.48)^2+0.05$	$0.05(X_{sat}-X)/(X_{sat}-0.48)$
κ_z^g	10E-10	10E-08	1	$1+(2X/X_{sat}-3)(X/X_{sat})^2$	
κ_r^l	10E-18	-	0	$0.95(X/0.48)^2$	$0.05(X-0.48)/(X_{sat}-0.48)+0.95$
κ_z^l	10E-14	-	0	$(X/X_{sat})^3$	

$$X^* = \frac{\langle \rho_b \rangle + \langle \rho_l \rangle}{\langle \rho_w \rangle} - X_{fsp}, \qquad X_{fsp} = 0.2, \qquad X_{sat} = \frac{\varepsilon_g^i \langle \rho_l \rangle^l}{\langle \rho_w^i \rangle} - X_{fsp}$$

REFERENCES

1 Larfeldt, J., Leckner, B. and Melaaen, M.C., "Modelling and measurements of the pyrolysis of large wood particles", Fuel, **79**, 1637-1643, 2000.

2 Suleiman, B.M., Larfeldt, J., Leckner, B., Gustavsson, M., "Thermal conductivity and diffusivity of wood", Wood Science and Technology, **6**, (33), 465-473, 1999.

3 Larfeldt, J., Leckner, B. and Melaaen, M.C., "Modelling and measurements of heat transfer in charcoal from pyrolysis of large wood particles", Biomass and Bioenergy, **18**, 507-514, 2000.

4 Green A. and Larfeldt J., "Results from measurements of drying and pyrolysis of wood cylinders", Report A98-222, Chalmers University of Technology, Göteborg 1998.

5 Melaaen M.C., "Numerical Analysis of Heat and Mass Transfer in Drying and Pyrolysis of Porous Media", Numerical Heat Transfer, Part A, **29**, 331-355, 1996.

6 Whitaker S., "Simultaneous Heat, Mass and Momentum Transfer in Porous Media: A Theory of Drying", Advances in Heat Transfer, **13**, 119-203, 1977.

7 Larfeldt J., "Drying and Pyrolysis of Logs of Wood", PhD Thesis, Department of Energy Conversion, Chalmers University of Technology, Göteborg, 2000.

8 Lai, W.-C., "Reaction Engineering of Heterogeneous Feeds: Municipal Solid Waste as a Model", PhD thesis, University of Washingtonm Seattle, 1991.

9 Grönli M., "A theoretical and experimental study of the thermal degradation of biomass", PhD Thesis, Division of Thermal Energy and Hydro Power, Trondheim, 1996.

10 Spolek G.A. and Plumb O.A., "Capillary Pressures in softwoods", Wood Sci Technol., 15, 189-199, 1981.

11 Di Blasi C., "Heat. Momentum and Mass Transport Through a Shrinking Biomass Particle Exposed to Thermal Radiation", Chem. Sci. Eng., **51**, 1121-1132, 1996.

12 Fyhr C., "Superheated Steam Drying of Wood Chips in Pneumatic Conveying Dryers", PhD Thesis, Department of Chemical Engineering Design, Chalmers University of Technology, Göteborg, 1996.

13 Petzold L. R., "A Description of DASSL: A Differential/Algebraic System Solver", Scientific Computing, Stepleman et al. (eds.), 65-68, IMACS, North-Holland, Amsterdam, 1983.
14 Siau J. F., "Transport Processes in Wood", Springer-Verlag, Series in Wood Science, 1984.
15 Roberts A.F., "Problems Associated with the Theoretical Analysis of the Burning of Wood", 13th Symp. (Int.) on Combustion, The Combustion Institute, Pittsburgh, 893-903, 1971.
16 Perré, P. And Degiovanni A., "Simulation par Volumes finis des Transferts Couple´s en Milieux Poreux Anisotropes: Séchage du Bois à Basse et à haute Temperature", Int. J. Heat Mass Transfer, 33, 2463, 1990.

Thermal Analysis and Kinetic Modelling of Wheat Straw Pyrolysis

M. Stenseng, A. Jensen & K. Dam-Johansen
Department of Chemical Engineering, Technical University of Denmark, Lyngby, Denmark

ABSTRACT
Simple kinetic models have been tested to fit the weight loss during pyrolysis of wheat straw under different temperature programs. The models include a single first order reaction, a three parameter nucleation model, a distributed activation energy model and a superposition model based on parallel first order reactions. Although the distributed activation energy model appears to give the best fit to data, it is still concluded that a biomass pyrolysis model valid over a wide range of process conditions is not available in the literature.

INTRODUCTION

The use of straw in power plants in Denmark has resulted in a need for research in the pyrolysis of straw. Very little regarding straw pyrolysis and straw pyrolysis kinetics are reported in the literature. A few thermogravimetric studies of the pyrolysis of biomass are reported in the literature [1,2,3,4,5], which include wheat straw. In those studies, the pyrolysis experiments were carried out at one set of experimental conditions only. No investigation of the influence of e.g. heating rate and final temperature has been found in the literature. In some of the studies [1,3,5] demineralized (washed) wheat straw was investigated as well. The washing was carried out in order to remove most of the salts and minerals in the wheat straw. Varhegyi et al. [5] washed the straw in order to determine the pyrolysis kinetics by a superposition model and wished to eliminate the influence of the salts. Jensen et al. [1] investigated the influence of washing the straw and adding salts to the washed straw in order to get a better understanding of the way the salts are bound in the straw. In practice, washing of straw may become a process used in the power plants in order to remove mainly potassium to reduce the deposition and corrosion in straw fired power plants [6].

Few straw pyrolysis models have been found in the literature. A few investigators [4,5] tested a superposition model for straw pyrolysis, in which the pyrolysis of straw is described by a number of parallel reactions, typically two or three, representing the pyrolysis of the main biomass constituents. The model was found to fit the weight loss data reasonably well when fitted to data at a single heating rate. Lanzetta and Di Blasi [7] made a model in which the pyrolysis of wheat straw is described by competing parallel and consecutive reactions. They tested their model for high heating rates (1500-

4200°C/min) and temperatures below 375°C, and found reasonable good agreement with data.

The main objective of this paper is to test simple kinetic models for the pyrolysis of wheat straw. In order to obtain well-defined conditions, especially regarding the sample temperature, low heating rate experiments in a thermogravimetric analyzer was preferred.

MATERIALS AND EXPERIMENTAL PROCEDURES

SAMPLES

The samples used in this study were Danish wheat straw pellets which were milled on a laboratory ultra centrifugal mill (Retzsch ZM 100) mounted with a 1 mm screen. In addition, washed wheat straw was used. The washing procedure was as follows: 1 g of milled straw was placed in 200 ml deionized water and stirred for 20 hours at room temperature, then filtered and washed with 300 ml deionized water, before drying [1,6,8].

EXPERIMENTAL SETUP AND PROCEDURES

The experiments were carried out in a Netzsch STA 409 C (Simultaneous Thermal Analysis - STA) in the TGA/DSC configuration. The STA has a vertical sample carrier with a reference and a sample crucible, and in order to account for buoyancy effects, a correction curve with empty crucibles was first conducted and then subtracted from the actual experiments. Platinum/Rhodium crucibles were used in order to get the best possible heat transfer. The thermocouple for each crucible was positioned just below and in contact with the crucible. The temperature obtained from the measurement is the temperature in the reference side. This temperature is converted to the temperature in the sample side by using the DSC-signal in μV and a temperature-voltage table for the thermocouple. The product gases were swept away by 100 Nml/min nitrogen which exited the top of the STA. The STA was calibrated for temperature and sensitivity (DSC) with metal standards at each heating rate.

Samples of 5 mg straw were pyrolyzed to a maximum temperature of 700°C. The influence of sample mass for the pyrolysis of wheat straw and washed wheat straw has previously been investigated for samples between 2 and 20 mg at a heating rate of 40°C/min [9], and it was found that the sample mass had no influence on the pyrolysis process. Heating rates of 5, 10 and 40°C/min was applied. Two types of temperature programs were used. In both types of temperature programs the sample was first heated to 110°C and kept at that temperature for 30 min in order to remove the moisture. In the first temperature program type, the sample was heated at a specified heating rate to 600°C. In the second temperature program the sample was heated at 10°C/min to 300°C, kept isothermal at that temperature for 60 min before heating at 10°C/min to 600°C. In both types of temperature programs the sample was then kept isothermal at the final temperature for 30 min before burning off the char in a nitrogen/oxygen mixture in order to determine the ash content.

RESULTS

The pyrolysis of wheat straw and washed wheat straw show different pyrolysis behavior. Besides a lower onset temperature of the pyrolysis and a higher char yield of the wheat straw, the main difference is the shape of the mass loss rate curves. The pyrolysis of wheat straw show an asymmetrical peak with a long tail, see Fig. 1. However, the pyrolysis of washed wheat straw shows two peaks, corresponding to hemicellulose and cellulose decomposition, and a tail, corresponding to lignin decomposition.

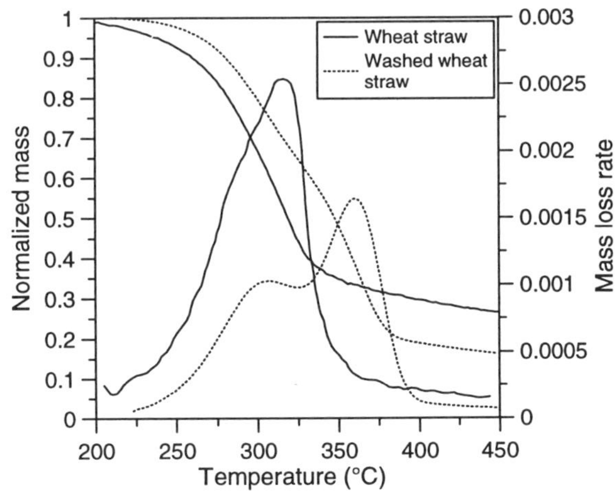

Fig. 1 The pyrolysis of wheat straw and washed wheat straw at 10°C/min.

The shape of the mass loss and mass loss rate curves determine the types of kinetic models that can be applied to fit the data. Simple kinetic models, which only include one reaction, may be applied to describe the pyrolysis of wheat straw, but cannot describe the two peaks observed in the pyrolysis of washed wheat straw.

Different kinetic models have been applied to fit mass loss data, including (with the parameters of each model in parenthesis):

1. a single first order reaction (A, E)
2. a three parameter nucleation model (A, E, m)
3. a distributed activation energy model using a natural logarithmic distribution (A, α, β)
4. a superposition model based on n first order reactions (c_i, A_i, E_i, i=1,n)

All four models were used to fit the pyrolysis of wheat straw, but only the superposition model was used to fit the pyrolysis of washed wheat straw. All four models have the flaw that they are not able to predict the char yield, which then has to be given as input to the models. Prediction of char yield can only be done by more complicated models.

The most simple model is the single first order reaction, which is described by:

$$\frac{dX}{dt} = k(1 - X) \tag{1}$$

in which X is the fractional weight loss normalized with the ultimate weight loss, and the rate constant (k) is given by the Arrhenius equation ($k = A\exp(-E/R_gT)$). An algorithm to fit the experimental data to Eqn. (1) was made in Fortran using a least squares method by Fletcher [10], in which the rate constant was expressed as follows:

$$k = k^*\exp\left(-\frac{E}{R_g}\left[\frac{1}{T} - \frac{1}{T^*}\right]\right) \tag{2}$$

This parameterization gives a lower coupling between the parameters E and k* than E and A in the normal Arrhenius equation. Note that k* is the rate constant at T*. The algorithm was used to fit kinetic constants to the pyrolysis of wheat straw at 5, 10 and 40°C/min (one data set per heating rate). The algorithm use the local temperature and does not rely on a constant heating rate. The data from an experiment were converted to dry ash free basis and the mass loss rate was normalized by the maximum mass loss rate. The data in the range where the normalized mass loss rate was above 0.1 was then used. This excludes the lignin tail from the data. The mass data were then converted to degree of conversion and normalized so the conversion of the final data point was 1. 300 points were used per data set. Kinetic parameters were fitted to the individual data sets as well as to all three data sets simultaneously . The kinetic values are listed in *Table 1*.

Table 1 Kinetics derived for individual heating rates as well as for all three heating rates for the pyrolysis of wheat straw using a single first order reaction model.

Heating rate (°C/min)	E (kJ/mol)	A (s^{-1})	$k_{277°C}$ (s^{-1})
All	137.9	$2.40 \cdot 10^{10}$	$1.92 \cdot 10^{-3}$
5	80.9	$6.11 \cdot 10^{4}$	$1.25 \cdot 10^{-3}$
10	78.5	$5.42 \cdot 10^{4}$	$1.91 \cdot 10^{-3}$
40	69.5	$2.37 \cdot 10^{4}$	$5.95 \cdot 10^{-3}$

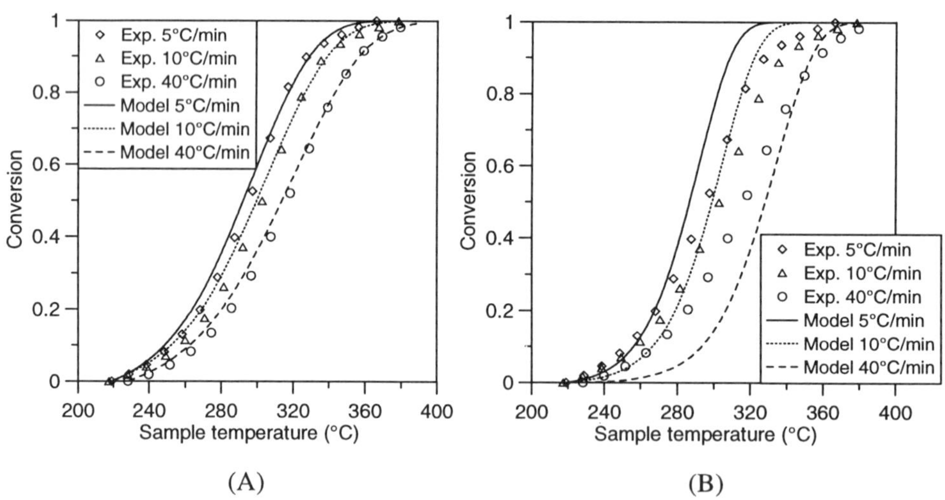

Fig. 2 The pyrolysis of wheat straw at 5°C/min, 10°C/min and 40°C/min. Single first order kinetics fitted to the individual curves (A) and simultaneously to all curves (B).

Table 1 shows that there is some difference between the kinetics fitted to the individual data sets. The activation energies for the pyrolysis at 5 and 10°C/min are close to 80 kJ/mol, but the activation energy for the pyrolysis at 40°C/min is approximately 10 kJ/mol lower. The conversion for all three data sets as well as the individual fits are presented in Fig. 2A. The model agrees reasonably well with the data. When the kinetics are fitted to all three data sets simultaneously, both the frequency factor and the activation energy increases dramatically. Fig. 2B shows the conversion for all three data sets as well as the kinetic from the simultaneous fit. The activation energy increases approximately by 60 kJ/mol to 140 kJ/mol and the frequency factor increases by a factor 10^5-10^6. The fit does not show as good agreement with the data as the individual fitted kinetics.

In order to test the applicability of the model over a wider range of conditions the kinetics derived at 10°C/min (from *Table 1*) was used to predict the pyrolysis of wheat straw heated to 300°C at 10°C/min, then kept isothermal for 60 min, before heating to 600°C at 10°C/min, see Fig. 3. The experimental data show that the addition of a 60 min hold time at 300°C results in somewhat different mass loss curve. The initial mass loss follows that of the data heated directly to 600°C, but when 300°C is reached, the mass loss rate becomes much lower, and after 60 min at 300°C, the conversion is only about 0.8. Full conversion is reached when the temperature is raised to 600°C. The char yield for the two experiments was approximately the same. The model, which fit the direct heating to 600°C reasonably well, does not work for the case with a 60 min isothermal segment at 300°C is inserted. The model predicted that the pyrolysis was far too fast, and a full conversion is predicted under the isothermal segment. This shows that first order kinetics only works well on the data on which it was fitted. A single first order model cannot predict the pyrolysis behavior under different conditions.

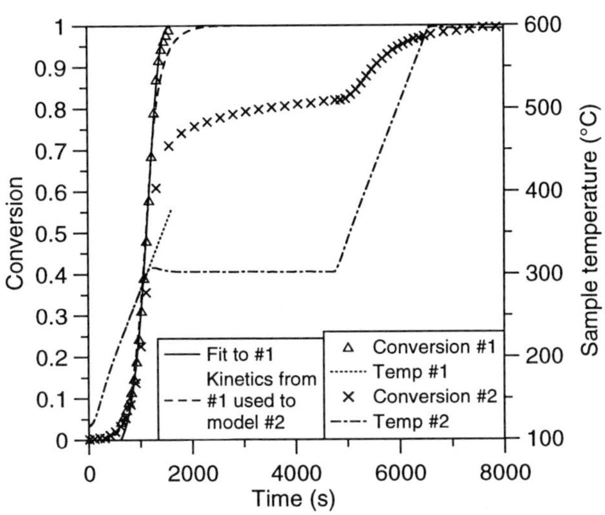

Fig. 3 The experimental data for the pyrolysis of wheat straw at 10°C/min to 600°C (#1) and to 300°C for 60 min before 600°C (#2). Model data fitted to #1 and used to model #2.

Reynolds and Burnham [11] found that a three parameter nucleation model fitted the pyrolysis of cellulose well. Here, the model was tested for the pyrolysis of wheat straw. The conversion rate was given by:

$$\frac{dX}{dt} = k(1-X)(1-0.99X)^m \qquad (3)$$

An algorithm similar to the one used for a single first order reaction was used to fit data to this model. The kinetics derived by Eqn. (3) is listed in *Table 2* for fits to the same data as used to fit the single first order reaction. Fig. 4A and 4B show the fit to the individual curves and the fit to the curves simultaneously. The fits are reasonable good, but not better than the fits using a single first order reaction. In fact, the fit to all three heating rates simultaneously gave a fit with a value of m close to zero, in which case the model reduces to a first order reaction.

Table 2 Kinetics derived for individual heating rates as well as for all three heating rates for the pyrolysis of wheat straw using the model by three parameter nucleation model.

Heating rate (°C/min)	E (kJ/mol)	A (s^{-1})	m	$k_{277°C}$ (s^{-1})
All	148.2	$2.22 \cdot 10^{11}$	$4.1 \cdot 10^{-8}$	$1.86 \cdot 10^{-3}$
5	86.4	$2.39 \cdot 10^{5}$	0.169	$1.48 \cdot 10^{-3}$
10	82.8	$1.69 \cdot 10^{5}$	0.177	$2.30 \cdot 10^{-3}$
40	76.6	$1.25 \cdot 10^{5}$	0.124	$6.67 \cdot 10^{-3}$

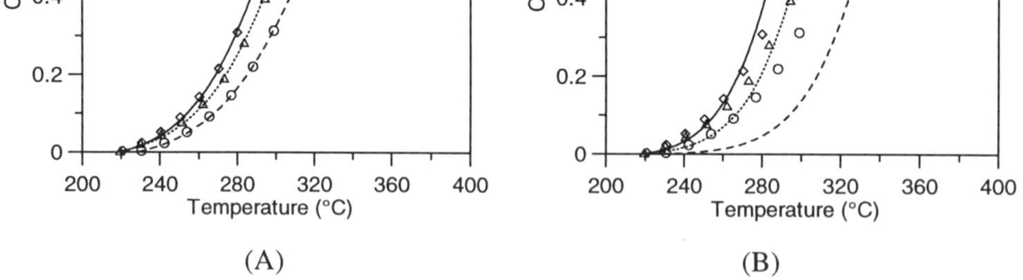

Fig. 4 The pyrolysis of wheat straw at 5°C/min, 10°C/min and 40°C/min. Three parameter nucleation kinetics fitted to the individual curves (A) and simultaneously to all curves (B).

Yet another model previously used to fit the pyrolysis of biomass is the distributed activation energy model. This model was originally used to model the pyrolysis of coal [12]. Niksa and Lau [13] found that the distributed activation energy model was the best for coal pyrolysis and that it should be included as an essential element in any pyrolysis model. The pyrolysis is assumed to occur via a number of parallel first order reactions which has the same frequency factor but varying activation energy. The pyrolysis rate for the reactions is:

$$\frac{d\alpha_i}{dt} = k_i(1 - \alpha_i) \qquad (4)$$

in which the rate constant (k_i) is given by an Arrhenius equation and the fraction of material which pyrolyses at a certain activation energy is given by a natural logarithmic distribution. In the original model of Anthony et al. [12] a normal distribution was used instead of the natural logarithmic distribution.

$$f(E) = \frac{1}{\sqrt{2\pi}}\frac{1}{\beta E}\exp\left(-\frac{1}{2}\left(\frac{\ln(E) - \alpha}{\beta}\right)^2\right) \qquad (5)$$

for which

$$\int_0^\infty f(E)\,dE = 1 \qquad (6)$$

This results in the following expression for the conversion:

$$X = \int_0^\infty \exp\left(-\int_0^t k(E)\,dt\right) f(E)\,dE \qquad (7)$$

An algorithm to fit the experimental data to this model was made in Fortran using a least square method by Fletcher [10]. In order to achieve the best result, the model was constructed to fit the frequency factor, A, and α, while the influence of β on the residual sum of squares had to be tested manually. As for the previous models, the model was used to fit data at a single heating rate and to fit data at three heating rates simultaneously. The best fit for the simultaneous fit to three heating rates was obtained with β = 0.055. This value was subsequently used to fit the data individually. *Table 3* lists the kinetic parameters and Fig. 5A and 5B shows the fit to the data curves for fits to individual curves and simultaneous fit to all three curves.

This model deviates from the previous described models by having frequency factors which are within a factor 10^3 between the kinetics derived for individual data sets and simultaneously to all three data sets. Another feature for this model is that the fit to all three data sets simultaneously gives a reasonable result. This shows that this model is better equipped to handle different experimental conditions.

Table 3. Kinetics derived for individual heating rates as well as simultaneously for all three heating rates for the pyrolysis of wheat straw using the distributed activation energy model with $\beta = 0.055$.

Heating rate (°C/min)	A (s^{-1})	α	E_{mean} (kJ/mol)	σ (kJ/mol)
all	$2.56 \cdot 10^{22}$	12.50	268.8	14766
5	$6.07 \cdot 10^{18}$	12.34	229.4	12627
10	$8.33 \cdot 10^{19}$	12.39	241.6	13298
40	$4.57 \cdot 10^{21}$	12.46	259.7	14296

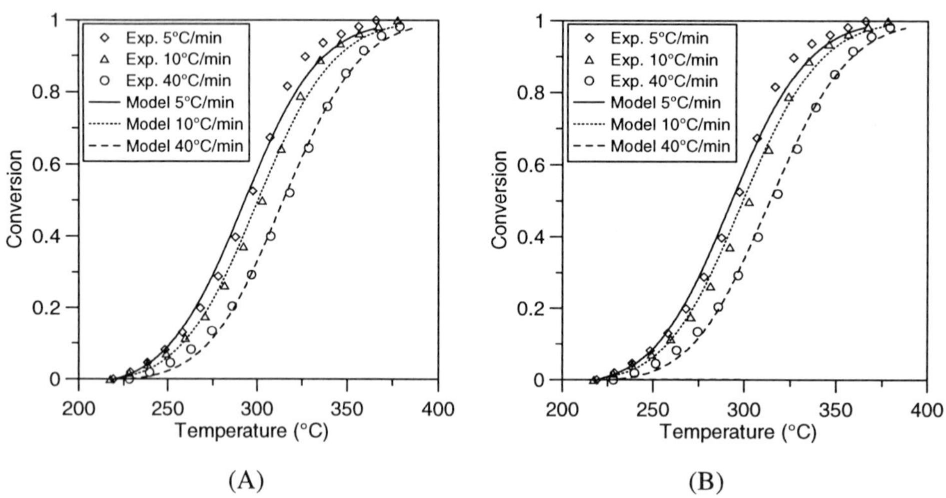

Fig. 5 Pyrolysis of wheat straw at 5, 10 and 40°C/min. Kinetics fitted to the individual curves (A) and simultaneously to three curves (B) by the distributed activation energy model.

To test the applicability of the model, the kinetics derived for the pyrolysis of wheat straw using the kinetics derived at 10°C/min (from *Table 3*) was used to predict the pyrolysis of wheat straw when the heating was delayed by inserting a 60 min isothermal segment at 300°C in the temperature program. Fig. 6 shows the experiments and the fits. Compared to the other models the distributed activation energy model performs better. It must, however, be remarked that given enough time at 300°C, also this model would predict full conversion.

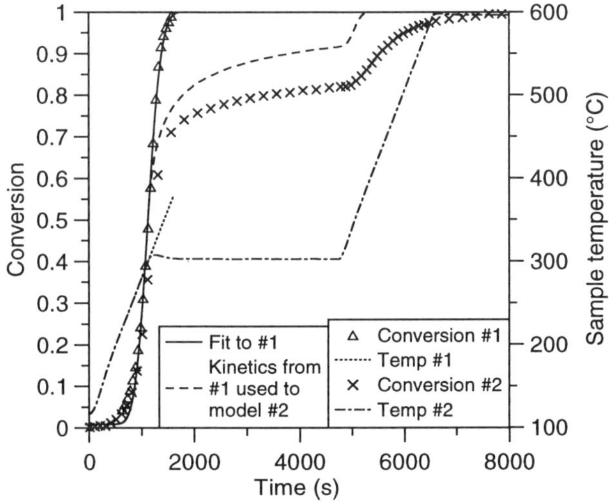

Fig. 6 The experimental data for the pyrolysis of wheat straw at 10°C/min to 600°C (#1) and to 300°C for 60 min before 600°C (#2). Model data using the distributed activation energy fitted to #1 and used to model #2.

The pyrolysis of wheat straw could be expected to be described by a superposition model, i.e. described by a number of parallel first order reactions, representing the decomposition of the biomass constituents (cellulose, hemicellulose and lignin) [4,5]. In this work, a superposition model has been used, in which the pyrolysis is assumed to be described by N independent first order reactions (for i=1, 2, ..., N):

$$X = \sum_i c_i \alpha_i \tag{8}$$

where X is the total conversion and the conversion rate is given by:

$$\frac{dX}{dt} = \sum_i c_i \frac{d\alpha_i}{dt} \tag{9}$$

The reacted fraction of component i is then given by:

$$\alpha_i = \frac{m_{0,i} - m_i}{m_{0,i} - m_{char,i}} \tag{10}$$

The components are assumed to decompose individually according to first order kinetics:

$$\frac{d\alpha_i}{dt} = k_i(1 - \alpha_i) \tag{11}$$

where the rate constant is given by the Arrhenius equation:

$$k_i = A_i \exp\left(-\frac{E_i}{R_g T}\right) \tag{12}$$

The coefficient c_i expresses the contribution of the partial process to the overall process:

$$c_i = \frac{m_{0,i} - m_{char,i}}{m_0} \tag{13}$$

The set of equations can be fitted to one data set (either TGA or DTG) by the least square method. An algorithm has been developed by Austegaard [14] and programmed in Matlab. Further information regarding the algorithm can be found in Grønli [15].

When the superposition model is used to model the pyrolysis of wheat straw, difficulties may arise because the mass loss rate curve shows an asymmetrical peak with a long tail. The absence of several distinct peaks makes the fit difficult. It is possible to achieve a good mathematical fit; however, the fit may not have a physical/chemical meaning. This may limit the fitted kinetics to be valid only at the conditions of the experiment they were fitted to. The pyrolysis of wheat straw at 10°C/min has been fitted to both mass loss data and mass loss rate data using both two and three parallel reactions. The kinetic parameters are listed in *Table 4*. All the fits describe the pyrolysis well, however, there is a large difference between the kinetic constants. Fig. 7 shows the fit to the mass loss rate using three reactions. The activation energies are very different depending on whether they are determined from mass loss or mass loss rate data. This indicates that the values of activation energy are not well determined by this method and the values of all the parameters may not have a physical meaning.

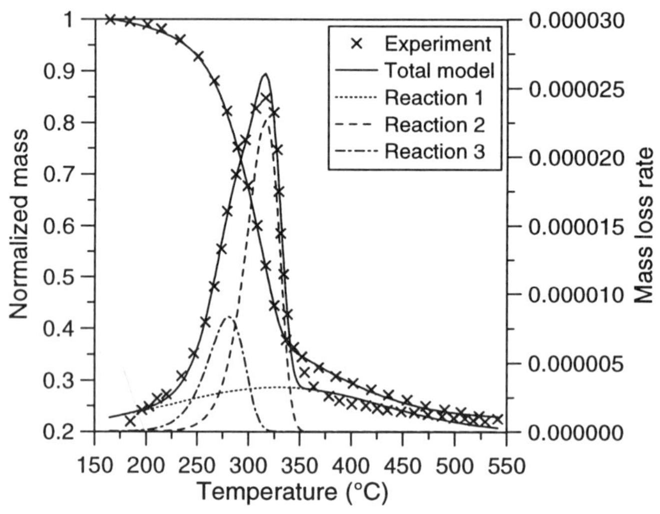

Fig. 7 The pyrolysis of wheat straw at 10°C/min fitted by three parallel reactions using the mass loss rate data.

The model contains many more parameters than the single first order reaction, and so it can only be justified by a wider range of applicability. To test this, the kinetics derived for the pyrolysis of wheat straw using three reactions and based on mass loss rate was used to predict the pyrolysis of wheat straw when the heating was delayed by inserting a 60 min isothermal segment at 300°C in the temperature program. Fig. 8 shows the experiments and the fits, and it is clear that the superposition model does not predict the case with the isothermal segment well. It is, however, slightly better than a single first order reaction (Fig. 3) due to the slower kinetics of the lignin tail.

Table 4 Kinetics for the pyrolysis of wheat straw at 10°C/min using 2 and 3 parallel reactions (for both mass and mass loss rate data).

	2 reactions (mass)	2 reactions (mass loss rate)	3 reactions (mass)	3 reactions (mass loss rate)
c_1 (%)	23.8	27	24.2	28.5
c_2 (%)	54.4	51.2	44.5	34.4
c_3 (%)	-	-	9.14	14.4
E_1 (kJ/mol)	22	20.6	23.8	24.5
E_2 (kJ/mol)	108.9	117.8	139.1	181.1
E_3 (kJ/mol)	-	-	114	139.8
A_1 (s^{-1})	$8.16 \cdot 10^{-2}$	$6.96 \cdot 10^{-2}$	$1.24 \cdot 10^{-1}$	$1.78 \cdot 10^{-1}$
A_2 (s^{-1})	$3.80 \cdot 10^{7}$	$4.13 \cdot 10^{8}$	$1.96 \cdot 10^{10}$	$1.16 \cdot 10^{14}$
A_3 (s^{-1})	-	-	$5.94 \cdot 10^{8}$	$1.34 \cdot 10^{11}$

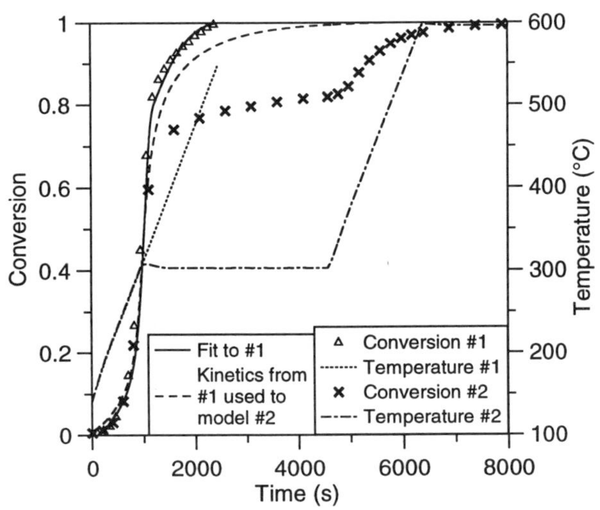

Fig. 8 The experimental data for the pyrolysis of wheat straw at 10°C/min to 600°C (#1) and to 300°C for 60 min before 600°C (#2). Model data fitted to #1 and used to model #2 using the kinetics derived from the superposition model with three reactions (based on mass loss rate data).

The superposition model was also applied for the pyrolysis of washed wheat straw, where a multi-reaction model is required to fit the data. Experiments show that the DTG-curves have two distinct peaks, which makes the kinetics determination easier than for raw straw. Fig. 9 shows the fit of three reactions to the mass loss rate data from the pyrolysis of washed wheat straw at 10°C/min. The fits show good agreement with the experimental data. The kinetic data are listed in *Table 5*. As for the pyrolysis of wheat straw, it may be expected that these kinetic values will fail to predict the pyrolysis at different conditions.

Table 5 Kinetics for the pyrolysis of washed wheat straw at 10°C/min 3 parallel reactions (for both mass and mass loss rate data).

	3 reactions (mass)	3 reactions (mass loss rate)
c_1 (%)	7.79	6.57
c_2 (%)	44.6	46.2
c_3 (%)	32.7	31.6
E_1 (kJ/mol)	68	90.4
E_2 (kJ/mol)	189.8	172.7
E_3 (kJ/mol)	114.4	111.6
A_1 (s^{-1})	$4.67 \cdot 10^2$	$3.33 \cdot 10^4$
A_2 (s^{-1})	$4.33 \cdot 10^{13}$	$1.59 \cdot 10^{12}$
A_3 (s^{-1})	$1.77 \cdot 10^8$	$1.03 \cdot 10^8$

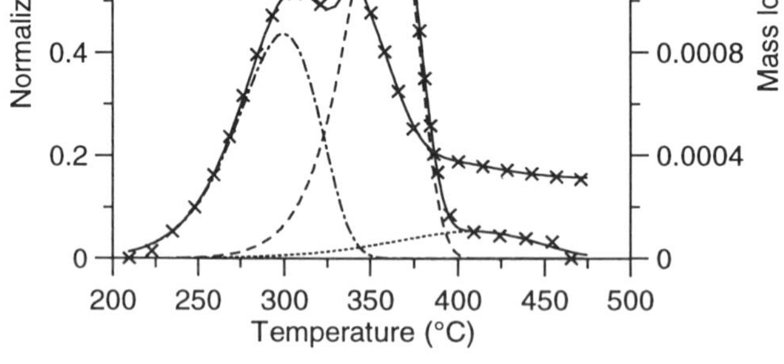

Fig. 9 The pyrolysis of washed wheat straw at 10°C/min fitted by three parallel reactions using the mass loss rate data.

The model for straw pyrolysis by Lanzetta and Di Blasi [7] was applied to the data in this study even though the conditions of the TGA-experiments are very different from the conditions used to fit the parameters in the model. The heating rate in the present study (5-40°C/min) is much lower than in the study by Lanzetta and Di Blasi [7], who used heating rates of 1500-4200°C/min, and the maximum final temperature was higher in this study (700°C) compared to 375°C by Lanzetta and Di Blasi [7]. The reaction scheme used by Lanzetta and Di Blasi [7] is given in Fig. 10.

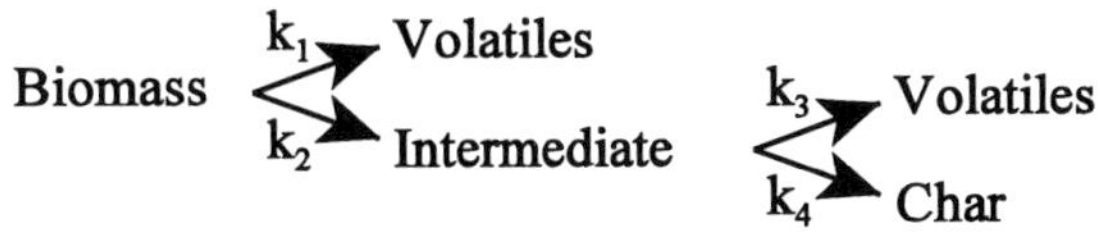

Fig. 10 Reaction scheme for the model by Lanzetta and Di Blasi [7].

Fig. 11 shows the experimental data from this study and results from simulations using the kinetic model by Lanzetta and Di Blasi [7] at 10°C/min. Only the normalized mass and mass loss rate are shown from the model. The model predicts that the pyrolysis starts at lower temperatures (below 150°C) than in the experiment and that the char yield is 65% compared to the experimentally determined char yield of 25%. This shows that the model cannot be extrapolated to work at TGA-conditions. This is an inherent problem of parallel reaction schemes.

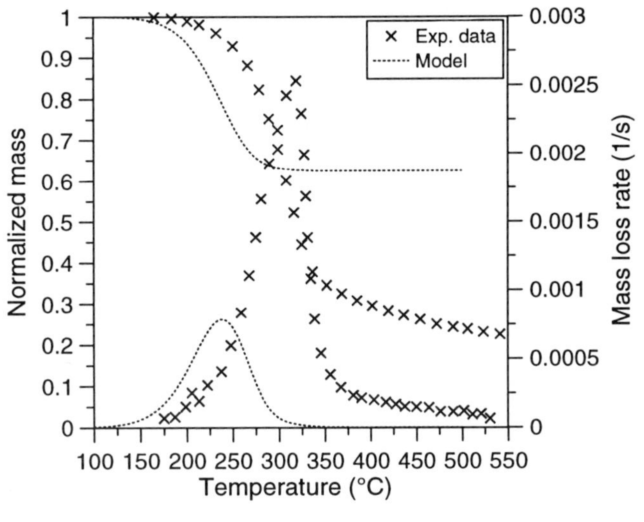

Fig. 11 Modelling of wheat straw pyrolysis at 10°C/min using the model by Lanzetta and Di Blasi [7].

CONCLUSION

Pyrolysis of wheat straw and washed wheat straw has been carried out by simultaneous thermogravimetric analysis and differential scanning calorimetry at heating rates of 5, 10 and 40°C/min.

Simple kinetic models have been tested against the weight loss and weight loss rate data from the pyrolysis experiments. The models include a single first order reaction, a three parameter nucleation model, a distributed activation energy model using a natural logarithm distribution and a superposition model based on first order reactions. All of the models were used to fit an individual set of experimental data, while some were also used to fit a set of experimental data with different heating rates. All the models could fit a single experimental data set reasonably well. However, a simultaneous fit to several experimental data sets at different heating rates did not give good results for a single first order reaction and the three parameter nucleation model. These models also failed to predict the pyrolysis behavior when an isothermal segment was introduced during the heating. The superposition model also failed to predict this. None of the models, however, could be extrapolated safely to conditions very different from those they were fitted to. The distributed activation energy model performs better for a set of experimental data with different heating rates. This model could also show the correct trend for the mass loss curve when an isothermal segment at 300°C was introduced during the pyrolysis of wheat straw at 10°C/min to 600°C. The distributed activation energy model gave the overall best results for the pyrolysis of wheat straw.

In addition a two-step model describing the pyrolysis of straw at high heating rates was extrapolated to predict the thermogravimetric data. The model did not work well at these heating rates.

A widely applicable biomass pyrolysis model is still not available in the literature.

ACKNOWLEDGEMENTS

This work is part of the CHEC research program (Combustion and Harmful Emission Control) funded by the Technical University of Denmark, the Danish Technical Research Council, Elsam, Elkraft, the Danish Energy Research Program, the Nordic Energy Research Program and the European Union.

NOMENCLATURE

A	frequency factor (s^{-1})
c_i	contribution of the partial process to the overall process (kg/kg)
E	activation energy (J/mol)
k	rate constant (s^{-1})
m	parameter in the three parameter nucleation model (-)
$m_{0,i}$	initial mass of component i (kg)
$m_{char,i}$	mass of char for component i (kg)
m_i	mass of component i (kg)
R_g	gas constant (8.314 J/molK)
t	time (min)
T	temperature (°C or K)
T_{peak}	temperature of maximum mass loss rate (°C or K)
X	overall conversion (kg/kg)

Greek

α parameter in the distributed activation energy model (-)

α_i degree of conversion for component i (kg/kg)

β parameter in the distributed activation energy model (-)

REFERENCES

1. Jensen, A., Dam-Johansen, K., Wójtowicz, M.A., Serio, M.A. (1997). *Energy & Fuels, 12*, 929-938.
2. Koós, M., Repás, M., Kosík, M., Reiser, V., Mihálov, V., Ciha, M. (1983). *Chem Zvesti, 39,* 399-408.
3. Raveendran, K., Ganesh, A., Khilar, K.C. (1995). *Fuel, 74,* 1812-1822.
4. Raveendran, K., Ganesh, A., Khilar, K.C. (1996). *Fuel, 75,* 987-998.
5. Varhegyi, G., Szabo, P., Antal Jr., M.J. (1994). In A.V. Bridgwater (Ed.), *Advances in thermochemical biomass conversion.* (pp. 760-770). Blackie Academic & Professional.
6. Knudsen, N.O., Jensen, P.A., Sander, B. & Dam-Johansen, K. (1998) Proceedings of the 10th European Conference and Technology Exhibition, Würzburg, Germany, June 8-11, pp. 224-228.
7. Lanzetta, M. and Di Blasi, C. (1998). *Journal of Analytical and Applied Pyrolysis, 44,* 181-192.
8. Stenseng, M., Jensen, A., Dam-Johansen, K. & Grønli, M. (1999) Proceedings to the 2nd Olle Lindström Symposium on renewable Energy, Bioenergy, Stockholm, June 8-11, pp. 97-104
9. Stenseng, M., Jensen, A. & Dam-Johansen, K. (2000) submitted to *Journal of Analytical and Applied Pyrolysis*
10. R. Fletcher (1971) A modified marquardt subroutine for non-linear least squares, Harwell Report, AERE R. 6799.
11. Reynolds, J.G. & Burnham, A.K. (1997) *Energy and Fuels, 11,* 88-97.
12. Anthony, D.B., Howard, J.B., Hotel, H.C. & Meissner, H.P. (1975) Fifteenth Symposium (International) on Combustion/The Combustion Institute, Pittsburgh, Pa, 1303-1317.
13. Niksa, S. & Lau, C.-W. (1993) *Combustion and Flame*, 94, 293-307.
14. Austegaard (1997) *Experimental and numerical study of a jetfire stop material and a new helical flow heat exchanger.* DPhil thesis, The Norwegian University of Science and Technology.
15. Grønli, M. (1996) *A theoretical and experimental study of the thermal degradation of biomass.* DPhil thesis, The Norwegian University of Science and Technology.

The potential of multivariate regression in determining formal kinetics of biomass pyrolysis

S. Völker, *42 Engineering, von-Behring-Str. 9, D-34260 Kaufungen, Germany*

Th. Rieckmann, *University of Applied Sciences Cologne, Department of Chemical Engineering and Plant Design, Betzdorfer Str. 2, D-50679 Cologne, Germany*

ABSTRACT: Multivariate regression has been applied to derive formal kinetic models for the mass loss of beech wood and wood components during pyrolysis. STA experiments have been performed with constant heating rates between 0.14 and 108 K/min. It was demonstrated that our experimental conditions provided (1) reliable process temperatures, (2) linear increase of temperature with time, and (3) negligible temperature gradients inside small samples. This allows the simultaneous fitting of a kinetic model with a single set of kinetic parameters to experiments at all investigated heating rates. Multivariate regression is an important tool for deriving reliable kinetic models with validity over the investigated parameter range. Using this approach, an appropriate model representing the best compromise between accuracy and number of parameters can be selected for the desired engineering application.

INTRODUCTION

The demand for sustainable development has led to growing interest in renewable energy and chemical feedstocks. Biomass is a source with global availability and its thermal conversion has attained a lot of attention during the past decades. The pyrolysis of biomass to yield valuable liquid and solid products is now on its way to commercialization. The development and optimization of technical processes bases on data from laboratory equipment as well as mini plants and scale-up is supported by process simulation. To be successful, this task needs reliable reaction models for the desired operation conditions. Since the molecular mechanisms during the thermal conversion of biomass are too complex to be modelled in full detail, reduced formal reaction models are used for engineering purposes which are characterized by lumping of compounds and mechanisms with limited numbers of reaction steps. Nevertheless, these formal reaction models have to comprise enough parameters to satisfy the requirement of validity over the whole technical parameter range to be evaluated and the reaction kinetics have to be separable from heat and mass transport influences.

Although the thermal decomposition reactions of biomass have been systematically investigated for many years [1-3], the biomass community is still debating the best formal reaction model and kinetic parameters for the primary thermal decomposition. Even for the chemically well defined component cellulose, the proposed models [4-10] include one-step first order models postulating or rejecting a change of the rate limiting reaction step going from low to high heating rates, as well as multi-step models with competitive and series reactions leading to volatiles and char.

Thermogravimetric analysis is a common tool for deriving formal reaction models for low heating rates which are relevant for e.g. carbonization processes. TGA experiments are usually performed in a dynamic mode with constant heating rates between around 100 K/min and less than 1 K/min. Although the simultaneous evaluation of results from experiments with different heating rates according to the technique of multivariate regression has been strongly recommended [11,12], this approach is still not commonly applied in publications on biomass pyrolysis. This might be due to the general assumption that TGA experiments with high heating rates cannot be performed without a thermal lag between the sample and the thermocouple measuring the process temperature [10]. Therefore in this study, conditions for TGA experiments have been investigated under which the process temperature is reliable and the temperature gradient inside the sample is negligible. The results of those experiments allow the fitting of formal kinetic models to the experimental mass loss curves according to the technique of multivariate regression. The intention was to work out guidelines for the development of formal reaction models with validity over a broad range of operation conditions which can be applied for design and scale-up of biomass reactors.

EXPERIMANTAL AND MODELLING

THERMOGRAVIMETRIC ANALYSIS

Experiments have been performed with a combined TGA/DTA (STA 503, *Bähr GmbH, Altendorfer Str. 12, D-32609 Hüllhorst*) applying constant heating rates β between 0.14 and 108 K/min. The maximum load of the STA is 1000 mg, and mass variations of $\pm$ 200 mg can be detected. The resolution is 1 µg. The STA apparatus has a horizontal weighing beam which generates an unusually low drag in the weighing direction and the effect of drag was further reduced by using high purity helium (99.999 %, *Linde*) as a low viscosity purge gas with a gas flow rate of 1.7 L/h. The thermocouples measuring the sample temperature and the reference temperature, respectively are integrated in the sample holder and are located directly beneath the platinum plates under the crucibles. This configuration provides an intense and reproducible thermal coupling to the sample.

A temperature calibration has been performed by analyzing the DTA signals from the melting peaks of the pure substances In (T_m = 429.8 K); $KClO_4$ (T_m = 572.6 K); and Ag_2SO_4 (T_m = 699.6 K). This temperature calibration has been performed for all heating rates, using open crucibles made from Al_2O_3 and the same conditions as in the experimental runs. Therefore, the corrected temperatures of the thermogravimetric analyses represent the sample surface temperatures well. The linearity of the T(t) curves was excellent for all heating rates. Linear regression of the T(t) curves to estimate the individual heating rate for each experiment resulted in correlation coefficients between 0.999997 for β = 0.14 K/min and 0.999832 for β = 108 K/min.

Samples from beech wood as well as from biomass components have been investigated using the same crucibles as for the temperature calibration. Microcrystalline cellulose (Avicel, *Merck*); lignin (Organosolv, *Aldrich*); and xylan from birchwood (extracted with NaOH, max. 5 % sulfate ash, *Roth*) have been chosen as biomass components. Samples from beech wood (Räuchergold, *J. Rettenmaier & Söhne*) were prepared by grinding the coarse particles and separating the fraction with particle diameters < 0.5 mm. Additionally, this fraction was extracted two times with hot water (335 K) to remove water soluble minerals. The beech wood contained 46.4 % cellulose; 22.3 % hemicellulose; 25.4 % lignin; 5.1 % water solubles; and 0.7 % ash. Initial sample masses m_0 between 1 and 20 mg have been used in the investigation of cellulose. The experiments with beech wood, xylan, and lignin were performed with $m_0 = 3$ mg. All samples were dried 1.5 h at 381 K in a helium stream at the beginning of the experiments.

MODELLING OF THE SAMPLE TEMPERATURE DISTRIBUTION

Derivation of kinetic models from TGA analyses bases on the assumption of a homogeneous temperature throughout the sample. To check this assumption, the spatial temperature distribution in the sample has been modelled by a simplified heat transport reaction model. The modelling has been performed for cellulose, because its heat demand per time interval is the highest of the investigated biomass samples due to its high reaction rate and endothermic reaction enthalpy.

The crucible contents of the TGA were modelled as a dynamic system with distributed parameters assuming plate geometry. The radius of the sample was taken as the characteristic length for the heat transport. The overall rate of reaction was approximated by an irreversible reaction of 1st order: biomass → solids + volatiles, assuming the validity of the Arrhenius expression for the temperature dependency of the rate constant.

$$r(\rho, T) = k_0 \exp\left(-\frac{E_A}{RT}\right)(\rho - \rho_F) \qquad (1)$$

The kinetic parameters were derived from the simultaneous evaluation of two TGA experiments with heating rates of 0.14 and 0.5 K/min, which can be assumed to be independent of heat transport influences. The material balance and the enthalpy balance resulted in a set of two meshed partial differential equations (PDE).

Material balance

$$\frac{\partial \rho}{\partial t} = -r(\rho, T) \qquad (2)$$

As a worst case scenario, the pyrolysis of cellulose was assumed to be endothermic during the entire decomposition with a reaction enthalpy of $\Delta H_R = 340$ kJ/mol. This is the most endothermic value measured for cellulose under vacuum by Mok and Antal [13] in experiments with open sample holders. The measured reaction enthalpy comprises the vaporization of volatiles and represents the total heat demand of the pyrolysis process.

Enthalpy balance

$$(1-\varepsilon_0)\rho_0 c_p \frac{\partial T}{\partial t} = \lambda \frac{\partial^2 T}{\partial x^2} + (-\Delta_R H)\ r(\rho, T) \qquad (3)$$

Initial conditions

$$T(t=0) = T_0 \qquad \rho(t=0) = \rho_0 \qquad (4), (5)$$

Boundary conditions

$$\left.\frac{\partial T}{\partial x}\right|_{x=0} = 0 \quad \text{(left boundary)} \qquad T\Big|_{x=h}(t) = \beta t + T_0 \qquad \text{(right boundary)} \quad (6), (7)$$

The temperature $T_{x=h}(t)$ at the right boundary is the corrected temperature representing the sample surface temperature (as described above). Therefore, external heat transfer limitations between STA oven and crucible had not to be considered.

The PDE system was transposed into a set of 2·n ordinary differential equations (ODE) by the finite differences technique (calculated with n = 128 finite differences).

$$\frac{\partial^2 T}{\partial x^2} \approx \frac{(T_{j-1} - 2T_j + T_{j+1})}{\Delta x^2} \qquad (8)$$

for $j = 2, n\text{-}1$ finite element.

At the left boundary (j = 1)

$$\frac{\partial^2 T}{\partial x^2} \approx \frac{(-T_j + T_{j+1})}{\Delta x^2} \qquad (9)$$

and at the right boundary (j = n)

$$\frac{\partial^2 T}{\partial x^2} \approx \frac{(T_{j-1} - 2T_j + T_{tc})}{\Delta x^2} \qquad (10)$$

The simultaneous numerical solution of the ODE system was performed using Matlab (*The MathWorks*). The kinetic and model parameters are summarized in Table 1.

Table 1 Parameters for the heat transport reaction model.

$\log_{10}(k_0/s^{-1})$	14.72	this work
E_A	197.5 kJ/mol	this work
$\Delta_R H$	340 kJ/kg	reference 13
c_p	2.5 kJ/kg/K	reference 14
λ	0.243 W/m/K	reference 14
ρ_0	1500 kg/m³	Merck
ε_0	0.7	Merck

FORMAL KINETIC MODELS

Formal kinetic models have been derived to describe the experimental mass loss of the biomass samples. The material balance of the crucible contents was described as a dynamic system with concentrated parameters which resulted in a single or a set of ordinary differential equations (ODE). The corrected sample temperature $T_{x=h}(t)$ was used for the time/temperature integration at a constant heating rate β which was calculated from the experimental data. The simultaneous numerical solution of the ODE system and the estimation of the kinetic parameters by the least squares (LSQ) method were performed using the software package Thermokinetics (*Netzsch*). The kinetic parameters according to different reaction models were calculated by fitting single experiments as well as by the multivariate regression technique [12]. TGA curves were used for the least-squares analysis because the TGA curve is less sensitive to noise and round-off error than the DTG curve. Additionally, this method uses the important data obtainable from the start and finish of the TGA curve which helps selecting an appropriate model [15].

RESULTS AND DISCUSSION

SAMPLE TEMPERATURE DISTRIBUTION

The modelling results of the temperature distribution in cellulose samples during TGA experiments are displayed in Figures 1 - 3. At a constant heating rate of 108 K/min, the temperature in the interior of a cellulose sample of 1 mg follows the temperature of the sample surface closely with a maximum temperature difference of less than 0.15 K between the different spatial layers of the cellulose sample (Fig. 1). Under the same conditions, the spatial layers of a cellulose sample of 3 mg exhibit a maximum temperature difference of 0.8 K (Fig. 2). This is mainly due to unsatisfied heat demands of the endothermic reaction, since a similar calculation with $\Delta_R H = 0$ kJ/mol yielded a maximum temperature difference of 0.2 K between the different spatial layers of the cellulose sample. In a cellulose sample of 20 mg at a constant heating rate of 108 K/min, a maximum temperature difference of nearly 25 K is predicted between the different spatial layers of the cellulose sample (Fig. 3). It is evident that such sample sizes are unsuitable for kinetic analyses applying the assumption of a homogeneous sample temperature, as stated in the literature before [10].

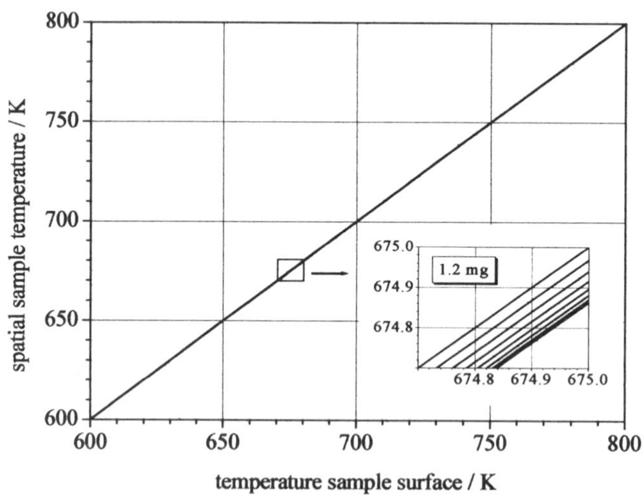

Fig. 1 Calculated temperature distribution for the different spatial layers in a cellulose sample of $m_0 = 1$ mg at $\beta = 108$ K/min according to the heat transport reaction model.

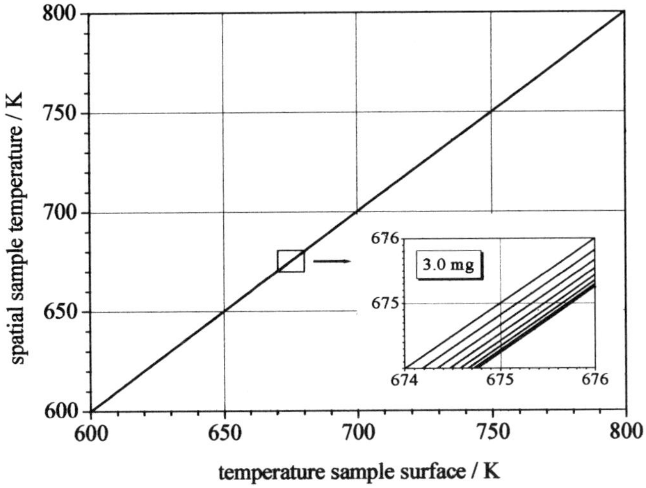

Fig. 2 Calculated temperature distribution for the different spatial layers in a cellulose sample of $m_0 = 3$ mg at $\beta = 108$ K/min according to the heat transport reaction model.

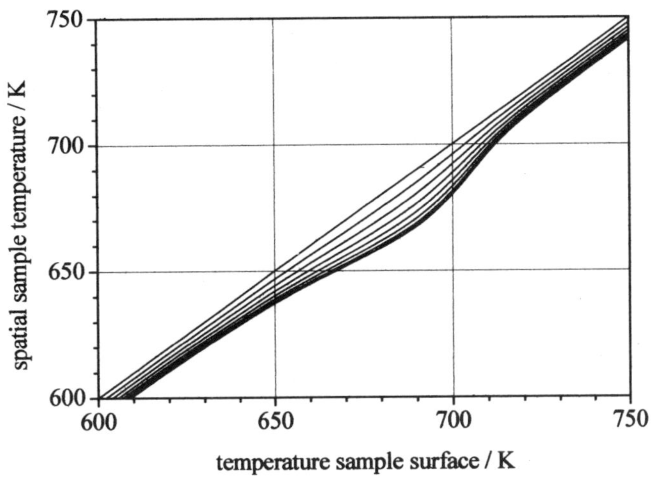

Fig. 3 Calculated temperature distribution for the different spatial layers in a cellulose sample of $m_0 = 20$ mg at $\beta = 108$ K/min according to the heat transport reaction model.

In Figure 4, the calculated mass losses for cellulose at different constant heating rates and initial sample masses are compared to experimental TGA results. The TGA curves at heating rates of 0.14 K/min and 0.5 K/min had been used to evaluate the kinetic parameters for the one step first order reaction model which was incorporated into the model to calculate the sample temperature distribution. Since the temperature gradients in those samples are nearly zero, the results of the heat transport reaction model represent simultaneously the best fit for the assumed reaction model. At a heating rate of 108 K/min, the initial sample mass influences the temperature at which a given mass loss is attained. Cellulose samples with $m_0 = 1 - 3$ mg are affected only to a minor extend by heat transport phenomena, and the heat transport reaction model represents the experiments very well. The temperature shift of the curve for 3 mg compared to the curve for 1 mg is less than 2 K. In a cellulose sample of 20 mg, the temperature inhomogeneity is pronounced and generates a significant temperature shift of the mass loss curve which is underestimated by our simplified model. This is due to mass transport effects which can not be neglected for a comprehensive evaluation [16] but are not included in our model.

The results confirm that TGA experiments are not significantly affected by heat transport phenomena if low initial sample masses as well as the described TGA configuration and experimental procedures are applied. The temperature gradients inside the samples are sufficiently small to allow the fitting of formal kinetic models to the experimental mass loss curves assuming a homogeneous sample temperature. Cellulose samples with initial sample masses of around 5 mg and higher can only be submitted to kinetic analyses under consideration of the enthalpy balance.

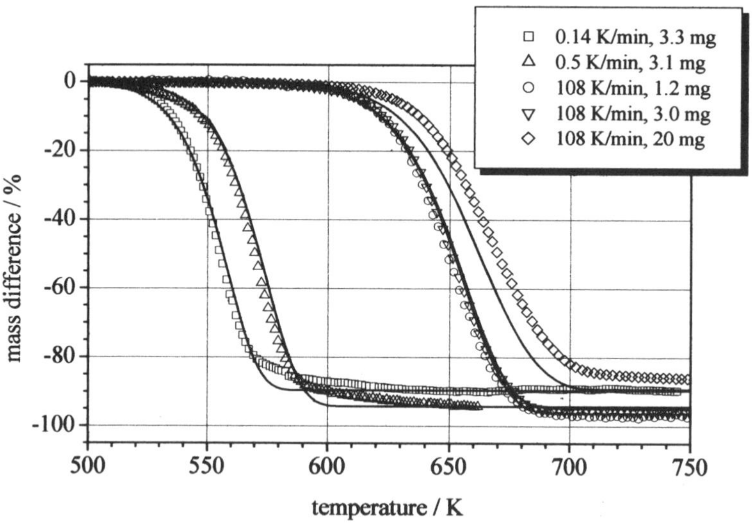

Fig. 4 Modelled mass losses (lines) for cellulose at different heating rates and initial sample masses and comparison to TGA results (symbols).

To minimize heat transport influences on our kinetic analyses, we have chosen an initial sample mass of 1 mg for cellulose at heating rates of 41 and 108 K/min. For cellulose at low heating rates as well as for all runs with xylan, lignin, and beech wood, initial sample masses of 3 mg have been chosen to improve the signal / noise ratio. The use of 3 mg samples at the high heating rates was justified by the observation of the DTA signals which showed that the endothermic heat of reaction for beech wood pyrolysis is at least four times smaller than that for cellulose pyrolysis. According to our heat transport reaction model, a beech wood sample with $m_0 = 3$ mg at $\beta = 108$ K/min would exhibit a maximum temperature difference of 0.4 K between the different spatial layers of the sample. The pyrolysis of lignin showed almost no overall reaction enthalpy whereas the pyrolysis of xylan was very slightly exothermic. The temperature gradients inside the samples can be neglected for kinetic analysis.

KINETIC ANAYLSIS

Avicel cellulose

The kinetic analysis of cellulose pyrolysis is extensively described in the literature with varying results regarding the proposed formal reaction mechanisms as well as kinetic parameters for identical reaction models [4-10]. One of the most widely used reaction models is an irreversible 1st order single step reaction. This model is grossly simplified with the major drawback, that it is unsuitable to describe the last stages of the pyrolysis process as well as the different char yields observed for different heating rates. Another disadvantage of this model is that evaluations of data sets from single experiments result in kinetic parameters varying over a broad range due to a compensation effect [16,17] which compensates influences of heat and mass transport and the inadequacy of the one step model.

A further problem with evaluating data sets of single experiments is that the formal reaction models contain enough parameters to achieve an equally good fit for a great number of model types. The choice of an appropriate reaction model is therefore complicated considerably. The technique of multivariate regression requires the identity of kinetic parameters for all data sets submitted to the analysis. This provides a very strong criterion and the number of applicable models as well as the range of resulting kinetic parameters narrows substantially. To demonstrate this, kinetic analyses basing on a single experiment with $\beta = 3$ K/min have been compared to kinetic analyses basing on the simultaneous evaluation of the data sets for all five heating rates. A single step reaction has been chosen for the mechanism and different reaction types have been tested. The results are summarized in Table 2.

The kinetic analyses of the single experiment yielded best fit models describing the experimental mass loss well with correlation coefficients between 0.99929 and 0.99993. Although each of the models could be used as formal reaction model for a process with a constant heating rate of 3 K/min, these models would predict widely differing behaviour for other heating rates due to their activation energies lying between 12 and 552 kJ/mol. On the other hand, the temperature shift of the mass loss with heating rate determines the activation energy resulting from the multivariate regression. This restriction reduces the number of formal reaction models from nine to six which describe the experimental mass loss well with correlation coefficients > 0.999. It is especially noteworthy that the kinetic parameters for those six models fall in a very narrow range with $\log_{10}(k_0/s^{-1})$ varying between 14.69 and 15.08 and E_A varying between 197.8 and 198.7 kJ/mol.

Kinetic parameters for the model with a single step first order reaction have also been calculated by a multivariate regression analysis of the DTG results for all heating rates. The best fit kinetic parameters are $\log_{10}(k_0/s^{-1}) = 14.75$ and $E_A = 196.7$ kJ/mol demonstrating that the result for the kinetic parameters is not influenced by choosing either the TGA or the DTG curves for the multivariate regression.

The comparison of the results of a round-robin study published by Grønli et al. [18] to our results for $\beta = 40$ K/min (the only heating rate investigated in both studies) shows that the correspondence between the experimental mass loss curves is very good. Nevertheless, the activation energy of 198.4 kJ/mol for the irreversible 1st order reaction from this work is considerably lower than the 211 - 232 kJ/mol found for the results from the different laboratories participating in the round-robin study. Instead, our results correspond better to an investigation performed by Reynolds and Burnham [7].

Table 2 Comparison of results from the evaluation of a single experiment with results from a multivariate regression for different formal kinetic models

Reaction type	*Evaluation of single experiment*			*Multivariate regression*		
	$\log_{10}$ (k_0/s^{-1})	E_A / kJ/mol	reaction order	$\log_{10}$ (k_0/s^{-1})	E_A / kJ/mol	reaction order
1st order reaction	16.87	220.6	1	14.85	198.4	1
	cc: 0.99968; mr: 0.02201			cc: 0.99927; mr: 0.03035		
nth order reaction	21.30	268.8	1.52	14.89	198.7	1.05
	cc: 0.99989; mr: 0.01256			cc: 0.99928; mr: 0.03012		
3D Jander's diffusion	31.39	396.7		15.37	217.7	
	cc: 0.99929; mr: 0.03156			cc: 0.98784; mr: 0.12352		
3D phase boundary reaction	13.90	193.6		14.11	196.2	
	cc: 0.99929; mr: 0.03161			cc: 0.99860; mr: 0.04193		
1st order autocatalytic reaction Prout-Tompkins	- 1.19	11.9	1	15.78	195.3	1
	cc: 0.99989; mr: 0.01230			cc: 0.96431; mr: 0.21006		
nth order autocatalytic reaction Prout-Tompkins	21.29	268.6	1.52	15.08	198.2	1.46
	cc: 0.99989; mr: 0.01255			cc: 0.99958; mr: 0.02302		
1st order autocatalytic reaction	16.87	220.6	1	14.77	197.8	1
	cc: 0.99968; mr: 0.02117			cc: 0.99924; mr: 0.03080		
nth order autocatalytic reaction	15.42	206.2	1.65	14.69	198.0	1.68
	cc: 0.99993; mr: 0.00982			cc: 0.99968; mr: 0.02009		
n dimensional nucleation Avrami-Erofeev	46.23	552.1	Dim.: 0.41	14.80	197.8	Dim.: 1.04
	cc: 0.99968; mr: 0.02114			cc: 0.99930; mr: 0.02970		

To improve the formal reaction model, multi-step models have been tested by multivariate regression with data sets from all heating rates. The pyrolysis of Avicel cellulose can best be represented by a formal reaction model comprising three reaction steps of nth order with two parallel reactions, one of them being a series reaction. Models with reactions (1) and (2) being competitive reactions yielded worse fits than models with parallel reactions. This might be due to different reaction pathways for crystalline and non-crystalline cellulose or an undetected formation of anhydrocellulose. However, for a mechanistic investigation a broader data base would be required.

The kinetic parameters for the best fit model

$$A \xrightarrow{1} B$$
$$C \xrightarrow{2} D \xrightarrow{3} E$$

are $\log_{10}(k_0/s^{-1})(1) = 14.97$, $E_A(1) = 198.7$ kJ/mol, $n(1) = 0.95$; $\log_{10}(k_0/s^{-1})(2) = 11.44$, $E_A(2) = 157.9$ kJ/mol, $n(2) = 0.95$; and $\log_{10}(k_0/s^{-1})(3) = 35.96$, $E_A(3) = 413.5$ kJ/mol, $n(3) = 4.60$. The total mass loss is the sum of the individual mass losses from the three reactions. Reaction (1) contributes 62.3 %, reaction (2) contributes 4.5 % and reaction (3) contributes 33.2 % to the total mass loss. The best fit model has a correlation coefficient of 0.99993, a mean of residues of 0.00966 and is displayed in Figure 5 together with the experimental results and the best fit for a single step 1st order reaction. The three step model fits nearly perfectly and could be interpreted as comprising a single tar forming reaction (1) and a serial char + gas forming reaction (2 and 3), as proposed by Várhegyi [5]. Reaction (1) has similar kinetic parameters as those resulting from the best fit of a single step mechanism, indicating that this reaction dominates the pyrolysis process of cellulose. The activation energy $E_A(1)$ matches the value given by Bradbury et al. for the volatilization step during cellulose pyrolysis [19]. Reaction (2) has an activation energy of 157.9 kJ/mol, comparing well to data from the literature [5,20]. The kinetic parameters for reaction (3) can be interpreted as covering a great number of parallel reactions taking place during the carbonization of the primary char.

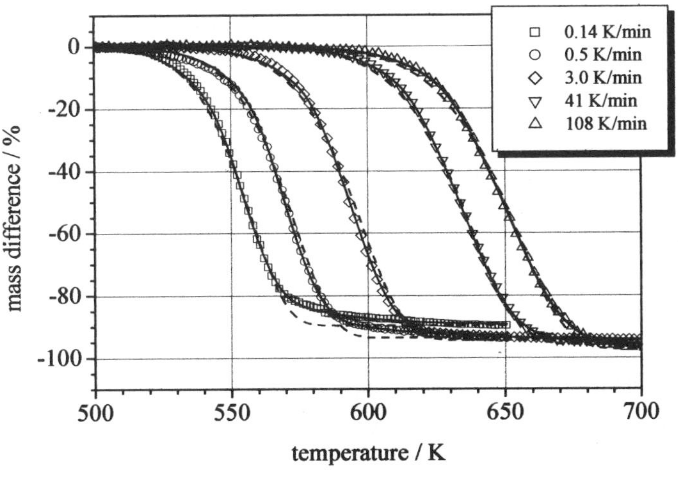

Fig. 5 TGA results for cellulose at different heating rates and best fit models for a single step mechanism (- - -) and a three step mechanism (——).

The reaction order of the first two reactions is 0.95, indicating that their reaction profile is narrower than that of a first order reaction. This could be a hint for a serial reaction or the presence of a preceding activation step [7], respectively. A multivariate regression fitting a four step mechanism to our experimental data did not improve the fit significantly due to the already high accuracy of the three step mechanism. To verify the presence or absence of an activation step, additional experiments under differing conditions have to be performed. The kinetic parameters from this work are valid under our investigated experimental conditions. They will probably change if the model is fitted to experiments with differing celluloses or experimental conditions due to influences of catalysts on reaction (1) and influences of mass transport on reactions (2) and (3). Especially mass transport plays a very important role in the pyrolysis of cellulose. This item demands further research to enable the evaluation of formal reaction models with improved transferability.

Xylan from birch wood

Xylan from birchwood decomposes over a broader temperature range than cellulose and the shape of the TGA curves suggests a mechanism with more than one reaction step. Consequently, one step formal models are not very successful. Applying multivariate regression, the best fit for this case is achieved with a single reaction of nth order with the kinetic parameters $\log_{10}(k_0/s^{-1}) = 19.63$, $E_A = 220.8$ kJ/mol and $n = 5.45$. A correlation coefficient of 0.99423 and a mean of residues of 0.06806 were calculated for this fit.

To represent xylan pyrolysis under our experimental conditions, a formal reaction model with three reactions in series was most successful:

$$A \xrightarrow{1} B \xrightarrow{2} C \xrightarrow{3} D$$

The following reaction types were used: A 2-dimensional diffusion reaction for reaction (1); a 3-dimensional diffusion reaction of the Jander's type for reaction (2); and a reaction of nth order for reaction (3). The best fit model has a correlation coefficient of 0.99955, a mean of residues of 0.01911, and the kinetic parameters from a multivariate regression are $\log_{10}(k_0/s^{-1})(1) = 19.14$, $E_A(1) = 203.5$ kJ/mol; $\log_{10}(k_0/s^{-1})(2) = 16.0$, $E_A(2) = 203.2$ kJ/mol; and $\log_{10}(k_0/s^{-1})(3) = 54.61$, $E_A(3) = 625.4$ kJ/mol, $n(3) = 15.02$. Reaction (1) contributes 10.6 %, reaction (2) contributes 61.6 % and reaction (3) contributes 27.8 % to the total mass loss.

The best fits for the one step and the three step formal reaction model are displayed in Figure 6. Formal reaction models with parallel reactions always yielded worse fits than equivalent models with series reactions. This confirms results from an investigation of Várhegyi et al. on the thermal decomposition of 4-methyl-D-glucurono-D-xylan [21].

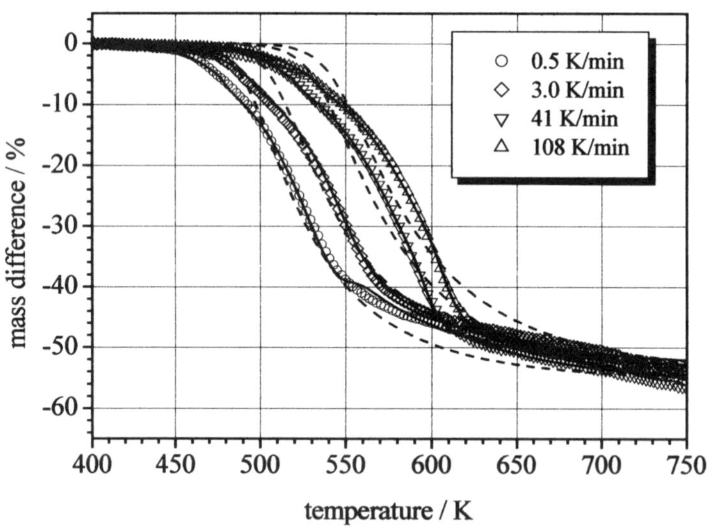

Fig. 6 TGA results for xylan at different heating rates and best fit models for a single step mechanism (- - -) and a three step mechanism (——).

Organosolv lignin

Organosolv lignin decomposes over an even broader temperature range than xylan. A formal reaction model with a single reaction step gives only a rough estimate for the pyrolysis behaviour of lignin. The best reaction type for a one step mechanism is the n-dimensional nucleation reaction (Avrami-Erofeev), especially for higher heating rates. Applying multivariate regression, the best fit is achieved with the kinetic parameters $\log_{10}(k_0/s^{-1}) = 11.99$, $E_A = 174.4$ kJ/mol and a dimension of 0.34. The correlation coefficient for this fit is 0.98833 and the mean of residues is 0.08904.
To represent lignin pyrolysis under our experimental conditions, a formal reaction model with two parallel series reactions had to be set up:

$$A \xrightarrow{1} B \xrightarrow{2} C$$
$$D \xrightarrow{3} E \xrightarrow{4} F$$

The following reaction types were used: A 3-dimensional diffusion reaction of the Jander's type for reactions (1); (2); and (3) and an n-dimensional nucleation reaction (Avrami-Erofeev) for reaction (4). The best fit model has a correlation coefficient of 0.99972, a mean of residues of 0.01362, and the kinetic parameters from multivariate regression are $\log_{10}(k_0/s^{-1})(1) = 5.49$, $E_A(1) = 95.4$ kJ/mol; $\log_{10}(k_0/s^{-1})(2) = 15.92$, $E_A(2) = 236.8$ kJ/mol; $\log_{10}(k_0/s^{-1})(3) = 8.44$, $E_A(3) = 138.5$ kJ/mol; and $\log_{10}(k_0/s^{-1})(4) = 46.22$, $E_A(4) = 635.1$ kJ/mol, dim. = 0.09. Reaction (1) contributes 14.6 %, reaction (2) contributes 13.5 %, reaction (3) contributes 23.2 % and reaction (4) contributes 48.7 % to the total mass loss. The best fits for the one step and the four step formal reaction model are displayed in Figure 7.

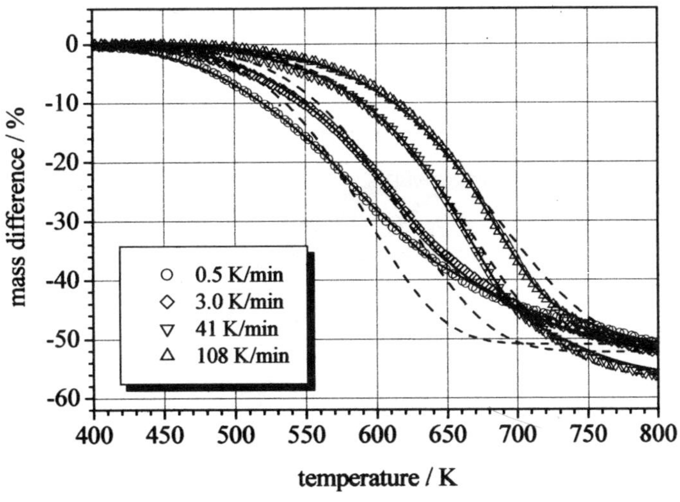

Fig. 7 TGA results for lignin at different heating rates and best fit models for a single step mechanism (- - -) and a four step mechanism (——).

Water washed beech wood

The pyrolysis of water washed beech wood can only be described by multi-step formal models. The most successful approach is a mechanism with three parallel reactions, as commonly used throughout the literature:

$$A \xrightarrow{1} B$$
$$C \xrightarrow{2} D$$
$$E \xrightarrow{3} F$$

A formal reaction mechanism with three 1st order reactions describes the experimental results very satisfyingly over the whole temperature range in case the experiments at different heating rates are evaluated separately. For each single heating rate, kinetic analysis results in kinetic parameters which can not be applied to describe the pyrolysis behaviour at any other heating rate. If multivariate regression is applied to TGA results from all heating rates simultaneously, the main parts of the pyrolysis process are still described very well, but the slow mass loss at the final stages of the decomposition can not be represented. The kinetic parameters for this formal reaction model are $\log_{10}(k_0/s^{-1})(1) = 15.62$, $E_A(1) = 188.4$ kJ/mol; $\log_{10}(k_0/s^{-1})(2) = 17.18$, $E_A(2) = 218.7$ kJ/mol; $\log_{10}(k_0/s^{-1})(3) = 14.99$, and $E_A(3) = 209.8$ kJ/mol. Reaction (1) contributes 17.2 %, reaction (2) contributes 18.9 % and reaction (3) contributes 64.0 % to the total mass loss. A correlation coefficient of 0.99856 and a mean of residues of 0.03802 were calculated for the best fit.

To improve the formal reaction model, a mechanism comprising three parallel nth order reactions was fitted to the experimental TGA curves from all heating rates. The best fit was achieved for the following kinetic parameters: $\log_{10}(k_0/s^{-1})(1) = 17.52$, $E_A(1) = 213.7$ kJ/mol, $n(1) = 4.61$; $\log_{10}(k_0/s^{-1})(2) = 14.90$, $E_A(2) = 208.3$ kJ/mol, $n(2) = 0.75$; and $\log_{10}(k_0/s^{-1})(3) = 15.29$, $E_A(3) = 169.0$ kJ/mol, $n(3) = 3.28$. Reaction (1) contributes 46.9 %, reaction (2) contributes 50.5 % and reaction (3) contributes 2.5 % to the total mass loss. A correlation coefficient of 0.99938 and a mean of residues of 0.02448 were calculated. The best fits for the formal reaction models with 1st order reactions and nth order reactions, respectively are displayed in Figure 8.

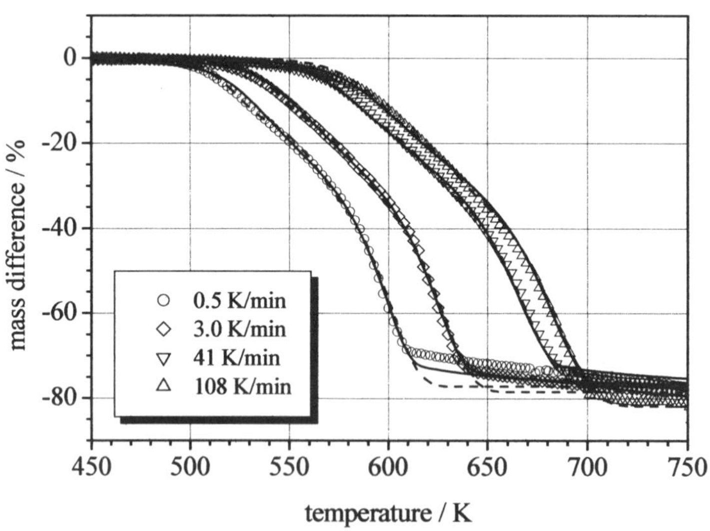

Fig. 8 TGA results for beech wood at different heating rates and best fit models for a three step mechanism with 1st order reactions (- - -) and nth order reactions (——).

CONCLUSIONS

Thermogravimetric analysis can be performed with heating rates up to 108 K/min without significant temperature gradients inside small samples. An STA apparatus enabling a temperature calibration under the exact experimental conditions will give reliable process temperatures. This allows kinetic modelling without consideration of the enthalpy balance.

Formal kinetic models have to be valid for different heating rates which will occur in all reactors with a distributed temperature. To obtain reliable kinetic models, experiments covering the desired range of heating rates have to be performed. A simultaneous evaluation of all experiments applying multivariate regression results in the description of the pyrolysis behaviour in the analyzed range of heating rates by a formal reaction model with one set of kinetic parameters. Multivariate regression is an important tool for deriving an appropriate model representing the best compromise between accuracy and number of parameters for the desired engineering application.

The results of our kinetic analyses demonstrate that formal kinetic models can be derived which describe mass loss during pyrolysis of biomass and its components quantitatively for heating rates varying over more than two decades. Nevertheless, care should be taken when transferring kinetic parameters from the literature. The exact chemical composition of the biomass samples is crucial to their pyrolysis behaviour. Additionally, mass transport phenomena influence the char forming reactions dramatically thus changing the observed macro kinetic parameters and the contribution of the individual reaction steps to the overall mass loss. To derive reliable reaction models for engineering applications, experiments with the desired technical particle sizes and bulk phase properties should be selected or mass transport models should be incorporated into the kinetic modelling.

NOTATION

c_p	heat capacity, J/(kg K)	t	time, s
cc	correlation coefficient	T, T_0	temperature, initial temperature, K
E_A	apparent activation energy, kJ/mol	$T_{x=h}$	temperature at sample surface, K
h	characteristic length of sample, m	x	spatial coordinate, m
$\Delta_R H$	enthalpy of reaction, J/kg	β	heating rate, K/min
k	rate constant, 1/s	ε_0	initial void fraction, $m^3_{gas}/m^3_{bulk\ phase}$
k_0	pre-exponential factor, 1/s	λ	heat conductivity, W/(K m)
mr	mean of residues	ρ	density, kg/m³
r	reaction rate, kg/(m³s)	$\rho_{F,}$	final density, kg/m³
R	ideal gas constant, J/(mol K)	ρ_0	initial density, kg/m³

REFERENCES

1. Antal M.J. (1982) Biomass pyrolysis: A review of the literature. Part 1 - carbohydrate pyrolysis. *Advances in Solar Energy*, **1**, 61-111
2. Antal M.J. (1985) Biomass pyrolysis: A review of the literature. Part 2 - lignocellulose pyrolysis. *Advances in Solar Energy*, **2**, 175-255
3. Hajaligol M., Howard J.B., Longwell J.P. and Peters W.A. (1982) *Ind. Eng. Chem. Process Des. Dev.*, **21(3)**, 457-465

4. Antal M.J. and Várhegyi G. (1995) Cellulose pyrolysis kinetics: The current state of knowledge. *Ind. Eng. Chem. Res.*, **34**, 703-17.
5. Várhegyi G., Jakab E. and Antal M.J. (1994) Is the Broido-Shafizadeh model for cellulose pyrolysis true? *Energy Fuels*, **8**, 1345-52.
6. Milosavljevic I. and Suuberg E.M. (1995) Cellulose thermal decomposition kinetics: Global mass loss kinetics. *Ind. Eng. Chem. Res.*, **34**, 1081-91.
7. Reynolds J.G. and Burnham A.K. (1997) Pyrolysis decomposition kinetics of cellulose-based materials by constant heating rate micropyrolysis. *Energy Fuels*, **11**, 88-97.
8. Antal M.J. and Várhegyi G. (1997) Impact of systematic errors on the determination of cellulose pyrolysis kinetics. *Energy Fuels*, **11**, 1309-10.
9. Di Blasi C. (1998) Comparison of semi-global mechanisms for primary pyrolysis of lignocellulosic fuels. *J. Anal. Appl. Pyrolysis*, **47**, 43-64.
10. Antal M.J., Várhegyi G. and Jakab E. (1998) Cellulose pyrolysis kinetics: Reviseted. *Ind. Eng. Chem. Res.*, **37**, 1267-75.
11. Conesa J.A., Marcilla A., Caballero J.A. and Font R. (2000) Comments on the validity and utility of the different methods for kinetic analysis of thermo-gravimetric data. Pyrolysis'2000, April 2-6, 2000, Seville, Spain.
12. Kaisersberger E. and Opfermann J. (1991) *Thermochim. Acta*, **187**, 151.
13. Mok W.S.-L. and Antal M.J. (1983) Effects of pressure on biomass pyrolysis. II. Heats of reaction of cellulose pyrolysis. *Thermochim. Acta*, **68**, 165-186
14. Curtis L.J. and Miller D.J. (1988) Transport model with radiative heat transfer for rapid cellulose pyrolysis. *Ind. Eng. Chem. Res.*, **27**, 1775-83.
15. Hoare I.C. and Hurst H.J. (1992) The evaluation of kinetic parameters from thermogravimetric curves. *Thermochim. Acta*, **203**, 127-35.
16. Völker S., Rieckmann Th. and Klose W. (2000) Thermokinetic investigation of cellulose pyrolysis – impact of final mass on kinetic results. Pyrolysis'2000, April 2-6, 2000, Seville, Spain.
17. Chornet E. and Roy C. (1980) Compensation effect in the thermal decomposition of cellulosic materials. *Thermochim. Acta*, **35**, 389-93.
18. Grønli M., Antal M.J and Várhegyi G. (1999) A round-robin study of cellulose pyrolysis kinetics by thermogravimetry. *Ind. Eng. Chem. Res.*, **38**, 2238-2244
19. Bradbury A.G.W., Sakai Y. and Shafizadeh F. (1979) A kinetic model for pyrolysis of cellulose. *J. Appl. Polym. Sci.*, **23**, 3271-80.
20. Lanzetta M., Di Blasi C. and Buonanno F. (1997) An experimental investigation of heat-transfer limitations in the flash pyrolysis of cellulose. *Ind. Eng. Chem. Res.*, **36**, 542-52.
21. Várhegyi G., Antal M.J., Jakab E. and Szabó P. (1997) Kinetic modeling of biomass pyrolysis. *J. Anal. Appl. Pyrolysis*, **42**, 73-87.

A Modeling Study On Cellulose Particle Pyrolysis Under Fluidized-Bed Conditions

Chunjiang Yu[1], Wennan Zhang[1] and Kefa Cen[2]

[1] *Department of Applied Science, Mid Sweden University, 871 88 Härnösand, Sweden*

[2] *Institute for Thermal Power Engineering, Zhejiang University, 310027 Hangzhou P.R.China*

ABSTRACT To study biomass pyrolysis processes, a numerical model is developed in this work focusing on the cellulose constituent. A single cellulose particle is supposed to experience pyrolysis at a fairly high heating rate in the case of typical fluidized bed conditions. The model involves the pyrolytic kinetic scheme and the detailed heat and mass transfers caused by radiation, conduction, diffusion and convection with respect to the solid, liquid and gas products from pyrolysis. From the solution of the model, the central role of the reaction heat versus a slow heat supply through the particle is identified. The dynamic characteristic of the pyrolysis and the evolutions of the internal pressure and the flow of the products are analyzed quantitatively. Computation of the model shows the importance of particle permeability, whereas the liquid phase of active cellulose and inter-particle secondary reaction of volatile play a negligible role during the pyrolysis under a typical fluidized bed conditions both for small and large particles. Various predictions by the model provide a good understanding of the complex process of biomass pyrolysis, which can be useful for applications of thermochemical conversions of biomass.

INTRODUCTION

Fluidized bed application is fairly popular in biomass thermochemical conversions. The one of reason is due to the high efficient heat and mass transfers between particles and gases in the reactor. The initial and essential step of biomass thermochemical conversions such as combustion and gasification in a fluidized bed is pyrolysis. Directly experimental study on the pyrolytic process is limited. Numerical simulation can be an effective way for understanding of the first step of biomass combustion and gasification in applications.

From literature, a great amount of work on biomass pyrolysis, including modeling of pyrolysis, has been found over the last two decades. However, the most of modeling works deals with low heat flux conditions or an extremely high heating rate (flash pyrolysis), or focus on the prediction of product distribution [1-3]. The comprehensive pyrolytic process inside a biomass particle with median intensive heating under fluidized bed conditions remains to be studied in detail for a better understanding of the process and optimal design of related industrial units.

The aim of the present work is to develop a mathematic model describing the behavior of a single cellulose particle when heated up under a fluidized bed condition. This may give an insight into the temperature profile of the particle in connection to the reaction heat, the characters of product gas and liquid flows, the secondary reaction of primary volatile and other important phenomena of biomass pyrolysis.

THE MATHEMATIC MODEL

The model is one-dimensional over a spherical, dry cellulose particle. The particle is supposed to expose to an inert, high-temperature environment of a fluidized-bed subject to an intense heating immediately. A typical constant overall heat transfer coefficient is adopted.

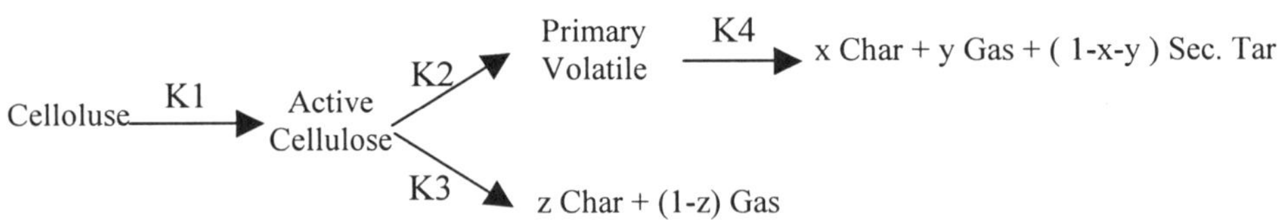

Fig. 1 Modified Broido-Shafizadeh Scheme [4].

Table 1 Kinetic data of the modified B-S scheme.

Reaction K_i	A_i [s^{-1}]	E_i [kj/mole]	ΔH_i[kj/Kg]	References
1	2.8e19	242.8	40	Bradbury, [4]
2	3.17e14	198.0	418	Bradbury, [4]
	1.26e18	238		Varhegyi et. al. [5]
3	1.32e10	150.7	418	Bradbury, [4]
	3.16e9	147		Varhegyi et. al. [5]
4	4.28e6	107.5	-42	Liden's [6]
	n=0.35			Colomba Di Blasi [7]
	i=0.35 ; j=0.3			Estimated

The kinetic scheme that has been chosen is basically the Broido-Shafizadeh [4] as shown in Fig.1. The most widely accepted Broido-Shafizadeh scheme provides a more reasonable description of the involved kinetics than other schemes. However, in the present model, some modifications are made. First, some more reliable kinetic data obtained by Varhegyi et. al. [5] is employed as seen in Table 1. Secondly, the intermediate product "active cellulose" is included, according to recent experimental study of O. Boutin et. al [8]. The experimental study not only proved the existence of such "active cellulose", but also showed the short-life intermediate product would experience a phase change from solid to liquid when temperature rises up to about 740K. A fast heating may raise the temperature of active cellulose over 740K before it consumed, which leads a liquid fluid present in the pyrolytic process. Third, the secondary reaction includes char and secondary tar as additional products to constitute a complete reaction.

The physical aspects involved are heat transfer and mass transfer. The external heat and mass transfers are free of resistance under the environment of fluidized beds. Only the internal heat and mass transfers are treated here. To develop the mathematic model, the following assumptions and statements are made:

1. There is a local thermal equilibrium among the solid phase, liquid phase and gaseous phase.
2. Negligible enthalpy flux due to fluid species diffusion.
3. Negligible kinetic and potential energy, negligible body forces.
4. No particle shrinkage and surface regression.
5. The inert gas and gaseous products resulted from reaction behave like ideal gas, with a constant mean molecular weight.
6. The secondary reaction quenches as soon as volatile products leave the solid surface.
7. The initial porous structure of cellulose is filled with inert gas before pyrolysis.
8. Physical properties (porosity, permeability, thermal conductivity, specific heat and fluid viscosity) depend on temperature and conversion extent.
9. The secondary reaction is homogeneous gas reactions.

Based on the above assumptions, The mass conservation through a differential control volume of a layer of spherical cellulose particle are thus formulated for cellulose (1), char (2), active cellulose (3), volatile (4), secondary tar (5), gas (6) and inert gas (7), respectively.

$$\frac{\partial}{\partial \tau}\langle \rho_c \rangle = iK4\langle \rho_v \rangle + nK3\langle \rho_a \rangle \tag{1}$$

$$\frac{\partial}{\partial \tau}\langle \rho_w \rangle = -K1\langle \rho_w \rangle \tag{2}$$

$$\frac{\partial}{\partial \tau}\langle \rho_a \rangle + \frac{1}{r^2}\frac{\partial}{\partial r}\left(r^2 \langle \rho_a \rangle^a \langle u_a \rangle\right) = K1\langle \rho_w \rangle - (K2 + K3)\langle \rho_a \rangle \tag{3}$$

$$\frac{\partial}{\partial \tau}\langle \rho_v \rangle + \frac{1}{r^2}\frac{\partial}{\partial r}\left(r^2 \langle \rho_v \rangle^f \langle u_f \rangle\right) = +K2\langle \rho_a \rangle - K4\langle \rho_v \rangle \tag{4}$$

$$\frac{\partial}{\partial \tau}\langle \rho_{t2} \rangle + \frac{1}{r^2}\frac{\partial}{\partial r}\left(r^2 \langle \rho_{t2} \rangle^f \langle u_f \rangle\right) = +(1 - i - j)K4\langle \rho_v \rangle \tag{5}$$

$$\frac{\partial}{\partial \tau}\langle \rho_g \rangle + \frac{1}{r^2}\frac{\partial}{\partial r}\left(r^2 \langle \rho_g \rangle^f \langle u_f \rangle\right) = +jK4\langle \rho_v \rangle + (1 - n)K3\langle \rho_a \rangle \tag{6}$$

$$\frac{\partial}{\partial \tau}\langle \rho_i \rangle + \frac{1}{r^2}\frac{\partial}{\partial r}\left(r^2 \langle \rho_i \rangle^f \langle u_f \rangle\right) = 0 \tag{7}$$

The variables in $\langle\rangle$ mean they are based on the whole control volume, in $\langle\rangle^f$ mean they are based on porous volume, in $\langle\rangle^a$ mean they are based on volume occupied by the active cellulose. The whole control volume is occupied by solid phase (cellulose and char), liquid phase (active cellulose) and gas phase (porous volume). Considering the flow of gas and liquid through the porous cellulose, Darcy's law (Eqs

8 and 9) and ideal gas law (Eq. 10) are introduced to govern the superficial velocity of the fluid with respect to the pressure and the particle permeability:

$$\langle u_a \rangle = -\frac{K_a}{\mu_a}\frac{\partial}{\partial r}\langle P \rangle \tag{8}$$

$$\langle u_f \rangle = -\frac{K_f}{\mu_f}\frac{\partial}{\partial r}\langle P \rangle \tag{9}$$

$$\langle P \rangle = \frac{\langle \rho_f \rangle^f RT}{M_f} \tag{10}$$

where K_a and K_f are intrinsic permeabilities of porous matrix for the liquid phase and the gas phase, respectively.

The energy conservation is written as Eq. 11.

$$\begin{aligned}
&\frac{\partial}{\partial \tau}\left(T\left(\langle \rho_w \rangle c_w + \langle \rho_c \rangle c_c + \langle \rho_a \rangle c_a + \langle \rho_v \rangle c_v + \langle \rho_{t2} \rangle c_{t2} + \langle \rho_g \rangle c_g\right)\right) \\
&+ \frac{1}{r^2}\frac{\partial}{\partial r}\left(r^2 T\left(\langle \rho_a \rangle^a c_a \langle u_a \rangle + \langle \rho_v \rangle^f c_v \langle u_f \rangle + \langle \rho_{t2} \rangle^f c_{t2} \langle u_f \rangle + \langle \rho_g \rangle^f c_g \langle u_f \rangle\right)\right) \\
&= \frac{1}{r^2}\frac{\partial}{\partial r}\left(r^2 K_{eff}\frac{\partial T}{\partial r}\right) + Q_r
\end{aligned} \tag{11}$$

where

$$\begin{aligned}
Q_r &= K1\langle \rho_w \rangle\left(\Delta H_1^0 + (T - T_0)(c_w - c_a)\right) + K2\langle \rho_a \rangle\left(\Delta H_2^0 + (T - T_0)(c_a - c_v)\right) \\
&+ K3\langle \rho_a \rangle\left(\Delta H_3^0 + (T - T_0)\left(c_a - (nc_c + (1-n)c_g)\right)\right) \\
&+ K4\langle \rho_v \rangle\left(\Delta H_4^0 + (T - T_0)\left(c_v - (ic_c + jc_g + (1-i-j)c_{t2})\right)\right)
\end{aligned}$$

describes the instantaneous energy absorption associated with chemical reactions.

K_{eff} is effective thermal conductivity of porous material, which takes into account thermal conductivity of the solid matrix and gas in pores and radiation in the pore system, as shown in Eq. 12:

$$K_{eff} = \varepsilon \times K_i + \eta \times K_w + (1-\eta) \times K_c + \frac{\varepsilon}{1-\varepsilon} \times \sigma \times d_p \times 4 \times T^3 \tag{12}$$

The following equations also hold true:

$$M_f = \left(\frac{\langle \rho_v \rangle}{\langle \rho_f \rangle M_v} + \frac{\langle \rho_{t2} \rangle}{\langle \rho_f \rangle M_{t2}} + \frac{\langle \rho_g \rangle}{\langle \rho_f \rangle M_g} + \frac{\langle \rho_k \rangle}{\langle \rho_f \rangle M_k}\right)^{-1} \tag{13}$$

$$\varepsilon_f = 1 - \frac{\langle \rho_w \rangle}{\langle \rho_w \rangle^w} - \frac{\langle \rho_a \rangle}{\langle \rho_a \rangle^a} - \frac{\langle \rho_c \rangle}{\langle \rho_c \rangle^c} \quad (14)$$

$$Ki = Ai \exp\left(-\frac{Ei}{RT}\right) \quad (i = 1,2,3,4) \quad (15)$$

The following statements are applied to the boundary conditions and initial conditions:

- The fluidized-bed is supposed to have atmospheric pressure and constant temperature of $T_r = 800°C$. The cellulose particle submerged in the fluidized bed is heated with an overall heat transfer coefficient: $H_{out} = 250\ Wm^{-2}K$. There is no diffusion and convection at the center of the particle. The concentrations of all released species are equal to 0 outside the particle. The initial temperature of the particle is 300K. The initial concentration of cellulose is 450 kgm^{-3}. The appearance density of inert gas in pores is 0.76 Kgm^{-3}.
- The solution of the 10 partial differential equations allows to determine temperature, composition and velocity as a function of both time and space. The system of equations is solved by the method of finite difference.

COMPUTATION RESULTS AND DISCUSSION

TEMPERATURE INSIDE THE PARTICLE

The spatial temperature profiles versus the time are shown in Fig. 2 for a 2 mm particle and in Fig. 3 for a 10 mm particle. In the figures, the particle is divided equally into 10 layers along the radius. At the first stage of temperature build-up (in 1 second in the case of 2 mm particle), the particle temperatures at different positions from the center to the surface rise in the shape of normal exponential temperature-rising profile similar to the field of a pure heat conduction object.

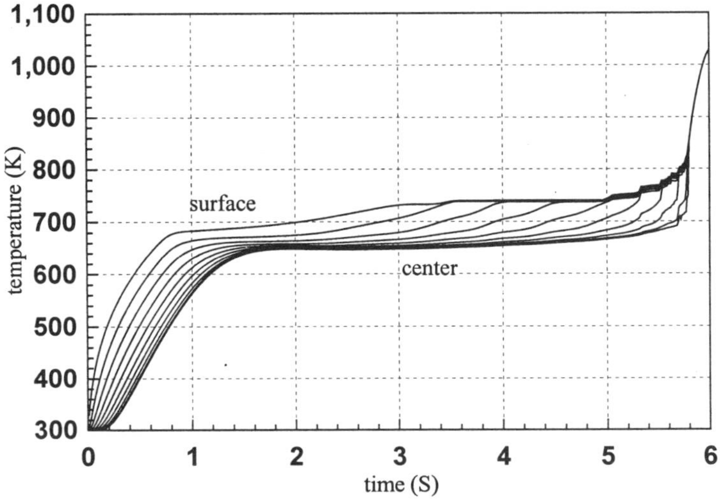

Fig. 2 Particle temperature vs. time (2mm particle).

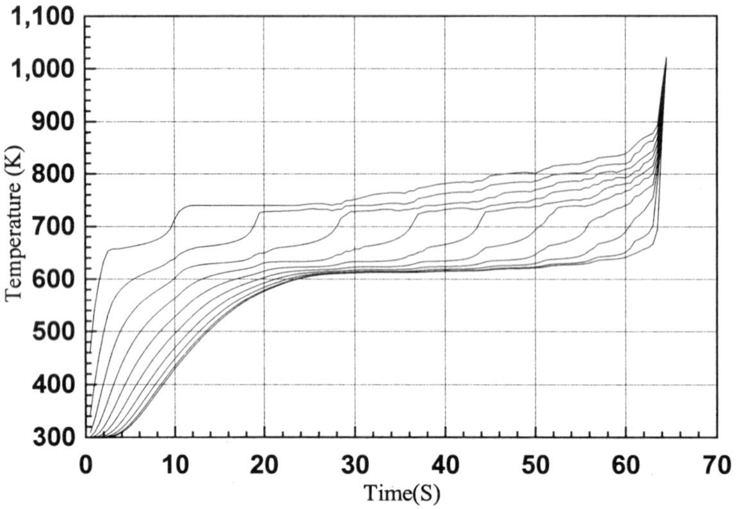

Fig. 3 Particle temperature vs. time (10 mm particle).

The heat from the environment is transferred into the particle layer by layer and become the sensible heat of the cellulose substance. This situation remains until the outside layer of particle reaches an enough high temperature of about 660K, at which, the chemical reactions start to come into effect. Due to the reaction heat needed for pyrolysis, the temperature increase is suppressed significantly, and is turned to the chemical reaction stage of rather flat temperature profiles. The temperature is held between 650 K to 750 K by endothermic pyrolytic reactions, which is much lower than the environmental temperature of 1073K.

The different stages of temperature build-up can clearly be seen to be corresponding to different zones in Fig. 4 where the temperature is plotted against the space variable. Zone A, drawn with the dashed lines, corresponds to the first stage of a pure heating of an object. Zone B, drawn with the solid lines, represents the endothermic chemical reaction stage. Zone C represents the fast temperature increasing stage (see Fig. 2) after the pyrolytic reaction is complete at the time of 5.8s. In this zone, only a char particle of developed porous structure is left. The temperature is uniform over the entire particle space and increases rapidly up to the environmental temperature due to a good access to the heat radiation and cease of endothermic reactions.

It should be emphasized that the endothermic reaction of decomposition plays the central role in determine the particle temperature profile. When pyrolysis begins to proceed intensively (after 1.6s in Fig 4), the reactions take place in out layer, absorb almost all the heat transferred from outer layers and cut down the energy flux to internal layers. This is the reason why the layers situated inside the reacting layers have an uniform and steady temperature variation over the time (see Fig. 2). Consequently, make the pyrolysis process can only proceed layer by layer.

Figure 3 shows the temperature profile of 10 mm size particle, similar to that of the 2 mm size particle. The shapes of these curves are very similar to those of small particle. However, the variation of temperature increasing rate for every layer is much more distinct than the case of 2 mm particle. This is because the high char yield of the larger particle reduces the radiative fraction of thermal conductivity, and increases the resistance of heat transfer.

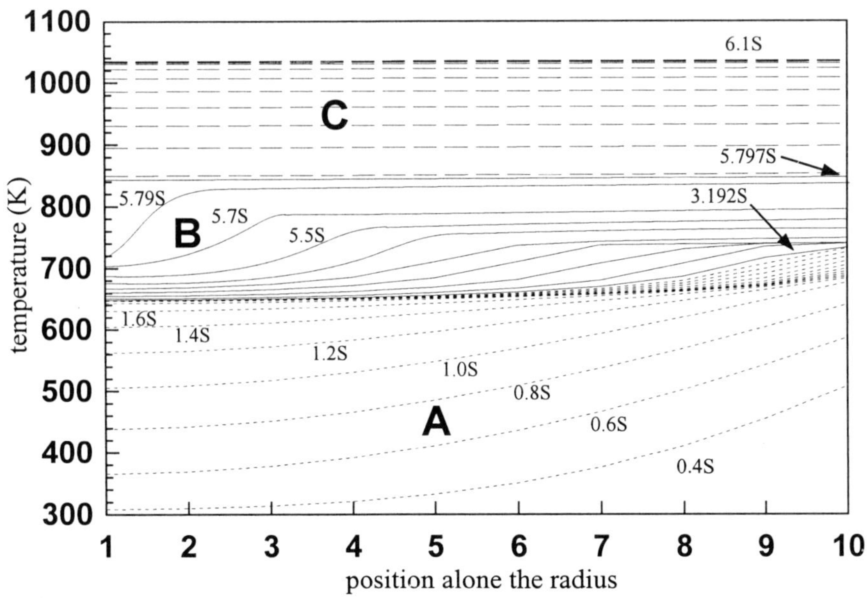

Fig. 4 Particle temperature vs. position (2mm particle)

PRESSURE INSIDE THE PARTICLE

Pressure is an important parameter of biomass particle pyrolysis, which governs the product gas flow through the particle, and consequently influences the mass transfer and heat transfer inside the particle. The results calculated by the model are shown in Figs 5 and 6 for particles of 2 mm and 10 mm in diameter. In Fig. 5, the numbers of '1' to '10' mark each layer from particle center to surface. Three peaks appear in the pressure profile over time, marked with 'A', 'B' and 'C'. The peak 'A' appears before chemical reactions, and is caused by expansion of inert gas in the pores when the temperature rises. The particle internal pressure rises fast to the highest value over the entire pyrolysis process due to high heating rate and comparably compact porous structure in the very beginning. The peak 'C' is caused by biomass devolatilization that proceeds layer by layer into the particle center and produces a large quantity of gaseous products. The small peak 'B' for the inside layers is induced from rapid devolatilization of the surface layer starting at about 0.8s. After peak 'C', the out layers become more porous char structure whose permeability is several order of magnitudes larger than raw cellulose, and gaseous products can flow out of particle with little resistance. Thus, the local pressure level off down to ambient pressure.

The general trend in the pressure profile for the small particle remains the same for the large particle of 10 mm as shown in Fig. 6. But some differences between them can be seen clearly. The peak 'B' disappears, the in-particle pressure has reached a much higher level and the pressure level difference between the peaks 'A' and 'C' is enlarged dramatically. These can be attributed to the low permeability of both reacting layers and char layers, leading to higher resistance inside the large particle against gas flow through the particle.

Unlike coal, the internal pressure of a biomass pyrolytic particle under fluidized bed condition is very low due to the high permeability of cellulose and char structure.

Even for the large particle of 10mm, the maximum internal relative pressure inside the particle is only a little more than 70Pa. Under such a low pressure, the physically cracking process caused by internal pressure can not be the case for biomass particles as it is for coal particles.

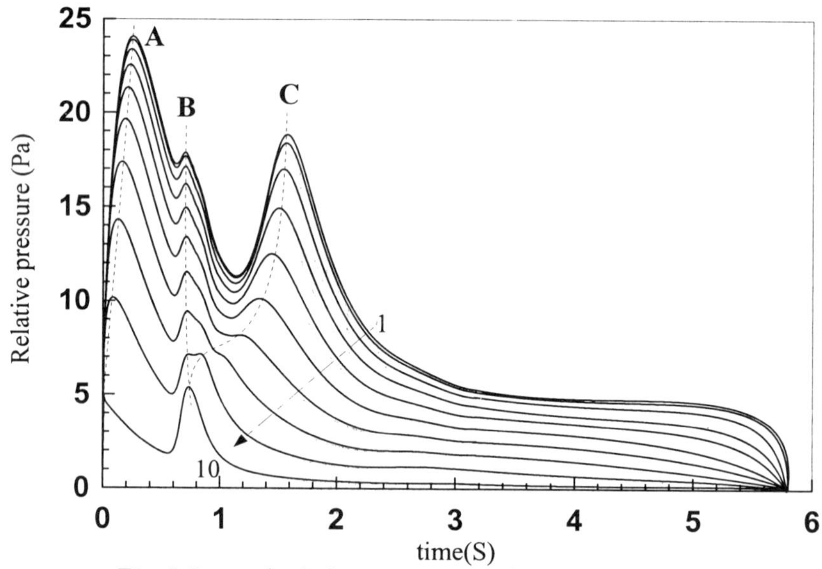

Fig. 5 Internal relative pressure vs. time (2mm particle)

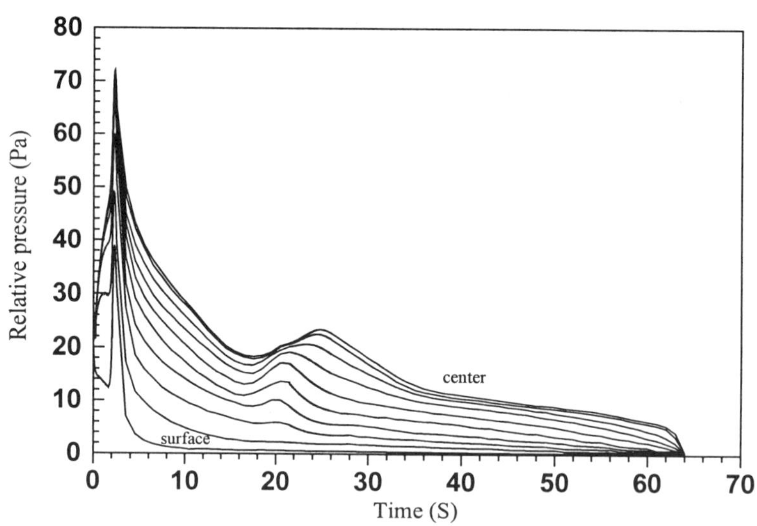

Fig. 6 Internal relative pressure vs. time (10mm particle)

GAS VELOCITY INSIDE THE PARTICLE

The gas phase contains inert gas from the original pores in cellulose, volatile from the primary decomposition, the second tar from the secondary reaction and permanent gas from both of the above two reactions. The flow of this gaseous mixture is driven by pressure gradient through the particle and is governed by many factors such as temperature, reaction rate, gas properties and porous structure of the particles.

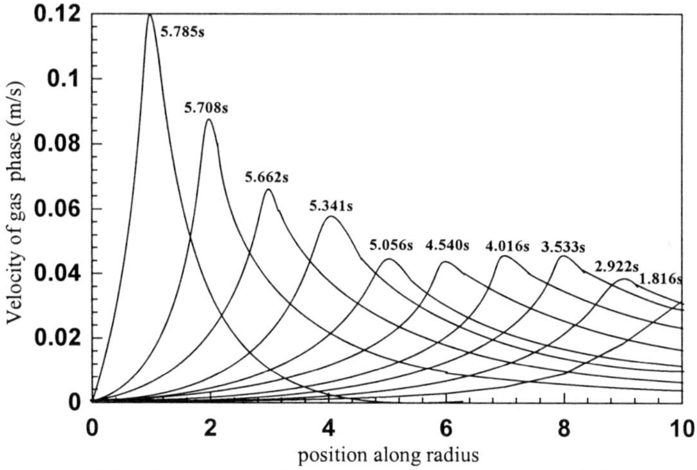

Fig. 7 Gas velocity vs. position (2mm particle)

The gas phase velocity distribution over the particle space is shown in Fig. 7. This figure consists of a serial of velocity distributions along the radius at different moments. Those moments are chosen for the reason that the gas velocities of local layers reach their maximum value at the moments. It should be mentioned that those moments just match the time when the active cellulose decomposition rates for corresponding layers reach the maximum values. It suggests that the gaseous product from active cellulose primary decomposition contributes the main part of the gas phase flow.

The maximum gas velocity at the moment of 5.787s is higher than others. This can be explained by particle size effect. A small particle has a shorter reaction period and a high reaction rate, which leads to a higher gas velocity.

LIQUID ACTIVE CELLULOSE

Concerning active cellulose, the most attention is given to its phase state. As recently proposed by Lede [9][3], the temperature of this intermediate substance is a key parameter. Active cellulose is believed to undergo a phase change at temperature around 740K. In slow pyrolysis as used for thermogravimetric analysis, active cellulose tends to be consumed in solid phase before it reaches the phase-change temperature, while in flash pyrolysis, visible liquid intermediate can be found at the high temperature frontier of reacting zone.

The flow of liquid active cellulose may have a significant influence on pyrolysis if it exists to a certain extent. The case of middle intensive pyrolysis is identified as shown in Figs 8 and 9. The existing zone of liquid active cellulose, in which the temperature is above 740K, can be found at the right-down corner of Fig. 8 for the small particle of 2 mm. The liquid active cellulose only appear in the internal layers of a limited volume at the very end of reaction period with quite low densities. The amount of liquid active cellulose is likely negligible. It is estimated that the mass of the active cellulose that flows out of the particle is less than 0.005% of origial cellulose mass. For the large cellulose particle of 10 mm, no liquid active cellulose at all can be found from the model calculation as shown in Fig. 9. Therefore, it can be concluded that the effect of active cellulose is negligible for the pyrolysis under middle intensive heating condition. For the case of flash pyrolysis, intensive heating supplies endothermic reactions with sufficent heat in time and raises the temperature of the solid active cellulose product above 740K immediately so that a large quantity of liquid active cellulose may emerge from the particle surface.

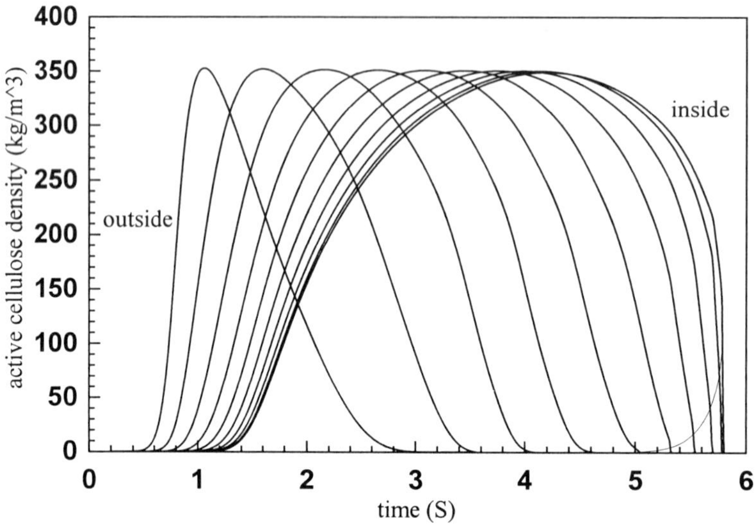

Fig. 8 Active cellulose density vs. time (2mm particle)

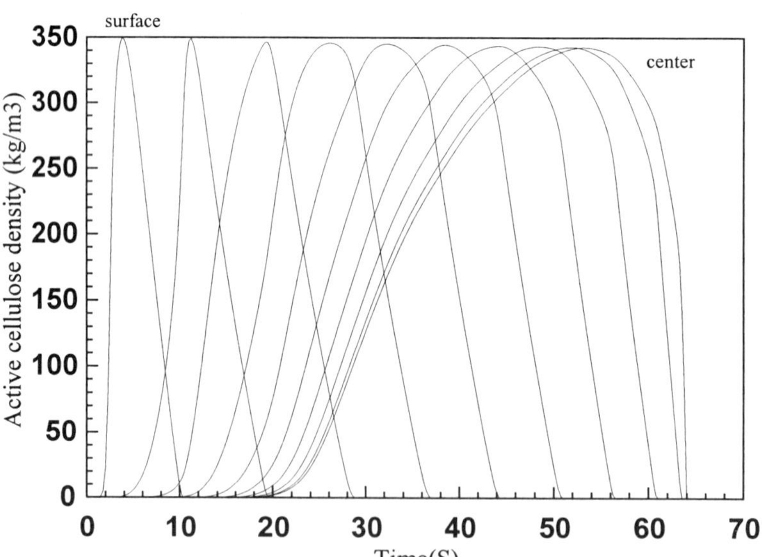

Fig. 9 Active cellulose density vs. time (10mm particle)

CELLULOSE DEVOLATILIZATION

Only cellulose is the feedstock of pyrolysis process treated in the model. Cellulose mass consumption represents the degree of the pyrolytic conversion. Figures 10 and 11 show the mass losing curves as pyrolysis proceeds for 2 mm and 10 mm particles respectively. As can be seen that cellulose density decreases all the way down to zero. The density in the outside surface layer decreases much faster than the center. It takes about 5s for the 2 mm cellulose particle to be devolatilized completely, and 60s for the 10mm particle. It also shown that devolatilization of cellulose particles proceeds layer by layer, which is more obvious for the outside layers and for the large particle.

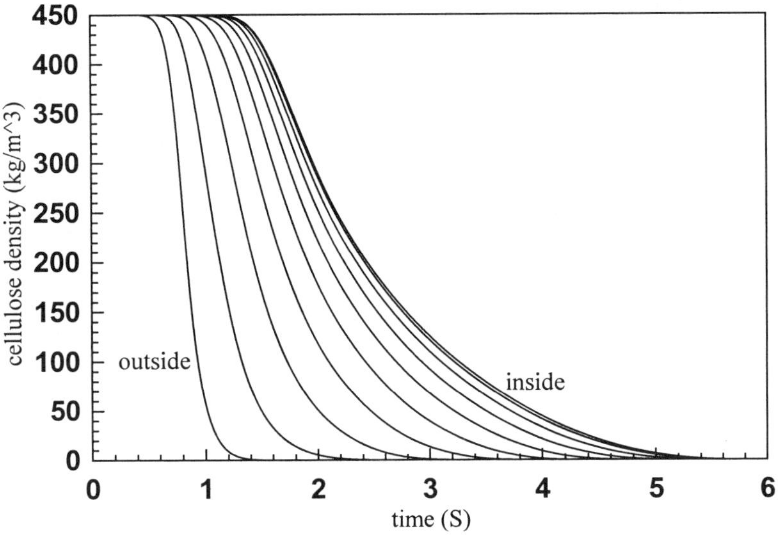

Fig. 10 Cellulose density vs. time (2 mm particle)

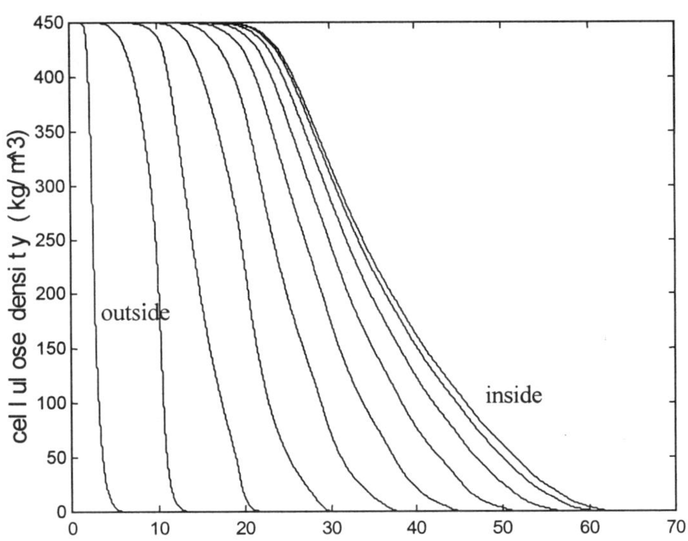

Fig. 11 Cellulose density vs. time (10 mm particle)

PRODUCT CHAR

Char is the only solid product of cellulose pyrolysis and constitutes the reaction-finished layer behind the reacting frontier. Figs 12 and 13 show the char density variation in solid curves and the density change rate in dashed curves (differential curves) as a function of time, i.e. char accumulation and production rate in each particle layer. The numbers of '1' to '10' again represent the particle layer from the center to the surface. Production of char can be divided into two stages. The first one is characterized by the peaks marked by numbers on the dashed curves, which correspond to the existing periods of active cellulose for every layer. Therefore, these peaks of char production are the results of primary decomposition. The second stage takes place after the end of pyrolysis reaction, which is marked by overlapping peak 'S' for every layer. This is because the remaining volatile in porous char structure stays still, and undergoes the secondary reaction at a high temperature. Before this stage, the secondary reaction is less important.

The two cases of different particle size show similarity of general variation trend and also some differences. The char yield for the large particle is higher due to a low heating rate during the pyrolysis process. The rate of increase in char density of the outer layers during the late stage of pyrolysis accelerates, which is associated to the secondary reaction of the primary volatiles.

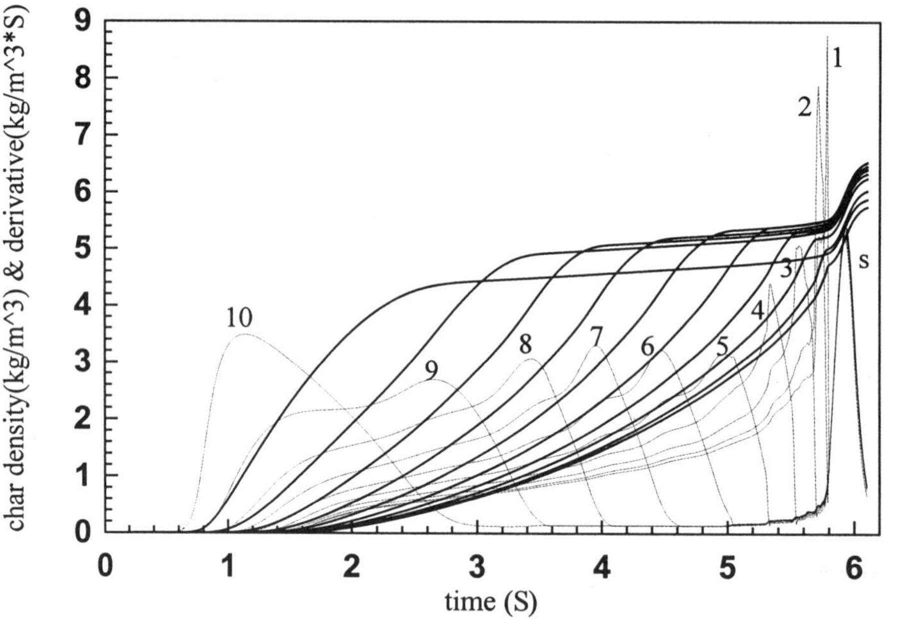

Fig.12 Char density vs. time (2mm Particle)

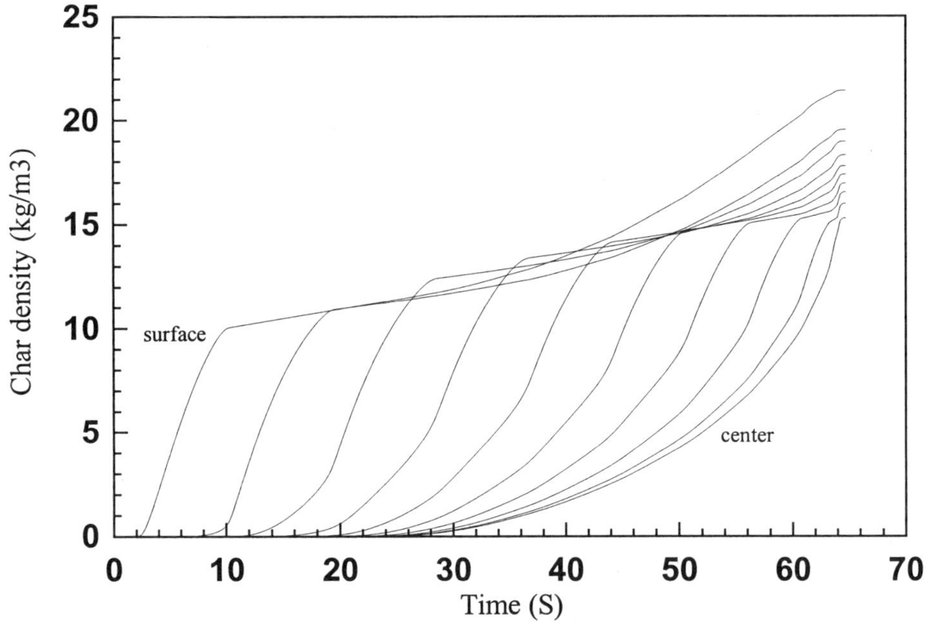

Fig.13 Char density vs. time (10mm particle)

PRODUCT VOLATILES

The volatile concentration is determined by the rate of volatile generation, the gaseous phase flow and cracking of the secondary reaction. The spatial distribution of volatile concentration over a 2 mm particle is shown in Fig. 14. The curves labeled with '1' to '10' correspond to the curves from 3.192s to 5.797s in Zone B of Fig. 4. It can be seen from Figs 4 and 14 that the volatile concentration in the reacting outer layers increases rapidly in the beginning of pyrolysis (at the moment of 0.6s). This trend holds until the surface layer is consumed at the moment of 3.192s (curve '1'). Afterwards, the improved permeability of outer layer enhances the outward product flow and suppresses the increasing of volatile concentration. As the pyrolysis frontier steps into the internal small layers, the quantity of produced volatile decreases. Consequently, the volatile concentration in outer layers begins to decrease from curve '5'. After curve '10' (moment 5.797s), the primary reaction stops and the outward flow becomes negligible. At this stage, the volatile inside particle char matrix is almost stagnant with an uniform distribution of the concentration.

Figure 15 shows the spatial distribution of volatile flux in a 2 mm particle. All the solid curves for time 3.92s to time 5.797s give a flat profile on the right hand side. For the moment specified by each curve, the primary decomposition has finished locally in the outer layers. This is also the case for a larger particle of 10 mm diameter. Thus, it is suggested that the secondary reaction plays a negligible role during this stage. In other word, volatile interior-particle crack during cellulose pyrolysis is not important under the fluidized-bed condition. From the model prediction, 0.088% of volatile undergoes the secondary crack inside a 2 mm particle, and about 2% for a 10 mm particle.

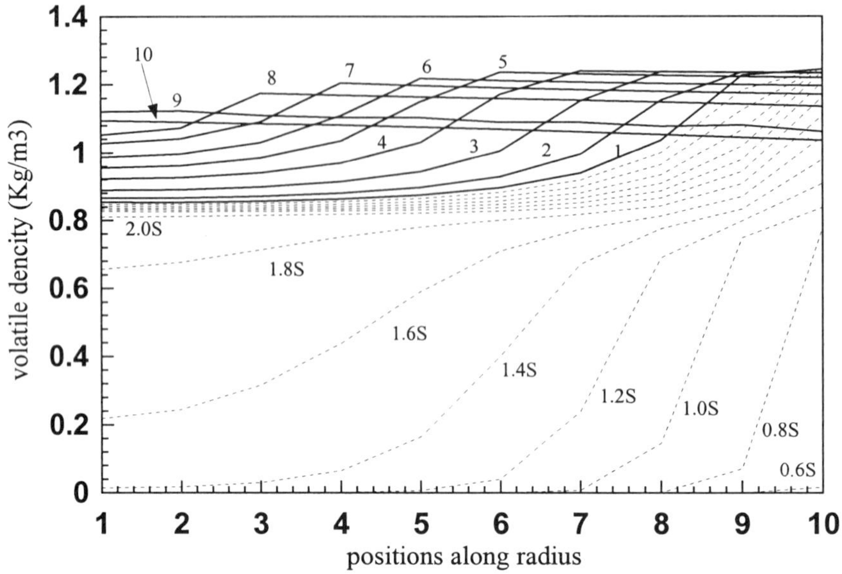

Fig. 14 Volatile concentration vs. position (2mm particle)

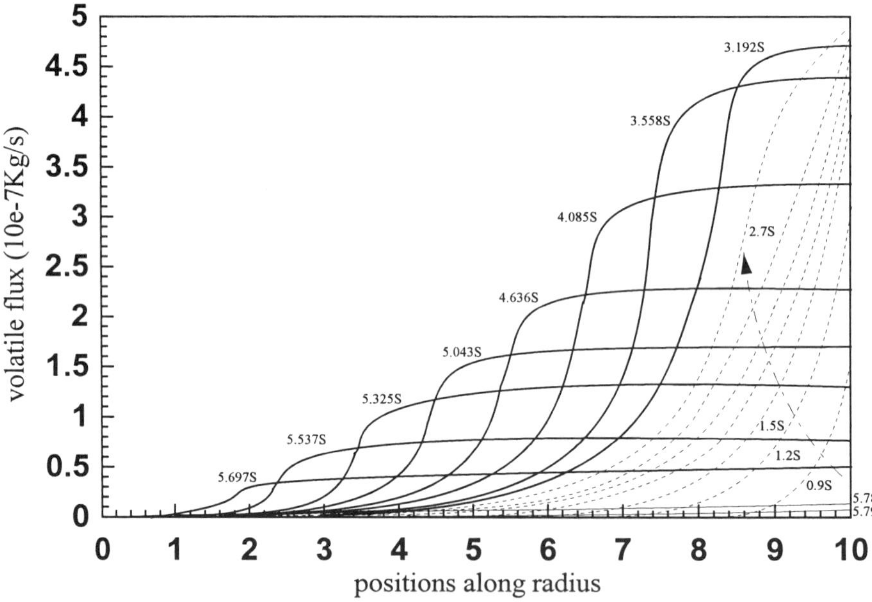

Fig. 15 Volatile flux vs. position (2 mm particle)

CONCLUSION

A comprehensive mathematic model describing the pyrolysis process of single cellulose particle under a fluidized bed condition is built up by formulating the transport phenomena and chemical reactions involved. The model addresses the most important parameters such as temperature, pressure, velocity and concentration over time and space, and gives reasonable predictions. The computation results show that the pyrolytic reactions take place in the form of layer by layer inside the particle. The reaction temperature is much lower than the environmental temperature. The endothermic primary decomposition reaction against a slow heat transfer through the particle plays the most important role in governing the biomass pyrolysis process. The particle permeability is the key parameter for the in-particle pressure and the heat and mass transfer through the particle. The appearance of liquid active cellulose is nearly impossible under the simulated pyrolytic condition even for small particles (2 mm). An extremely small part of volatile can be consumed by internal-particle secondary cracking.

ACKNOWLEDGEMENTS

The authors would like to acknowledge the financial support of EU under Project Contract: JOR3CT980306, for the realization of this work. The co-operation of the partners involved in the EU-project is also greatly appreciated. Thanks should also go to Swedish National Energy Administration for a part of the financial support

NOMENCLATURE

μ	dynamic viscosity
ρ	density
σ	Stefan-Bolzman constant
K_i (i=1,2,3,4)	Arrhenius rate
E_i(i=1,2,3,4)	activation energy
A_i(i=1,2,3,4)	pre-exponential factor
u	velocity
P	pressure
R	universal gas constant
M	molecular weight
C	heat capacity
T	temperature
ΔH_i(i=1,2,3,4)	heat of reaction
K_{eff}	effective thermal conductivity
ε	porosity
d_p	pore diameter
τ	time
η	level of conversion $(\langle\rho_a\rangle+\langle\rho_w\rangle)/\langle\rho_w\rangle^0$

Subscripts

w	cellulose
c	char
a	active cellulose
v	volatile
t_2	secondary tar
g	permanent gas

i	inert gas
f	total gas phase ($v+t_2+g+i$)

Superscript

0	initial state
$<>^{a}$	variable based on volume occupied by active cellulose
$<>^{f}$	variable based on volume occupied by total gas phase
$<>^{w}$	variable based on volume occupied by cellulose
$<>^{c}$	variable based on volume occupied by char

REFERENCES

1. Jalan, R.K. and V.K. Srivastava, (1999) Studies on pyrolysis of a single biomass cylindrical pellet – kinetic and heat transfer effects. *Energy Conversion & Management*, Vol.40, 467-494
2. Di Blasi, C., (1996) Heat, momentum and mass transport through a shrinking biomass particle exposed to thermal radiation, *Chemical Engineering Science*, Vol.5, No.7, 1121-1132
3. Lede, J., (1994) Reaction temperature of solid particles undergoing an endothermal volatilization. Application to the fast pyrolysis of biomass. *Biomass and Bioenergy*, Vol.7, No.1, 49-60
4. Bradbury, A. G. W. et. al. (1979) A kinetic model for pyrolysis of cellulose. *Journal of applied polymer science,* Vol.23: 3271-3280
5. Varhegyi G.P. et. al. (1994) Kinetics of the thermal decomposition of cellulose under the experimental conditions of thermal analysis. Theoretical extrapolations to higher heating rates. *Biomass and Bioenergy*, Vol.7 No. 1, 69-74
6. Liden, A.G. et. al. (1988) A kinetic model for the production of liquid from the flash pyrolysis of biomass, *Chem. Eng. Comm.* Vol.65, 207-221
7 Di Blasi, C.m (1996) Kinetic and heat transfer control in the slow and flash pyrolysis of solid. *Ind. Eng. Chem. Res*, Vol.35, 37-46
8. Boutin, O. et. al. (1998) Radiant flash pyrolysis of cellulose — evidence for the formation of short life time intermediate liquid species. Journal of Analytical and applied pyrolysis, Vol.47, 13-31
9. Lede, J. et al., (1996) The nature and properties of intermediate and unvaporized biomass pyrolysis martierial, *Development in thermochemical biomass conversion*, IEA Bioenrgy, Editor: A.V. Bridgwater, D.G.B. Boocock, p. 27-41

Modeling Potassium Release In Biomass Pyrolysis

Chunjiang Yu [a,b] and Wennan Zhang [a]

[a] *Department of Applied Science, Mid Sweden University, 871 88 Härnösand, Sweden*

[b] *Institute for Thermal Power Engineering, Zhejiang University, 310027 Hangzhou, P.R.China*

ABSTRACT: Alkali compound emission during biomass thermochemical conversion gives rise to a number of problems such as agglomeration, slag, fouling and metal corrosion in the conversion process system. Understanding the behaviour of alkali emission from biomass fuels is important to solve these problems. In this work, a study focused on the potassium compound release in biomass pyrolysis is carried out. The transformation of the element, K, is interpreted based on literature and the present study. A mathematic model of the potassium compounds emitting from biomass during pyrolysis is proposed. The different existing forms of potassium in biomass, the chemical equilibrium of the compounds at the pyrolytic environment and the potassium compound release due to vaporization at high temperature are taken into account in the model. The result of the mathematic model provides an understanding of the alkali compound release in biomass pyrolysis and the effects of the pyrolytic temperature, the fuel composition and residence time on the release process.

INTRODUCTION

Biomass particles undergo a rapid heating once fed into a reactor of thermal conversions. The thermal degradation occurs in the first stage, which destroys the matrix structure of the main constituents in biomass: cellulose, hemicellulose and lignin. In the same time, the trace inorganic elements that are bonded in biomass fuel in different forms are released and transformed into various compounds. Some of them may form volatile compounds and are vaporized easily during the process, while others reside in char matrix as a part of ash. The alkali compounds released give rise to a number of problems in thermochemical conversion systems, such as bed material agglomeration and sintering of fluidized beds, slagging on refractory-lined walls in the furnace, fouling on heat transfer tubes, deposit-related corrosion of superheaters, filters, gas turbines and engines [1][2]. Alkalis enriched micron-size char particles in pyrolytic bio-oil present a potential problem to the combustion in engines.

Pyrolysis is the first and essential step in biomass thermochemical conversions, which is supposed to determine, basically, the quantity and forms in which alkalis are distributed into gas, liquid and solid phases. In the applications, for instance, the fuel gas from pyrolysis or gasification can be a good quality fuel with less alkali for reburning, while the remaining char and ash are alkalis-enriched and can be considered for fertilizer. In order to find solutions to the alkalis-related problems, the pyrolysis of biomass particle with respect to the alkali compound release needs to be studied in detail.

Potassium in biomass exists as dissolved salts in the moisture or as a cation attached to carboxylic and other functional groups, which is highly mobile. It is most possibly bonded to chlorine as KCI if chlorine is available in biomass fuels [3]. The most possible potassium compounds released from biofuels are KCI, K_2SO_4 and KOH [1]. A measurement by Surface Ionization Mass Spectrometry technique [4] showed that potassium release to the vapour phase in pyrolysis takes place in two stages. The first stage is in the temperature range of 200 C° to 500 C°. The potassium release can be well coupled to the biomass devolatilization, and insensitive to the chlorine content in the fuel. In the second stage of temperature above 500 C°, the potassium release is at its full speed, depending on the chlorine content in the fuel.

Based on reported experimental studies, it can reasonably be assumed that the potassium release completes in two stages of low and high temperatures. In the first stage of low temperature, the organic bonded potassium releases following devolatilization of biomass constituents, cellulose, hemicellulose and lignin. In the second stage of high temperature, the inorganic compounds of potassium such as KCI and KOH are vaporized into the vapour phase. The present study will give a quantitative description of potassium release in the two stages, respectively.

EXPERIMENTAL

Alkalis contained in biofuels vary in a wide range depending on biomass sources. Agricultural residues such as straw are on the top in the alkalis content, followed by grass, short rotation coppices and bark of trees. Woody biomass has the least content of alkalis. In the present work, a straw is used as a typical fuel for potassium release study. The elementary analysis of the straw is given in Table 1.

Table 1 Composition of the straw used in the experiment (w% on dry basis).

C	H	O	N	S	Cl	Si	Ca	K	others
43.2	4.96	37.7	1.22	0.05	0.56	7.61	0.29	2.22	2.19

In the experiment, the straw was heated up in a lab-scale pyrolyzer with external heating element. The temperature in the pyrolyzer was increased gradually and kept at about 800 C° for a period of time. The temperature of the fuel was registered as a function of time. The temperature history of the straw pyrolysis is shown in Fig. 1.

The residual char was weighted and analyzed for determination of inorganic content. 0.39 kg char per 1 kg straw was obtained at the end of the test. The inorganic content in the residual char is given in Table 2.

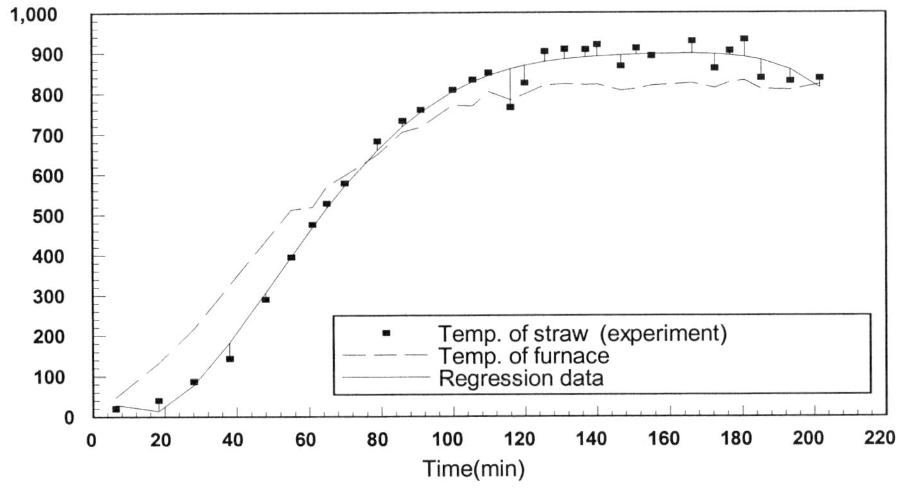

Fig. 1 Temperature history during the experimental test of biomass pyrolysis.

Table 2 Inorganic content in the residual char after pyrolysis (%w)

K_2O	MgO	CaO	P_2O_5
4.44	2.98	2.5	0.5

Thus, the potassium release to the vapor phase in the test can be calculated from the data of Table 2 as expressed below:

Potassium release = (potassium in fuel – potassium in residual char)/potassium in fuel
= 35.3 %

MATHEMATIC MODEL

The potassium release is divided into two parts, the organic bonded potassium release and the inorganic potassium compound vaporization. The amount of the organic potassium in biomass can be determined experimentally with the chemical fractionation method. The chemical fractionation technique distinguishes different types of inorganic material according to their solubility in a series of increasingly aggressive solvents. Those materials soluble in water are soluble salts and loosely bonded material. Those soluble in ammonium acetate solvent are the organic bonded materials (i.e. ion exchangeable by ammonium acetate). Those soluble in hydrochloric acid are typically carbonates or sulfates. Those not soluble in any of these solvents are commonly in the form of oxides, silicates or sulfides. The literature data for straw are shown in Table 3, which gives the organic bonded potassium of 13%.

Table 3 Chemical fractionation results of straw (from [5])

Element	Water Soluble %	Ion Exchangeable %	Acid Soluble %	Residual %
Si	0	0	0	100
Na	93	7	0	0
K	87	13	1	0
S	16	0	0	84
Cl	100	0	0	0

ORGANIC BONDED POTASSIUM RELEASE

The modeling of the organic potassium release is based on the following assumptions:

1) There is no big difference in distribution of organic potassium between three main biomass components of cellulose, hemicellulose and lignin.
2) The influence of other inorganic materials and moisture content in biomass on organic bonded potassium release is negligible due to slow reactions at low temperature and short devolatilization time of biomass. .
3) The organic potassium release follows devolatilization of cellulose, hemicellulose and lignin, respectively.
4) All organic potassium is released to the vapor phase through carry-over of sub-micro particles with gas flow.

The kinetic model of biomass devolatilization is employed to describe the potassium release process as given in Eq. 1.

$$\frac{d\alpha_i}{dt} = A_i \exp\left(-\frac{E_i}{RT}\right)(1-\alpha_i) \qquad (i=1,2,3) \tag{1}$$

where i=1, 2, 3 denote cellulose, hemicellulose and lignin. α_i is the fraction of potassium released, A_i and E_i are the pre-exponential factor and the activation energy. The data are introduced from literature as given in Table 4.

Table 4 Parameters of potassium release model

Parameters	Value, unit
Activation energy E_1[6]	154000 kj/mol
Activation energy E_2[6]	199000 kj/mol
Activation energy E_3[6]	34000 kj/mol
Preexponential factor log A_1[6]	13.9 min^{-1}
Preexponential factor log A_2[6]	16.6 min^{-1}
Preexponential factor log A_3[6]	1.7 min^{-1}
Content of hemicellulose[7]	18%
Content of cellulose [7]	36%
Content of lignin [7]	16%
Constant in Antoine Eq. for KCl[8]	A=4.782 B=7440.691 C=-122.709
Constant in Antoine Eq. for KOH[8]	A=4.028 B=5854.906 C=-144.113

Incorporating the relationship of temperature verse time used in the experimental test, the organic potassium release is calculated as shown in Fig. 2. It can been seen from the figure that the organic potassium release starts at 180 C° as lignin devolatilizes first. The rapid release appears at temperature about 300 C° up to 400 C° due to fast devolatilization of cellulose and hemicellulose. Afterwards, the potassium release is slowed down by slow devolatilization of lignin. Up to the temperature of 500 C°, the organic bonded potassium is released completely.

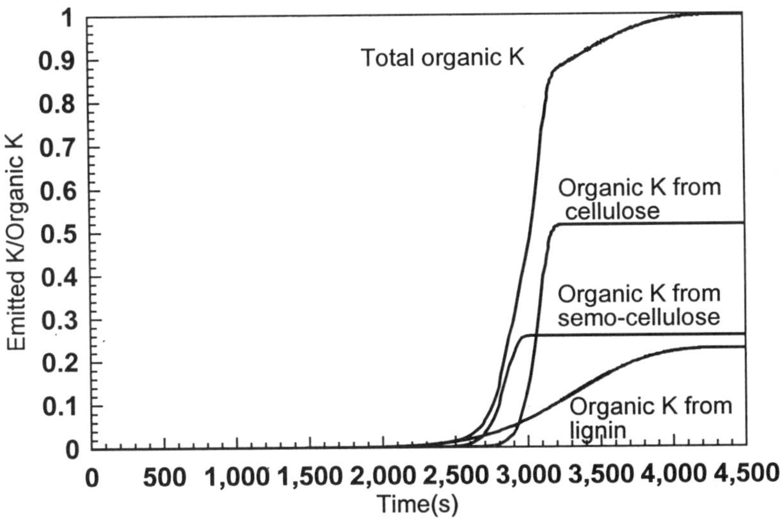

Fig. 2 Organic potassium release during biomass pyrolysis.

INORGANIC POTASSIUM COMPOUND VAPORIZATION

The modeling of inorganic potassium release is based on the following assumptions:

1) 30% of potassium bonded in KCl reacts with silica oxide to form the eutectic compund into solid phase at a low temperature [3].
2) In addition to KCl and K_2SO_2, KOH is regarded as the only potassium compound produced from the reactions of potassium salts and moisture. The reactions are complete before vaporization.
3) KCl and KOH become condensed small particles in biomass after drying, and vaporize into the gas phase by molecule dispersion.
4) The vaporization takes place under a stagnant environment, on resistance-free particle surface and at zero partial pressure of the potassium compounds far from the particles.
5) K_2SO_4 is supposed to remain in the solid phase in comparison with the volatilities of KCl and KOH.

The above assumptions define the potassium release process as a mass transfer from the particle surface into the vapor phase. The vapor partial pressure of KCl and KOH vary with temperature, which can be calculated by the well-known Antoine equation:

$$\log_{10}(P) = A - (B/(T + C)) \tag{2}$$

in which, P is the saturation vapor pressure (bar), T the compound temperature and A, B, C are constants. The vapor dispersion process can be described with *Sherwood number*,

i.e. $$Sh = k_A D / \Gamma_A \tag{3}$$

where k_A is the mass transfer coefficient of A in a medium (m/s), D the particle diameter and Γ_A the mass diffusivity of A in a medium (m^2/s). Assumption 4. leads to Sh = 2. Thus, from Eq. 3, the release rate of A on the particle surface, $\Phi_{mol\square A}$ (mol/s), is expressed as the following equation,

$$\Phi_{mol\square A} = k_A\, A\, C_A = 2\ \Gamma_A/D\ \pi D^2\ P/RT \tag{4}$$

where A is the particle surface area (m^2), C_A mole concentration of A on the particle surface (mol/m^3) and R the ideal gas constant.

The KCl and KOH diffusivities can be estimated with Hirschfelder's equation [9]:

$$\Gamma_A = \frac{1.8583 \times 10^{-3} T^{1.5}}{p_0 (\sigma_{AB})^2 \Omega_D} \left(\sqrt{\frac{1}{M_A} + \frac{1}{M_B}} \right)^{0.5} \tag{5}$$

The above equation expresses the diffusion of vapor A through B medium, in which, M_A and M_B are the molecular weight (kg/kmol), P_0 the absolute pressure, σ_{AB} and Ω_D are the molecule collision diameter and the molecule collision integral which are functions of substance properties and can be determined by empirical equations [10].

Assume Q_A (kg) to be the total A vapor mass released, the following relation between Q_A and D holds true:

$$D = \sqrt[3]{D_0^3 - 6Q_A / \rho\pi} \tag{6}$$

where D_0 is the initial diameter of the particles and $\rho$the density of A substance. Putting Eq. 6 into Eq. 4 leads to the expression of A substance release rate by the vaporization of the substance:

$$\frac{dQ_A}{dt} = 2\pi\Gamma_A(t)\sqrt[3]{D_0^3 - 6Q_A / \rho\pi}\,\frac{P(t)}{RT(t)} M_A \tag{7}$$

Thus, KCl and KOH releases can be calculated from Integration of Eq. 7 by incorporating the fuel temperature history measured in the experimental test. The results are shown in Fig. 3.

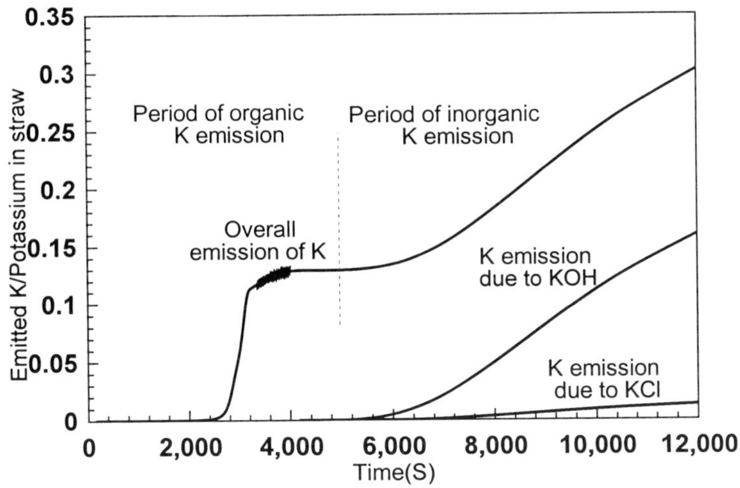

Fig. 3 Potassium release during biomass pyrolysis.

DISCUSSION

The potassium releases from straw pyrolysis in different ways are shown in Fig. 3 for comparison. The sum of the organic potassium release and the inorganic potassium compound vaporization gives the overall potassium release. At low temperature of 200 to 500 C° indicated on the left of the dashed line, the inorganic compound vapor pressures are kept at a too low level to give any contribution to the total potassium release. In this stage of pyrolysis, the potassium release is entirely attributed to the organic bonded potassium release, following the biomass devolatilization process. At the temperature of 500 C°, the all organic potassium in the fuel is released (13% of total K in the present work). After this, the release process is switched to the inorganic potassium release indicated on the right of the dashed line. The KOH vaporization starts at a temperature of about 800 C°, and hold a high vaporization rate as the temperature is kept about 900 C°. The KCl vaporization, on the other hand, starts at a higher temperature of about 850 C° and hold a low vaporization rate under the same condition.

From the model computation, 15.89% of total fuel potassium is released in the form of KOH, While 1.29% of total fuel potassium is released in the form of KCl. The total potassium release (plus the organic potassium release) is calculated from the model to be 30.18%. This value is fairly in agreement with the experimental measurement (35.3%) in the present work.

After vaporization, KOH may be converted into KCl by reaction with HCl depending on the chlorine content in the fuel. Chlorine contained in the fuel is released already, to a great degree, as HCl in the biomass devolatilization stage.

In summary, a small amount of potassium is released in the biomass devolatilization stage due to the limited content of organic bonded potassium in the fuel. The main potassium release occurs at a high temperature by means of vaporization of potassium compounds. The pyrolytic temperature, fuel properties and the residence time have an significant influence on the potassium release to the gas phase. In the case of a lower temperature of 800 C° being maintained instead of 900C° in the present study, the inorganic potassium release will be reduced to 6.26% as calculated from the model. Thus, 80.7% of potassium can be kept in the solid residue of pyrolysis. Therefore, low temperature pyrolysis and gasification show a great advantage when a high alkali content biofuel is treated.

CONCLUSION

Alkalis-related problem in the thermochemical conversion of biofuels is one of the main limits to the increasingly utilization of biomass. A theoretical study focusing on the potassium release from biomass pyrolysis is carried out. The mathematic model built up in the study gives computation of potassium release in biomass pyrolysis in a good agreement with the experimental measurement. According to the model, the potassium release in the biomass pyrolysis can reasonably be explained as two parts, organic bonded potassium release and inorganic potassium compound vaporization. The latter part plays a more important role in determining the amount of alkalis released to the vapor phase. The alkalis release depends on fuel properties, pyrolytic residence time and the pyrolytic temperature. Keeping a lower temperature from 900C° to 800 C° will capture alkalis in the solid residue by more than 80% under the conditions of the present study.

ACKNOWLEDGEMENTS

The authors would like to acknowledge the financial support of EU under Project Contract: JOR3CT980306, for the realization of this work. The co-operation of the partners involved in the EU-project is also greatly appreciated. Thanks should also go to Swedish National Energy Administration for a part of the financial support.

REFERENCES

1. Bryers, R. W. (1996) Fireside Slagging, Fouling, and High-temperature Corrosion of Heat-transfer Surface Due to Impurities in Steam-raising Fuels. *Prog. Energy Combust. Sci.* Vol.22 22-120.
2. Valmari, T., Lind, T. M., Kauppinen, E.I., Sfiris, G., Nilsson, K. and Maenhaut, W. (1999) Field study on ash behavior during Circulating Fluidized-Bed Combustion of Biomass.2.Ash Deposition and Alkali Vapor Condensation. *Energy & Fuels,* 13, 390-395.
3. Björkman, E. and Strömberg, B. (1997) Release of Chlorine from Biomass at Pyrolysis and Gasification Conditions, *Energy & Fuel,* 11,1026-1032.
4. Olsson, J. G., Jäglid, U. and Pettersson, J. (1997) Alkali Metal Emission during Pyrolysis of Biomass. *Energy & Fuel*, 11, 779-784.
5. (1996) Alkali Deposits Found in Biomass Boilers: The behavior of inorganic material in biomass–fired power boilers – field and laboratory experiences, NREL/TP-433-8142, SAND96-8225 VII.
6. Hsisheng, T., and Wei, Y.C. (1998) Thermogravimetric Study on the Kinetics of Rice Hull Pyrolysis and Influence of Water Treatment. *Ind. Eng. Chem. Res.*, 37,3806-3811
7. Orfao, J. J. M., F.J.A. Antunes and J.L. Figueiredo (1999) Pyrolysis kinetics of lignocellulosic material- three independent reactions model. *Fuel*, 78, 349-358

8. Stull, D.R. (1947) Vapor Pressure of Pure Substances Organic Compounds. *Ind. Eng. Chem.*, 39, 517-540
9. Hirschfelder, J. O., R. B. Bird and E. L. Spotz (1949) *Chem. Rev.* 44. 205
10. Brokaw, R.S. (1969) *Ind. Eng. Chem.* Proc.Des.Develop, 8, 240

Comparative study on char properties and pyrolysis kinetics of different lignocellulosic wastes

P.R. Bonelli, P.A. Della Rocca, G.E. Cerrella, A.L. Cukierman*
Programa de Investigacion y Desarrollo de Fuentes Alternativas de Materias Primas y Energia (PINMATE) - Departamento de Industrias - Facultad de Ciencias Exactas y Naturales. Universidad de Buenos Aires. Intendente Guiraldes 2620. Ciudad Universitaria - (1428) Buenos Aires, Argentina.

ABSTRACT: Chemical composition, heating values and surface properties of the char samples generated from slow pyrolysis of different abundant lignocellulosic wastes under identical operating conditions were examined to assess comparatively their suitability for potential use as biofuel and/or for further processing to produce activated carbons. Sawdust from two different wood species (*Aspidosperma australe, Populus deltoide)*, peanut husks, and olive stones were selected. Proximate and ultimate analyses as well as physical adsorption measurements of nitrogen and carbon dioxide were performed. Heating values were estimated from a multivariate correlation in terms of samples' elemental composition. All the chars obtained are found potentially useful as relatively pollution-free solid biofuels, the char arising from sawdust of *Aspidosperma* wood having the greatest potential. Comparison of surface properties with those determined for the olive stones-char, taken as a reference, suggests that the chars from peanut husks and from sawdust of *Populus* wood could be potential adequate candidates for further conversion into activated carbons. Kinetic measurements of the wastes pyrolysis were additionally conducted by non-isothermal thermogravimetry over the temperature range 25°C~900°C. Kinetic parameters were estimated by applying a published model, which assumes a steadily increasing variation in the activation energy with the course of pyrolysis and allows effective representation of experimental data for all the selected wastes over the whole range of temperatures. Appreciable differences in the estimated kinetic parameters are found.

INTRODUCTION

Large amounts of lignocellulosic wastes generated in processing of wood and agricultural products constitute attractive potential raw materials for production of charcoal and activated carbons, especially in those countries where economies are strongly based on agriculture and forestry. Conversion may additionally contribute to an overall waste reduction strategy in order to diminish environmental pollution arising from accumulation, landfilling and/or open field burning (1-3).

---

* Corresponding author. Phone: 54-11-45763383. Fax: 54-11-45763366. E-mail: analea@di.fcen.uba.ar

Slow pyrolysis has been recognised as a convenient process to attain high yields of the solid product (char). The char has potential as a solid fuel. In addition, activation of the char to produce activated carbons may represent another economically interesting option. Commercially available activated carbons are highly porous carbonaceous materials with high BET surface areas typically in the range from 400 to 1500 m^2/g. These properties are responsible for the large adsorption capacities characterising activated carbons which are commonly used to remove pollutants from gaseous and liquid streams.

Char properties are important to assess its potential use for specific applications. They depend strongly on the raw material employed as feedstock and on pyrolysis conditions. Despite the significant progress that has been achieved, it is still difficult to predict with confidence char properties for a given lignocellulosic waste from available information in the literature. In particular, char surface properties are relevant to activated carbon production by the "physical" activation two-stage method. This method involves pyrolysis of the precursor and subsequent activation, i.e partial gasification, of the resulting char employing steam, CO_2 or flue gases as the most common activating agents (4).

On the other hand, efficient conversion of lignocellulosic wastes into char requires knowledge of pyrolysis kinetics of each waste. Regardless of the final char application, pyrolysis kinetics plays a key role for the design of commercial reactors, depending on the reactivity of each waste (5-6). Pyrolysis of lignocellulosic wastes is complex due to the different degradation behaviour of major constituents, i.e. cellulose, hemicellulose and lignin, interactions between constituents, and effects of minute amounts of mineral matter naturally present in whole biomass samples (7-8). As a consequence, a generalized knowledge of pyrolysis kinetics of any lignocellulosic material has not been yet achieved.

In this context, the present work aims at examining comparatively: (1) chemical compositions, heating values and surface properties of char samples obtained from slow pyrolysis of different lignocellulosic wastes, in relation to their potential as biofuel and/or for further processing to produce activated carbons; (2) pyrolysis kinetics of the selected wastes, necessary for the design of the reactors for char production.

EXPERIMENTAL SECTION

Different lignocellulosic wastes extensively generated in wood and agricultural products processing in Argentina were selected. The wastes employed were sawdust from two different wood species, Guatambú (*Aspidosperma australe)*, a native species, and Poplar *(Populus deltoide),* husks of peanut (*Arachis hypogaea),* and olive (*Olea L. europaea*) stones free from pulp. They are respectively denoted in the text as AA, PD, PH, and OS.

The wastes were milled and screen-sieved. Wastes composition was determined in terms of the major constituent biopolymers, holocellulose (cellulose + hemicellulose) and lignin, and extractive components that are soluble in ethanol-benzene. Lignin and extractives were isolated according to TAPPI standard methods. Holocellulose was obtained following the procedure described elsewhere (9).

For charcoal production, fractions of 1.3 x 10^{-3}m (1.3 mm) average particle diameter were used. The wastes were thermally treated under flowing nitrogen in a fixed bed reactor heated by an electrical furnace. The samples were subjected to a heating rate of 0.25 °C s^{-1} up to 850°C, and held at this temperature for 1 hour. Afterwards, the resulting chars were cooled under nitrogen to room temperature, weighted, and carefully stored for characterisation. Typical yields of the chars ranged between 15 and 40%, depending on each waste.

Proximate analyses of the wastes and char samples were conducted following the ASTM standard methods. An elemental analyser (Carlo Erba EA 1108) was used to determine elemental compositions.

Surface properties of all the samples were determined from physical adsorption measurements by volumetric techniques, using N_2 at -196 °C and CO_2 at 25 °C. The complementary use of these adsorbates allows characterisation of complex networks of pores of different sizes (5). A Micromeritics Gemini 2360 surface area analyser was employed for N_2 adsorption experiments, whereas adsorption measurements performed with CO_2 were carried out in a Micromeritics Accusorb 2100E sorption instrument. All the samples were outgassed overnight at 120°C at a final pressure of 1.33 x 10^{-4} Pa (10^{-6} mm Hg) prior to adsorption measurements.

Kinetic measurements of pyrolysis of the different wastes were performed by non-isothermal thermogravimetric (TG) analysis, from ambient temperature up to 900°C. A thermogravimetric balance Netzsch STA 409 coupled with a nitrogen flow device and a data acquisition system was used for the kinetics measurements. From a preliminary set of experiments, sample mass, particle diameter, and gas flow rate, for which diffusional effects are negligible, were determined. Nitrogen flow rate of 2.5 x 10^{-3} L s^{-1}, sample masses of 10±0.2 mg, and 37-44 μm particle diameter were employed for all the experimental runs. Experiments at different heating rates, in the range 0.17-1.7°C s^{-1} were also carried out. No appreciable differences in mass loss against temperature curves were found for the heating rates examined, pointing to negligible heat transfer effects (10).

RESULTS AND DISCUSSION

CHEMICAL COMPOSITION, HEATING VALUES AND SURFACE PROPERTIES OF THE SAMPLES

Contents of holocellulose and lignin, free from extractives, are shown in Table 1.

Table 1 Content of main constituents of the lignocellulosic wastes. #

Raw Material	% Holocellulose	% Lignin
AA sawdust	73	26.6
PD sawdust	-	25.4
Peanut husks	65	28.8
Olive stones	74	44.3

Expressed on an extractive-free basis

As already mentioned, holocellulose and lignin percentages, reported in Table 1, were determined independently by different experimental methods. Thus, values may not add to 100%. Olive stones possess a considerably larger percentage of lignin compared to the other selected wastes.

Tables 2 and 3 report proximate and ultimate analyses of the wastes and derived char samples, respectively. Heating values (HV) of the wastes and chars were estimated by an own multivariate correlation in terms of samples' elemental compositions, previously reported (5). Estimated heating values on a dry-ash free basis are listed in Table 3.

Table 2 Proximate analyses (dry basis) of the lignocellulosic wastes and char samples.

Sample	Volatile Matter (%)	Ash (%)	Fixed Carbon* (%)
AA sawdust	72.7	1.0	26.3
AA-char	8.1	7.2	84.7
PD sawdust	71.4	0.5	28.1
PD-char	7.4	4.6	88.0
Peanut husks	84.9	1.7	13.4
PH-char	38.8	4.2	57.0
Olive stones	74.8	0.4	24.8
OS-char	22.1	1.6	76.3

*Estimated by difference.

Table 3 Ultimate analyses and estimated heating values (HV) of the lignocellulosic wastes and char samples. #

Sample	%C	%H	%N	%O*	HV (MJ/kg)
AA sawdust	48.2	5.6	0.9	45.3	19.01
AA-Char	92.8	1.3	0.8	5.1	29.65
PD sawdust	48.3	5.8	0.8	45.1	19.14
PD-char	89.6	0.8	0.3	9.3	27.40
Peanut husks	47.4	6.1	2.1	44.4	19.39
PH-char	84.6	0.8	0.9	13.7	25.80
Olive stones	46.6	6.0	-	47.4	18.81
OS-char	94.4	0.5	0.8	4.3	27.67

All the results are expressed on a dry-ash free basis.
*Estimated by difference.

Pyrolysis of the wastes promotes pronounced changes in chemical compositions, as reflected in the results obtained for the char samples that are strongly dependent on each waste. Release of volatile matter leads to enhance the inorganic material of the samples, the char from AA sawdust showing the highest ash content (Table 2). All the wastes are enriched as fuels during pyrolysis. Increases in %C and decreases in %O may be appreciated in Table 3. Variations in chemical compositions of the selected lignocellulosic wastes due to pyrolysis are in accordance with well-known general trends (5, 11). Char samples possess high carbon contents (84 - 95%), negligible contents of sulphur, and nitrogen percents lower than 1%.

Similar heating values are found for the different wastes. As expected, all char samples show higher heating values compared to the parent wastes owing to the enhanced %C, the differences in HV among the chars becoming larger. Heating values are comprised between 25 MJ/kg and 30 MJ/kg. Char from AA sawdust shows the highest HV, followed by the char samples produced from olive stones and PD sawdust. The PH-char presents the lowest HV. Heating values of the wastes and chars are within the range reported in the literature for this kind of biomass and derived chars (11-13).

Surface properties of the samples were examined from nitrogen and carbon dioxide adsorption measurements.

Figure 1 illustrates typical N_2 adsorption isotherms determined on the char samples produced from the different wastes, providing information about samples larger pores, mainly macropores, mesopores and larger micropores. Nitrogen adsorbed volumes expressed in standard conditions of temperature and pressure (STP) per sample mass unit, V_a, as a function of the relative pressure (p/p_0) are shown in the figure.

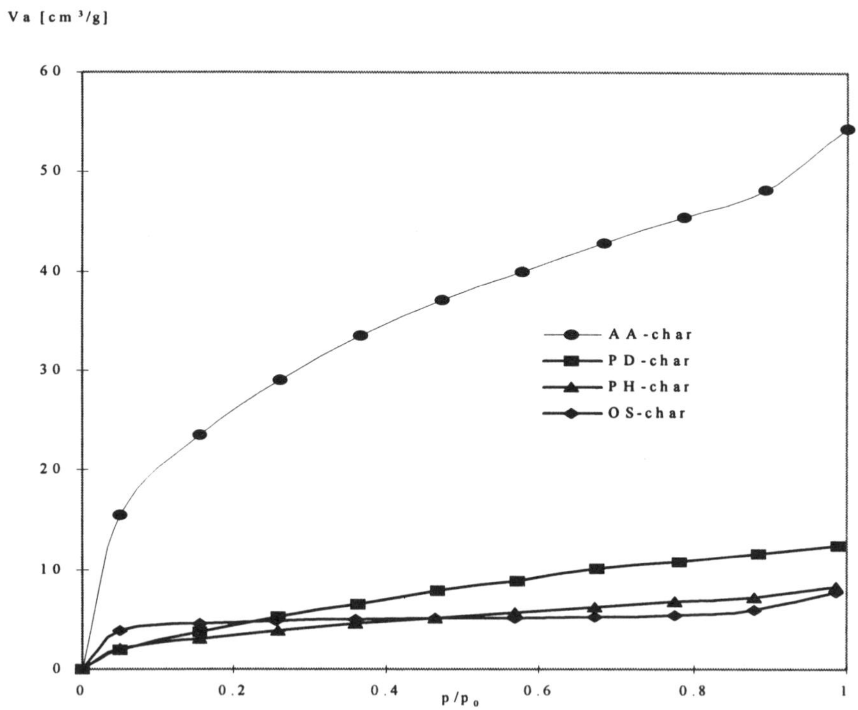

Figure 1 Nitrogen adsorption isotherms for char samples derived from the different lignocellulosic wastes.

Differences in the shapes of the isotherms are found (Figure 1). The OS-char shows a flat plateau which characterises type I-isotherms according to IUPAC classification, pointing to a predominantly microporous structure. N_2 volumes adsorbed on the AA-char are considerably larger compared to those determined on the other wastes. Results indicate that char samples possess different pore networks, that seem to depend on the original botanical structure inherent to the type of waste used for char production (10).

The BET model was applied to fit N_2 isotherms and evaluate the specific surface area of the samples (S_{BET}), following the conventional procedure (14). Estimated BET surface areas are detailed in Table 4.

Table 4 Surface properties of the lignocellulosic wastes and derived chars, as estimated from N_2 and CO_2 adsorption data applying BET and Dubinin-Radushkevich equations.

Sample	S_{BET} (m^2/g)	S_{CO2} (m^2/g)	S_{CO2} / S_{BET}
AA sawdust	1.6	140	87.5
AA-char	100.0	320	3.2
PD sawdust	1.3	82	63.1
PD-char	21.0	490	23.3
Peanut husks	1.0	107	107.0
PH-char	13.0	534	41.1
Olive stones	0.6	53	88.3
OS-char	17.0	717	42.2

Micropores in the lignocellulosic wastes and resulting chars were analysed from CO_2 adsorption isotherms by applying the Dubinin-Radushkevich (DR) equation (14). DR plots

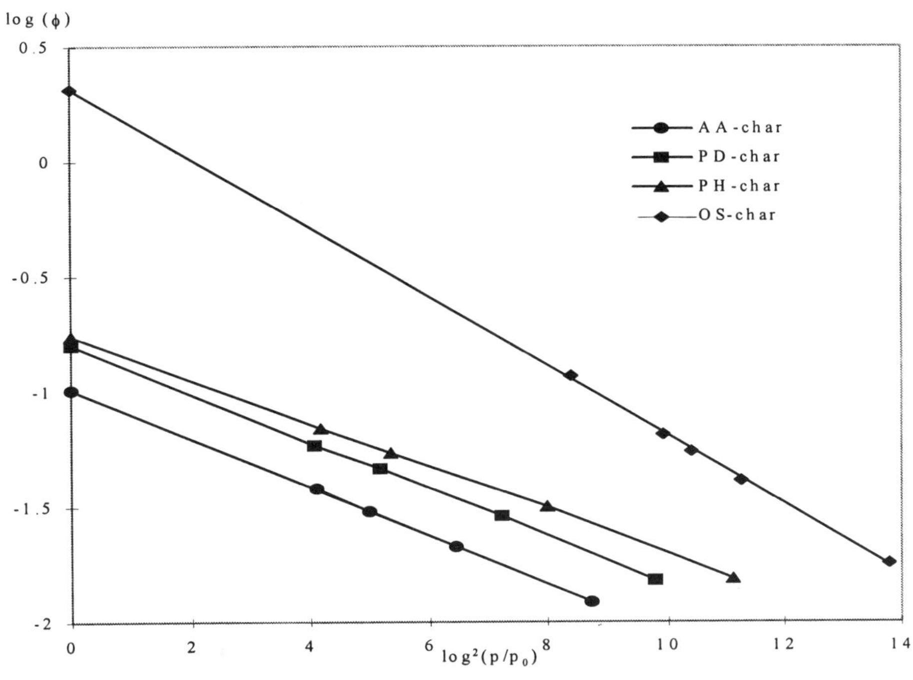

Figure 2 Dubinin-Radushkevich (DR) plots for char samples resulting from the different lignocellulosic wastes, as calculated from CO_2 adsorption data.

for the char samples are illustrated in Figure 2. In this figure, the amounts of CO_2 adsorbed (ϕ) expressed as a liquid volume per sample mass unit are plotted against $\log^2 (p/p_0)$.

As can be seen from Figure 2, CO_2 adsorption data for all the samples are well described by the DR- equation. Consequently, the total micropore volume (ϕ_o) for each sample was obtained from the ordinate at the origin of the DR plots. ϕ_o values were then used to calculate the apparent surface areas (S_{CO2}) according to the evaluation procedure described in detail elsewhere (14).

The estimated CO_2 surface area for each sample is listed in Table 4. The ratios between the nitrogen and carbon dioxide surface areas (S_{CO2}/S_{BET}) are also included in the same table, for the sake of comparison.

Char samples show larger S_{BET} and S_{CO2} areas than the wastes. Release of volatile matter during pyrolysis of the wastes, as a result of the complex network of reactions that takes place, promotes the development of new pores and/or favours widening of pre-existent pores. Some restructuring in the char matrix may cause additional increases in microporosity leading to larger S_{CO2} values (10).

All the samples possess BET surface areas considerably smaller than those calculated from CO_2 adsorption data, pointing to very narrow micropores or obstructions of pore entrances that restrict N_2 diffusion. Thus, differences between values of S_{BET} and S_{CO2} suggest prevalence of micropores over meso and macropores in the lignocellulosic wastes and chars. The relative contribution of micropores to the samples' structures may be additionally inferred from the S_{CO2}/S_{BET} ratio. Char samples present smaller ratios than the wastes, indicating decreases in the relative proportions of micropores owing to the thermal treatment.

Results indicate that the AA-char has the largest BET surface area and smallest amount of micropores compared to the char samples from the other wastes, pointing to a more open structure. The S_{CO2}/S_{BET} ratio for this char is one order of magnitude smaller. On the contrary, the char from olive stones shows the largest amount of micropores and the highest S_{CO2}/S_{BET} ratio suggesting a relatively close, compact structure. Similar, intermediate proportions of micropores, as judged from S_{CO2} areas, characterise the chars produced from peanut husks and PD sawdust, although the S_{CO2}/S_{BET} ratio for the latter is smaller.

The OS-char obtained under similar pyrolysis conditions to those employed in the present study has proved suitable for production of high-quality activated carbons, leading to BET surface areas larger than 1100 m^2/g by subsequent partial gasification using carbon dioxide as activating agent (15). Activated carbons of microporous character were obtained. Hence, properties of the char samples from the other selected wastes are compared with those exhibited by the OS-char in an attempt to predict potential conversion into activated carbons of reasonably good quality.

Although S_{BET} and S_{CO2} areas for the OS-char differ from those determined for the rest of the char samples, the PH-char shows the closest S_{CO2}/S_{BET} ratio to that of the OS-char, pointing to a similar relative contribution of pores of different sizes to the structure, followed by the PD-char. The AA-char possesses an interesting large BET surface area, but the S_{CO2}/S_{BET} ratio is considerably smaller than that corresponding to the OS-char. Large BET surface areas and sorption capacities of activated carbons arise from the presence of significant amounts of micropores (4). Thus, apparently, the AA-char should not lead to a pronounced development of porosity through further activation due to its relatively more open structure. Heschel and Klose (16) reported that coarse-cellular structures, such as those characterising high porosity woods, are disadvantageous as activated carbon feedstock. Nevertheless, the AA-char might be useful for preparation of activated carbons presumably with a smaller BET surface area but with a larger average pore size that could be appropriate for the removal of large size compounds in liquid-phase.

The ash content of char samples is another feature that has to be taken into account since activated carbons with low ash contents are normally desired. The low ash content of the olive stone-char should be emphasised. From this viewpoint, the relatively higher ash content of the AA-char compared to char samples from peanut husks and PD sawdust is disadvantageous.

KINETICS OF THE WASTES PYROLYSIS

Figures 3 and 4 illustrate typical thermogravimetric (TG) curves obtained for pyrolysis of the different lignocellulosic wastes at a heating rate of 0.83 °C s^{-1}. The TG curves show the recorded mass losses, expressed as weight fraction (w), against the temperature (T). Weight fractions are given by $w=m/m_o$, where m, m_0 are the instantaneous and initial mass samples, respectively. Pyrolysis rates (-dw/dt) for the different wastes were estimated by differentiation of the weight fraction-time curves. In Figure 5, the pyrolysis rates are represented as a function of temperature.

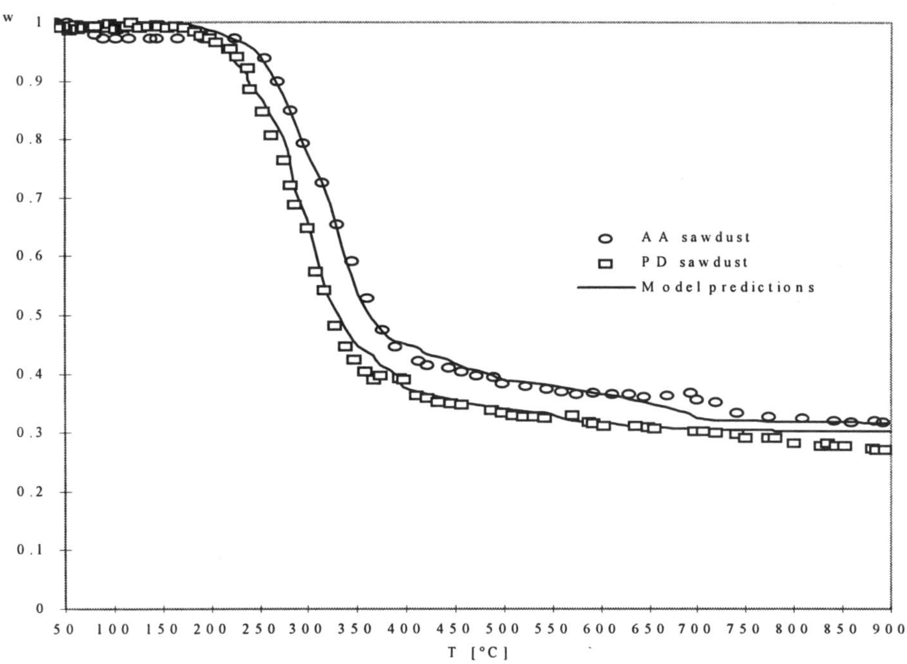

Figure 3 Thermogravimetric curves of pyrolysis of sawdust from both wood species, *Aspidosperma australe* (AA) and *Populus deltoide* (PD): weight fraction (w) against temperature (T). Comparison between experimental data (symbols) and model predictions (solid lines).

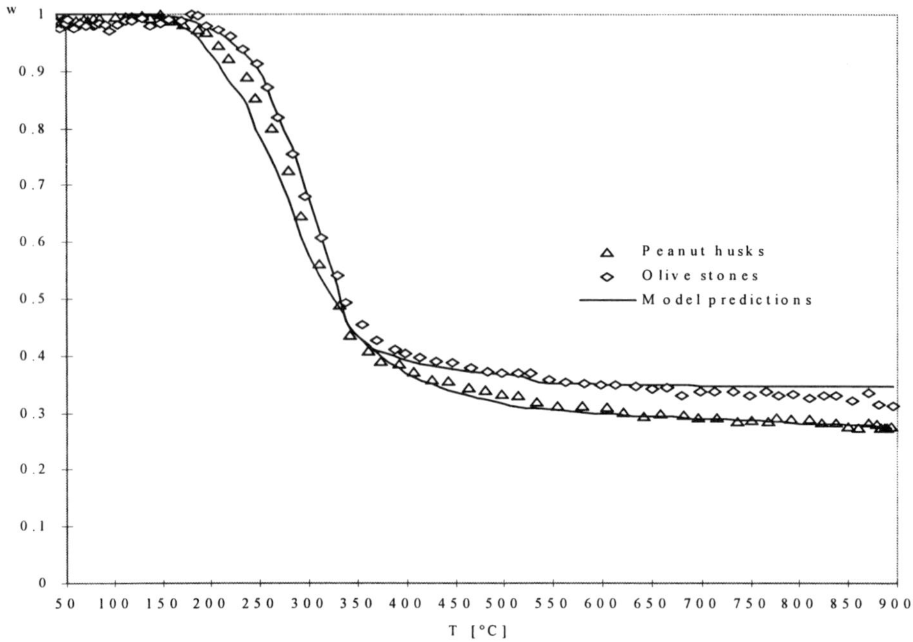

Figure 4 Thermogravimetric curves of pyrolysis of peanut husks and olive stones: weight fraction (w) against temperature (T). Comparison between experimental data (symbols) and model predictions (solid lines).

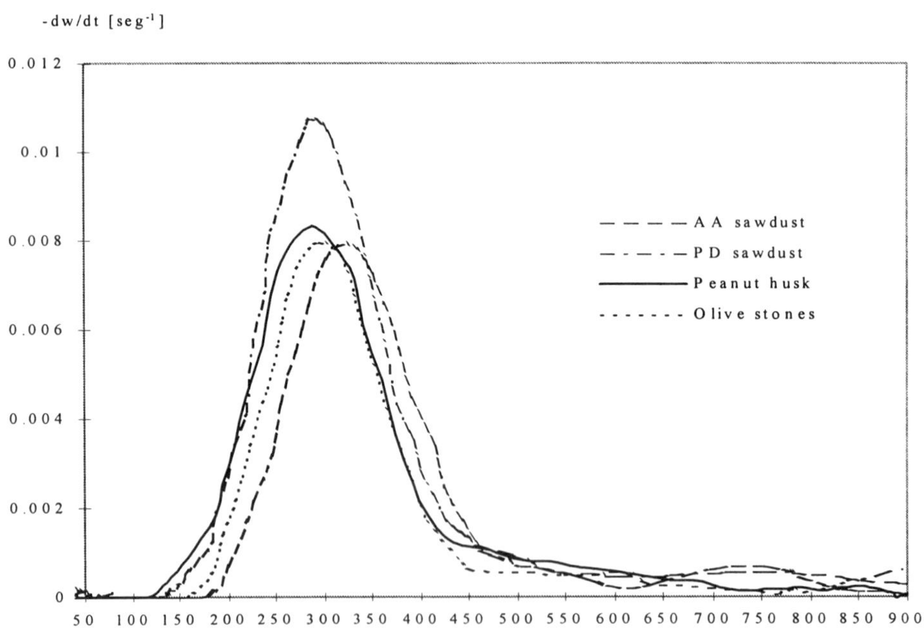

Figure 5 Influence of the temperature on the pyrolysis rates for the different lignocellulosic wastes.

Pyrolysis of the wastes becomes substantial above 150°C. Maximum pyrolysis rates for the selected wastes are attained in an interval of temperatures ranging between 290°C and 320°C. At higher temperatures, pyrolysis slows down. This behaviour may be attributed to the thermal degradation of major biopolymers constituting the lignocellulosic wastes. At low temperatures, degradation of cellulose and hemicellulose having a polysaccharide structure takes place. As temperature increases, degradation of lignin, which has a predominantly aromatic structure and is known to show more resistance to pyrolysis than holocellulose, seems to become progressively more prevalent. Mineral matter composing the wastes also appears to play a role affecting the pyrolytic behaviour (8, 17). The highest maximum rate is found for pyrolysis of sawdust from Poplar wood, while lower values are determined for the other wastes (Figure 5).

A model published in the literature (18) was used in order to fit experimental data and evaluate kinetic parameters over the whole range of degradation temperatures investigated, up to 900°C.

The model assumes that pyrolysis of the wastes occurs as a first-order overall decomposition. It considers that the physical and chemical changes which take place within the solid as pyrolysis proceeds cause solid deactivation. The reaction rate is given by:

$$(-dw/dt) = k_{app}\,(w - w_{\infty})$$

Changes within the solid are assumed to affect the reaction rate constant (k_{app}) and are taken into account through an increase of the activation energy with the temperature and the solid conversion according to:

$$k_{app} = k_0 \exp\left[-E_{A0}\,(1 + \beta T X^{\gamma}) / RT\right]$$

where X is the normalised fractional conversion, given by:

$$X = (1-w) / (1-w_{\infty})$$

w, w_{∞}, are the instantaneous and residual weight fractions, respectively. k_0 is the pre-exponential factor, E_{A0}, the initial activation energy, for X=0, and β, γ, fitting parameters (β, the deactivation rate, γ, the order with respect to X). R, T, are the universal gas constant and the absolute temperature, respectively.

For model application, fitting of TG curves (w vs T) was carried out taking into account the measured linear relationship between temperature and time. As mentioned in the experimental section, from preliminary experiments it was verified that different heating rates in the range 0.17-1.7°C s^{-1} led to similar w against T curves.

Model parameters were estimated by non-linear regression analysis. The sum of squares between the experimental and calculated instantaneous weight fractions was minimised. Estimated values of the model parameters together with standard deviations (SD) are detailed in Table 5.

Model fit to experimental results for sawdust from both wood species and agricultural wastes is shown in Figures 3 and 4, respectively. As can be seen in these Figures, the model appropriately describes experimental data over the wide range of temperatures for all the selected wastes. Other models were also applied. However, they were unable to describe experimental results for pyrolysis of all the wastes, led to larger SD, or adequately represented the data only for restricted ranges of temperature (10).

Table 5 Estimated characteristic parameters of the kinetic model.

Raw Material	E_{Ao} (kJ/mol)	k_o x 10^{-4} (min^{-1})	β (K^{-1})	γ	SD (%)
AA sawdust	59.4	12.4	0.0022	8.0	1.30
Poplar sawdust	59.1	3.70	0.0023	8.4	1.28
Peanut husks	35.0	0.17	0.00089	5.0	0.58
Olive stones	67.0	340	0.00095	7.0	0.56

Pyrolysis of sawdust from the two wood species is characterised by similar values of initial activation energy. In contrast, larger differences in the estimated E_{A0} values for pyrolysis of the agricultural wastes are found. The degradation of olive stones shows a relatively higher value of initial activation energy, presumably due to its larger lignin content (Table 1).

CONCLUSIONS

Characterisation of chars produced from selected lignocellulosic wastes by slow pyrolysis under identical operating conditions is carried out. Distinctive compositions, heating values and surface properties of the chars are determined.

All the chars obtained are found potentially useful as relatively pollution-free solid biofuels. Char samples possess high carbon contents (84-95%), negligible contents of sulphur, and nitrogen percents lower than 1%. Ash contents are also sufficiently low, with the exception of the AA-char which has a relatively higher ash percent (≈7%). Heating values lay between 25 MJ/kg and 30 MJ/kg. The char from sawdust of the native wood species (*Aspidosperma australe)* shows the greatest potential as biofuel, followed in a decreasing order by the chars arising from olive stones, Poplar wood sawdust, and peanut husks. The higher ash content of the AA-char might be an inconvenient for this application.

Comparison of surface properties of char samples with those characterising the OS-char, which has proved suitable for preparing high quality activated carbons through subsequent activation, suggests that chars from peanut husks and Poplar sawdust could be adequate candidates for further conversion into activated carbons with well-developed surface areas. Nevertheless, other factors that are relevant to manufacture and applications of activated carbons should be further examined.

Pyrolysis kinetics of all the selected lignocellulosic wastes is properly described over the wide thermal degradation range 25°C~900°C by a model that considers an increasing dependence of the activation energy on the temperature and waste conversion with the process course. Appreciable differences in the estimated kinetic parameters are found.

ACKNOWLEDGMENTS

The authors gratefully acknowledge Consejo Nacional de Investigaciones Científicas y Técnicas (CONICET), Universidad de Buenos Aires (UBA) and Agencia Nacional de Promoción Científica y Tecnológica (ANPCYT) from Argentina for financial support.

REFERENCES

1. Bridgwater, A.V., Bridge, S.A. (1991). A review of biomass pyrolysis and pyrolysis technologies. In: *Biomass Pyrolysis Liquids Upgrading and Utilisation.* (Ed. by A. V. Bridgwater & G. Grassi), pp. 11-92. Elsevier Applied Science, London and New York.
2. Rashid Khan, M., Gorsuch, C. (1996). Utilisation of waste materials: examples of business successes. In: *Conversion and Utilisation of Waste Materials.* (Ed. by M. Rashid Khan.), pp 3-13.Taylor and Francis, Washington DC.
3. Karaosmanoglu, F., Isigigur-Ergùdenler, A., Sever, A. (2000). Biochar from the straw- stalk of rapeseed plant. *Energy and Fuels* **14**, 336-339.
4. Patrick, J.W. (Ed.) (1995). *Porosity in Carbons: Characterisation and Applications.* Halsted Press, New York and Toronto.
5. Cukierman, A.L., Della Rocca, P.A., Bonelli, P.R., Cassanello, M.C. (1996). On the study of thermochemical biomass conversion. *Trends in Chemical Engineering* **3**, 129-144.
6. Della Rocca, P.A., Cerrella, E.G., Bonelli, P.R., Cukierman, A.L. (1999). Pyrolysis of hardwoods residues: on kinetics and chars characterisation. *Biomass and Bioenergy* **16**, 79-88.
7. Antal Jr, M.J., Varhegyi, G. (1995). Cellulose pyrolysis kinetics: the current state of knowledge. *Ind. Eng. Chem. Res.* **34**, 703-717.
8. Caballero, J.A., Conesa, J.A., Font, R., Marcilla, A. (1997). Pyrolysis kinetics of almond shells and olive stones considering their organic fractions. *J Anal. Appl Pyrolysis* **42**, 159-175.
9. Browning, B. I. (1970). *Methods of Wood Chemistry,* Volume 2, Wiley, J. (Ed) New York (Chapter 19).
10. Della Rocca, P.A. (1998) *Study on biomass thermal conversion processes.* DPhil thesis, Universidad de Buenos Aires.
11. Minkova, V., Marinov, S.P., Zanzi, R., Björnbom, E., Budinova, T., Stefanova, M., Lakov, L. (2000). Thermochemical treatment of biomass in a flow of steam or in a mixture of steam and carbon dioxide. *Fuel Processing Technology* **62**, 45-52.
12. Raveendran, K., Ganesh, A. (1996). Heating value of biomass and biomass pyrolysis products. *Fuel* **75**, 1715-1720.
13. Di Blassi, C., Signorelli, G., Di Russo, C., Rea, G. (1999). Product distribution from pyrolysis of wood and agricultural residues. *Ind. Eng. Chem. Res.* **38**, 2216-2224.
14. Gregg, S.J., Sing, K.S.W. (1982). *Adsorption, Surface Area and Porosity.* Academic Press Inc., London.
15. Horowitz, G.I., Lombardi, G., Cassanello, M.C., Cukierman, A.L. (1997). Gasification of an agricultural by-product: a study towards activated carbons

preparation. In: *Proc. 4th Asian Pacific International Symp. on Combustion and Energy Utilisation.* Chulalongkorn University, Bangkok, pp 757-761.
16. Heschel, W., Klose, E. (1995). On the suitability of agricultural by-products for the manufacture of granular activated carbon. *Fuel* **74**, 1786-1791.
17. Raveendran, K., Ganesh, A., Khilar, K. C. (1995). Influence of mineral matter on biomass pyrolysis characteristics. *Fuel* **74**, 1812-1822.
18. Balci, S., Dogu, T., Yücel, H. (1993). Pyrolysis kinetics of lignocellulosic materials. *Ind. Eng. Chem. Res.* **32**, 2573-2579.

The Pyrolysis Kinetics of a Single Wood Particle

Davidsson, K. O.[1] Pettersson, J. B. C, [1] Bellais, M. [2], Liliedahl, T. [2] and Sjöström, K. [2]

[1] *Department of Chemistry, Physical Chemistry, Göteborg University, SE-412 96 Göteborg, Sweden*

[2] *Department of Chemical Engineering and Technology, Royal Institute of Technology, SE-100 44 Stockholm, Sweden*

ABSTRACT: Experimental results from birchwood and pinewood pyrolysis in a new single particle reactor are presented. Apparent kinetic parameters for the mass-loss of wood particles (5-800 mg) at temperatures from 300 to 860 °C are determined. Kinetic parameters for the evolution of CO, CO_2, H_2O, H_2 and CH_4 are also established. The drying process was examined and it was found that drying and pyrolysis increasingly overlap in time as temperature rises and that the overlap is substantial above 450 °C.

INTRODUCTION

Increasing environmental concern, as to the global warming, has made renewable energy sources for combustion and gasification an advantageous option. If the use of coal were shifted towards increased use of biomass the CO_2-emission would decrease. To perform such a shift, existing and future power plants have to be (re)constructed for biomass conversion. Although biomass was mans first fuel the large-scale energy production from biomass is rather a new issue, especially compared to coal, where one can rely on decades of experience. Hence there is a need for knowledge about biomass behaviour in large-scale conversion.

In the present paper wood pyrolysis will be considered, since wood is the most important biomass in many areas. The first step in the combustion process is pyrolysis, which is the production of combustible volatiles as the temperature of the solid fuel increases. Wood pyrolysis is a complex phenomenon due to the composition and structure of the fuel. Despite this, many models have been proposed for the pyrolysis of a single wood particle. Bamford [1] coupled the kinetics of the pyrolysis with the heat conduction within the wood. Kung [2] included internal transfer by heat convection of the volatiles and reaction endothermicity. Kansa *et al.* [3] included Darcy's law. Chan *et al.* [4], Saastamoinen and Richard [5], Bilbao *et al.* [6] and Melaaen and Grønli [7] have considered the drying process. Other effects such as shrinking [8,9] and exterior conditions [10] have been considered. Di Blasi [11] made a two-dimensional model of the wood pyrolysis. Although these models may be regarded crude on a molecular level they may be too complicated to be used for modelling of a power plant reactor. Wood pyrolysis experiments have been performed with TGA (thermogravimetric analysis) applying temperature ramps [12-14]. An exception is Petek *et al.* [15] who pyrolysed wood particles by injection feeding to a thermobalance preheated to 500-950 °C. This way of heating resembles the heating of a wood chip in large-scale conversion. A few experiments have been made where the time-resolved gas evolution is treated [12,14,15, 17-19].

It is not evident to us that the behaviour of a thermally thick wood particle, upon entering a hot reactor, is predictable from standard TGA-studies because of the differences in heating processes. In a TGA-study one typically finds that a heating rate of a few to a few hundred degrees per minute has been applied. This is to be compared with the heating of a particle rapidly moved from 20 °C to a reactor of 400 or even up to 800 °C. We have chosen wood, rather than cellulose or lignin, because it is unclear whether wood can be described by simple combination of the pyrolysis behaviour of its components. Antal [16] clearly states that this is not the case.

To contribute to the knowledge of the behaviour of a single wood particle upon entering and pyrolysing in a large-scale reactor, experiments in a new single particle reactor were performed. The wood particle (0.6-800 mg, cubic or cylindrical) falls through a cooled tube from room temperature onto a plate inside the reactor surrounded with gases of 300-850 °C. The mass is recorded and the light gases emitted are measured with mass spectrometry yielding time resolved data. The single particle pyrolysis in a preheated environment resembles the situation for a particle in a large-scale reactor, even though the heat transfer and attrition due to interaction with other particles and bed material are not accounted for. The purpose of the present paper is hence to obtain apparent kinetic parameters for mass-loss and gas emission for a wood particle under a range of conditions during pyrolysis.

EXPERIMENTAL

The experimental system consists of a single particle reactor, a mass spectrometer (Baltzers QMG 421C) and a balance (Sartorius BP 211 D) and is shown schematically in Fig. 1. In the middle of the reactor is an Al_2O_3-plate on which the sample is pyrolysed. It is surrounded by five U-shaped heaters, which can heat the reactor up to 1000 °C. The temperature of the reactor is measured with thermocouples (TC type K) close to the plate, which is connected to the balance. Above the plate is a cooled insertion tube, through which the sample is brought. Level with and a few mm from the plate is situated the capillary, through which pyrolysis gases are sampled to the mass spectrometer. The capillary consists of an Al_2O_3-tube, which is inside the reactor, and a heated deactivated silica tube (diameter = 0.23 mm) between the reactor and the mass spectrometer. Visual observation and other measuring techniques are also possible. Preheated inert gas at atmospheric pressure enters through holes in the bottom of the reactor. Balance- and mass spectrometer data are recorded and represent time resolved mass and emission of light gases respectively. In the present case tars were not collected.

The Al_2O_3-plate is heated from the inert gas flow and no large temperature gradients occur over the distance between the plate and the TC. Therefore it is reasonable to assume that the gas and the plate are of the same temperature as the TC. At higher temperatures radiation becomes increasingly important so that the TC and the plate may become warmer than the gas but at these temperatures radiation is also more important than convection and conduction for the pyrolysis.

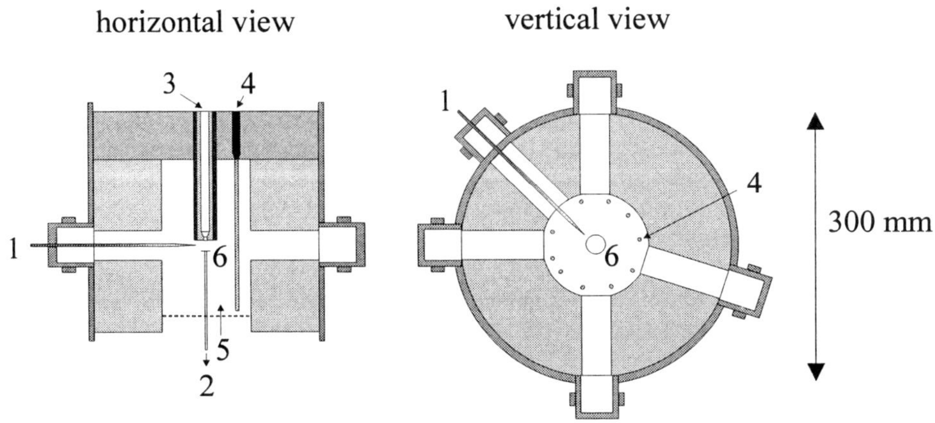

Fig. 1 Horizontal and vertical views of the experimental set-up. 1: mass spectrometry-capillary, 2: to the balance, 3: insertion tube, 4: heater, 5: pre-heated gas, 6: Al_2O_3-plate (for the sample); The figure is not entirely to scale.

At 400 °C heat transfer to the particle was calculated. The energy required to evaporate 85% of the water in moist particles corresponds to an average heat transfer of 69 kWm^{-2}.

Birchwood (*Betula verrucosa*) was mainly used in the experiments but a few experiments were performed with pinewood (*Pinus silvestris*). Both these trees are common in northern Europe. Samples of cubic and cylindrical forms and masses from 0.6 to 800 mg were used. The masses have been divided into four groups and the shape of the particles in each group is given in Table 1.

Table 1 Description of sample sizes and shapes

mass	1 mg	5 mg	60-100 mg	600-800 mg
shape	cubic 1mm	cubic 2mm	cylindrical 5×5mm and cubic 5mm	cylindrical ϕ 10 mm×15 mm

The samples were usually dried but a series of experiments was carried out with moist samples. Elemental analysis showed that both birch and pine contained 50% C, 6% H, 44% O and 0.1% N with respect to dry mass.

A typical experiment was carried out in the following manner. The sample was cut, weighed and dried at 105 °C until it had constant mass. It was brought onto the plate by falling through the insertion tube. The reactor was held at a constant temperature and had been evacuated with respect to oxygen (<0.05%). The balance recording continued until the particle mass was constant and the mass spectrometer recording continued until background levels of the selected gases were reached. The sample was taken out, weighed and stored.

The retention time for CO_2 in the capillary leading to the mass spectrometer is 1-2 s and we conclude that for CO_2, as well as for other light gases of interest, the retention time in the tube is negligibly small. Heavier substances, like tar, are not detected due to longer or infinite retention time. The balance data shown here are in most cases raw data implying that the balance responds very quickly to a mass change.

Mass spectra were analysed and it was found that no mass to charge ratio (m/z) above 84 was registered. Further analysis also showed that m/z = 2, 15, 18, 28 and 44, to a major part, correspond to H_2, CH_4, H_2O, CO and CO_2 respectively.

To examine the effect of the sample plate on the results, other plates than the Al_2O_3-plate were used. A three mm thick Cu-plate with relatively large heat capacity, maximising the heat transfer between the plate and the sample, and a plate consisting of a Pt-net suspended on thin Al_2O_3-supports, so that the sample rested on the net with close to free convection around, were used. No significant effect of the sample holder was found.

To examine the effect of the ambient atmosphere a few experiments were carried out in N_2. No difference compared with experiments in Ar could be found.

RESULTS

The results were obtained under Ar-atmosphere with dried birchwood samples, if not otherwise stated. The evaluation of data can be subdivided into that concerning sample mass and that concerning gas emission.

SAMPLE MASS

The pyrolysis of a wood particle entering a hot zone from room temperature can roughly be said to proceed in two phases. During the first phase, the heating phase, the particle is heated so that water, if any, evaporates and the temperature rises to about 200 °C at which devolatilisation of wood is known to start. Above this temperature, the mass-loss phase starts, where devolatilisation and consequently mass-loss occur. The rate of the heating phase is governed by external and internal heat transfer. The mass-loss phase is kinetically or heat transfer controlled. It should be noted that the temperature at which pyrolysis occurs may differ from the ambient. Both phases show strong temperature dependence, which can clearly be seen in Fig. 2, where mass as a function of time at different temperatures is shown. The initial mass of the birchwood sample was 80-100 mg cubic. The temperature dependence of the pyrolysis process is evident from the leftmost diagram, where it can be seen that the pyrolysis decreases from hours to less than ten minutes as the temperature increases from 315 to 400 °C.

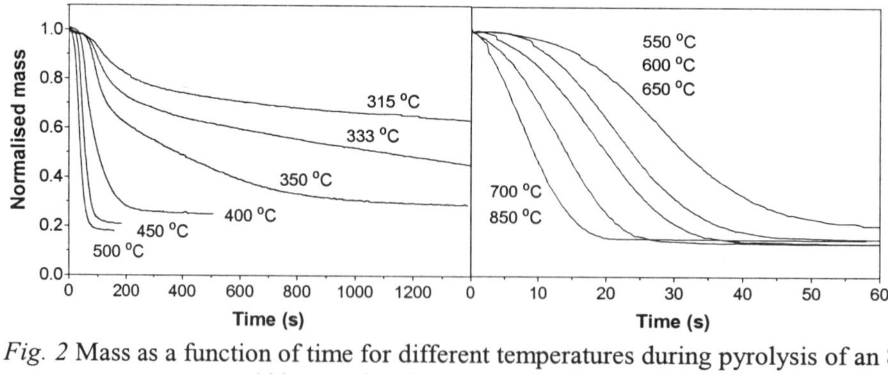

Fig. 2 Mass as a function of time for different temperatures during pyrolysis of an 80-100 mg cubic dried birchwood sample.

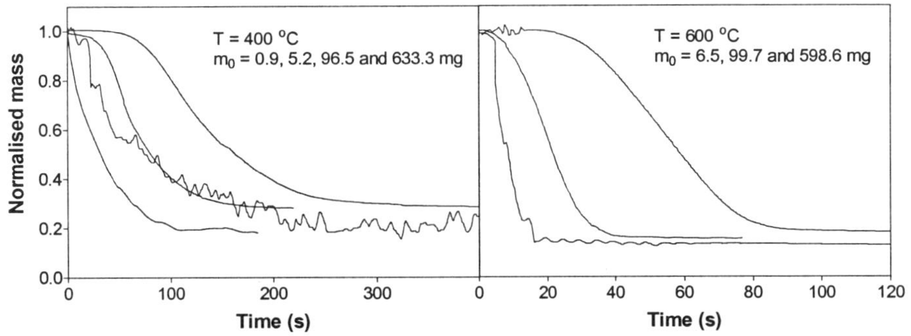

Fig. 3 Mass as a function of time for different initial sample masses at 400 and 600 °C during pyrolysis of a cubic dried birchwood sample. In the leftmost diagram the curve for $m_0 = 0.9$ mg has been smoothed.

Fig. 3 shows mass as a function of time for different initial sample masses. A small sample is heated faster than a large one, implying a shorter heating phase but the mass-loss phase is also shorter. The char yield for different samples is shown in Fig. 4. The yield ranges from almost 40% to less than 10%. An increase in reactor temperature (in this temperature range) obviously causes a decrease in char yield. A higher reactor temperature means a higher heating rate and a higher final temperature. Both these parameters have been shown to have a reducing effect on the char yield [17,20,21], with final temperature apparently being the most important. Also the size of the sample affects the char yield so that a smaller size yields less char. Two reasons may be given for this. Smaller samples experience larger heating rates and therefore also higher average temperatures. The other explanation would be that in a small particle, tars flow through a thin char layer compared with a large particle, reducing the importance of secondary reactions [22], i. e. tar cracking into volatiles and char re-building. The effect of particle size on char yield is less pronounced at higher temperature.

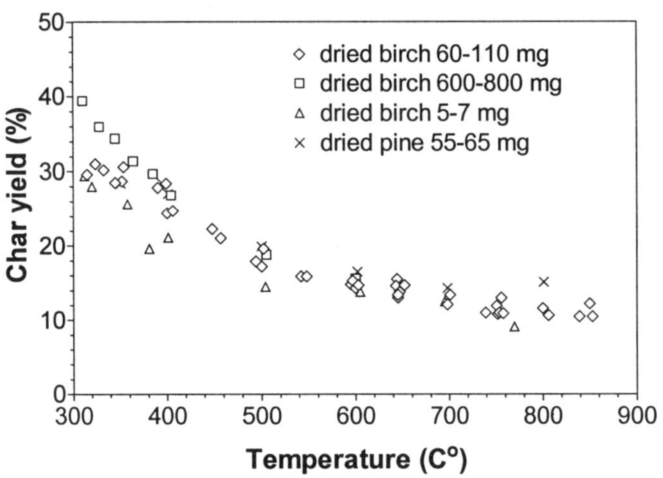

Fig. 4 Char yield vs. temperature for different sizes of dried birchwood samples and one size of dried pinewood samples.

In order to investigate the mass-loss process of the particle, apparent rate coefficients from a first-order expression were obtained. The following expression was applied:

$$m = (m_0 - m_c)\, e^{-kt} + m_c, \qquad (1)$$

where m, m_0 and m_c are mass, initial mass and char mass (mass at infinite time) respectively and t is time. These parameters were obtained from experiments, while the rate coefficient, k, was obtained by fitting of Eq. 1 to experimental data. Assuming that the rate coefficient can be described by the Arrhenius expression

$$k = Ae^{-E/RT}, \qquad (2)$$

where A, E and T are pre-exponential factor, activation energy and temperature respectively, the activation energy can be found as can be seen in Fig. 5. Due to the heating phase preceding the mass-loss, a kinetic first order expression cannot comprise the complete process. Despite this, the value of the rate coefficient at different temperatures gives an idea of the rate-limiting step and the time it will take to pyrolyse the particle at different temperatures. The activation energy below 370 °C is in agreement with those found for the pyrolysis of the constituents of wood, i. e. cellulose, hemicellulose and lignin [20,23-26] indicating that the pyrolysis kinetics are rate determining. The explanation for this is that, at low temperatures, the heating phase is very fast compared to the mass-loss phase, which may take hours. Above 370 °C, the activation energy is about 30 kJmol^{-1} for all sample sizes except the smallest ones. This low activation energy cannot correspond to a chemical reaction. We contribute this to external heat transfer limitation. The activation energy in this temperature range can be said to reflect the temperature dependence of the heat transfer rate.

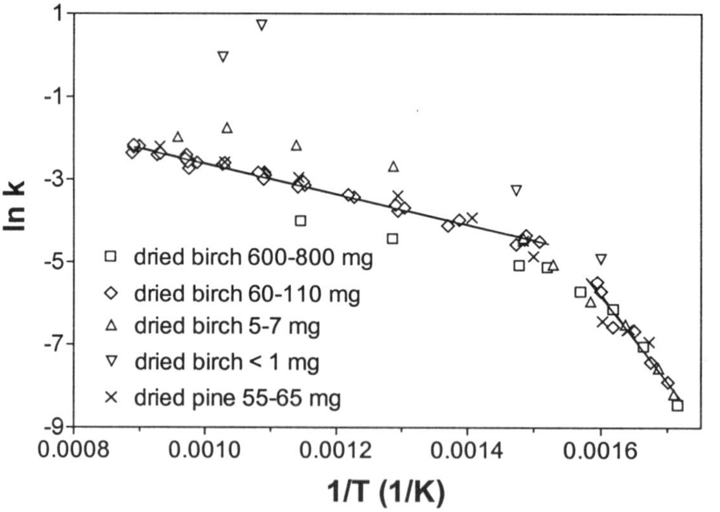

Fig. 5 Arrhenius plot according to a first-order mass-loss expression for pyrolysis of dried birchwood particles. Least square fits to dried birchwood samples of 60-110 mg are indicated in the figure. Above 370 °C, $E = 31$ kJmol^{-1} and $A = 3$ s^{-1}. Below 370 °C, $E = 177$ kJmol^{-1} and $A = 1.9 \cdot 10^{12}$ s^{-1}.

Figure 5 indicates that in the range 5-800 mg the size of the particle does not affect the apparent activation energies, implying that the temperature dependence of the limiting process is the same within these ranges of temperature and size.

A series of experiments was made with pinewood, and a comparison of the pyrolysis of pine and birch is made in Fig. 6. The qualitative differences of the mass-loss are small but reproducible. Birch loses weight a little faster during the heating phase. This may be attributed to higher content of bound water or some other easily gasified substance like hemicellulose. The compositions of birch and pine are listed in Table 2.

Table 2 Characteristics of birch and pine by mass-% [26].

substance	Scandinavian birch	Scandinavian pine
cellulose	40	44
hemicellulose	39	26
lignin	21	29

Hemicellulose is a thermally unstable compound and is known to start to pyrolyse at 200°C [23,27,28]. During a short period of the mass-loss phase the mass of birch is higher than that of pine but the char yield of pine is slightly higher. The differences are small and may be attributed to different heat transfer but because of the independence of temperature (in the range 400-700 °C) it is more likely that different structure and composition explains the difference in mass history. The evaluation of the kinetic parameters, conducted as above, shows no difference from birch (see Fig. 5).

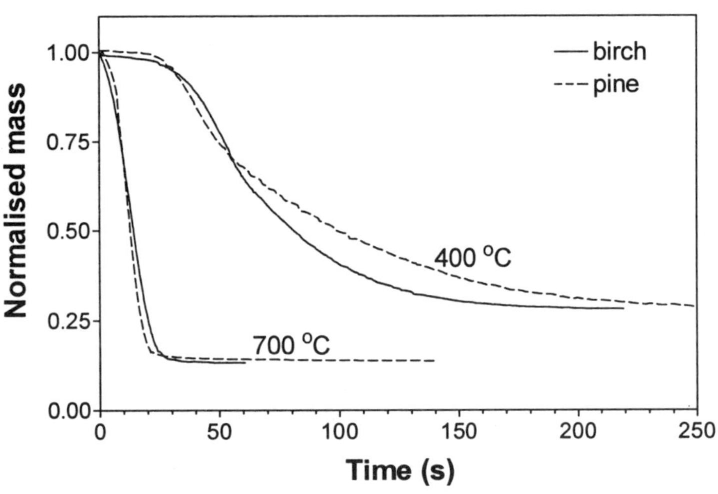

Fig. 6 Mass vs. time during pyrolysis of dried pine- and birchwood samples of equal sizes (55-100 mg) at two different temperatures.

GAS EMISSION

The gas emission during the pyrolysis of birchwood was analysed by use of mass spectrometry. Mass-loss and the emission of selected gases (mass spectrometer ion current) during pyrolysis at 500 °C and 800 °C are shown in Fig. 7. The gas emission curves correlate in time, as expected, with the mass-loss curve. This observation is evidence of the short response-time for the experimental set-up. As the temperature increases, more CO and H_2 are evolved compared with the other gases in accord with Beaumont and Schwob [17]. The reason for this may be a change in pyrolysis chemistry inside the particle with temperature. This agrees with the observation that the char yield decreases with temperature [17,20,21]. Antal [16] presented a scheme for the thermal decomposition of lignocellulosic materials. This scheme included a high activation energy fragmentation of the solid and it would occur at very high heating rates and yield CO and H_2. In the present work at high reactor temperatures at least the outer layers of the particle will be heated very fast and may undergo this reaction. Another explanation would be that reactions outside the particle change with temperature. For example H_2O and CO may react to give H_2 and CO_2 and its equilibrium is shifted towards higher concentrations of H_2 and CO_2 as temperature rises. This can though not fully explain our observations since we do not observe an increase in CO_2. However, the reaction between CO and H_2 to produce CH_4 and H_2O is shifted towards CO and H_2 with increasing temperature. This could partly explain the observations even though we see no decrease in CH_4-concentration.

A first-order expression was applied to the gas emission data:

$$g_i = 1 - e^{-kt}, \qquad (3)$$

where g_i is the relative amount of gas i released at the time t (amount of gas released at time t divided by amount of gas released at infinite time). Rate coefficients were obtained by fitting of Eq.3 to experimental data. Arrhenius diagrams for the emission of all the selected gases showed that the activation energy is around 30 $kJmol^{-1}$, i. e. close to the one obtained for the mass-loss above 370 °C. The Arrhenius-diagram for CO is given as an example in Fig. 8. The low activation energy, similar to the activation energy of the mass-loss, also implies that the measured gas emission is a product from within the particle and not the result of gas phase reactions outside. This means that the gas phase reactions discussed above do not have a major effect on the observations.

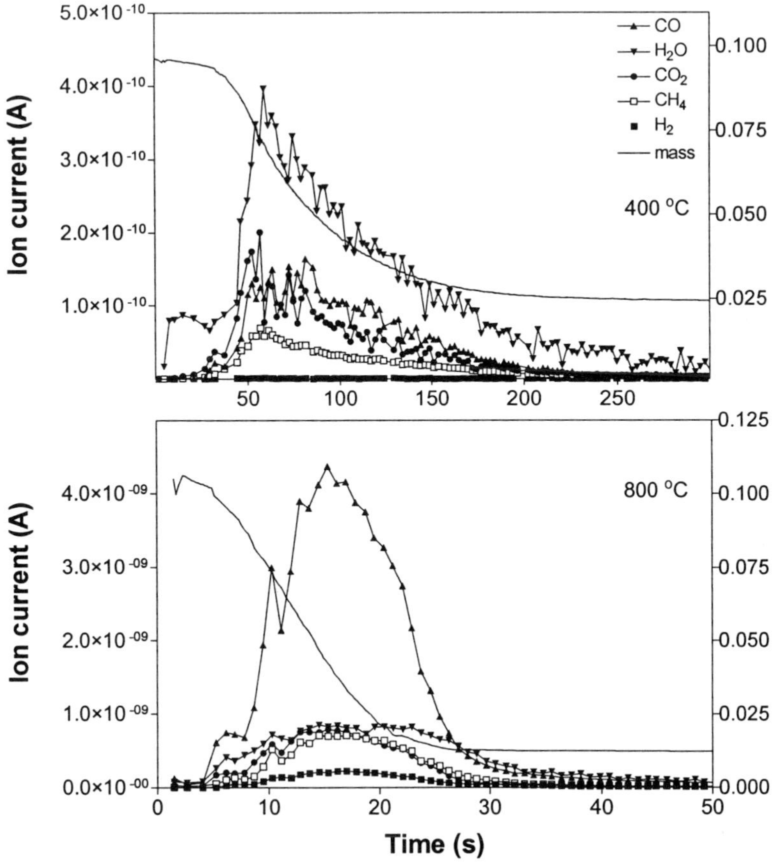

Fig. 7 Mass and MS signals as functions of time during pyrolysis of dried birchwood particles at 400 and 800 °C.

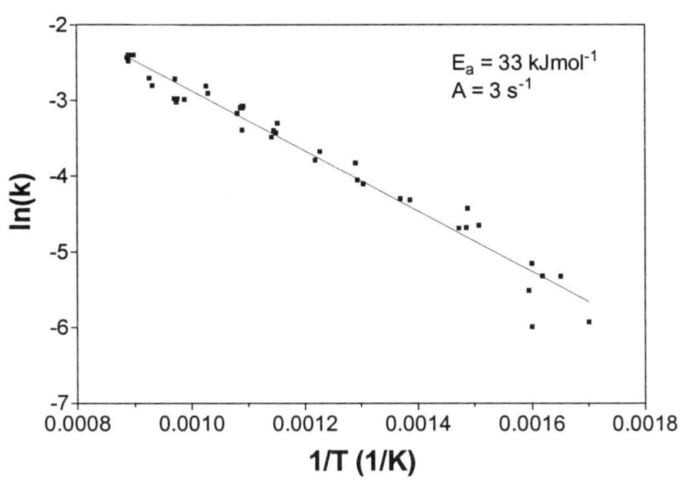

Fig. 8 Arrhenius-plot of CO emission during pyrolysis of cubic dried birchwood particles of approximately 100 mg. The line corresponds to a least square fit. Kinetic parameters are given in the figure.

Drying

Experiments were performed where the birchwood particle was dried and weighed and thereafter immersed in water until it had a moisture content of 50-70 % on dry basis. It was then pyrolysed as described above. A comparison between pyrolysis at 400 °C of a moist and a dried birchwood sample is given in Fig. 9, where the mass has been normalised to initial dry mass. This figure illustrates the main features of pyrolysis of moist wood. It is clear that the conversion takes longer time due to the longer drying period. The mass-loss phase (due to pyrolysis) is also somewhat slower for the moist particle but the char yield is very similar in the two cases. At this temperature and below, a distinct drying phase is observed but at higher ambient temperatures the drying and pyrolysis tend to overlap as illustrated in Fig. 10, where the mass spectrometer signal for H_2O is given for the pyrolysis of samples at different temperatures. Much of these findings agree well with Di Blasi et al [29]. At 400 °C one clearly sees two peaks at about 50 seconds and 130 seconds. These peaks correspond to drying and devolatilisation respectively. At 500 °C one can see a small peak at 85 seconds corresponding to pyrolysis but this peak is strongly overlapping with the drying peak. At 600 °C there is practically only one peak.

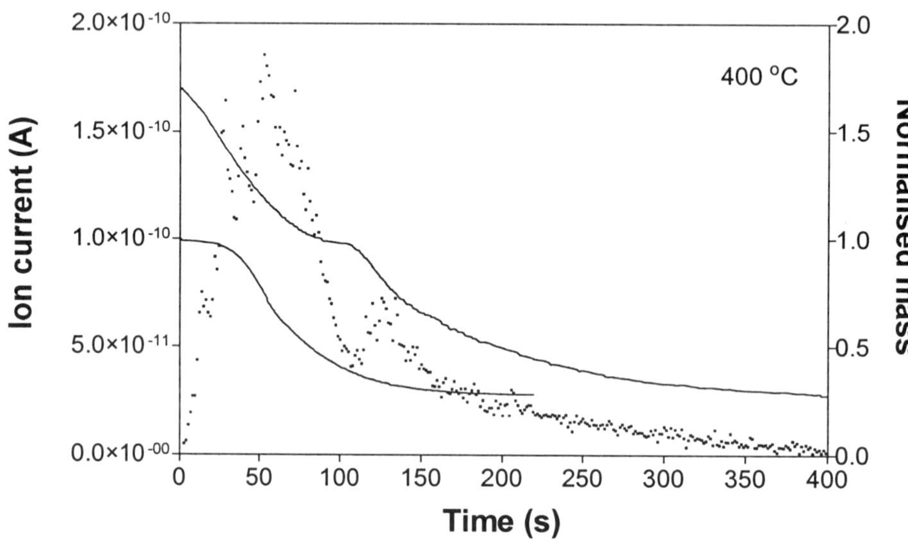

Fig. 9 Comparison of mass as a function of time for a dried and a moist birchwood particle (m_0(dry) = 63 and 96 mg) during pyrolysis at 400°C. The MS signal from water is given for the moist particle. Initial dry mass = 1.

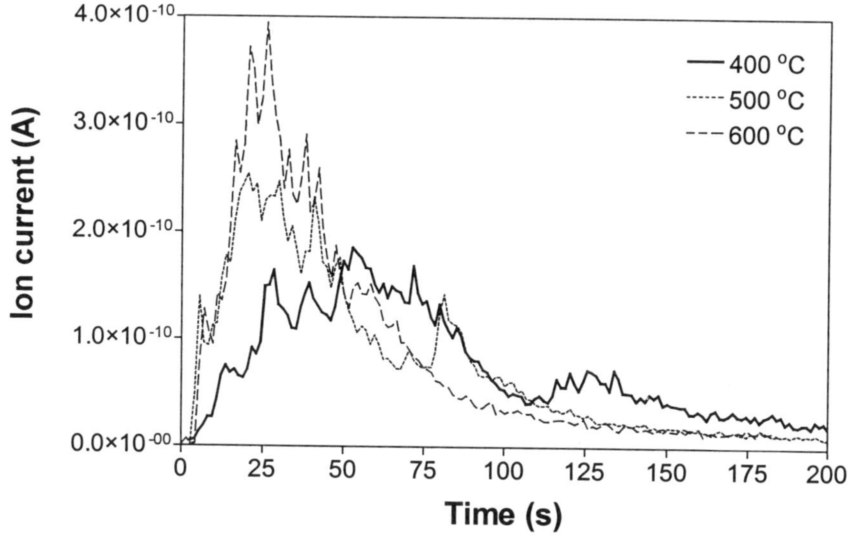

Fig. 10 MS-ion current as a function of time for H_2O at different temperatures during pyrolysis of moist birch particles of 60-100 mg initial dry weight.

CONCLUSIONS AND FUTURE WORK

The main results and conclusions from this work can be summarised as follows:

(1) Apparent kinetic parameters have been obtained for mass-loss and non-condensable gas evolution. The activation energy, above an ambient temperature of 370 °C, was found to be about 30 kJmol^{-1} under the present circumstances. The activation energy for the mass-loss below 370 °C agrees with those found in the literature for the kinetics of the main wood constituents.

(2) Dried Birch and pine behaves somewhat differently and the difference is qualitatively independent of ambient temperature. It is likely that differences in structure or composition, rather than differences in heat transfer, explains this.

(3) The evolution of CO and H_2 is enhanced by increased temperature and may be attributed to the fragmentation pathway of lignocellulosic materials mentioned by Antal [16]. The trends for the other gases are not as clear but evolution of CH_4 shows an increase whereas a slight decrease can be observed for CO_2 and water.

(4) Water is emitted in two phases. We conclude that the first phase is evaporation of free water and that the second phase is water produced in the pyrolysis reactions. These phases are fairly separated up to 400 °C, above, which they increasingly overlap. Increasing the temperature increases the importance of the internal heat transfer limitation of the processes. Thus, at high temperatures, a drying wave precedes a pyrolysis wave. If the particle is large, both phenomena occur at the same time. This is important from a simulation point of view: for high temperature, the overall pyrolysis cannot be simulated by a two-step model, namely a drying model and a pyrolysis model. A global model featuring both sets of equations has to be solved.

As shown in the preceding parts, kinetic parameters cannot be directly calculated when internal heat transfer limits pyrolysis. A model taking into account both kinetic scheme and heat- mass transfers becomes necessary. A one-dimension model has already been implemented and solved. It features a classical set of equations for heat and mass transfers in porous media, i.e. heat transfer through convection, conduction, radiation and mass transfer due to pressure gradient (Darcy's law) and binary diffusion. Different kinetic schemes from the literature are and will be tested: mass-loss as lumped first order reaction, formation of volatiles, tars and char from decomposition of cellulose, hemicellulose and lignin [26], the Broido-Shafizadeh model [30] and the one proposed by Di Blasi [31]. None of them can describe the composition of the volatiles and supplementary schemes have to be found.

Drying could be simulated in a separated model. However, since it was shown that drying and pyrolysis occur at the same time for large particles and high temperatures, the model should solve simultaneously an equation set for drying and one for pyrolysis. The model also needs to be extended to two dimensions due to the wood structure and the particle form.

ACKNOWLEDGMENT

We would like to thank Mr. Benny Lönn for support and construction work. This project was financed by The Swedish National Energy Administration, project No P10444.

REFERENCES

1. Bamford C. H., Crank J. and Malan D. H. (1946) The combustion of wood. *Proc. Camb. Phil. Soc.* **42**, 166-82.
2. Kung H. C. (1972) A mathematical model of wood pyrolysis. *Combustion and Flame* **18**, 185-95
3. Kansa E. J., Perlee H. E. and Chaiken R. F. (1977) Mathematical model of wood pyrolysis including internal forced convection. *Combustion and Flame* **29**, 311-24
4. Chan W.-C. R., Kelbon M. and Krieger-Brocket B. (1985) Modelling and experimental verification of physical and chemical processes during pyrolysis of a large biomass particle. *Fuel*, **64**, 1505-13
5. Saastamoinen J. and Richard J. R.(1996) Simultaneous drying and pyrolysis of solid fuel particles. *Combuston and Flame,* **106**, 288-300
6. Bilbao R. Mastral J. F. Ceamanos J and Aldea M. E. (1996) Modelling of the pyrolysis of wet wood. *J. Anal. Appl. Pyrolysis,* **36**, 81-97
7. Melaaen M. C and Grønli M. G. (1997) Modelling and simulation of moist wood drying and pyrolysis. proc. *Dev. Thermochem. Biomass Conversion,* **1**, 132-46
8. Villermaux J., Antoine B., Lede J. and Soulignac F. (1986) A new model for thermal volatilization of solid particles undergoing fast pyrolysis. *Chemical Engineering Science,* **40**, No. 1, 151-7
9. Di Blasi C. (1996) Heat momentum and mass transport through a shrinking biomass particle exposed to thermal radiation. *Chemical Engineering Science*, **51**, No. 7, 1121-32

10. Miller R. S. and Bellan J. (1996) Analysis of reaction products and conversion time in the pyrolysis of cellulose and wood particles. *Combust. Sci. and Tech.*, **119**, 331-73
11. Di Blasi C. (1997) A transient, two-dimensional model of biomass pyrolysis. Proc. *Dev. in Thermochemical Biomass Conversion* **1**, 147-60
12. Bilbao R., Arauzo J., Murillo M. B. and Salvador M. L. (1997) Gas formation in the thermal decomposition of large spherical wood particles. *J. Anal. Appl. Pyrolysis*, **43**, 27-39
13. Varhegyi G. and Antal M. J., Jr. (1989) Kinetics of the thermal decomposition of cellulose, hemicellulose, and sugar cane bagasse. *Energy & Fuels,* **3**, 329-35
14. Szabó P., Várhegyi G., Till F. and Faix O. (1996) Thermogravimetric/mass spectrometric characterization of two energy crops, Arundo donax and Miscanthus sinensis. *J. Anal. Appl. Pyrolysis*, **36**, 179-90
15. Petek J., Rummer B., Seebauer V., Steiner G. and Staudinger G. (1999) Thermal decomposition of large wood particles in inert and reactive atmosphere: mathematical simulation and experiments with rapid heat up. *6th international conference on circulating fluidized beds.* In: *Circulating Fluidized Bed Technology VI,* (Ed. by J. Werther), pp. 469-474, Dechema e. V.
16. Antal M. J. (1985) Biomass pyrolysis: A review of the literature, part 2- lignocellulosic pyrolysis. in *Advances in Solar Energy* (Ed. by Boer K. W. and Duffie J. A) **2**, 175-255. Plenum Press, New York
17. Beaumont O. and Schwob Y. (1984) Influence of physical and chemical parameters on wood pyrolysis. *Ind. Eng. Chem. Process Des. Dev.*, **23**, 637-41
18. Ohlemiller T. J., Kashiwagi T. and Werner K. (1987) Wood gasification at fire level heat fluxes. *Combustion and Flame*, **69**, 155-70
19. Varhegyi G., Antal M. J. Jr., Szekely T., Till F., Jakab E. and Szabó P. Simultaneous Thermogravimetric-mass spectrometric studies of the thermal decomposition of biopolymers. 2. sugar cane bagasse in the presence and absence of catalysts. *Energy & Fuels*, **2**, 273-7
20. Min K. (1977) Vapor-phase thermal analysis of pyrolysis products from cellulosic materials. *Combustion and Flame*, **30**, 285-94
21. Antal J. A. Jr., Mok W. S. L., Varhegyi G. and Szekely T. (1990) Review of methods for improving the yield of charcoal from biomass. *Energy & Fuels*, **4**, No. 3, 221-5
22. Connor M. A. and Salazar C. M. (1988) Factors influencing the decomposition processes in wood particles during low temperature pyrolysis. In: *Research in Thermochemical Biomass Conversion*, pp. 164-78 (Ed. by A. V. Bridgwater and J. L. Kuester). Elsevier, London.
23. Bar-Gadda R. (1980) The kinetics of xylan pyrolysis. *Thermochimica Acta*, **42**, 153-63
24. Nguyen T., Zavarin E. and E. M. Barall (1981) Thermal analysis of lignocellulosic materals. *J. Macromol. Sci.-Rev. Macromol. Chem.* C20, 1-65
25. Kanury M. and Blackshear Jr. P. L. (1970) Some considerations pertaining to the problem of wood-burning. *Combustion Science and Technology*, **1**, 339-55 24.
26. Grønli M. (1996) A theoretical and experimental study of the thermal degradation of biomass. PhD Thesis, Norges teknisk-naturvitenskaplige universitet Tronheim
27. Rajeswara T. R. and Sharma A. (1998) Pyrolysis rates of biomass materials. *Energy*, **23**, No. 11, 973-8
28. Roberts A. F. (1970) A review of kinetics data for the pyrolysis of wood and related substances. *Combustion and Flame*, **14**, 261-72

29. Di Blasi C., Hernandez E. G. and Santoro A. (2000) Radiative pyrolysis of single moist wood particles. *Ind. Eng. Chem. Res.* **39**, 873-882
30. Shafizadeh F. (1982) Chemistry of pyrolysis and combustion of wood. *Progress in Biomass Conversion*, **3**, 51-76
31. Di Blasi, C. and Russo, G. (1994), Modeling of transport phenomena and kinetics of biomass pyrolysis. In: *Advances in Thermochemical Biomass Conversion.* (Ed. by A.V. Bridgwater) pp. 906-921. Blackie Academic & Professional, London

Dynamics and products of wood pyrolysis

Colomba Di Blasi, Carmen Branca, Antonio Santoro,
Elier Gonzalez Hernandez and Raul Alberto Perez Bermudez
Dipartimento di Ingegneria Chimica, Università degli Studi di Napoli "Federico II" - P.le V. Tecchio, 80125 Napoli - Italy

ABSTRACT The radiative pyrolysis of wood (thick cylinders and chip beds) has been investigated experimentally for external radiative heat fluxes in the range 28-80kW/m^2, resulting in maximum sample temperatures of 600-950K. Radial temperature profiles, product yields and composition, and devolatilization rates have been measured. The influences of wood variety (hardwoods and softwoods) on the pyrolysis characteristics are discussed and comparisons are made with biomass (agricultural residues).

INTRODUCTION

A high number of studies is available on the pyrolysis behaviour of wood particles [1,11]. For large part of the investigations, measurements are concerned with the time history of temperature at different radii or stations along the axis of the sample, while in a few cases, mass loss curves, profiles of solid density and gas overpressure have also been measured. For complete process characterization all these measurements are necessary and, for applications in thermochemical conversion, especially yields (and composition) of the main product classes, i.e. gases, char and liquids. However, these have been measured only in a very few cases [7,10,11]. In addition, the measurements produced by the different laboratories are not directly comparable because of the differences in reactor configurations (heat transfer mechanism from reactor to particle), wood variety (different physical properties and chemical kinetics) and sample characteristics (size, shape and moisture content). Therefore, fundamental information is not currently available for optimal feedstock pretreatment and selection, in order to guarantee efficient operation of conversion units and to reduce the emission of noxious substances. Also, the applicability of mathematical models [12,13] for process design and development is highly dependent on the validation stage, which requires sets of data measured over wide temperature intervals.

In this study results of an extensive experimental activity are presented on the pyrolysis of wood (chips and thick cylinders) for a wide range of external heating conditions and several varieties, namely, two hard-woods (beech, chestnut) and three soft-woods (Douglas fir, redwood and pine).

EXPERIMENTAL SYSTEM

The experimental system applied for the pyrolysis of wood cylinders is the same presented in [11,15,16] and a schematic representation is reported in Fig.1, which also includes details of the heater and the reactor. The furnace, manufactured by Research Inc., presents four tubular quartz infrared lamps with a tungsten wire filament that emits radiant energy in proportion to the applied voltage. Elliptical, polished aluminium, water-cooled reflectors focus the high-density infrared energy, emitted by lamps, onto a cylindrical shaped target area (diameter 6.5×10^{-2} m). The furnace, equipped with a PID controller and a transducer (SCR), allows a constant radiative heat flux to be emitted. To avoid interaction between the volatile pyrolysis products and the lamps, a quartz tube (ID 6×10^{-2} m), transparent to infrared radiation, is located inside the furnace and used as a reaction chamber.

A wood cylinder is vertically positioned in the uniformly heated zone of the reactor, through a suspension system, which is connected to a precision balance. The sample is exposed to the same radiative heat flux along the lateral surface. For each chosen radiation intensity, steady temperatures of the radiant heater are achieved within a couple of minutes (maximum heating rates of about 750K/s) but, given the thick sample, pyrolysis takes place under heat transfer control.

A nitrogen flow is applied at the top of the quartz reactor in order to reduce the extra-particle residence time of volatile pyrolysis products and to establish the proper reaction environment. A nominal (ambient conditions and absence of sample) gas velocity of 0.0125m/s has been employed for all the tests, which corresponds to volatile residence times along the heated section of about 6s.

Two separate tests are made, one for the continuous recording of the sample weight loss and another for temperature measurement, product collection and gas analysis. Temperatures along the particle radius, r, at the median section are continuously monitored (0.5×10^{-3} m bead chromel-alumel thermocouples) at five positions starting from the center. The composition of the gas is analyzed, at selected times, through a gas-chromatograph equipped with a TCD and a packed column. These measurements are also applied to evaluate the yields of non condensable gaseous components (indicated as "gas"), through integration of the concentration of each species over the time of the experiments. After complete conversion, the power is turned off and the solid residue is left under a nitrogen flow until its temperature lowers to ambient values. This residue is indicated as "char". The liquids are collected through a condenser train, consisting of water-cooled traps, cotton wool demisters and a silica gel bed. All the condensable products collected and weighed from the traps (organic compounds (tars) and product water formed) are indicated as "liquids".

RESULTS

The tests have been carried out for two hardwoods, beech (Fagus sylvatica) and chestnut (Castanea sativa), and three softwoods, Douglas fir (Pseudotsuga menziesii), redwood (Sequoia sempervirens) and pine (Pinus pinea). Chemical analyses are reported in Table 1. Extractives are determined by means of the extractor Soxhtec HT2. Acetone (60ml) is used as solvent (1g of biomass) with residence times for the boiling and rising stages equal to 90 and 20min, respectively (temperature equal to 363K). Lignin is determined according to the Klason method and holocellulose is obtained by difference.

The chemical composition of Douglas fir and redwood is about the same (29-33% lignin, 65-56% holocellulose and 6-11% extractives, respectively), but differences are larger with pine wood, which presents much lower lignin contents (24%) with 69% holocellulose and 7% extractives. Differences exist also between the two hardwoods, with the chemical composition of beech consisting of 20% lignin, 78% holocellulose and 2% extractives. Chestnut is characterized by a high content of extractives (16%) associated with 18% lignin and 66% holocellulose. Differences are also observed in the initial wood density.

In a first set of experiments, pyrolysis has been carried out of wood cylinders (grain direction is parallel to the longitudinal axis) with 4cm diameter and length, after drying at 373K for 8-10h. The applied radiative heat flux has been varied in the range 28-80kW/m^2, so as to achieve steady surface temperatures, T_s, varying from about 600K to 950K. Measurements have been made of radial profiles of temperature, yields of the main product classes and composition of the gas. For beech, chestnut and pine, weight loss curves have also been recorded.

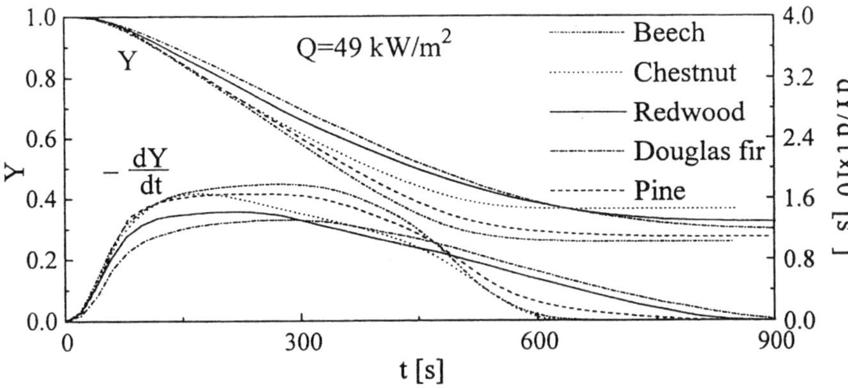

Fig.2A - Solid mass fraction and time derivative of the solid mass fraction for a heat flux of Q=31kW/m^2 and different wood chips varieties (beds with 4cm diameter).

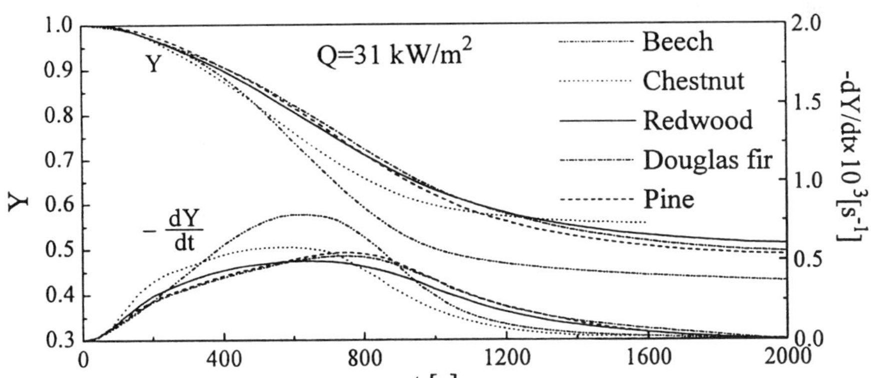

Fig.2B - Solid mass fraction and time derivative of the solid mass fraction for a heat flux of Q=49kW/m^2 and different wood chips varieties (beds with 4cm diameter).

Variety	Beech	Chestnut	Douglas fir	Redwood	Pine
extractives %	2	16	6	11	7
lignin %	20	18	29	33	24
holocellulose %	78	66	65	56	69

Table 1 Chemical composition of wood samples.

The reproducibility of the tests is good, as maximum variations in the steady temperatures of 10-20K are observed only for applied heat fluxes above 40kW/m^2, probably as a consequence of structural failures and volume variations, while variations in the weight loss dynamics are always negligible.

Weight loss curves and temperatures profiles have also been measured for wood chips beds (after drying) for low (31kW/m^2) and high (80kW/m^2) applied radiation intensities. In order to privilege the influence of the chemical composition, similarity in the physical behaviour of the packed beds has been accomplished by working with small particles (0.5-1mm) and the same bed density (0.212g/cm^3) . The sample consists of about 8g arranged to give a bed with 4cm diameter and 3cm height. It is worth noting that, as found in previous studies (for instance [14]), for a chosen bed density, particle sizes do not influence the pyrolysis characteristics, indicating that the heat transfer resistance across the bed predominates over that across the particles.

WEIGTH LOSS DYNAMICS OF WOOD CHIPS BEDS

From the qualitative point of view, the behaviour of the different wood chip beds is the same, but large quantitative differences exist, as shown by comparing weight loss dynamics in Figs. 2A-2B. For low heat fluxes (Fig.2A), maximum devolatilization rates are observed for the two hardwoods, whose degradation process also terminates earlier. More precisely, for times shorter than 300s, the degradation rate of chestnut is faster, but for longer times it is beech wood which attains the fastest rate. The degradation rates are comparable for all the softwoods.

For intermediate (Fig.2B) and high heat fluxes, apart from very short times, when the degradation dynamics of the different varieties are very similar, it is still beech wood which attains the highest rates. The most interesting result is the similar behaviour of redwood and Douglas fir, which degrade with much slower rates than the hardwoods, while the degradation behavior of pine wood is, in some way, comprised between that of beech wood and chestnut. In other words, close similarity exists between beech and pine, on one side (fast devolatilization rates), and redwood and Douglas fir, on the other (slow devolatilization rates).

Figures 2A-2B also show that the lowest residues are detected for beech, whereas the highest solid residues are obtained for the other hardwood variety, chestnut. Though at low temperatures the solid residue is high for all the softwoods, as soon as temperatures become sufficiently high, pine wood residues are significantly lower (and practically coincide with those of beech wood for very high temperatures) than those of the other two softwoods. Finally, the conversion times, defined as the time when the devolatilization rate lowers to 1/10 of the maximum value, show the shortest values for the two hardwoods. Times are significantly longer for all the softwoods. Hence, contrary to the trends shown by the solid residues, close similarity exists between the hardwoods, on one side, and the softwoods, on the other.

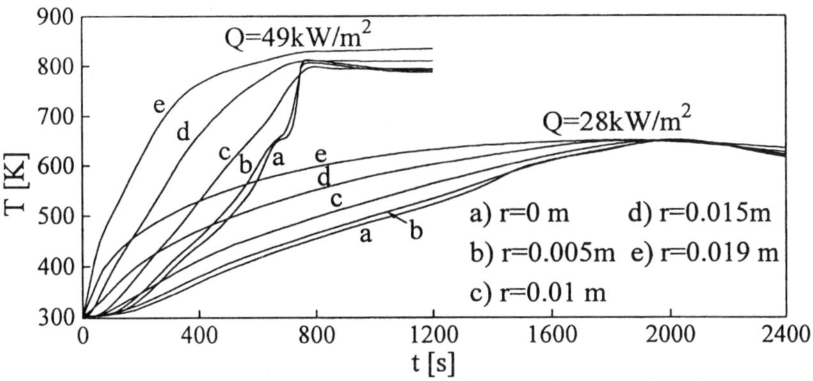

Fig.3 - Time-temperature profiles at several radial positions for beech wood cylinders exposed to Q=28 and 49kW/m^2.

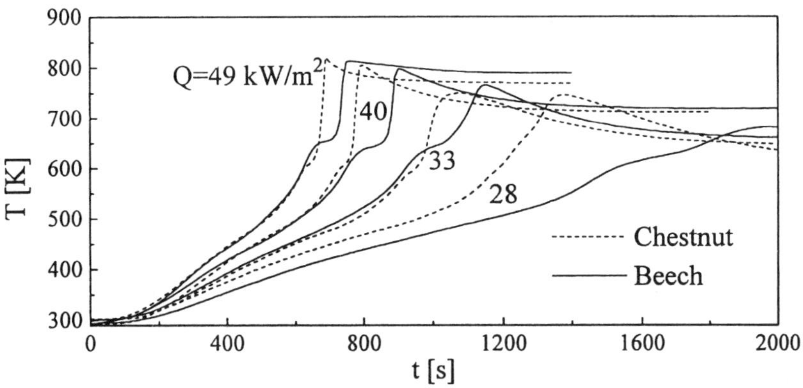

Fig.4 - Time-temperature profiles at the sample centerline for chestnut (dashed lines) and beech (solid lines) wood cylinders exposed to several radiation intensities.

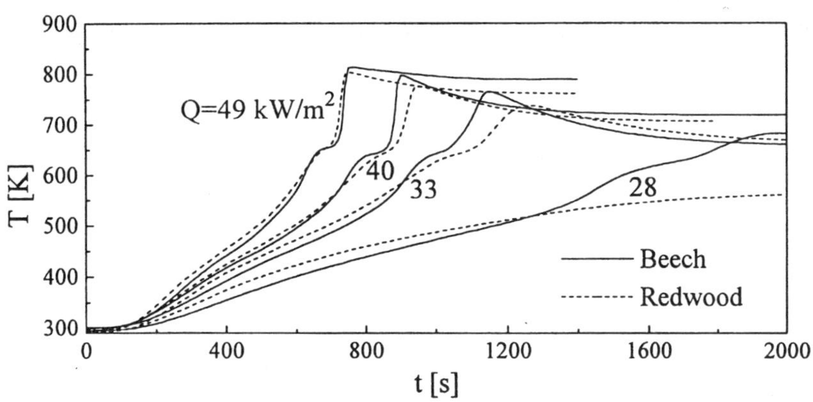

Fig.5 - Time-temperature profiles at the sample centerline for redwood (dashed lines) and beech (solid lines) wood cylinders exposed to several radiation intensities.

Given the same heating conditions established for the different varieties of wood chips, differences in the degradation dynamics are mainly due to the variable amounts (Table 1) and nature of the main components, though other factors, such as interactions between the component fractions, catalytic actions by the mineral matter content, may play an important role. Beech, on one side, and Douglas fir and redwood, on the other, can be considered as classical examples of hardwoods (low lignin contents) and softwoods (high lignin contents), respectively. The high extractive content of chestnut and the relatively low lignin content of pine introduce significant deviations with respect to the basic characteristics of the two wood categories. Compared with beech wood, chestnut degradation starts at shorter times, global devolatilization rates are lower and solid residues significantly higher. These features can be, for large part, attributed to extractive degradation [10], whereas the higher temperatures needed by the softwood categories to attain fast degradation rates can be due to the higher degradation temperature of their lignin fraction [10].

HEATING AND WEIGHT LOSS DYNAMICS OF WOOD CYLINDERS

The macroscopic behavior of degrading wood cylinders is again qualitatively similar for the different varieties and examples are given for beech wood at low ($28kW/m^2$) and high ($49kW/m^2$) radiative heat fluxes (Fig.3). The temperature profiles show the existence of two main regions: the most external is essentially determined by heat transfer effects and the second by reaction energetics. For the first region, heating rates become slower as the distance from the external surface increases, because of the increasing internal heat transfer resistance, but the degradation of the main wood components takes place at about the same time. For the inner core of the sample, due to the slower heating rates of the final conversion stage, the endothermic degradation of the low-temperature components (extractives and holocellulose) is completed before the high-temperature and exothermic degradation of lignin [15,16]. Indeed, there is an almost flat region which corresponds to the first endothermic stage. The subsequent exothermic lignin degradation gives rise to local maxima in the time derivative of temperature.

The temperature dynamics remain qualitatively the same but the size of the second region becomes successively smaller as the externally applied radiation intensity is increased. In other words, as the spatial temperature gradients increase, endothermic and exothermic processes become more coupled, given that they take place at the same time for a successively larger portion of the sample. The simultaneous degradation of all the wood components results in a local process nearly isothermal, dominated by transport phenomena.

Given the high contribution of the low temperature components in the wood compositions (67-82%, Table 1), their initial degradation temperature, as reported by the centerline thermocouple, can be considered as representative of the pyrolysis process, T_p . On the other hand, lignin degradation takes place over a wide range of thermal conditions [17,18] and its exothermicity does not allow a clear and univocal definition of a reaction temperature. The time needed to attain the temperature T_p can be considered as the characteristic heating time, t_h, and can be used to evaluate the average heating rate as $\beta=(T_p-T_0)/t_h$, where T_0 is the initial temperature.

Though, from the qualitative point of view, the dynamics of sample heating are the same for the different wood varieties, significant differences exist for the pyrolysis temperature and the average heating rate and, consequently, for the minimum heat flux needed to achieve significant wood degradation (i.e. volatile yields at least equal to 50% of the initial dry solid mass). Figures 4-5 compare examples of the time histories of temperature at the sample centerline, while Figure 6 reports T_p and β as functions of the applied radiation intensity. For both temperature dynamics and parameter values, heat fluxes of about 40kW/m^2 separate a region of strong dependence from a region of weak dependence. On the whole, it can be observed that the average heating rates increase with the external radiative heat flux and roughly reproduce those of thermal analysis. The average heating rates are, for large part, determined by pyrolysis temperatures which, after a rapid increase (Q=40kW/m^2), approach constant values. With reference to maxima, the lowest T_p values (about 600K) are observed for chestnut, as a consequence of the high content of extractives and hemicellulose, which degrade at temperatures lower than cellulose. The highest T_p values are measured for beech wood (about 650K). Softwoods, in particular redwood and Douglas fir, present pyrolysis temperatures slightly lower than beech wood. The higher levels of extractives and the complete degradation of their hemicellulose content within a narrow temperature range [10] may be the factors responsible for these features.

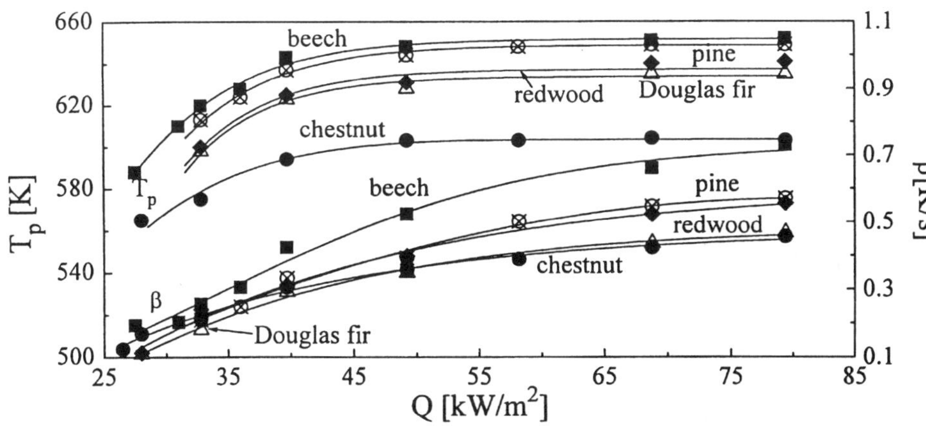

Fig. 6 Pyrolysis temperature, T_p, and average heating rate, β, as functions of the applied radiative heat flux for different wood varieties (cylinders).

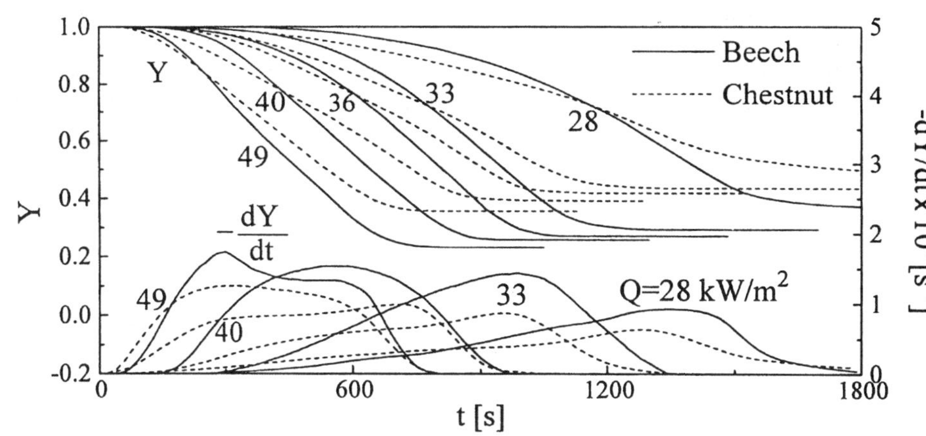

Fig. 7 Time profiles of the solid mass fractions and time derivatives of the solid mass fraction for chestnut (dashed lines) and beech (solid lines) wood cylinders exposed to several radiation intensities.

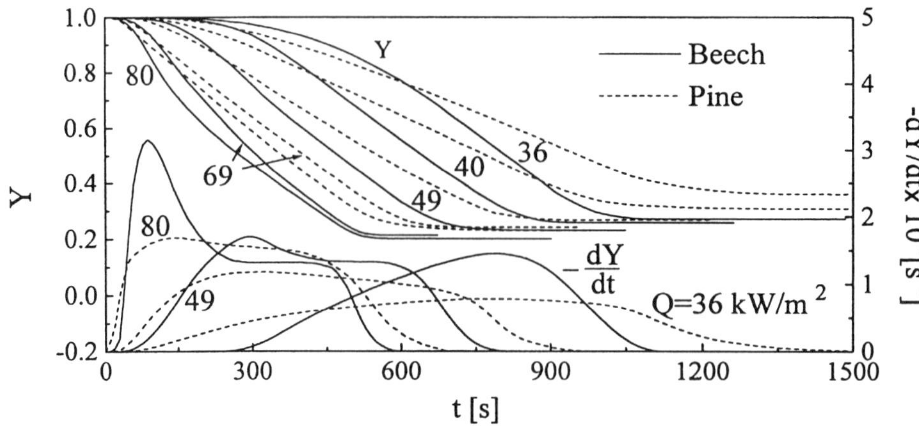

Fig. 8 Time profiles of the solid mass fractions and time derivatives of the solid mass fraction for pine (dashed lines) and beech (solid lines) wood cylinders exposed to several radiation intensities.

Differences in the temperature dynamics among varieties are high especially for Q=28kW/m^2, when heating rates are slow and thermal conditions barely adequate for degradation. The effects of variable contribution and/or reactivity of the components are enhanced also in relation to process energetics (exothermic lignin degradation). Thus large differences appear between the two hardwoods (Fig. 4) and between hardwoods and softwoods (Fig. 5). The very low pyrolysis temperature of chestnut causes that, for a given heat flux, the region controlled by reaction energetics is always larger than for the other wood varieties. In particular, for the lowest heat flux investigated, it regards the whole sample radius, as the degradation of extractive and holocellulose components, on one side, and lignin, on the other, take place into two separate stages. Therefore, despite of the low lignin content, its separate degradation is able to cause a significant temperature increase above the steady (final) value (Fig. 4), with effects on both product distribution and weight loss dynamics. From Fig. 5, where a classical hardwood (beech) and a classical softwood (redwood) are compared, it appears that the second exothermic stage is completely absent for the latter when Q=28kW/m^2 . In accordance with component analysis [10], this is due to higher degradation temperatures of softwood lignin. As a consequence, higher applied radiation intensities are needed in order to cause a softwood conversion above 50%.

A comparison between the weight loss dynamics for the two hardwoods considered in this study can be made through Fig.7. For beech wood, the apparent kinetics of weight loss can be described as a one-stage process, for Q=40kW/m^2 . Indeed, the temperature gradients are small, the size of the degrading region is large and maximum degradation rates are attained for heat-up conditions. Devolatilization characteristics are, on the whole, qualitatively similar to those discussed for woodchips beds. However, the complete or significant separation between the endothermic and exothermic degradation stages results in two different regions in the devolatilization rate curve. Indeed, a first wide region of low values (low-temperature component degradation) is followed by a narrow second zone of higher values (lignin degradation). Due to the significant exothermicity of the second stage, large part of the devolatilization is completed at times shorter than for beech.

For Q> $40kW/m^2$ and both hardwood varieties, the process tends to become globally two stage, with the first due to surface degradation and the second to sample heat up. Indeed, the fast rise of the surface temperature causes the prompt devolatilization of a thin layer, below the irradiated surface, while the increase in the local heating rates and the spatial temperature gradients cause the simultaneous degradation of the different wood components. Again, chestnut degradation starts at shorter times (so the appearance of a first peak associated with surface degradation is displaced towards higher Q values), global devolatilization rates are significantly lower, but conversion time is about the same as for beech wood. That is, degradation dynamics of the two hardwoods tend to become the same as the reaction conditions are made severe.

Figure 8 compares the global degradation characteristics of beech and pine woods. On the whole, it appears that pine devolatilization starts at shorter times and ends at times longer than beech (again with the same trends as for wood chips). The higher content of extractives, on one side, and the lower reactivity of softwood lignin, on the other, may account for these features (including longer conversion times). Devolatilization rates are, on the average, lower and the process is characterized by slightly higher solid residues. Also, given the lower density of pine wood, spatial temperature gradients are lower and the devolatilization rates remain relatively constant for large part of the degradation process. Again, differences between the two different wood varieties are reduced for high external radiation intensities.

An evaluation of the importance of internal heat conduction time versus external heat transfer time, based on the external radiation intensity, can be made by means of the Biot number, Bi, and the use of average medium properties. Bi values are comprised between about 10 and 40 [11,15] for the range of heating conditions considered, indicating that internal heat transfer is controlling and it also becomes successively more important as Q is increased (higher values for softwoods). The Biot number is defined as $Bi=(\varpi QR)/(\lambda \Delta T)$, where average values are used for the medium properties ($\varpi=0.8$, $\lambda=1.4x10^{-1}$ W/mK (hardwoods [15]) and $\lambda=0.93x10^{-1}$ W/mK (softwoods [15]) and temperature difference, ΔT, referred to the pyrolysis temperature). Also, the characteristic time associated with internal heat transfer can be compared with the characteristic time of chemical reaction, through the thermal Thiele number, Th, and average devolatilization rates [11,15] and medium properties. The thermal Thiele number is defined as $Th=(k\rho_s c_s R^2)/\lambda)$ (again average values for the medium properties are used with $\rho_s=450kg/m^3$ for hardwoods and $\rho_s=310kg/m^3$ for softwoods, $c_s=1.25$ kJ/kg K and, k, the average devolatilization rate). Computed values of Th vary from 0.4 to 2.3 and, for all the woods, applied heat fluxes of about $40kW/m^2$ roughly separate a region where chemical reaction times are slow (sample temperatures are hardly sufficient for wood degradation) from another where internal heat transfer becomes relatively more important.

Heat and mass transfer processes are affected by volume variation and structural changes undergone by wood while degrading, though these effects on the conversion characteristics are difficult to quantify. Sample shrinkage, mainly along the radial direction, appears as a reduction in the sample diameter by factors of 15% for beech wood and Q<$40kW/m^2$. For higher heat fluxes, structural failure occurs at a large extent and precise evaluations are not possible. A small reduction in the sample diameter (about 2-3%) is also observed for the other varieties and Q=$40kW/m^2$. For higher values, successively larger swelling occurs. For chestnut, it corresponds to an increase in the sample diameter by factors of 3-9% and it is also accompanied by formation of cracks. These are completely absent for softwoods, which present swelling factors of 3-10% (pine and Douglas fir) and 10-20% (redwood).

PYROLYSIS PRODUCTS

Product (char, liquids and gas) yields from wood cylinder pyrolysis are reported, as percent of the initial dry solid mass (%, wt), as functions of the radiation intensity in Figs. 9-11. The mass closure is good (95.5-99%) and, from the qualitative point of view, all the wood varieties show the same trend. Again two regions are shown with distributions highly dependent on the heating conditions for external heat fluxes below 40kW/m^2 (sample temperatures below 750K). For this region, as the external heat flux is increased, the final char yield decreases whereas volatile (gas and liquid) yields increase, as a result of a competition between charring and devolatilization reactions, which become successively more favored [18,19]. As secondary reaction activity is considered to be negligible for temperatures below 750K ([11, 15]), product distribution is essentially the result of the selectivity of primary wood decomposition reactions.

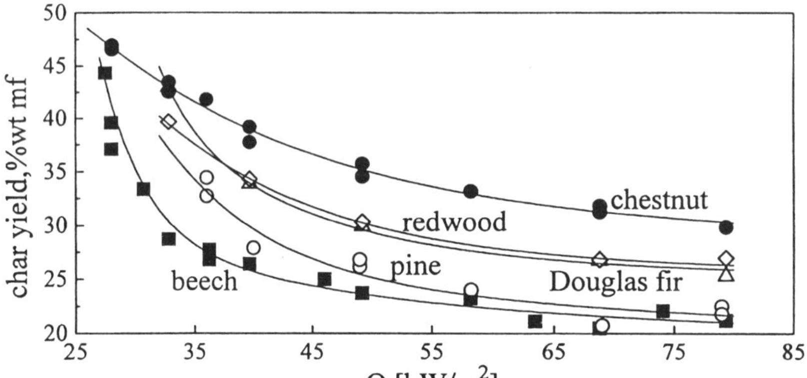

Fig.9 - Char yields from wood cylinders pyrolysis as functions of the radiative heat flux.

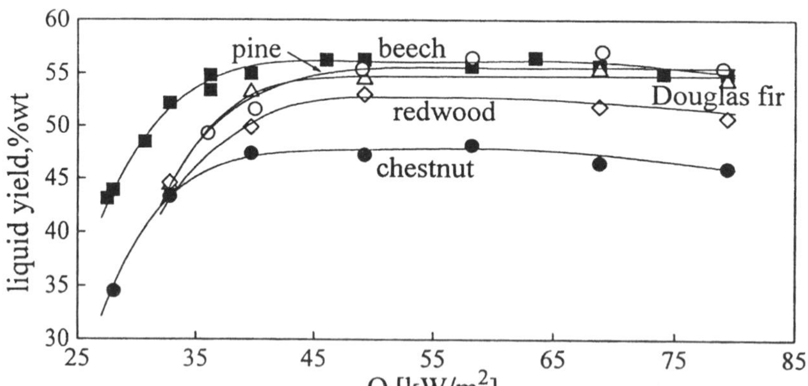

Fig.10 - Liquid yields from wood cylinders pyrolysis as functions of the radiative heat flux.

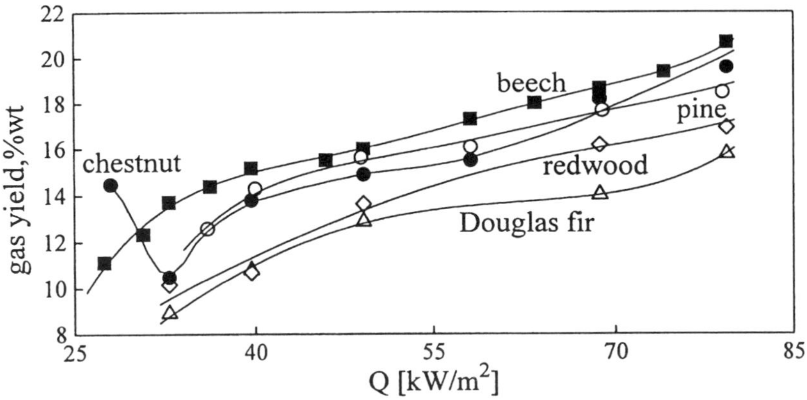

Fig.11 - Gas yields from wood pyrolysis cylinders as functions of the radiative heat flux.

For radiative heat fluxes above 40kW/m^2, liquid yields present a barely visible maximum, due to secondary tar reactions [19] as, for the experimental conditions of this study, these effects are small also for temperatures above 750K (small activity only within the pores of the degrading solid). Indeed, intra- and extra-particle residence times of volatile specie are short, as a consequence of the relatively fast devolatilization rates and the negligible intra-particle resistance to mass transfer, on one hand, and to the forced nitrogen flow, on the other. In the second place, the volatile temperatures are much lower (values always below 550K [11]) than those reported by thermocouples in strict contact with the solid phase. On the whole, all the wood varieties show the same qualitative dependence on the heating conditions, which, as pointed out in [11], is also in agreement with previous experimental and modelling analyses.

As for the solid residue (char), it should be observed that the high yields reported at low heat fluxes also include some unreacted material.

In particular, sample temperatures are not sufficiently high to cause complete degradation of the ligninic components, especially for softwoods. For the whole range of thermal conditions investigated, the highest char yields are measured for chestnut and the lowest for beech wood, whereas softwoods attain intermediate values. For the most severe conditions, when the volatile content of the char is at a minimum [20], yields vary from about 21% (beech and pine) to 27% (redwood and Douglas fir) and 33% (chestnut). Consequently wood variety causes variations in the char yield up to 12% with respect to the initial dry wood mass.

An inspection of the liquid yields as functions of the radiative heat flux reveals that the higher the char yields are the lower the liquids. The maximum liquid yields, attained for applied heat fluxes of 49-69kW/m^2, vary from about 56-57% (beech, Douglas fir and pine) to 52% (redwood) and 47% for chestnut. Hence, the wood variety may cause variations in the optimal liquid yields up to 10% of the initial dry wood mass.

Variations in the gas yields with the wood variety are lower than for the other two product classes. Indeed, for the highest heat flux (and maximum yields) they are about 5% of the initial dry solid wood. The highest yields are measured for beech, pine and chestnut (21-18.5%), with lower values for redwood and Douglas fir (17-16%). The local maximum observed for chestnut and Q=28kW/m^2 is due to the separate degradation of the low-temperature components and lignin for the whole sample, resulting in a second high exothermic stage with maximum temperatures well above the final steady value (Fig. 4), causing significant solid devolatilization.

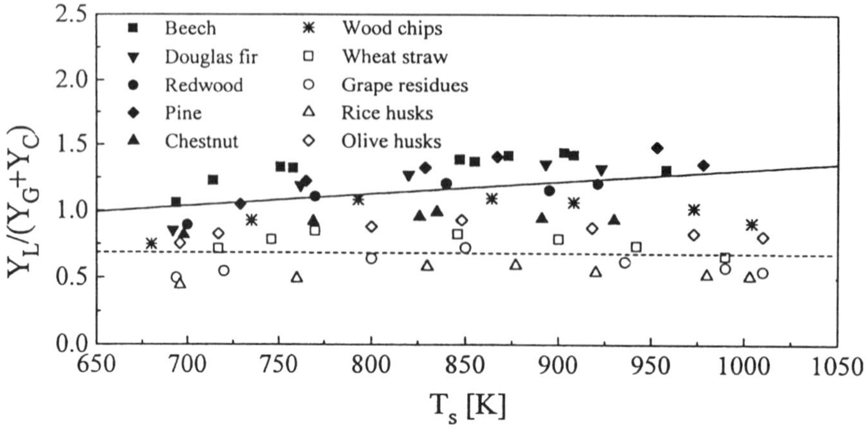

Fig.12 - Ratio $Y_L/(Y_C+Y_G)$ (ration of liquid yields to char and gas yields) as a function of the surface temperature in the pyrolysis of wood cylinders and packed biomass beds (for both cases a diameter of 4cm) exposed to radiative heating.

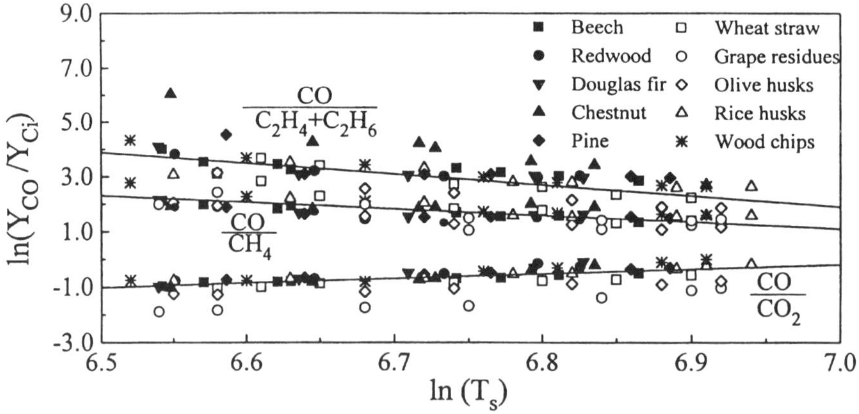

Fig.13 - Logarithms of the ratio CO/CO_2, CO/CH_4 and CO/C_2 yields as functions of the logarithm of T_s (steady surface temperature) for different wood varieties (cylinders) and packed biomass beds (for both cases a diameter of 4cm) exposed to radiative heating.

Chemical composition may explain, for large part, the differences in product distribution among the different wood varieties. Previous studies [21] report that extractive free wood produces up to 10% higher liquid yields than untreated wood. It was postulated [22] that the layer of extractives which surrounds the wood fibers acts as a barrier in the mass transfer of volatile pyrolysis products, out from the reacting sample. The increase in the residence time of tar vapors inside the high temperature pores of the degrading wood was thus considered responsible for an enhanced activity of secondary tar degradation reactions, with higher char or gas yields [21]. The results of this study are in qualitative agreement with such findings given that the lowest liquid yields are observed for the varieties with the highest extractive content (chestnut and redwood).

Also this feature is associated with high char (chestnut) or both char and gas (redwood with respect to Douglas fir, both belonging to the same variety and with very similar composition) yields. However, the explanation based on secondary reaction activity could be consistent only for high temperatures, while the above characteristics are observed for the whole range of heat fluxes. Hence, it is plausible that the products from the degradation of extractives, especially if these are present in high amounts, may contribute significantly to product distribution, that is, to the formation of char and/or gas, at the expense of liquid products. Specific features may depend on the chemical pattern of extractives which vary greatly between the different wood species and also between different parts of the same tree. Finally, in accordance with thermogravimetric results (for instance [10]), lignin is the main responsible for char formation, as redwood and Douglas fir produce higher yields of this product than pine and beech.

Figure 12 shows that, for reaction temperature above 650K, the ratio between the liquid yields and the yields of the other two product classes slightly increases with temperature (woods) or remains roughly constant for biomass. Indeed, for comparison purposes, results previously obtained for several agricultural residues [14] (packed beds of the same size as wood cylinders) have been examined. Hence, once the process becomes controlled by internal heat transfer, increasing the external temperature has a certain effect only on the relative amounts of gases and char (lignin degradation is more favored). Ratio values become successively lower with holocellulose content, with significant differences mainly between chestnut wood and the other woods, and also between the different biomass species.

The pyrolysis gas consists mainly of CO_2, CO, CH_4 and lower amounts of H_2 and C_2 hydrocarbons (details reported in [14] for biomass and in [15] for wood). At low temperatures, the evolution of CO and CO_2 (and water vapor) is due mainly to the degradation of extractives, hemicellulose and the activity of the first path in cellulose degradation, leading to gas and char formation. As temperature increases, the formation of tar vapor from cellulose becomes predominant, while lignin degradation also attains fast rates. The last process is largely responsible for char formation and also for the evolution of CO_2, CO, CH_4 and H_2 . From the quantitative point of view, the two hardwoods and pine (all including low lignin contents) are characterized by CO_2 yields higher than redwood and Douglas fir. Yields of CO and CH_4 are comparable for the different woods, apart from the slightly lower CH_4 and C_2 yields for chestnut, caused by the low lignin content.

Previous pyrolysis studies, for instance [20], report very general, approximate relationships for the ratios of the major gas phase products, CO, CH_4, CO_2 and C_2 hydrocarbons, over wide ranges of experimental conditions (temperatures). These are due to comparable values of the activation energies for the formation of the different species, with reaction mechanisms representative of apparent kinetics, taking into account both primary and secondary reactions. As apparent kinetics are also important in relation to large wood samples, a plot is constructed of the logarithms of the ratio of yields of major gas-phase species on the logarithm of the steady surface temperature, T_s, and comparisons are again made with agricultural residues [14]. Examples for CO/CO_2, CO/CH_4 and CO/C_2 are shown in Fig.13. It can be seen that a linear dependence is roughly established, confirming that major gas species present comparable kinetic constants (in particular activation energies) for both the different wood and biomass varieties.

CONCLUSIONS

The influence of wood variety (beech, chestnut, Douglas fir, redwood and pine) on the conventional pyrolysis characteristics has been investigated for a wide range of applied radiation intensities (from 28 to 80kW/m^2), which correspond to final sample temperatures of 600-950K.

The first difference between the pyrolytic behaviour of hardwoods and softwoods is the higher initial (minimum) heat flux required by the latter category for achieving conversion levels above 50% (about 33kW/m^2 against 28kW/m^2). This is a consequence of the higher degradation temperature of softwood lignin and the failed contribution of its reaction exothermicity for thermal enhancement. The lower reactivity of this component also results in slightly longer conversion times for softwoods.

For all the wood varieties, the temperature field shows an outer and an inner region, along the cylinder radius, determined by the simultaneous and the sequential degradation of the components, respectively. Though the details of process dynamics depend on the external heat flux and the wood category, the differences tend to disappear for heat fluxes above 40kW/m^2, when a shift from chemical reaction to internal heat transfer control occurs.

The wood components, independently of their nature, attain maximum degradation rates in a relatively narrow temperature range, i. e. 500-650K. Consequently, global parameters, such as pyrolysis temperature, average devolatilization rate, product yields and gas composition are affected mainly by the relative amounts rather than by the different reactivity (hardwood or softwood category) of the components. In general, despite of the same variety (hardwoods), the differences in product yields between beech and chestnut are very large. On the other hand, the differences are significantly lower between beech and pine (softwood), given the comparable contributions of the main components.

Finally, softwoods with nearly equal chemical composition (redwood and Douglas fir) also present nearly equal product distribution. Thus, in terms of product distribution, the percentage of the different components in the chemical composition of wood is more important than the wood variety (or the different reactivity of the main components). In particular, the share between solid and liquid products from the degradation process is highly dependent on the holocellulose contribution in the wood chemical composition, with successively higher solid residues as lignin and/or extractive contents increase. Differences between pyrolysis products from wood and biomass are high for liquid and char yields, whereas the gas yields are comparable and so are the apparent kinetics of single species formation.

Finally, this study provides an extensive set of data on thick wood pyrolysis which can be better interpreted and generalized by the use of mathematical models taking into account the effects of transport phenomena and chemical reactions. Models including such features are already available in the literature (for instance, see References 23,24) and have proven to give quantitative predictions of temperature dynamics, but product yield predictions are still unacceptable, mainly because of unreliable kinetic constants. Therefore, this issue deserves further investigation before extensive computer simulation and/or development of more advanced physical models of thick wood pyrolysis are proposed.

RFERENCES

1. Roberts A. F., Clough G., Ninth Symp. (Int.) on Combustion, The Combustion Institute, Pittsburgh, 1963, pp. 158-66.
2. Tinney, E. R., Tenth Symposium (Int.) on Combustion, The Combustion Institute, Pittsburgh, 1965, pp.925-30.
3. Kanury, M. A., Combustion and Flame, 18, 75-83 (1972).
4. Lee C. K., Chaiken R. F., Singer J. M., Sixteenth Symposium (Int.) on Combustion, The Combustion Institute, Pittsburgh, 1976, 1459-70.
5. Pyle D. L., Zaror C. A., Chemical Engineering Science 19, 147-158 (1984).
6. Ohlemiller T. J., Kashiwagi T., Werner K., Combustion and Flame, 69, 155-170 (1987).
7. Chan W. R., Kelbon. M., Krieger-Brockett B., Ind. Eng. Chem. Res., 27:2261-75 (1988).
8. Fredlund B., Fire Safety Journal, 20, 39-69 (1993).
9. Bilbao R., Millera A., Murillo M. B., Ind. Eng. Chem. Res., 32, 1811-17 (1993).
10. Gronli M. G., A theoretical and experimental study of the thermal degradation of biomass, PhD. Thesis, NTNU, Trondheim, Norway, 1996.
11. Di Blasi C., Gonzalez Hernandez E., Santoro A., Ind. Eng. Chem. Res., 39, 873-82, (2000).
12. Di Blasi C., Chemical Engineering Science, 55, 2931-44, (2000).
13. Di Blasi C., Modelling the fast pyrolysis of cellulosic particles in fluidized bed reactors, Chemical Engineering Science, in press, 2000.
14. Di Blasi C., Signorelli G., Portoricco G., Ind. Eng. Chem. Res., 38: 2571-81, 1999.
15. Di Blasi C., Branca C., Santoro A., Gonzalez Hernandez E., Pyrolytic behaviour and products of some wood varieties, Combustion and Flame, in press 2000.
16. Di Blasi C., Branca C., Santoro A., Perez Bermudez R. A., Weight loss dynamics of wood chips under fast radiative heating, J. of Analytical and Applied Pyrolysis, in press 2000.
17. Antal M. J., Varhegyi G., Ind. Eng. Chem. Res. 34, 703-17 (1995).
18. Shafizadeh F., in Fundamentals of Biomass Thermochemical Conversion, (R. P. Overend, T. A. Milne, L. K. Mudge, Eds.), Elsevier, London,, pp. 183-218 (1985).
19. Di Blasi C., Progress in Energy and Combustion Science, 19, 71-104 (1993).
20. Scott D. S., Piskorz J., Radlein D., Ind. Eng. Chem. Res. 24, 581-88 (1985).
21. Roy C., Pakdel H., Brouillard D., Journal of Applied Polymer Science 41, 337-48 (1990).
22. Ahmed A., Pakdel H., Roy C., Kaliaguine S., Journal of Analytical and Applied Pyrolysis 14, 281-94 (1989).
23. Di Blasi C., Chemical Engineering Science, 51, 1121-1132 (1996).
24. Di Blasi C., Int. J. of Heat and Mass Transfer, 41, 4139-4150 (1998).

The Mathematical Modeling of Biomass Pyrolysis in a Fixed Bed and Experimental Verification

G. Chen[1], J. Andries[1] and D. Y. C. Leung[2]

[1] *Section of Thermal Power Engineering, Department of Mechanical Engineering and Marine Technology, Delft University of Technology, Mekelweg 2, 2628 CD Delft, the Netherlands*

[2] *Department of Mechanical Engineering, The University of Hong Kong, Pokfulam Road, Hong Kong, China*

ABSTRACT: A modelling study of biomass pyrolysis in a fixed bed has been investigated using a novel kinetic scheme, which comprises all the basic Phenomena occurring at biomass pyrolysis. The model considers transient behaviour of biomass particle's physical parameters which are a function of the conversion rate. Mass transfer covering the term of convective flow, heat transfer through conduction, convection and radiation, as well as the change in the structure occurring in the biomass particle are accounted for. The prediction based on this mathematical model shows good consistence when compared with those experimental data on conversion, gas and tar yield, gas and tar evolution, and temperature over a wide range of operating conditions executed in the fixed bed. The model can be also applicable to other cases like biomass or coal gasification.

INTRODUCTION

Biomass is a renewable resource of plenty of storage covering all the world land surface particularly in rural districts, its conversion to synthetic fuels and chemical products appears attractive, especially in developing countries where the imported fuel and chemical products are much unbearable in price. In addition, biomass can be utilized together with coal in varied proportion with an additional advantage of synergistic effects with regard to char reactivity, tar formation and emission of harmful components[1]. Currently, among all kinds of conversion routes, thermo-chemical conversion, especially pyrolysis is being paid more attention in consideration of economy combined with technology. The pyrolysis process can be carried out very effectively in a fixed bed in conventional way with additional advantages of relatively easy design, operation and optimization. However, determining a kinetic approach which could describe well pyrolysis mechanism involving in this case is much difficult due to the varying composition of biomass materials used and the complexity of the chemical reactions involved in the pyrolysis. Furthermore, the mathematic modelling coupling knowledge of kinetics with heat and mass transfer description appears much more difficult, but this task is really mandatory to be undertaken in that case when precise prediction by model are expected.

Since the beginning several studies have been carried out. Bamford et al. proposed a model by the equation of heat transfer coupled with the equation for heat changes due to chemical reactions[2]. Afterwards, this model was adopted and adjusted by some researchers in order to incorporate the effects of internal convection and variable transport properties[3,4]. Using dimensionless groups to Bamford's model, Pyle and Zaror defined the relative importance of internal and external heat transfer and of the intrinsic pyrolysis kinetics[5]. A general pyrolysis model was specified by Kothari and Antal with a coupled set of two ordinary and three partial differential equations, as well as approximate boundary conditions[6]. This model concerned three levels of problems covering chemical reaction, external heat transfer, internal heat transfer and internal mass transfer. Based on the assumption of pseudo-steady-state, a transport model applicable for radiative heat transfer was presented by Curtis and Miller[7]. This model concerns only convection flow and reaction terms but neglects diffusion and pressure-driven flow. Recently, Hastaoglu and Hassam applied a general gas-solid reaction to flash pyrolysis of wood, where convective mass transfer and diffusion through the porous medium were considered but pyrolysis mechanism was depicted using simple two-parallel primary reactions followed by a secondary cracking reaction[8]. More recently, Batra and Rao introduced into the model a heat transfer equation combined with the pyrolysis reaction rate for biomass pyrolyzed in an annular finned pyrolyser, still using simple two order decomposition model[9].

The following basic steps can be used to describe in mechanism biomass pyrolysis process: step 1, the degradation of the virgin biomass materials into primary products (tar, gas and semi-char); step 2, the decomposition of the primary tar into secondary products (tar, gas and char), and step 3, the accompanied reaction between primary gas and semi-char. Unfortunately, almost those above-mentioned mathematical models on the one hand did not reflect step 2 or step 3, even if step 2 and 3 together, on the other hand majority of them did not incorporate convective terms of heat and mass transfer by volatile products inside reactor.

Although under certain experimental condition, step 2 or step 3 may be ignored due to its non-conspicuous influence to the final pyrolysis product distribution, however, a good mathematical model for biomass pyrolysis should be versatile applicable to other pyrolysis conditions, and thus it should be involved the above-mentioned three steps of process, of course heat and mass transfer equations should be included also. This paper presents this kind of mathematical model. Although the model is constructed based on sawdust pyrolysis, it is quite straightforward to apply the same approach to other cases such as straw and municipal solid waste pyrolysis even if to biomass or coal gasification or metal ore reduction.

MODELLING

KINETIC MODELLING

The basic principle of the overall kinetic pyrolysis model (OKP model) proposed by us is shown in *Fig. 1*. Under our experimental conditions aimed at the maximum production of medium calorific gas, step 2 and step 3 exert significant influences to final products yielded, and thus attention should be given to them. Our kinetic scheme obviously differs from those mechanisms used previously[5,10-12]. The OKP model reveals the significant secondary cracking of the primary pyrolysis tar product and in particular interactions between semi-char and gaseous phase which always exist irrespective of reaction temperature, residence time of the volatile phase. Those important phenomena of adsorption, deposition, char-gas reaction, cracking and repolymerization reactions are all incorporated in this model. Of course, the OKP model can be simplified to those existed in particular case where a certain secondary reaction occurs only to a minimum degree and thus may be neglected.

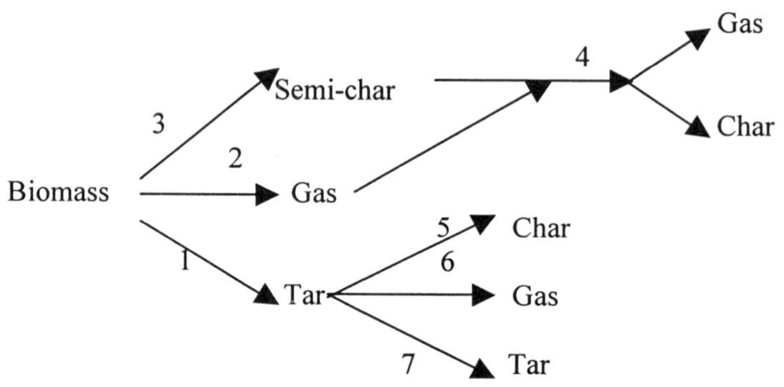

Fig. 1. Kinetic scheme proposed by us for biomass pyrolysis.

Kinetic scheme is typically described by the following partial differential equations, which are in closure physically.

$$\frac{\partial W_d}{\partial t} = -(k_1 + k_2 + k_3)W_d$$

$$\frac{\partial C_1}{\partial t} = k_3 W_d^{n3} - k_4 C_1$$

$$\frac{\partial C_{21}}{\partial t} = \delta k_4 C_1$$

$$\frac{\partial C_{22}}{\partial t} = k_5 T_{a1}$$

$$\frac{\partial T_{a1}}{\partial t} = k_1 W_d^{n1} - (k_5 + k_6 + k_7) T_{a1}$$

Here $k_j = A_j \exp(-E_j / LT)$; the initial conditions is as follows:

t=0: $W_d = W_0$, $C_1 = C_{21} = C_{22} = T_{a1} = 0$

Where, W_d is degradable portion of biomass weight at time t; A_j is pre-exponential factor at reaction j; C_1 and C_{21}, C_{22} are primary and secondary char weight, respectively; E_j is apparent activation energy at reaction j; T_{a1} is primary tar weight; $k_1 - k_7$ are kinetic rate constants; T is temperature and L is gas constant.

MATERIAL CONSERVATION

For tar and total volatile phases, the following two equations are applicable.

$$\frac{\partial(\varepsilon\rho_T)}{\partial t} + \frac{\frac{1}{r}\partial(r\rho_T U)}{\partial r} = k_1\rho_d - \varepsilon(k_5 + k_6)\rho_T$$

$$\frac{\partial(\varepsilon\rho_V)}{\partial t} + \frac{\frac{1}{r}\partial(r\rho_V U)}{\partial r} = k_1\rho_d + \varepsilon\delta k_4\rho_{G1} - \varepsilon k_5\rho_{Ta1}$$

Where, r is radial distance of reaction front; ε is void fraction; δ is deposition coefficient; U is radial diffusion velocity; ρ_T , ρ_d , ρ_V , ρ_{Ta1} and ρ_{G1} are concentrations of total tar, degradable biomass, total volatile phase, primary tar and primary gas, respectively. This material balance equations contain terms only for convective flow and reaction, neglecting both diffusion and pressure-driven flow. This is due to the fact that early simulation showed diffusion and pressure-driven flow in the material balance several orders of magnitude smaller than the convection and reaction terms[7,13]. In our material conservation equations, a term with respect to the transient term is introduced. A pesudo-steady-state assumption is not made for tar and total volatile phase flow because residence time of volatile phase in the reactor is comparable in magnitude to the time necessary for temperature profile to change significantly.

ENERGY CONSERVATION

The energy balance in the fixed bed includes convective, conductive and radiative heat transfer and heats of reactions terms. The equation is described below:

$$[\varepsilon\rho_V C_{V,p} + (1-\varepsilon)\rho_s C_{s,p})]\frac{\partial T}{\partial t} + \rho_V U C_{V,p}\frac{1}{r}(r\frac{\partial T}{\partial r}) = K_e\frac{1}{r}\frac{\partial}{\partial r}(r\frac{\partial T}{\partial r}) + \sum_{j=1}^{7} k_j \Delta h_j$$

Where K_e is the effective thermal conductivity, $C_{V,p}$, $C_{s,p}$ are specific heat of total volatile phase and solid phase, respectively; ρ_s is the concentration of solid phase and Δh is reaction heat. The first term of the left-hand side in the equation is the generation term related to internal enthalpy of solid and volatile phases. The second term is the convective flow of the volatile components. The first term of the right-hand side is the effective conductive flow and the second term is the reaction heat with regard to chemical reactions and phase changes.

INITIAL AND BOUNDARY CONDITIONS

In order to solve mass and energy balance equations, initial and boundary conditions should be given. Initially, the biomass material is in a quiescent environment at ambient conditions and thus is specified as uniform temperature and solid compositions. For t>0, the spatial conditions at the centerline (r=0) are specified by symmetry, the two sides (r=R) are exposed to radiation by a constant temperature of heater. At these two sides, radiant heat transfer from the surface takes places. Further conditions of constant ambient pressure and zero gradient of tar and gas mass at centerline are also used.

At $t = 0$: $T = T_o$; $\rho_s = \rho_0$; $\rho_T = \rho_V = \rho_{Ta1} = \rho_{G1} = 0$; $U = 0$

At $t > 0$, $r = 0$: $\frac{\partial T}{\partial r} = 0$;

$r = R$: $K_e\frac{T}{r} = e\sigma(T_r^4 - T^4)$; $U = 0$

Where: e is emissivity of pyrolyzing biomass particle; σ is Stefan-Boltzmann coefficient; T_r is pre-set temperature.

SOLUTIONS OF THE MODELLING EQUATIONS

Assumptions

(1) The radiative resistance and conductive resistance outside inner surface of the reactor are advertently omitted and thus the pre-fixed temperature of the heater can be assumed as the initial value of inner wall of the reactor.
(2) Biomass particle is taken as non-shrinkable during pyrolysis process and its influence may be accounted for by the change in void fraction of the bed.
(3) The temperature is symmetric along the centerline and thus only radial temperature profile exists as the ratio of height of reactor to its diameter is larger than 5.

Parameters and physical properties

Table 1 presents the values of physical properties and simulation parameters used, which were derived from data given in our experimental results and literature. These values are regarded as overall values with regard to not only the corresponding values of the biomass material itself but also of its derivatives produced. It is considered that the local values of these properties in particular heat conduction coefficient K_{con} and specific heat $C_{s,p}$ are functions of the pyrolysis temperature and the conversion rate of biomass. K_{con}, $C_{s,p}$ and B (permeability) were estimated as the sum of the value corresponding to the virgin biomass material and of the char at a certain temperature. Each of the values of three parameters contributes to the sum in a way proportional to the conversion rate. The expression is as follows:

$$X = (1-\eta)X_b + \eta X_c$$

Table 1. Parameters and properties used.

Parameters and properties	Source
$\Delta h_1 = \Delta h_2 = \Delta h_3 = -418 kJ/kg$	Chan[15]
$\Delta h_5 = \Delta h_6 = 42 kJ/kg$	Koufopanous[2]
$\Delta h_4 = -492 kJ/kg$	estimated
$\Delta h_7 = 42 kJ/kg$	estimated
$\rho_{s,0} = 166.5 kg/m^3$	Chen[14]
$C_b = 2.3 kJ/kg \cdot K$	Kung[4]
$C_c = 1.1 kJ/kg \cdot K$	Kansa[16]
$d = 4 \times 10^{-5} m$	Chan[15]
$K_b = 0.11 W/m \cdot K$	Chen[14]
$K_c = 10.46 \times 10^{-2} W/m \cdot K$	Lee[17]
$\mu = 3 \times 10^{-5} kg/m \cdot s$	Kansa[16]
$e = 0.95$	Pyle[5]
$\varepsilon_0 = 0.70$	Chen[14]
$B_b = 1.0 \times 10^{-5}$ Darcys	Di Blasi[13]
$B_c = 0.32$ Darcys	estimated

Where X is to represent K_{con}, $C_{s,p}$ and B; $\eta = {(W_0 - W)}/{W_0}$, is the conversion rate of biomass material in the bed. Advertently, the contribution of conduction of volatile phase is not accounted for by K_{con} due to its value more than one order in magnitude smaller than the values of virgin biomass or of char. Dummy thermal conductivity of K_{rad} (also named as equivalent thermal conductivity) was calculated as follows:

$$K_{rad} = 13.5T^3 \frac{\sigma}{e} \frac{d}{\varepsilon}$$

The effective thermal conductivity K_e, defined by the following equation, which accounts for the combined contribution of conductive and radiative heat transfer can be thus applicable.

$$K_e = K_{con} + K_{rad}$$

The specific heat of total volatile phase, $C_{V,p}$, is taken as the molar-weighted average of tar-phase and gas-phase, and the tar as well as gas phases specific heats are also taken as the molar-weighted average of all its components, respectively. The values of the specific heats of tar and gas phases are define by equations below:

$$C_{T,p} = -286.721 + 1549.749 \frac{T}{T_0}$$

$$C_{G,p} = 41.934 + 1.572 \frac{T}{T_o}$$

Radial velocity U can be derived from the assumption of ideal gas and the void fraction of the bed can be calculated below:

$$\varepsilon = 1 - (1 - \varepsilon_0) \frac{W}{W_0} \frac{\rho_{s,o}}{\rho_s}$$

Where the density of solid in the bed, ρ_s, can be deduced from general pyrolysis rate expression with a modified coefficient, $1/(1-\varepsilon)$.

Mathematical manipulation

Considering the following work is to be carried out in the fixed bed, where reaction temperature is not constant whilst the constant heating rate (Φ) may be assumed within accuracy, therefore here only a certain heating rate is concerned. In this case, kinetic equations should be re-constructed by expression of $T = \Phi t + T_0$ (T_0 is initial temperature). The best values of $k_1 \sim k_7$ were obtained, using the least square analysis where the sum of square of deviation between theoretically calculated residual mass values and experimentally measured results by thermogravimetry. Detailed method for obtaining kinetic parameters is available elsewhere[14] .

The method of zonation was applied to the energy and material conservation equations. Based on centered finite difference approximations, this method can transform three partial differential equations in radial distance and time to ordinary differential equations in time only. Following this, the ordinary differential equations were solved by using Crank-Nicholson algorithm. On the basis of this, the volumetric fluxes of those tar-phase and total volatile phase components were integrated with time by using improved Euler method to evaluate overall pyrolysis product yields, and afterwards the gas yield can be deduced.

EXPERIMENTS

The verified experiments were carried out in the fixed bed, which was set up by us in particular for the purpose of investigating pyrolysis behaviors of biomass materials.

The detailed description of that apparatus is shown elsewhere[14]. The following parameters and operating conditions were used: for fixed bed, 80mm i. d., 450mm in height, bed temperature 500~900 $^{\circ}C$; for pine sawdust particle: 0~1mm in diameter, apparent density 166.5 kg/m^3, initial porosity 0.70, energy density 3007.7 kJ/ m^3, apparent char density 95 kg/m^3; initial sawdust weight 0.375kg. In addition, the following estimated properties were also used: initial sawdust permeability 1.0E-5 darcys, char permeability 0.2 darcys; molar molecular weight of tar, gas, initial sawdust and char 145, 31, 145 and 12, respectively.

Those kinetic experiments were carried out using a thermogravimetry. The biomass sample first was kept in a differential compartment and then introduced into the reaction zone by an inert gas flow of nitrogen. The temperature and the sample weight are continuously and automatically monitored. The experiments were based on constant heating rate of 50 $^{\circ}C/\min$ while not on other heating rates in order to keep close simulation of those following experiments executed in fixed bed.

RESULTS AND DISCUSSIONS

The relation of weight-loss and temperature variation at pine sawdust pyrolysis by thermogravimetry is shown in *Fig. 2*, where the continuous line is predicted result by model and the symbols are the experimental values. As constant heating rate is considered, the correlation of weight loss and temperature variation represents weight-loss evolution history. As can be seen from this figure, the kinetic model predicts well the weight-loss and temperature variation.

Fig. 3 shows temperature histories at the wall surface and center of the reactor, respectively. In general, the simulated results agree well with those experimentally monitored data. The divergence appearing in the figure may be contributed to the negligible moisture evaporation terms in the model. The evaporation of moisture contained in the virgin biomass particles is an endothermic process, and thus monitored temperatures are lower than those predicted at the initial stage of pyrolysis. Another possible explanation to that divergence is due to term Δh_4 incorporated in energy balance equation. Up to now, there is still no accurate description to the reaction scheme related to k_4.

Fig. 4. shows the evolution history of volumetric flow of gas-phase yielded. The experimental dots if regressed by a curve whether in shape or magnitude is well similar to that theoretically predicted. The maximum gas-phase flow is achieved at time of 240s or so. Theoretically, this point of time is exactly corresponding to the beginning of fast decomposition of cellulose, main component of biomass. Also the phenomena represented by this figure reflects the same trend by *Fig. 3*, where the predicted temperature is normally a little higher than of experimentally measured. In this figure, it is easily to be observed that the predicted volumetric flow curve is also put forward a little. The area beneath the predicted volumetric flow curve is corresponding to the simulated non-condensable gas phase yield.

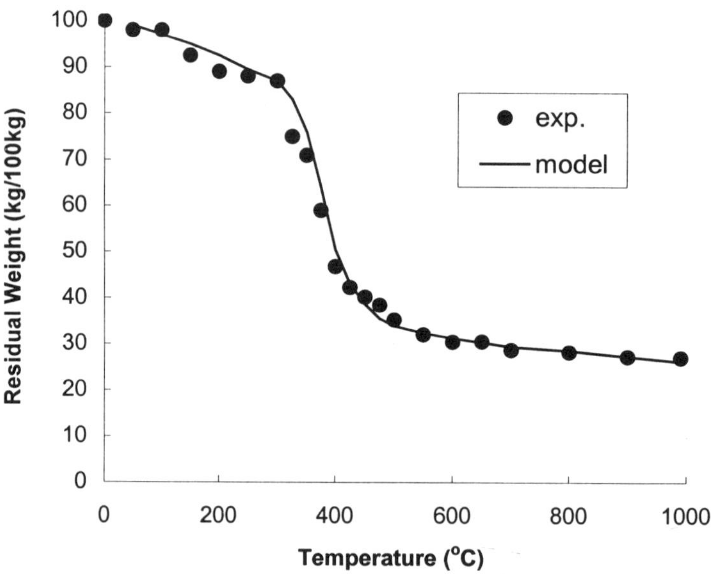

Fig. 2. Weight-loss versus temperature variations for pine sawdust pyrolysis at constant heating rate of 50 $°C/\min$. Dots, experimental data; solid line, model predictions.

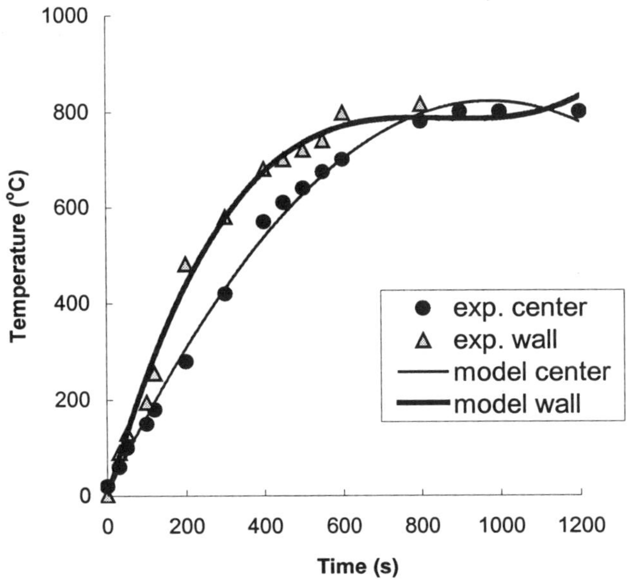

Fig. 3. Temperature profile at the center-line and inner wall surface of the reactor versus time during pyrolysis at temperature of 800 $^{\circ}C$. Symbols, experimental data; line, model predictions.

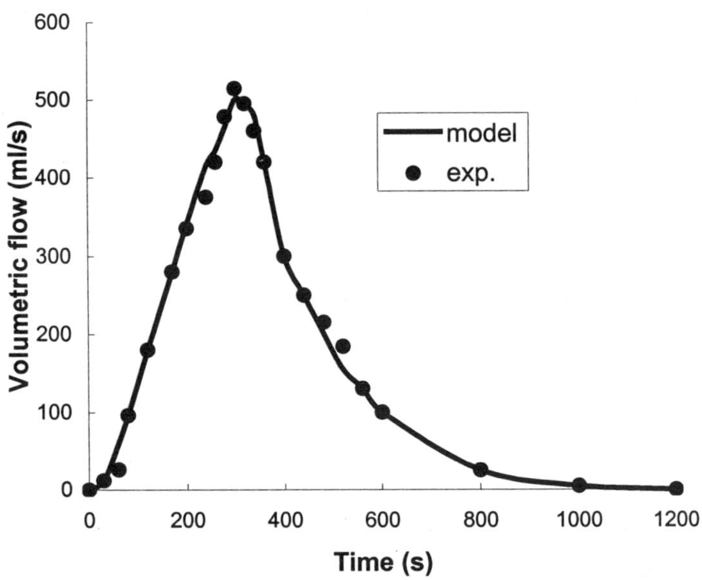

Fig. 4. The evolution history of volumetric flow of non-condensable gas-phase yielded during pyrolysis at temperature of 900 $^{\circ}C$. Dots, experimental data; solid line, model prediction.

The final product yields at different temperatures are presented in *Fig. 5*. It can be seen an increase in temperature results in increasing value of gas yield defined by kg of per kg of biomass feed, while reducing value of tar phase yield. In the latter case, the decrease observed could be related to secondary reactions (tar cracking reactions) where tar components of heavy molecular mass decompose into that of light molecular mass and accompanied gases. Also, when the reaction temperature increases further from 700 $^\circ C$, the decrease observed in the value of tar yield is becoming weaker. This may be attributed to the fact that tar cracking is easy to process at temperature of 500 $^\circ C$ above and furthermore approximately accesses final point as temperature reaches 700 $^\circ C$. It can be also observed that the increase of gas yield is sufficient to compensate for loss of tar yield. Char-gas interactions occurring at temperature of 700 $^\circ C$ higher probably contribute to this observation. When compared with the data obtained experimentally, the values of gas and tar yields estimated by the model are shown to be ideal, though an over-estimation of gas yield and an under-estimation of tar yield are presented. The over-estimation may be due to the fact that char-gas reactions of char and H_2O, char and CO_2 incorporated in the model are not significantly influential to products yields under our experimental conditions where heating-time is comparable to devolatilization-time of biomass particles. The under-estimation of tar yield, of course, is accompanied by an assumption of secondary reactions occurring in all yielded tars.

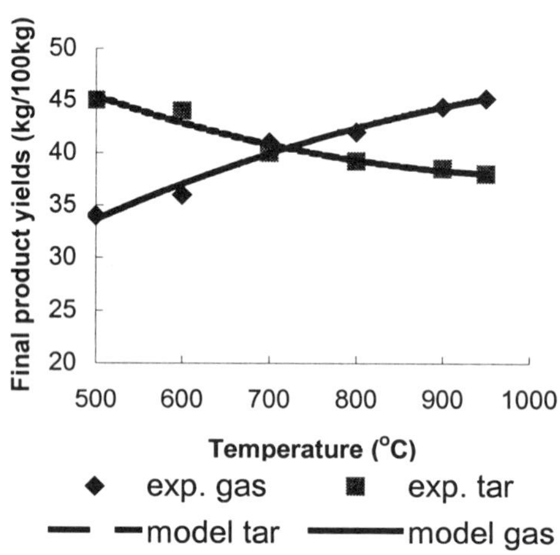

Fig. 5. The final gas and tar yields at different temperatures. Symbols, experimental data; solid line, predicted gas yield; dash line, predicted tar yield.

Fig. 6 presents the evolution of tar and gas yields measured by weight percent. Also the figure shows the corresponding simulated results. The profile is sharp at initial stage of the process (0~4minutes), followed by a relatively sharp stage (4~7minutes) and then by a flat period (7minutes~end). The gas and tar yields at the time period 0~7 minutes account for 90% and 95% of the total yields, respectively. Such considerations are important when the economics of the process are to be taken into account. The simulated results are well corresponding to the measured data, considering the complexity of sawdust pyrolysis process. From this figure, conversion rate η can be obtained, however, this part of work will be discussed in the near future publication.

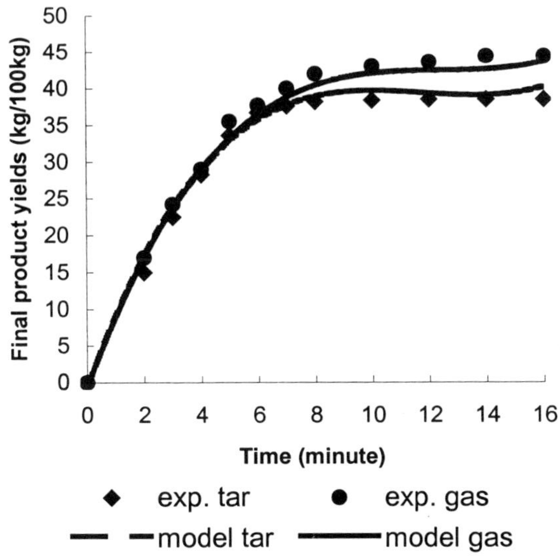

Fig. 6. Evolution of gas and tar yields during pyrolysis at temperature of 900 $^{\circ}C$. Symbols, experimental data; solid line, predicted gas yield; dash line, predicted tar yield.

CONCLUSIONS

It has proven that our kinetic pyrolysis scheme covering three basic steps can be well applied to the analysis of sawdust pyrolysis in a fixed bed. Further to this, the self-constructed mathematical model incorporating heat and mass transfer equations combined with our OKP model can be used to successfully predict the behaviour of biomass particles pyrolyzing in the fixed bed and the comparison between experimentally obtained values and predicted results by model shows good agreement. The modelling also considers the structural changes of biomass particles inside the bed through the equation of void fraction. In addition, the model incorporates convective flow terms in material conservation and radiative heat transfer in energy conservation. With the aid of the model, it is completely possible to optimize various parameters for any required yield.

REFERENCES

1. Andries J. & Hoppesteyn P. D. J. (1997) Pressurized combustion of biomass-derived low calorific value, fuel gas. In: Biomass, Gasification and Pyrolysis, (Ed. by M. Kaltschmitt & A. V. Bridgwater), pp. 1282-291.
2. Bamford C. H. & Crank J. et al. (1946) The combustion of wood. Proceed. Cambr. Phil Soc., 42, 166-82.

3. Capart R. & Fagbemi L. et al (1986) Wood pyrolysis: a model including thermal effects on the reaction, energy from biomass. (Ed. By W. Palz), pp. 842-846. Elsevier Applied Science Publication.
4. Kung K. C. (1972) A mathematical model of wood pyrolysis. Combustion and Flame, 18, 185-95.
5. Pyle D. L. & Zaror C. A. (1984) Heat transfer and kinetics in the lower temperature pyrolysis of solids. Chem. Eng. Sci., 39, 147-58.
6. Kothari V. & Antal M. J. (1985) Numerical studies of the flash pyrolysis of cellulose. Fuel, 64, 1487-494.
7. Curtis L. J. & Miller D. J. (1988) Transport model with radiative heat transfer for rapid cellulose pyrolysis. Ind. Eng. Chem. Res., 27, 1775-783.
8. Hastaoglu M. A. & Hassam M. S. (1995) Application of a general gas-solid reaction model to flash pyrolysis of wood in a circulating fluidized bed. Fuel, 74, 697-703.
9. Batra D. & Rao T. R. (2000) Analysis of an annular finned pyrolyzer-II. Energy Convers. Manage, 41, 573-83.
10. Koufopanos C. A. & Papayanakos N. (1991) Modelling of the pyrolysis of biomass particles: studies on kinetics, thermal and heat transfer effects. Can. J. Chem. Eng., 69, 907-15.
11. Ward S. M. & Braslaw J. (1985) Experimental weight loss kinetics of wood pyrolysis under vacuum. Combustion and Flame, 61, 261-69.
12. Di Blasi C. (1996) Kinetic and heat transfer control in the slow and flash pyrolysis of solids. Ind. Eng. Chem.Res., 35, 27-46.
13. Di Blasi C. (1996) Heat, momentum and mass transport through a shrinking biomass particle exposed to thermal radiation. Chem. Eng. Sci., 51, 1121-132.
14. Chen G. (1998) Study on mechanism for biomass pyrolysis and its experimental verification. DPhil thesis, Zhejiang University.
15. Chan W. R. & Krieger B. B. (1985) Modelling and experimental verification of physical and chemical process during pyrolysis of a large biomass particle. Fuel, 64, 1505-513.
16. Kansa E. J. & Perlee H. E. (1977) Mathematical model of wood pyrolysis including internal forced convection. Combustion and Flame, 29, 311-24.
17. Lee C. K. & Chaiken R. F. (1976) Charring pyrolysis of wood in fire by laser simulation. Sixteenth Symposium on Combustion, the Combustion Institute, pp. 1459-470.

Origin and Nature of Paramagnetic Moieties in Pyrolysis Oils

T. Dizhbite, G. Dobele, N. Mironova, G. Telysheva,
Latvian State Institute of Wood Chemistry,
27 Dzerbenes St. , LV-1006, Riga, Latvia
D. Meier, O. Faix
Federal Research Centre for Forestry and Forest Products, Institute for Wood Chemistry and Chemical Technology of Wood, Leuschnerstr. 91, D-21031 Hamburg, Germany

ABSTRACT: Paramagnetic properties of pyrolysis bio-oils produced on different fast pyrolysis pilot plants have been studied using ESR spectroscopy. A narrow (0.3-0.4 mT) ESR singlet detected in the oil samples was assigned to stable paramagnetic centers of expanded π-conjugation systems, which involve 6 – 12 benzene rings. The content of paramagnetic species varied significantly (3 – 4000 ppm) for oil samples from different plants. It has been shown that paramagnetic domains in oils are related to both char microparticles admixtures and polyarylene structures present in the tar (pitch) fraction of oil, rich in "pyrolytic lignin". Based on ESR and polarization microscopy data it is assumed that formation of anisotropy associates of π-conjugation paramagnetic domains, forming an ordered mesophase, took place. The possibilities of ESR for monitoring phase separation of pyrolysis liquids during storage is discussed.

INTRODUCTION

Proposed application of bio-oils as fuel for gas turbines or diesel engine generators requires stability of their properties in conformity with the standard specifications. However, it has been recognized for a long time that bio-oils properties tend to change during storage resulting in an increase in molecular weight and viscosity, phase separation and fine char sedimentation [1, 2]. The aging of bio-oils depends on a combination of physico-chemical processes and chemical reactions including association and agglomeration, polymerization and condensation as well as redox processes.

Bio-oils can be considered a multicomponent system, which often contain impurities such as char fines and ash elements which have paramagnetic properties [3]. However, there are no published data about paramagnetic properties of bio-oils obtained by fast pyrolysis of wood, although many research groups have reported paramagnetic properties of charcoal and petroleum pitch. It is well known that the various paramagnetic impurities can play a key role in promoting ageing as they act as catalysts for hydrolytic and/or oxidative reactions. Therefore, study of paramagnetic properties, their nature and origin seems to be necessary to forecast the stability of bio-oils, and to assess of the role of the different paramagnetic centers in the ageing process of bio-oils. Hence, this study aimed at monitoring pyrolysis liquids by ESR and analyzing the results.

MATERIALS AND METHODS

Bio-oils were produced in 1995-1998 from various wood species on fast pyrolysis pilot plants, which differed by their construction features: NREL (ablative pyrolysis with hot gas filter, poplar wood, 1996), Aston FB (fluidized bed, poplar wood, 1995), IWC, TP 29 (fluidized bed, bamboo, beginning of 1998) BTG (rotating cone, pine wood, 1995), VTT (circulating fluidized bed, pine wood, 1997), IWC, E-filter (electrostatic precipitator, beech wood, end of 1998) and IWC K1-F10 from different coolers (beech wood, end of 1998).

ESR spectra at X-band were measured with a RE-1006 spectrometer operating at 9.6 GHz (empty cavity at ambient temperature) and 50-kHz magnetic field modulation. Spectra were recorded at a microwave power of 1 mW and a modulation amplitude 0.02 mT to avoid line shape distortions, which could arise from experimental conditions such as microwave saturation and overmodulation. Scan range was 150-420 mT. Calibration of g-values is based on diphenylpicrylhydracyl (DPPH, g=2.0036) and Cr^{3+} (g=1.9796) standards. The amounts of the paramagnetic species (PMS) were calculated by double integration of the resonance line areas.

Oil samples for ESR measurements were introduced in the ESR sample tube for ESR measurements in a layer of 5 mm thickness. Typically ESR measurements were carried out on samples in open tubes. To avoid oxygen influence air from the tube was carefully removed by repeated thawing and freezing under high vacuum (ca. $1*10^{-4}$ Torr) and then sealed off. All measurements were carried out at ambient temperature.

An automatic microscopic image analyzer "Morphoquant" (Karl Zeiss) with a high resolution optical microscope was applied to measure the sizes of fibres. The accuracy of measuring was 0.1 μm. About 1000 microparticles were analyzed in each oil sample.

Ageing of oil samples was carried out under the following conditions: 80°C for 24 hours [2].

Oil viscosity was measured using a capillary viscometer prior and after ageing.

RESULTS AND DISCUSSION

ESR measurements have shown that all bio-oils studied contain measurable quantities of paramagnetic species, excluding the sample from NREL (Table 1). For the latter only trace amounts of PMS, < 0.1 ppm, was observed.

The ESR spectra of the oils exhibit a symmetric singlet signal at g-value about 2.0025 without hyperfine structure. Line width values are within a narrow range of 0.3-0.4 mT. On the contrary, PMS content varies considearbly from sample to sample, being 3 ppm and 4000 ppm for oils from VTT and BTG plants, respectively (Table 1). The ESR signals from ashes were not detected.

Table 1 Parameters of ESR spectra of bio-oils.

Oil sample	ESR line width* (mT)	g-value**	Concentration of paramagnetic species: spins/g	ppm
NREL poplar	-	-	$< 1.10^{15}$	<0.1
Aston FB poplar	3.4	2.0027	$(6.5\pm0.5)* 10^{17}$	130
TP 29 bamboos	2.9	2.0024	$(5.1\pm0.5) *10^{17}$	100
Twente V BTG pine	3.7	2.0028	$(1.4\pm0.2) *10^{17}$	30
VTT ITP pine	4.1	2.0028	$(1.5\pm0.4) *10^{16}$	3
E-Filter LDR	3.1	2.0024	$(1.9\pm0.1) *10^{19}$	4000
K1-F10-LDR	3.4	2.0025	$(6.6\pm0.5) *10^{17}$	130

* The accuracy of line width measuring was ±0.1

**The accuracy of g-value measuring was ±0.0001

In the case of relatively fresh bio-oil samples, IWC E-Filter LDR and K1-F10-LDR, produced a couple of weeks before ESR monitoring started, a prominent tendency of a decrease in PMS amount within the first three months of storage was observed. After that the PMS content was stable (Table 2).

Table 2 Dynamics of PMS content in the bio-oils during storage in open ESR measuring tubes at ambient temperature.

Content of paramagnetic species, $*10^{-17}$ spins/g, after storage for:	Oil sample					
	Aston FB poplar	IWC bamboos	BTG pine	VTT pine	IWC beech E-Filter	IWC beech K1-F10
Initial sample	6.5±0.5	5.1±0.5	1.4±0.2	0.15±0.04	190±10	6.6±0.5
7 days	6.6±0.4	4.6±0.5	1.2±0.1	0.19±0.05	180±15	6.4±0.5
14 days	6.2±0.6	4.8±0.5	1.4±0.2	0.18±0.04	160±10	5.7±0.5
21 days	6.4±0.5	4.5±0.5	1.2±0.2	0.17±0.04	140±10	5.3±0.2
30 days	6.5±0.5	5.2±0.5	1.6±0.3	0.19±0.05	130±7	4.8±0.5
45 days	6.5±0.5	4.8±0.3	1.4±0.2	0.16±0.03	130±10	4.8±0.3
60 days	6.9±0.6	5.3±0.6	1.4±0.1	0.15±0.03	130±10	4.6±0.5
80 days	6.3±0.5	4.6±0.5	1.4±0.3	0162±0.04	120±10	4.4±0.2
200 days	6.1±0.5	5.0±0.5	1.5±0.2	0.15±0.04	120±15	4.4±0.1
300 days	6.2±0.3	5.2±0.2	1.2±0.3	0.15±0.02	120±10	4.4±0.3

Paramagnetic centers of all other oil samples exhibit a high stability: for most samples the PMS content and other ESR spectral parameters practically did not change during storage neither under vacuum nor in the presence of air oxygen at ambient temperature (Table 2).

Analysis of the detected parameters reveals that ESR spectra of the bio-oils are characterized by paramagnetic centers, whose unpaired electron is delocalized in the π-conjugation system (π-polyconjugation paramagnetic domains). The presence of expanded π-conjugation systems, bearing paramagnetic species, is a characteristic feature for carbonized materials, i.e. chars, coals and coal pitches and conducting synthetic polymers, "synthetic metals". Up to now the origin of paramagnetic centers in polyconjugated systems is under discussion. Nowadays, the most accepted model is the concept of solitons for polyacetylenes and polarons for polyaromatic polymers [4, 5].].

According to the ESR theory polyconjugated systems are characterized by the degree of spin density delocalization, which can be estimated on the basis of the ESR line width [6]. The parameter **n** corresponding to the number of equivalent protons interacting with an unpaired electron, can be estimated from the following expression:

$$\mathbf{n} \geq 1 + (\Delta H_0/\Delta H_{eff})^2, \qquad (1)$$

where ΔH_0=2.25 mT is a splitting corresponding to interaction of a p-electron with a proton on a CH-fragment and ΔH_{eff} is the experimentally observed ESR spectrum line width. For the oils studied **n** is in the range 25-30. According to the well-known characteristics of thermal biomass conversion, the backbone of π-conjugation systems of bio-oil components consists of aromatic rings. So the delocalisation region of unpaired electron can derive from 6 to 12 benzene rings, which corresponds to the paramagnetic domains size of ≥ 30-50 Å.

The above-mentioned decrease in the PMS content for the fresh samples from the BTG plant could be due to the decay of unstable radicals formed by breaking of chemical bonds during pyrolysis. However, a persistence of the ESR spectrum line width and shape suggests that the paring of a part of PMS through the π-conjugation system or by intermolecular interaction may also take place [7].

Based on the concepts proposed in literature of structural peculiarities of biomass derived pyrolysis oils and pitches [8,9], which have about 50% of aromatic and olefinic carbons, these materials can contain high-developed π-conjugation systems as structural constituents. At the same time, ESR characteristic observed for the samples studied are close to the same for charcoals obtained from pyrolysis at temperature about 500°C [10]. It remains uncertain, whether the paramagnetic centres in the oils are associated with π-conjugation systems coming from high molecular weight lignin fragments, or whether they are associated with fine char impurities.

Taking into account the absence of detectable amounts of paramagnetic centers in the NREL oil (the plant is equipped with a hot vapour filter), it seems that the ESR spectra of all other oils are due mainly to the presence of char microparticles.

However, careful study of the spectra prior and after oil samples microfiltration revealed the presence of one or more polyconjugated species depending on pyrolysis plant construction features. It has been shown that microfiltration (pores size 2 μm) of oil samples decreased sufficiently both the quantity of microparticles and their size. In the case of the pyrolysis oil obtained from IWC plant, an ESR signal was not detected after filtration, showing that fine char particles caused the detection of paramagnetic moieties in this sample. However, the PMS content in oil samples from Aston FB, BTG and VTT plants remained constant after filtration.(Table 3).

Table 3 Changes in parameters of ESR spectra of bio-oils after microfiltration and ageing.

Oil sample	Concentration of paramagnetic species, spins/g		
	Prior filtration	After filtration	After ageing of filtrated samples
NREL poplar	$\leq 10^{15}$	$\leq 10^{15}$	$\leq 10^{15}$
Aston FB poplar	$(6.5\pm0.5)*10^{17}$	$(7.7\pm0.8)*10^{17}$	$(7.2\pm0.4)*10^{17}$ $(1.6\pm0.3)*10^{17}$* $(46\pm5)*10^{17}$**
IWC bamboos	$(5.1\pm0.5)*10^{17}$	0.00	0.00
BTG pine	$(1.4\pm0.2)*10^{17}$	$(1.5\pm0.2)*10^{17}$	$(1.6\pm0.2)*10^{17}$
VTT pine	$(1.5\pm0.4)*10^{16}$	$(1.6\pm0.4)*10^{16}$	$(1.5\pm0.2)*10^{16}$
IWCE-Filter beech	$(1.9\pm0.1)*10^{19}$	$(0.8\pm0.1)*10^{19}$	$(1.0\pm0.1)*10^{19}$
IWC K1-F10 beech	$(6.6\pm0.5)*10^{17}$	$(6.8\pm0.5)*10^{17}$	$(6.9\pm0.5)*10^{17}$ $(0.33\pm0.05)*10^{17}$* $(290\pm10)*10^{17}$**

* The probe was not stored in an ESR measuring tube but was taken from the top of the vessel with the bulk oil sample

** The probe was taken from the bottom of the vessel with the bulk oil sample

Filtration of the IWC E-Filter sample led to decreasing average effective diameter of microparticles from 4.5 μm to 2.5 μm (Fig. 1). Simultaneously PMS concentration decreased about two times (Table 3). For this sample narrowing of the ESR spectrum line was observed after ageing from 30.32 mT to 2.5 mT accompanied with a slight increase in the PMS content (Table 3). In accordance with expression (1) narrowing of the ESR line reflects an increase in delocalization region of unpaired electron. The cause of this phenomenon could be chemical and physico-chemical processes. The former is connected with condensation/polymerization reactions and the latter with densification of intermolecular packing of coplanar aromatic cycles resulting in increasing electron exchange interaction between them [11].

An increase in the PMS content after ageing confirms that the main reason of the ESR line narrowing is physico-chemical intermolecular interaction resulting in formation of more dense microstructures, e.g. aggregates consisting of π-conjugation domains. Owing to co-planarity of π-conjugation regions, such aggregates have more ordered structures than bulk oil and to some extent can be considered as ordered mesophase. In accordance with the assumption of the presence of ordered entities, which rotate a plane of light, polarization was observed by polarization microscopy in the IWC E-Filter oil sample after ageing.

Ageing of this sample at higher temperature (105°C) led to anisotropy in its ESR spectrum with an amplitude of anisotropy of 0.9 mT [12]. The average g-value of the anisotropy spectrum is equal to that of isotropic spectrum. Anisotropy observation under conditions of intermolecular exchange of unpaired electrons is possible only in the case of mutual orientation of the main molecular axes of paramagnetic domains located in high ordered mesophase structures [13]. This leads to a preferential orientation of PMS with respect to the magnetic fields.

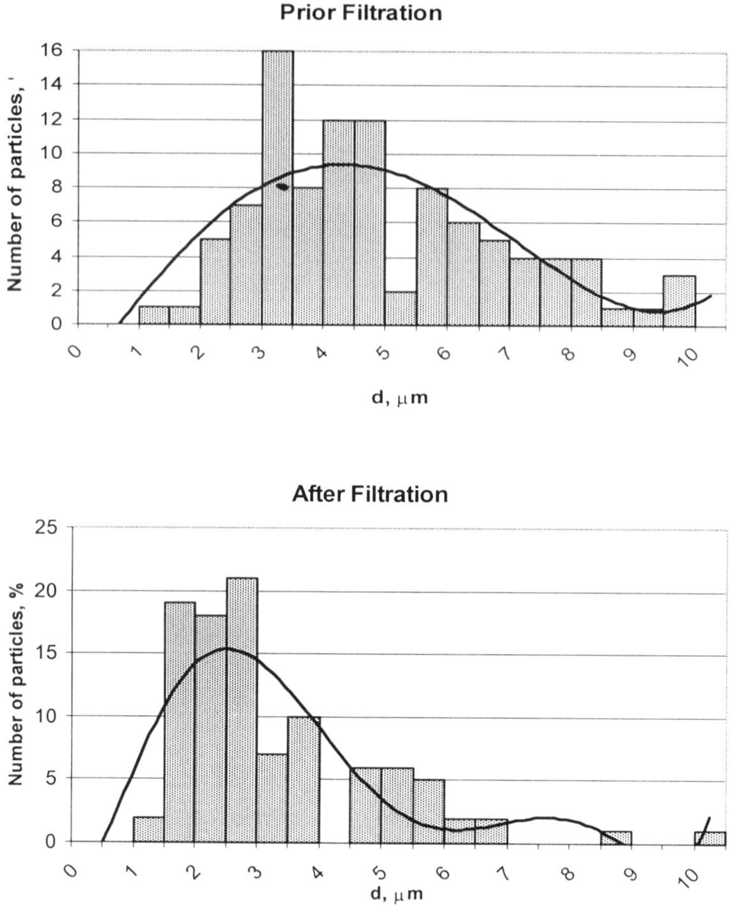

Fig. 1 Histograms of the distribution of char particles before and after microfiltration of BTG oil.

Results obtained so far allow to consider paramagnetic species of bio-oils as natural spin probes, which can provide an information about changes in the oil's microstructure. With this respect, testing of bio-oils phase separation during ageing could be an important application of the ESR method.

Comparison of the results of viscosity measurements (using a capillary viscometer) before and after ageing (Table 4) showed that for a series of bio-oils viscosity decreased or practically did not change after ageing.

Table 4 Changes in viscosity of bio-oils after microfiltration and ageing.

Oil sample	before filtration		after filtration		after ageing of filtered samples	
	$\nu_{20°C}$ (cSt)	$\nu_{50°C}$ (cSt)	$\nu_{20°C}$ (cSt)	$\nu_{50°C}$ (cSt)	$\nu_{20°C}$ (cSt)	$\nu_{50°C}$ (cSt)
NREL poplar	370.53	41.46	339.89	43.07	222.44	29.44
Aston FB poplar	690.61	64.68	726.43	66.43	752.08	63.86
IWC bamboos	5.72	2.31	6.24	2.40	5.72	2.22
BTG pine	1758.14	105.19	1728.42	107.75	2321.78	133.25
VTT pine	1258.21	89.54	1256.06	88.17	1416.84	97.93
IWC E-Filter beech	-	-	265.73	31.60	372.77	42.65
IWC K1-F10 beech	-	-	30.38	11.11	14.93	5.02

Obviously a viscosity index recommended for characterization of bio-oil stability [2], "using a viscometer type available in a participating laboratory", cannot be correctly applied due to tendency of bio-oils, non-Newtonian liquids, to phase separation into thin oil, thick tar and solid admixtures [1]. However, the ESR method allows to reveal considerable difference between the properties of the samples prior and after ageing. It was shown (Table 3) that the concentrations of paramagnetic centers in samples taken after ageing from the top and the bottom of oil storage vessel differed significantly, e.g. in the cases of oils from IWC and Aston installations, by 10 and 800 times, respectively.

CONCLUSIONS

ESR measurements indicated the presence of highly stable paramagnetic species in bio-oils, which are associated with paramagnetic centers (PMS), whose unpaired electron is delocalised in the π-conjugation system (π-conjugation paramagnetic domains). The delocalisation region of unpaired electron comprises 6 to 12 benzene rings that corresponds to domains size of 30-50 Å.

Partially these PMS can be assigned to the presence of fine char impurities in bio-oils. An increase in the PMS content after ageing is connected with physico-chemical association of coplanar aromatic rings resulting in the formation of more ordered microstructures with a higher density.

ESR could be considered as a monitoring procedure of oil's susceptibility to the phase separation during the storage.

REFERENCES

1. Diebold J. P., Milne T. A., Czernik S., Oasmaa A., Bridgwater A.V., Cuevas A., Gust S., Huffman D., Piskorz J. (1997) Proposed specification for various grades of pyrolysis oils. In: *Developments in Thermochemical Biomass Conversion* (Ed. by A.V.Bridgwater, D.G.B.Boocock), **1**, pp. 433-447. Blackie Academic and Professional.
2. Czernik S. (1999) Workshop on stabilisation /upgrading of bio-oils. *Pyrolysis Network. Vacuum Pyrolysis*, 8, 3.
3. Scahill J., Diebold J.P., Feik C. Removal of residual char fines from pyrolysis vapors by hot gas filtration (1997) Proposed specification for various grades of pyrolysis oils. In: *Developments in Thermochemical Biomass Conversion* (Ed. A.V.Bridgwater, D.G.B.Boocock) **1**, pp. 253-266. Blackie Academic and Professional.
4. Kume K. and Mizoguchi, K. (1993) Magnetic resonance and resistivity studies on polyacetylene. *Progr. Theor. Phys. Suppl.*, **113**, 91-96.
5. MacDiamid A.G., Epstein A.J. (1991) "Synthetic metals": a novel role for organic polymers. *Makromol. Chem., Macromol. Symp.*, **51**, 11-28.
6. Blumenfeld L., Voevodsky V., Semenov A. (1962) *Application of ESR in Chemistry*. Novosibirsk.
7. Hayashi H. and Yamamoto T. (1998) New π-conjugated polymers derived from a benzimidazole unit. Preparation, solvatochromism, and oxidation of poly(aryleneethynlene)s composed of 2-(3,5-di-tert-butyl-4-hydroxyphenyl)benzimidazole bearing a hindered phenolic substituent. *Macromolecules*, **31**, 6063-6070.
8. Scott D.S., Legge R.L., Piskorz J., Majerski F., Radlein D. (1997) Fast pyrolysis of biomass for recovery of specialty chemicals. In: *Developments in Thermochemical Biomass Conversion* (Ed. by A.V.Bridgwater, D.G.B.Boocock), **1**, pp. 523-535. Blackie Academic and Professional.
9. Pasa V.M.D., Carazza F., Otani C. (1997) Wood tar pitch: analysis and conceptual model of its structure. In: *Developments in Thermochemical Biomass Conversion* (Ed. by A.V.Bridgwater, D.G.B.Boocock), **1**, pp. 448-444617. Blackie Academic and Professional.
10. Mrozovski S., Gutsze A. (1977) From char to carbon: a study of the electron spin resonance. *Carbon,*15,335-342.
11. Kovarski A. L., Lopatin S.I., Kasparov V.V., Tikhonov N. A. and Oleinik E.F. (1999) *Visokomolekularnye soedineniya*, **A 40** , 1803-1808.
12. Barta P. S., Nniziol M., Pron A. and Pacyna A. (1993) Low temperature magnetic properties for poly(3-alkylthiophenes) and poly(4,4'-dialkyl-2,2'-bithiophenes). *Synthetic Metals,* **55-57**,5003-5007.
13. Shklyaev A.A., Miloshenko T.P., Lukovnikov A.F.(!988) ESR spectra and orientational order in semi-coke anisotropic structure. *Khimiya Tverdogo Topliva,* N1, 70-75.

New Prospects for Biocarbons

Michael Jerry Antal, Jr., Xiangfeng Dai, Brent Shimizu, Man S. Tam, Hawaii Natural Energy Institute, University of Hawaii, 2540 Dole St., Holmes Hall 246, Honolulu, HI, 96822, USA; Morten Grønli, SINTEF Energy Research, 7034 Trondheim, NORWAY.

ABSTRACT: We present a concise review of recent advances in technology for producing biocarbons, and new results concerning the electrical conductivity of biocarbons. In light of the fact that biocarbons can be produced in yields which attain the theoretical limit, we recommend that their potential as fuels, reductants, adsorbents, and electrodes be more fully explored. In particular, we recommend research aimed at the development of a biocarbon fuel cell.

INTRODUCTION

In 1956 M. King Hubbert used a mathematical model to predict that oil production in the U.S.A. would peak around 1969, and that world oil production would peak early in the 21st century (1). Oil production in the 48 states peaked according to the model, and there is now widespread agreement that world oil production will peak sometime during the next ten years (2). Most projections of future energy supplies indicate an important role for biomass and other renewables as a substitute for fossil fuels. To fulfill this role, biomass must be converted into electricity or a suitable transportation fuel. Conversion methods include gasification for the production of a low to medium Btu gas or hydrogen, and "liquefaction" methods to produce biocrude "oils" (by fast pyrolysis) or ethanol (by fermentation). Electric power can be produced by combustion of the low Btu gas or the biocrude in boilers, gas turbines, or diesel engines. Likewise, electric power can be produced from hydrogen by electrochemical combustion in a fuel cell. In the U.S.A. ethanol is currently the renewable fuel of choice for transportation, but hydrogen may become competitive if less costly methods can be discovered to produce and store the fuel. These technologies have been the focus of worldwide research and development efforts since the "energy crisis" in 1973.

In addition to gaseous and liquid fuels, biomass can be converted to charcoal and carbonized charcoal (a relatively pure carbon). We call these products "biocarbons". Unlike the gaseous and liquid fuels discussed above, biocarbons have attracted little interest. One reason for the disinterest is the assumption that the properties of biocarbons are similar to coal, and the world has plenty of coal. But in fact, the properties of biocarbons are quite different than those of coal (see below). Another reason for disinterest is the low efficiency of traditional methods for converting biomass to charcoal. Simple metal kilns used in the developing world often realize charcoal yields of 20 to 25 wt% on a dry basis (3). Such inefficient processes are the principal cause of the deforestation of many tropical countries (including Thailand, Haiti, and Madagascar). Moreover, because of pollution associated with charcoal production, the charcoal fuel cycle is among the most greenhouse-gas intensive energy sources employed by mankind (4). In the U.S.A. a typical yield of charcoal manufactured from hardwoods in a Missouri kiln operated on a 7 to 12 day cycle is about 28 wt% (5) on a dry basis. This yield is also low: simple stoichiometric arguments show that charcoal yields in excess of 40 wt% should be possible, and thermochemical equilibrium calculations indicate that a fixed carbon yield of about 30 wt% should be achieved when equilibrium is reached in a pyrolytic reactor operating at 300 °C. In 1986 our research led us to realize that high yields of charcoal can be obtained when pyrolysis is conducted at elevated pressure wherein the vapors are held in contact with solid products of pyrolysis. A grant from the State of Hawaii enabled us to build a pilot plant, which demonstrated charcoal yields of 42 to 62 wt% with fixed carbon contents of 70% or higher on a 1 hour operating cycle (6). In this paper we report measurements of the fixed carbon yield that result from pyrolysis of a variety of biomass feeds at elevated pressure. We compare these values to values predicted by thermochemical equilibrium calculations. For some of the feeds considered, the yield of fixed carbon delivered by our equipment attains the theoretical limit.

Another assumption of the biomass community is that woody materials will be the principal biomass feedstock. Recent studies by Hiromi Yamamoto and his co-workers (7) call this assumption into question. Their work shows that by 2100 most arable land will be dedicated to crop production in order to feed ten billion hungry people. Consequently, there will not be much land available for energy plantations. Instead, the residues of cereal crops will represent the largest available biomass resource. In this work we show that grain residues (e.g. rice and oat hulls) can be converted to biocarbons with yields comparable to those of hardwoods. Moreover, nut shells, some food processing wastes (e.g. garlic wastes), and tropical species (e.g. bamboo and leucaena wood) offer higher yields of biocarbons than traditional hardwood feeds. In light of these findings, we suggest that there are many new and exciting prospects for biocarbons as fuels, reductants for metal ores, and adsorbents for water treatment.

EXPERIMENTAL

At SINTEF charcoals from birch and oak wood were produced at atmospheric pressure in a 3.7 L stainless steel retort placed in a Nabertherm muffle oven. The muffle oven is equipped with a programmable temperature controller, which provides control of the heating rate, temperature, and reaction time. The retort has a capacity of approximately 1 kg of dry wood per batch. The typical dimensions of the feedstock pieces were 3 cm x 3 cm x 5 cm. During heating and cooling, the retort was purged with N_2 to prevent air leakage into the system.

At HNEI high-yield biomass charcoals were produced in two reactors with similar configurations and operating procedures. The process development unit (PDU) has an internal volume of 80 L and can produce as much as 10 kg of charcoal per cycle (6). Because the pyrolytic reactions which transform biomass to charcoal are exothermic (8), the pilot plant requires little heat input. The laboratory reactor (9) is a pressure vessel with an internal volume of 7.2 L. At the bottom of this reactor there are two internal 4 kW cartridge heaters that are controlled by a temperature controller. Also, a rod heater is installed down the central axis of the reactor. The reactor's outer wall is heated by four band heaters totaling 4.4 kW. The pressure within the reactor is controlled by an on-off valve. Combustible gases released by the valve are burned in a flare. More details concerning the design and fabrication of these two reactors are available in the literature (6, 10).

RESULTS AND DISCUSSION

Table 1 displays the results of rapid, and slow pyrolyses of birch and oak wood at 0.1 MPa. The charcoal yield for rapid heating of birch is 29.5%, and the fixed carbon yield is 21.4%. An increase in the heating time from about 90 min to about 4 h followed by a 4 h soak at 450 °C increases the fixed carbon yield to 22.8%. The yields of charcoal (31.2%) and fixed carbon (24.0%) from oak wood are somewhat higher than those obtained by American industry with this feed in a Missouri kiln.

Table 1 Charcoal and fixed-carbon yields realized at atmospheric pressure (0.1 MPa).

Feed	Proximate Analyses[1]			Average Charcoal Yield[2] (wt%)	Average Fixed C Yield[2] (wt%)
	VM (wt%)	Fix-C (wt%)	ash (wt%)		
Birch Wood[3]	26.4	72.5	1.15	29.5	21.4
Birch Wood[4]	21.5	77.2	1.23	28.8	22.3
Birch Wood[5]	19.9	79.1	1.04	28.8	22.8
Oak Wood[4]	21.7	76.9	1.42	31.2	24.0

[1]% of dried charcoal.
[2]% of dried feed material. Fixed-C yield = Charcoal yield * (100 – % Volatile Matter – %Car Ash) / (100 – %Feed Ash).
[3]rapid heating from 20–450°C (within 90 min), followed by 10 min soak at 450°C.
[4]slow heating from 20–450°C (within 240 min), followed by 60 min soak at 450°C.
[5]slow heating from 20–450°C (within 240 min), followed by 240 min soak at 450°C.

Table 2 displays values of the charcoal and fixed-carbon yields obtained from a variety of biomass feedstocks by pyrolysis at elevated pressures. The ± values associated with oak wood represent the sample standard deviation of four identical runs on the pilot plant. In Table 2, charcoal yields range from 34.6% (birch wood) to 58.6% (kukui nut shell). The average fixed-carbon yields for these two feedstocks are 27.2% and 41.4% (respectively). A comparison of the elevated pressure fixed-carbon yields reported in Table 2 with those obtained by SINTEF at atmospheric pressure and low heating rates (see Table 1) shows that elevated pressures significantly improve the fixed-carbon yields for each of the wood species studied by both groups. A more comprehensive study involving 5 different wood feedstocks, reported by the authors in reference (10), corroborates the beneficial effects of elevated pressure for charcoal production. Remarkably, carbonization at elevated pressure realizes high yields without a prolonged reaction time. When the pressure vessel is already hot (as would be the case in a continuously operating commercial reactor), the feed is completely carbonized after about 1 h of reaction time.

Table 2 Charcoal and fixed-carbon yields realized at 1.0 MPa

Feed	Proximate Analyses VM (wt%)	Proximate Analyses Fix-C (wt%)	Proximate Analyses ash (wt%)	Average Charcoal Yield[1] (wt%)	Average Fixed C Yield[2] (wt%)	Theor. Fixed C Yield (wt%)
Bamboo	27.0	67.1	5.9	45.9	32.1	33.1
Birch Wood	20.9	78.5	0.6	34.6	27.2	32.4
Garlic Waste	27.3	49.2	23.5	42.9	25.5	27.6
Kukui Nut Shell	26.9	69.6	3.5	58.6	41.4	42.0
Leucaena Wood	27.0	69.4	3.6	44.2	32.3	32.9
Oak Wood[3]	24.0 ± 2.0	75.0	1.0	39.8 ± 1.3	29.9 ± 1.1	34.2
Oat Hull	24.9	62.3	12.8	40.2	26.3	32.1
Rice Hull	14.9	47.2	37.9	46.2	27.6	32.1

[1]% of dried charcoal.
[2]% of dried feed material. Fixed-C Yield = Charcoal yield * (100 – %Volatile Matter – %Char Ash) / (100 – %Feed Ash).
[3] Results from process development unit (PDU).

The thermochemical equilibrium value of the fixed-carbon yield constitutes a benchmark against which the experimental values can be compared. We refer to this value as the "theoretical" yield of carbon, which we hypothesize to be the upper limit attainable by thermal processes. We used the thermochemical equilibrium software STANJAN (11) to calculate the mass fraction of solid carbon present at equilibrium when the elements C, H, and O with molar ratios dictated by the composition of each feedstock (see Table 3) are mixed at 1.0 MPa and 300 °C. We neglect the presence of nitrogen and sulfur in the STANJAN calculations because these two elements compose only a small fraction of the mass of the biomass feed. The ash content of the feed is also neglected. In general, STANJAN predicts that solid C, and the gases CO_2, H_2O and CH_4 should be the only significant products present in equilibrium, and that the distribution of these products is not strongly dependent upon either the assumed pyrolysis temperature or the assumed pressure. The predicted thermochemical equilibrium values for the fixed-carbon yields are displayed together with the experimental values in Table 2. The measured values of the fixed-carbon yields for kukui nut shell, leucaena wood, and bamboo effectively equal the theoretical yields within the margin of error of the data.

Table 3 Elemental analyses of feedstocks.

	C (wt%)	H (wt%)	O (wt%)	N (wt%)	S (wt%)	ash (wt%)
Bamboo	47.65	5.77	44.23	0.27	0.11	3.91
Birch Wood	48.07	6.00	45.56	0.17	<0.05	0.20
Garlic Waste	37.85	4.97	43.12	0.49	0.22	17.07
Kukui Nut Shell	55.76	5.60	37.99	0.34	0.03	1.45
Leucaena Wood	45.95	6.06	41.23	2.42	0.27	5.11
Oak Wood	50.13	5.98	44.76	0.08	0.03	0.19
Oat Hull	46.00	5.91	43.49	1.13	0.15	4.91
Rice Hull	38.86	4.86	37.15	0.42	0.06	20.97

Remarkably, agricultural wastes offer yields of charcoal and carbon that are comparable to hardwoods. Rice hulls are a particularly interesting example because of their widespread availability and difficulties associated with their disposal. Based on our measured fixed-carbon yields, we estimate that 370,000 mt of fixed carbon could be produced from the 1,700,000 mt of rice hulls that are generated annually in the USA. High-yield charcoal produced from these rice hulls could satisfy the Norwegian ferrosilicon industry's current and future demand for carbon. Because rice hull charcoal consists of small particles, industry might not consider it to be suitable for use as a reductant. But this charcoal can be formed into briquettes, which resist crushing and are durable. Inexpensive charcoal briquettes made from agricultural wastes could cause a paradigm shift in the way metal ore is refined around the world.

When a proximate analysis of charcoal is conducted, the charcoal is heated in a covered crucible to 950 °C and held at this temperature for about 6 min. The carbonized charcoal product of this process possesses interesting properties. It is a relatively pure carbon. For example, the ultimate analysis of carbonized oak wood charcoal is 92.8% C, 1.09% H, 3.49% O, 0.24% N, 0.04% S, and 1.46% ash. Table 4 presents measured values of the resistivity of #40 mesh oat hull charcoal carbonized at different temperatures. The resistivity decreases by six orders of magnitude and reaches a value of 3.7 Ω - cm for carbonization at 900 °C. For comparison sake, we used the same apparatus to measure the resistivity of a packed bed of commercial grade graphite powder. As expected, the value we obtained, 0.16 Ω - cm, was somewhat smaller than that of the carbonized charcoal. Nevertheless, the low electrical resistivity of carbonized biomass charcoal opens some interesting doors. In their classic text on fuel cells, Bockris and Srinivasen (12) conclude that carbon fuel cells are impractical because (i) coal is not an electrical conductor, and (ii) graphite is too scarce and expensive to be used as a fuel. In light of the good electrical conductivity of carbonized biomass charcoal, we believe that the research community should give attention to the development of a biocarbon fuel cell.

Table 4 Measured electrical resistivity of a packed bed of carbonized oat hull charcoal (#40 mesh) compressed at 846 kPa.

Carbonization Temperature (°C)	Resistivity (Ω - cm)
600	7.4×10^6
700	9.9×10^4
800	35
900	3.7

Carbonized charcoal is also suitable for the production of very high quality activated carbons. The conventional method for physical activation of charcoal involves gasification in steam or carbon dioxide at high temperatures. The yield of activated carbon from dry biomass for this process is typically about 8 wt%. To improve this yield, and to develop a less costly method for activation, we initiated research on a novel activation process involving (i) carbonization of the charcoal at temperatures below 750 °C, (ii) air oxidation in boiling water (AOBW) of the carbonized charcoal at temperatures below 240 °C, and (iii) activation (a second carbonization) of the oxygenated carbon. High surface area (>900 m^2/g) activated carbons have been produced in our laboratory from various agricultural wastes by the AOBW process (13). In step (ii) the boiling water controls the oxidation temperature and holds the reaction rate at a relatively low value; thereby precluding mass transfer limitations that quickly appear at higher temperatures. Under these conditions dissolved oxygen in the water is able to diffuse into the micropores and efficiently remove carbon. Note that the equilibrium concentration of dissolved oxygen in boiling water at 240 °C exceeds the concentration of oxygen in air at room conditions. Yields of activated carbons produced by the AOBW process are about 100% higher than those reported by industry for the synthesis of coconut shell activated carbon.

CONCLUSIONS

(1) In the case of oak wood used to manufacture charcoal in Missouri kilns, pyrolysis at 1.0 MPa increases the fixed-carbon yield from ≤24 wt% in an externally heated retort (or the Missouri kiln) to 29.9 wt% in the pressure vessel. The time requirement is reduced from ≈10,000 min for the Missouri kiln to about 70 min in the pressure vessel.

(2) High yields of fixed-carbon are realized at 1.0 MPa from a wide variety of agricultural wastes. For example, rice and oat hulls offer fixed-carbon yields of 27.6 and 26.3 wt% (respectively). These yields are higher than those obtained from traditional hardwood feeds used by industry to manufacture charcoal in Missouri kilns.

(3) More attention should be given to biomass species that are prone to the formation of carbon by pyrolytic methods. Likewise, industry should give more attention to the production of carbon from agricultural wastes. Briquettes manufactured from rice hull charcoal could satisfy the Norwegian ferrosilicon industry's current and future demand for renewable carbons. The substitution of renewable carbon from biomass for coal in

the metallurgical industry and elsewhere can reduce greenhouse gas emissions while creating new jobs in the developing world and enhancing the profitability of agribusiness.

(4) Carbonized biomass charcoals are good conductors of electricity. Consequently, carbonized biomass charcoal could be used to fuel a biocarbon fuel cell. We urge the research community to give attention to the development of a biocarbon fuel cell.

(5) Very high quality activated carbons can be produced from carbonized biomass charcoals by conventional gasification processes, but the yield is low. We urge the research community to give attention to improving the yields of activated carbons from biomass.

REFERENCES

1. M. K. Hubbert, *Transactions of the American Society of Mechanical Engineers* **101**, 16–30 (1979).
2. C. J. Campbell, J. H. Laherrere, *Scientific American*, 78–83 (1998).
3. N. Shah, P. Girard, C. Mezerette, A. M. Vergnet, *FUEL* **71**, 955–962 (1992).
4. K. R. Smith, *Energy for Sustainable Development* **1**, 23–29 (1995).
5. "Charcoal: Production, Marketing, and Use" *U.S. Dept. of Agriculture Forest Service Report #2213* (1961).
6. M. J. Antal, Jr. *et al.*, *Energy Fuels* **10**, 652–658 (1996).
7. H. Yamamoto, K. Yamaji, J. Fujino, *Applied Energy* **63**, 101–113 (1999).
8. W. S. L. Mok, M. J. Antal, P. Szabo, G. Varhegyi, B. Zelei, *Industrial and Engineering Chemistry Research* **31**, 1162–1166 (1992); W. S. L. Mok, M. J. Antal, *Thermochimica Acta* **68**, 165–186 (1983).
9. X. Dai, Ph.D. Thesis, University of Hawaii at Manoa (1998).
10. M. J. Antal *et al.*, *submitted for publication in Ind. Eng. Chem. Res.*.
11. W. C. Reynolds, *STANJAN Thermochemical Equilibrium Software* (Stanford University, Stanford, CA, ed. 3.91, 1987).
12. J. O. M. Bockris, S. Srinivasan, *Fuel Cells: Their Electrochemistry* (McGraw-Hill, New York, 1969).
13. M. S. Tam, M. J. Antal, E. Jakab, G. Varhegyi, *Submitted to Ind. Eng. Chem. Res.*

Issues in Value-Added Products from Biomass

D. C. Elliott
Pacific Northwest National Laboratory, Richland, Washington, USA

ABSTRACT: While methods for converting biomass to alternative energy products has been developing over the past 30 years, process development for chemical products is an area of increasing interest and effort. Although the need for renewable resources in the fuels market is still likely to have the most economic impact worldwide, chemical products derived from biomass can also be more than just niche market applications. However, the specific chemical processing required for refined products requires improved chemical handling methods for separations and purifications, as well as improved catalyst systems. Development of these unit operations has lagged behind the process research focused on the finished products. This paper describes the critical processing issues that must be addressed to more effectively use biomass feedstocks for producing value-added chemicals. Specific areas include biomass component separation, catalyst development for aqueous processing, and trace component effects in catalytic processing of biomass feedstocks. The paper also describes some of the process research which has been performed or is now underway in our laboratory and others.

THERMOCHEMICAL CONVERSION TO VALUE-ADDED PRODUCTS

Thermochemical processes for producing value-added chemicals from biomass have followed similar paths as processes for generating alternative fuel products, namely gasification and liquefaction methods. These methods are discussed below for both fuels and chemicals production to give an overview of the current technology.

PROGRESS IN FUELS PRODUCTION

The impetus for using thermochemical conversion of biomass to produce value-added products began with the need for new fuels to displace limited petroleum resources. Since the oil embargo in the mid-1970s, work has been conducted worldwide to develop new processing methods for alternative fuel products. The bulk of the research and development has focused on generating liquid and gaseous fuel products, with only limited work on direct combustion of solid biomass.

Liquefaction processes for biomass conversion can be categorized as slurry-phase, high-pressure processing, usually with catalysts; or as dry pyrolysis at low or reduced pressure[1]. The research has concentrated on the fuel properties and has often overlooked the detailed chemical analysis of the biomass-derived products. Both process types produce a complex mix of oxygenated organic components. The mixture has a lower heating value than petroleum liquids and is more chemically reactive.

Gasification systems have typically been either pyrolysis systems, often with added steam, or partial oxidation systems, using either air or oxygen[2]. Direct addition of catalysts into these systems has usually not been successful. Analysis of only the low molecular weight components in the gas product has been the typical limitation of these process development efforts, with the residual high-molecular weight product classified as "tar".

PROGRESS IN CHEMICALS PRODUCTION

The recent interest in chemical production is based on a higher return expected for chemical products versus fuels. For example, biomass gasification can be used to produce a synthesis gas of hydrogen and carbon monoxide. This gas can be used in catalytic synthesis of a range of chemicals, from methanol and formaldehyde to higher hydrocarbons, in the same way that synthesis gas derived from natural gas can be used[3]. However, by breaking down the biomass to the basic building blocks all product differentiation relative to fossil fuels is lost.

Biomass liquefaction methods can also be used to produce chemical products. With the more mild treatment at liquefaction conditions, the intrinsic properties of biomass composition are retained to a greater degree. Both pyrolysis and high-pressure process conditions can be used to generate value-added chemical products. Pyrolysis of biomass produces a complex product slate of oxygenated chemicals, only a few of which are present in more than 1% concentration. In earlier times, most of the methanol and acetic acid was produced from wood pyrolysis, but such processing is not economical in the current petroleum-based economy. Today, liquid smoke, a food flavoring containing a mixture of pyrolysis products, is a well-known chemical product from wood pyrolysis. New processing methods and new products are now being developed, such as resins[4], levoglucosan[5], and specialty chemicals[6], that use flash pyrolysis-derived bio-oil. Pressurized catalytic processing may also provide future opportunities for chemicals production. Such systems have been developed for methane[7] and hydrogen[8] production. Commercialized recovery of liquid chemicals requires advancing the technology of separation and concentration for these processes, however.

More recent chemical production methods using biomass seem to be utilizing pretreatment technologies from the past. Whereas pretreatment in thermochemical conversion is geared to material handling issues such as densification and feeding[9], pretreatment in bioconversion involves concentration and purification of the glucose or carbohydrate components to maximize the processing efficiency of the microbes[10]. The preconcentration of the useful biochemical feedstock components and removal of deleterious components, as done in biochemical conversion, is also beneficial in thermochemical processing.

Direct conversion of cellulose in an unpurified state has been shown in the case of levulinic acid production. Production of levulinic acid as a high-yield direct product of controlled acid hydrolysis of cellulose is an exception to the complex product slate typically produced from biomass feedstocks. While the levulinic acid is only derived from the cellulose portion of the biomass and other components end up as byproducts, technology has now been demonstrated for recovery of the levulinic acid product at yields and purity sufficient to generate market interest[11]. Based on related research developments, levulinic acid may prove to be an important building block for chemical products derived from biomass[12]. In order for new technologies to be developed for chemical production from biomass, issues related to the feedstock and aqueous processing environment need to be addressed.

BIOMASS FEEDSTOCK SOURCES AND PRETREATMENT PROCESSING

Biomass feedstocks usually contain a wide variety of chemical functional types within the biopolymer structure. Since thermochemical conversion has typically focused on the use of the "whole" biomass, separating the chemical functional types has been of less interest. Using the "whole" has been viewed as the most cost-effective approach while the thermochemical processes were considered robust enough to handle the range of chemical functional types. As a result, the products were a complex mixture of chemical entities useful primarily for fuel applications. The costs of collecting individual chemical products could not be justified in most cases.

A new processing paradigm is now being developed based on using cleaner, simpler biomass-derived feedstocks to address the issue of biomass complexity. The complexity issue has two facets, the mixture of organic functional types and the inclusion of inorganic components. Although wood is "clean," in terms of being low in sulfur and nitrogen (two major pollution sources found in coal and petroleum), it is still complex structurally, containing aromatic (lignin) components - like coal - but also both polymeric C_6 and C_5 structure. It also has significant levels of mineral content - much lower than coal but much higher than petroleum. Other biomasses can contain very high levels of mineral content (grass straws) or nitrogen (all types of green foliage or seeds). Complication of the organic structure with protein components is commonly found, and the presence of lipids is a complication in some other cases. The different specific components are useful as feedstock for different chemical products. In order to effectively use biomass for chemical production, the various components should be used separately for feedstock.

PLANT-DERIVED SUGAR/STARCH FEEDSTOCKS

There is no directly recoverable glucose feedstock from biomass. Even the sugar feedstock produced directly by plants in some cases, such as beet or cane sugar (sucrose), is actually a disaccharide. It must be processed through hydrolysis (in this case known as "inversion") to be useful as a glucose/fructose mixture for chemical production. However, these sugars are generally too expensive except for specialty chemical production, such as sorbitol or mannitol, or in cases of special economic circumstances, such as a captive market or high transportation costs.

Corn-derived starch is a readily available feedstock in the U.S. It is produced efficiently in large quantities in corn wet mills and provides the basis for much of the biomass-derived chemical production presently in the market. The starch is enzymatically hydrolyzed on a large scale to produce glucose, much of which is isomerized to fructose for use as food sweetener, and the rest mostly goes into ethanol production for automotive fuel. The several separation processing steps in the corn wet mill are summarized in Figure 1. Equivalent processing could be applied to wheat and potato or other starch when the feedstock is available at a suitably low price.

Some of the corn-derived glucose goes into various fermentation processes that lead to other chemical products. For example, large-scale production was recently announced for lactic acid and polylactide plastic derived from it[13]. Citric acid and lysine are two other important fermentation products.

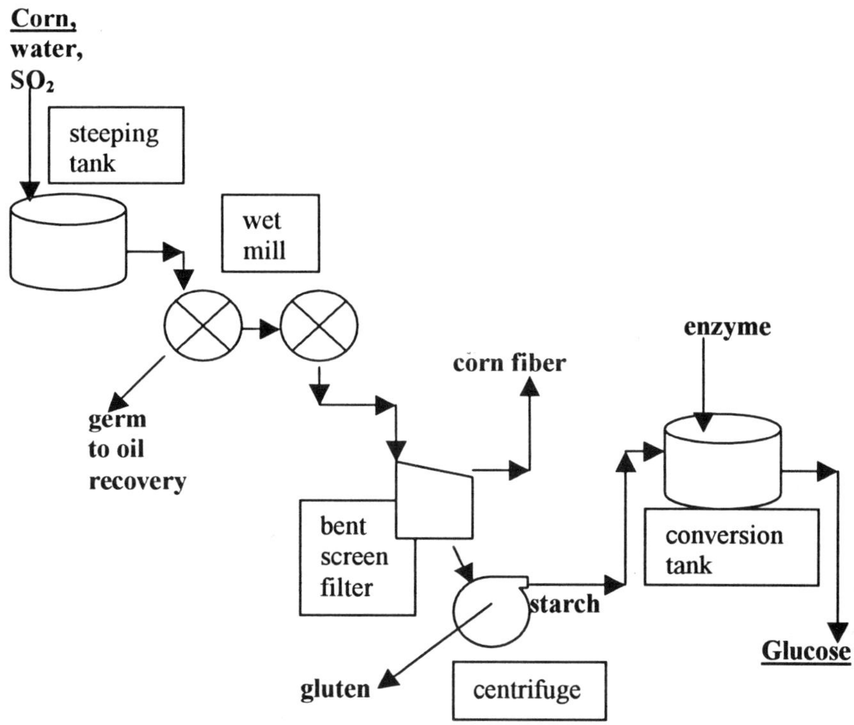

Figure 1 Glucose Recovery from Corn

HYDROLYZATE FEEDSTOCKS AND DERIVED FEEDSTOCKS

Hydrolysis of biomass-derived carbohydrate could produce inexpensive feedstock for processing to chemicals. [The technical aspects of hydrolysis are discussed in detail in the next section.] The major structural components of biomass, cellulose and hemicellulose, can be broken down to the individual sugar monomers which can be used as feedstock for chemical production. The monomers can be processed directly by thermochemical means, such as catalytic hydrogenation, or can be bioprocessed into to various fermentation-derived feedstocks. Processes are being developed for fermentation products, including lactic acid (mentioned above) and succinic acid. The succinic acid has potential application for polyester monomer (1,4-butanediol) and solvent (tetrahydrofuran) production[14].

PULPING ANALOGS DERIVED FEEDSTOCKS

Various pulping technologies for wood and other biomass have been used for centuries to produce fiber, primarily for paper production. These processes effectively separate and clean the cellulose for use as fiber. However, the hemicellulose and lignin byproducts are not typically used for chemical products. The depolymerized hemicellulose in pulping liquor is a large source of biomass carbohydrate that could potentially be used for chemical production.

Another related processing byproduct, available in large quantities, is the unrecovered fiber disposed as paper mill sludge. This cellulosic feedstock is usually contaminated with various paper-making chemicals, such as clay and lime which would likely require separation before the cellulose could be used. However, processes are being developed that directly use the carbohydrate, including the levulinic acid process mentioned above.

New pulping technology is in development that could lead to more effective recovery of the biomass components in a higher purity form. Such "Clean Fractionation" technology is expected to produce high purity streams of cellulose, lignin (in solution), and hemicellulose (as dissolved sugars) by a specially developed solvent pulping process[15]. The cellulose and lignin would be particularly clean products, while the hemicellulose stream would contain residual dissolved mineral matter.

ANIMAL WASTE FEEDSTOCKS

The use of animal livestock waste for chemical production would provide a solution for this otherwise problematic material. Because of increased large-scale livestock operations, the waste has become locally concentrated to levels that often exceed environmentally desirable disposal methods. The composition of animal waste is not adequately documented to determine which chemical manufacturing processes would benefit from its use as a feedstock. The limited information on the composition of carbohydrates, protein structure, and other components needs to be expanded. The secondary issue is the need to develop appropriate separation and cleanup processes to produce useful and consistent feedstock for chemical production. It is clear that there are economically interesting amounts of waste available for processing if sufficient amounts of feedstock can be economically derived from it.

BIOMASS UTILIZATION ISSUES

Processes for depolymerization of the biopolymer, separation of the various chemical types, and aqueous phase processing all must be established for biomass conversion to chemical products to become widespread.

DEPOLYMERIZATION

Depolymerization of the biopolymer structure is an issue in almost all utilization concepts. The established chemical technology using cellulose as a base material is a significant exception. Recovery of proteins is proving to be a second major exception. For the most, part the carbohydrate structure needs to be broken down into useful monomers like glucose or xylose. Over the years, the assessment of hydrolysis with catalysts, both mineral and biological, has been evaluated with a range of biomass feedstocks.

Acid/base hydrolysis

Autohydrolysis is the hydrolysis without added catalyst. In fact, it requires the presence of water and heat that likely generate the initial acid sites (acetate in the case of biomass), which then further catalyze continuing hydrolysis. Hydrolysis is generally practiced in a staged method, using different concentrations of acid and processing temperatures to maximize the yield of desired sugar products and minimize the formation of degradation products.

Acid hydrolysis of wood is a well-known procedure, but it has never been considered an economical process for sugar production. Most of the technology developed for wood has broader application to most biomass. In Europe initial efforts focused on high concentration acid (40% H_2SO_4 or HCl) at high temperatures (150-250°C). Later efforts focused on a low-concentration acid first-step for hemicellulose removal, followed by a higher concentration acid step to break down the more refractory cellulose. Refinements of this type of processing have proceeded unabated. A recent search of the U.S. patent database shows over 200 patents describing acid hydrolysis of biomass just since 1976. Dilute acid hydrolysis has also been developed as a pretreatment for enzymatic hydrolysis[16]

Base hydrolysis has also been practiced for depolymerization of biomass. Alkali pulping of biomass is an industrial example of a well-known way to break down the hemicellulose without destroying the cellulose. In other cases, base hydrolysis has been used as a pretreatment to facilitate microorganism access to the carbohydrates for bioprocessing of biomass. In some cases, the base is added in the form of ammonia which has the advantage of providing nitrogen nutrient value for the microorganisms as well. More intensive base treatment has been used to attack cellulose for the production of organic acid products. Krochta's group has evaluated a range of base treatments that result in the production of acids like formic or acetic, as well as several hydroacids and diacids[17].

Enzymatic hydrolysis

Enzymatic hydrolysis has received great attention in recent years. In combination with fermentation, it has been found particularly attractive. The value of the enzymatic process is its specificity for the reaction and resulting high product selectivity. This also means that the enzymes have limited application, and often a suite of enzymes must be developed to achieve an effective depolymerization of the complex structure in a typical biomass material.

Wood hydrolysis has been limited to cellulose degradation by cellulase enzymes. These enzymes are typically low activity and highly inhibited by the glucose product. As mentioned above, mild acid pretreatment has been found to be an important first step in the biomass utilization process. The pretreatment is used both to break down the hemicellulose to sugars and to disrupt the lignocellulosic structure and the crystallinity of the cellulose.

Depolymerization of hemicellulosic structure is still being developed. Because of its more complicated structure involving several kinds of sugar units in variously linked conformations, more than one specific enzyme is needed. More simple plant starches can be effectively broken down to sugars with combinations of enzymes, including amylases and glucoamylases.

COMPLEX MIXTURE SEPARATION

Following the depolymerization of the carbohydrate structure of the biomass, numerous separations will likely be required. These complex mixture separations are at the forefront of the process development effort for biomass utilization for chemicals production. Several technologies are available for other applications and are being developed for the specific needs of biomass processing.

An interesting model can be found in the existing and developing technology for cheese whey utilization. Cheese whey, the residual aqueous product from cheese production, is a dilute solution (6% dry solids), with a major component of carbohydrate (75% lactose, dry solids) and lesser amounts of protein, and minerals as well as trace amounts of lipids and vitamins. Separations have been based on precipitation of lactose and minerals following concentration to produce food-grade products. A more recent development is the recovery of a protein concentrate by ultrafiltration. The resulting permeate still contains a significant amount of lactose with minerals, vitamins, and non-protein nitrogenous compounds. The mineral components may be removed by ion exchange or precipitation. Lactose recovery may either be by utilization of the solution or by concentration, precipitation, and recrystallization to a purified form.

Recovery of the sugars from solution may or may not be an important processing step. Separation of individual sugars may allow production of higher valued specific products. Such separation is possible using simulated moving bed continuous chromatography processing. This type of processing can also be used for final purification by removing residual mineral and protein from the carbohydrate stream. However, the several sugars derived from biomass may be processed to final products before separation in some cases. This option is particularly applicable to thermochemical processing, in which the catalyst and heat are less specific in the reaction mechanism than enzymatic processing, in which the biocatalyst differentiates between various isomers.

The use of the sugars in solution or concentration and recovery for final use is also an important question. The aqueous processing environment raises important processing issues as will be described below. Use of the dilute solution requires considering the tradeoff of additional reactor volume versus effects of concentration on thermodynamic equilibriums of the reactions involved. The solution can be concentrated through straightforward evaporation, but due to the high heat of vaporization of water, there is a significant energy cost. While water is an excellent solvent for the sugars, it is low cost as well. Because of the good solvent properties of water for the sugars, concentration by solvent extraction is not an attractive or effective alternative, even disregarding the fact that such an option almost undoubtedly introduces a less environmentally friendly and more expensive solvent to the process. The most cost-effective approach appears to be to <u>not</u> recover the sugars from the water but to proceed with the processing to the point that a product can be more efficiently recovered.

AQUEOUS PHASE CATALYTIC PROCESSING

Use of catalysts in an aqueous processing environment has received much less attention than more traditional catalytic processing. Most heterogeneous catalyst development work is focused on gas phase processing or on hydrocarbon (petroleum) processing. In either case, there is only a limited partial pressure of water in the reactor and no liquid phase water. Use of the aqueous phase to process biomass-derived sugars requires understanding the high-pressure liquid water phase as the reaction environment. In this environment, conventional catalyst formulations have limited lifetimes because of catalyst metal oxidation or sintering and/or catalyst support hydrolysis or dissolution. The use of biomass-derived feedstock can also lead to the contamination of the catalyst with dissolved mineral species and other inorganic components such as sulfur.

Catalyst material stability in the aqueous phase

Chemical process development underway in our laboratory is using high-pressure liquid water as the processing medium. Catalysts would be incorporated in many of the processes being studied. To date, commercial catalyst formulations tested have been unacceptable for chemical processing in the high-pressure liquid water environment. We have reported earlier that conventional support materials, including γ-alumina (and other high-surface area aluminas), cements like calcium aluminate, silica, silica-aluminas, hydrous titanates, and natural minerals like kieselguhr, have not been usable at 350°C and 200 atm[18]. Pelletized forms of these materials soften, and actual dissolution has occurred in some cases. Reaction of the aluminas to a hydroxide form (böhmite) has also been confirmed.

Catalytic metal species are often unstable in high-pressure liquid water environments. Of the first row Group VIII metals, only nickel remains reduced at 350°C and 200 atm; both iron and cobalt are readily oxidized. Of the other transition metals tested to date, copper and Group VIII noble metals remain reduced, but zinc, chromium, tungsten, and molybdenum are oxidized. Even though nickel remains reduced at operating conditions, it is not stable with respect to crystallite size. Based on x-ray diffraction approximations, average crystallite size increase can account for dramatic reduction of nickel surface area. In addition, nickel metal is subject to dissolution into the aqueous processing medium leading to loss of catalyst material and contamination of the products.

We have found, however, that a ruthenium metal catalyst on titania support has high activity in these systems. Because the titania is stable in the aqueous processing environment, the catalyst remains physically intact. The ruthenium is also stable in this environment and does not oxidize or sinter/agglomerate or dissolve into the water. In addition, because of its high specific activity, the ruthenium can be used at low metal loadings of only 3 wt% compared to nickel metal catalysts of 50 wt%, or nearly 100% in the case of Raney nickels. Also because of its high activity, this catalyst can be used at significantly lower processing temperatures and still attain high processing effectiveness.

The form of the titania is also an important issue. We have found that the rutile form is the preferred component for use in the aqueous processing environment. The anatase form is readily converted to the rutile in the aqueous environment over a period of hours to days depending on the operating temperature. As a result, ruthenium catalysts formulated on anatase support lose activity over time, while those formulated on rutile maintain their activity at operating conditions.

Catalyst processing results with improved formulations

We have demonstrated that the ruthenium on titania catalyst is useful for the specific applications of lactose, glucose, and xylose hydrogenation to lactitol, sorbitol, and xylitol, respectively. In these applications, the use of the new catalyst allows the process to be accomplished at lower processing temperature while maintaining high processing rates and high conversion and product selectivity. As an example, some typical processing results are presented in Table 1 for the new ruthenium on titania catalyst. This high activity is maintained from startup, while the nickel catalysts lose activity over a period of hours to arrive at a typical operating activity.

Table 1 Hydrogenations of lactose, glucose, and xylose.

Feedstock (20 wt%)	Temp. (°C)	Pressure (psig)	LHSV L/L/hr	Feedstock conversion (%)	Selectivity to sugar alcohol
Lactose	100	1900	2	100	95
Lactose	100	1900	3	99.7	96
Lactose	100	1900	4	99.2	98
Glucose	100	1900	4	100	97
Glucose	100	1900	6	99.95	99.8
Xylose	100	1900	4	100	99

Comparative tests have been performed in the semi-batch reactor system to evaluate the Ru/TiO_2 catalyst versus a more conventional nickel-based catalyst. These tests show that ruthenium at only 3% metal loading has about the same activity as nickel at 50% metal loading. This comparison is only for short-term activity of the catalyst. As demonstrated in the continuous flow tests, the nickel catalyst loses activity readily in the first hours on stream, while the ruthenium maintains its activity.

Table 2 Semi-batch hydrogenations of lactose, glucose, and xylose.

Feedstock (12 wt%)	Temp. (°C)	Pressure (psig)	Time at Temp. (hr)	Feedstock conversion (%)	Selectivity to sugar alcohol
lactose/Ru	100	1500	3	98.8	97
lactose/Ni	100	1500	3	98.8	88
lactose/Ru	140	1500	1	95.3	85
lactose/Ni	140	1500	1	98.6	64

Poisoning of catalysts in aqueous phase processing of biomass-derived feedstocks

The processing of mineral-loaded, biomass-derived sugar feedstocks can be problematic. We have reported in tests with cheese whey and brewer's spent grain the calcium and magnesium precipitate on the catalyst as various phosphate minerals, e.g., whitlockite or durapatite[19]. Removing these elements from the feedstock may be required even at the lower temperatures used for chemical production, e.g., catalytic hydrogenation at 150-250°C, as opposed to the higher temperature tested in the reference. However, avoiding the near-critical temperature processing condition may facilitate the use of the contaminated feedstocks from biomass.

The sulfur content of biomass can also be important when metal catalysts are involved in the processing. Sulfur poisoning of nickel metal catalysts is well known, particularly as a result of petroleum processing. The difficulty led to the development of nickel modified molybdenum catalysts, which are used in the sulfided form at relatively high temperatures for some hydrogenation reactions in treatment of petroleum distillates. In the case of biomass, sulfur content is mostly contained in protein components or as sulfates. Removing the protein can likely reduce the sulfur content to an insignificantly low level. Sulfate removal may require anion exchange.

CONCLUSIONS

Effective utilization of biomass for value-added chemical product synthesis will require development of new applications of important unit operations. Carbohydrate recovery from the biomass is the key near-term application for production of commodity chemicals. Protein recovery will continue to have an important niche market in the purified form as food and a larger low-value market in the crude form as animal feed. Important processing information for carbohydrate depolymerization can be found in the literature from biochemical conversion of biomass. New process applications of separation technologies are just now being developed and refined for use with biomass-derived carbohydrate and protein streams. The use of an aqueous processing environment for carbohydrates will require careful consideration of the differences that type of environment entails, such as the effect on catalyst formulations.

REFERENCES

1. Elliott, D. C., Beckman, D., Bridgwater, A. V., Diebold, J. P., Gevert, S. B., and Solantausta, Y. (1991) Developments in Direct Thermochemical Liquefaction of Biomass: 1983-1990. *Energy & Fuels*, **5**(3), 399-410.
2. Beenackers, A.A.C.M. and Maniatis, K. (1996) Gasification Technologies for Heat and Power from Biomass. In: *Biomass for Energy and the Environment*, (Ed. P. Chartier, G.L. Ferrero, U.M. Henius, S. Hultberg, J. Sachau, and M. Wiinblad) pp. 228-259, Pergamon/Elsevier, Oxford, United Kingdom.
3. Anikeev, V. (1999) Modern Technologies of the Biomass Conversion for Chemicals, Carbon Sorbents, Energy, Heat and Hydrocarbon Fuels Production. In: *Biomass: A Growth Opportunity in Green Energy and Value-Added Products.* (Ed. R. P. Overend and E. Chornet), Volume 1, pp. 563-570. Elsevier Science, Ltd., Kidlington, United Kingdom.
4. Roy, C., Calve, L., Lu, X., Pakdel, H., and Amen-Chen, C. (1999) Wood Composite Adhesives from Softwood Bark-Derived Vacuum Pyrolysis Oils.In: *Biomass: A Growth Opportunity in Green Energy and Value-Added Products.* (Ed. R. P. Overend and E. Chornet), Volume 1, pp. 521-526. Elsevier Science, Ltd., Kidlington, United Kingdom.
5. Pernikis, R., Zandersons, J., and Lazdina, B. (1997) Obtaining of Levoglucosan by Fast Pyrolysis of Lignocellulose. Pathways of Levoglucosan Use. In: *Developments in Thermochemical Biomass Conversion*, (Ed. by A.V. Bridgwater and D.G.B. Boocock), pp. 536-548. Blackie Academic & Professional, London.
6. Scott, D.S., Legge, R.L., Piskorz, J., Majerski, P., and Radlein, D. (1997) Fast Pyrolysis of Biomass for Recovery of Specialty Chemicals. In: *Developments in Thermochemical Biomass Conversion*, (Ed. by A.V. Bridgwater and D.G.B. Boocock), pp. 523-535. Blackie Academic & Professional, London.
7. Elliott, D.C., Baker, E.G., Butner, R.S., and Sealock, L.J., Jr. (1993) Bench-Scale Reactor Tests of Low Temperature, Catalytic Gasification of Wet Industrial Wastes. *Journal of Solar Energy Engineering*, **115**, 52-56.
8. Minowa, T., Ogi, T., Yokoyama, S. (1997) Hydrogen Production from Lignocellulose Materials by Steam Gasification Using a Reduced Nickel Catalyst.

In: *Developments in Thermochemical Biomass Conversion*, (Ed. by A.V. Bridgwater and D.G.B. Boocock), pp. 932-944. Blackie Academic & Professional, London.

9. Miles, T.R. (1984) Biomass Preparation for Thermochemical Conversion. In: *Thermochemical Processing of Biomass.* (Ed. A.V. Bridgwater), pp.69-90, Butterworths, London.
10. Robinson, J.S. (1980) Primary Biochemical Conversions, In: *Fuels from Biomass: Technology and Feasibility.* pp. 167-183, Noyes Data Corporation, Park Ridge, New Jersey.
11. Denn, J. (1997) Test Plant Aims to Make Something of Nothing. *Albany Times Union*, p. E1 & E4, November 29, 1997, Albany, New York.
12. Elliott, D. C.,Fitzpatrick, S. W., Bozell, J. J., Jarnefeld, J. L., Bilski, R. J., Moens, L., Frye, J. G., Jr., Wang, Y. and Neuenschwander, G. G. (1999) Production of Levulinic Acid and Use as a Platform Chemical for Derived Products. In: *Biomass: A Growth Opportunity in Green Energy and Value-Added Products.* (Ed. R. P. Overend and E. Chornet), Volume 1, pp. 595-600. Elsevier Science, Ltd., Kidlington, United Kingdom.
13. Brown, R. (2000) CDP Will Build Huge PLA Unit at Cargill's Site. *Chemical Marketing Reporter*, **257**(3), p.1.
14. Werpy, T.A., Frye, J.G., Jr. et al. (2000) Production of Succinic Acid Based Renewable Chemicals. In: ACS Symposium Series, to be published.
15. Black, S.K., Hames, B.R., and Myers, M.D. (1998) Method of Separating Lignocellulosic Material into Lignin, Cellulose and Dissolved Sugars. *U.S. Patent No. 5,730,837.*
16. Torget, R., Himmel, M., Wright, J.D., and Grohmann, K. (1988) Initial Design of a Dilute Sulfuric Acid Pretreatment Process for Aspen Wood Chips. *Applied Biochemistry and Biotechnology*, **17**, 89-104.
17. Krochta, J.M., Tillin, S.J. and Hudson, J.S. (1988) Thermochemical Conversion of Polysaccharides in Concentrated Alkali to Glycolic Acid. *Applied Biochemistry and Biotechnology*, **17**, 23-32.
18. Elliott, D.C., Sealock, L.J., Jr., and Baker, E.G. (1993) Chemical Processing in High-Pressure Aqueous Environments. 2. Development of Catalysts for Gasification. *Industrial and Engineering Chemistry Research*, **32**, 1542-1548.
19. Elliott, D.C., Phelps, M.R., Sealock, L.J., Jr., and Baker, E.G. (1993) Chemical Processing in High-Pressure Aqueous Environments. 4. Continuous-Flow Reactor Process Development Experiments for Organics Destruction. *Industrial and Engineering Chemistry Research*, **33**, 566-574.

Combined Chemicals and Energy Production From Biomass Pyrolysis

D.A. Himmelblau
Biocarbons Corporation, 71 Cummings Park, Woburn, MA 01801, USA and R.W. Beck, P.O. Box 9344, Framingham, MA 01701, USA

ABSTRACT: While the emphasis on biomass pyrolysis liquids has been fuel production, higher or added value can be obtained by chemical production. Pyrolysis oil yield and composition are functions of reactor design and operating conditions, especially temperature, and the raw pyrolysis oil composition can be modified to improve yield and selectivity for higher value chemical products in order to produce a better return than from fuel use alone.

The pyrolysis products that are currently commercial chemicals will have to compete with conventional petrochemical or agrichemical production, especially in terms of product purity and scale of production. Unfortunately, ready markets do not exist for many of the potential chemical products from biomass pyrolysis such as levoglucosan, because they have not been previously available in large quantities at reasonable cost. The markets cannot be developed without additional application research and distribution channels. Chemical company partners will be needed.

This paper uses pyrolysis products from an air-blown, fluidized-bed reactor designed to produce liquids for use in adhesives as an illustration of what might be produced, separated and sold commercially for higher value. Guidelines for chemicals production and product recovery are suggested for pyrolysis processes in general. Recommended research and development topics to aid commercialization and increase chemical product recovery and project cash flow are presented.

INTRODUCTION

The purpose of this paper is to guide or at least stimulate thinking about obtaining as much value as possible from the chemicals produced from biomass pyrolysis. At optimum, all solid and liquid products could be used for chemicals, rather than fuel. At minimum, additional cash flow could be produced by recovery or production of any high value chemicals in the liquid product, leaving the balance of the products to be used as fuel.

The data in this paper is specific to the air-blown, fluidized-bed reactor system used by Biocarbons Corporation (5,6) that produces a water-insoluble liquid yield of up to 25 weight percent (of dry feed), a char yield of up to 15 weight percent and the balance as gas, mostly CO_2, H_2, CO, H_2O and CH_4. The reactor system was specifically chosen to produce a liquid stream that could be used (as it turns out in its entirety) as a lower cost substitute for phenol in the phenol-formaldehyde adhesive resins used for making water-resistant wood panels such as plywood and oriented strandboard. Recovery and/or production of higher value materials from the original products increases the net revenue for this process.

The first part of this paper will present the chemical recovery options for the product spectrum produced by Biocarbons Corporation's reactor system. This is followed by a general discussion and guidelines for chemical recovery from pyrolysis. This paper will also present suggestions for additional research and development to create new high value chemical products and recovery methods.

BIOCARBONS CORPORATION'S CHEMICAL PRODUCTS

By operating an air-blown, fluidized-bed reactor at about 590 to 600°C, Biocarbons Corporation has been able to produce a water-insoluble liquid product that is phenolic in nature. While product yields are lower than for fast pyrolysis processes designed to maximize liquid fuel production from biomass, the selectivity is 100%; no further product separation is required before the oil can be used to make adhesive.

Like other pyrolysis oil processes, Biocarbons Corporation's reactor produces a large number of oil compounds. For mixed hardwood (maple, birch and beech) pyrolyzed at typical operating conditions, 69 peaks were found by GC/MS analysis. Of these, the 14 peaks present at above 2 mole percent, represented 45 mole percent of the product that came through the GC. The 27 peaks between 1 and 2 mole percent, represented an additional 37 mole percent of the product. These compounds that were identified are listed in Table 1, in order of appearance (time). Several of the #4-position groups could also be occurring at the #3 position. <u>All are reactable to make a phenol-formaldehyde type adhesive.</u> Pyrolysis oil from pine that was made at the same operating condition (but has not yet been tested for adhesive use) had essentially the same compounds present at >1 mole percent, but at different relative concentrations. Some lower concentration compounds such as fatty acids are only produced from pine, but these compounds are specific to softwoods and the composition of softwoods. A comparison between the mixed hardwoods and pine products is shown in Table 2.

Although some have three or more, most of the compounds in Table 1 have two available positions for methylene linkages, compared to three for phenol. Some have only one site, and are polymer chain breakers. Most of the aldehydes are capable of linking at positions in addition to the HC=O group. The overall "linkability" of the compounds present - an average of 2.0 positions per molecule - helps explain why complete substitution of pyrolysis oil for phenol does not produce a suitable thermoset, and why 50 percent or less phenol is still needed to provide an adequate network of methylene linkages. Adhesive production and testing results are reported elsewhere (3,4).

Operating the reactor at about 500°C increases the liquid product yield by about 10 weight percent, but changes the product composition, making it less suitable for use in an adhesive (see Table 3). Most of the compounds produced are 2,4 substituted phenols that only have one position available for polymerization.

Table 1 Compounds Identified in Air-Blown, Bubbling Bed Pyrolysis Oil At > 1 Mole Percent In Order of Appearance From the Column (Time).

Phenol	1H-Inden-1-one, 2,3-dihydro
1,2-Cyclopentanedione, 3-methyl	1,2-Benzenediol, 4-methyl
Phenol, 2-methyl	2-Methoxy-4-vinylphenol
Phenol, 4-methyl	Phenol, 2-methoxy-3-(2-propenyl)
Phenol, 2-methoxy	4-Ethylcatechol
Phenol, 2,4-dimethyl	Vanillin
Phenol, 4-ethyl	Phenol, 2-methyl-6-(2-propenyl)
Phenol, 2-ethyl	Phenol, 4-ethyl-2-methoxy
Phenol, 2-methoxy-4-methyl	Phenol, 2-methoxy-4-(2-propenyl)
1,2-Benzenediol	1,3-Benzenediol, 4-propyl
Phenol, 2-(1-methylethyl)	Ethanone, 1-(4-hydroxy-3-methoxyphenol)
or Phenol, 2-ethyl-4-methyl	Benzaldehyde, 4-hydroxy-3,5-dimethyl
1,2-Benzenediol, 3-methyl	4-Hydroxy-2-methoxycinnamaldehyde
1,4-Benzenedicarboxaldehyde-2-methyl	

Table 2 Comparison Between Mixed Hardwood and Pine Pyrolysis Oils Made At 590°C Using Selected Compounds.

Compound	Hardwood GC/MS Area	Pine GC/MS Area	Ratio
Phenol	2.18%	1.84%	1.18
Acetophenone	----	0.60	----
Phenol, 2-methyl	2.31	1.66	1.39
1,2-Benzenediol	3.86	2.87	1.34
1,2-Benzendiol-3-methyl	2.93	2.71	1.08
4-Ethylcatechol	5.26	4.22	1.25
2-Methoxy-4-vinylphenol	1.04	1.67	0.62
Vanillin	1.54	1.46	1.05
n-Hexadecanoic acid	----	0.61	----
Furan, 3-phenyl	0.74	----	----
4-Hydroxy-2-methoxycinnamaldehyde			

Compounds in the water-insoluble liquid produced at around 600°C with current high value (greater than $2.00/kg) and commercial uses include vanillin and catechol (1,2-benzenediol, pyrocatechol). However these compounds are difficult to remove economically as discussed below and do have adhesive value. It is more feasible to remove catechol from the scrubber water as discussed below.

To recover the liquid product from the reactor product gas stream, Biocarbons Corporation uses a venturi scrubber with a closed water recirculating loop that (for commercial operation) is kept above the water dew point of the gas to avoid net water production (An electrostatic precipitator would be used downstream of the scrubber

Table 3 Major (>2% of Area) GC/MS Compound Peaks Identified In 11/99 Mixed Hardwood Oils.

Compounds Identified At 500°C	Compounds Identified At 590°C
Phenol, 2-methoxy	Phenol
Phenol, 2-methoxy-4-methyl	Phenol, 2-methyl
2-methoxy-4-vinylphenol	Phenol, 4-methyl
Phenol, 2,6-dimethoxy	Phenol, 2-methoxy
Phenol, 2-methoxy-4-(1-propenyl)	Phenol, 2,4-dimethyl
Eugenol	Phenol, 2-ethyl-6-methyl
Phenol, 2,6-dimethoxy-4-(2-propenyl)	1,2-benzenediol, 4-methyl
Benzaldehyde, 4-hydroxy-3,5-dimethyl	2-Methoxy-4-vinylphenol
3,5-Dimethoxy-4-hydroxycinnamaldehyde	Phenol, 2,6-dimethoxy
	Eugenol
	2-Methoxy-6-methylphenol
	Vanillin
	Phenol, 2-methoxy-4-(1-propenyl)
	5 additional peaks
	3,5-Dimethoxy-4-hydroxycinnamaldehyde

to collect the remaining oil fume.). As the various compounds reach saturation in the water, they deposit into the water-insoluble oil phase that is heavier than water and separates from the water phase by gravity. To prevent water soluble acids from reaching saturation and entering the water insoluble phase (if a third phase is not formed), the scrubber water would be treated with metal hydroxide to form precipitating salts after the treatment shown in Figure 1. Compounds typically found in the 600°C reaction scrubber water at high concentrations and their relative concentrations are shown in Table 4. Of note are catechol (and levoglucosan (1,6-anhydro-β-d-glucopyranose).Catechol is present in the water at a higher concentration than in the oil phase, because of a high solubility in water.

The catechol can be recovered by proprietary chemical reactions used to convert commercially produced catechol into even higher value products. An alternative method is shown in Figure 1. A side stream of the scrubber water at about 40 weight percent "solids" is removed, heated and flashed to separate the higher boiling point catechol and sugars from the water and organic acids, aldehydes and ketones (an additional drying step may be needed). If not separable by melting point difference, the catechol can be selectively washed out from the sugars with methanol. The water, hydroxyacetaldehyde and formic acid are distilled overhead from the higher boiling compounds -acetic acid, hydoxyacetone, etc. The formic and other organic acids can be removed from water by adding a metallic hydroxide (sodium, calcium, copper, iron, etc.) to produce the corresponding lower solubility formate, etc. Catechol yield from dry wood seems to be in range of 0.65 to 1.35 weight percent.

Levoglucosan could also be recovered from the scrubber water, and is of special note since it is a major high temperature decomposition product of cellulose for all reactor systems. Levoglucosan has the potential to be used as a precursor for a wide

Table 4 Selected Relative Concentrations In Scrubber Water of Compounds Made At 590°C From Mixed Hardwoods.

1-Hydroxy, 2-propanone (acetol)	1.35
Acetaldehyde	1.95
Phenol	1.00
1,2-Benzenediol (catechol)	4.42
1,2-Benzendiol, 3-methyl	1.12
1,2-Benzendiol, 3-methoxy	1.20
1,2-Benzenediol, 4-methyl	2.07
Levoglucosan	12.04

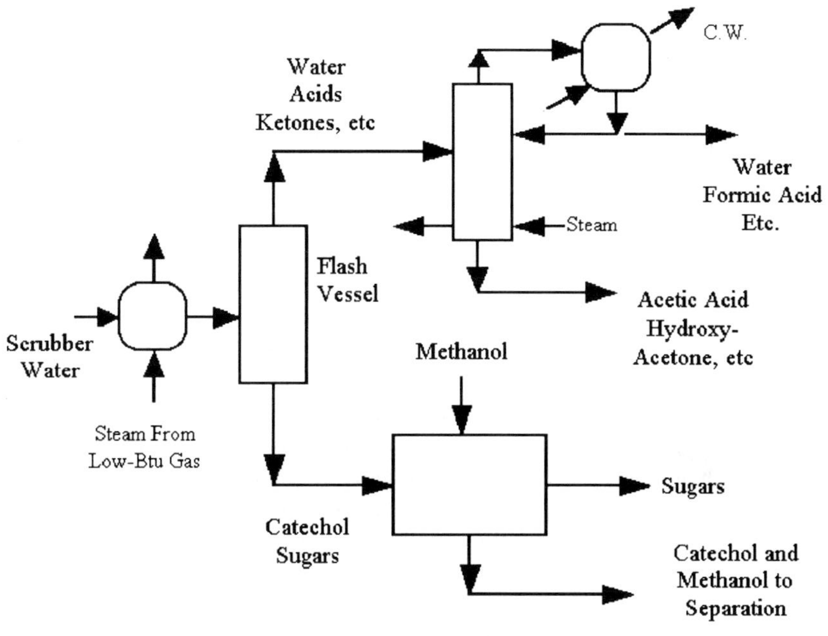

Figure 1 Product Separation And Catechol Recovery.

range of industrial chemicals including levoglucosenone, linear dextrans, vitamin H, polyurethanes, epoxies, surfactants and chiral ligands (8).

Finally, while the char produced from the pyrolysis process can be sold for making charcoal, depending on the cost of transporting the char to a charcoal-producing facility, the char can also be used to make activated carbon. Because the char is what remains after compounds have rapidly volatilized out of the original structure, the removal of the remaining hydrogen and oxygen, and the opening of the pore structure is rapid and can be done below typical steam activation temperatures and residence times. The yields are approximately 50% by weight with surface areas of at least 500 m^2/g and Iodine numbers over 500. The optimization of the activated carbon grades and yields is ongoing.

REVENUE BENEFITS

Recovering (a 0.6 weight percent yield of) catechol for sale at about 75% ($1.85/kg) of the large volume/contract sales price, and selling charcoal at $77/tonne, F.O.B. the plant would increase the gross cash flow from a 180 dry tonne of wood per day plant making 14,750,000 kg per year of pyrolysis oil for adhesive use (at $0.077/kg) by about $1,140,000 per year or about 20%. However, the net cash flow increase, exclusive of any additional cash flow for other by-products is about $700,000 per year or about 30%. A catechol yield of 1.2 weight percent would increase the net cash flow by an additional $600,000 per year. Converting the char to activated carbon with a yield of 50% and a sales price of $0.80/kg, F.O.B. the plant, would increase the net cash flow by about another $1,000,000.

COMMERCIAL CONSTRAINTS ON CHEMICAL RECOVERY ECONOMICS

Three strong warnings need to be attached to this type of economic analysis. First, there is not an infinite market for any chemical product. For many of the chemicals that potentially can be produced from biomass, the markets are small relative to commodity petrochemicals. Most of the chemicals that can be produced from biomass should be considered specialty chemicals. Unless new large uses can be found for them, e.g. using levoglucosan as an intermediate building block, large scale production of any of these currently high-value chemicals will drop the price. Second, attention must be paid to the current purity and grades of product being sold, and the costs required to achieve these specifications. Even if absolute purity specifications are achieved, the types of impurities remaining will be different than those in the conventional chemical. The effects of these impurities on current uses will have to be evaluated. If the amount used in any most applications is small, this could become very expensive. It is also unlikely that small-volume users of any chemicals recovered from pyrolysis oil will be willing to change their manufacturing processes to accommodate new impurities.

Finally, because of the small yields and scale relative to even many specialty petrochemicals, one should not automatically assume that the recovery "production" costs of chemicals from biomass pyrolysis are necessarily competitive with the actual production costs from petrochemicals; there is usually a large mark-up over costs, because the markets are small, the number of producers is small and there are technical barriers to entry, sales and application development. It would be wise to co-develop recovery with a current manufacturer, just to assure distribution channels. If recovery costs are lower than conventional production, and the market is growing, there will likely be interest, because expansion capital costs can be avoided, and incremental additional production becomes feasible.

GUIDELINES FOR CHEMICALS PRODUCTION AND RECOVERY

USE PROCESS AND FEEDSTOCK SELECTIVITY IN PROLYSIS OIL PRODUCTION

While all pyrolysis oil production reactor systems produce similar materials, each reactor produces a unique compound slate. The first decision, especially for a potential chemical or fuel producer, rather than a reactor developer, is to determine what products to make and which reactor system to use. The operating parameters of any reactor system designed to produce pyrolysis oil, especially temperature, can be altered to change the pyrolysis oil product composition and yield. Different feedstocks will produce different pyrolysis oil compositions and by-products, e.g. amorphous silica from rice hulls or rice straw, fatty acids from pine. Finally, feedstock pretreatment and/or catalysis, or reactor-bed catalysis can be used to improve specific product yields (7). Reactor system developers need to examine what they can produce and make this information available to chemical manufacturers and suppliers/owners of biomass feedstocks. **This assumes that analysis of the entire liquid product from thermal conversion can be made, including quantitative analysis for any compounds that are being considered for recovery**. Physical characterization - pH, viscosity, solids content, etc.is also needed. However, what can be produced is of no value, if it cannot be recovered or used economically. This involves examining the trade-offs between yield and current commercial value, recovery costs, and potential commercial value.

SEPARATION AND RECOVERY HURDLES

Recovery of specific compounds from pyrolysis oil is made difficult by the thermal sensitivity of most pyrolysis oils. The aldehydes produced can react with the phenols, furans, ketones, etc. present, forming viscous polymerized compounds that have no value as distinct chemicals (but still may be usable as adhesive feedstock). These reactions are accelerated by temperature. Many of the compounds produced are photosensitive or are oxygen sensitive, again polymerizing when exposed. Addition of bases to adjust pH by organic acid neutralization will catalyze most of these polymerizations. Divalent cations can polymerize some of the vinyl groups present from the lignin similar to styrene polymerization. Some of these reactions are useful for making adhesives, but make thermal separation for chemical product recovery difficult without using low temperature vacuum distillation.

Initial separation of water-soluble material from water-insoluble material is usually easy, because most pyrolysis oils are really oil-in-water emulsions, with the organic acids and/or compounds with hydroxyl groups usually serving as the emulsification agent. The water-insoluble compounds - phenolics, etc., can be separated from the water-soluble products (organic acids, etc.), by adding water, changing the pH, heating (the riskiest) or some combination of all three, similar to water separation from crude petroleum.

Solvent extraction can be used to remove groups of compounds (1) from water soluble or water-insoluble fractions, but solvent extraction is not particularly selective, and recovery of the solvent can be expensive.

WHAT CAN BE PRODUCED ECONOMICALLY

While multiple products can be produced from biomass pyrolysis, pyrolysis and product recovery is not really a refinery, because thermal processing, the backbone of petroleum processing is not usually a viable option for product separation. There are also large differences in the economies of scale between a petroleum refinery and a biomass conversion plant, because of the differences in cost for transporting low density biomass by truck or rail versus transporting petroleum by ship or pipeline. The economic radius for a petroleum refinery feedstock can be worldwide, versus about 100 kilometers for a biomass plant, unless the biomass is the gathered residue from an agricultural processing plant. Storage requirements and feedstock availability are also major differences. Consequently, biomass processing plants, even paper mills utilize feedstock on the order of 1000 tonnes per day or less, rather than 40,000 tonnes per day or more for a economic world-scale refinery. The by-products of petroleum refining that are the building blocks of the petrochemical industry are produced in larger quantities than a typical biomass-to-energy conversion facility. For example, a single fluidized catalytic cracker for upgrading heavy oil fractions can produce over 1000 tonnes per day of propylene for making polypropylene. Finally, biomass and biomass compounds contain much more oxygen than petroleum products. Duplicating petrochemical building blocks from biomass can require a large amount of expensive hydrogen and processing to remove this oxygen.

Consequently, making most conventional petrochemicals from thermal conversion of biomass is not going to be economic until petroleum prices actually reach the dizzy heights predicted in the early 1970's, i.e., not in your lifetime. Other than fuels, the products that make economic sense to try and produce and then use as either end products or synthesis intermediates are the chemicals that can actually be produced and recovered from biomass pyrolysis without a high recovery costs - materials that can be recovered in an acceptable form for $0.20 to 0.50 per kg or less. This means using the solid by-products, organic acids, aldehydes, ketones, phenols, furans, sugars, pentanones, etc. that are "currently" produced, trying to improve specific product yields and selectivity as much as possible. This rule of thumb might be ignorable for low concentration pharmaceutical precursors such as steroids or fragrance precursors that have been found in pyrolysis oil, depending on the value and recovery and purification costs.

The pyrolysis products that can be produced fall into two categories- compounds that are or have already been used, and compounds that are newly available in bulk and at lower cost such as levoglucosan. For the former there is a wealth of literature that exists prior to the widespread of use of petrochemicals (for example reference 2) that is usually ignored by lazy investigators that end up reinventing or ignoring prior research. Much of the use of the chemicals that used to be made from biomass and coal was cast aside when cheaper petrochemical synthesis pathways became available. However, many of these forgotten applications and syntheses should be newly competitive with higher volume production and therefore lower cost biomass chemicals. This also may apply to biomass chemicals newly available in "bulk" as specialty chemicals for which bench studies were done, but whose commercial prospects were dim, because of the actual or perceived cost of making the chemicals. The new "bulk" intermediaries should also be evaluated for use via all the chemistry, especially for making polymers, that has been developed since the advent of the petrochemical age in the 1930's.

PRODUCT RECOVERY TECHNIQUES

Product recovery is some combination of rote methods of using processing tools and inspired application of new and conventional separation tools. For example, if aldehydes make distillation infeasible because of polymerization products and/or vacuum distillation costs, then one can consider utilizing the Cannizzaro reaction to make organic acid salts and ketones from the aldehydes before proceeding. Many bench-scale processing techniques such as thin-film evaporation have been scaled up to handle larger quantities of temperature sensitive materials. Assistance in utilization of these techniques can usually be obtained from the equipment vendors.

Other suggested product recovery tools that may not be readily apparent are:

(1) Organic acid removal by resin separation in a chromatographic simulated moving bed
(2) Water removal by membrane separation to promote selective precipitation of lower solubility compounds, followed by solvent separation of the precipitates
(3) Melting point separation of thermally stable solids
(4) Adding or seeding compounds to cause precipitation of specific chemicals
(5) In-situ chemical reactions such as oxygen initiated reactions, e.g., aldehydes to acids

SUGGESTED RESEARCH AND DEVELOPMENT

In order to effectively produce chemicals from biomass pyrolysis, developers need to think beyond the mindset of maximizing production of liquids and stuffing these liquids into a combustion device. The proper thought process should be that of a chemical engineer - what can I make from this and what do I then do with it (to become rich and retire early). For a few specific chemicals that are identified through analysis and are easy to recover, no additional research and development are needed other than scale-up; the commercialization problems are market related. For most potential products, some additional help is needed. The following are suggested research and development topics:

(1) Polymerization of the sugars such as levoglucosan into thermoplastics
(2) Utilization (beyond polymerization) of any oxirane sugars via ring opening
(3) Fluidized bed catalysis for better selectivity and removal of compound oxygen
(4) Conversion of "new" bulk five and six carbon sugars to furans and pyrans, respectively
(5) Organic acid and other oxygenated compound separation by resins using a simulated moving bed
(6) Identification of specific solvents for compound classes, i.e., aldehydes, ketones,etc.
(7) Fragrance and flavors from the aldehydes produced
(8) Increased steroid precursor yields and better selectivity
(9) Pyrolysis oil mixture properties and quantitative analysis techniques
(10 Recovery and usage of pentanones and quinones

(11) Increased yields of polyhydric phenols
(12) For any reactor system, testing of feedstocks known to be rich in valuable, extractable components
(13) Production of silicons and silicates from amorphous silica from rice hulls and straw
(14) Increased selectivity and/or recovery of vinyl and propenyl phenols for polymerization ,etc.

ACKNOWLEDGEMENTS

The author would like to thank Carrie Davis and Lynn Gennaro of Northeastern University and Dietrich Meier of the Institute for Wood Chemistry, Hamburg, for the analytical assistance provided.

REFERENCES

1. Chum H. et al. (1990) Process for Fractionating Fast-Pyrolysis Oils and Products Derived Therefrom. U.S. Patent 4,942.269.
2. Ellis C. (1935) The Chemistry of Synthetic Resins, Reinhold Publishing, New York.
3. Himmelblau D.A., Grozdits G. A. and Gibson M. A. (2000) Performance of Wood Composite Adhesives Made With Biomass Pyrolysis Oil. To be published as part of *Wood Adhesives 2000*, Forest Products Society, Madison, WI.
4. Himmelblau D.A. and Grozdits G.A. (1999) Production and Performance of Wood Composite Adhesives with Air-Blown, Fluidized-Bed Pyrolysis Oil. In *Proceedings, Fourth Biomass Conference of the Americas*, Elsevier, New York.
5. Himmelblau D. A. and Grozdits G.A. (1999) Production of Wood Composite Adhesives with Air-Blown, Fluidized-Bed Pyrolysis Oil. In *International Contributions to Wood Adhesion Research*, Forest Products Society, Madison, WI.
6. Himmelblau D.A. (1996) Phenol-Formaldehyde Resin Substitutes from Biomass Tars. In *Wood Adhesives 1995*, Forest Products Society, Madison, WI.
7. Piskorz J, Radlein R, et.al. (1995) Fast Pyrolysis of Pretreated Wood. In *Proceedings of the Second Biomass Conference of the Americas*, National Renewable Energy Lab, Golden, CO.
8. Resource Transforms. Levoglucosan - A Chiral Raw Material, Waterloo, Ontario.

Sibunit Supported Catalysts for Hydrogenolysis of a C-O Bond in 'Bio-Crude-Oil' Components

D.G. Aksenov[1], A.N. Startsev[1] and B.N. Kuznetsov[2]

[1] *Boreskov Institute of Catalysis, 630090 Novosibirsk, Russia*

[2] *Institute of Chemistry and Chemical Technology, 660049 Krasnoyarsk, Russia*

ABSTRACT: New catalysts for hydrogenolysis of C-O bond are proposed. The catalysts were prepared by anchoring of Mo, (Ni,Mo) and (Co,Mo) complexes to the surface of a new carbon support Sibunit. Two types of the active component were prepared – oxide and sulfide forms. The catalysts were tested in a model reaction of tetrahydrofuran hydrogenolysis. As shown, the catalysts are active in the purposeful reaction of C-O- bond hydrogenolysis and do not catalyze the side reactions – polymerization and dehydration of tetrahydrofuran.

INTRODUCTION

Utilization of the liquid synthetic raw materials produced on thermal conversion of biomass has received much attention in recent years [1]. "Bio-crude-oil" possesses high viscosity, contains solid resin precipitates, and a large number of oxygen containing contamination's. For this reason "bio-crude-oil" cannot be used directly as a fuel. Hydrorefining of "bio-crude-oil" allows one to decrease the concentration of oxygen to the required level, to decrease viscosity, and to remove solid resin materials. Hydrogenolysis of alcohol and ether groups proceeds easily. By contrast, hydrogenolysis of tetrahydrofuran (THF) and its homologues runs with great difficulty, though their concentration in "bio-oil" is rather high [2].

The traditional catalysts have usually a short period of operation because of high rates of deactivation processes, such as polymerization, carbonization, and decomposition of the active component. The catalysts are routinely prepared by precipitation, which results in formation of a wide range of surface species, providing a number of side reactions. The side reactions may also occur on alumina, the main support for the hydrogenolysis catalysts. Alumina is an active catalyst for dehydration processes as well.

In our work, Sibunit, an artificial carbon material, was used as a support. The methods of Sibunit synthesis permit one to vary surface and size of the support pores within a wide range. Because of high thermal stability, mechanical strength, and inertness to many chemical reactions, Sibunit is very promising as a support for the hydrofining catalysts operating under severe conditions.

EXPERIMENTAL

The catalysts were prepared by supporting oxalate Mo and Mo-Ni (Co) complexes [3]. After drying the catalysts were calcined at 500°C in air (CatMo, CatNiMo, and CatCoMo). Some catalysts were sulfurized with hydrogen sulfide (CatMoS, CatNiMoS, and CatCoMoS) at 400°C. The concentration of Mo in all samples was about 10%. For bimetallic catalysts, the ratio Mo : Ni(Co) is close to 2.

The catalysts were tested on hydrogenolysis of tetrahydrofuran (THF). The reaction was performed in a flow setup at the hydrogen pressure of 20 atm and 250-400°C. After the reactor we installed a vessel, filled with glass fiber, to trap the non-volatile reaction products.

RESULTS AND DISCUSSION

Hydrogenolysis of THF may follow different schemes to produce either butane (scheme 1), or butenes (scheme 2) or butadiene (scheme 4). In addition, polymerization, yielding crown ethers (scheme 3), may also occur.

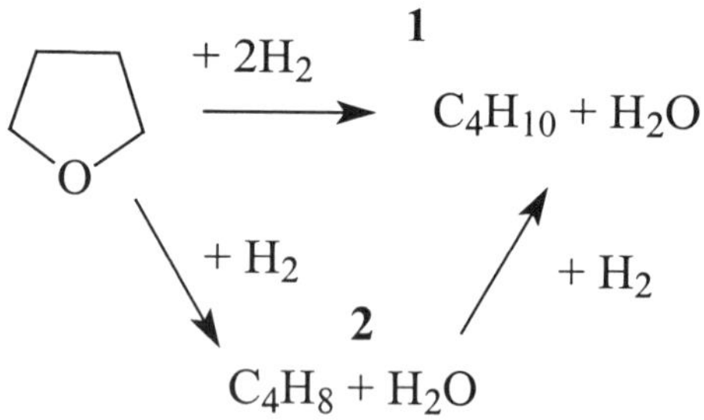

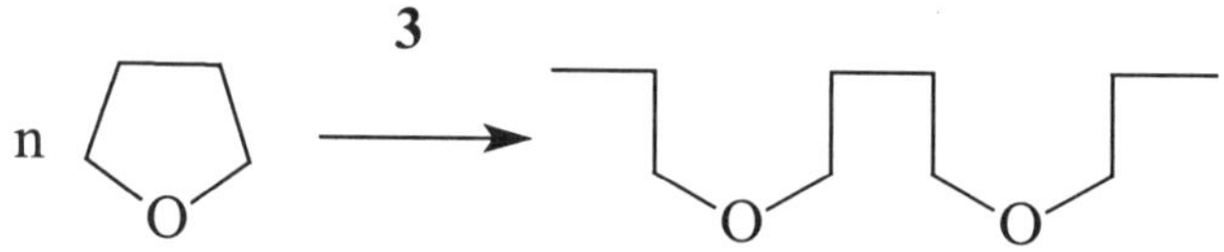

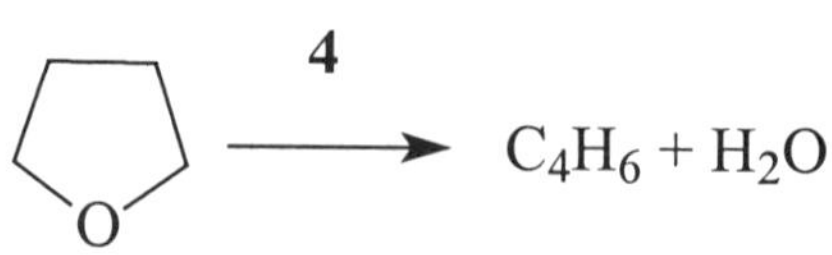

In contrast to alumina, Sibunit is not active on hydrogenolysis of THF (Table 1). The products of polymerization were not observed even at high temperature (Table 2). The monometallic molybdenum catalysts exhibit low activity. The oxide catalysts are more active than sulfides by a factor of 3. At high temperature, their activity is comparable (Table 1).

Table 1. THF hydrogenolysis over the oxide and sulfide catalysts.

Catalysts	Reaction temperature, °C	Activity, mol(THF)/(molΣMet*h)
Sibunit	300	–
	350	–
	400	–
KMo	300	5,6
	400	31,0
KMoS	300	1,8
	350	11,9
	400	34,8
KNi	250	3,4
	300	31,4
KNiS	300	8,5
	350	13,2
	400	16,6
KCoS	300	2,7
	350	13,3
	400	26,3

Both catalyst types preserve activity for several hours (Fig. 1). The activity and stability of the bimetallic catalysts strongly depend on their type. The oxide catalysts exhibit higher activity and stability than the sulfide catalysts. In three hours, the activity of sulfide catalysts decreases by ~20%, and that of the sulfide catalysts, by a factor of 3. For bimetallic catalysts, there is the synergetic effect.

At the initial reaction stage, the activity of oxide Mo-Ni catalysts is higher than that of Mo catalysts by a factor of 5. In three hours, this value decreases to 4 (Fig. 1a). For the sulfide samples, this effect is less pronounced (Fig. 1b). In the next run, the catalyst activity was determined at 400°C, then the temperature was decreased to 300°C. This procedure was repeated for several times. For both oxide and sulfide molybdenum catalysts, activity and the product composition are reproduced with high accuracy in a number of experiments (Fig. 2). At high temperature, the bimetallic catalysts are quickly deactivated during the reaction (and the experimental cycles are not reproducible).

Butadiene was not observed in the reaction products on all studied catalysts, including the pure support. The composition of the hydrogenolysis reaction products depends primarily on the reaction temperature (Table 2).

For both oxide and sulfide molybdenum catalysts, the ratio between butane and butenes is much higher than 1. As temperature rises, this ratio decreases for the oxide forms and increases for the sulfide forms. Polymerization begins at $T > 300°C$ on both catalyst forms. For all bimetallic catalysts, excepting Mo-Ni oxide catalysts, the ratio butane : butenes is higher than 1 at 300°C. The amount of resin materials holds with time only for Mo catalysts. For bimetallic catalysts, the conversion of THF into resin materials increases with time by ~10-15%. For all catalysts, excepting Mo-Ni sulfide catalysts, the yield of resin complexes increases with increasing temperature.

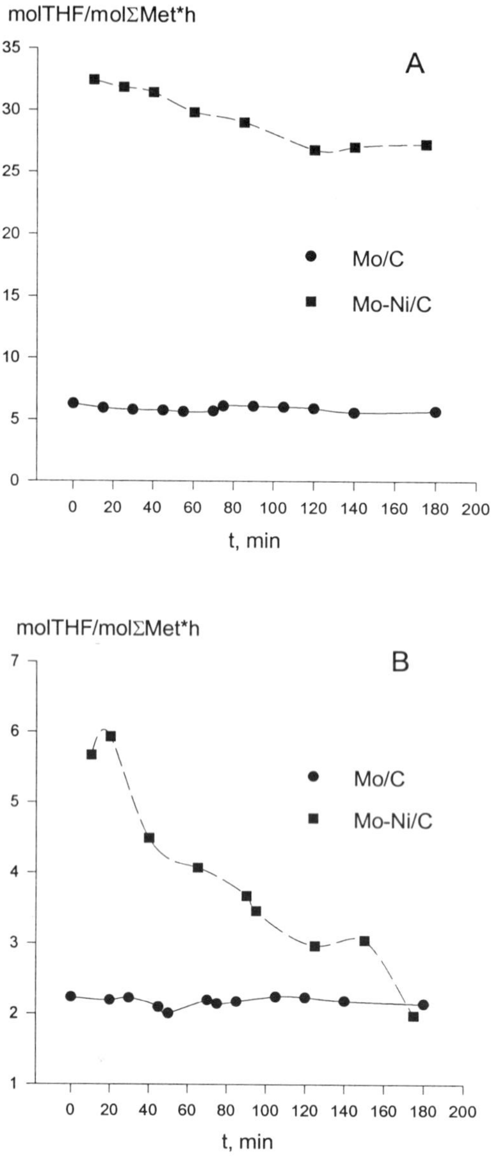

Fig. 1 THF hydrogenolysis over the oxide (A) and sulfide (B) catalysts. Reaction temperature is 300°C.

At high temperature and in the presence of Mo complexes, THF is polymerized to yield crown ethers (scheme 3) [4]. Probably, this reaction is responsible for deactivation of the bimetallic catalysts supported on Sibunit. The monometallic catalysts are not deactivated even at high yields of resin complexes and high temperature. It is likely that the polymers do not deposit on the active component and the support surface.

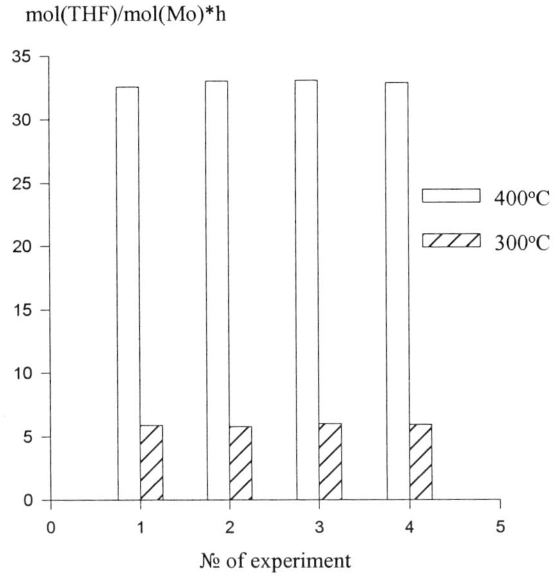

Fig.2 Changes in activities over monometallic catalysts at different temperatures in the consecutive cycles of the reaction temperature

Table 2 Product composition of THF hydrogenolysis after 30 min of beginning of reaction.

Catalysts	Reaction temperature, °C.	Butane/butenes ratio	α^*
Sibunit	300	–	–
	400	–	–
CatMo	300	30,1	–
	400	17,4	45,2
CatMoS	300	19,2	–
	350	52,7	21,3
	400	43,5	53,5
CatNiMo	250	1,2	30,2
	300	0,6	47,6
CatNiMoS	300	2,5	25,5
	350	6,3	13,9
	400	24,8	8,1
CatCoMoS	300	10,8	6,2
	350	4,1	34,5
	400	41,9	48,3

* - THF conversion to polymer compound.

For bimetallic catalysts, strong bonds between oxygen of the formed polymers and nickel are formed. According to the structure of the active component of the sulfide catalysts, electron density on the nickel atoms is reduced, which provides formation of strong bonds between nickel and the molecules possessing lone electron pairs [5]. Because of this phenomenon, the number of active sites in the sulfide bimetallic catalysts quickly decreases, which results in a sharp decrease in the catalyst activity. In oxide catalysts these bonds are less strong. As a result, a major part of the formed polymers is removed from the reactor, and the catalyst deactivation proceeds slower, though the number of resin complexes is large (Table 2).

The fact that butane is not found in the reaction products indicates that THF is not dehydrated (scheme 3) on the study catalysts and oxygen is removed from THF by hydrogenolysis (schemes 1 and 2). On the oxide and sulfide molybdenum catalysts, butane is the main reaction product, that is why hydrogenolysis predominantly follows scheme 1. This is also true for bimetallic sulfide catalysts, if the reaction occurs at high temperature. For the rest catalysts, a part of butenes in the reaction products is rather high. This suggests that hydrogenolysis of the most part of THF follows scheme 2.

CONCLUSIONS:

- Sibunit is the most suitable support for the catalysts of hydrogenolysis of the C-O bond in THF as a model of the "bio-crude-oilcomponent;
- dehydration does not occur on the study catalysts;
- hydrogenolysis of the C-O bond follows scheme 1 on the sulfide and oxide molybdenum catalysts and bimetallic sulfide catalysts, and schemes 1 and 2 on the other catalysts;
- polymerization as a side reaction does not proceed on the monometallic catalysts at low temperature;
- bimetallic catalysts are quickly deactivated because of strong bonds formed between the polymers and the active component.

REFERENCES

1. Bridgwater A.V. (1994) Catalysis in thermal biomass conversion. *Appl. Catal. A: General,* **116,** 5-47.
2. Veldsink J.W., Bouma M.J., Schoon N.S.-H., Beenackers A.C.M. (1997) Heterogeneous Hydrogenation of Vegetable Oils: A Literature Review. *Catal. Rev.- Sci. Eng.,* **39,** 253-318.
3. Aksenov D.G., Klimov O.V., Startsev A.N. (1997) Alumina supported sulfide catalysts. VI. Hydrogenolysis of tetrahydrothiophen. *Kinet. Katal.* **38,** 903-7.
4. Yukelson I.I., (1969) Methods of the Basic Organic Synthesis, Nauka, Moscow.
5. Startsev A.N. (1995) The Mechanism of HDS Catalysis. *Catal. Rev.-Sci. Eng.,* **37,** 353-423.

Multi-parameter Assessment of Sunflower Husk-sawdust Layer Hydraulic Resistance

Gubynskyy M., Shishko Y., Usenko A., Vvedenskaya T.
National Metallurgical Academy of Ukraine
4 Gagarin Avenue, Dniepropetrovsk 49005, Ukraine

ABSTRACT Modelling and computing of thermal biomass treatment processes involving thick layers requires that the gas permeability and hydraulic resistance of such layers be known. The paper presents the results of experimental research into the hydraulic resistance of the thick layer comprised of two biomass varieties: sunflower husks and sawdust. The research was conducted within the range of Re alteration from 5-70, and layer porosity change 0.2-0.7.

INTRODUCTION

The utilization of agricultural and wood biomass waste for power generation requires knowledge of the physical properties of the wastes.

These include low filling density and significant compressibility of waste during transportation in the channel. The last factor is efficiently used in the technology of sawdust briquetting by biomass friction against the channel walls. Biomass compression diminishes unit dimensions and intensifies the process of thermal treatment, which is especially important in conditions of excess pressure. Gas permeability of the layer and its hydraulic resistance are generally considered to be the major parameters determining the thermal treatment processes in the layer.

The regime of gas motion is one of the essential factors influencing the process. As the gas is moving along the channel the particles of the layer are streamlined, the laminar regime giving way to the turbulent one at Re=4. In this phase the actual value of gas filtration velocity may significantly exceed the conventional gas velocity related to the channel cross-section. The movement of the gas in the layer depends on dimensions, amount, and shape of the channels formed by the particles. These parameters stand in functional relation to the granulometric composition and porosity of the layer, as well as particle dimensions and shape.

Porosity is greatly determined by the particle shape. Ball shape allows for the minimum void volume in an irregular filling. However, the layer porosity characteristics obtained for the simplest case of ball-shaped evenly-sized particles cannot be applied for determining the type of the filling, or the shape of the channels formed by the particles and, consequently, for estimating the layer resistance.

Literature on the issue provides a sufficient number of equations for computing the hydraulic resistance of the layer [1-6]. In processing raw data the authors departed from two theoretical approaches modelling the pattern of particle motion in a gas:

- A capillary model presenting a system of channels with developed roughness;
- A model of the particle ensemble, with gas streamlining particular elements (particles) of the layer.

Hydraulic channel diameter d_e was assumed to be a defining parameter for the first model, while equivalent particle diameter d_k was chosen for the second model. The conventional equation for computing hydraulic resistance is known to be:

$$\Delta p = \xi_l \frac{\rho w_i^2}{2},$$

Where $\xi_l = f(Re, E)$ - coefficient of hydraulic resistance;

$$E = \frac{V_{njh}}{V_l}$$ - layer porosity;

w_i - gas velocity related to the layer cross-section area.

Experimental relations resulted in different expressions for computing Re and ζ_l. The value of ξ_l is as a rule inversely proportional to Re first power and E first or second power, i.e.

$$\xi_l \sim \frac{1}{Re}, \frac{1}{E^n}$$

The above-mentioned relations are valid only for certain shapes and dimensions of particles as well as for a limited range of porosity values. At E=0.4 the range of variation does not exceed ±0.08. These relations are rather generalized which tells on the calculation error and calls for further experimental proof of their validity.

A series of experiments has been undertaken with the view to determine hydraulic resistance of sunflower husks and sawdust of coniferous wood.

RESEARCH METHOD

The scheme of experimental unit is presented in Figure 1. Biomass layer under analysis is placed in the cylindrical container 200mmØ. The layer is enclosed from the top and the bottom by the specially designed grates whose hydraulic resistance is extremely low in comparison with the layer hydraulic resistance, the fact that was taken into account in the processing of experimental data.

The grates prevent the material from being carried away and ensure that the biomass layer be compressed and held in the experimental chamber. The layer height (250mm) remains constant in the process of the experiment as extra portions of the material are added into the cylinder. Then the layer is reduced by the hydraulic press to the initial volume. The amount of additionally-charged material was measured by weighing the material after the experiment. The absolute error in weighing did not exceed 0.5g. Porosity and density of the layer varied from 0.73 to 0.21, and from $100 kg/m^3$ to $400 kg/m^3$ respectively.

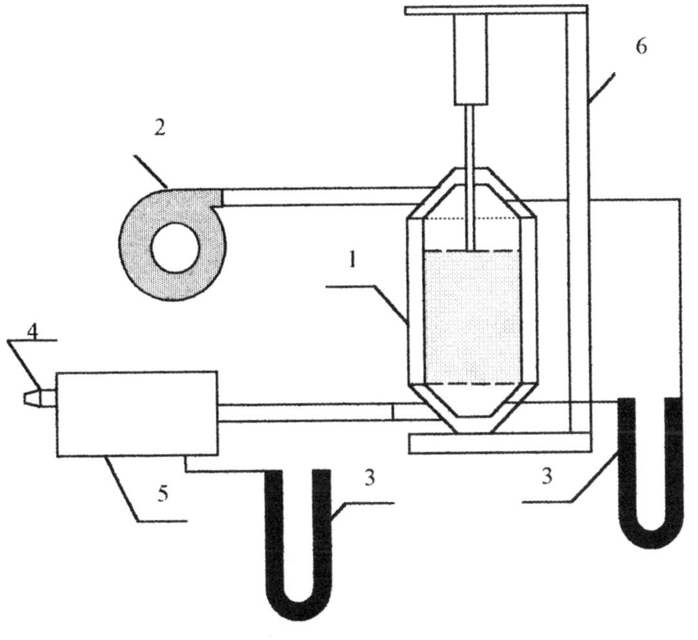

Figure 1 Experimental unit.
1-container with material under investigation
2-blower;3-U-shaped manometer;4-calibrated nozzle;
5-chamber for the flow velocity levelling ;6-hydraulic press

TB-1-25 blower blew the air through the material. The regulating valve in the blower's inlet maintained the air consumption. The amount of the consumed air was measured by a calibrated round-edged nozzle (length-diameter ratio: l/d = 4.3) which refers to the consumption coefficient μ=0.97. The nozzle was placed in the chamber for gas velocity levelling. The air pressure near the nozzle was measured by a liquid U-shaped manometer. The amount of air passing through the layer was calculated by:

$$Q = \mu \frac{\pi}{4} d^2 \sqrt{2gH}$$

Where H – air pressure near the nozzle, mm of water column.

The difference of pressure in the layer was measured by the manometer, the absolute error being 0.5 mm of water column.

ANALYSIS OF EXPERIMENTAL RESULTS

The results of experimental research into husks and sawdust hydraulic resistance as related to air velocity and layer density are presented in *Figure 2* and *Figure 3*.

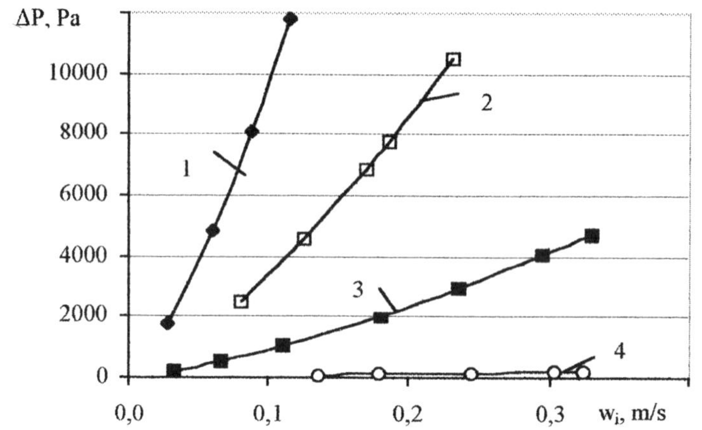

1 - E=0,73 ; 2 – E= 0,60; 3 – E=0,47; 4 – E=0,20

Figure 2. Experimental data on the sunflower husks layer hydraulic resistance

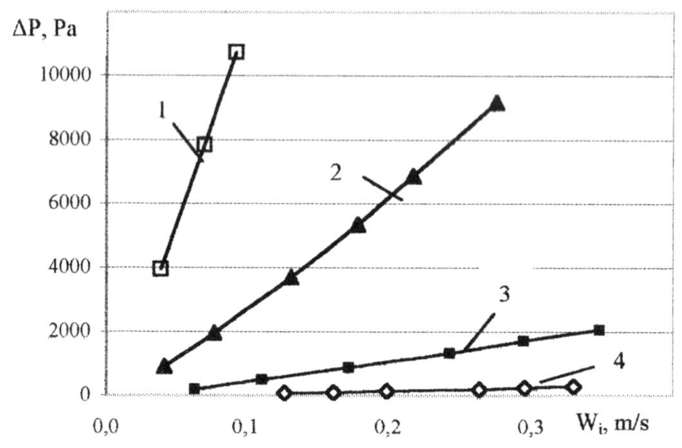

1- E=0,72; 2-E=0,55; 3 – E= 0,38; 4 – E= 0,21

Figure 3 Experimental data on the sawdust layer hydraulic resistance

Conventional empirical equations for computing hydraulic resistance in the thick layer [1,2] were used for evaluating pressure loss ΔP. Computations were based on the actual geometric characteristics of the particles and the data received from studying the layer granulometric composition as obtained by direct measurements [7] alongside with the sieve analysis of sawdust. The comparison of the computed and experimental values (*Figure 4*) testify that the obtained relations ensure the required accuracy of calculations only within the range of high biomass porosity (without external layer compression).

The value of error grows significantly with the increase of the air velocity and particles packing density. That is why it seems incorrect to use conventional computational instruments in the given situation.

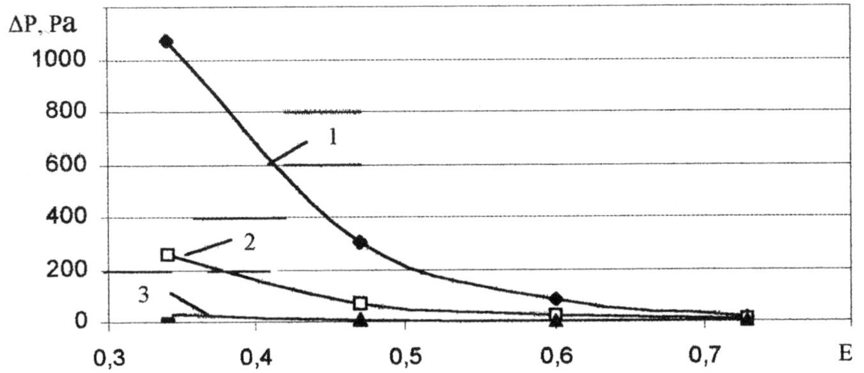

1-experimental values; 2, 3 –computed values [1,2].
Figure 4. Pressure losses in sunflower husk layer at air velocity W=0,2 m/s

Depending on the type of relations between resistance coefficient and Re the value of ξ can be computed by the power equation:

$$\xi = a \cdot Re^{b},$$

where coefficients a and b are functions of the layer porosity. Computational equations for coefficients a and b are presented in *Table 1*.

Table 1. Computed equations for calculating empirical coefficients a and b

Biomass type	Coefficient a	Coefficient b
Sunflower husks	$a = 10^5(-9,69E^3+19.34E^2 - -12,97E + 2,94)$	$b = 1,48E^2 - 1,06E + 0,80$
Sawdust	$a = 4 \cdot 10^6 EXP(-12,7E)$	$b = -7,56E^3 + 10,69E^2 - - 3,88E - 0,44$

CONCLUSIONS

The empirical equations obtained allow to evaluate biomass layer hydraulic resistance with an error up to ± 10%. The resulting computational instruments will be further applied for fixed bed pyrolysis units designing as well as for constructing chargers providing continuous feeding of material and pyrolizer's hermetic sealing by creating the necessary hydraulic resistance of the biomass layer.

REFERENCES

1. Aerov M.E.(1998) *Hydraulic and thermal fundamentals of the apparatus operation for fixed and fluidized beds.* /Russian/, Moscow, Chymyia Publishing House.
2. Kutateladze S.S. (1995) *Reference manual on heat exchange.* /Russian/, Moscow, Energia.
3. Sethonenko B.D., Orlik V.N., Alekseenko V.V.(1998) *Matematicheskoe modelirovanie teploperedachi i gidravlicheskogo soprotivlenia v regenerativnom teploobmennike s sharovoi nasadkoi* /Russian/, Ekolotechnologei i resursosberegenie , #5.
4. Telegin A.S. (1993) *Teplotechnicheskie rascheti metallurgicheskich pechei* /Russian/,. Moscow , Metallurgia
5. Agafonova M. E., Toritsin L. N., Lekomtseva E. D. (1990) *Gidravlichiskoe soprotivlenie sloevoi nasadki djpduchonagrevatelia* / Metallurgicheskaia ntpljtechnika./ /Russian/,. Moscow , Metallurgia
6. Kiet F., Blek Y.(1993) Osnovi teploperedachi. /Russian/,. Moscow , Mir.
7. Gubinsky M. (1999) Biomass pyrolysis of organic waste in a rarefied layer. *Proceedings of the 4th Biomass Conference of Americas*, Oakland, CA.

Organic composition of liquidized model kitchen garbage

S. Inoue, T. Minowa, S. Sawayama and T. Ogi
National Institute for Resources and Environment,
16-3 Onogawa, Tsukuba, Ibaraki 305-8569, Japan

ABSTRACT: Model kitchen garbage (87.1% moisture content, 11.3% organic content) was liquidized at 175 °C for 1h. The liquidized garbage was filtered into a liquid phase and a solid phase and then the organic compositions of the liquidized garbage and its centrifuged phases were analyzed and compared to mechanically disrupted garbage.

The total carbon amounts in 100 g of the mechanically disrupted garbage and liquidized garbage in the liquid phase were 2168.1 and 2297.0 mg, respectively. The total carbon amount was almost similar value. The solubilization of solid and/or insoluble material was not caused through the liquidization process. The liquidization phenomenon is mainly the thermal rupture of the cells.

During liquidization, low molecular organic acids were produced and their amounts increased compared with that in the mechanically disrupted garbage. The increment of the low molecular organic acids might be due to the degradation of the proteins and sugars. The total sugar content in the liquidized garbage, including simple sugars, oligosaccharides, polysaccharides, and their derivatives, was lower than that in the mechanically disrupted garbage. Considering that low molecular organic acids can be easily decomposed by anaerobic digestion, these results show that the anaerobic digestion of the liquid phase could contribute to a higher digestion rate compared with that of the mechanically disrupted garbage.

Introduction

The kitchen garbage can be converted from a solid to a liquid slurry by thermochemical liquidization at 150-250 °C.[1] The liquidization of kitchen garbage has wide range of applications, such as pretreatment for transportation using pipelines and anaerobic digestion. By applying the liquidization process, we examined the anaerobic digestion of model kitchen garbage and the filtrate of the liquidized model kitchen garbage. The anaerobic treatment of the filtrate indicated fast digestion compared with the mechanically broken up original model kitchen garbage.[2] Upflow anaerobic sludge blanket (UASB) systems have been widely used for the high-rate treatment of various organic wastewaters since UASB reactors have exhibited superior performance compared to other types of anaerobic reactors when utilized with high loading rates. By using this system, the filtrate of the liquidized model kitchen garbage was continuously digested for 82 days.[3] The anaerobic digestion of a liquid phase of the liquidized model kitchen garbage exhibits fast digestion rates. An analysis of the organic constitute of the liquidized food waste during the liquidization process is important for providing preliminary information and optimum conditions about the potential for the pretreatment of the anaerobic digestion.

Experimental

Model kitchen garbage

Garbage discharged from homes or food industries has various properties, depending on the season or region. In this work, model garbage[4-6] was used as a starting material. The components of the model garbage were sliced cabbage (92.4%), boiled rice (5.3%), dried fish (0.6%), shell (1.1%) and butter (0.6%). Table 1 shows the properties of model kitchen garbage. The moisture content was determined by weighing the garbage after drying at 105 °C for 24 h. The ash content was determined by weighing the residue after heating at 600 °C for 1 h, and the organic content was obtained by difference.

Table 1 Properties of model kitchen garbage

	Model garbage (%)
Moisture content	87.1
Organic	11.3
Ash	1.6

Mechanically disrupted model kitchen garbage

Mechanically disrupted model kitchen garbage was obtained by the addition of the

Table 2 Properties of mechanically disrupted garbage and its filtered phases

	Mechanically disrupted garbage (%)	Liquid phase (%)	Solid phase (%)
Moisture content	93.8	96.5	83.6
Organic	5.9	3.4	15.7
Ash	0.3	0.1	0.7

same weight of distilled water to the model food waste followed by disruption using a mixer for 1 min.[2] The product was filtered into a liquid phase (81.0 wt%) and a solid phase (19.0 wt%) using a 20 mm pore size nylon mesh. Table 2 shows the properties of mechanically disrupted garbage and its filtered phases.

Liquidization of model kitchen garbage

The model kitchen garbage was placed into a 1 L autoclave made of stainless steel. After purging with nitrogen gas, it was charged to 2 MPa. The reaction was started by heating the autoclave with an electric furnace. After heating the autoclave to 175 ℃, the temperature was maintained for 1 h and then the autoclave was cooled to room temperature. Separation of the liquidized model kitchen garbage into solid and liquid phases was conducted by filtration using a 20 μm pore size nylon mesh. Table 3 shows the properties of liquidized model kitchen garbage and its filtered phases.

Table 3 Properties of liquidized model kitchen garbage and its filtered phases

	Liquidized garbage (%)	Liquid phase (%)	Solid phase (%)
Moisture content	89.7	93.2	74.0
Organic	9.7	6.0	25.1
Ash	0.6	0.8	0.9

Analysis of water solubles in the liquid phases

The carbon amount in the liquid phase was measured using a total organic carbon (TOC) meter (Shimazu, TOC-5000A). Organic acids were analyzed with a Dionex DX-500 ion chromatograph. The total sugar content was measured by the method of Dubois *et al.*.[7] Sugars were analyzed by using a high performance liquid chromatography (HPLC) (Shimazu, LC-10AT).

Results and Discussion

Liquidization

The TOC in the liquid phase of mechanically disrupted garbage and liquidized garbage in the liquid phase were 12803 and 31479 mg/l, respectively. The total organic carbon (TOC) for the liquid garbage was 2.5 times as that for the mechanically disrupted garbage. However, the addition of water of same weight as the model garbage was needed to obtained the mechanically disrupted garbage. The total carbon amount in the liquid phase of mechanically disrupted model garbage and liquidized garbage on the basis of 100 g-model garbage were 2168.1 and 2297.0 mg, respectively (shown in Table 4). The water soluble organic carbon amount was similar value in spite of the liquidization; the solubilization of solid and/or insoluble material did not occur by the liquidization process. Minowa *et al.* reported that liquidization phenomenon of garbage is due to the thermal rupture of cells; thus releasing intra-cell water from the cells.[8]

Table 4 TOC in the liquid phases (in 100 g-model garbage)

Organics	Model garbage (liquid phase)	Liquidized garbage (liquid phase)
TOC	2168.1 mg-C	2297.0 mg -C

Organic acids in the liquid phases

Figure 1 shows the concentration of organic acids in the liquid phases as measured by ion chromatography. The low molecular organic acids concentration in the liquidized model kitchen garbage is higher than that in the mechanically disrupted garbage. It has been reported that protein denatures at 150 °C[9] and proteins in sewage sludge hydrolyze during the liquidization process at 175 °C.[10] Krochta *et al.* reported the thermochemical degradation of cellulose into organic acids.[11] Therefore, organic acids could be obtained by the hydrolysis of protein and the thermochemical degradation of sugar. The major phenomenon of liquidization is rupture of cells. The decomposition of soluble material might have occurred by minor reaction, such as hydrolysis.

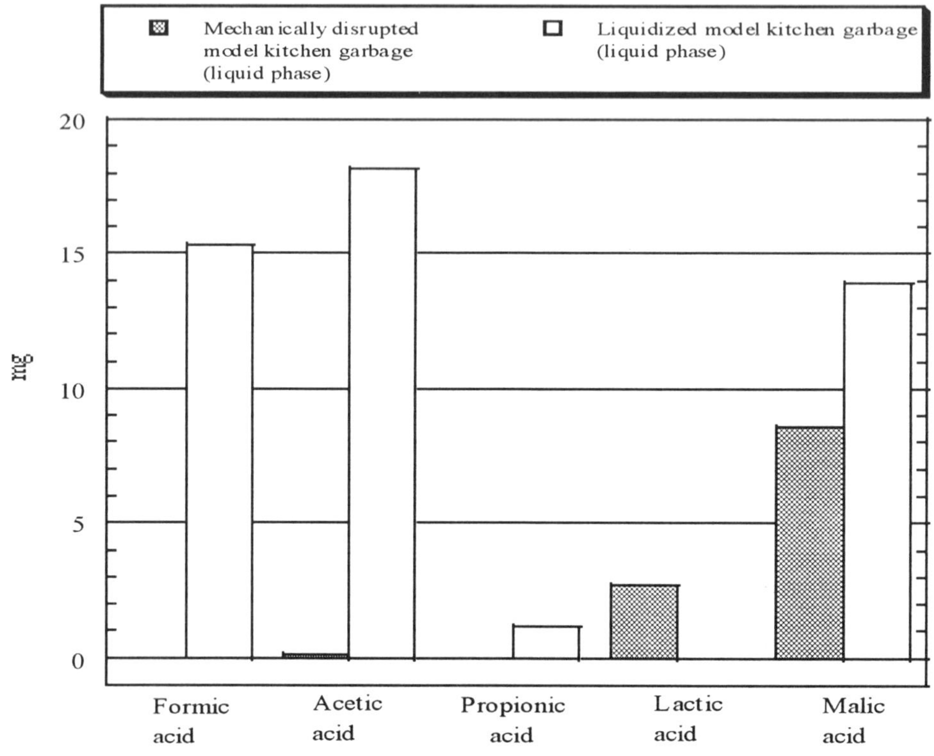

Figure 1 Organic acids concentration in the liquid phases (in 100 g-model garbage)

Sugar in the liquid phases

Table 5 shows the total sugar content in the liquid phases. The total sugar amount in the liquid phase for liquidized model kitchen garbage, including simple sugars, oligosaccharides, polysaccharides, and their derivatives, is lower than that for the mechanically disrupted garbage. The results of the HPLC showed that the fructose and glucose contents decreased during the process. (shown to Figure 2) They were decomposed to other compounds, such as sugar derivatives and acids. The decrease of sugar amount consists with the increase of organic acids amount as mentioned above.

Table 5 Total sugar content in the liquid phases (in 100 g-model garbage)

Organics	Model garbage (liquid phase)	Liquidized garbage (liquid phase)
Sugar amount	89.9 mg	59.9 mg

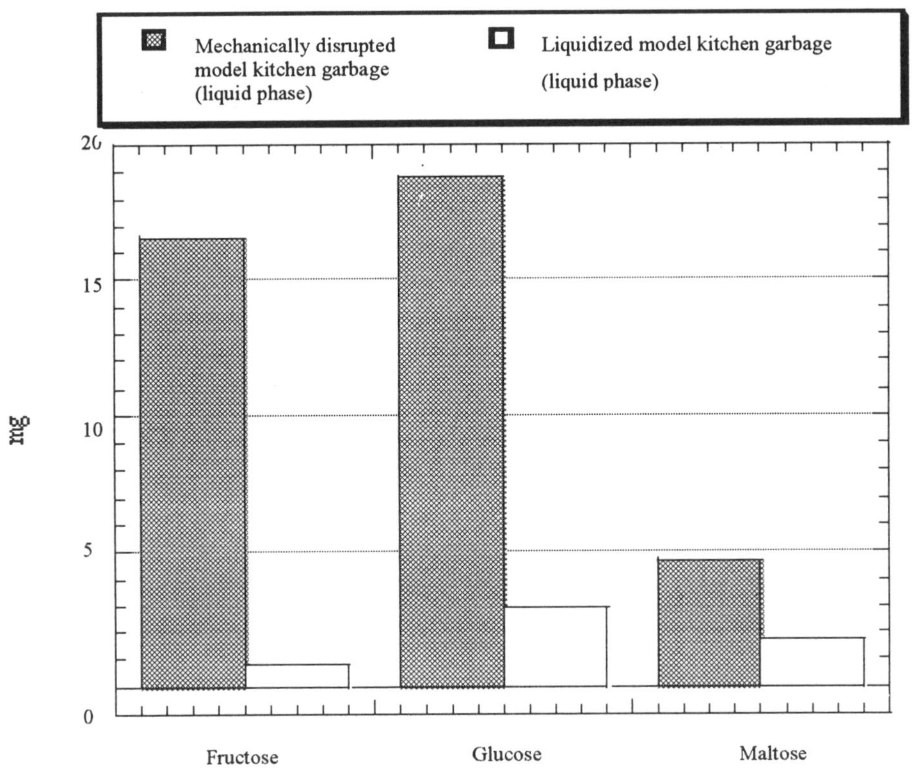

Figure 2 Sugar concentration in the liquid phases (in 100 g-model garbage)

Application of liquidized garbage

As mentioned above, additional water is needed for mechanical disrupting. One of the advantages of the liquidization is that the model garbage is changed to liquid slurry without an increase of volume.

The liquid phase of liquidized model kitchen garbage contained mainly organic acids and sugars and their derivatives. In our previous study, anaerobic digestion of liquid phases obtained from liquidization of the model kitchen garbage was reported. The biogas yield from the liquid phase obtained from the liquidized model kitchen garbage was higher than that from the mechanically disrupted garbage.[2] Furthermore, the liquid phase was continuously anaerobically digested using a UASB method.[3] The decomposition and solubilization of solid organic materials is one of the key steps for anaerobic digestion. These results suggest that the anaerobic digestion of liquid phase could contribute to a higher digestion rate compared with that of mechanically disrupted garbage.

Conclusion

Model kitchen garbage was liquidized by a thermochemical process. The organic composition of liquidized model kitchen garbage and its separated phases were analyzed. The following conclusions were obtained:

(1) The liquidization phenomenon is due to the thermal rupture of cells; the intra-cell water could then be released from the cell.

(2) The low molecular organic acids concentration in the liquidized model kitchen garbage is higher than that in the mechanically disrupted garbage. The increment of the low molecular organic acids might be due to the degradation of the proteins and sugars.

(3) The sugar concentration in the liquidized model kitchen garbage is lower than that in the mechanically disrupted garbage.

Acknowledgements

We thank Ms. Yukiko Fukuda, an assistant staff member, for her help during the experiments.

References

1. Minowa T., Murakami M., Dote Y., Ogi T. and Yokoyama S. (1995) Oil production from garbage by thermochemical liquefaction. *Biomass Bioenergy*, **8** (2) 117-120
2. Sawayama S., Inoue S., Minowa T., Tsukahara K. and Ogi T. (1997) Thermochemical liquidization and anaerobic treatment of kitchen garbage. *J. Ferment. Bioeng.*, **83** (5) 451-455

3. Tsukahara K., Yagishita T., Ogi T. and Sawayama S. (1999) Treatment of liquid fraction separated from liquidized food waste in an upflow anaerobic sludge blanket reactor. *J. Biosci. Bioeng.*, **87**, (4) 554-556
4. Koyama K, Tanaka N. and Inoue Y. (1983) Study on anaerobic decomposition of municipal solid waste. *Proc. Environ. Sani. Engng Res.* **19**, 136-145
5. Tanaka E., and Morishita K.(1985) Citizens and waste disposal. *Environ. Res. Quart.* **56**, 79-88
6. Tanigawa N., Wakabayashi M. and Noguchi T. (1989) 1988 FY research of waste composition. *Report of Tokyo Metropolitan Cleansing Laboratory*, pp.1-42
7. Dubois M., Gilles K. A., Hamilton J. K., Rebers P. A. and Smith F. (1956) Colormetric method for determination of sugars and related substrances. *Anal. Chem.*, **28**, (3), 350-356
8. Minowa T., Dote Y., Sawayama S., Yokoyama S. and Murakami M. (1995) Phase changing of garbage from solid to liquid slurry by thermal liquidization. *J. Chem. Eng. of Japan*, **28** (6) 727-731
9. Dote Y., Inoue S., Ogi T. and Yokoyama S. (1996) Studies on the direct liquefaction of protein-contained biomass: The distribution of nitrogen in the products. *Biomass Bioenergy*, **11** (6) 491-498
10. Inoue S., Sawayama S., Ogi T. and Yokoyama S. (1996) Organic composition of liquidized sewage sludge. *Biomass Bioenergy*, **10** (1) 37-40
11. Krochta J. M., Hudson J. S. and Drake C. W. (1984) Alkaline thermochemical degradation of cellulose to organic acids. *Biotechnol. Bioeng. Symp.*, **14,** 37-54

The volatility of tars from pyrolysis of biomass materials

V. Oja[1] and M. R. Hajaligol
Research Center, Philip Morris USA, Richmond, Virginia, USA.
[1] *Current Address: Chemical Engineering Department, Tallinn Technical University, Tallinn, Estonia.*

ABSTRACT: The volatility of tars from pyrolysis of biomass and biomass-derived materials is of special interest. This interest is related to the question whether or not biomass tar evaporation has a significant effect on the total rate escape of tar from pyrolyzing substance and therefore on pyrolysis kinetics. In fact, in many practical applications (especially in fossil fuel thermal conversion processes), tar vaporization is known to be an important step during pyrolysis that influences both yield and composition of pyrolysis products.

As part of the answer lies in ability to describe the volatility of biomass tars, the work was primarly undertaken to get some insights into this area. Due to the very complex nature of pyrolysis tars, there are not suitable estimation methods available - experimental data are needed for further development of these techniques. The experimental data are needed to increase our understanding of this phenomenon. In this paper, we present results from tobacco, hemicellulose and lignin tars.

Our preliminary results indicate that the heats of vaporization of the "low volatility" fractions studied here fall somewhere between 110 and 140 kJ/mol. Therefore, incorporation of the heats of vaporization into the global activation energy should be relevant for an accurate assessment of pyrolysis kinetics.

INTRODUCTION

Understanding the volatility of tars from pyrolysis of biomass materials should be an important aspect in the design of bimass thermal conversion processes, including gasification and combustion. So far, the effect of external transport limitation due to tar vaporization has been quantitatively described for coal pyrolysis and this transport limitation has been shown to cause changes in pyrolysis product yields, compositions and even in pyrolysis kinetics. Coal (as well as biomass) is a macromolecular network in which thermal degradation in a non-oxidizing environment includes, among other processes, formation of volatile (semi-volatile) fragments or species. Smaller and higher volatile fragments vaporize, bigger and less volatile tend to form char as a result of longer residence times. It could be said that there is a competition between vaporization and char forming reactions. In some cases (especially in softening coals) the impact on product yield and kinetics due to vaporization is more significant than in others (non-softening coals), but qualitatively the external mass transport limitation is correct for all fossil fuels including lignites [Suuberg *et al.,* 1978]. Therefore, advanced pyrolysis modelds originally developed for coal pyrolysis [Serio *et al.*, 1987; Niksa and Kerstein, 1987; Grant et al., 1989] include tar vaporization as one relevant step in the modeling schemes and use vapor pressure correlations to predict volatilities of pyrolysis tars. Recently, some of these advanced models have been applied to pyrolysis of biomass materials [Serio *et al.*, 1994; Chen *et al.*, 1998]. One modeling code to be improved for the advanced pyrolysis models is the ability to predict tar volatilities. The complex nature of pyrolysis tars makes estimation of the effects of tar volatility unpredictable without experimental evidence for each individual biomass tar.

There is evidence that external transport limitations due tar vaporization contribute to the overall kinetics of bimass pyrolysis. Suuberg *et al.* [1996, 1997] first addressed this problem by measuring cellulose tar volatility. They hypothesized that tar escape via vaporization was an important parameter in cellulose pyrolysis and therefore in determining the global mass loss kinetics at high heating rates. However, in this respect, it should be remembered that opposite to the other type of biomasses, cellulose tars, depending on the pyrolysis conditions, can contain a significant amount of low volatility levoglucosan and its condensation products (up to 60 %). In present context it may be stated that the volatile species from biomass pyrolysis have much smaller molecular weights and therefore are much more volatile than coal tars. On the other hand, biomass tars generally have much higher heteroatom content, which is known to lower the volatility significantly, especially when hydrogen bonds forming functional groups are present. The subject of tar vaporization in biomass pyrolysis should be given more attention.

EXPERIMENTAL

SAMPLES

The tars studied were generated from lignin, xylan, and tobacco starting materials. These afforded representative samples of a wide variety of tar chemical and physical properties including a cellulose-like material, hemicellulose, a sample with a high phenolic content (lignin) and a material with a large nitrogen content – tobacco. Tobacco used in this study was a low alkaloid tobacco (to minimize the contribution of the nicotine to tar volatility). Birch wood xylan is representative of hemicellulose, and was purchased from Fluka. Alkali lignin was obtained from Aldrich.

TAR PREPARATION

This study was primarily focused on the volatility of tars that were prepared under conditions that minimize the extent of secondary reactions. Tars were produced using a fast pyrolysis technique (heating rate greater than 5 °C/sec). About 2 or 3 grams of sample on a wire mesh holder was pushed into the pre-heated tubular pyrolysis reactor. The sample temperature was measured by a thermocouple and the holder was pulled out from the heated region after the sample reached desired temperature. The final temperature was selected based on thermogravimetric experiments at heating rate of 20 °C /min for each material studied and are shown in Table 1. The inert gas (N_2) flow was 300 ml/min at room temperature. That corresponds to the gas phase tar maximum residence time of 3 seconds in the hot zone of the reactor. This approach was selected in order to produce enough material to analyze, but at the same time to minimize the extent of secondary reactions.

Volatile room temperature condensibles pyrolysis products were collected at cold parts within the reactor and in a collecting system made of a liquid trap followed by a Cambridge filter. The liquid trap contained inhibitor free tetrahydrofuran (THF) at 0°C. The filter and the cold parts of reactor were washed with inhibitor free THF. Dry tars were prepared by evaporating the solvent in a vacuum oven at about 45°C.

DEFINITION OF TAR

The definition of tar is arbitrary and operationally defined. The tar preparation technique used resulted in loss of light materials (low molecular weight) or the "high volatility" fraction of room temperature condensable organic pyrolysis products -- the total tar. The material remaining, or "low volatility fraction", was defined here as the tar to be analyzed. The relative amount of "low volatility" and "high volatility" fractions in the total tar depends mostly on the starting materials and pyrolysis conditions. For example, in coal pyrolysis, the "high volatility" fraction is no more than 5-10 weight percent of the total tar, but in the case of biomass, it can contribute up to 80% by weight. Table 1 shows the yields of pyrolysis products based on dry and ash free starting material.

Table 1. Yields of biomass pyrolysis products (dry and ash free bases).

Material	T_{final} [°C]	TGA	Tubular reactor		
		Char yield [wt%]	Char yield [wt%]	Gas + "high volatility" tar [wt%]	Collected "low volatility" tar [wt%]
tobacco	550	24	24	66.4	9.6
lignin	525	49	50	36	14
xylan	450	30	31	58	11

To get a better estimate of the amount of "low volatility fraction" in total room temperature condensable products, crude semi-quantification of evolved gases (only CO, CO_2, H_2O) was carried out in a thermogravimetric analyzer (TGA) coupled with a mass spectrometry. This estimation was based on mass spectrometric data of pyrolysis volatile products at a heating rate of 20 °C/min. We are aware that the gas yield during pyrolysis varies somewhat with heating rates, but here our aim is to obtain an approximation. Based on our estimations the amount of gases (CO, CO_2 and H_2O) were 40% for tobacco, 25% for lignin and 10% for xylan of the total weight. This suggests that the amount of "low volatility" tar in the total tar would be about 30% for tobacco, 55% for lignin and 20% for xylan by weight.

It should be mentioned that a sharp borderline between "low volatility" and "high volatility" fractions does not exist. There is a continuous change that is sensitive to temperature and vacuum fluctuations/uncertainties in a vacuum oven. Therefore, based on our observations, the amount of "low volatile" fractions might be up to 15 to 20% greater than the values shown if conditions were such that the maximum amount of "low volatility" tars was produced.

RESULTS AND DISCUSSION

Biomass tars are known to be very complex mixtures which contain significant amounts of polar compounds that can potentially form hydroxyl bonds. This means that current vapor pressure estimation techniques are not only difficult to apply, but are inaccurate for mixtures with significant oxygen functionalities. Therefore we have chosen to approach this problem from an experimental standpoint.

BIOMASS TAR VOLATILITY

It is known that volatility (vapor pressure) of tars depends to a great extent on molecular weight (on sum of van der Waals forces). Therefore, one of the objectives of our volatility concept was the volatility relation to the molecular weight distribution of the volatilizing material. Field Ionazation Mass Spectrometric technique was adopted for this purpose. This technique is especially useful because in addition to volatility data, it provides molecular weight distribution information. As our main interest was vaporization data, the prepared tars were reheated in the Field Ionization Mass Spectrometry (FIMS) inlet probe at a heating rate of 3°C/min. This differs somewhat from traditional molecular weight distribution determination experiments by FIMS, where samples usually are pyrolysed within the inlet of the mass spectrometer.
Figure 1 shows vaporization behavior for tars at pressure 10^{-4} torr at a heating rate of 3 °C/min. These are relative vaporization data that relate volatilities of tars to the number average molecular weights under these specific experimental conditions.

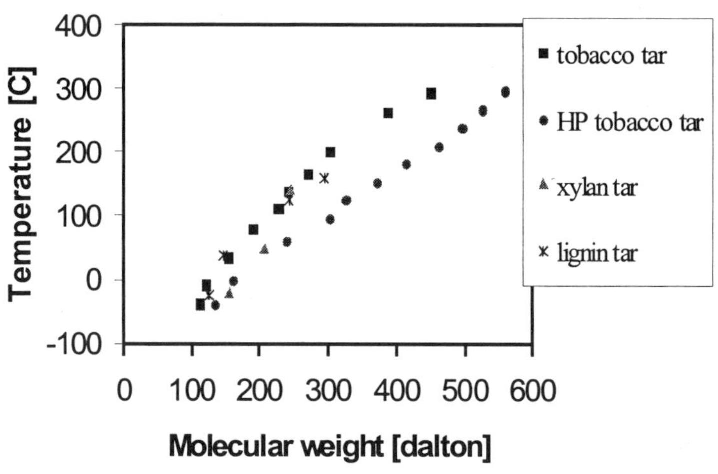

Fig 1. Temperatures of volatility as a function of molecular weight.

Some general outcomes can be drawn from the data:

(1) First, volatility depends to a significant extent upon the molecular weight of the tar species. As a first approximation, it can be assumed that all compounds with the same molecular weight generate similar vapor pressures. From this it fallows that for any tar, it is reasonable to say that an average molecule with higher molecular

weight has lower volatility, that is vapor pressure decreases as molecular weight increases.

(2) Second, the overall tar vaporization processes (positions of the vaporization curves in *Fig. 1*) is determined by the chemical composition of the tar. The comparison between tobacco tar (solid squares) and its hexane phase partitioned fraction (referred as "HP tobacco tar"; solid circles) indicates clearly this important effect of chemical composition and functional groups. Note that the hexane phase partitioned fraction represents the less polar fraction of tobacco tar obtained by partitioning tar species from methanol solution into a hexane phase.

These data indicate that one cannot describe volatility without obtaining information on molecular weight distribution and chemical composition of the vaporizing species. In order to more conveniently interpret the data in the *Fig. 1*, following are some characteristics of tars studied. Molecular weight distributions fitted form FIMS data on the "low volatility" fraction of tars are shown in *Fig. 2*.

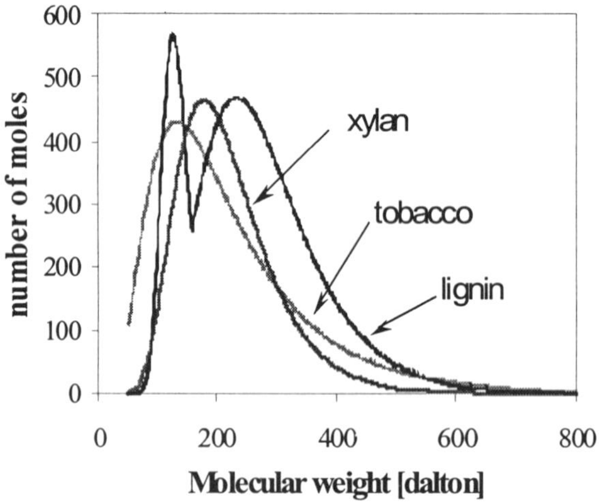

Fig 2. Molecular weight distributions fitted from FIMS data.

This figure shows that all three tars contain a quite broad range of pyrolysis products with very different molecular weights and the distribution curve shape and broadness varies from tar to tar. At this point it is worth noting that presenting the FIMS data as a molecular weight distribution is only a rough approximation. Although practically useful, the assumption of continuous spectra of compounds may not be adequate for hemicellulose and lignin tars -- these tars contain strong signals of individual compounds. Table 2 shows elemental composition for each tar studied here.

Table 2. Elemental analysis results for tars studied (dry bases).

tar	MW	C	H	O	N	S	H/C	O/C	N/C
tobacco	219	71.71	7.03	11.3	9.95		1.18	0.12	0.119
lignin	246	75.93	6.34	15.36	0.78	1.59	1.00	0.15	0.009
xylan	207	69.99	6.56	23.45	0		1.12	0.25	

The oxygen content indicates that these tars are quite polar in nature. For example, an average molecule of tobacco tar would have 3, lignin tar 2.5 and xylan tar 3 polar atoms, which are likely to form hydrogen bonds. H/C ratios between 1 and 1.2 indicate cyclic and saturated or aromatic structures. This might suggest that some secondary reactions had occurred despite our efforts to avoid these. Therefore, the work currently in progress is expected to overcome this problem by using a wire mesh reactor (also referred as a heated grid reactor) instead of the previously used tubular reactor.

HEATS OF VAPORIZATION AND BOILING POINT ANALYSIS

The non-isothermal Knudsen effusion technique was used to study vapor pressures of tars. The experimental details have been described previously [Oja and Suuberg 1997, 1998]. About 10 mg of dry tar was placed into a hermetic effusion cell with a small orifice, from which the saturated vapor effuses out into the vacuum outside the cell. To overcome effects caused by changes in tar composition during effusion, the experiment involved first a continuous cool-down followed by a continuous heat-up of the sample. From the mass loss data (by taking into account both cool-down and heat-up as a whole cycle) the vapor pressure was calculated using the Knudsen equation.

The goal was to determine vapor pressures of tars cycle-by-cycle, where between each cycle, certain amounts of higher volatility species were evaporated. As the effusion technique built in this laboratory was best suited for measuring vapor pressures from 10^{-5} to 10^{-3} torr, the temperature had to be continuously increased as more and more volatile species were lost in the process. The process is further complicated by the fact that pyrolysis tars generally have a tendency to "age" with time, especially, at higher temperatures. It is known that condensation/polymerization type reactions become considerably facile at temperatures above 100 °C.

Therefore, the main problem related to the experimental determination of vapor pressure was the thermal instability of tars. Significant signs of thermal instabilities for all three tars were observed around 140 –150 °C. Because of the tar thermal instabilities, we present here only results from vapor pressure measurements performed below 110°C and between 110 and 140 °C.

Table 3 shows heats of vaporization and boiling points determined from vapor pressure measurements below 110 °C. It is reasonable to assume that tars under these conditions can be considered "fresh" i.e. no significant condensation type reactions had taken place.

Table 3. Results from vapor pressure measurements below 110 °C.

sample	Mass loss [wt.%]	Heat of vaporization [kJ/mol]	boiling point at 760 torr [°C]	boiling point at 10 torr [°C]
tobacco	6%	132	234	170
lignin	6%	119	292	210
xylan	4%	121	307	220

Note that the values correspond to the tars that had lost 4-6% of their initial mass. The boiling point was estimated assuming that the heat of vaporization is constant.

Table 4 shows heats of vaporization and estimated boiling point values from vapor pressure measurements between 110 and 140 °C.

Table 4. Results from vapor pressure measurements between 110 and 140°C.

sample	Mass loss [wt %]	Heat of vaporization [kJ/mol]	boiling point at 760 torr [°C]	Boiling point at 10 torr [°C]
tobacco	24%	129	345	250
lignin	19%	138	335	250
xylan	15%	111	420	290

As evident from above data, the boiling points increase as molecular weight increases as a result of losing higher volatility spieces. In order to more conveniently interpret the data shown above, Table 5 shows pyrolysis decomposition temperatures and differential mass loss peak temperatures for the materials studied. The temperatures were obtained by thermo-gravimetric analysis at low heating rates.

Table 5. Differential weight loss results from biomass pyrolysis.

sample	Start of weight loss [°C]	Major peaks [°C]
tobacco	150	275 and 315
lignin	200	335
xylan	190	240 and 290

At this point, it is worth of noting that typically vaporization enthalpies tend to increase as the vapor pressure decreases. However, the heat of vaporization of possibly "aged" tar (Table 4) was similar to the "fresh" tar (Table 3) in the case of tobacco or lower in the case of xylan, even if the vapor pressure lowered significantly (by a factor of ca. 10). This indicates changes in chemical structure (mainly in polar interactions), which are due to loss of volatile species, as well as from condensation type reactions.

On the bases of the present work the heats of vaporization of the "low volatility" fractions fall somewhere between 110 and 140 kJ/mol. It should be emphasized here that Milosavljevic and Suuberg [1995] suggested that primary tar forming reactions were nearly thermo-neutral in the case of cellulose pyrolysis. The results of our co-workers from the Combustion Research Laboratory indicate that the pyrolysis of other biomasses is less endothermic than cellulose. Therefore, incorporation of heats of vaporization into the global activation energy should be relevant for an accurate assessment of pyrolysis kinetics.

SUMMARY

The situation regarding the biomass tar vaporization is more complex, and the work above was an attempt to bring some insights into biomass tar volatility. We believe that the question of biomass tar volatility should be given more attention in the discussion of yields of biomass pyrolysis products and even pyrolysis kinetics.

More specific analysis of tars and correlations of vapor pressure data as a function of molecular weight or/and some chemical characteristics were beyond the scope of this presentation, although these are important considerations. Detailed pyrolysis models [Chen et al., 1998; Serio et al., 1994] include a tar vaporization step and use correlations for predictions of vapor pressures of primary pyrolysis tars. Unfortunately, very few experimental data to develop such correlations are available.

REFERENCES

Chen, Y., Charpenay, S., Jensen, A., Wotjowicz, M.A., and Serio, M.A. (1998) *27th Symposium (Int.) on Combustion,* p. 1327, The Combustion Institute.

Grant D.M., Pugmire, R.J., Fletcher, T.H., and Kerstein, A.R. (1989) *Energy Fuels* **3**, 175.

Milosavljevic, I. and Suuberg, E.M. (1995) *Ind. Eng. Chem. Res.* **34**, 1081.

Niksa, S. and Kerstein, A.R. (1987) *Fuel* **66**, 1389.

Oja, V. and Suuberg, E.M.(1997) *Anal. Chem.* **69**, 4619.

Oja, V. and Suuberg, E.M.(1998) *Energy and Fuels* **12**, 1313.

Serio, M.A., Hamblen, D.G., Markham, J.R., and Solomon, P.R. (1987) *Energy Fuels* **1**, 138.

Serio, M.A., Charpenay, S., Bassilakis, R., and Solomon, P.R. (1994) *J. Biomass Bioenergy* 7, 107.

Suubereg, E.M., Milosavljevic, I., and Oja, V. (1996) 26th Symposium (Int.) on Combustion, p. 1515, The Combustion Institute.

Suuberg, E.M., Peters, W.A., and Howard, J.B. (1978) *Ind. Eng. Chem. Process Des. Dev.* **17**, 37.

Release of Chlorine From Biomass and Model Compounds at Pyrolysis and Gasification Conditions

B. Strömberg and F.Zintl,
TPS Termiska Processer AB, Studsvik S-611 82 Nyköping; Sweden

ABSTRACT Experiments with KCl-pretreated biomass and different organic model compounds have shown that they release chlorine when they are heated to moderate temperature levels (300-500°C). The experiments included both pyrolysis and gasification conditions. Both low-molecular model compounds and macromolecular compounds were included, of both biological and technical origin. The results allowed a ranking of the functional groups concerning their contribution to the release of Cl from the materials. The following functional groups contribute significantly to the release of chlorine when heating the samples to 300 °C:

At pyrolysis conditions:

Free carboxylic acid > ester in lactone ring ≅ hemiacetal > acetal ≅ open ester ≅ hydroxymethyl on ring or chain >> terminal hydroxymethyl

At gasification conditions:

Hemiacetal > free carboxylic acid ≅ ester in lactone ring ≅ acetal > open ester.

It must be noted that all common biomasses contain biopolymers (carbohydrates or others) which are rich in these functional groups. Some of them as for example crops which have been fertilised with KCl, can contain elevated amounts of chlorine. The release of significant parts of this chlorine from chlorine containing fuels can probably not be suppressed by pre-treatment under pyrolysis or gasification conditions at low temperature.

INTRODUCTION

Emission of hydrogen chloride is the third most important contribution to the global acidification from human activities. The two first are SO_2 and NO_x. The HCl is a local pollutant, contrary to the other two. It is soluble in water and easily dissolved in rain droplets and, therefore, usually falls down near the emission source. The hydrogen chlorine emissions from combustion and gasification processes has been calculated to 3.5 Mt./year. The major part of the estimated global contribution of HCl to the atmosphere is evaporation from the oceans. Even with a redeponation of 90-% HCl to the oceans, the estimated emission will reach approximately 120 Mt./year. The majority of the emitted chlorine from a combustion process will leave as HCl in the gas phase which may cause problems like corrosion and formation of dioxins.

Both the amount and the origin of the chlorine varies between different types of fuels. In coal, the concentration varies normally between 50-2000 mg/kg. The origin is mainly groundwater which has been incorporated into the coal after its formation.

In biomass the chlorine content can vary from less than 100 mg/kg up to 7000 mg/kg [1]. Biomass fuels show comparably low ash content. The uncombustible fuel components which form the ash are dominated by compounds of alkaline and alkaline earth metals (K,Na,Ca,Mg) and non-metallic anion-forming elements (Cl,S,P,Si).

Woody biomass is low in ash content, sulphur, chlorine and silica, and the ash itself is dominated by compounds of Ca and K. Biomass fuels from crops are higher in ash content, sulphur, chlorine and silica. The ash is dominated by oxides of Ca, K and Si. The fuels based on the growing parts of trees (bark+debries) show ash compositions between the two extremes, wood and acre crops. Therefore the largest chlorine emissions are expected from energy crops (grasses) and from agricultural by products (straw etc).

Little is known about the bonding and chemical forms of the ash-constituting elements in the plant materials. The alkaline metal ions can easily be leached from the fuels and are probably present as inorganic or organic salts, or associated to parts of the organic macromolecular structures.

A first hypothesis in this project has been that the main part of the chlorine in the biomass fuels is present as simple chloride ions, forming some type of salts with inorganic or organic cations. The high degree of water-leachability of Cl from the acre crops suggests this type of weak bonding in water-soluble salts. Since K is the ash constituent with the highest affinity to chlorine, easily forming KCl, it was supposed that Cl was present in the fuels mainly as KCl, or that this salt was formed by reaction after degradation of other forms of potassium and chlorine when the fuel was burnt or pyrolysed. It would then be expected that the release of chlorine or chlorine compounds (even HCl) to the gaseous phase during thermal treatment of biomass would be limited by the equilibrium vapour pressure of the potassium chloride. Very little volatilisation of Cl was therefore expected for temperature lower than the melting point of KCl (770°C), since the vapour pressure of solid KCl is negligible.

A number of chlorine-rich biomass fuels were studied in small-scale laboratory pyrolysis experiments. Contrary to the above mentioned hypothesis it was shown that during pyrolysis of biomass 20 to 50% of the total chlorine evaporated already at 400°C.[2]. See Figure 1.

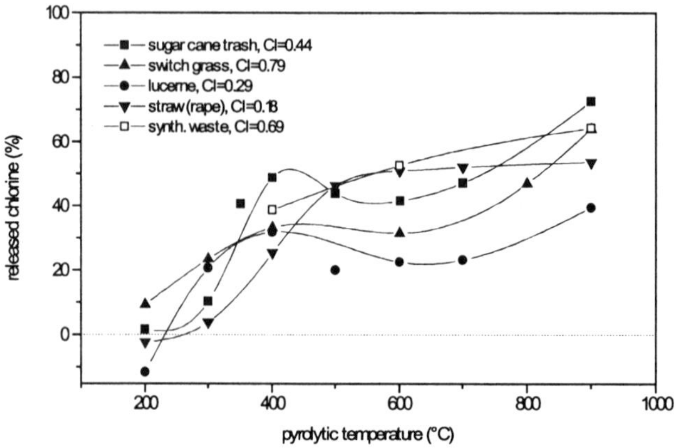

Fig. 1 Percentage of released chlorine as a function of the pyrolytic temperature for different biomasses.

To explain the release of chlorine at low temperature, a mechanism has been suggested in which the chlorine in the fuel during pyrolysis can be transformed to the gaseous and volatile hydrogen chloride.[2] This mechanism suggests that:

(1) The organic matter in the fuel contains or develops under pyrolysis a number of carboxylic acid (-COOH) groups.
(2) These carboxylic groups react with KCl in an ion exchange reaction which sets free HCl:

$$R\text{-}COOH(s) + KCl(s) \rightarrow R\text{-}COOK(s) + HCl(g)$$

(3) HCl is then evaporated to the gas phase. The potassium carboxylate is more stable but can be deteriorated in a later step e.g.

$$R\text{-}COOK \rightarrow R + CO_2(g) + K(g)$$

The hypothesis is that the pyrolytic thermal degradation of the biofuels leads to the formation of chain-bound carboxylic groups which then react with KCl or NaCl under formation of alkaline carboxylate and gaseous HCl. This mechanism would then explain:

(1) The early release of a significant part of the fuel chlorine during biomass pyrolysis.
(2) The observation that the early release of Cl not requires any hetero atoms (e.g. N,S,P) in the structure.
(3) Why the early release of Cl needs a threshold temperature when the Cl and the alkaline metal ions are present in an organic material without initially accessible

carboxylic acid groups. These groups are then formed on heating and thermal degradation.

The suggested mechanism was confirmed in series of experiments with three model compounds cellulose, oxicellulose, and xylan [3]. In the present work a great number of new model compounds were chosen to study the activity of different functional groups more in detail. The aim with the work was to verify the suggested mechanism and to investigate the influence from other functional groups on the early release of chlorine from organic materials.

EXPERIMENTAL

METHOD

The model compounds were all pretreated with a saturated and filtrated solution of potassium chloride KCl (2%) in methanol. A solution in methanol was chosen to establish a good contact of the materials with the KCl. Water solution was avoided, since the methanol can more easily be removed from the samples without heating.

Approximately 2 g. of the desired model compounds were heated in a stream of dry nitrogen (pyrolysis) or of a mixture of 92 % nitrogen and 8% oxygen (gasification). The sample was placed in a combustion boat of porcelain. The boat was heated in a quartz tube which was placed in an electrically heated furnace. The furnace was heated with 50°C/min to the desired temperature which was kept for 30 minutes, and the sample was cooled under nitrogen and weighed. The fraction of solid residue left in the vessel was calculated and analysed on its chlorine content.

Whenever possible, the three temperature levels 300, 400, and 500 °C were studied for all materials and both gas atmospheres. In some cases the experiments failed, or were impossible. This was the case for e.g. adipic acid at temperatures above 300 °C, since the vapor pressure of adipic acid is too high

MODEL COMPOUNDS

The model compounds were chosen so that they contained structural elements and functional groups which are usual in the natural materials that constitute the biomasses. Both macromolecular (polymeric) and low-molecular compounds were chosen. Some of the model compounds were natural materials or extracted from them, some were synthetic, some were natural materials after chemical treatment. Table 1 gives a list over all model compounds, and which functional groups they contain.

Table 1: Model compounds and their functional groups.

Model Compound	Type	Functional groups
Cellulose	BM	exocyclic hydroxymethyl groups, acetal
Oxicellulose	BM-CT	free carboxylic acid groups, acetal
Xylan	BM	acetal
Cellulose triacetate [CTA]	BM-CT	acetal, open ester
Pectin	BM	free carboxylic acid, open ester, acetal
Poly(acrylic acid) [PAA]	SM	free carboxylic acid on polymer chain
Poly(methyl methacrylate) [PMMA]	SM	open ester, attached to polymer chain
Copolymer of ethylene and acrylic acid [PE-PAA]	SM	free carboxylic acid attached to polymer chain
Polyesterpolyol [Poly-poly]	SM	hydroxymethyl groups attached to polymer chain
ester groups building the chain		
Polyvinyl butyral [PVB]	SM	acetal groups (rings) attached to polymer chain
Glucose	BM	exocyclic hydroxymethyl group, hemiacetal
Sucrose	BM	exocyclic hydroxymethyl groups, acetal,
hemiacetal		
Maltose	BM	exocyclic hydroxymethyl groups, acetal,
hemiacetal		
Lactose	BM	exocyclic hydroxymethyl groups, acetal,
hemiacetal		
Cellobiose	BM	exocyclic hydroxymethyl groups, acetal,
hemiacetal		
Sorbitol	BM-CT	terminal hydroxymethyl groups
Gluconolacton	BM-CT	ring-constituting ester (lacton), hydroxymethyl
group attached to lacton ring		
Mucic acid	BM-CT	free carboxylic acid
Adipic acid	SM	free carboxylic acid

BM: *biological material or extracted from such material*
BM-CT: *biological material, modified by chemical treatment*
SM: *synthetic material*
[ ] *abbreviation used in figures etc.*

RESULTS AND DISCUSSION

In figure 2 results from the model compounds cellulose, oxicellulose, and xylan, are shown.

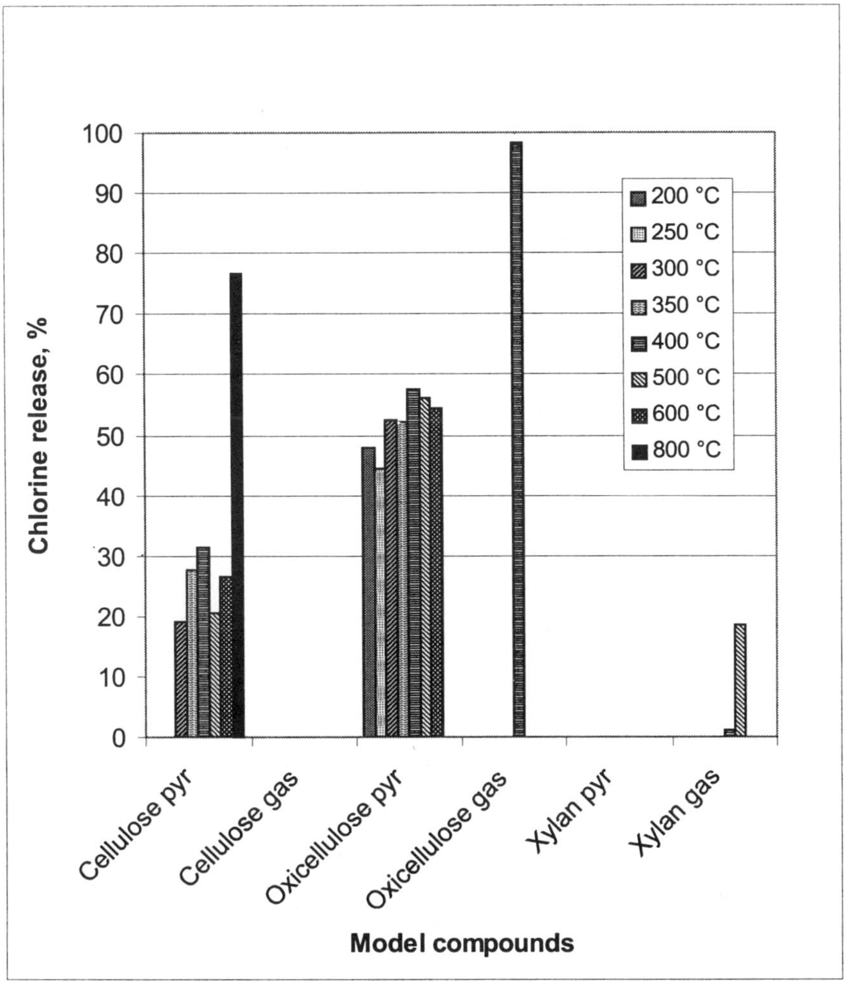

Fig. 2 Release of chlorine (%) from KCl-pretreated model compounds cellulose, oxicellulose, and xylan, during pyrolysis (N_2) and gasification ($92\%N_2 + 8\%O_2$) experiments.

Figure 2 shows the previously presented results from pyrolysis [1] together with the new results from gasification of cellulose, oxicellulose and xylan. Chlorine is released to a significant degree from both cellulose and oxicellulose at 300°C, but not from xylan, during pyrolysing conditions. No results from gasification of cellulose could be obtained since there was to small amounts left after the experiments to analyse. For xylan a small release of chlorine can be observed at 500°C but more significant release of chlorine do not occur until 600°C. This result confirms the suggested reaction mechanism. The details have been discussed previously [3].

Figures 3 and 4 show all the results from the chlorine release obtained from experiments with model compounds.

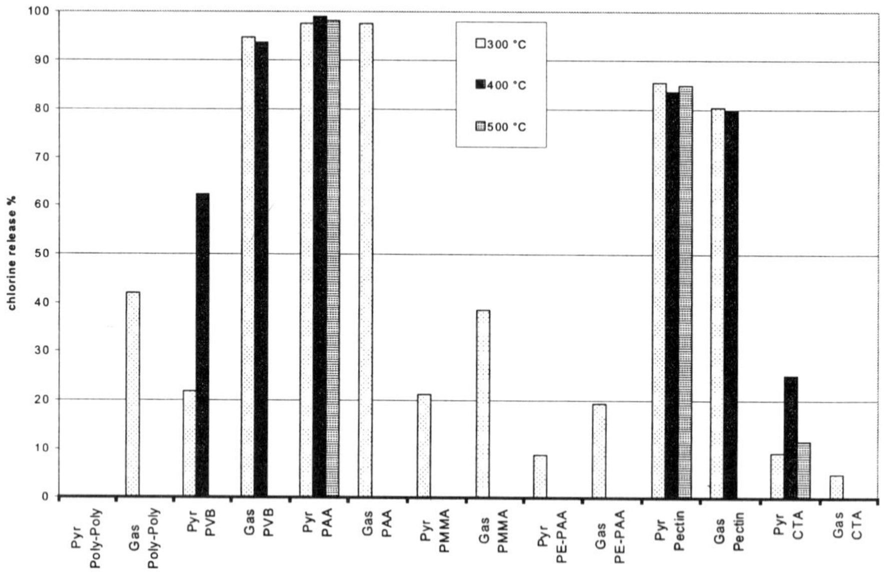

Fig. 3 Release of chlorine (%) from KCl-pretreated macro-molecular model compounds during pyrolysis (N_2) and gasification (92%N_2 + 8%O_2) experiments. Abbreviations explained in table 1.

From the results the strong effect of the free carboxylic groups on the chlorine release is verified. The chlorine release from the compounds with free acid groups showed no or little dependence from temperature.

The Polyvinylbutyral PVB with its high concentration of acetal groups shows a significant chlorine release during pyrolysis, and a very high release for gasification. The latter shows that the acetal during gasification probably is rapidly oxidized under formation of free acid groups. Even the polyesterpolyol shows a significant activity, but only for gasification, not for pyrolysis. The reason is unclear.

Also the model compounds with esterified carboxylic acid groups showed significant chlorine release. Their chlorine release was strongly temperature dependent, since higher temperature supports the cleavage of the ester bond under formation of free acid.

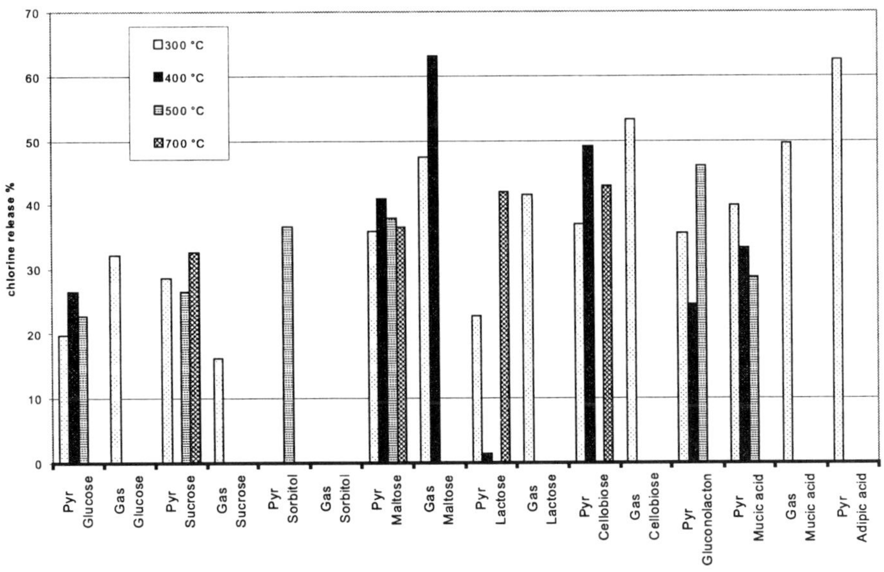

Fig. 4 Release of chlorine (%) from KCl-pretreated molecular model compounds during pyrolysis (N_2) and gasification ($92\%N_2 + 8\%O_2$).

Figure 4 shows the results for sugars and sugar derivatives. The sorbitol, which lacks the hemiacetal and acetal groups of the sugars, shows weak chlorine release only during pyrolysis at 500 °C.

In some model compounds, especially in the sugars in figure 4 , hydroxymethyl groups and groups like acetal and hemiacetal have been the dominating functional groups. For hydroxymethyl it was necessary to make a distinction between the hydroxymethyl in terminal positions (at the end of open chains) and those which are attached to rings, chains or rings which constitute chains. The gasification conditions could not be studied for some of these substances at higher temperature levels since they gave too little solid residue.

It can clearly be seen that at 300°C more chlorine is released during gasification than during pyrolysis for most of the model substances.

The results from all model compounds show that, with a few exceptions, a significant release of chlorine at pyrolysis conditions occurs already in the temperature range 300-500 °C, which is too early for a gasification of undissociated KCl. The exceptions have been: xylan, sorbitol, and polyesterpolyol.

This strongly confirms the ion exchange mechanism which initially had been suggested. The model compounds with free carboxylic groups show the strongest effect, which is consistent with the theory. The other compounds show weaker, but temperature-dependent effect. This means that even other functional groups, such as chemically bound carboxylic acid (open or ring-shaped esters) or chemical precursors for such carboxylic acid (hydroxymethyl groups, acetals, hemiacetals) can be activated at elevated temperatures, probably by chemical rearrangements, in the case of gasification additionally supported by partial oxidation.

From the results it can be concluded that the following types of functional groups contribute more or less strongly to the early release of chlorine:

Free carboxylic acid	(strongest effect, temperature-independent)
Carboxylic acid in ester	(open ester or lacton ring)
Acetal	
Hemiacetal	
Hydroxymethyl	(when attached onto chains or rings)
Hydroxymethyl	(when at the end of a chain, “terminal”)

RANKING

Based on this conclusion, a method has been developed to determine the relative activity of the functional groups. This was done based on the assumption that the observed effect (chlorine release in %) is caused by the simple sum of the contributions from the different functional groups, which are present in the model compounds.

For all model compounds where reliable values for the chlorine release had been obtained , the concentrations $C_{functional\ group}$ of all functional groups from the materials “as used” were determined.

A value S was defined for each model compound, which was calculated from these concentrations:

$$S = (a * C_{free\ acid}) + (b * C_{lactonic\ ester}) + (c * C_{open\ ester}) + (d * C_{hemiacetal}) + (e * C_{acetal}) + (f * C_{hydroxymethyl\ on\ cain\ or\ ring}) + (g * C_{terminal\ hydroxymethyl})$$

These S-values for the model compounds were plotted against the results for chlorine release for the temperatures 300, 400, and 500 °C (pyrolysis) and for 300 °C (gasification). For gasification at higher temperatures there were too few reliable results.

Using linear regression the relative influence from each functional group was calculated. The factors (a,b,c,d,e,f,g) were chosen in that way that their total amount equals 100 (normalized results). These factors give a crude ranking of the activity of the functional groups. The results are presented in Figure 5.

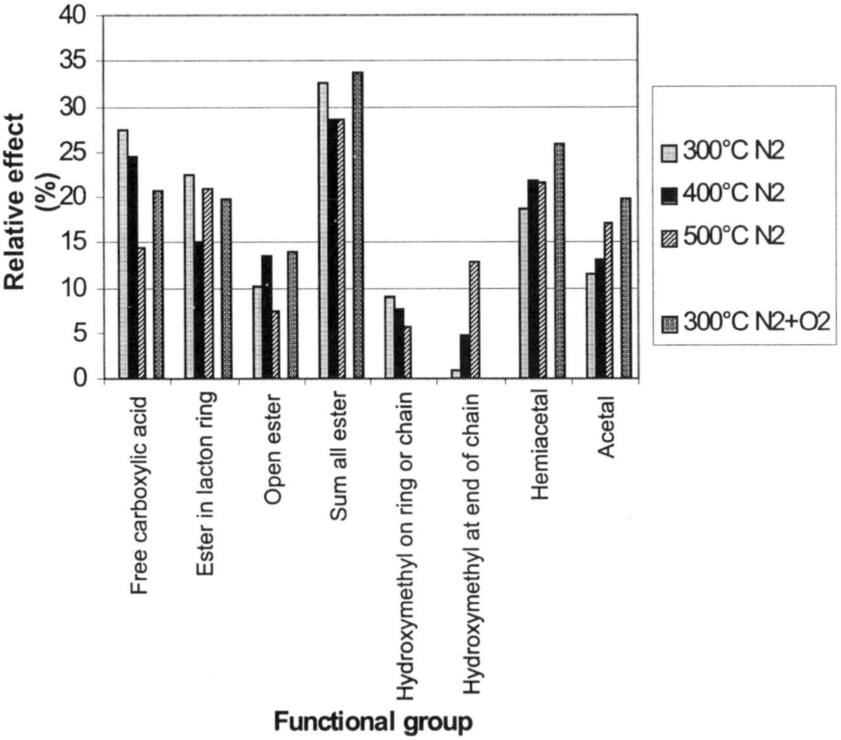

Figure 5. The relative effect of the presence of certain functional groups in the model compounds on the chlorine release during pyrolysis or gasification in the temperature range 300-500 °C.

SUMMARY

The earlier results from pyrolysis of cellulose, oxicellulose and xylan together with the new results from gasification of these substances and all results from experiments with organic model compounds confirm the previously suggested mechanism (ion exchange with carboxylic acid).

(1) The release of chlorine at pyrolysis or gasification of chlorine-containing biomasses or similar model compounds is significant already at temperature levels which are several hundreds of degrees lower than what would be expected for the volatilisation of KCl. In some cases, less than 300 °C can be sufficient.

(2) Many model compounds which only consist of chemically bounded C, H, and O, and pretreated with KCl, can show the early chlorine release when pyrolysed or gasified at 400 °C. No heteroatoms such as S, N, or others, are required for the effect.

(3) In most cases, the release at gasification conditions is higher than at pyrolysis conditions.

(4) As expected, the model compounds with significant amounts of free carboxylic acid groups showed the strongest chlorine release.

(5) Other model or natural compounds which contained no free carboxylic acid groups, showed significant and temperature-dependent chlorine release at pyrolysis or gasification between 300 °C and 500 °C when at least one of the following functional groups was frequent in the substance:

 Ester in open or ring-bound (lactonic) form
 Acetal
 Hemiacetal
 Hydroxymethyl (-CH_2-OH), when attached to a ring or onto the side of a chain.

 Hydroxymethyl was found to be inactive at 300 °C, when it formed the end of a straight unbranched molecule (or polymer chain), but its activity increases strongly with increasing temperature.

(6) A simple ranking of the relative activity of these functional groups showed that the relative activity of free carboxylic groups is strongest at 300 °C at both pyrolysis and gasification conditions. This effect is more or less "wiped out" at higher temperatures at pyrolysis.

(7) The effect of the other (non-carboxylic acid) functional groups can be explained by their ability to rearrange under formation of active forms like carboxylic acid at elevated temperatures.

CONCLUSIONS

In all biomass fuels, several of the functional groups active in the low temperature release of chlorine are present in significant amounts in one or several of the biopolymers, which constitute the biomass.

The consequence for practical applications is that the effect of chlorine release at the pyrolysis or gasification of chlorine-containing fuels will be significant as soon as a temperature level of about 300 °C is exceeded.

The ability to release a part of their chlorine concentration already at comparably "moderate" temperatures seems to be an inherent property of these fuels. This seems to be caused by the chemical properties of the natural substances that constitute the main part of these fuels.

ACKNOWLEDGEMENT

This work has been performed within the program "FB förbränning/förgasning" under contract no P1286. The support from STEM, Swedish National Energy Administration, is gratefully acknowledged.

REFERENCES

1. Mojtahedi,W. & Backman,R. (1989) *The fate of sodium and potassium in the pressurised-bed combustion and gasification of peat.* J.Inst.Energy, **1989**, 189-196.

2. Björkman,E.& Strömberg,B. (1997) *Release of chlorine from biomass at pyrolysis and gasification conditions.* Energy&Fuels, **11,** 1026-1032.

3. F. Zintl, B. Strömberg, and E. Björkman (1998) *Release of chlorine from biomass at gasification conditions.* In: "Biomass for Energy and Industry", 10th European Conference and Technology Exhibition, Würzburg, Germany 8-11 June 1998, pp. 1608-1611.

The Char Residues from Pyrolysis of Biomass – some Physical Properties of Importance

E.M. Suuberg, I. Aarna, I. Milosavljevic
Brown University, Division of Engineering, Providence, R.I. 02912 USA

Abstract: The pyrolysis of pure cellulose and actual biomass materials has been studied with an eye towards characterization of the char products formed during pyrolysis. Very few data exist regarding the thermal properties of biomass or cellulose chars, particularly at temperatures above ambient. Those engaged in modeling the pyrolysis behavior of these materials have had to guess at many of the key parameters. Several of these parameters have been here experimentally determined as part of an examination of heat transfer controlled pyrolysis of pure cellulosic material. It is likewise the case that the development of porosity in biomass materials during pyrolysis is not yet well understood, despite the significant commercial interest in producing activated carbons from these materials. This study also provides experimental data on this topic.

INTRODUCTION

THE THERMAL PROPERTIES OF A MODEL CELLULOSIC MATERIAL

The pyrolysis and combustion of cellulosic solids have been extensively studied phenomena. There have been many reviews of different aspects of the problem (e.g., biomass pyrolysis[1-3], modeling of combustion [4] and smoldering [5]). There remains some controversy regarding even the most basic aspects of the pyrolysis process, including even the apparent kinetics of pyrolysis of pure cellulose [6-10]. It is clear that transport limitations of various kinds are responsible for some of the disagreement in the literature, though the relative importance of heat and mass transport limitations have yet to be fully sorted out. The process of evaporation of pyrolysis tars (or partially decomposed sugars) has been seen to be an important facet of both the heat and mass transport processes. In a study on the thermal effects of cellulose pyrolysis [11], it was concluded that the endothermicity of pyrolysis arises mainly from the latent heat of tar evaporation, and is counterbalanced by exothermic char formation due to tar cracking. The magnitude of the pyrolysis endotherm is comparable to the sensible enthalpy requirements, under certain conditions. Thus the thermal requirements of the pyrolysis process are themselves quite complicated. In addition, while the processes governing the transport of heat to the reaction zone may be qualitatively understood, there remain significant uncertainties regarding the rates of heat transport. In this paper, some of the key thermal and transport properties governing the rate of the pyrolysis process in cellulosic solids are examined.

The original motivation for studying the thermal properties of cellulosic chars came from our study on bulk cellulose pyrolysis under conditions simulating those existing in a fire. In such a situation, the flame over the surface of the solid supplies heat to the pyrolyzing solid. In our work, the radiative and conductive feedback of heat from the flame to the surface was simulated using radiant heaters. The experiments were carried out in an inert gas environment, to maintain as well-defined a heat transfer environment as possible, free from complications due to actual combustion heat sources. A convective flow of the inert gas was used to sweep away volatiles from the vicinity of the surface, and the heat transfer effects of the sweep gas were also taken into account.

The material selected for study was pure cellulose. This material was selected in an effort to produce "realistic", but highly reproducible samples. Samples were pressed to densities typical of those encountered with wood. Such samples had the advantage that a large amount of kinetic work on cellulose could be brought to bear in understanding the results of the experiments. The work was concerned with the behavior of bulk material, in which sample dimensions are at least of order centimeters in all directions. There have been many other studies of cellulosic pyrolysis on this scale [e.g., 12-21]. Some have noted a well-defined pyrolysis front [12-16] and a period of relatively constant mass loss rate [15,16,19,21], particularly with high imposed radiative fluxes. Others have seen a sharp rise followed by a slow decrease in mass loss rate [17]. Generally, the data revealed a significant sensitivity to char thermal conductivity [20]. The effects of oxygen [18] and retardants [21] can be significant in affecting observed behavior. In summary, experiments with bulk cellulosic samples have provided evidence of a conduction limited process, often characterized by a very sharp boundary between char and unreacted cellulose, particularly at high incident radiative fluxes. The pyrolysis wave is often seen to move at near-constant velocity through a significant portion of the process, as our own results have also shown [22].

In attempting to model the heat transfer-controlled behavior, we were surprised to find, in the literature, a significant variability in the key thermal properties that such a model required. Quite different predictions could result from using different assumptions regarding the thermal properties. To address this obviously significant impediment to reliable model development, an experimental program focusing on thermal property measurements was initiated. The results are summarized below.

THE POROUS PROPERTIES OF BIOMASS CHARS

The interest in the properties of the chars derived from cellulosic or biomass solids extends beyond those associated with thermal transport in the char. Insofar as the char residue from a pyrolysis process must typically be burned, gasified, or put to use as an activated carbon product, there is also a need to examine the porous nature of the char. In activated carbons, the pore structure is key to adsorption performance. In combustion or gasification, the porosity can play a role in determining conversion kinetics in the intrinsic rate controlled or pore diffusion controlled regimes.

There exists significant empirical understanding of how to prepare useful activated carbon products from biomass materials. Still, there is relatively little information available on this topic to guide those whose main interest is in the pyrolysis of biomass to produce oils or combustion energy. In these types of processes, the conditions are often far removed from those used to produce activated carbons. This paper therefore also outlines some of the main characteristics of the porous structure developed in biomass materials during pyrolysis and during subsequent oxidative gasification.

EXPERIMENTAL

MATERIALS

The cellulose studied in this work was Whatman CF-11 powder, with an ash content of 0.009% and a moisture content of 7.68% (as-received). Powder samples were formed into pellets using an ordinary laboratory press, without any additives. Three different density samples were prepared by pressing for varying lengths of time at various pressures. The final pellet densities were: 0.965±0.041 g/cc, 0.691±0.035 g/cc and 0.458±0.026 g/cc. These densities are comparable to those of woods ranging from softwoods to hardwoods. All samples were 38 mm in diameter and had a thickness of about 10 mm. The pellets were generally quite hard; only the lowest density material showed some propensity to lose powder from near its edges. These pellets were used in the simulated fire apparatus mentioned above, in order to examine the kinetics of volatile release in situations in which the screening of oxygen from the solid surface (by the volatiles efflux) is simulated by performing the experiments in inert gas [22]. These experiments will not be described further here, except as needed to explain the thermal property results.

It was observed that in pressing the samples, the cellulose fibers tended to mat down in the plane perpendicular to the pressing direction. This gave samples a somewhat banded appearance. Experiments were conducted with the banded structure (grain structure) both perpendicular to and parallel to the incident radiative flux. This seemed to make little difference in the results that were obtained. Generally, the simulated fire experiments were conducted with the samples having the bands perpendicular to the incident heat flux, and this was the orientation for the thermal property measurements as well. Again, for these pure cellulosic materials, the orientation seemed to be of little significance.

In the case of the porosity measurements, in addition to the same cellulose samples, samples of oak and white pine were also examined. These samples of wood were ordinary commercial lumber samples (the same pine having been previously used in fire experiments at the Center for Fire Research of the National Institute for Standards and Technology). The pine had an initial dry density of 0.377 g/cc, and the oak 0.734 g/cc. Both of these wood samples were pyrolyzed for two hours at 1273 K, in helium, in order to prepare the chars for testing. The char yields were in the neighborhood of 28% by mass of the starting oak and 26% by mass of the starting pine.

RADIATION ABSORPTIVITY MEASUREMENTS

Virgin cellulose pellets and cellulose chars produced in the simulated fire apparatus were both examined. Two different measurements were made. One involved measuring the reflected radiation in the mid-infrared from 2.5 to 25 μm (4000 to 400 cm^{-1}). These measurements were performed in a diffuse reflectance cell within an FTIR spectrometer. These experiments revealed some wavelength dependence of reflectivity. Reflectance was also measured in-situ in the simulated fire apparatus, by arranging the samples, a fluxmeter, and the heating lamps such that surface reflection of the incident radiation could be determined over a fairly wide range of angles around the specular angle. This arrangement did not assure the collection of radiation over all angles, as did the FTIR measurement. There will be a dominance of a specular component in the reported values below. The results from this technique were, however, compared with independently calculated values from a surface energy balance on a sample instrumented with near-surface thermocouples. The agreement appeared to be reasonable, implying that the diffuse reflectance component did not contribute much to the energy balance. Measurements of the reflectivity were performed with the cellulose or char surface still cold, such that re-radiation from the pellet surface was not significant.

THERMAL DIFFUSIVITY MEASUREMENTS

Thermal diffusivity measurements were performed on pellets by the use of a transient response technique. Pellets of cellulose or char were instrumented with thermocouples at known distances from the surface. Samples were then brought into contact with a hotplate whose surface temperature varied in an approximately sinusoidal manner. Samples of virgin cellulose could only be tested up to surface temperatures of about 500 K, to avoid pyrolysis. Char samples could be tested to temperatures up to those seen during pyrolysis (in excess of 750 K). Samples were covered by a bell jar, purged with nitrogen, in order to avoid reactions with oxygen.

The pseudo-one dimensional conduction problem defined by the experimental system has a well-known solution [23]. While it proved difficult in practice to satisfy perfectly the boundary conditions for the ideal solution, a good enough approximation was possible such that the phase lag could be used to calculate the diffusivity directly [24].

Together with information on the sample pellet density (from measurements of its mass and volume) and heat capacity (see below), the thermal conductivity of the sample could be calculated.

HEAT CAPACITY

Heat capacities were determined by using a differential scanning calorimeter (DSC). A sapphire standard was used for DSC cell calibration. Samples were cut directly from the samples used in the thermal diffusivity measurements. Heat capacity was determined, as usual, from the heat requirement of the sample in response to a particular change in temperature. Multiple scans were performed to verify that sample degradation did not influence the results. Char samples could be heated up to 850 K in these experiments. The DSC cell was continuously purged with nitrogen, in order to avoid oxidation of the samples.

POROSITY MEASUREMENTS

Adsorption isotherms were determined in an automated volumetric gas adsorption apparatus (Autosorb 1, Quantachrome Co.). Adsorption of nitrogen was performed at 77 K. Before measurements, samples were outgassed at 672 K for at least 8 hr in vacuum. Where some burnoff of the char was desired, the reactions were performed in an Online Instruments TG-plus thermogravimetric analyzer. The reactions were performed in a mixture of helium and oxygen, flowing at a rate of about 220 cc/min. Samples of 30-50 mg were dispersed on a circular platinum pan with a large flat surface and raised sides, resulting in a particle beds of about 1 mm thickness. Temperatures between 573-748 K were used for gasification. The partial pressures of oxygen was 2.02 kPa. Char samples were outgassed at 1173 K for 30 minutes prior to reactivity measurements. Samples were then reacted to the appropriate level of burnoff and then quenched. Burnoff is expressed on a dry,ash-free basis.

RESULTS AND DISCUSSION

CHAR YIELD AND DENSITY

The char yield is defined here as the fractional mass left after the mass loss ceases under given conditions. The char yield was not observed to be a very strong function of initial cellulose sample pellet density, over the range examined here. The char yield did appear to be a function of the temperature that any particular point in the sample achieved. The asymptotic char yield for these bulk samples was in the neighborhood of 10%. Unfortunately, the role of mass transfer cannot be disregarded in predicting what the char yield should be under any particular conditions. It is possible that deep within a sample, the char yield is higher than at its periphery due to a greater opportunity for secondary cracking of tar volatiles to take place due to their hindered escape from the hot solid [7]. Our earlier work clearly established the importance of the interplay of thermal history and mass transport limitations in determining char yields. Slower heating and more resistance to escape of volatiles clearly promotes higher char yields [7].

The above naturally raises a question as to what exactly is meant by the term "char" in the context of bulk pyrolysis experiments. The definition can clearly only be operational, and is simply the mass that is left when an apparently steady state condition is achieved, regardless of how incomplete the pyrolysis is or however much carbon has resulted from processes involving secondary carbonization reactions of tars and their precursors. Thus all reports of "properties" of cellulosic chars must be viewed in this light.

Figure 1 shows the variation in actual char density, as a function of char yield. These data were obtained from various experiments in which bulk samples were pyrolyzed in the simulated fire apparatus to a surface temperature of around 800 K. Different ultimate char yields were observed as a function of position in the samples. Different ultimate char yields were observed as a function of position in the samples.

It should be carefully noted that the variation in final *char* density with char yield is not inconsistent with the above observations concerning the lack of effect of initial *cellulose* density on the ultimate fractional char yield.

The important conclusion is that char density is a nearly linear function of char yield, as might be expected. Such an assumption has been commonly employed in modeling of the pyrolysis behavior. The slope is, however, not what would be expected from a simple extrapolation to zero char yield. The true density would be underpredicted by such a common extrapolation. The reason is that shrinkage during pyrolysis is significant. Such shrinkage would tend to increase the density of the char in comparison with a simple extrapolation to zero char yield. The error associated with neglect of shrinkage can be seen by a comparison of the dashed line and the solid line in Figure 1. The shrinkage is observed during all experiments, and makes the calculation of actual temperature gradients in a char problematic.

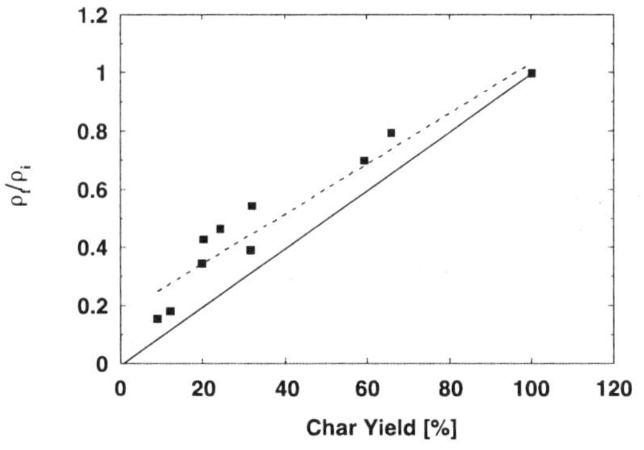

Figure 1. The ratio of final to initial cellulose sample densities for various char yields.

HEAT CAPACITIES OF CELLULOSE AND CHAR

Values of the heat capacity of the cellulose and chars are shown in Figure 2. The different curves for "chars" reflect the values obtained for cellulose pyrolyzed between 300 and 600°C.

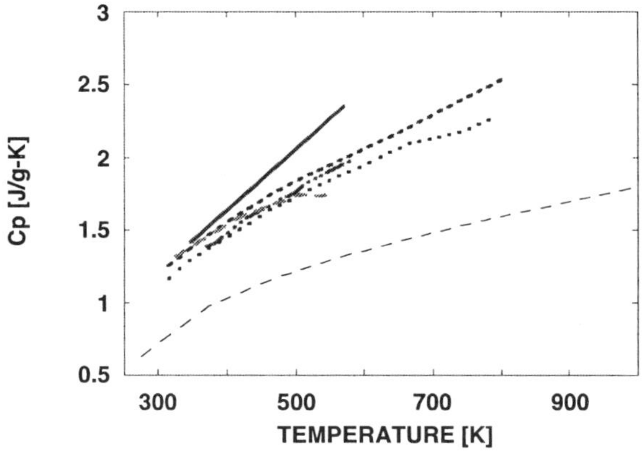

Figure 2. The heat capacities of virgin cellulose (solid line) and cellulose chars (heavy broken lines). Graphite is also shown for reference (thin dashed line).

There was, within the uncertainty of the heat capacity data, no variation with preparation conditions (neither initial pellet density nor pyrolysis temperature made much difference). The values in Figure 2 are at considerable variance with some values used in modeling of wood and biomass pyrolysis, depending upon temperature. For example, values for wood and biomass chars have been taken to be 1.38 [25], 2.5 [26], 1.0 [27-29], 0.71 [30], 0.67

[31], 2.0 [32], and 2.3 [33], all in J/g-K.

The measured values of Figure 2 show that the true heat capacity values, for the cellulose studied here, are in the range from 1.5 to 2.5 J/g-°C, irrespective of initial density. The results show that the variation in heat capacity with temperature should not be neglected, as it often is in modeling studies. Uncertainties in the magnitude of the char heat capacity have, however, been judged to be of only modest importance in determining pyrolysis rates [34,35], so it is not necessarily surprising that this parameter has not been earlier identified as a potential source of significant error. Still,where there is concern with the accuracy of net sensible enthalpy effects, it appears that there could be difficulties encountered in using many commonly assumed values. This problem is exacerbated by large uncertainties in enthalpies of pyrolysis [11], which may be significantly endo- or exo-thermic, depending upon conditions and the position of the pyrolysis front.

THERMAL CONDUCTIVITIES OF CELLULOSE AND CHARS

Values of the thermal conductivity of different chars are shown in Fig. 3, together with the conductivities of the cellulose pellets, and for reference, gaseous nitrogen. The chars tested for thermal diffusivity (conductivity) had fairly uniform properties because they were prepared in a pyrolysis furnace, and not in the simulated fire apparatus. This tended to minimize temperature gradients, but there was no assurance of absolutely uniform density, for the reasons noted above. Preparation of the chars followed a temperature history designed to simulate that in the simulated fire apparatus.

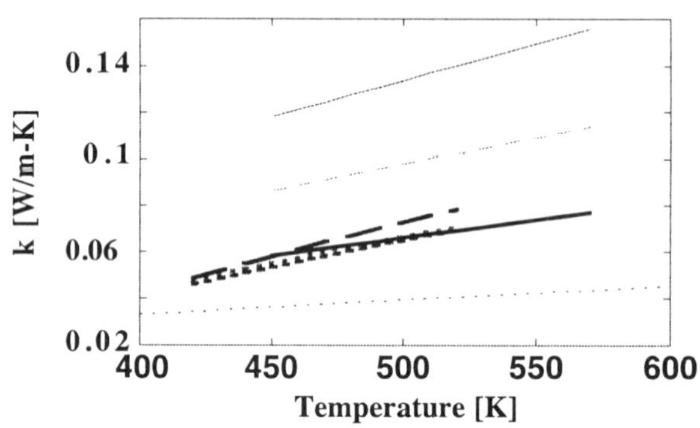

Figure 3. Thermal conductivity of virgin cellulose (0.458 g/cc - heavy solid line; 0.678 g/cc - thin dashed line; 0.928 g/cc - thin solid line), cellulose chars (heavy dotted lines) and nitrogen gas (thin dotted line).

The thermal diffusivities of different density cellulose samples were all comparable, meaning that their thermal conductivities scaled roughly with density. The conductivity values obtained for chars from different density samples were all rather close, because it was observed that the final char densities of samples prepared for these tests did not vary significantly, or in a consistent manner, with initial pellet density. The final char densities (0.33 to 0.52 g/cc) often bracketed that of the low density unpyrolyzed cellulose sample, thus it is little surprise that these samples exhibited conductivities comparable to that of the low density unpyrolyzed cellulose pellets. This implies that it is the porosity of the sample that is the primary factor in determining the thermal conductivity, since the true solid densities of the char and cellulose are somewhat comparable.

The char thermal conductivity values obtained here are again in good agreement with some and at variance with others that have been utilized or reported in the literature, e.g., 0.19 [27], 0.05 [30], 0.04 [25], 0.03 [32], 0.11 [33], of order 0.1 [36], around 0.15 [29] and $0.08 \cdot 10^{-4}T$ (°C) [37], all in W/m-K.

SURFACE ABSORPTIVITY

Surface absorptivity (emissivity) was initially determined using an FTIR spectrometer operating in the mid-infrared. This wavelength was not entirely commensurate with the wavelength of the radiative heaters, but little attention was paid to this point initially, because the measurements were performed to "confirm" the black nature of the char surface. Results are shown in Figure 4, as integral reflectance through the mid-IR, as a function of pyrolysis time. The cellulose itself is seen to have a reflectivity of only about 10% to begin with, and this rapidly drops as pyrolysis begins. This is what would normally be expected as the surface blackens, and becomes more absorbing (and less reflective) during pyrolysis. Interestingly, the reflectance value began to climb again as pyrolysis continued, implying a lowering of absorptivity with carbonization.

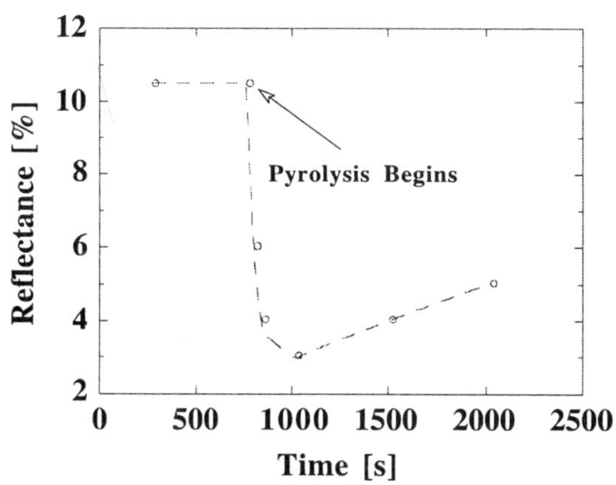

Figure 4. Surface reflectance of high density cellulose pellets as measured by FTIR in the mid-infrared. The times refer to the time of heating under an incident flux of 40 kW/m^2.

Difficulties in closing a calculational energy balance in the simulated fire experiments prompted review of the FTIR absorptivity measurements (absorptivity being taken as one minus reflectivity, since the sample transmits no radiation through its interior). The alternate procedure described above was therefore employed for obtaining absorptivity in the wavelength range of the actual heat lamps. Again, this involved the use of the simulated fire apparatus and its own heaters as a radiative source, and using a radiation fluxmeter looking towards the surface for reflected radiation. The result of these experiments was that the cellulose gave an average absorptivity over the wavelength range of heater operation (about 0.5 to 8 μm) of 0.53 (±0.05). Various chars were tested at various flux levels and gave an average absorptivity of 0.77 (±0.09). These latter two values may be viewed as maximum values for the respective materials since, again, they came from measurements near the specular angle. These values gave energy balances that were in considerably better agreement than those using the higher absorptivity values from the mid-IR measurements. The difference is presumably associated with the fact that the bulk of the radiation from the heat lamps is of considerably shorter wavelength than mid-infrared. There must be a rather significant wavelength dependence of absorptivity in this range.

There is obviously an important consequence of using char absorptivitites of 0.95-0.98 as implied in Figure 4, as opposed to using the value of 0.77 which better describes the real situation in a fire. The former will not result in as rapid heating of the sample. Rarely has much concern been shown for such important departures from near-black char surface behavior in the modeling of pyrolysis in cellulosic systems.

Another area of concern related to such radiative heating calculations concerns the "screening" of the surface (during active pyrolysis) by what might be described as "smoke" or an aerosol of tar droplets. In our work with the simulated fire apparatus, it appeared that the degree of screening depended upon the inert purge gas flowrate over the sample surface. Higher flowrates, not unexpectedly, helped clear away the smoke, and an optical pyrometer trained on the surface revealed an enhanced reflected light signal with increase in purge flowrate. It was difficult to obtain quantitative results from the particular experiments that were performed as part of this study. There was strong experimental evidence, though, that the extent of surface screening was at least 10%, in experiments involving heat fluxes of around 40kW/m^2. This evidence was gathered in experiments in which a fluxmeter was placed into the center of a pyrolyzing sample. Taken together with the lower than expected surface absorptivity, the calculation of heat absorption rate is seen to be potentially subject to some rather severe overestimates.

POROSITY OF CELLULOSE AND BIOMASS MATERIALS

The porosity of the cellulose chars was studied using nitrogen adsorption at 77K. The results for the fresh cellulose char and the char burned off to differing extents in oxygen are shown in Figure 5.

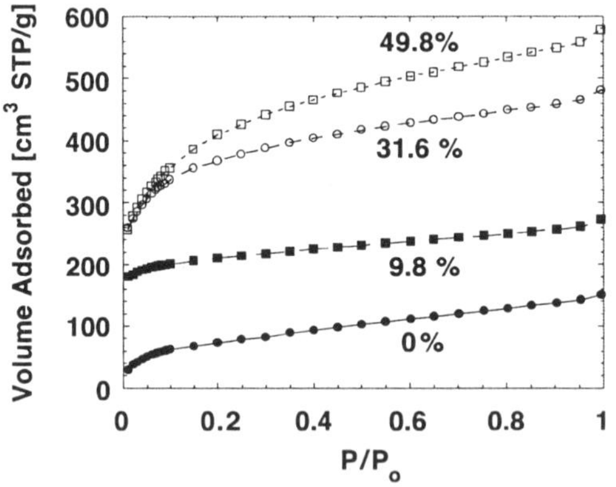

Figure 5. Nitrogen isotherms at 77 K on cellulose chars. The values on the graph refer to the degree of burnoff in 2% oxygen at 723 K.

The results in Figure 5 display a very typical pattern of behavior for microporous carbons. The definitions of pore sizes recommended by IUPAC are micropores (<20Å), mesopores (20Å-500Å) and macropores (>500Å), and the micropores will adsorb nitrogen at the lowest values of relative pressure, P/P_0 [38]. The apparent non-zero intercept on the ordinate implies an important role of the smallest micropores. The dramatic increase in the value of the intercept implies that a large amount of new microporous volume is being opened up by burnoff. It is of course precisely this sort of porosity development that is desired during the "physical activation" processes for producing activated carbons.The so-called BET surface area [38] of the cellulose char starts at a value of around 265 m^2/g and grows to 1374 m^2/g at 49.8% burnoff. The latter value is typical of good commercial activated carbons. These carbons have mesopore volumes of between 0.1 and 0.3 cc/g.

Figure 6 shows a comparison of the nitrogen isotherms obtained on the pine and oak samples, in comparison to the isotherm on cellulose. All are for the raw chars, with no burnoff. This figure illustrates that the initial microporosity in many biomass materials is much lower than it is from pure cellulose. It is believed that this is associated with the chemical differences in the starting materials; the hemicellulose and lignin components tend to promote filling of the porosity during pyrolysis. Because the fresh wood chars have much smaller amounts of accessible microporosity, they also exhibit very low BET surface areas (37 m^2/g for pine and 27m^2/g for oak char).

Figure 7 shows that the development of microporosity is very significant in the wood chars, just as it was in the cellulose char. In fact, a very similar pattern of surface area development is seen in the wood chars as in the cellulose chars.

The implication of the above results is that cellulosic and biomass materials tend to form microporous chars. This is already well known in the activated carbon community. The tendency to form microporous materials has to do with the forming of a rigid, highly crosslinked structure during pyrolysis. Materials that tend to melt during pyrolysis generally give rise to ordered, mesophasic carbon materials with low microporosity. These results also tend to emphasize the fact that it is inappropriate to characterize biomass char porosity solely on the basis of fresh char porosity. While the fresh chars have very low microporosity, they are quickly activated, or "opened up" upon exposure to oxidizing gases at high temperatures (oxygen in this case, but carbon dioxide, nitric oxide and steam also are effective).

The rapid opening of porosity with burnoff at low burnoffs carries with it another important consideration. Because the reactivity of the char to oxidizing gases depends upon the ability of the gases to reach reactive surface, the development of porosity at low burnoffs carries with it an important increase in apparent reactivity of the char with burnoff. Thus it is important in characterizations of reactivity of pyrolysis chars to recognize this possibility, and to conduct reactivity experiments at a range of burnoffs, under which conditions this trend will become more apparent. This topic will be explored in a forthcoming paper by our group.

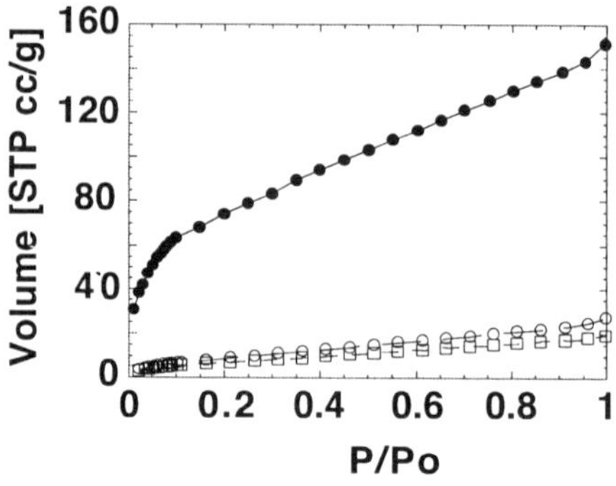

Figure 6. The nitrogen 77 K isotherms on raw cellulose char (filled circles), pine char (open circles) and oak char (open squares).

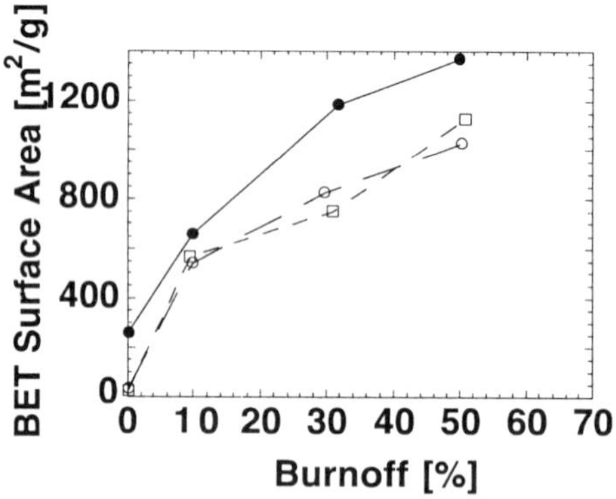

Figure 7. The nitrogen BET surface areas for raw cellulose char (filled circles), pine char (open circles) and oak char (open squares).

CONCLUSIONS

The literature shows a wide variation in values of key thermal properties of cellulosics and their chars. Values of thermal conductivity, heat capacity, enthalpy of pyrolysis and surface emissivity apparently cannot yet be safely taken from one study and applied in another. The values of these properties obtained in this study suggest that the specific heat capacity of cellulose char varies significantly with temperature between about 1.3 J/g-K at ambient temperature to about 2.5 J/g-K at 800 K. The heat capacity of the raw cellulose is comparable to that of the char at low temperatures, but quite a bit higher at temperatures just below the onset of pyrolysis (around 550 K).

The thermal conductivity of cellulose chars is a strong function of final sample density. Thus the sample porosity (on a macroscopic scale) plays a key role in determining the conductivity, which was typically found to be in the range from 0.04 to 0.08 W/m-K at temperatures from ambient to 575 K. The cellulose samples shrink during pyrolysis, and achieve a density in the range from 0.33 to 0.53 g/cc, regardless of initial density.

The surface absorptivity or emissivity of cellulose chars cannot be safely assumed to be near unity. Near unity values are found for wavelengths in the mid-infrared, but at the shorter wavelengths characteristic of thermal radiation in combustion environments, the emissivity may be closer to 0.8. Significant energy balance errors may be made in assuming higher values.

Whole biomass samples are characterized by rather low microporosity and surface areas when freshly prepared. This is in contrast to pure cellulose, which exhibits significant microporosity initially. The biomass samples do, however, develop a significant amount of microporosity (activate) with modest amounts of burnoff in oxygen. In this regard, cellulose is a good model for whole biomass behavior.

Acknowledgment

The financial support of the U.S Department of Energy (under grant DE-FG2699-FT40582) and the National Science Foundation (under grant BES-9523794) is gratefully acknowledged.

REFERENCES

1. Antal, M.J. Jr. (1982) in *Advances in Solar Energy*, (K. Boer and J. Duffie, Eds.), American Solar Energy Society, Vol.1, p. 61.
2. Antal, M.J. Jr. (1985) in *Advances in Solar Energy*, (K. Boer and J. Duffie, Eds.), American Solar Energy Society, Vol.2, p. 175.
3. Antal, M.J., Jr. and Varhegyi, G. (1995), *I&EC Res.*, **34**, 703-717.
4. Di Blasi, C. (1993) *Prog. Energy Comb. Sci*, **19**, 71-104.
5. Ohlemiller, T. (1985) *Prog. Energy Comb. Sci*, **11**, 277-310.
6. Milosavljevic, I. and Suuberg, E.M. (1995) *I&EC Res.*, **34**, 1081-1091.
7. Suuberg, E.M., Milosavljevic, I. and Oja, V. (1996) *26th Symp. (Int.) on Combustion,*, The Combustion Institute, Pittsburgh, pp 1515-1521.
8. Antal M. J., Jr., Varhegyi G. and Jakab E. (1998) *I&EC Res.*, **37**, 1267-1275.
9. Reynolds, J.G. and Burnham, A.K. (1997)*Energy and Fuels*, **11**, 88-97.
10. Antal, M.J.,Jr. and Varhegyi, G. (1997), *Energy and Fuels*, **11**, 1309-1310.
11. Milosavljevic, I. and Suuberg, E.M. (1996) *I&EC Res.*, 35:653-662.
12. Kanury, M. A. and Blackshear, P.L.(1966) *Pyrodynamics* **4,** 285-298.
13. Blackshear, P.L., Jr., and Kanury, M.A. (1965) *10th Symp. (Int.) on Combustion*, The Combustion Institute, Pittsburgh, pp. 911-923.
14. Kanury, M.A. and Blackshear, P.L., Jr. (1970) *Comb. Sci. Tech.*, **1,** 339-355.
15. Lee, C.K., Chaiken, R.F. and Singer, J.M. (1976) *16th Symp. (Int.) on Combustion,* The Combustion Institute, Pittsburgh, pp. 1459-1470.
16. Lee. C.K. and Diehl, J.R. (1981) *Comb. Flame*, **42,** 123-138.
17. Ohlemiller, T.J., Kashiwagi, T. and Werner, K. (1987) *Comb. Flame*,**69,** 155-170.
18. Kashiwagi, T., Ohlemiller, T.J. and Werner, K. (1987) *Comb. Flame*,**69,** 331-345.

19. Becker, H.A. and Phillips, A.M., *Comb. Flame* (1984) **58,** 255-271.
20. Chan, W.-C., Kelbon, M., and Krieger, B.B. (1985)*Fuel*, **64**, 1505-1513.
21. Chen, Y., Frendi, A., Tewari, S.S., and Sibulkin, M. (1991) *Comb. Flame*, **84**, 121-140.
22. Milosavljevic, I. and Suuberg, E.M. (1992) *ACS Div. Fuel Chem. Prepr.*, **37**, 1567-1574.
23. Carslaw, H.S. and Jaeger, J.C. (1959) *Conduction of Heat in Solids*, Clarendon Press, Oxford.
24. Milosavljevic, I. (1994) Ph.D. Thesis, Division of Engineering, Brown University.
25. Atreya, A. and Wichman, I.S. (1989) *J. Heat Transfer*, **111**, 719-725.
26. Kung, H-C. (1972) *Comb. Flame*, **18**, 185-195.
27.Kung, H.-C. (1975) *15th Symp. (Int.) on Combustion,* The Combustion Institute, Pittsburgh, pp. 243-253.
28. Holve, D.J. and Kanury, A.M. (1982) *J. Heat Transfer*, **104**, 344-350.
29. Evans, D.D. and Emmons, H. (1977) *Fire Res.,* **1,** 57-66.
30. Kansa, E.J., Perlee, H.E., and Chaiken,R.F. (1977) *Comb. Flame*, **29,** 311-324.
31. Kung, H.-C. and Kalekar, A.S. (1973) *Comb. Flame*, **20,** 91- 103.
32. Kanury, A.M. and Holve, D.J. (1982) *J. Heat Transfer*, **104,** 338-343.
33. Weatherford, W.D. and Sheppard, D.(1965) *10th Symp. (Int.) on Combustion,* The Combustion Institute, Pittsburgh, pp. 897-910.
34. Desrosiers, R.E. and Lin, R.J. (1984) *Solar Energy*, **33,** 187-196.
35. Chen, Y., Delichatsios, M., and Motevalli, V. (1993) *Comb. Sci. Tech.*, **88,** 309-328.
36. Parker, W. J. (1988) D.Sc. Thesis, Dept. of Mechanical Engineering, George Washington University.
37. Koufopanos, C.A., Papayannakos, N., Maschio, G., Lucchesi, A. (1991) *Can. J. Chem. Eng.*, **69,** 907-915.
38. Gregg, S.J., Sing, K.S.W. (1982) Adsorption, Surface Area and Porosity, Academic Press, New York.

Biomass Fast Pyrolysis in an Air-blown Circulating Fluidized Bed Reactor

I. Boukis, M.E. Gyftopoulou and I. Papamichael
Centre for Renewable Energy Sources (CRES)
19th km Marathonos Ave., 190 09 Pikermi, GREECE

ABSTRACT: A novel circulating fluidised bed reactor for biomass fast pyrolysis integrally utilizing the energy content of the by-product char in the pyrolysis reactor has been operated and further amended. This paper is aiming to examine the amendments adopted for the reliable operation of the downstream processing as well as other system components, to analyse the results and to explore scale-up possibilities. In addition, potential end-use applications are discussed and first operational results of a bio-oil fuelled Stirling engine are presented.

INTRODUCTION

Biomass valorisation to high value added energy products is of growing interest, since bioenergy has a series of socioeconomic advantages, such as:

(1) Direct reduction of fossil CO_2 emissions.
(2) Energy independence on a regional scale via the rational exploitation of forestry and agricultural residues.
(3) Reduction of agricultural surpluses to comply with the Common Agricultural Policy.
(4) Income increase in rural areas followed by a reduction in urbanisation rates.

Several potential routes have been developed to evolve the vast bioenergy potential. As such, the production of pyrolysis liquids (also known as bio-oil or bio-crude oil) via biomass fast pyrolysis is approaching market introduction and pre-commercialization in Europe. Bio-oil, as a liquid fuel with the strategic advantage that it can be stored until required or readily transported to where it can be most effectively utilised for power and/or heat applications. Besides the remarkable progress accomplished regarding biomass fast pyrolysis, further challenges to be faced exist in improving the technology and adapting applications to cope with the unusual behaviour and characteristics of the liquid product (1).

On this background, a Circulating Fluidised Bed (CFB) reactor for biomass fast pyrolysis was extensively tested by CRES in the framework of previous JOULE I and II Programmes. The CFB reactor configuration conceived is differentiated from any other reactor system considered for biomass fast pyrolysis, since it incorporates the integral utilisation of the energy content of the by-product char in the same reactor vessel. Char derived by biomass fast pyrolysis is captured in the solids recovery system and recycled in the lower part of the CFB reactor, where it is combusted with air, thus providing the necessary energy requirements for the subsequent biomass pyrolysis carried out in the riser section (2).

IMPROVEMENTS IN DOWNSTREAM PROCESSING

Design aspects and operational problems faced during the development of the CFB reactor have been discussed and reviewed in the past (3). During the first period of the CFB reactor operation (2) a number of experimental runs were successfully performed. At that time, the condensible vapors recovery system consisted of an indirectly cooled shell-and-tube heat exchanger (STHE) and a cotton wool filter (in the following, Downstream Configuration I, DC-I). The relatively lower liquids yields obtained, in comparison to those obtained by other researchers (4), were mainly attributed to this, DC-I, configuration and particularly to the indirect pyrolysis vapors recovery system practiced, since:

(1) Deposits of heavy components formed in the tubes, due to indirect cooling of gas/vapors in the STHE, act as the forefront in polymerisation reactions, converting primary organic vapors to secondary and tertiary tars.
(2) The layer, formed by the deposits created in the heat exchanger tubes, reduces the heat transfer coefficient leading to ineffective cooling of the gas/vapor products, prolonging their residence time at higher temperatures and consequently reducing the product yield.
(3) The deposits created, reduced the free area and initiated tube plugging and excess pressure build-up, while resulted to limited system availability.

It has been reported (1) that the recovery of pyrolysis liquids is not a simple condensation process, because their rapid indirect cooling is leading to the formation of stable aerosols and micron-size droplets, which are usually entrained in the gaseous stream, thus avoiding capture. Therefore, impingement and coalescence of the pyrolysis vapors is considered an essential feature in any liquids recovery process.

On the aforementioned grounds, a scrubbing system for the recovery of pyrolysis vapors was designed, constructed, installed and tested. This system, further referred to as Downstream Configuration II, DC-II (5), is actually based on the existing STHE, and is consisted of:

(1) A scrubbing section, virtually a plate-and-ring disk system arrangement for the proper distribution of the scrubbing medium and the maximisation of heat transfer area.
(2) The previously used STHE for the indirect cooling of the scrubbing medium.
(3) A properly designed bottom vessel for disengagement of liquids and pyrolysis gas, through which the pyrolysis gas coalesces hence enhancing mass transfer and improving vapors capture.
(4) A recirculating Moyno-type (screw) pump with independent flow control.

Pyrolysis experiments with wheat straw were conducted to commission DC-II. Although the experiments exhibited substantially improved system availability, they suffered lower total liquids yields (loss of char and heavy liquid fractions, see further on the experimental results section).

An improved configuration of the DC-II gas/vapors recovery system integrated an electrostatic precipitator (EP) replacing the cotton wool filter (denoted as DC-III). Gas enters the bottom of the precipitator and leaves the top, while a liquid outlet situated at the lower part of the precipitator serves for the collection of the captured pyrolysis liquids.

More experiments conducted in order to validate the performance of the new downstream components, revealed improved availability and higher mass balance closures (above 95%). The final configuration of the biomass fast pyrolysis experimental set up is depicted in Figure 1 (5).

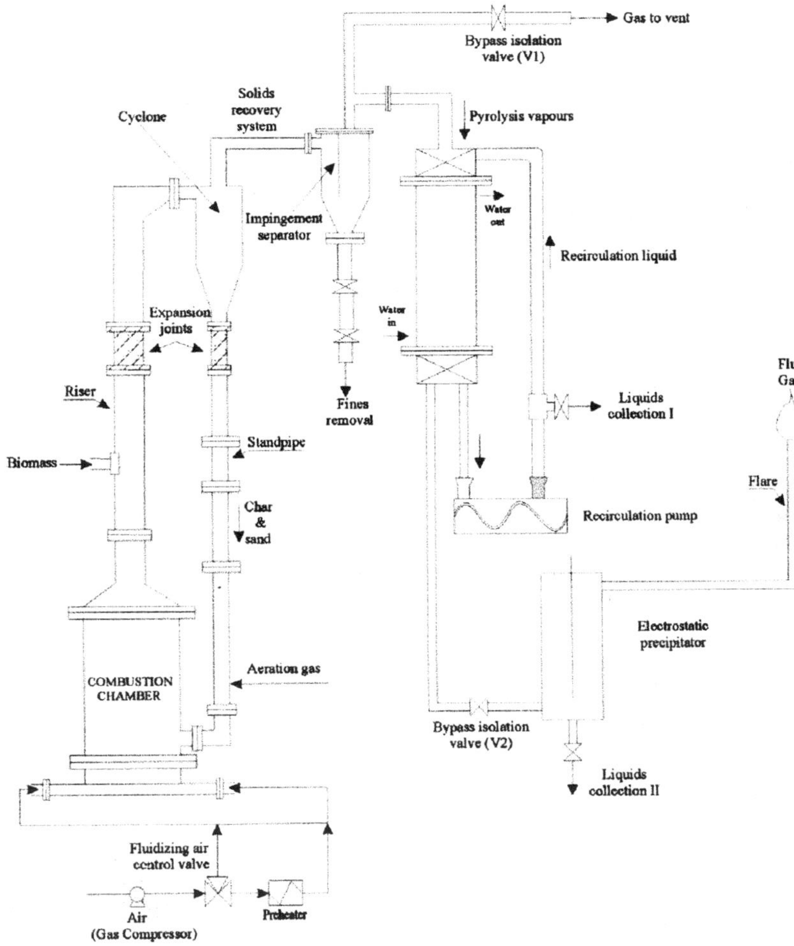

Fig. 1 CFB reactor with DC-III.

EXPERIMENTAL RESULTS

Five experimental runs (R7-R11) were performed with DC-I and pine of 1-2 mm as a feedstock with acceptable mass balance closures (91.99 - 97.52 %) and liquid yields up to 61.5 % wt on moisture and ash free (maf) feed basis.

Two experiments (RS1, RS2) were performed with DC-II and Swedish wheat straw with particle size in the range of 1 to 2 mm. During these experiments the solids recovery system (cyclone and impinger, Figure 1) did not perform satisfactorily with the very fragile straw char. The char, exposed to the very harsh environment (fast moving sand particles) of the CFB reactor and atritted to very fine particles, almost submicron powder, could not be collected by conventional solids recovery systems. Consequently, the char was either adhered with the liquids heavy fraction to the STHE inner tubes or readilly accumulated in the liquid recovery system.

In addition, sintering of the sand occured, due to the straw ash melting at significantly lower temperatures than the respectives for wood ash. It was observed in form of white spots at the char combustor walls.

After the installation of the DC-III, i.e. scrubber, STHE and EP, two more runs (R12, R13) were conducted with a mixture of softwood with a particle size in the range of 1.5 to 2 mm.

During run R12, increase of the EP's supply voltage resulted in improving the collection efficiency, as indicated by the change of colour in the gaseous stream leaving the EP from brownish to transparent. Short-circuiting, attributed to droplets-assisted bridging between the two electrodes, occurred during the EP operation and was followed by severe instability problems to all the nearby electronic devices, as well as significant reduction of the EP's voltage. After the completion of R12 the scrubber was dismantled and cleaned. It was noticed that the solids loading was much less than the respective of the straw experiments.

In run R13, the EP performed satisfactorily and the pyrolysis liquids recovered from this point had very different physical appearance (higher density and viscosity and lower moisture content) compared to the liquids recovered from the scrubber. However, short-circuiting occurred again in the EP approximately after 30 minutes of operation, leading to the thought of installing a pump for the continuous removal of the collected liquids.

In total, five pyrolysis experiments were conducted with the latest configuration (DCIII, Figure 1) and a mixture of hardwood in order to achieve longer and more reliable operation of the system. The particle size of the feedstock was within the range if 1.5 to 2 mm. Operating conditions for all runs are presented in Table 1.

Operation of the biomass fast pyrolysis pilot plant with DC-III and woody feedstocks, specifically mixed softwoods (mainly pine) and hardwoods, revealed the following:

(1) The reactor operability is easily sustained once steady-state operation is reached.
(2) The woody feedstocks produced much less fine char than straw, and the gas entrained solids were more efficiently recovered by the solids recovery equipment.
(3) The modified vapor recovery system (DC III) resulted in the higher pyrolysis liquids yields obtained (63.4% on a maf feedstock basis) for the feedstock size

(1.5<d_p<2.0 mm) tested. Nevertheless, these yields are still lower than those achieved by others researchers due to reasons referred to above.

(4) The quality of the bio-oil was significantly improved with respect to the liquids water content. During the earlier DC-I and II experiments, the loss of the heavy fraction of the pyrolysis liquids (retained in the cotton-wool filter) resulted in a wattery liquid (> 35% wt water), while during DC-III experiments the water content was reduced to approx. 20% wt.

(5) A significant problem remains the relatively high solids (char) content in the pyrolysis liquids, due to the fragmentation of char and wood fines in the harsh environment which are subsequently entrained in the gaseous stream, an inherent characteristic encountered in all CFB reactor systems, (ineffective solids recovery system).

Table 1 Operating conditions and total liquid yields for the experiments conducted.

Run number	T Riser (^{0}C)	T Com-bustor (^{0}C)	Biomass flow-rate (kg maf/h)	Fluidizing air flow-rate (kg/h)	Vapor residence time (ms)	Total liquids yield (%wt on maf feedstock)	Mass balance closures (%)
R7 (DC-I)	581	678	8.44	16.00	374	40.12	94.00
R8 (DC-I)	605	700	10.45	14.70	388	54.46	95.36
R9 (DC-I)	579	645	10.20	13.90	396	57.43	97.52
R10 (DC-I)	496	748	8.96	12.10	458	61.50	94.18
R11(DC-I)	550	740	10.22	13.00	373	49.91	91.99
RS-1 (DC-II)	630	700	13.20	17.00	360	43.00	89.78
RS-2 (DC-II)	670	720	9.90	13.80	477	38.00	91.42
R12 (DC-III)	470	610	8.50	18.40	557	61.00	97.29
R13(DC-III)	520	700	8.50	17.50	557	56.00	97.61
R14 (DC-III)	450	520	10.30	14.70	459	63.40	95.14
R15 (DC-III)	500	580	9.64	13.88	491	62.80	94.91
R16 (DC-III)	535	605	10.17	14.78	465	61.90	95.96

FURTHER IMPROVEMENTS IN THE PILOT PLANT

After the second period of the CFB reactor operation (DC-II, DC-III) further amendments in the plant were decided to solve respective problems.

Firstly, the existing feeding system incorporated in the pyrolysis pilot plant has been replaced, due to the following reasons:

(1) Excessive wear of motors, axes and gears.

(2) Not very reliable feeding (variations of up to 10-15% from average have been observed during calibration tests).

(3) Insensitivity of feeder to deliver reliably over a wide span of flowrates.

A new feeding system has been designed, constructed and installed in the plant. It comprised of:

(1) A feedstock hopper (capacity 40-45 kg of biomass) with a stirring device.
(2) A system of three screws, a horizontal one, a vertical one and another horizontal one and their relevant motors. The horizontal screw in the hopper is the main screw that regulates the biomass flow rate. The other two screws are just transporting the biomass into the reactor. The second horizontal screw is connected directly to the reactor.

The feeding system has been calibrated with feedstock in the range of 1 to 12 kg/h on a wet feedstock basis. It must be noted that during the calibration of the system the repeatability and the reproducibility of the results were proved to be very accurate, especially when they were compared with the respective ones of the previous feeding system (2). The biomass flow remains relatively steady, even while the hopper is discharging to the dosing screw. A significant deviation in the results appears only when less than 1 kg of biomass is left inside the hopper. This feeding system is expected to result in better control of the process and more reliable and accurate mass balances.

Secondly, aiming to avoid the short-circuiting occurring in the EP, due to droplets-assisted bridging between the two electrodes, a peristaltic pump was installed in the liquid outlet situated at the lower part of the precipitator. During commissioning of the EP connected with the peristaltic pump apparent improvement was noticed. Although the nearby situated electronic devices failed again at approximately 30 minutes of steady state operation, the voltage provided by the power supply unit to the EP remained relatively constant (10% drop compared to 50-70% drop without the pump), and the EP continued to clean the pyrolysis vapours. In addition the quantity of pyrolysis liquids that were captured by the EP increased significantly.

The failure of the electronic devices is attributed to the electromagnetic field created by the high voltage of the EP. A possible, easy way to resolve this problem is the isolation of the EP with a metalic construction that will absorb the main part of the electromagnetic field. Further experiments should be performed to prove the effectiveness of this solution.

INVESTIGATION OF SCALE-UP POSSIBILITIES

During the implementation of the JOR3-CT96-0099 Contract (5), the potential possibilities for scaling-up the CFB biomass fast pyrolysis reactor technology have been studied. Provision of technical information to the Technical University of Vienna led to the construction of three Cold Flow Models (CFMs) with 10, 135 and 1000 kg/h nominal biomass feed rate, in order to derive scale-up guidelines for the CRES Circulating Fluidised Bed (CFB) biomass pyrolysis reactor. The methodology followed was based on the similarity rules theory (6) and the consideration of non-dimensional numbers derived from the fundamental equations of two-phase flow.

Extensive experiments to study the fluid mechanic behaviour have been carried out at these CFMs corresponding to CFB fast pyrolysers (5). Satisfactory agreement was reached between the 10 kg/h CRES CFB pyrolyser and the CFMs, leading to the conclusion that appropriately scaled and designed CFMs are well suited for the investigation of three basic fluid mechanic scale-up criteria. These criteria can be used for the design of a commercial scale CFB biomass fast pyrolysis reactor (5).

The first two criteria are:

(1) Constant ratio of the specific circulation rate of bed material to the specific feed rate.
(2) Constant gas residence time in the riser.

The third criterion - constant rate of feed rate to bed inventory - is not that strict.

All models tested retain the minimum necessary specific circulation rate of bed material ($m_{rec,mat}/m_{biom}>10$) to ensure adequate heat transfer to biomass particles and high feedstock ablation rates.

Investigations with different geometric and operating parameters have aided to optimize reactor design and process operating conditions. The measurements obtained indicate sufficient specific solids circulation rates to achieve adequate heat tranfer to biomass incoming feed and close to the optimum required gas residence time in the riser of the corresponding fast pyrolyser even with a 1000 kg/h biomass feed (7).

According to the acquired results, a pilot scale (approx. 100 kg maf biomass/h) CFB pyrolyser would have a riser diameter of approximately 0.16 m, while a demonstration scale one (approx. 1000 kg maf biomass/h), a riser diameter of 0.40 m. The specific feed rates for these two units correspond to specific feed rates of 1.3 and 2.6 kg/m^2s respectively, which are the highest for all units investigated so far in the literature (2).

POTENTIAL APPLICATIONS

Biomass fast pyrolysis is apprehended to offer significant economic advantages over other thermal conversion processes, as it decouples fuel production from energy generation, since the end products are liquids. Interesting challenges in further developing and modifying biomass fast pyrolysis technology, include the liquids upgrading and the adaptation of applications to accept the unusual behaviour and characteristics of the liquid products (suspended char, alkali metals, low pH, high viscosity, etc.)(1,8).

Bio-oil end use applications have been extensively reported in the literature (8). Among the different end-use applications considered, small and medium scale Stirling CHP systems fuelled with bio-oil have been reported to achieve higher thermal and electric efficiencies than any other options with the same fuel in similar power range (9). In the framework of JOR3-CT98-0310 Contract, co-ordinated by CRES, very promising results have been initially obtained from the operation of a Stirling engine (30 kW_{th}) equipped with a newly developed Flox® (flameless oxidation) burner utilizing bio-oil as a fuel. The special features of the Flox® burner consist of internal premixing of combustion air and the exhaust gases and avoidance of high temperature peaks in the flame. The measured particulate emissions were extremely low, at least 10 times lower than the relevant of a diesel engine for the same power range (10), due to prolonged fuel residence time in the Stirling engine external combustion chamber provided by the internal mixing of combustion air and exhaust gas. Internal mixing also resulted in the efficient burn-out of suspended fuel-contained char particles, which constitutes a major remedy for high-solids content bio-oils. During the operation of the Stirling CHP system, very low CO and NO_x emissions have been measured (CO < 6ppm, NO_x <70 ppm)(10). Besides, very good fuel atomisation performance has been reported with an air pressure atomiser, but since air preheating was not possible, the engine efficiency suffered (11). It is expected that these problems will be successfully overcome within the continuation of this contract.

CONCLUSIONS AND RECOMMENDATIONS

The CRES CFB biomass fast pyrolysis pilot plant performance and availability was improved after amendments to the pyrolysis vapors collection and recovery train. The latest configuration adopted (DC-III), namely a scrubbing-type device followed by an electrostatic precipitator, resulted in:

(1) Improved collection efficiency.
(2) Improved quality of the pyrolysis liquids collected.
(3) Prolonged pyrolysis plant operation.
(4) Significantly simpler cleaning and maintenance procedures.
(5) Better process integration.

The CFB reactor operation is currently performed on a fully-autothermal mode, i.e. all additional heat supply is suspended, as soon as steady-state conditions are reached. To the authors' knowledge, this constitutes a major breakthrough, since no other single-bed reactor system has been operated on a similar mode. This achievement is far more significant since it was performed on a low-throughput, pilot plant (approx. 10 kg/h), where no internal optimisation of energy supply was practised. The results obtained also verify in practice the long known postulation that no external heat supply for fast pyrolysis is overall required (12).

Scale-up potential has been investigated through the study of the hydrodynamics for three cold flow models. The scale-up is not considered to create any severe problems in reactor operation, since it will be associated with design features not easily attainable to small, bench-scale equipment. Therefore, both the operability (adoption of suitable automation measures) and the performance (rational design) are expected to be substantially improved.

The only significant opearational problem remains the relatively high solids (char) content in the liquids, due to the excessive fragmentation of char and wood fines in the harsh environment incorporated in CFB reactors. It is well-known that it is difficult to properly size and manufacture small-scale (< 1") cyclones with a high performance and adequate solids removal from gaseous (pyrolysis vapors) streams. However, this is expected to be resolved during scale-up, where larger dimensions and manufacturing capabilities will allow a more precise close-up of design calculations and manufacturing specifications.

The potential utilization routes for the final products have been extended with the integration of small CHP systems, namely a Stirling engine, successfully operated on bio-oil for sufficient time.

Additional actions should be taken in order to achieve reliable performance of the scrubber under any conditions. A more efficient solids recovery system is essential to be incorporated in the plant. Scale-up at pilot (approx. 100 kg/h) or demo-scale (approx. 1000 kg/h) should also be considered.

ACKNOWLEDGMENTS

Financial support from the European Commission, through JOR3-CT96-0099 and JOR3-CT98-0310 Contracts of the JOULE III Programme, is appreciated.

REFERENCES

1. Bridgwater A.V. (1999) Principles and practice of biomass fast pyrolysis processes for liquids. In: *J. Anal. Appl. Pyrolysis*, 51, 3-22.
2. Boukis I (1997) *Fast pyrolysis of biomass in a circulating fluidised bed reactor.* PhD thesis, Aston University.
3. Boukis I. (1995) Practical implications during operation of a CFB air-blown pyrolyser. In: *Bio-oil Production and Utilisation*, Proceedings of the 2nd EU-Canada Workshop on Thermal Biomass Processing, (Ed. By A.V. bridgwater &E.N. Hogan), pp. 49-65, CPL Press.
4. Bridgwater A.V.,Peacocke G.V.C. (2000) Fast pyrolysis processes for biomass. In: *Renewable and Sustainable Energy Reviews*, 4, 1-73.
5. Final report (1999), Contract JOR3-CT96-0099, *A novel approach for the integration of biomass pyrolytic conversion processes in existing markets of liquid fuels and chemicals.*
6. Glicksman J. (1984). In: *Chem.Eng.Sci.*, Vol. 39, No 9, p. 1373.
7. Haslinger W., Hofbauer H., Gavriil L., Boukis I.(1999) Scale-up guidelines for a circulating fluidized bed biomass pyrolyzer. In: *Proceedings of 6th International Conference On Circulating Fluidized Beds*, August 22-27, Wurzburg, Section: Process design and scale-up.
8. Bridgwater A.V., Meier D., Radlein D. (1999) An Overview of Fast Pyrolysis of Biomass. In: *Organic Geochemistry,* 30, 1479-1493.
9. Bandi A., Personal Comunication on 25-5-2000.
10. Bandi A., and F. Baumgart (2000), Fast pyrolysis liquid feed to a FLOX® burner. In: *Proceedings of Progress in Thermo-chemical Biomass Conversion*, Tyrol, Austria, 17-20 September.
11. Third progress report (2000), Contract JOR3-CT98-0310, *Small - scale combined heat and power from bio-crude oil fuelled to a Stirling engine.*
12. Bogley W.J., et.al., (1977) *Solid Waste Utilisation-Pyrolysis*, Oak Ridge National Laboratory, Oak Ridge, TN.

Rotating Cone Bio-Oil Production and Applications

B.M. Wagenaar[1], R.H. Venderbosch[1], J. Carrasco[2], R. Strenziok[3], B.J. van der Aa[4]

[1] *BTG Biomass Technology Group BV, Enschede, The Netherlands.*

[2] *Centro de Investigaciones Energéticas Medioambientales y Technológicas, Madrid, España.*

[3] *Universität Rostock, Rostock, Germany.*

[4] *KARA Engineering Almelo B.V., Almelo, The Netherlands.*

ABSTRACT: A biomass pyrolysis processing chain is considered which starts with the raw feedstock acquisition and ends with electricity production. The following results are discussed:

*Biomass pretreatment:*Poplar and straw have been selected as typical feedstock types. First the energy use of drying has been investigated as well as the possibilities to obtain the heat by energy integration. Milling of these feedstocks to various particulate sizes has also been examined, together with the energy requirements. The energy requirements were obtained by monitoring the power consumption of the milling engines.

Development of a 200 kg/hr pyrolysis plant: The pyrolysis process includes the biomass feeding section, the pyrolysis reactor and the liquid collection system. Various tests are discussed, demonstrating the technical feasibility of the process. For example, the yields of bio-oil, char and gas are typically 70, 15 and 15 weight percent. Finally, an economic evaluation of the pyrolysis process is presented in terms of the investment and bio-oil production costs.

Prime movers on pyrolysis oil: Results of test concerned with the combustion of pyrolysis oils in a boiler are reported. And finally, preliminary attempts of combustion experiments with a gas turbine will be described.

INTRODUCTION

Fast pyrolysis is a technology by which biomass is decomposed into bio-oil, char and gas. When wood is used as a feedstock, bio-oil is the major product (70 wt.%) The gaseous by-product (15 wt.%) can be fired in a boiler or in a gas engine. Finally, the char (15 wt.%) may be combusted in the pyrolysis unit to drive the process auto-thermally. Only the ash (up to 1 wt.%) is left as a waste stream.

Fast pyrolysis produces a clean liquid biomass energy carrier. Advantages of the technology are:

(1) Bio-oil is cheaper to transport than biomass.
It has a volumetric energy density of 20 GJ/m3.
Wood chips have an energy density of only 4 GJ/m3.

(2) Bio-oil is cleaner than biomass.
The ash content in bio-oil is a factor 100 lower than in biomass. Possible contaminants like K, Cr, and Cu remain in the char.

(3) The cost of bio-oil production is relatively low due to the mild conditions.
Fast pyrolysis occurs at 500 °C and at atmospheric pressure.

While demonstrating the pyrolysis technology chain (including feedstock pre-treatment, bio-oil production and electricity production), problems which have to be overcome are:

(1) the selection of proper feedstock pre-treatment systems which result in low specific costs;
(2) the reliability of the pyrolysis pilot plant at the scale of 200 kg per hour;
(3) retrofitting of prime movers for bio-oil utilization and demonstration of the total system.

This paper will evaluate the conditions to produce bio-oil and electricity on basis of biomass fast pyrolysis. The objective is then to report, is a sequential form, the results obtained with respect to the three topics mentioned above.

BIOMASS PRE-TREATMENT

The objective of this task is to consider, for each biomass type selected in a previous sub-task, the following aspects:

(1) Which pre-treatment steps are required to produce a suitable feedstock for the rotating cone reactor.
(2) Which type of pre-treatment equipment should be used.
(3) What are the energy requirements and the associated costs to pre-treatment operations.

The particle size reduction tests for various selected biomass types have been executed in the pilot installation of CIEMAT. A flow diagram of the installation is shown in Figure 1. Basically, the equipment consists of two hammer mills connected in series, the first one has a nominal power of 11 kW and is used for pre-grinding (crushing). The second hammer mill has a nominal power of 15 kW and is used for refining the pre-ground biomass.

Figure 2 shows the effect of hammer milling with sieves of different mesh sizes. The solid curve 1 depicts the size distribution of the original poplar chips. The chips have a size varying from 5 to 30 mm. Grinding of the poplar chips with a coarse sieve of 8 mm requires 44 kWh per oven dry ton (odt) as shown by the curve: ground (8 mm).

Grinding of the poplar chips first with the 8 mm sieve and then with a 2.5 mm sieve requires a total of 132 kWh per oven dry ton. Finally, grinding of the chips first with the 8 mm sieve and then with the 2.0 mm sieve requires a total of 179 kWh per oven dry ton as shown by the dotted curve: ground (2 mm).

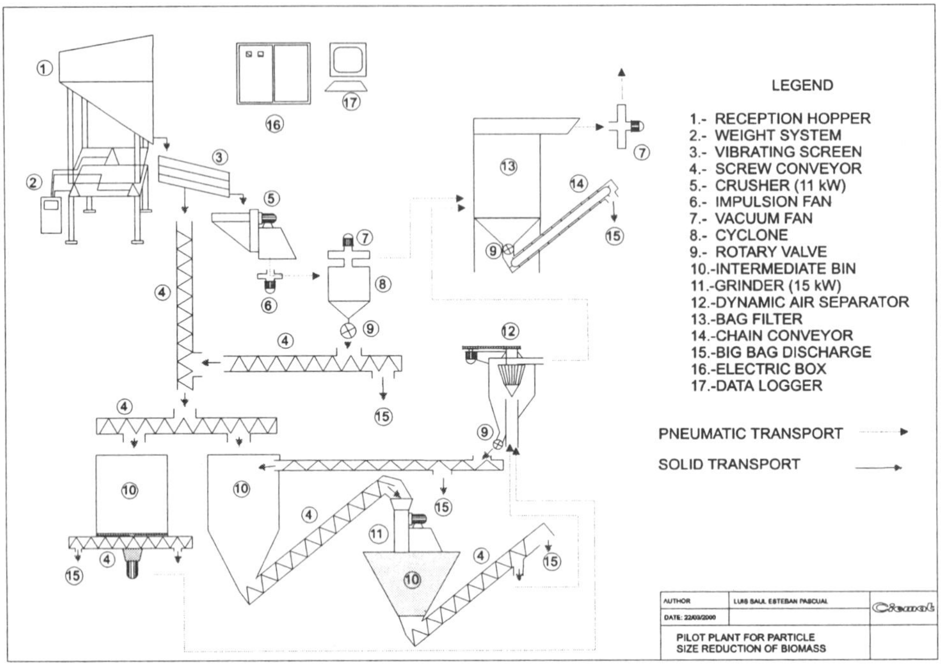

Figure 1: The biomass pre-treatment plant.

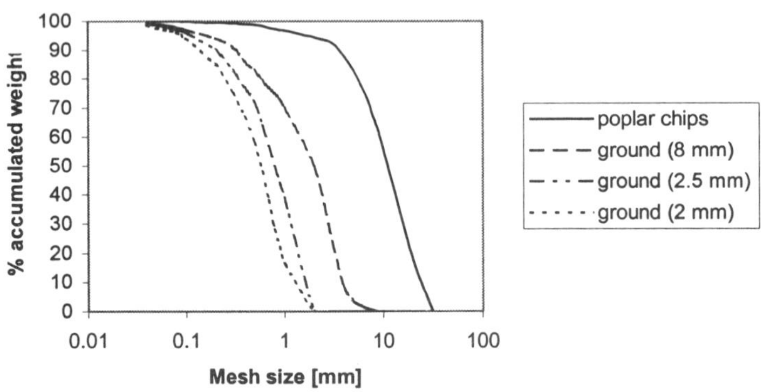

Figure 2: Weight distribution versus the mesh size.

These results allow an assessment of the pre-treatment costs of biomass on a larger scale. The installation envisioned is a pre-treatment station as shown in Figure 1 with a throughput of 4 ton per hour. It contains a biomass dryer, a coarse grinder with an 8 mm sieve and a fine grinder with a 2 mm sieve. The following basic assumptions have been introduced in the economic evaluation:

(1) Heat of the pyrolysis process is used for drying.
(2) Cost of electricity: 0.066 ECU/kWh for Spain.
(3) Electricity consumption in drying: 4.7 kWh/ton.
(4) Moisture content of biomass after drying: 12 wt%

Table I presents the investment costs for a 4 ton per hour pre-treatment station. The variable costs are presented in Table II. The electricity cost of grinding accounts for more than 80% of the variable costs.

Table I: The investment cost of a 4 ton/hour raw materials pre-treatment station.

Investment cost [Thousand Euros]			
Item	Biomass type		
	Short rotation poplar	Wheat straw	Rice husks
Rotary dryer	360	0	0
Storage silo (400 m3)	60	60	60
Pre-grinding	165	206	0
Refining	282	279	78
Storage silo (300 m3)	42	42	42
Bale chopper	0	30	0
Installation cost (20%)	182	117	36
Subtotal	*1092*	*734*	*216*
Site preparation	120	120	120
Buildings	30	30	30
Subtotal	*150*	*150*	*150*
Total cost	1242	884	367

As can be observed from Table II, the pretreatment cost varies between 7 and 20 Euro's/ton for a fully equipped pretreatment station which can grind to less than 2 mm.

Table II: Added costs for pre-treatment.

Pre-treatment costs [Euro/ton]			
Item	Biomass type		
	Poplar	Wheat straw	Rice husks
Fixed cost			
Depreciation (15 yr)	2,5	1,8	0,7
Labour costs	4,3	4,3	4,3
Total fixed costs	6,8	6,1	5,0
Variable cost			
Electricity	10.8	6.4	1.5
Maintenance	1.5	1.7	0.3
Total variable costs	12,4	8,1	1,8
Overheads (3% of fixed costs)	0,2	0,2	0,2
PRE-TREATMENT COSTS	19,4	14,3	7,0

Recent operational data from the pyrolysis plant from April 2000 show that biomass with a particulate size of 3 mm can be easily processed with could lead to a pre-treatment station with only 1 coarse grinder equipped with a sieve of 8 mm mesh size. Such a stripped down pretreatment station would lead to pre-treatment costs, which vary from 5 to 10 Euro/ton depending on the type raw material, which is treated.

DEVELOPMENT OF A 200 KG/HR PLANT

The development of the rotating cone reactor started a decade ago at the University of Twente. Since then the technology has evolved from laboratory scale to a pilot scale unit which is characterized by a biomass throughput of 200 kg per hour. In this section first the pyrolysis reactor is described as the core of the technology. Then the entire pilot plant is described and finally, characterization results of the plant are presented.

THE ROTATING CONE REACTOR

The core of the pyrolysis pilot plant is the rotating cone reactor which is a compact high intensity reactor in which biomass of ambient temperature is mixed with hot sand. Upon mixing with the hot sand of 550 degrees Celsius biomass decomposes into 70 weight percent condensable vapours, 15 weight percent non-condensable gases and 15 weight percent char. An important characteristic of this reactor type is the absence of carrier gas since it is the rotating action of the cone which propels the solids from the reactor entrance to its exit. Because of the absence of carrier gas, the vapour products are not diluted and their flow is minimal. An undiluted and concentrated product flow from the reactor leads to small downstream equipment with related minimal investment costs. Finally, the 200 kg/hr pilot plant contains a single cone with a top diameter of 0.65 m and a bottom diameter of 0.21 m. The top angle of the cone is pi/2 radians, as is shown in Figure 3.

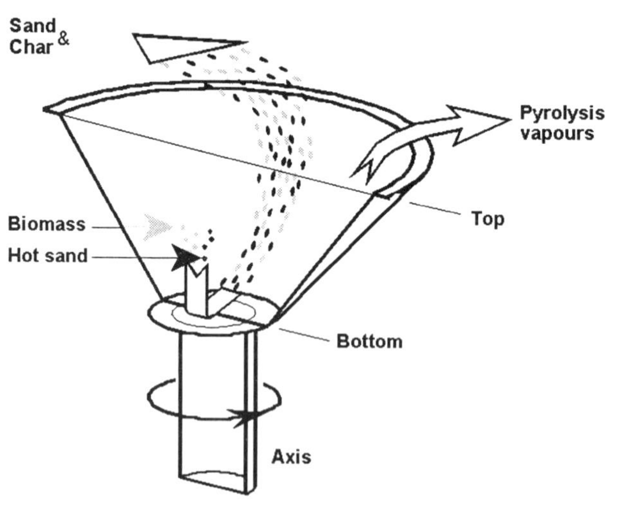

Figure 3: The rotating cone reactor.

In mechanical terms, the reactor technology is remarkably simple and robust. The rotational speed of the cone is only 300 rpm and after more than 1000 hours of operation signs of abrasion or wear are absent. This is due to the combination of a relatively slow moving solids phase, directed parallel to the wall, and the selection of SS310 as the cone wall material. The extended axis, which supports the cone, leads to minimal heat leakage and allows the bearings to operate at 30 degrees Celsius (90 °F). For reasons of robustness, heavy-duty truck bearings have been installed, which operate at 30% of their design speed in a vibration free environment. They are expected to last longer than the depreciation period of the plant. Scaling-up of the rotating cone reactor is possible by increasing its diameter. For capacities which require a cone diameter larger than 2 meters, stacking of multiple cones on a single axis leads to the lowest investment costs. This conventional approach is also encountered in centrifugal disk separators or rotating disk contactors. With these options all pilot plant capacities between 2 and 100 ton/hr can be served.

DESCRIPTION OF THE PILOT PLANT BTG 200P

The starting point of the design of the pilot plant was a specification of its three major design goals:

(1) Design and operational data of the pilot plant must provide a sufficient amount of know how to keep the investor risk for a commercial size plant at an acceptable level.
(2) The pilot plant know how must allow for scaling up in a single step to a commercial size plant of 2 t/h.
(3) All systems and sub-systems of an anticipated commercial plant must be present in the pilot plant.

These considerations did lead to a plant with minimal infrastructure requirements. In thermal energy terms, the pilot plant is self sustained and is a net producer of heat when the biomass throughput is larger than 70 kg per hour. Inert gases are not required, and as a consequence they are absent. The electricity demand of the BTG 200P is 23 kW (3 phase, 25 A, cos $\varphi = 0.80$). Finally, the cooling tower requires 0.6 m3 water per ton of biomass processed, to compensate for evaporation losses. In the present pilot plant BTG 200P all char is combusted to heat the process and the pyrolysis gas is burned in a stack. This leaves as emissions to the environment a flue gas stream and an ash stream. A process flow sheet of the pyrolysis process is presented in Figure 4.

The mass and energy balances, which are related to the flowsheet of Figure 4, are presented in Table III. The following assumptions have been used in the calculations:

(1) Wood is processed with 6 wt% moisture and 1 wt% ash.
(2) Bio-fuel oil consists of a single phase mixture of water and organics.

Because the data in Table III originate from mass and heat balance calculations, the balance closure is 100% for the mass and heat balance.

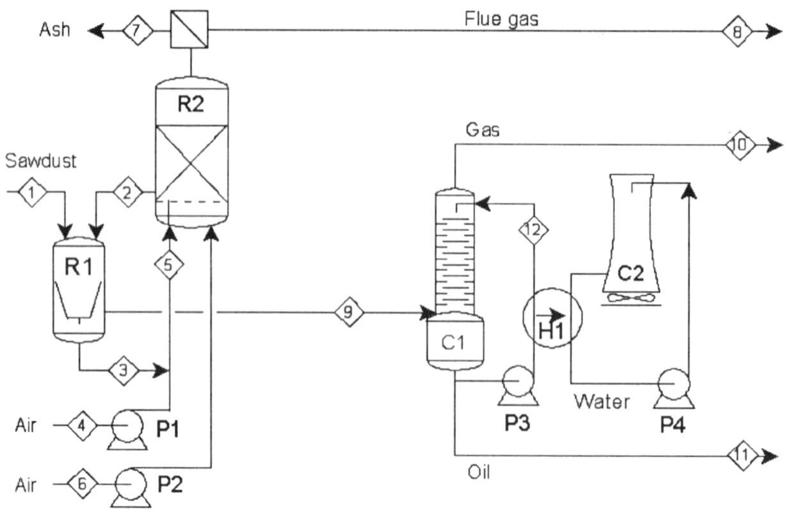

Figure 4: The flowsheet of the pyrolysis process.

Table III: Results from the mass and energy balance calculations for each flow of the process flowsheet.

Flow nr.	1	2	3	4	5	6
Stream	feed kg/h	cut kg/h	rcr exit kg/h	riser air kg/h	riser kg/h	air kg/h
Wood	200	-	-	-	-	-
N_2	-	-	-	88	88	184
O_2	-	-	-	22	22	46
CO_2	-	-	-	-	-	-
Pyro gas	-	-	-	-	-	-
H_2O	12	-	-	-	-	108
Organics	-	-	-	-	-	-
Char	-	-	30	-	30	-
Ash	2	-	2	-	2	-
Sand	-	4500	4500	-	4500	-
Total	214	4500	4532	110	4642	338
T [K]	291	829	761	291	747	291
P[bar a]	1.03	1.03	1.03	1.10	1.10	1.15

Table III continued

Flow nr.	7	8	9	10	11	12
Stream	Ash kg/h	Flue gas kg/h	gas+ oil kg/h	pyro gas kg/h	bio-oil kg/h	bio-oil kg/h
Wood	-	-	-	-	-	-
N_2	-	272	-	-	-	-
O_2	-	-	-	-	-	-
CO_2	-	91	-	-	-	-
Pyro gas	-	-	30	30	-	-
H_2O	-	115	47	-	47	3384
Organics	-	-	105	-	105	7560
Char	-	-	-	-	-	-
Ash	2	-	-	-	-	-
Sand	-	-	-	-	-	-
Total	2	478	182	30	152	10944
T [K]	829	829	761	302	310	302
P[bar a]	1.00	1.00	1.03	1.00	1.00	3.00

A composed photograph of the plant is shown in Figure 5. Not visible in the photograph are the two cyclones in series between the pyrolysis reactor and the condenser and bio-fuel oil recycle loop of the condenser.

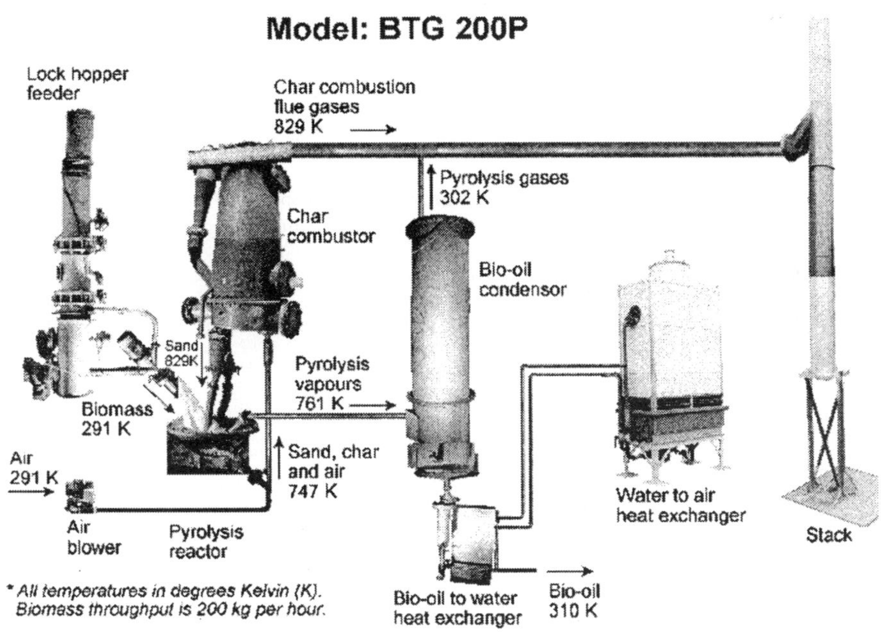

Figure 5: A photograph of the pilot plant.

The lock hopper feeding is located most on the left hand side. This feeder injects the biomass in the pyrolysis reactor. Hot sand is fed into the pyrolysis reactor from the char combustor, which is located above the reactor. With means of a riser, the char containing sand from the pyrolysis reactor exit is recycled into the char combustor. All gaseous pyrolysis products are directed to the bio-fuel oil condenser. Cooling of the pyrolysis vapours releases heat which is removed from the bio-fuel oil by means of a water circuit. This water circuit is air cooled.

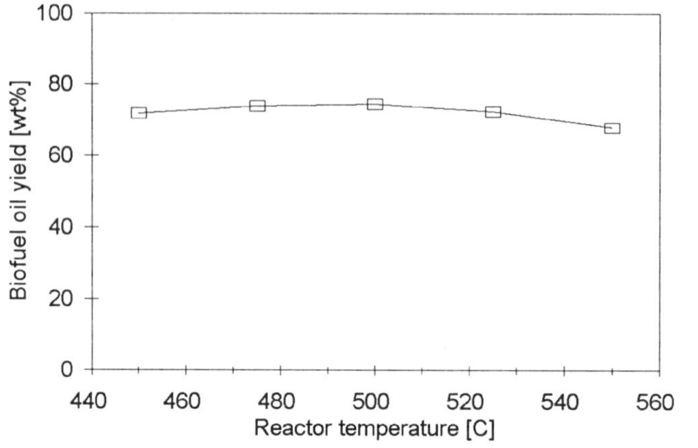

Figure 6: The oil yield versus the reactor temperature.
Mixed wood sawdust, 1-3 mm; vapour residence time 1.0 s.

A typical operational characteristic of the BTG 200P is shown in Figure 6. A maximum bio-oil yield of 74 wt% has been measured when processing clean wood waste from the wood processing industry. Since this type of sawdust is available in abundance, BTG envisions to contract this stream for large scale future operations. The first step of such grand scheme is to demonstrate the capability of pyrolysis plants to process this commercially available feedstock. Figure 6 was constructed from measurement data obtained during a long duration run of 4 days.

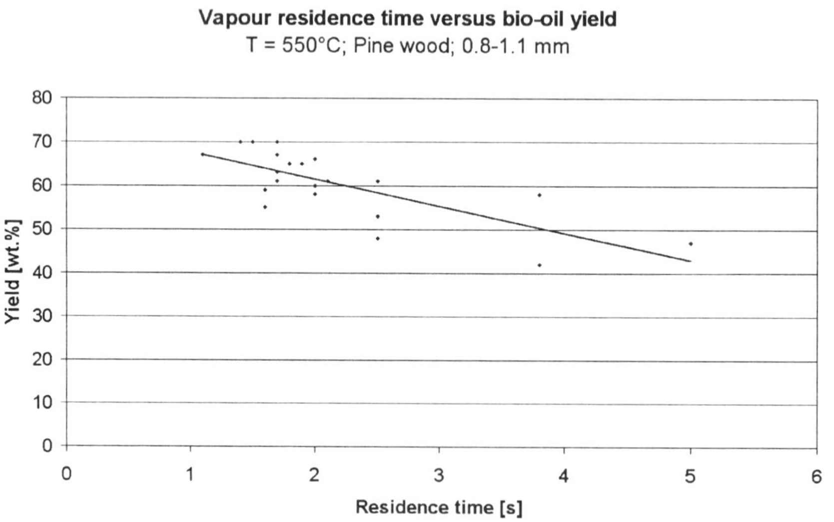

Figure 7. The bio-fuel oil yield versus the vapour residence time.

Wood decomposes almost completely at pyrolysis conditions and so do the bio-fuel oil vapours. Therefore, a proper plant design is aimed at minimizing the residence time of the pyrolysis vapours in the hot enclosure. A measurement campaign has been conducted on the BTG 200P to obtain the relationship between the bio-fuel oil yield and the vapour residence time in the hot enclosure. These data are presented in Figure 7. For these measurements, pine sawdust of a well defined size has been used. Figure 7 shows that 5 weight percent of vapour loss can only be achieved for a vapour residence time which is less than 2 seconds. For a residence time of 5 seconds, more than 20 weight percent of the bio-fuel oil is lost due to gas phase vapour cracking. The BTG 200P usually operates at a gas phase residence time which is less than 1.5 seconds.

Current achievements of the pilot plant:

(1) the plant capacity of 260 kg/hr exceeds its design specifications by 30%.
(2) a total of 20 ton of bio-fuel oil has been produced for various clients.
(3) a bio-oil yield of 70 wt% on dry basis.
(4) the longest continuous run lasted for 4 days.
(5) 3 mm pine wood, sawdust residues from a wood waste supplier, poplar, beech and straw have been successfully converted to bio-fuel oil.

PYROLYSIS ECONOMICS

The pilot plant generated a sufficient amount of know how to satisfy the design goals which have been stated at the start of paragraph 3.2. The economics can be summarized in a nutshell: BTG anticipates selling bio-oil at a price of 6 euro/GJ (100 euro/ton) for quantities larger than 30 kton. The bio-fuel oil is produced from clean wood residues and qualifies for a green label.

PRIME MOVERS ON PYROLYSIS OIL

Rostock University used the bio-fuel oil from BTG in various end-user applications like a combustion furnace and a turbine. Results from both applications are presented hereafter.

COMBUSTION DEVELOPMENT

The combustion characteristics of the bio-fuel oil from BTG has been assessed in a 300 kW(th) furnace at Rostock University. This combustion facility is depicted in Figure 8 and consists of a burner head, an burn out chamber immediately under the burner. Connected to the flame chamber is a horizontal heat exchanger segment and finally the exhaust flue gases are diluted with air to be released by the stack.

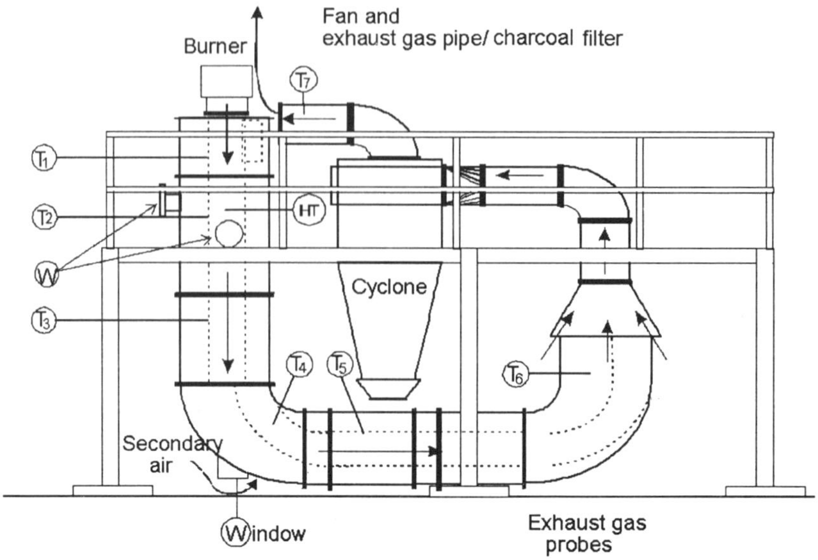

Figure 8: The 300 kW(th) combustor setup of the University of Rostock.

This flame tunnel set-up was originally equipped with a modified oil burner RZ 3.3 from MAN, Germany. The burner was then fitted with a pilot burner for ignition of the bio-oil. A fuel pump delivers the fuel flow rate and pressure for load levels varying between 70 % and 100 % via electrically controlled valves.

In the original configuration, a malfunctioning of the fuel pump did occur when switching from diesel to bio-oil and vice-versa. It was assumed that residues of the different fuels mix and clog the pump and fuel delivery system. In order to avoid this clogging problem, an alcohol fuel is burnt immediately before switching over. After this modification a satisfactory and stable start-up of the combustion facility was achieved.

When burning bio-oil is has been established that it is necessary to preheat the bio-oil to 60 - 70 °C. The pre-heating of the bio-oil leads to a decreased viscosity and enhances the atomisation of the oil. As a consequence, a stable flame was obtained. The condition of the flame tube after 20 minutes of bio-oil operation is shown in Figure 9. No coking residues were found and only a fine, light greyish layer was formed by deposition.

Figure 9: The burner head when viewed from within the combustion chamber. Two spark ignition electrodes can be observed on the right hand side.

Flame tunnel emissions which have been measured with the exhaust gas probes are presented in Table IV.

Table IV. Combustion tunnel emissions.

	Components					
Fuel	O2 %	CO2 %	CO ppm	NO ppm	NO2 ppm	SO2 ppm
Diesel	6.8	12.0	0	65	0	8
Bio-oil	6.0	16.0	9	137	1	7

From Table IV can be observed that the emissions from bio-fuel oil combustion are slightly higher compared to the emissions from diesel combustion.

TURBINE DEVELOPMENT

A Deutz gas turbine has been retrofitted to allow for the feeding and combustion of bio-fuel oil in the turbine. This 80 kW(el) gas turbine is characterized by a nominal rotational speed of 50000 rpm and a pressure ratio of 2,5. A number of test runs have been executed with the turbine rig and a steady operation could be sustained for hours. A summary of the emission measurements of the gas turbine is presented in Table V.

Table V. Turbine emission results.

	Components					
Fuel	O2 %	CO2 %	CO ppm	NO ppm	NO2 ppm	SO2 ppm
Diesel	5	9.7	405	141	9	n.a.
Bio-oil	5	10.5	3434	30	42	n.a.

FUTURE PROSPECTS

Results from the development activities lead to the following route of commercialisation:

(1) At present, bio-fuel oil for co-firing purposes can be produced at commercial terms.
(2) Power generation from bio-fuel oil in turbines or diesel engines is expected within 2 years.

ACKNOWLEDGEMENT

The development program of the rotating cone technology has been supported amongst others by:
- EC DGXII, FAIR program,
- NOVEM, EWAB program.

CFD for the Modelling of Entrainment in Fluidised Bed Fast Pyrolysis of Biomass

H.Gerhauser, S.C. Generalis, R.A. Hague, A.V. Bridgwater
Bio-Energy Research Group, Aston University, Birmingham, B4 7ET, United Kingdom

ABSTRACT Bio-oil generated by fluidised bed fast pyrolysis of biomass holds considerable promise towards reducing dependence on fossil fuels. One problem still holding back progress is the efficient removal of char by entrainment from the reactor, which is impossible to model with traditional empirical correlations. A new model combining computational fluid dynamics (CFD), implemented in the commercial software package CFX, and particle entrainment theory has therefore been developed and found to give good qualitative agreement with experimental entrainment data. The CFD model uses a multiphase Eulerian-Eulerian technique for the bulk of the fluidised bed and individual particle tracking in the freeboard. Furthermore, the predictions of the CFD calculations for a number of important fluidised bed characteristics were checked, namely bubble diameter and velocity, terminal particle velocity and bed expansion. These were found to correspond closely to observations. Finally, several design modifications were tested on a laboratory cold flow model, which showed a significant improvement in char entrainment, and successfully modelled by computational simulation.

INTRODUCTION

Biomass is a significant renewable resource, which will contribute towards replacing fossil fuels as they become depleted and less environmentally acceptable. It is particularly important to find alternatives to crude oil, which still enjoys a virtual monopoly as a transportation fuel.

Presently, there are two main routes for obtaining biomass based liquid substitutes, biological (ethanol, rape oil) and thermochemical conversion (pyrolysis oil and hydrocarbons by Fischer-Tropsch synthesis from gasification). Biological methods are established technology, but suffer from several drawbacks, notably a low overall conversion efficiency and the need to divert food production. Fast pyrolysis on the other hand can deal with a larger variety of organic feedstocks allowing far greater volumes and the use of materials that are now wasted. The technology, however, is still young and there are a number of areas, such as liquid stability, heating value and viscosity, where improvements and further research are desirable.

Aston University is one of the institutions engaged in that work and is investigating fast pyrolysis in fluidised bed reactors. The specific sub-task dealt with in this paper is the removal of char, which is an unavoidable by-product that has been found to catalyse reactions that lower the pyrolysis oil yield. It is achieved by selective and enhanced entrainment from the bed followed by separation in a series of cyclones.

Entrainment may be defined as the carryover of ejected particles, while selective entrainment of finer or less dense particles is often referred to as elutriation. In most industrial processes, neither entrainment nor elutriation are desirable[1], which is in sharp contrast to this particular application. Consequently, there is very little research aimed specifically at enhancing the selective removal of less dense material from fluidised beds. Most research on entrainment is based on dimensional analysis applied to experimental data either with no or very limited consideration of the underlying physics[2]. Predictions made from these correlations are limited to very simple geometries. They may vary widely even for reactor arrangements close to the experimental conditions they are based on, and are often completely unreliable when conditions are markedly different. In several internal studies they have been found inadequate for entrainment and elutriation predictions in the fluidised bed system under investigation. The problem is too complex to be adequately represented by a small number of ordinary equations that would simply require substitution of a few parameters to obtain the rates of entrainment of the different particle size fractions.

Hence more elaborate ways of modelling were considered, namely CFD and the discrete element method (DEM)[3,4,5]. Computational fluid dynamics was chosen, as it can be applied to systems with a very large number of particles, where DEM becomes far too computationally expensive. In CFD the flow domain is discretised into a number of grid points for which the fluid flow equations (Navier-Stokes) are then solved numerically. There are two ways to deal with solid particles in CFD. In particle tracking Newton's second law is applied to the particles and the interaction with the fluid is obtained via drag terms. Interactions between the particles are usually neglected rendering this approach inappropriate for modelling high particle concentrations such as occur inside fluidised beds. To deal with large particle concentrations a two-phase approach is normally taken that considers locally averaged values for the particle phase, rather than individual particles.

CFD AND ENTRAINMENT THEORY

INTRODUCTION

The computational fluid mechanics calculations were implemented using the commercial software package CFX, which allows considerable flexibility through User Fortran routines. The bulk of the particles in the fluidised bed was modelled using the standard Eulerian-Eulerian two-phase model that treats the particles not as single entities but rather as a locally averaged second phase. Ejected particles, however, were tracked individually. The starting positions and velocities of the particles were obtained by assuming a bubble ejection mechanism based on experimental observations, Eulerian CFD results and the entrainment literature[1].

TWO-PHASE FLUID BED MODEL

A Eulerian-Eulerian method is applied that uses locally averaged variables for the particle and fluid phases. The optimum scale of averaging was obtained by running the same simulation with successively better grid and time step resolutions until the numerical solution became sufficiently insensitive to further increases in the number of grid points and time steps.

The equations resulting from the averaging process are of a similar form to the standard Navier-Stokes equations. Each phase has its own set of equations that are solved individually, but linked through inter-phase transfer terms. They were first derived by Gidaspow[6]. The equations for continuity and momentum are as follows.

Continuity:

$$\frac{\partial}{\partial t}(r_\alpha \rho_\alpha) + \nabla \bullet (r_\alpha \rho_\alpha \mathbf{U}_\alpha) = \sum_{\beta=1}^{N_P} (\dot{m}_{\alpha\beta} - \dot{m}_{\beta\alpha})$$

Momentum:

$$\frac{\partial}{\partial t}(r_\alpha \rho_\alpha \mathbf{U}_\alpha) + \nabla \bullet (r_\alpha (\rho_\alpha \mathbf{U}_\alpha \otimes \mathbf{U}_\alpha - \mu_\alpha (\nabla \mathbf{U}_\alpha + (\nabla \mathbf{U}_\alpha)^T)))$$

$$= r_\alpha (\mathbf{B} - \nabla p_\alpha) + \sum_{\beta=1}^{N_P} c_{\alpha\beta}^{(d)} (\mathbf{U}_\beta - \mathbf{U}_\alpha) + \mathbf{F}_\alpha + \sum_{\beta=1}^{N_P} (\dot{m}_{\alpha\beta} \mathbf{U}_\beta - \dot{m}_{\beta\alpha} \mathbf{U}_\alpha)$$

where r refers to the volume fraction, the subscripts α and β to the respective phases and **B** to body forces

In order to obtain the interphase momentum transfer term in regions with a gas volume fraction less than 0.8 the Ergun equation[7] is adapted:

$$c_{\alpha\beta}^{(d)} = 150 \frac{(1-r_\alpha)^2 \mu_\alpha}{r_\alpha d^2} + 1.75 \frac{(1-r_\alpha) \rho_\alpha |\mathbf{U}_\beta - \mathbf{U}_\alpha|}{d}$$

In regions with lower particle volume fractions a modified form of the single particle drag correlation is used instead:

$$c_{\alpha\beta}^{(d)} = \frac{3}{4} \frac{C_D}{d} r_\beta \rho_\alpha |\mathbf{U}_\beta - \mathbf{U}_\alpha|$$

The drag coefficient C_d is obtained from the particle Reynolds number:

$$\mathrm{Re} = \frac{\rho_\alpha U d}{\mu_\alpha}$$

C_d	$= 24\ \mathrm{Re}^{-1}(1+0.15\mathrm{Re}^{0.687})$	for Re<1000
or	$= 0.44$	for Re>1000

The equations are solved numerically using a pressure correction technique, commonly known as SIMPLE [8], and a semi-implicit method to handle the strong coupling between the two phases similar to the IPSA algorithm by Spalding[9].

The inlet boundaries are specified using the standard Dirichlet boundary condition with a constant velocity. The outlets are pressure boundaries. The initial condition is no flow and particles filling the bed up to a specified height (that is the volume fraction of particles is specified).

PARTICLE TRACKING

Particle tracking in the freeboard is performed with the help of Newton's second law taking account of the drag force exerted by the gas on the particles:

$$\mathbf{F}_D = 0.125\pi d^2 \rho C_D |\mathbf{v}_R| \mathbf{v}_R$$

The effect of the particle movement on the fluid phase, and interactions between particles are neglected. This is justified because the number of particles in the freeboard is low. Particles are allowed to bounce off walls, where a coefficient of restitution of 0.8 is applied, which is based on the elasticity of sand and metal collisions but can be varied considerably (from 0.5 to 0.9) without affecting the entrainment results noticeably.

ENTRAINMENT MECHANISM

When a fluid is allowed to rise through a loose bed of particles, there will be a pressure drop across the bed that acts as a lifting force on the particles. At the point of incipient fluidisation this pressure drop has become large enough to balance the weight of the particles forming the bed. Further fluid flow then percolates through the bed in the form of bubbles.

The eruption of these bubbles at the bed surface is responsible for the ejection of particles of all size classes into the freeboard. Very fine particles may even be entrained without the assistance of a bubble, if their terminal falling velocity is below the superficial gas velocity in both the bed and the freeboard.

Linking particle tracking and the two-phase fluidised bed model is not straightforward, because individual particle velocities are subsumed in the local averages and therefore not directly available from the fluidised bed model. Consequently, the initial particle velocities for particle tracking have to be obtained from what the fluidised bed model does give, namely the extent of bed expansion, the position and velocity of the bubble, gas velocities and local averages for particle velocities.

Several models in the literature[1] use the bubble velocity to estimate the initial particle velocity without distinguishing between different particle sizes and densities, that is heavier and bigger particles are given the same starting velocity as lighter and smaller ones. This wrong assumption implemented in particle tracking would indicate that large particles reach a greater height than small ones, completely contrary to experimental observation.

The particle tracking is therefore divided into two phases. In the first the acceleration within the fast gas jets that are found in and just above the erupting bubbles is simulated and in the second the accelerated particles are followed through the freeboard. In practice this means an average gas jet velocity and the time and distance the particle spends within that jet are estimated from the fluidised bed model, and then put into the particle tracking model. This gives an initial particle velocity for tracking the particles through the freeboard. The initial position for this second phase of particle tracking is also obtained from the fluidised bed model.

A final question that needs to be answered is the number of particles drawn into the gas jets during bubble eruptions. Based on the literature[1] and experimental observations undertaken during the course of this study, it is assumed that a layer with a thickness equal to the mean particle diameter in the bed is involved in the ejection process. From the surface exposed to a particular gas jet the total mass of involved particles is then calculated. Size classes for char and sand particles are allocated the same percentage of the ejected particle mass as in the bed as a whole.

The total entrainment is calculated by dividing it up into small classes (for example char between 180 and 300 microns, ejected from a bubble between 2 and 3 cm in diameter) and then summing the individual contributions.

At this point it may also be clarified that most two-phase Eulerian calculations were performed in two dimensions, while most particle tracking operations were done in three dimensions. This was due to excessive computing times being required by the two-phase calculations.

SUMMARY OF PROCEDURE FOR OBTAINING ENTRAINMENT

In order to make the procedure used for obtaining entrainment clearer, it is helpful to restate it in a simplified and shortened form:

1. A Eulerian two phase calculation is performed in two dimensions giving locally averaged velocities and volume fractions for the particle and gas/vapour phases. Furthermore, bed expansion and bubble characteristics are obtained.
2. It is assumed that particles are predominantly entrained when a bubble hits the surface of the bed. Furthermore, it is assumed that the volume of particles ejected at that point is equal to the volume of a layer of average particle size thickness. The size distribution of the ejected particles is assumed to be the same as in the bed.
3. Individual particles are tracked through a constant flow field obtained by extrapolating the 2-D field at the moment of bubble eruption into 3-D. The number of particles leaving the flow field are counted by the computer. A very large number of particles needs to be tracked (of the order of 5000) to obtain acceptable results.

EXPERIMENTAL WORK

Entrainment and important fluidised bed parameters were measured for a 1 kg/h and a 5 kg/h fast pyrolysis reactor. In addition, a cold-flow model of the 1kg/h rig was built to study fluidisation aspects that were difficult to obtain from the pyrolysis reactor. The cold-flow model was subsequently modified to validate the model's capability to deal with changes in the reactor geometry. The cold flow rig is illustrated in Figure 1.

PARTICLE SIZE ANALYSIS

The bed material is sand, which is mixed with char produced by the pyrolysis of wood. Three methods were used for size analysis: sieving, microscopic examination and laser diffraction. All the methods gave similar values for the particle diameters of sand and char, though microscopic examination showed that many char particles had a needle form and therefore one significantly longer dimension that could not be captured by the other two methods. The different particles are illustrated in Figure 2 and the initial size distributions for sand and char in Figures 3 and 4.

Fig. 1 Cold fluid bed model

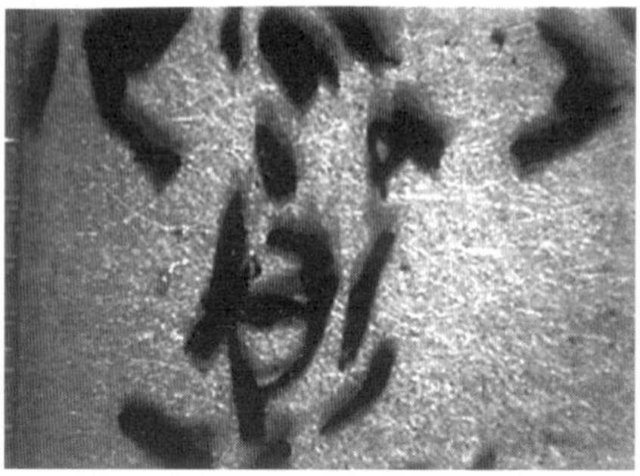

Sieved char less than 351 microns

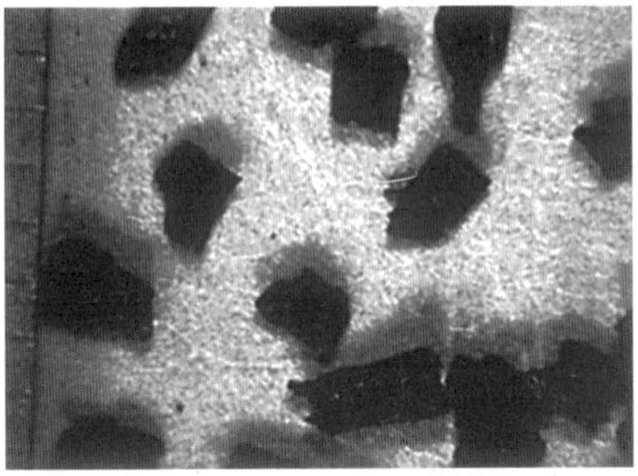

Sieved char greater than 1003 microns

Unsieved wood nominal size 3 mm.

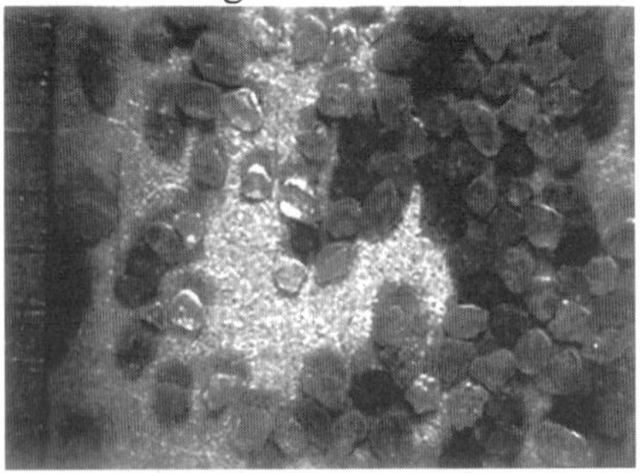

Sieved sand between 500 and 600 microns

Fig. 2 Char, wood and sand particles (Scale to the left of each picture, units in mm).

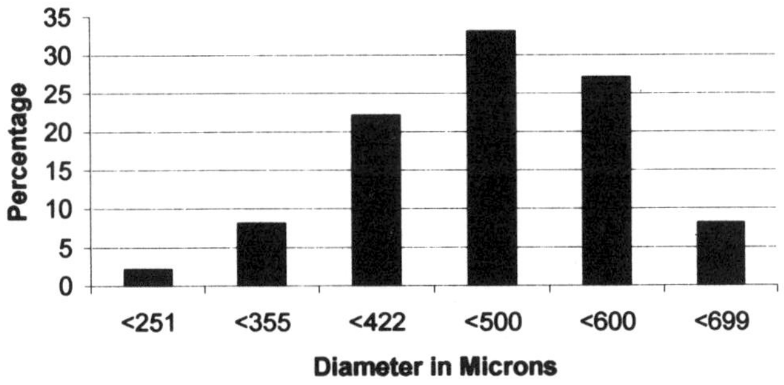

Fig. 3 Initial size distribution for sand (size classes based on British Standard).

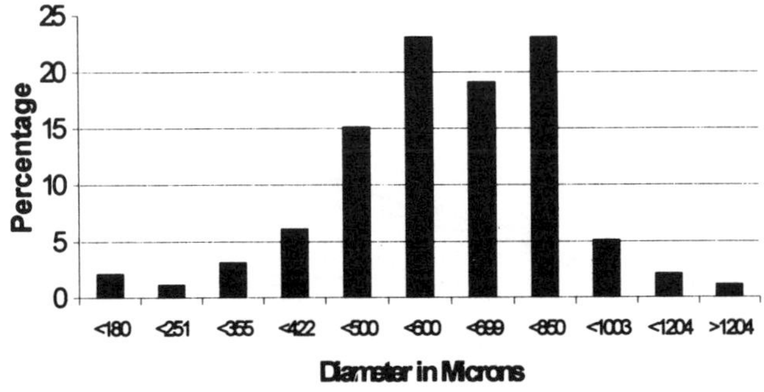

Fig. 4 Initial size distribution for char (also based on BS for sieving).

CFD RESULTS COMPARED WITH EXPERIMENT

The qualitative fit of the CFD work and experiment is illustrated in Figure 5. As can be seen clearly, that fit is very good.

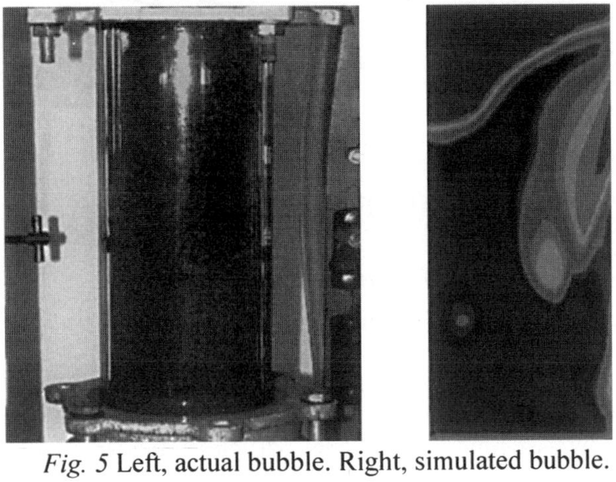

Fig. 5 Left, actual bubble. Right, simulated bubble.

TERMINAL FALLING VELOCITY

The terminal velocity that particles reach when in free fall is an important particle characteristic. It is also fairly straightforward to measure and was therefore used to check the accuracy and validity of the particle tracking model. The measurement technique is entrainment of a small sample of a particular size class in a narrow tube taking the average of the band within which 90% of the small sample are entrained. In the computational modelling of the particle tracks the upper end of the class size is used as the particle diameter and the particles are started from rest and from a falling velocity greater than their terminal velocity.

For sand the fit is very good (see Figure 6). The measured velocity is within the band indicated by particle tracking. For char (see Figure 7) the picture is slightly more complicated. For smaller diameters, the measured falling velocity is above and not within the range given by the particle tracking. The microscopic examination explains the reason.

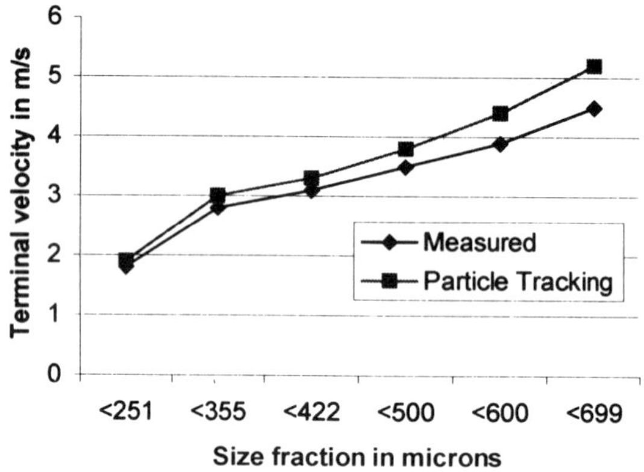

Fig. 6 Terminal velocity of sand.

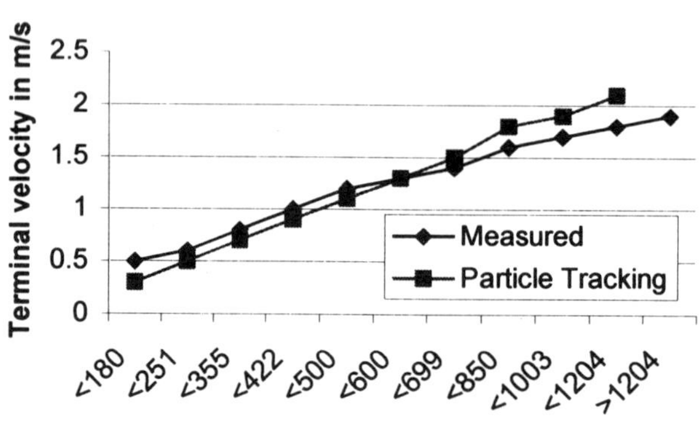

Fig. 7 Terminal velocity of char.

While sand particles approximate spheres fairly closely, char particles are long and thin cylinders. Furthermore, the smaller the char particle the greater the aspect ratio. This may be explained by assuming that the char particles fall apart along the grain rather than across it. As sieving measures the diameter of the thin cylinders and not their length, the effective diameter is underestimated, particularly so for smaller char particles. Consequently, the diameters used in the particle tracking undertaken to estimate entrainment had to be adjusted. This was done by shifting the measured curve to the right until the measured and computed value for the size class corresponded.

BUBBLE VELOCITY

This variable was measured experimentally using a video camera. Several bubbles were analysed for each velocity and an average taken. Likewise a number of bubbles generated by the CFD software for the same conditions were examined. These two values are also compared (see Figure 8) to a widely employed correlation for bubble velocity by Davidson and Harrison[10]:

$$u_B = (u-u_{mf}) + 0.711\,(gD)^{0.5}$$

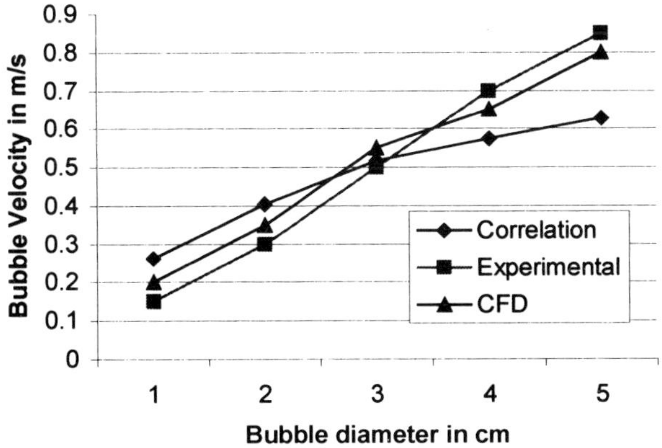

Fig. 8 Bubble velocity.

The CFD results are in much closer agreement with the experimental values than the above correlation, in particular for larger bubble diameters, where the bubble velocity is significantly underpredicted.

There is quite a degree of variation between different bubbles. Even for exactly the same conditions they may vary considerably in terms of velocity, size and shape. This dynamic and irregular behaviour is very well captured in the CFD analysis and cannot possibly be obtained by a simple correlation that is only capable of giving an average.

BUBBLE DIAMETER

In comparison to the bubble velocity the diameter shows a much greater range of variation. Correlations therefore usually only give the maximum diameter rather than an average that would be difficult to apply. A correlation by Mori and Wen[11] has been chosen for purposes of comparison (see Figure 9) with experimental and CFD results:

$$d_B = (u-u_{mf})^{0.4} * [1.49D^2 - (1.49D^2 - 1.38g^{-0.2} A_B^{\ 0.4}) e^{-0.3h/D}$$

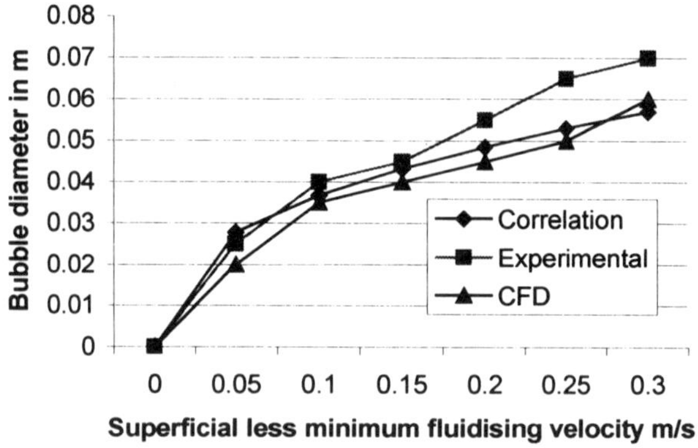

Fig. 9 Maximum bubble diameter.

The correlation by Mori and Wen predicts the actual maximum diameter fairly well, though it underpredicts for higher minimum fluidising velocities. The CFD calculations give lower values for the maximum at all velocities. One reason is the small number of bubbles that could be modelled within the available computing time in comparison to the number available for experimental study. Furthermore, at higher velocities convergence problems were encountered again reducing the number of bubbles available for comparison.

BED EXPANSION

The surface of the fluidised bed is not static, but rather subject to dynamic movement. For validation of the model the maximum height reached by the surface is therefore plotted (Figure 10).

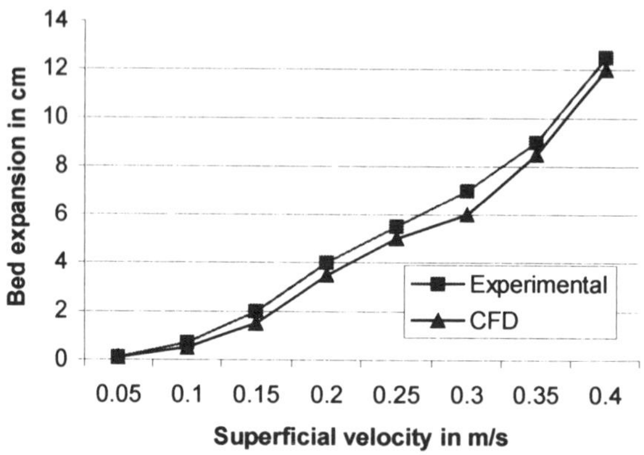

Fig. 10 Bed expansion.

ENTRAINMENT

From Figures 6 and 7 (shown earlier), which illustrate the terminal falling velocities of the sand and char particles, it can be seen that virtually all char particles have a lower terminal velocity than the vast majority of sand particles. Based on this consideration, separation can therefore be effected within a gas stream that has a superficial velocity in between those of sand and char, that is about 1.5 to 2 m/s. Based on this consideration a number of modifications was implemented both in the CFD modelling work (see Figure 11-14) and experimentally.

Figure 15 illustrates sand entrainment on the original cold flow rig. While the trend for sand entrainment estimated from the CFD calculations is correct, on a quantitative basis the under-prediction is significant, namely by a factor of up to three for the lower superficial velocities. The main source of error is the entrainment mechanism.

The next graph (Figure 16) shows the percentage of char entrained out of the original and modified designs. At the start of the experiment the bed consists of 95% sand and 5% char. The superficial velocity is 0.45 m/s.

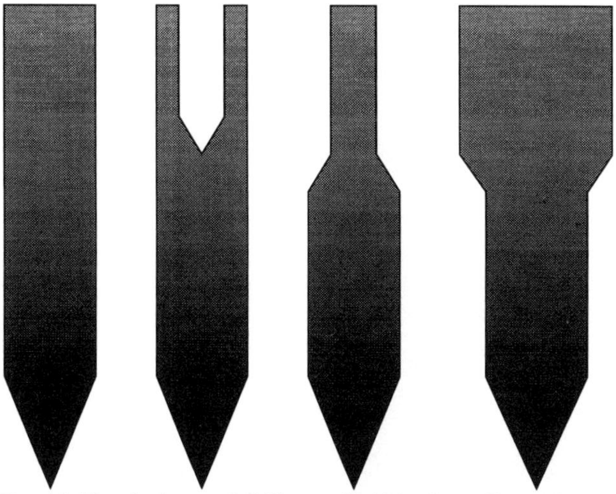

Fig. 11 Simulations of different fluid bed configurations

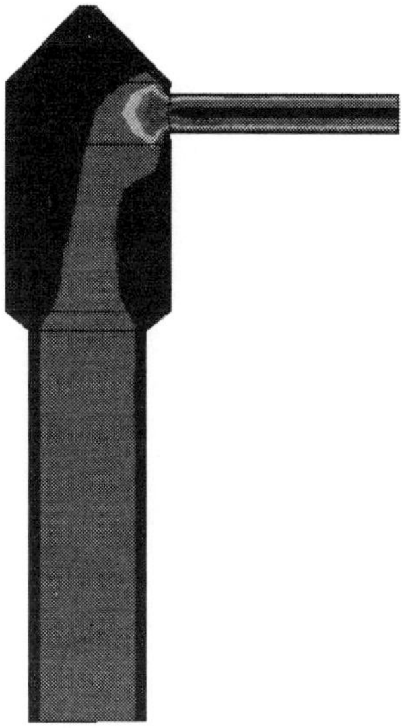

Fig. 12 Simulation of original 1 kg/h cold rig showing a contour plot of vapour speed.

Fig. 13 Simulation of the modified 5 kg/h reactor

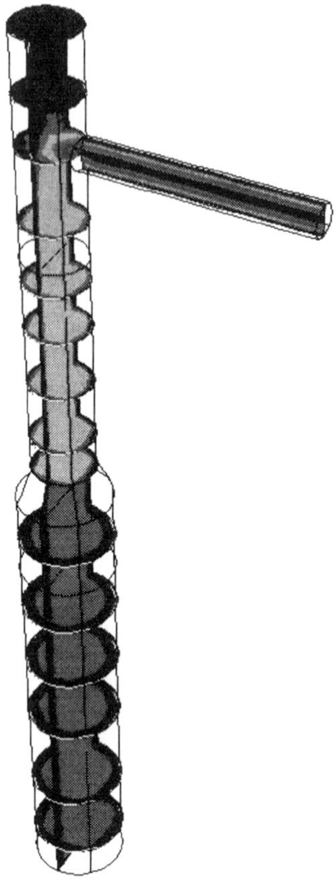

Fig. 14 Modified cold flow rig showing contour plots of speed along cross sections of the reactor

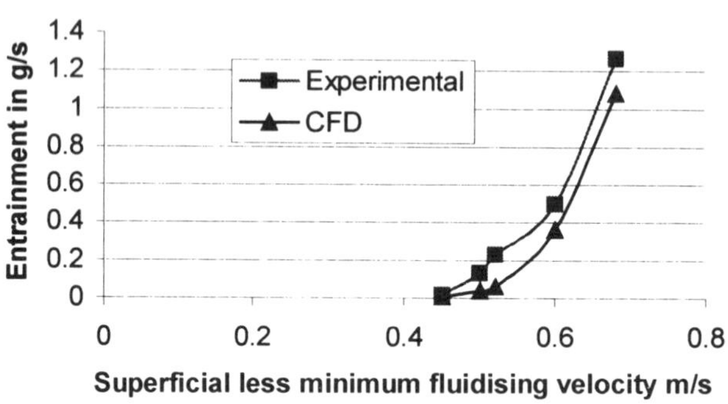

Fig.15 Sand entrainment on the original cold flow rig.

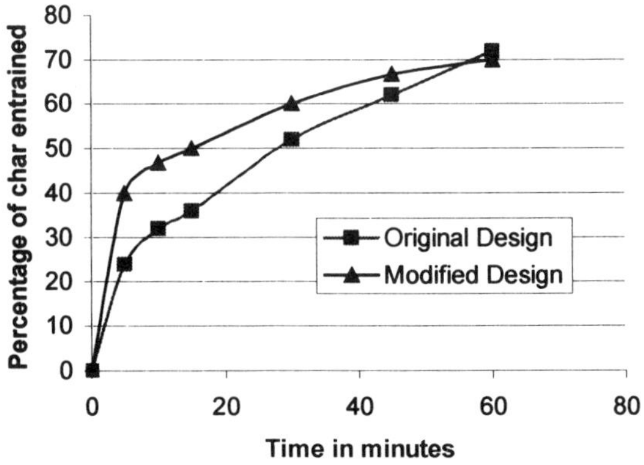

Fig. 16 Comparison of mixed sand and char entrainment in original and modified cold flow rig design.

The graph indicates that the initial entrainment rate for the modified design is much greater than for the original design. The analysis of the entrained material also showed that in the improved design ten times fewer sand particles were entrained and that their distribution was skewed with no sand particles greater than 500 microns. On a qualitative basis that is exactly what the CFD calculations predict (see Figure 17). The figure shows that the majority of large char particles and nearly all smaller char particles that are ejected are entrained, while sand particles with a diameter of 800 microns fall back.

Finally, a comment on the different geometries should be made. For the two reactor sizes investigated so far, a simple freeboard widening was found to reduce both overall entrainment and the selectivity of that entrainment with respect to char, while a narrowing (either directly, or by using an insert) proved beneficial on both counts. It should be stressed, however, that for different scales, bed materials or other changes a new calculation is required.

CONCLUSIONS

Classical empirical correlations have been found to be incapable of satisfactorily modelling entrainment in fluidised bed fast pyrolysis reactors.

A new model that combines CFD (for Eulerian two-phase modelling and individual particle tracking) with an ejection mechanism has been developed. This has been successfully validated against experimental data for a number of different geometries with agreements within 10% at higher entrainment rates and within 50% at low entrainment rates. The poorer agreement at low entrainment rates is believed to be due to the fact that some of the less frequent bed conditions are not captured in the CFD model.

For predictions relating to fluidised bed characteristics, the agreement is much better than for entrainment, and is typically within 5% of the measured variables.

Using the cold flow rig an improvement in the initial char entrainment rate by a factor of two and enhanced separation from sand by a factor of ten have been achieved and correctly predicted by the entrainment model.

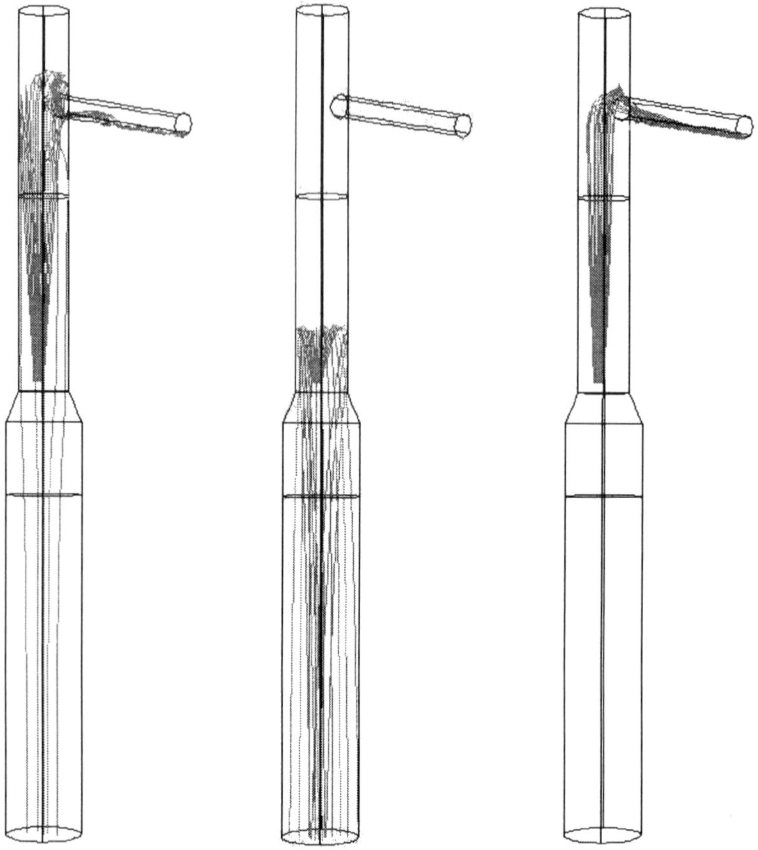

Fig. 17 Particle Tracks,
Left - large char particles (800 microns); Centre - sand (800 microns); Right – small char particles (400 microns)
All particles close to maximum ejection height and velocity.

ACKNOWLEDGEMENTS

The work reported in this paper is based on contributions to EC Contract JOR3-CT97-0197 and the authors would like to acknowledge the support of the European Commission JOULE Programme.

NOMENCLATURE

A_B	distributor area over number of orifices	m^2
$\mathbf{B}$	body forces	N/m^3
$c_{\alpha\beta}$	interphase momentum transfer term	kg/m^3s
C_D	drag coefficient	
d_B	bubble diameter	m
d	particle diameter	m
D	bed diameter	m
$\mathbf{F}$	interphase non drag forces	N/m^3
$\mathbf{F_D}$	drag force	N
g	gravitational acceleration	m/s^2

$\dot{m}$	mass flow between phases	kg/m^3s
h	bed height	m
N_P	number of phases	
p	pressure	Pa
r	volume fraction	
Re	Reynolds Number	
t	time	s
U	speed	m/s
$\mathbf{U}$	velocity	m/s
u	superficial velocity	m/s
u_{mf}	minimum fluidising velocity	m/s
u_B	bubble velocity	m/s
$\mathbf{v_R}$	fluid velocity relative to particle	m/s
α,β	subscripts for the two phases	
ρ	density	kg/m^3
μ	viscosity	kg/ms

REFERENCES

1. Fung A.S., Hamdullahpur F. (1993) Effect of bubble coalescence on entrainment in gas fluidized beds. *Powder Technology,* **77**, pp 251-265
2. Zenz F.A. et al (1958) A Theoretical-Empirical Approach to the Mechanism of Particle Entrainment from Fluidized Beds. *American Institute of Chemical Engineering Journal*, **4**, pp. 472-479
3. Witt P.J. et al (1998) A numerical model for predicting bubble formation in a 3D fluidised bed. *Applied Mathematical Modelling*, **22**, pp. 1071-1080
4. Tsuji Y. et al (1993) Discrete particle simulation of two-dimensional fluidised bed. *Powder Technology*, **77**, pp. 79-87
5. Hoomans B.P.B et al (1996) Discrete particle simulation of bubble and slug formation in a two-dimensional gas-fluidised bed: A hard-sphere approach. *Chemical Engineering Science*, **51**, pp. 99-118
6. Gidaspow D. (1994) *Multiphase Flow and Fluidization.* Academic Press
7. Ergun S. (1952) Fluid Flow through packed columns. *Chem. Eng. Prog.*, **48**, pp. 89-94
8. Patankar S.V., Spalding D.B.(1972) A calculation procedure for heat, mass and momentum transfer in three-dimensional parabolic flows. *International Journal of Heat and Mass Transfer*, **15**, pp. 1787-1806
9. Spalding D.B. (1980) Numerical computation of multiphase flow and heat transfer. In: Recent Advances in Numerical Methods in Fluid Mechanics (Ed. by C. Taylor & K. Morgan), Pineridge, Swansea, pp. 139-168
10. Davidson J.F., Harrison D. (1963) *Fluidised Particles.* Cambridge University Press
11. More S. and Wen C.Y. (1975) *American Institute of Chemical Engineering Journal*, **21**, p. 117

Modelling, Scale-up and Demonstration of a Vacuum Pyrolysis Reactor

J. Yang, D. Blanchette, B. de Caumia and C. Roy
Institut Pyrovac Inc., 333 Franquet Street, Sainte-Foy, Quebec G1P 4C7, Canada

ABSTRACT A novel reactor configuration has been developed in our laboratory which addresses the heat transfer limitations usually encountered in vacuum pyrolysis technology. In order to scale-up this reactor to an industrial scale, a systematic study on the heat transfer, the chemical reactions and the movement of the bed of particles inside the reactor has been carried out over the last ten years. Two different configurations of moving and stirred bed pilot units have been used to scale-up a continuous feed vacuum pyrolysis reactor, in accordance with the principle of similarity. A dynamic model for the reactor scale-up was developed, which includes heat transfer, chemical kinetics and particle flow mechanisms. Based on the results of the experimental and theoretical studies, an industrial vacuum pyrolysis reactor, 14.6 m long and 2.2 m in diameter, has been constructed and operated. The operation of the pyrolysis reactor has been successful, with the reactor capacity reaching the predicted feed rate of 3000 kg/h on a biomass feedstock anhydrous basis.

INTRODUCTION

Over the last ten years, the vacuum pyrolysis process has been developed at the bench and pilot scales at the Université Laval and Pyrovac Institute Inc. in Québec, Canada. This thermal decomposition process enables a large variety of solid and semi-liquid wastes to be transformed into useful products. Vacuum pyrolysis is typically carried out at a temperature of 400-500°C and a total pressure of 2-20 kPa [1]. These vacuum conditions allow the pyrolysis products to be rapidly withdrawn from the hot reaction chamber, thus preserving the primary fragments originating from the thermal decomposition reactions.

A limitation of vacuum pyrolysis technology is heat transfer. Previous studies have shown that the rate of heat transfer is essentially the rate limiting step for pyrolysis reactions [2]. Conventional pyrolysis reactors such as multiple hearth furnaces, rotary kilns and screw type reactors exhibit overall heat transfer coefficients ranging from 10 to 60 $W \cdot m^{-2}K^{-1}$ [3], depending on the type of feedstock handled. The low thermal conductivity of the feedstock materials partially explains why the heat transfer fluxes are low.

The new vacuum pyrolysis industrial scale reactor, described herein, uses an indirect heating system with eutectic molten salts flowing inside the heating plates. The feedstock moving on top of these plates is heated both by conduction and radiation mechanisms. In addition, the novel transport system in the reactor efficiently agitates the feedstock, thus greatly improving the heat transfer. The process is commercialized under the trademark of Pyrocycling^{TM}. It is an owned technology [4].

In a properly designed industrial scale reactor, feedstock conversion is achieved at a certain throughput capacity. In order to scale-up the reactor, heat and mass transport phenomena must be studied. This includes heat transfer phenomena, feedstock conversion kinetics and the movement of particles inside the reactor. In this work, both experimental and theoretical studies were carried out to investigate these phenomena. Two different configurations of moving and stirred bed reactors, the batch scale rotative and a continuous feed Process Development Unit (PDU), have been used to generate the data in accordance with the principle of similarity. A dynamic model to scale-up the reactor was then tested.

DESCRIPTION OF THE PYROCYCLING^{TM} PROCESS

The process flow chart of the new vacuum pyrolysis industrial scale reactor is presented in Figure 1. From the feedstock storage facility, separated from the plant building, the feedstock is loaded in a hopper and carried along by extracting screws into a pneumatic conveying system up to a cyclone, located on the roof of the plant building. The feedstock flows down by gravity from the cyclone into a vibrating screen separator, providing a granulometry range between US Sieves 0.5 and 40 mesh. The feedstock with the specified particle size then sluices out through the vacuum feeding system, consisting of two rotary valves set in line and between which the auxiliary liquid-ring vacuum pump suction line is connected. A positive pressure drop is maintained over the second rotary valve (between the vacuum pump suction and the reactor) in order to avoid the escape of pyrolysis gas from the reactor. The feedstock is then fed into the 14.6 m long and 2.2 m diameter reactor by screw conveyors.

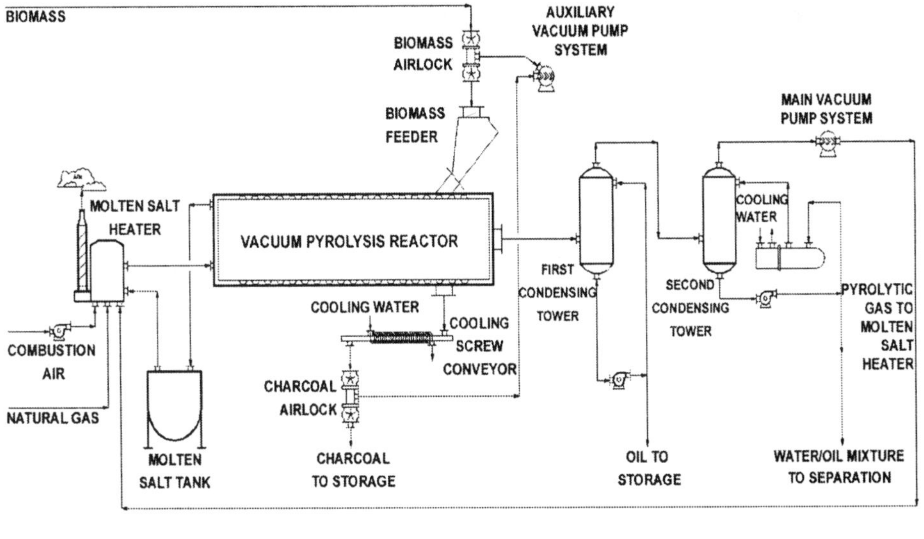

Figure 1 Process Flow Chart of the Vacuum Pyrolysis Pyrocycling^{TM} Technology

Inside the reactor, the biomass is pyrolyzed on two heating plates, while being carried along and mixed by a raking system. The 10.4 m long by 1.2 m wide heating plates are composed of tubes in which molten salts circulate as a heat carrier medium. The reactor is also heated by electric tracers installed on the exterior shell. The total pressure in the reactor is maintained at 20 kPa.

At the outlet of the reactor, the solid product is moved out of the reactor by a three-stage water-cooled screw conveyor system. The pyrolytic vapours and gaseous products are withdrawn from the reactor via a 600 mm header to a two-stage condensation system, where the heavy bio-oils are condensed in the first packed tower while a mixture of light bio-oil and pyrolytic water is condensed in the second packed tower. The bio-oil/water mixture is afterwards separated. The bio-oil fractions from both packed towers are mixed together and stored (either directly in 1 m^3 polyethylene containers or into a 50 000 L tank). The aqueous phase is neutralized with liquid caustic prior to being disposed of (temporary solution) into the municipal sewer system. The non condensable gas drawn up from the second packed tower by the main vacuum pump is further cooled past the pump, compressed to 170 kPa pressure and fired in combination with natural gas into a molten salt heater to provide the energy required for the pyrolysis reaction.

From the molten salt 9 m^3 storage tank, which is equipped with heating elements as well as the piping loop circuit, the molten salt is pumped to the furnace to be heated, circulates then into the reactor heating plates and flows back into the tank. The total surface area occupied by the process equipment, excluding the biomass and product storage rooms, is equivalent to 850 m^2 split into two floors.

HEAT TRANSFER FUNDAMENTALS

When the biomass feedstock particles enter the vacuum pyrolysis reactor, they are immediately heated by the heating plate and pushed forward by the blades. The temperature of the feedstock increases quickly while the feedstock is moving towards the outlet of the reactor. It thus produces a temperature distribution along the longitudinal direction of the reactor, with the lowest temperature at the entrance position and the highest temperature at the reactor outlet. At a certain position in the reactor the feedstock temperature reaches a chemical decomposition temperature. While the feedstock flows forward and its temperature further increases, the pyrolysis reactions extend and convert the feedstock into vapours, gases and solid residues. At the outlet of the reactor, the maximum temperature is reached and the final conversion is achieved.

During the pyrolysis process, the final conversion mainly depends on three phenomena: the heat transfer from the reactor to the feedstock, the feedstock movement in the reactor and the kinetics of pyrolysis reactions. The heat transfer rate determines the rate of temperature increase of the feedstock. The feedstock movement behaviour determines the residence time of the feedstock particles in the reactor. In turn the heating rate and the residence time control the quantity of energy transferred and thus the temperature distribution throughout the feedstock in the reactor. Once the temperature distribution is known, the kinetic behaviour of the feedstock determines the final conversion at the reactor outlet.

A systematic study has been carried out in our laboratories over the last decade on the above three phenomena with the objective to design industrial reactors (See Figure 2). Three theoretical and semi-theoretical models have been developed. The heat transfer model, namely the "Surface Renewal Model" has been developed to predict the heat transfer coefficient in the reactor. The feedstock movement model, known as the "Single Blade Volume Output Model" was developed to predict the average residence time of the feedstock in the reactor. The kinetic model, here a "Parallel Kinetic Model", predicts the conversion of the feedstock as a function of temperature. The combination of these three models gives birth to a dynamic model to scale-up the reactor by allowing the prediction of the final conversion of the feedstock.

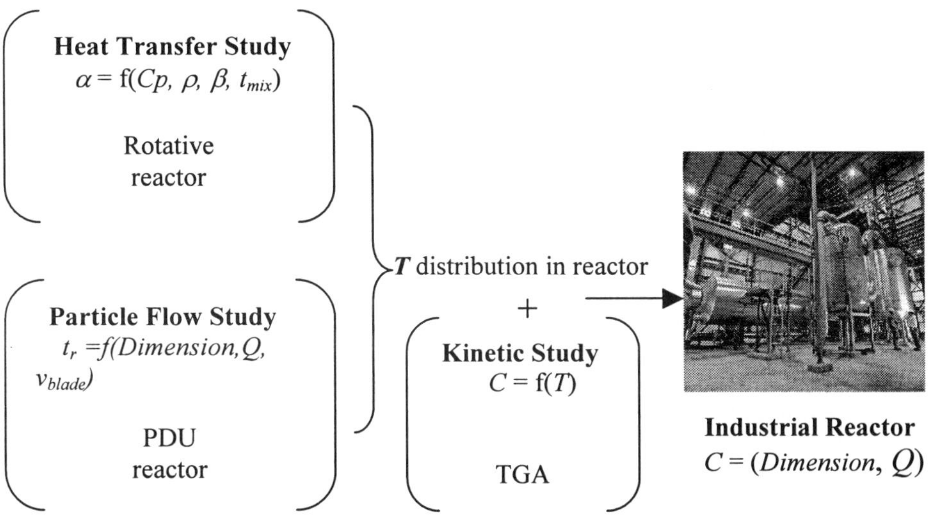

Figure 2 Chemical and Transport Phenomena Occurring in a Vacuum Pyrolysis Reactor and Their Role in Reactor Design.

THE "SURFACE RENEWAL MODEL"

The proposed Surface Renewal Model is based on the heat transfer model that Schlünder [5] developed for the indirect contact vacuum drying of solid particles. In Schlünder's model, particle movement due to the agitation was described as a series of time periods during which the bed of particles is assumed to be static, followed by a rapid mixing of the particles. Thus the average heat transfer coefficient during the static period can be calculated using the Penetration Theory, using the following equation:

$$\alpha_{bed} = \sum_{k=1}^{\infty} B_k \alpha_k \qquad (1)$$

where α_{bed} is the overall heat transfer coefficient in the bed, and B_k and α_k represent the amount and the heat transfer coefficient of fraction k. The variables, B_k and α_k, are calculated by the following equations:

$$B_k = \beta \cdot (1-\beta)^{k-1} \qquad (2)$$

$$\alpha_k = \frac{\alpha_{ws}}{1 + \frac{\sqrt{\pi}}{2}\sqrt{\frac{\alpha_{ws}^2}{(\rho C_p \lambda)_{bed}} k t_{mix}}} \tag{3}$$

In equation (3), ρ, C_p and λ are thermophysical properties linked to the nature of the feedstock. Thus, the main parameters which govern the reactor design are the surface-to-particle contact heat transfer coefficient, α_{ws}, and the bottom layer renewal efficiency, β. Since these two parameters are directly influenced by the reactor design, they allow a relationship to be established between the reactor design and the overall heat transfer throughout the bed of particles.

THE "SINGLE BLADE VOLUME OUTPUT" MODEL

In a moving and stirred bed reactor, the most important parameters representing the feedstock flowing behavior are the residence time t_r, the velocity of feedstock movement v_{bed} and the feedstock bed thickness in the reactor h_{bed}. The "Single Blade Volume Output" model proposed in this work is based on the assumption that at the outlet of the reactor, the amount of feedstock accumulating in front of each agitation blade is determined by the value of h_{bed} using a power expression.

$$V_{blade} = c \cdot (h_{bed})^n \tag{4}$$

where c and n are constants. Their value depends on the frictional properties of the material and the configuration of the agitation blades. The value of c and n can be determined experimentally. If a steady flow in the reactor is assumed, the mass flow in the reactor, (Q_{in}), must be equal to the mass flow out of the reactor:

$$Q_{in} = \rho_{bed} \cdot V_{blade} \cdot n_{blade} \cdot \frac{v_{blade}}{L_{chain}} \tag{5}$$

where Q_{in} is the feedstock feed rate, ρ_{bed} is the bulk density of the feedstock, n_{blade} and v_{blade} are the number and the velocity of the agitation blades and L_{chain} is the length of the chain to which the agitating blades are attached. V_{blade} and h_{bed} are determined by combining Eq. (5) and Eq.(6). The velocity of feedstock movement v_{bed} can then be calculated by the following expression which is derived from a mass balance around the reactor:

$$v_{bed} = \frac{V_{blade} \cdot v_{blade} \cdot n_{blade}}{h_{bed} \cdot W_{bed} \cdot L_{chain}} \tag{6}$$

Using equation (6), the residence time of the feedstock in the reactor can then be determined by equation (7):

$$t_r = \frac{L_{bed}}{v_{bed}} \tag{7}$$

where L_{bed} is the total length of the two heating plates.

THE "PARALLEL KINETIC MODEL"

For many materials that can be treated in the vacuum pyrolysis reactor, the weight loss of the pyrolyzed material can be expressed by the sum of the conversion of each group component in the material:

$$\frac{dC}{dt} = \sum_{j=1}^{N} x_j \frac{dCj}{dt} \qquad (8)$$

where m is expressed as $(m\text{-}m_0)/(m_{fin}\text{-}m_0)$. x_i is the ratio of each component. For example, in the case of wood bark, the major components are assumed to be lignin, cellulose, hemicellulose and water. The conversion of each component can be described by the Arrhenius Equation:

$$\frac{dC_j}{dt} = Z_j e^{-E_j/RT}(1 - C_j)^{n_j} \qquad (9)$$

where E, Z and n are the activation energy, the pre-exponential factor and the reaction order. In this work, the parameters were determined by thermogravimetric analysis.

A DYNAMIC MODEL TO SCALE-UP THE VACUUM PYROCYCLING™ PYROLYSIS REACTOR

Based on the above three models, a dynamic model to scale-up the vacuum pyrolysis process was developed, which correlates the temperature and the mass of feedstock at any position on top of the heating plates inside the reactor, as a function of heat transfer, particle flow and pyrolysis kinetics phenomena. The energy conservation in the reactor is the foundation of the model. It assumes i) steady flow, ii) one dimensional temperature variation and, iii) feedstock thermal properties vary as a function of temperature T.

$$\frac{(C_P m T)_{i+1} - (C_P m T)_i}{\Delta t} = \alpha_{bed}(T_{ws} - T_i) dA + dA\, \sigma(T_{shell}^4 - T_i^4) + m_i q_p \qquad (10)$$

The term on the left side of the equation represents the flow of internal energy in and out of the system, where m is determined by the kinetic Equations (1-9). Δt is determined by the total number of nodes generated, N, and the feedstock residence time in the reactor, t_r which can be calculated by equations 4-7. The first term on the right side represents the heat transfer from the bottom heating plates; T_{ws} is the temperature of the heating plate and α_{bed} is the heat transfer coefficient which is determined by the heat transfer equations (Eq. 1-3). The second term is the radiation heat transfer contribution from the reactor wall. The last term represents the kinetic energy released during the pyrolysis reaction, which is assumed to be proportional to the rate of pyrolysis reaction (Eq.8-9).

EXPERIMENTAL

Two different configurations of moving and stirred bed reactors, the batch scale rotative and PDU reactors, have been used for the tests in accordance with the principle of similarity.

THE "BATCH SCALE ROTATIVE REACTOR"

The batch scale rotative reactor was developed and used as a tool to validate the new heat transfer model [4, 6]. The batch reactor consists of a well insulated tank which contains molten salt and is equipped with heating elements in order to be able to heat the salt to the set point temperature. A second well insulated tank can be placed in the salt bath. The feedstock enters the reactor through the feedpipe. The same agitation blades as those found in the industrial reactor are used. The stirring mechanism transports and agitates the feedstock in a circular manner. The center of the reservoir is kept free of feedstock by a scraping mechanism. The diameter of the feedstock tank is 107 cm and the effective heat transfer area is 0.82 m^2 [6].

Both cold and hot tests were carried out in the batch scale rotative reactor. In the cold tests, the bottom layer renewal efficiency, β (see Eq. 1-3), has been measured as a function of the agitation speed [7]. During the hot tests, a certain quantity of feedstock is dropped into the reactor when the reactor has been first heated up to 520°C by the molten salt. Several thermocouples record the temperature change in the molten salt and in the bed of particles. The heat transfer coefficient from the molten salt and the feedstock in the reactor is determined as a function of the temperature variation [6].

THE PDU VACUUM PYROLYSIS REACTOR

The PDU vacuum pyrolysis reactor is a semi-continuous horizontal pilot plant reactor 3 m long with a diameter of 0.6 m and a throughput capacity of about 50 - 200 kg/h, depending on the feedstock treated. The configuration of the PDU reactor is almost the same as that of the industrial reactor, except that the PDU has smaller agitation blades. Two types of tests have been conducted with this reactor, the cold and the hot runs. In the cold tests, the particle flow behaviour is studied by a stimulus-response technique, under different agitation speeds and feed rates. The hot tests enable the conversion to be determined as a function of the feed rate and the agitation speed.

THERMOGRAVIMETRIC ANALYSIS

Thermogravimetric analysis (TGA) has been used to study the kinetics of feedstock pyrolysis. In this work, a Seiko 220 TG/DTA thermal balance system was used. The sample mass was 10.0 ±0.1 mg. The heating rate was 10 °C/min. For each test, the nitrogen gas was first introduced in the furnace to remove the air. Then the pressure in the furnace was lowered by a vacuum pump to 13 kPa absolute.

THE FEEDSTOCK

The feedstock was air-dry softwood bark obtained from a wood shredding plant. It was composed of approximately 31% v/v of balsam fir (*Abies balsamea*), 55% v/v white spruce (*Picea glauca*) and 14% v/v black spruce (*Picea mariana*). The moisture content of the feedstock at the inlet of the reactor averaged 10.0% by wt. The proximate analysis gave: 74.8 volatile matter (V.M.), 22.3% fixed carbon and 2.9% ash. The sample particle size is comprised between 0.5 and 40 mesh US sieves.

RESULTS AND DISCUSSION

THE TGA RESULTS

The furnace of a TGA instrument can be considered as a small scale reactor, in which a linear heating rate and a uniform temperature and pressure environment exist. The weight loss measured under such conditions is only controlled by the kinetics of the feedstock pyrolysis. Figure 3 shows the weight loss (the TG curve) and the rate of weight loss (the DTG curve) of softwood bark thermal decomposition measured by TGA. As shown in Figure 3, the softwood bark pyrolysis starts at 160 °C. At 358°C, it reaches the maximum rate of weight loss. At 400 °C, one observes a turning point, as the extensive pyrolysis phase is terminated and a slow pyrolysis process continues until 900°C. At 900°C, the pyrolysis is completely terminated. The weight loss then reaches 74% of the initial weight.

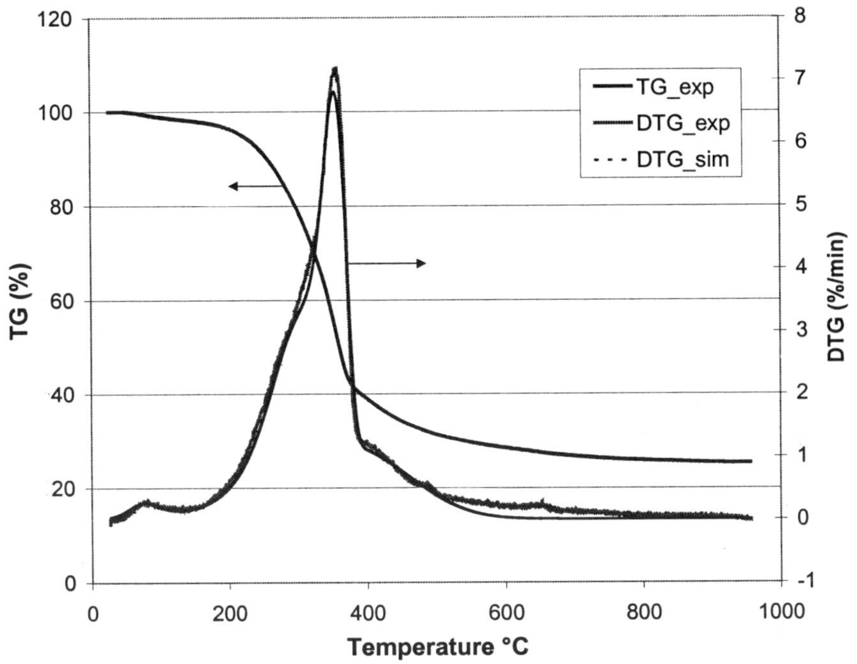

Figure 3 Weight Loss (the TG curve) and the Rate of Weight Loss (the DTG curve) of Softwood Bark as Measured by TGA

Similar to most lignocellulosic materials, the softwood bark used in this work basically contains three organic chemical components: cellulose, hemicellulose and lignin. Using the "Parallel Kinetic Model" (Eq. 8-9) to fit the DTG curve in Figure 3 (the solid line), the activation energy E, the reactor order n and the ratio of each of the three components can be determined. The simulated DTG curve is drawn in Figure 3 (the dashed line). It shows a very good agreement with the measured DTG for the temperature range of 30°C to 520°C. As the temperature in the vacuum pyrolysis reactor is less than 520°C, the simulation is satisfying. The kinetic parameters obtained from the DTG curve simulation are presented in Table 1. The sum of the ratio of each component gives 95%, while the other 5% weight loss occurs in the temperature range from 520°C to 900°C, which is not of interest in this study.

Table 1. Kinetic Parameters Obtained by Fitting the TGA Results

Components	E (kJ/mol)	Z (1/min)	n	Component Content (%)
Cellulose	180	1.0×10^{15}	1	35
Hemicellulose	80	1.58×10^{7}	1	20
Lignin	30	2.4×10^{2}	1	38
Water	60	5.0×10^{8}	2	2

THE HEAT TRANSFER COEFFICIENT IN THE ROTATIVE REACTOR

The heat transfer coefficient in the batch rotative reactor has been determined both theoretically and experimentally. As indicated by the heat transfer model (Eq.1-3), the heat transfer coefficient between the reactor and the feedstock is a function of the thermophysical properties of the feedstock C_p , ρ and λ, the contact heat transfer coefficient α_{ws} and the bottom layer renewal efficiency β. When the feedstock material is known, the thermal properties and α_{ws} can be determined [5]. The heat transfer coefficient becomes a function of β and t_{mix}.

The thermal properties of the softwood bark are given in Table 3. The values were determined experimentally in this work [8].

Table 2. Physical and Thermal Properties of the Softwood Bark Feedstock

Bulk density ρ, kg/m^3	250
Specific heat capacity C_p, J/kg K	2720
Thermal conductivity λ, W/m K	0.074
Contact heat transfer coefficient α_{ws}, W/m² K	640

The main parameter influencing the value of β is the speed of the agitation blades [7]. Through cold tests in the rotative reactor, the experimental values of β can be determined. The results are shown in Table 3.

Table 3. Bottom Layer Renewal Efficiencies for the Circular Agitation Device in the Batch Rotative Reactor

Turning speed (rpm)	Characteristic time, t_{mix} (s)	Bottom layer renewal efficiency, β
10	6.1	0.56
19	3.1	0.51
26	2.0	0.46
38	1.6	0.47

When the thermal properties and the β values listed in Tables 2 and 3 are introduced into the heat transfer equations, the heat transfer coefficients can be determined as a function of the agitation speed (see Figure 4, the solid line). Figure 4 shows a quick α increase with the agitation speed. The maximum α is 105 W/m²K which is reached at t_{mix}=2.1.

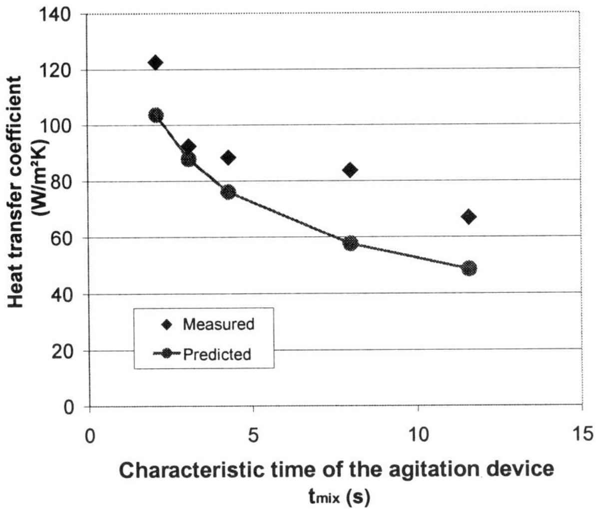

Figure 4 Predicted and Measured Heat Transfer Coefficients in the Rotative Batch Reactor.

The heat transfer coefficient in the rotative reactor has been determined experimentally. The tests measured the temperatures of the feedstock in the reactor and that of the heating plate. Then the heat transfer coefficient is calculated based on the variation of the feedstock internal energy and the temperature difference between the feedstock and the heating plate [6]. The maximum α is reached at t_{mix}=2.1, and corresponds to 180 W/m²K. This value incorporates both the heat transfer from the heating plate and from the reactor wall. It has been estimated that the heat transfer by radiation from the reactor wall provides approximately 1/3 of the overall heat transfer. Figure 4 presents the measured heat transfer from the heating plate to the feedstock as a function of t_{mix}, which is equal to 2/3 of the measured overall heat transfer coefficient (see the diamond marks). Comparison between the experimental and the theoretical results indicates that the experimental measurements and the model predictions are in good agreement. Even though the predicted α seems to be closer to the experimental value at high turning speeds, the difference between the experimental and theoretical values is less than 21%. This difference may be due to the fact that the model only considers the heat transfer from the heating plate to the feedstock. In reality, there is also some convective heat transfer caused by the evolving pyrolysis vapor products.

THE MOVEMENT OF PARTICLES IN THE PDU REACTOR

As shown by the "Single Blade Volume Output Model", the correlation between the single blade volume V_{bed} and the bed height h_{bed} is the key to determine the feedstock flowing behaviour in the vacuum pyrolysis reactor. Figure 5 presents the V_{bed} measured as a function of h_{bed}. The square marks in Figure 5 are obtained using the large blades designed for the industrial reactor. The round marks are linked to the small blades in the PDU reactor. Through curve fitting, the c and n in equation (5) were determined to be 110 and 1.2 (the solid line) and 23 and 1.2 (the dashed line) for the large and the small blades, respectively. When c and n are known, the feedstock residence time t_r can be calculated using the Equations 4-7. Figure 6 shows the predicted residence time in the industrial reactor. The design parameters used for the calculations are presented in Table 4. Figure 4 shows that t_r significantly varies with the speed of the agitation blades and moderately varies with the feedstock feed rates.

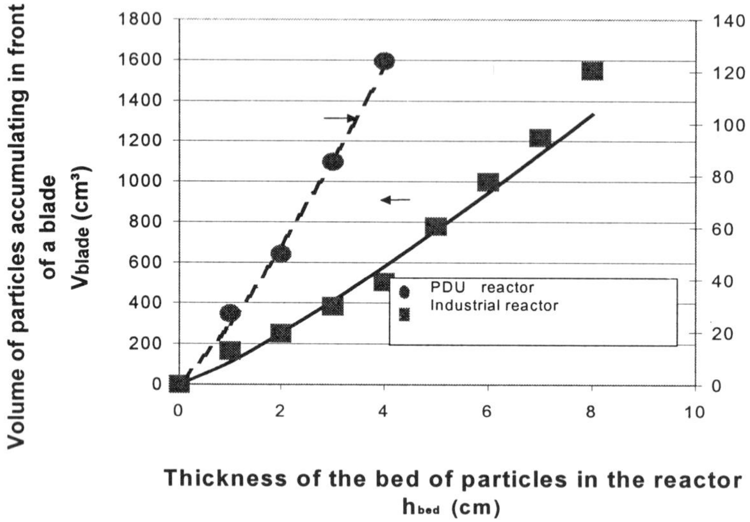

Figure 5 Observed Values of V_{bed} as a Function of h_{bed}

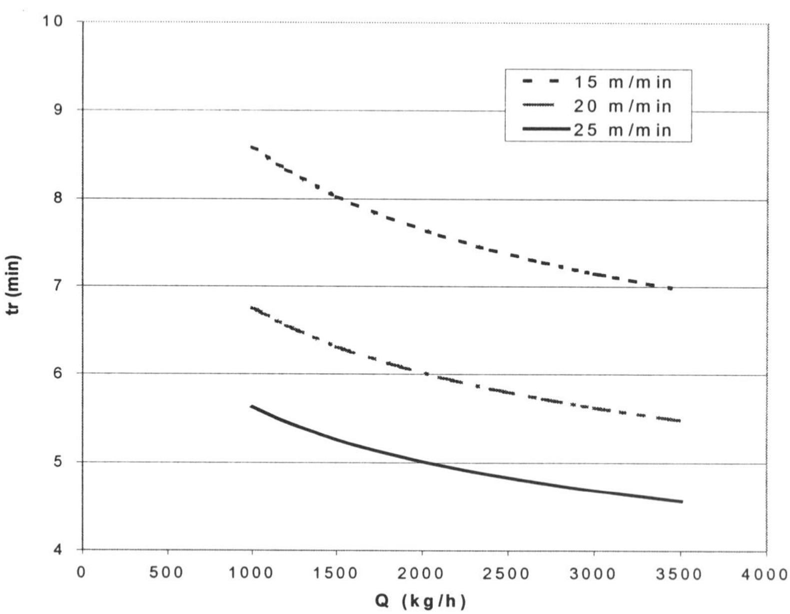

Figure 6 Predicted Residence Time in the Industrial Reactor as a Function of the Velocity of the Agitation Blades.

Table 4. Design Parameters for the PDU Industrial Scale Reactors

Parameters	PDU	Industrial
W_{bed}	0.4 m	0.96 m
W_{blade}	6.4×10^{-3} m	0.038 m
L_{chain}	5.24 m	26.7 m
L_{bed}	4.04 m	20 m
n_{blade}	96	385

To validate the "Single Blade Volume Output Model", the feedstock residence time was also studied experimentally by using the stimulus-response technique. Using scrap tires as feedstock, the residence time in the PDU reactor has been measured, which showed a very good agreement with the model prediction [8]. The experiments for measuring the t_r of wood bark feedstock in the PDU reactor is in preparation. The results will be presented later.

BIOMASS CONVERSION IN AN INDUSTRIAL SCALE PYROLYSIS REACTOR

The mathematical model to scale-up the vacuum pyrolysis reactor predicts the temperature distribution in the reactor and the final conversion of the feedstock under varying feed rates and agitation speeds, once the reactor size and configuration are fixed. The dimension of the industrial scale reactor is presented in Table 4. The values of W_{bed} and L_{bed} presented are the effective dimension for heat transfer study, which are smaller than the true values (the true values of W_{bed} and L_{bed} are 2m and 20.8m, respectively). When one uses as inputs the thermal properties (Table 3), the flow of particles characteristics (Table 2 and Figure 5) and kinetic (Table 1) parameters in the mathematical model (equations 1-10), the temperature and the conversion distributions can be predicted. The speed of the agitation blades is assumed to be 25 m/min in the calculation. The temperature of the heating plate is 500°C. The average t_{mix} in the reactor is about 11 s. The heat transfer coefficient is thus estimated to be 100 W/m² K, which includes both the heat transfer from the heating plate and from the reactor wall. The feedstock moisture is assumed to be 10%. To calculate the conversion behaviour, it is assumed that the conversion is 100% when the volatile matter content in the residual wood charcoal is 35%.

As depicted in Figure 7, the temperature in the reactor increases more rapidly at the beginning due to the large temperature difference between the reactor and the feedstock. The maximum temperature is reached at the end of the reactor, before the product flowing out of the reactor. In Figure 7 the temperature distribution changes greatly with different feed rates: the lower the feed rate, the higher the final temperature of the feedstock. Figure 8 illustrates the predicted feedstock conversion in the reactor as a function of the reactor zone and the feed rate. The conversion is low at $X < 2$ m. In the case of a feed rate of 1000 kg/h, a rapid conversion is observed between $2 < X < 10$. Here the conversion reaches 100% even before reaching the end of the reactor. At feed rates of 2000-3500 kg/h, the conversion occurs much later, at X=5 m. The final conversion is lower and reaches 100%, 96% and 92% at the reactor outlet, for feed rates of 2000, 3000 and 3500 kg/h, respectively. Figure 9 illustrates the predicted final temperature (dashed line) and the final conversion (solid line) in the reactor, as a function of the feed rate. It shows that when the feed rate is ≤ 2600 kg/h, the final temperature is higher than 450°C, and the final conversion reaches 100%.

When the feed rate is higher, the final conversion decreases due to a decrease of the charcoal outlet temperature.

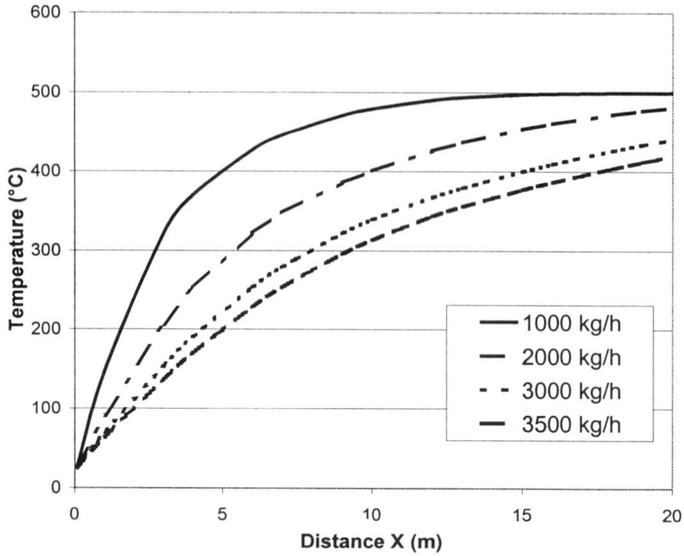

Figure 7 Predicted Temperature Distributions in the Reactor as a Function of the Distance along the Reactor and the Feed Rate.

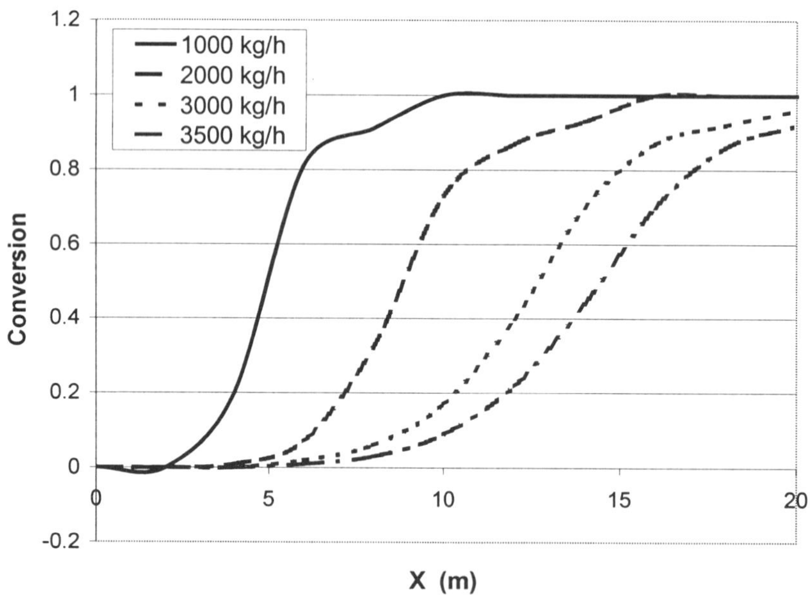

Figure 8 : Predicted Feedstock Conversion in the Reactor as a Function of the Distance along the Reactor and the Feed Rate.

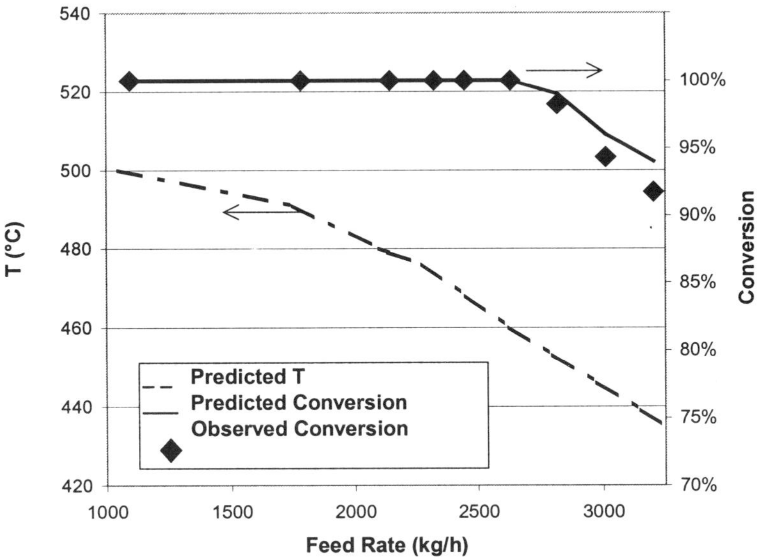

Figure 9 The Predicted Feedstock Temperature Outlet and Ultimate Conversion, and the Observed Conversion in the Industrial Reactor, as a Function of Various Feed Rates.

Figure 9 shows the final conversion measured as a function of the feed rate during experimental runs in the industrial Pyrocycling^{TM} reactor described earlier (diamond marks). When the feed rates are less than or equal to 2600 kg/h, the final conversion is always higher than 100%. When feed rate reaches 3200 kg/h, the conversion is 92%. Comparing the measured and the predicted conversions shown in Fig.9, a very similar trend can be observed for the conversion. For most of the points in Fig.9, the predicted conversion is very close to the experimental results, although the predicted values are a little bit higher. The most likely reasons for the slight discrepancies are: i) the mathematical model considers no mass transfer resistance when the pyrolysis vapour or gas products are evolving in the reactor and ii), there may be vapour products re-condensing in the reactor particularly when the vapour products are flowing from the hot zone to the colder zone, a factor which is not taken into account in the mathematical model.

CONCLUSIONS

The objective of scaling-up a vacuum pyrolysis reactor is to achieve the desired capacity and conversion of feedstock by determining the dimension of the reactor. The feedstock conversion in a vacuum pyrolysis reactor mainly depends on three factors: the heat transfer coefficient from the reactor to the feedstock; the residence time of the feedstock in the reactor and the kinetics of the feedstock pyrolysis reactions. The heat transfer coefficient and the residence time determine the quantity of energy transferred and thus the temperature distribution throughout the feedstock in the reactor. The temperature distribution and the kinetics determine the final conversion achieved at the reactor outlet.

A heat transfer model, namely the "Surface Renewal Model" has been developed. It predicted that the heat transfer coefficient a vacuum pyrolysis reactor fed with biomass is 80-140 W/m^2 K, depending on the process conditions. A model for the movement of particles, known as the "Single Blade Volume Output Model" was developed to predict the average residence time of the feedstock in the reactor. It showed that the residence time of the biomass in a 20 m long reactor ranges between 5 and 9 min, with a speed of the agitation blades varying from 15-25 m/min. The biomass conversion starts at 180 °C. At 520 °C, 95% conversion is achieved. A kinetic model, called the "Parallel Kinetic Model", simulated the TGA results and enabled the determination of the activation energy and the fraction of each component in the feedstock. Finally the combination of these three models gives birth to a dynamic model to scale-up the reactor by allowing the prediction of the final conversion of the feedstock.

An industrial vacuum pyrolysis reactor, 14.6 m long and 2.2 m in diameter, has been constructed and operated. The operation of the pyrolysis reactor has been successful, with the reactor capacity reaching the predicted feed rate of 3000 kg/h on a biomass feedstock anhydrous basis. Using the dynamic model, the conversion of the biomass feedstock in the reactor has been predicted as a function of the biomass feed rates. Comparison of the predicted and measured conversions in the reactor shows a fairly good agreement.

NOTATION

A	surface area, m^2
B	the amount of the agitated fraction k of the first layer of particles
C	conversion
C_p	specific heat capacity, $J \cdot kg^{-1}K^{-1}$
d	diameter, m
E	activation energy kJ mol^{-1}
h_{bed}	thickness of the feedstock bed of particles in the reactor, m
i	grid location on the longitude axis
j	index of each component in the pyrolysis material
k	index for the first layer of particles having different residence times
L_{bed}	the length of the heating plate, m
L_{chain}	the length of the chain to which the agitation blades are installed, m
m	mass, kg
n	the reaction order
n_{blade}	number of agitation blades installed in the reactor
N	the total number of nodes generated in numerical calculations
Q_{in}	the feedstock feed rate, $kg \cdot h^{-1}$
q_p	heat of reaction, W kg^{-1}
R	ideal gas constant, J K^{-1} mol^{-1}
T	temperature, K
t	time, s
t_{mix}	characteristic time of the agitation device, s
t_r	residence time of feedstock in the reactor, s
Δt	the time interval for the feedstock flowing from one grid to the next, s
v_{bed}	velocity of feedstock in the reactor, m s^{-1}
v_{blade}	the velocity of the agitation blade, m s^{-1}

V_{blade} the volume of the particle accumulating in front of an agitation blade, m^3
W_{bed} the width of the heating plate, m
W_{blade} the width of the agitation blade, m
X longitudinal distance along the reactor, m
Z pre-exponential factor, min^{-1}

GREEK LETTERS

α heat transfer coefficient, $W \cdot m^{-2}K^{-1}$
β bottom layer renewal efficiency
λ thermal conductivity, $W \cdot m^{-1}K^{-1}$
ρ density, $kg \cdot m^{-3}$
σ Stephan-Boltzmann constant, $W \cdot m^{-2}K^{-4}$

REFERENCES

1. Roy, C., Labrecque, B. and de Caumia, B. (1990) Recycling of scrap tires to oil and carbon black by vacuum pyrolysis. *Resour. Conserv. Recycl.*, Vol. 4, pp. 203-213.
2. Yang, J., Tanguy, P. A. and Roy, C. (1995) Heat transfer, mass transfer and kinetics study of the vacuum pyrolysis of a large used tire particle. *Chem. Eng. Sci.* Vol. 50, No. 12, pp. 1909-1922.
3. Perry, R.H. and Green, D. (1984) Perry's Chemical Engineers' Handbook, Sixth edition, McGraw-Hill Book Co., New York, pp. 11.48 - 11.49.
4. Roy, C., D. Blanchette and B. de Caumia. Horizontal Moving and Stirred Bed Reactor. Canadian Patent Claim Number 2,196,841. US Patent Number 8,811,172. International Patent Number 98,902,153.0.
5. Schlünder, E. U. (1984) Heat transfer to packed and stirred beds from immersed bodies. *Chem. Eng. Process.*, Vol. 18, pp. 31-53.
6. C. Malendoma, Yang, J. and C. Roy. (2000) Determination of the Overall Heat Transfer Coefficient in a Vacuum Pyrolysis Moving and Stirred Bed Reactor. Chem. Eng. Res. Des. Vol. 78, pp. 633-642.
7. Roy, C., D. Blanchette, L. Korving, J. Yang and B. de Caumia. Developments in Thermochemical Biomass Conversion. A.V. Bridgwater and D.G.B. Boocock, Eds. Blackie Academic and Professional, London, UK, 1997 pp. 351-367.
8. M. Gupta, J. Yang, and C. Roy, Density of Softwood Bark and Softwood Char: Procedural Calibration and Measurement by Water Soaking and Kerosene Immersion Method. Manuscript in preparation.
9. Yang, J., C. Roy, M. Gupta and X. Roy, Study of Tire Particle Flowing Behaviour in an Agitated and Moving Bed Reactor Manuscript in preparation.

Thermal efficiency of the HTU® Process for Biomass Liquefaction

F. Goudriaan[a], B. van de Beld[d], F.R. Boerefijn[b], G.M. Bos[c], J.E. Naber[a], S. van der Wal[b], and J.A. Zeevalkink[c]

[a] *Biofuel B.V., Rendorppark 30, 1963 AM Heemskerk, The Netherlands*

[b] *Jacobs Comprimo Nederland, Postbox 58026, 1040 HA Amsterdam,, The Netherlands*

[c] *TNO-MEP , Postbox 342, 7300 AH Apeldoorn, The Netherlands*

[d] *Biomass Technology Group (BTG) B.V. , Postbox 217, 7500 AE Enschede, The Netherlands*

ABSTRACT The development of the HTU® Process is now well under way. A pilot plant with an intake capacity of some 10 kg/h (dry basis) is in operation. Biomass is converted by treatment in liquid water at temperatures from 300 to 350°C and pressures from 100 to 180 bar. The product is 'Biocrude', a heavy organic liquid with a lower heating value of 30-35 MJ/kg.

A case study is presented for the HTU conversion of sugar beet pulp (130 kton/a on a dry basis). A process description is given, and the heat economy is discussed. For the generation of process heat some external fuel is required (2% of the heating value of the feedstock).

The *thermal efficiency* is defined as the ratio of heating values of biocrude product and feedstock plus external heat input It has a theoretical maximum of 79%. The process designed here has a thermal efficiency of 75%.

The total capital expediture is 30 M$. At a zero cost of feedstock the price of the biocrude product is $ 153/ton, or $ 4.6/GJ.

A number of items for further improvement of the thermal efficiency is discussed. It requires a systematic exergy analysis. However, any further gain in heat economy has to be traded off against increased capital expenditure and reduced operability.

INTRODUCTION

Hydrothermal liquefaction is a process for the conversion of biomass into an organic product oil. It has been widely studied in the early 1980's (see the recent extensive literature survey in [1]). Starting in 1983 Shell Research performed process development based on experiments in autoclaves and a continuous bench scale unit. This resulted in a conceptual design for the HTU process including cost estimate [2,3]. The work was discontinued in 1988.

In 1997 Stork Engineers and Contractors (now Jacobs Comprimo Nederland), Shell Netherlands, TNO-MEP, BTG and Biofuel formed a consortium for further development and commercialisation of the HTU process. With financial support from the Dutch Government (the EET programme) a 5.5 M$ project is under way from 1 November 1997 to 1 September 2000. It encompasses, experiments in autoclaves, product research, process design, and the construction and operation of a pilot plant with a throughput of 10-20 kg/h (db), [See list of abbreviations at the end of this paper] see Figure 1. The business plan shows favourable commercial prospects [4]. A feasibility study is being carried out which aims at the construction of a demonstration plant, capacity 10-20 kton/a (db), in 2002.

Fig. 1 HTU pilot plant, located at TNO-MEP, Apeldoorn, The Netherlands.

The HTU process

In the HTU process the feedstock is treated in *liquid* water at temperatures ranging from 300 to 360°C. The pressure is 100-180 bar and the residence time 5-20 minutes. Oxygen is removed from the biomass, mainly as CO_2. This results in a product designated as "biocrude" with an oxygen content as low as 10-18 %w. The biocrude is not miscible with water and it has a relatively high heating value (LHV is 30-35 MJ/kg). Its main application as such is power generation, e.g., co-combustion in a coal-fired power station. Alternatively it can be upgraded by catalytic hydrodeoxygenation. Scouting experiments have demonstrated that in this way a diesel fuel with excellent properties can be produced. By its nature, the HTU process is very well suited for the conversion of *wet* biomass feedstocks. These are generally low in price, such as residues from agriculture, forestry, food processing industry, etc. However, at a longer term, large-scale processing of crops from energy plantations is considered as well.

Thermal efficiency

Based on the results of the process development work the design of a commercial-scale HTU process unit is now under way. Heat integration and energy economy are of prime importance to keep the processing cost low. A useful criterion for the energy economy is the *thermal efficiency*. This paper discusses a case study of such a process design.

HTU PROCESS DESCRIPTION

CHEMISTRY OF OXYGEN REMOVAL

In chemical terms the key to biomass liquefaction is the removal of oxygen. Biomass contains typically 40-45%w (DAF basis) of oxygen. Oxygen removal increases the heating value and it leads to a product with more hydrocarbon-like properties causing it to be immiscible with water. In thermochemical liquefaction the oxygen is either removed as water or as carbon dioxide. Removal as water leads ultimately to carbon as the remaining product (examples: charcoal manufacture, pyrolysis). Removal of carbon dioxide tends to leave a product with a higher H/C ratio and therefore a higher LHV.

At HTU reaction conditions a significant proportion of the oxygen is removed as CO_2. The decarboxylation selectivity is defined as the ratio (oxygen removed as CO_2) : (total of oxygen removed as CO_2 and H_2O). Table 1 shows decarboxylation selectivities for a number of biomass HTU conversion cases with different feedstocks and reaction conditions. It appears that the selectivity is almost constant.

Table 1 Deoxygenation selectivity.

	% of oxygen ending up in product fraction			
Product fraction	case 1	case 2	case 3	case 4
Biocrude	14	21	15	12
Light organic product (OD)	20	12	9	9
CO2	35	36	37	40
H2O	31	31	38	40
Decarboxylation selectivity	0,53	0,54	0,49	0,50

The chemical reaction mechanism is not quite understood [5]. Apart from water and carbon dioxide the reaction products are:

Biocrude. This is the main product. It has an oxygen content of 12-20 %w and the atomic H/C ratio is 1.0-1.3. Average molecular weight is about 300.

Organics dissolved (OD). This is the organic fraction dissolved in the process water after cooling the reaction mixture to ambient temperature. It consists mainly of acetic acid, acetone, and components like methyl-cyclopentenone and hydroxypyridine.

Gases. Apart from carbon dioxide, some carbon monoxide is formed, and traces of methane and hydrogen.

As a feedstock for the case under study sugar beet pulp is selected. Its properties are given in Table 2.

Table 2 Sugar beet pulp feedstock composition

SBP Feed Composition, %w		DAF	dry	wet
DAF feed,	C	45.75	42.32	9.31
	H	6.42	5.94	1.31
	O	46.17	42.71	9.40
	N	1.66	1.54	0.34
Minerals			7.49	1.65
Water				78.00
Total		100.00	100.00	100.00
%w of wet feed		20.35	22.00	100.00

A typical product distribution is shown in Table 3. With the SBP feedstock less biocrude and more OD are produced than with wood [3].

Table 3 Product distribution and LHV balance

Product		%w on feed (DAF)	LHV, MJ/kg	LHV, MJ/kg feed
DAF feed		100.00	17.5	17.5
Products				
Biocrude		41.27	33.32	13.75
Gas,	CO_2	24.25	0	0.00
	CO	1.20	10	0.12
	CH_4	0.26	50	0.13
	H_2	0.03	121	0.03
Organics Dissolved (OD)		12.35	26	3.21
Water		20.64	0	0.00
Total products		100.00		17.25

PHASE EQUILIBRIA

Phase equilibrium data are essential for process design. Prof. J. de Swaan Arons and Dr. Chongli Zhong (Presently Professor at Beijing University of Chemical Technology.) of Delft University of Technology are developing a thermodynamic model, which also predicts phase equilibria. Main uncertainties are in the modeling of the biocrude properties. Experimental verification of phase equilibrium data with autoclave and pilot plant equipment is ongoing.

An alternative method is to realise that carbon dioxide and water constitute over 90 mole % of the reaction mixture. Literature data [6] for the CO_2-H_2O phase system are used, and for the distribution of biocrude and OD over the phases, distribution coefficients are derived from experiments and literature.

Thirdly, flowsheeting programmes like AspenPlus® also provide data on thermodynamics and phase equilibria. Here again, biocrude and OD have to be introduced, e.g., as pseudo-components.

THERMOCHEMICAL DATA

Specific heat

The specific heats of various biomass samples were measured. The specific heat of biocrude was estimated from a group contribution method.

LHV data

Table 3 gives for each of the products its LHV and its contribution to the total LHV of the product mixture. The LHV of biocrude is given as 33.32 MJ/kg. This value is derived from experimental determination and from calculation on the basis of elemental composition. The two methods show a good agreement. The Table shows that the LHV of the biocrude is 79% of the LHV input of the feedstock. The LHV of the combustible by-products (gases and OD) is some 20% of the LHV input. It is, in principle, available for the generation of process heat.

Standard enthalpy of reaction

In theory the reaction enthalpy at 298 K can be derived from the LHV balance as given in Table 3. This method gives $\Delta H_{r,298}$ = 17.25 - 17.50 = - 0.25 MJ/kg feed (DAF). However, the accuracy of the most important LHV data (feed and biocrude) is considered to be at best ±10%. So, this would yield $\Delta H_{r,298}$ = -0.25 ± 3 MJ/kg (DAF). On this basis, it cannot be determined whether the reaction is endothermic or exothermic.

For the present purpose we derive the reaction enthalpy from autoclave experiments. In a standard experiment, a 2 litre autoclave is filled with some 350 g of water and heated to 330°C. Then an amount of SBP feed is injected with a specially constructed "high-pressure syringe". The injection of the (cold) feed makes the temperature drop sharply, and then it increases again, the main driving force being the heat content of the autoclave wall. The standard experiment was then repeated with the injection of water in stead of SBP feed. By comparing the course of the temperature of the contents of the autoclave for the blank and the standard experiments, one can estimate the reaction enthalpy. The temperature curves are shown in Figure 2. Calculations take into account the heat capacity of the wall and the contents of the autoclave, enthalpy of vaporisation, and other relevant factors. The resulting values of the reaction enthalpy are: Initially endothermic, $\Delta H_{r,298}$ = + 0.40 MJ/kg (db), and then exothermic, $\Delta H_{r,298}$ = -1.00 MJ/kg.

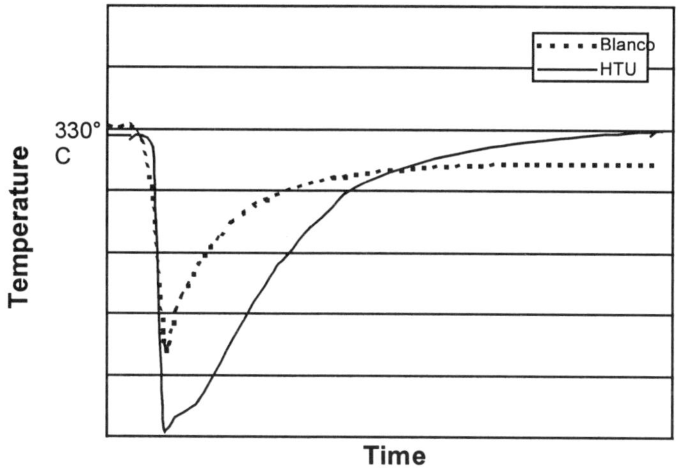

Fig. 2 Temperature recordings in blank and HTU experiments.

Indications for the reaction enthalpy are also derived from the continuous pilot installations, both at Shell Research (1984-1986) and the present one at TNO-MEP. Here one measures the steady state electric energy supply required for heating the reaction mixture to the desired temperature. By comparing this with a blank experiment where only water is heated one concludes that the overall reaction heat effect is exothermic, and its value is consistent with the result from the autoclave work.

CONSIDERATIONS FOR PROCESS DESIGN

In this section we discuss a case study for the design of a commercial scale HTU installation, based on the data given above. The intake capacity of SBP feed is 130 kton/a, or 4.51 kg/s db assuming 8000h/y. A simplified process flow scheme for the main process loop is shown in Figure 3.

Reactor conditions

On the basis of autoclave and pilot plant experiments a HTU reactor temperature of 330°C is selected, with a liquid residence time of 10 minutes. The product distribution in Table 3 pertains to these conditions. The reaction pressure does not have much of an effect, provided the water is in the liquid phase. The decisive factor in choosing the reactor pressure is the extent of vapour phase formation in the reactor. A lower pressure causes a larger proportion of the reaction mixture to evaporate. As a consequence more heat has to be supplied in **E1** to maintain the reactor temperature at the desired level. This tends to be disadvantageous in view of the temperature level of this heat supply. A concomitant advantage of the lower pressure is the lower equipment cost in all of the high-pressure section of the process. These factors, along with the possible advantage to have a larger part of the reaction mixture in the vapour phase, have to be traded off. In the present case a pressure of 165 bar was chosen as the optimum.

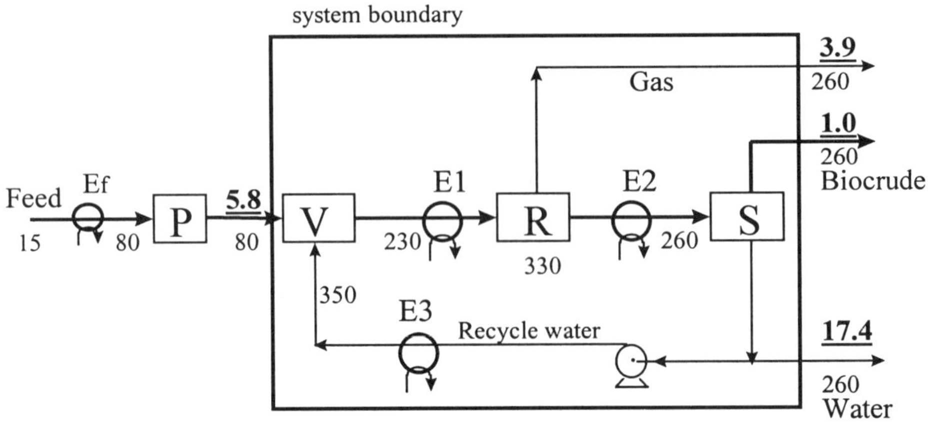

Legend: 80 = temperature in °C
5.8 = Enthalpy, MW

Fig. 3 Flow scheme for main HTU process loop.

Feed pump

A major accomplishment of the present project is the cooperation with a pump manufacturer who has constructed a pump for the pilot plant installation, for bringing the feedstock to the HTU process pressure of 100-200 bar *in one step.* On the basis of this, and their experience with commercial-size pumps, the manufacturer has made a design and cost quotation for a commercial-scale pump.

In the process scheme of Fig. 3 the SBP feedstock is first heated to 80°C, using heat available in the process (see below). For heat economy reasons it would be desirable to choose this temperature considerably higher. For the time being this is not done because it would require pressurized equipment upstream of the feed pump. The pump(s) **P** bring the feedstock to the HTU pressure.

Preheating of the feedstock

At the outlet of the pump the feedstock flows through a duct with a diameter of, say, 200 mm. Supply of heat through the wall of the duct would require virtually all of the heat transport to occur by conduction. The penetration of heat to the core of the "solid" cylinder of feedstock would require many hours. It is therefore not a practical solution. Hence, alternative ways of preheating are called for.

Almost all feedstocks of interest have such properties that they do not flow readily through either pipes or shell of a conventional heat exchanger without distribution problems. After a number of alternatives had been considered it was concluded that the most adequate way of preheating is *direct injection of hot water.* In Fig. 3 this is done in vessel **V** where a stream of hot recycle water is mixed with the feed. A special way of contacting achieves efficient heat and mass transfer. It has been verified in our experiments that at 200-250°C biomass feedstocks of different origin, including wood chips, are softened and form a paste-like substance [7,8]. The residence time required for this is 2-100 minutes, depending on the temperature. Some carbon dioxide forms as well, but owing to the combination of high pressure and relatively low temperature, it remains in solution and does not form a separate vapour phase.

In our case study it was decided to have the softener vessel **V** at a temperature of 230°C. The temperature of the recycle water has to be high in order to have the amount of recycle water as small as possible. This is essential to limit the size of all equipment in the high-pressure section. For this reason the recycle water is heated to 350°C in heater **E3**.

HTU reactor

In the reactor **R** the HTU reactions take place. At the prevailing temperature and pressure a vapour phase forms. The design of the reactor vessel is such that it allows for sufficient stageing to prevent backmixing, and for efficient separation of the vapour phase. Its volume is in the order of 40 m^3.

Product cooling and separation

The liquid effluent of the reactor is cooled in cooler **E2** to 260°C. This is sufficiently low to stop any undesired reactions leading to degradation of the biocrude. In separator section **S** the two liquid phases, water and biocrude, are separated. Experiments in autoclaves have shown that at this temperature the two liquid phases separate well and that their mutual solubility is sufficiently low to avoid excessive loss or contamination of the product. From the water stream from the separator the recycle water is routed to the recycle pump and then to heater **E3**. The remainder is the stream "water" leaving the main section in Figure 3.

Downstream process sections

The products leaving the main process loop shown in Figure 3 are further handled in sections not shown in the Figure. The *gas* phase stream is expanded. The possibility of recuperation of work by means of an expansion turbine is under study. Then, the stream is further cooled and treated in such a way that biocrude and water are routed to the liquid streams as much as possible. For the OD contained in the gas stream, it is desired to leave them in the gaseous phase.

The handling of the *biocrude* stream depends on the eventual application of the product. In general it will run off at some 80°C to storage and transport.

The *water* stream is passed through flasher(s), coolers etc. with the aim of having as much as possible of the OD stripped to the gas phase. The remaining liquid water stream is treated by anaerobic digestion. Here biogas is formed which contains the major part of the LHV originally contained in the OD present in the waste water stream. An after-treatment by aerobic digestion is required in most cases to comply with water discharge regulations. Also, the minerals dissolved in the effluent of the water treater are removed to a level corresponding to effluent regulations.

Furnace

The gas streams originating from all downstream process section are combined. They contain all of the non condensable gases CO, CH_4 and H_2, and part of the OD. This gas is combined with the biogas and combusted in the process furnace. High-temperature heat is generated and transferred to a hot-oil or high-pressure steam circuit to provide heat to the process. If necessary (see below) additional external fuel is burnt in the furnace so as to match the heat requirement.

If regulations require so, a catalytic deNOxer will be installed to reduce the NO_x content of the flue gases. For reasons of heat economy the flue gas temperature should be as low as possible and preheating of the combustion air is applied.

HEAT ECONOMY AND THERMAL EFFICIENCY

HEAT BALANCE

Basis of calculations

For the purpose of a scouting study like the one under review it is convenient to use a simplified method of calculating the enthalpy of process streams. Only the enthalpy of water (liquid and/or vapour), biomass and biocrude are taken into account. The required data are readily available, the calculation is simple and the error made by this simplification does not exceed some 10%. The reference level for H=0 is liquid water at 0°C. The enthalpies of the streams entering and leaving the main process loop have been calculated in this way. They are indicated in Fig. 3.

The mass balance for the main process loop is obtained in a straightforward fashion from the product distribution given in Table 3, the recycle water flow rate being obtained from an enthalpy balance over vessel **V** (see below).

Main heaters and coolers

The enthalpy balance around softener **V** considers the sensible heats and the endothermic reaction enthalpy of 4.51 kg/s(db) ∗ 0.4 MJ/kg(db) = 1.80 MW. The flow rate of the recycle water stream is found to be 1.20 ∗ the feed stream (on the basis of water). The enthalpy to be supplied in heater **E3** is 10.4 MW.

The enthalpy balance around the reactor **R** deals with the exothermic reaction enthalpy of 4.51 kg/s(db) ∗ -1.0 MJ/kg(db) = - 4.51 MW. The duty of heater **E1** is 17.5 MW.

Cooler E2 serves to rapidly cool down the liquid product mixture and to recover high-temperature enthalpy. The outlet temperature of **E2** is important for the overall enthalpy economy since it determines the enthalpy content of the streams exiting at the system boundary. By that it determines the required overall heat input to the main process loop shown in Figure 3. For overall enthalpy economy reasons it should be as low as possible. This temperature is found from optimisation of the hot oil (or high-pressure steam) system that transmits heat from the process furnace to the main process loop. The temperature of the hot oil leaving **E1** cannot be lower than 240°C. This stream picks up the heat from **E2**, where it is heated to at most 320°C. With these data the enthalpy balance over **E2** gives the outlet temperature (process side) to be 260°C. The duty of **E2** is 14.0 MW.

Enthalpy balance, main process loop

From the optimisation of the hot oil system in the preceding paragraph it is found that the process furnace has to supply 13.9 MW to the hot oil. The overall enthalpy balance over the main process loop then reads (see Fig. 3 for system boundary):

Enthalpy In:		
Stream out of **P**:	5.8 MW	
Reaction enthalpy	2.7	= (-0.4 + 1.0) * 4.51
Net from furnace	13.9	
Total In	22.3 MW	

Enthalpy Out:	
Gas stream from **R**	3.9 MW
Biocrude from **S**	1.0
Waste water	17.4
Total Out	22.3 MW

Overall enthalpy balance, entire process

The enthalpy of the streams leaving the main loop is 22.3 MW. Part of it is utilised for heating the feedstock in $\mathbf{E_f}$, duty 4.7 MW, and for providing heat to flashers and strippers in the downstream process sections. The remainder is some 10-15 MW. It is cooled away in air or water coolers.

In the downstream process sections some 40% of the OD ends up in the gas to the furnace. The remainder of OD has to be removed in the waste water treating section. Here it is assumed that 80% of the LHV is conserved as LHV of the biogas. In downstream processing some 3% of the biocrude product is lost to gas and water streams. It is assumed that 2% of the biocrude LHV becomes available in the furnace. So, the LHV input to the furnace from the various process streams is (in MJ/kg feed DAF, see Table 3):

Combustible gases, 100%	0.28
OD via gas, 40% of 3.21	1.28
Biogas, 80% of 60% of 3.21	1.54
Biocrude lost, 2 % of 13.75	0.28
Total	3.38 MJ/kg DAF feed

This corresponds to 3.38 (MJ/kg DAF feed) * 4.17 (kg DAF feed/s) = 14.09 MW

The required enthalpy *output* of the furnace into the hot oil is 13.9 MW. If we assume the furnace to have an efficiency of 90% the required enthalpy *input* to the furnace is 15.4 MW. Availablility from process streams is 14.1 MW. So an external input of (15.4 - 14.1) = 1.3 MW is required from additional fuel.

THERMAL EFFICIENCY

Definition

For the present purpose we define "thermal efficiency" as:

$$\eta_{th} = \frac{\text{(LHV of biocrude output)}}{\text{(LHV of feed)} + \text{(LHV from external fuel)}} * 100\%$$

Input of electric energy is disregarded on the presumption that electricity will be purchased, and as such it figures in the economic analysis. From the figures in Table 3 it follows that the theoretical maximum of the thermal efficiency for the present case is (13.75/17.5) = 78.6 %.

Result

In the process design presented above the (LHV from external fuel) is 1.3 MW.

The feedstock LHV is 17.5 (MJ/kg DAF feed) * 4.17 (kg DAF feed/s) = 72.98 MW.

The LHV of biocrude output is 97% of 13.75 (MJ/kg DAF feed) * 4.17 (kg DAF feed/s) = 55.62 MW. So the thermal efficiency found here is:

$$\eta_{th} = \frac{55.62}{72.98 + 1.3} * 100\% = 74.9\ \%$$

When electricity consumption is considered (as fuel consumed in generating the required electricity with an efficiency of 50%), thermal efficiency drops to 72.5 %.

PROCESS ECONOMY

For the HTU installation described above an estimate has been made for the total capital expenditure, including, a.o., a 15% development contingency. For the intake capacity of 130 kton/a (db) the capital expenditure is 30 M$. This is significantly lower than what would have obtained from cost estimates based on an earlier process design [3]. This reflects the cost efficiency of the process improvements applied since 1994.

In Table 4 an economic analysis of the process is given. A capital charge of 12% per year is assumed. Maintenance, overhead, insurance, etc. are covered with a 6% per year on the capital. Labour costs are based on two operators per shift plus the appropriate day shift personnel. Electricity is purchased for 0.04 $/kWh. External fuel costs 5 $/GJ. In this analysis it is assumed that the feedstock cost is zero. This may be true for certain residues from agriculture, food industry, etc. Equally well the feedstock price may be negative or positive. It is beyond the scope of this paper to discuss this any further.

DISCUSSION

By successful heat integration in the above process setup a thermal efficiency of 75% has been achieved, which is quite satisfactory. Further improvement may be obtained by having the outlet temperature of the main process loop at a value lower than 260°C. This requires a different heat integration in the hot oil system. Furthermore a different choice of the temperature of softener vessel **V** might lead to a higher efficiency. A more detailed exergy analysis would seem required for this.

Table 4 Process economy

Capacity intake, kton/a (db)	130		
Net production biocrude, kton/a	48.1		
Installed capital, M$	30		
Yearly cost	M$/a	$/t product	%
Capital charge (12%)	3.60	75	49
Maintenance, overhead. etc (6%)	1.80	37	24
Labour	1.40	29	19
Feedstock	0.00	0	0
Electricity	0.39	8	5
External fuel	0.19	4	3
Total	7.38	153	100
$/GJ		4.6	

However, an improved heat efficiency leads, in general, to more pieces of equipment. Table 4 has shown that about 75% of the process cost is related to the capital investment. Therefore a gain in thermal efficiency has to be carefully traded off against a higher capital investment and a more complicated operability.

A fundamental improvement of the process must be sought in the chemistry. If biocrude could be obtained in a higher yield and with a lower oxygen content, and if the yield of OD would be lower this would raise the thermal efficiency. Supporting studies into the chemistry of the HTU reactions are about to be started.

A lower water content of the feedstock would obviously be beneficial since less heat has to be supplied and the equipment size diminishes. This has to be balanced against the cost of the feedstock.

In the process scheme attention should be paid to lower loss of biocrude and a larger proportion of OD routed to the gas stream. These factors lead directly to a higher thermal efficiency and a smaller waste water treatment section.

It must be stressed that the present case study is to a large extent a simplified one. Examples of simplifications are:

- The build-up of minerals, OD, CO_2 etc. in the recycle loop has not been taken into account.
- The method of calculation of the enthalpy of the process streams is a simplified one.
- The estimation of capital expenditure was done on the basis of a conceptual design. A more thorough basic engineering study is presently being conducted.

CONCLUSIONS

A case study has been made for a conceptual process desin for the HTU conversion of sugar beet pulp with an intake capacity of 130 kton/a (db). The main conclusions are:

- Optimisation of the heat economy is essential.
- In the present case some additional external fuel is required for process heat. It amounts to about 2% of the LHV of the feedstock.
- Thermal efficiency as defined here has a theoretical maximum of 79 % for the present conversion. The process designed here has a thermal efficiency of 75 %.
- The total capital expediture is 30 M$. At a zero cost of feedstock the cost of producing the biocrude product is $ 153/ton, or $ 4.6/GJ.
- A number of items for further improvement of the thermal efficiency is discussed. It requires a systematic exergy analysis. However, any further gain in heat economy has to be traded off against increased capital expenditure and reduced operability.

ABBREVIATIONS

DAF	Dry and Ash Free
db	dry basis
LHV	Lower Heating Value at 298 K
OD	Organics Dissolved in process water phase.
SBP	Sugar beet Pulp

REFERENCES

1. Venderbosch R.H. & Sander C. (2000) *Hydroconversion of wet biomass: a review.* GAVE rapport 9919 (order at publicatiecentrum@novem.nl).
2. Annee J.H.J. & Ruyter H.P. (1986) *Process for producing hydrocarbon-containing liquids from biomass.* European Patent 0204354, to Shell Inernationale Research Maatschappij B.V.
3. Goudriaan F., Grinsven P.F.A. van & Naber J.E. (1994) Electricity generation from biomass via the HTU process, In: *DGMK Conference "Energetische und stoffliche Nutzung von Reststoffen und nachwachsenden Rohstoffen"*, Velen; DGMK Tagungsbericht 9401 (ISBN 3-928164-70-8) 149-162
4. Naber J.E., Goudriaan F., Wal S. van der, Zeevalkink J. & Beld B. van de (1999) The HTU process for biomass liquefaction: R&D strategy and potential business development. In: *Proc. of the 4th Biomass Conference of theAmericas, Oakland (Calif.),* (Ed. by R.P. Overend & E. Chornet, ISBN 0 08 043019 8), **1**, 789-95.
5. Luijkx G.C.A. (1994) *Hydrothermal conversion of carbohydrates and related compounds.* Ph.D. Thesis, Delft University of Technology.
6. Tödheide K. & Franck E.U. (1963) *Z. Physik. Chemie (Frankfurt),* **394,** 387-401; also in: Gmelin *Handbuch der anorganischen Chemie,* 6th edn. (1973) **3** pp 54-7. Verlag Chemie, Weinheim/Bergstr.

7. Mok W.S-L. & Antal M.J. (1992) Uncatalysed solvolysis of whole biomass hemicellulose by hot compressed liquid water. *Ind. Eng. Chem. Res.* , **31**, 1157-61.
8. Jollez P., Chornet E. & Overend R.P. (1993) Steam-aqueous fractionation of sugar cane bagasse: an optimisation study of process conditions at the pilot plant level. In: *Advances in Thermochemical biomass conversion* (Ed. by A.V. Bridgwater) pp. 1659-69. Blackie, London.

Thermochemical Treatment of Radiata Pine using Hot Compressed Water

T. OGI, S. INOUE
National Institute for Resources and Environment,
16-3 Onogawa, Tsukuba, Ibaraki 305-8569, Japan
Y. YAZAKI
Monash University, Dep. Chem. Eng., Clayton, Victoria 3800, Australia

ABSTRACT: Radiata pine bark contains a large amount of polyflavanoids (radiata tannin) and other valuable materials suitable for feedstock of adhesives and resins. We tried to extract radiata tannin effectively and to partially degrade the bark to water soluble materials using hot compressed water.

Radiata pine bark was extracted using hot compressed water either with or without 1% NaOH at temperatures of 100°C and higher under different pressures for various holding times. When 1% NaOH was added to the compressed water system, the yields of the extractives and polyflavanoids increased remarkably, and the highest yield (31.3%) of the extractives was obtained at 140°C and at 10 atm. The hot compressed water extraction requires less than 1/6 the heating energy required for the conventional extraction.

In order to obtain partially degraded water-soluble materials for producing wood adhesive and/or resin feedstocks, Radiata pine bark was treated with 0-15% H_2O_2 solutions at temperatures higher than 100°C. The highest yields of water-soluble materials (47.4 and 47.9%) were obtained with 15% H_2O_2 solution at 120°C. The water-soluble fraction was analyzed by GC-MS and other methods. It was found that the water-soluble fraction mainly consisted of low molecular weight acidic organic compounds, and succinic acid and malonic acid were major compounds. These compounds can be used as feedstocks for isocyanate adhesives.

1. INTRODUCTION

In Australia, Radiata pine (Pinus Radiata) is widely silvicultured and a large amount of bark is generated unused. This bark contains large amounts of polyphenols called "polyflavanoids" (radiata tannin) which react with formaldehyde to give high-quality water-proof wood adhesives. An effective tannin extraction is one of the most important key technologies to produce water-proof and structural wood adhesives. It is well-known that the extractives yield increases with increase of water temperature for bark extraction up to 100°C. Furthermore, extraction of radiata pine bark with aqueous sodium hydroxide (NaOH) gives a higher yield of extractives than extraction with water (1). However, the amount of NaOH used for extraction should be very carefully monitored, because highly alkaline conditions cause the rearrangement of the A-rings structure of polyflavanoids, resulting in the loss of reactivity of polyflavanoids with formaldehyde (2, 3).

In our laboratory we have studied thermochemical treatment of biomass using hot compressed water and found solid-like biomass wastes were converted to liquidized materials by the release of water soluble compounds inside cells due to thermal rupture of cell around 175°C at 40 atm (4). Applying this liquidization process, we tried to extract radiata tannin from the bark.

The bark also contains other materials suitable for resin feedstock. We tried to partially degrade the bark to water-soluble materials using hydrogen peroxide (H_2O_2). Hydrogen peroxide (H_2O_2) yields only water and/or oxygen upon decomposition, so it is an ecologically and environmentally desirable pollution control agent. Alkaline H_2O_2 solutions are also used to chemimechanically bleach ground wood, and to chemically treat pulps. The use of H_2O_2 for pulping and bleaching is based on delignification and further degradation of compounds produced during delignification (5).

In this paper, we report the results of effective tannin extraction from the bark and degradation of the bark for resin feedstocks using a hot compressed water process (6).

2. EXPERIMENTAL

Radiata pine bark sample

The bark samples were collected from a plywood and lumber mill that was using logs from 35-year-old Radiata pine trees at Mt. Gambier in South Australia. The bark samples were dried to 12% moisture content and then ground to pass a 6.2 mm screen in a Wiley mill. The resulting particle size distribution was: 6.2 mm to 1.4 mm (31.6%), 1.4 mm to 1.0 mm (25.2%), 1.0 mm to 0.6 mm (19.8%) and less than 0.6 mm (23.4%). Furthermore, for a comparative study, some of the bark samples were also ground to pass through a 0.6 mm screen. The chemical composition of the Radiata pine bark sample was analyzed according to the Japan Industrial Standard method (7). Elemental and chemical composition of Radiata pine bark are as follows;

Composition of Radiata Pine Bark

Moisture: 9.3 wt%	Organics: 89.9 wt%		Ash: 0.8 wt%
Elemental Composition (wt%)	C: 51.9,	H: 5.7,	O: 42.1*
Chemical Composition (wt%)			
Lignin (polyflavanoids)	56.6		
Cellulose	12.4		
Hemicellulose	17.8		*by difference

Hot compressed water and 1% NaOH aqueous extractions at temperatures higher than 100°C

All experiments were carried out using a stainless steel autoclave of 300 mL capacity. The bark samples (20 g) and water or 1% aqueous NaOH (based on the weight of the oven-dried bark) solution (70 mL), in which the ratio of bark to solvent was 1:4 (w/w), were added to the autoclave. The autoclave was flushed and then pressurized (4 atm or 10 atm) with nitrogen gas. The temperature was increased to 100°C (or 120, 140, 160, 180°C) and either immediately air-cooled (0 hour holding time) or maintained at a designated temperature for 30 minutes (0.5 hour holding time). After heating, the autoclave was cooled to room temperature and then opened. The bark mixtures were separated into water soluble fraction and insoluble residues by filtering and washing with water.

Extractives yield, Stiasny values and polyflavanoids yield

The yield of extractives was calculated on the basis of the weight loss of the oven-dried extracted bark, compared with the weight of the oven-dried unextracted bark.

The Stiasny reaction was carried out using the freeze-dried extractives. The dried sample (100 mg) was dissolved in water (10 mL), to which 36% formaldehyde solution (2 mL) and 10M HCl (1 mL) were added, and heated under reflux for 30 minutes. The Stiasny precipitates were collected using a no. 3 sintered glass filter and were dried at 105°C for 16 hours. The Stiasny value (original) was calculated on the following equation:

$$\text{Stiasny value (original)} = \frac{\text{Weight of Stiasny precipitates}}{\text{Weight of original bark}}$$

The adjusted Stiasny value was then obtained by dividing the Stiasny value (original) by 1.112, since the value of 111.2 is the Stiasny value for catechin being used as a model compound for radiata tannin.

Treatment of bark using H_2O_2 at temperatures higher than 100°C

The ground bark (Particle size 1.25 mm to 0.65 mm; 3.0 g) and 0, 5, 10, or 15% H_2O_2 Solution (30 mL) was added to an electrically heated 100 mL stainless steel autoclave. After this was purged with nitrogen gas, a pressure of 10 atm nitrogen gas was applied and then the autoclave was heated. The temperature was increased to 100°C (or 110, 120, 130, 140°C), maintained at the designated temperature for 30 min, and then reduced to room temperature.

Fractionation and analyses of products after treatment with H_2O_2

Fig. 1 is a schematic diagram showing the procedures for the fractionation of products after the bark was treated with H_2O_2. The evolved gas was collected in a sample bag and then analyzed using gas chromatography (GC). The GC analysis was done using a WG-100 column (Shimadzu GC-8A). The temperatures of the oven, injector and detector were 50°C and 70°C, respectively. The TCD current was 120 mA. The carrier gas was He at a flow rate of 33 mL min^{-1}.

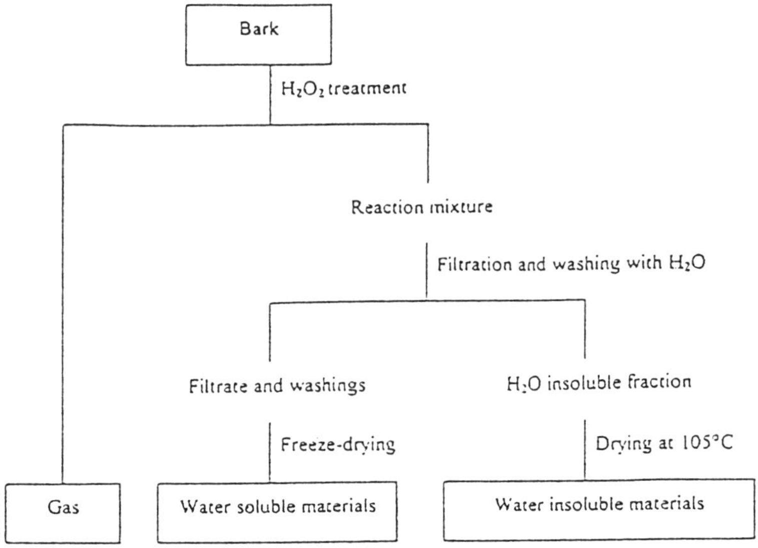

Fig. 1 Procedures for the fractionation of products from Radiata pine bark

The products obtained from the treatment of the bark with H_20_2 were separated into water-soluble and insoluble fractions by filtering and washing with water (60 mL) on a No.3 glass filter. The water-soluble fraction was diluted with water exactly to 200 mL and was measured for Total Organic Carbon (TOC), pH (using Duotest pH indicator paper), and residual H_20_2 (titrimetric permanganate method), then freeze-dried. The TOC was measured using a TOC meter (Shimadzu TOC-5000A). Elemental composition was analyzed by a NA-1500 Elemental Analyzer (Carlow Erba NA-1500). GC-MS analysis was done using a Hewlett-Packard 5890 Series 2GC with HP 5971 MS Detector. GC was done on a DB-5 column using He carrier gas at 0.9 mL min^{-1}. Injector and detector temperatures were 320 and 280°C, respectively. The sample was injected into the GC column at 40°C, left for 3 min, and then heated to 320°C at 5°C min^{-1}.

3. RESULTS AND DISCUSSION

Hot compressed water and 1% NaOH aqueous extractions

Table 1 shows the results of extraction of Radiata pine under hot compressed water system in two cases; that is, the case using compressed water only and the case with 1% NaOH aqueous solution, together with the conventional reflux extraction case.

The extraction using hot compressed water at temperatures higher than 100°C gave lower yields of extractives and polyflavanoids than those extracted using conventional hot water extraction at 100°C under ambient pressure.

However, when a small amount of NaOH (1% of the weight) was added to the hot compressed water for the extraction of Radiata pine bark, the extractives yields increased remarkably. The highest extractives yield (31.3%) was obtained from the extraction at 140°C and 10 atm with no time at temperature. The second highest one (25.9%) was obtained also at 140°C but

Table 1. Extractives yield, Stiasny value and polyflavanoids yield from the bark

Extraction condition			Extractives		Stiasny value		Polyflavanoids
Temp. (°C)	Pres. (atm)	Holding time (h)	Yield (%)	pH	Original (%)	Adjusted (%)	Yield (%)
Compressed water only							
100	4	0	14.4	3.8	85.0	76.4	11.0
100	4	0.5	12.9	3.7	73.4	66.0	8.5
100	10	0	14.8	3.8	82.0	73.7	10.9
100	10	0.5	15.6	3.8	70.2	63.1	9.8
140	4	0	14.4	3.8	74.6	67.1	9.7
140	4	0.5	11.7	3.8	32.2	29.0	3.4
140	10	0	12.2	3.8	48.4	43.5	5.3
140	10	0.5	9.9	3.8	27.3	24.6	2.4
180	4	0	13.3	3.8	43.3	38.9	5.2
180	4	0.5	5.2	3.7	22.5	20.2	1.1
180	10	0	6.9	3.7	15.1	13.6	0.9
180	10	0.5	3.8	3.7	27.4	24.6	0.9
Compressed 1% NaOH aqueous solution							
100	4	0	23.4	5.6	85.6	77.0	18.0
100	4	0.5	22.3	5.3	83.9	75.4	16.8
100	10	0	21.0	5.9	84.2	75.7	15.9
100	10	0.5	21.8	5.0	80.3	72.2	15.8
120	10	0	23.8	5.6	84.0	75.5	18.0
140	4	0	25.9	5.9	77.4	69.6	18.0
140	4	0.5	23.9	5.3	72.3	65.0	15.8
140	10	0	31.3	5.9	77.2	69.4	21.7
160	10	0.5	13.7	5.9	13.9	12.5	1.7
180	10	0	22.9	5.9	71.1	63.9	14.6
180	4	0	20.6	5.9	68.9	62.0	12.8
180	4	0.5	13.7	4.7	8.4	7.6	1.0
180	10	0	16.8	5.9	25.5	22.9	3.8
180	10	0.5	14.9	4.7	11.1	10.0	1.5

Conventional extraction at ambient pressure

Extraction condition				Extractives		Stiasny value		Polyflavanoids
Temp. (°C)	Reflux (h)	time time	Ext. Solvent H_2O	Yield (%)	pH	Original (%)	Adjusted (%)	Yield (%)
100	0	1	1% NaOH	13.3	4.4	88.0	79.1	10.5
100	0	1	H_2O	17.4	6.5	83.3	74.9	13.0
100	0.5	1	1% NaOH	18.2	4.4	88.8	79.9	14.5
100	0.5	1	H_2O	22.6	6.2	84.1	75.6	17.1
100	0.5	2	H_2O	19.6	4.4	88.7	79.8	15.6
100	0.5	2	1% NaOH	30.5	6.2	82.8	74.5	22.7

at 4 atm pressure, while the lowest (13.7%) was obtained after extraction for 30 min at 140°C and 10 atm pressure and after extraction for 30 min at 180°C and 4 atm pressure. When treated at lower pressure (4 atm) with no time at temperature, the extractives yields were very similar (23.4% and 25.9%) at 100°C and 140°C and slightly decreased to 20.6% at 180°C.

When the bark was treated for 30 min at temperature, the extractives yields were less than those of the bark extracted with no time at temperature. The greatest decrease (17.6%) from 31.3% to 13.7% was found in the bark extracted at 10 atm and 140°C between with no time at temperature and with 30 min holding time at temperature.

The extractives obtained from the extraction at 4 atm pressure with no time at temperature gave the highest adjusted Stiasny value (77.0%) at 100°C which decreased to 69.6% and 62.0% at 140°C and 180°C, respectively.

The highest polyflavanoids yield (21.7%) was obtained from the bark extracted at 10 atm and 140°C with no time at temperature, followed by the yield of 18.0% from the extractions at 100°C and 4 atm, 140°C and 4 atm, and 120°C and 10 atm pressure with no time at temperature. On the other hand, the smallest polyflavanoids yield (1.0%) was obtained after treated for 30 min at 180°C under 4 atm pressure.

Comparisons between the conventional and the hot compressed extraction process

Ambient pressure has been used for the extraction of Radiata pine bark, although different conditions have been used for the tannin extraction of wattle (*Acacia mearnsii*) bark. This is largely due to our previous experience that the application of temperatures higher than 100°C decreases the extracted yield from the bark.

The yields and Stiasny values of extractives obtained from the bark samples using the conventional extraction method at 100°C are shown in Table 1. At ambient pressure, the extractives yields increased with the increase of extraction time (reflux time) and also with the addition of NaOH.

The extractives yields (14.4% and 14.8%) obtained from the extraction with hot compressed water at 4 and 10 atm were slightly higher than that (13.3%) obtained from the conventional extraction method at ambient pressure with no time at temperature. However, after extraction for 30 min, the reverse result was obtained in which the yields (12.9% and 15.6%) from the extraction with hot compressed water were much lower than that (18.2%) for the conventional extraction. In addition, the extractives obtained from the conventional extraction method gave much higher Stiasny values than those of the extractives obtained from the extraction with hot compressed water. When using water only, there is not much advantage in hot compressed water extraction.

However, when using 1% NaOH hot compressed solution, the highest extractives yield (31.3%) was obtained at 140°C, 10 atm, with no time at temperature. This extractive had a high Stiasny value, therefore high polyflavanoids yield (21.7%). Since the highest extractives yield was obtained at 140°C under 10 atm pressure with no time at temperature, two additional extractions were carried out at 120°C and 160°C. The results are included in Table 1. The extractives yields were 23.8% and 22.9% and their adjusted Stiasny values were 75.5% and 63.9% at 120°C and 160°C, respectively. Therefore 140°C was the best temperature which gave the highest yield. It is interesting to note that 140°C is the temperature, at which lignin has been reported to soften (8, 9, 10).

When the polyflavanoids are subjected to highly alkaline conditions, rearrangement of the A-ring structure of the polyflavanoids takes place so that the polyflavanoids are likely to have lost their reactivity resulting in poor glue bond quality (11, 12). Accordingly, the pH of the extractives solution obtained from the 1% aqueous NaOH solution at higher temperatures and pressures is a very critical factor for obtaining wood adhesives of good quality. However, the experimental results showed that the pH's of the 1% NaOH extractives were between 4.7 and 5.9, which being acidic are suitable pH's for the formulation of wood adhesives.

The conventional extraction using 1% aqueous NaOH solution gave almost the same yields of extractives (30.5%) and polyflavanoids (22.7%) as the hot compressed extraction. Concerning extractive and polyflavanoids yields, there was not so much difference between conventional and hot compressed water process. However, there is much difference for the energy consumed for extraction process.

Energy consumption and CO2 emission from the extractions

The conventional process is based on the extraction with hot water, which is 20 times the bark weight, heated under reflux for 30 min and repeated once. However, the hot compressed water extraction uses water amounting to 4 times the bark weight, heated up to 140°C, cooled immediately, and the extraction is not repeated.

Based on these conditions and assuming that the extractives yields are 30.5% and 31.3% from the conventional (1% NaOH) and the hot compressed water (1% NaOH) extractions, respectively, the energy required for heating the water from room temperature (20°C) to the designated temperatures (either 100°C or 140°C) was calculated on the basis of the production of 1t of the tannin extracts from the bark.

The specific heat for water is 4.18 [MJ t^{-1}. C^{-1}] and the amount of bark to obtain ton of tannin extracts was 1/0.305 [t bark / t tannin extracts] for the conventional and 1/0.313 [t bark / t tannin extracts] for the hot compressed water extractions, the energy requirement for heating the water to either 100°C or 140°C under these conditions can be calculated according to the following equations;

Conventional hot water extraction

$$1/0.305 \times 20 \times 4.18 \times (100 - 20) \times 2 = 43{,}900 \text{ MJ t}^{-1}$$

Hot compressed water extraction

$$1/0.303 \times 4 \times 4.18 \times (140 - 20) = 6{,}600 \text{ MJ t}^{-1}$$

Consequently, the hot compressed water extraction required less than 1/6 the heating energy required for the conventional extraction method. Furthermore, since the conventional extraction method requires additional energy to keep the water at 100°C for 30 minutes twice, the new hot compressed water extraction may be a preferred method.

Assuming that in order to generate 1[kWh] of electricity, 0.547[kg-CO_2] is emitted (13) and 1 [J] is equivalent to 2.8 x 10^{-7} [kWh], the amounts of CO_2 emission for these methods would be;

Conventional hot water extraction

43,900 MJ = 12,300 [kWh] = 6,700 [kg-CO_2]

Hot compressed water extraction

6,600 MJ = 1,900 [kWh] = 1,000 [kg-CO_2]

Therefore, assuming that 7,000 t of tannin extracts are produced using the hot compressed water method in a year, the CO_2 emission would be 7,000 t/year, whilst that from the conventional method would be 46,900 t/year.

Consequently, since the calculations described above did not consider the energy to keep the temperature at 100 °C during the extraction time (2 x 0.5 hours), the conventional extraction method requires at least six times the amount of energy required for the extraction with hot compressed water at 140 °C with no time at temperature. At the same time, the conventional extraction method emits at least six times the amount of CO_2 than does the hot compressed water extraction method.

Decomposition of Radiata pine bark treated with H2O2.

The yields of the water-soluble and insoluble materials obtained from Radiata pine bark after treatment with H_2O_2 solutions at temperatures higher than 100 °C are summarized in Fig. 2.

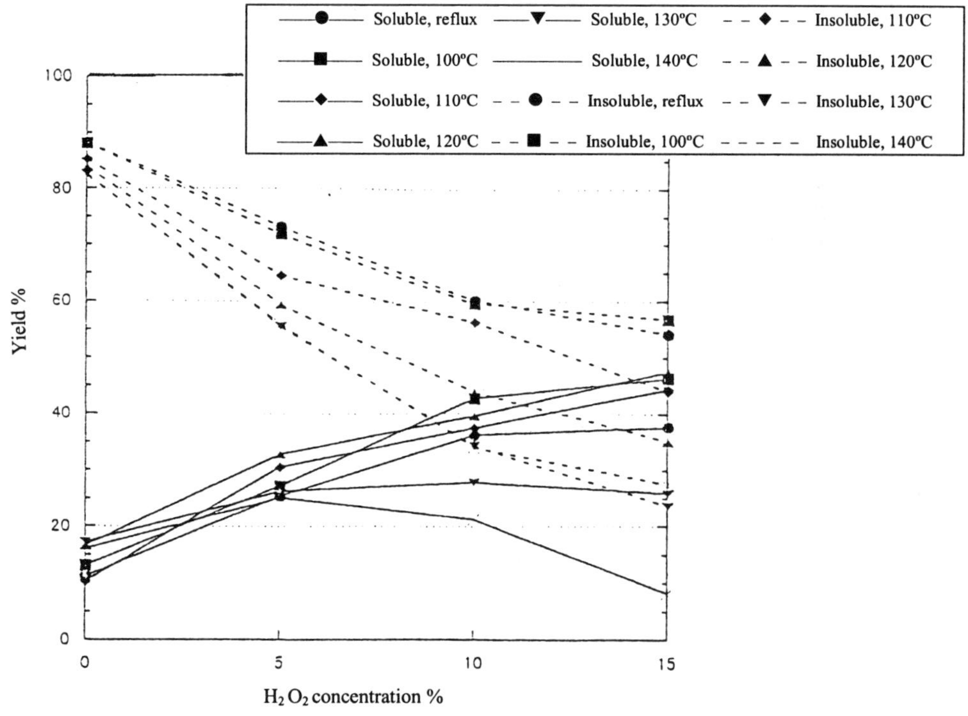

Fig. 2 The yields of water soluble and insoluble materials

At temperatures higher than 100 °C, the yield of water-soluble materials increased with increasing H_2O_2 concentration from 0 to 15%. The highest yield of water-soluble material (47.4%) was obtained when the bark was treated with 15% H_2O_2 solution at 120 °C for 30 min. Very similar yields were obtained with the same H_2O_2 charge at 100 °C and 110 °C. Over 130 °C, the yield of the water-soluble materials decreased due to the production of gases such as CO_2 and CO.

Various amounts of H_2O_2 remained in the aqueous phase after the bark treatments. The highest concentration of residual H_2O_2 (80%) was obtained when the bark was treated with 15% H_2O_2 under reflux conditions. The decomposition of H_2O_2 was dependent on the temperature applied for the treatments, and there was no H_2O_2 after treatment at 140 °C. Residual H_2O_2 limits the use of the water-soluble materials, so it is important to find conditions in which there is little residual H_2O_2 in the aqueous phase. When the bark with H_2O_2 solution was heated higher than 140 °C, an exothermic reaction took place, which was to vigorous to control, and during the reaction all H_2O_2 was decomposed and consumed. The suitable degradation condition was the following: temperature 120 °C, 30 min, 10-15% H_2O_2.

Chemical characterization of products

More than half of the bark was obtained as water soluble materials by the hot-compressed H_2O_2 solution process at 120 °C. The water soluble materials were fractionated further into ethyl acetate and aqueous fractions and analyzed by NMR and GC-MS.

In Fig.3, the GC-MS spectrum of ethyl acetate fractions and the compounds identified are shown. The compounds with asterisk (*) were also identified in aqueous fractions.

The results of the GC-MS for water-soluble materials showed the presence of many low molecular weight organic acidic compounds. The pHs of the water layers after treatments were between 2.8 and 5.0. Kadla et al. studied the reactions of lignin model compounds treated with H_2O_2 at 90 °C. In addition to the reactions of the phenolic representatives, the non-phenolic compounds are shown to react at these higher temperatures via an S_N2 mechanism to ultimately give the corresponding benzoic acids and dimeric ether compound. (14) Krochta et al. reported the thermochemical degradation of cellulose into organic acids (15). Consequently, it is expected that lignin and / or cellulose contained in the bark react to give acids during the treatments with H_2O_2 solutions at temperatures higher than 100 °C. Radiata pine bark mainly consists of cellulose, hemicellulose, and lignin (polyflavanoids). It is a complex feedstock so that many organic compounds, other than acids, might be contained in the water soluble materials. Russel et al., recognized the formation of aromatic compounds, when cellulose was converted by a thermochemical treatment. (16) The results of NMR and GC-MS analysis also showed phenolic and non-phenolic compounds were contained in the water soluble materials, which maintain the structure of lignin and polyflavanoids after the treatment. The main compounds in the water soluble fraction were succinic acid and malonic acid. It is presumed that these compounds are found by degradation and oxidation of lignin. The solubulization process of the bark consists of a very complicated combination of various reactions such as oxidation, dehydration, and cyclization. The organic acidic compounds contained in the water-soluble materials can be used in isocyanate adhesives.

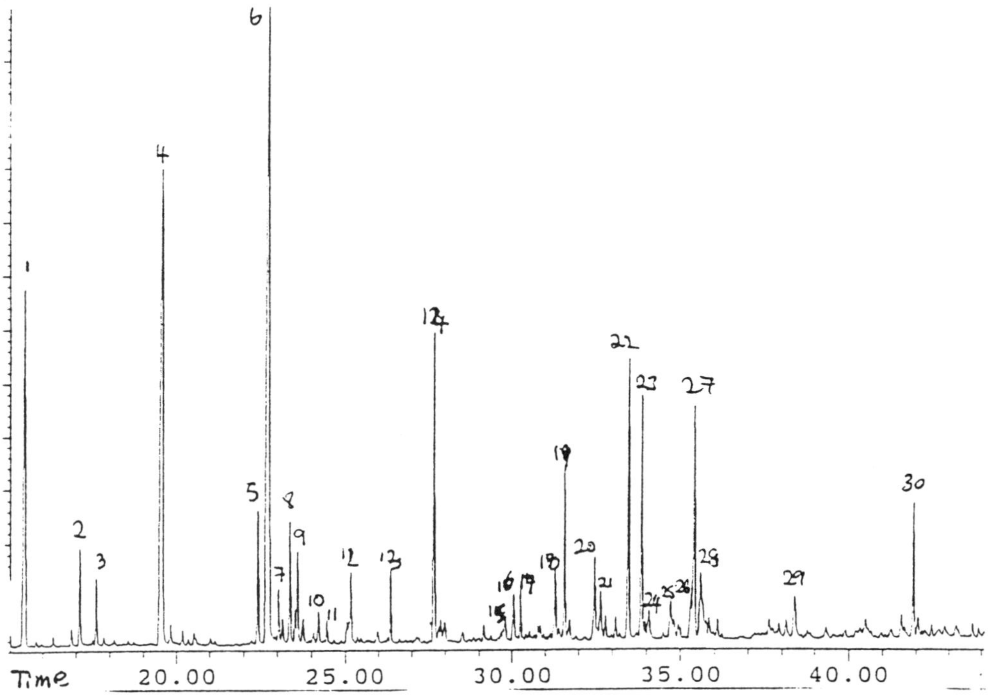

Peak #	Compounds	Peak #	Compounds
1	Glycolic acid	14	DL-malic acid *
2	Levulinic acid	15	2-Hydroxypentanedioic acid
3	3-Hydroxy propanoic acid *	17	*meso*-Tartaric acid *
4	Malonic acid *	18	Xylonic acid *
5	Maleic acid *	19	2,3-Dihydroxy succinic acid *
6	Succinic acid *	22	Tricarballyic acid
7	Methyl succinic acid	23	Aconitic acid
8	2,3-Dihydroxy propanoic acid *	24	Vannilic acid
9	Fumaric acid	27	3,4-Dihydroxy benzoic acid
12	Pentanedioic acid		
13	2- Pentanedioic acid		

Fig. 3 GC-MS spectrum of water soluble fraction and compounds identified

4. CONCLUSION

The extraction of Radiata pine bark using hot compressed water at temperatures higher than 100 °C gave lower yields of extractives and polyflavanoids than those extracted using the conventional hot water extraction at 100 °C under ambient pressure. However, when a small amount of NaOH was added to the compressed water, the extractives yields from the bark increased remarkably. The extraction of the Radiata pine bark at 140 °C and 10 atm pressure with no time at temperature provided the highest yields of extractives (31.3%) as well as polyflavanoids (21.7%). The hot compressed water extraction requires less the 1/6 the heating energy for the conventional extraction.

Treatment using hot compressed H_2O_2 solution:
When Radiata pine bark was treated with 0-15% H_2O_2 solution, the highest yield (47.4%) of water soluble materials was obtained at 120 °C with 15% H_2O_2.
The water-soluble materials contain many organic acidic compounds in which succinic acid and malonic acid were major compounds. These obtained compounds could be used in wood adhesives, because they contain hydroxyl groups.

ACKNOWLEDGEMENTS:

This project was financially supported by the Agency of Industrial Science and Technology (AIST) in Japan and also by the MFP Branch, Department of Industry, Science and Technology (DIST) in Australia. We thank Ms. Y.Fukuda, an assistant staff member at NIRE, for her help during the experiments and Dr.A.F.A.Wallis and Mr.R.Wearne at CSIRO for their GC-MS analysis.

REFERENCES:

1. Yazaki Y. (1985) , *Holzforshung*, **39**, 267
2. Sears K.D. Casebier R.L. Hergert H.L. Stout G. H. and McCandish L.E. (1974) *J. Org. Chem.*, **39**, 3244
3. Yazaki Y. and Aung T. (1989) *Holzforshung,* 43, 281
4. Sawayama S. Inoue S. Yagishita T. Ogi T. and Yokoyama S. (1995) *J. Fermentation and Bioengineering*, 79, 300
5. Hosoya S. (1992) *Kami-parupu gijutu kyoukaishi*, **46**, 1344
6. Some parts of this paper have been published in the following papers; Inoue S. Asaga M. Ogi T. and Yazaki Y. (1998) *Holzforshung,* **52**,139, and Inoue S.Ogi T. and Yazaki Y. (1999) *Bull. Chem. Soc. Jpn.*, **72**, 2135
7. Migita N. Yonezawa Y. and Kondo T. (1981) Wood Analysis (Chapter 10), Vol 2, Kyoritu Shuppan Inc. Tokyo, Japan 1
8. Atack D. (1972) *Svensk Papperstidn*, **75**, 89
9. Goring D.A.I. (1963) *Pulp Paper Mag. Can*. **64**, 517
10. Hillis W.E. and Rozsa A.N. (1978) *Holzforshung*, **32**, 68
11. Yazaki Y. and Aung T. (1989) *Holzforshung,* 43, 281
12. Yazaki Y. and Collins P. (1994) *Holz Roh-Werkstoff* **52**, 307
13. NEDO (New Energy Development Organization) report (NEDO-GET 9410-1) (1995) Chap.11, 28p

14. Kadla J. Chang H. and Jameel H. (1997) *Holzforshung*, **51**, 428
15. Krochta J.M. Hudson J.S. and Drake C.W. (1984) *Biotechnol. Bioeng. Symp.*, **14**, 37
16. Russel J.A. Miller R.K. and Molton P.M. (1983) *Biomass*, **3**, 43

Chemical conversion of biomass resources to useful chemicals and fuels by supercritical water treatment

S. Saka · R. Konishi
Department of Socio-Environmental Energy Science, Graduate School of Energy Science, Kyoto University, Yoshida Honmachi, Sakyo-ku, Kyoto 606-8501, Japan

ABSTRACT : A batch-type supercritical water system was used to study the chemical conversion of sugi (*Cryptomeria japonica* D. Don) and buna (*Fagus crenata* Blume) woods to useful chemicals and fuels. Cellulose, hemicelluloses and milled wood lignin (MWL) as their cell wall components were also studied to understand the reaction behaviors of wood in supercritical water. High performance liquid chromatography (HPLC) and gel permeation chromatography (GPC) have been performed on the reaction mixtures of these samples. The results clearly indicated that the woods can be almost fully liquefied in supercritical water for 5 sec. In addition, the water-soluble fraction was mainly derived from cellulose and hemicelluloses with average molecular weights in a range between 100 and 200, whereas the methanol-soluble fraction derived mainly from the lignin and hydrophobic in nature with an average molecular weight of 530. From these fractions, useful chemicals such as glucose, levoglucosan, 5-hydroxymethyl furfural, furfural and some phenolic compounds could be obtained.

INTORODUCTION

Biomass resources will become more important in the future as alternatives to fossil resources, which will be exhausted sooner or later. The features of biomass are renewable, carbon-neutral, and abundant. However, these resources have not been utilized efficiently and their unused portions are wasted in the world. Human beings are, thus, urged to develop efficient utilization technology of biomass, especially cellulosic biomass resources.

For the conversion of biomass resources into useful chemicals and bio-energy, three major process types, direct combustion, gasification and liquefaction, are effective. Among them, liquefaction for liquid transpiration fuels is, in particular, important because it is difficult to replace them with other energy forms. Thus, a study of the liquefaction of biomass is of interest.[1,2]

There exist two major conversion methods, biochemical and thermochemical, of wood into useful chemicals and liquid fuel. Enzymatic saccarification is one of the biochemical conversion method[3], whereas acid-, base-, and metal-catalyzed hydrolyses are the thermochemical conversion method.[4,5] However, the hydrolysis reaction does not proceed at a sufficiently high reaction rate. In addition, corrosion of the reactor may occur in the latter and it requires purification process of the waste water.

Recently, some researchers have suggested that ether or ester linkages are hydrolyzed in supercritical water even without any catalyst.[6] Sakaki et al.[7-9] have investigated non-catalytic decomposition of cellulose in subcritical water, and cellulose was found to be decomposed to water solubles, which were decomposed further after their yield reached nearly 80%. On entering the second decomposition process, the water solubles were converted into gaseous products and methanol-soluble products, and solid products like char were formed from the liquid phase. The hydrolysate of cellulose obtained in this process was subjected to a fermentation test, and the formed glucose was successfully converted into ethanol.

In our previous study,[10] a supercritical biomass conversion system was developed with a bath-type reaction vessel to hydrolyze woody biomass resources and to convert them into useful chemicals and fuel. This system can cover a range of up to 280MPa in pressure and up to 500℃ in temperature, measured with the pressure gauge and thermocouple installed.

Water used in this supercritical state behaves very differently from water under normal pressure and temperare.[11] In such a supercritical state, the water can be expected to act as an acid or base,[12] but by returning the system to ordinary conditions before pyrolysis occurs, glucose and its derivatives could be obtained in water from cellulose. Therefore, supercritical water treatment can be superior to enzymatic saccharification or ordinary acid hydrolysis mentioned above, for the chemical conversion of biomass to useful chemicals.

Sasaki et al. have also studied cellulose hydrolysis in subcritical and supercritical water[13] and cellulose was found to be rapidly hydrolyzed in supercritical water to sugars. This experiment was conducted with a flow-type reactor in a range of temperature from 290 to 400℃ at 25MPa with a high pressure slurry feeder developed to feed the cellulose-water surries. As a result, hydrolysis products of about 75% in their yield were found in supercritical water, much higher than those in subcritical water. This is because around the critical point of water, the hydrolysis rate of cellulose jumps over one order of magnitude larger in supercritical condition than the subcritical state of water. In addition, Townsend et al.[6] reported that the aromatic model compounds of coals were hydrolyzed in supercritical water and that ether bonds of the compounds can be cleaved during the treatment. It can be, therefore, expected that biomass components can be hydrolyzed in supercritical water.

In this study, thus, we selected sugi as softwood and buna as hardwood species. Just for comparison, cellulose and hemicellulose (glucomannan, xylan) and lignin (milled wood lignin) were also studied for their reaction in supercritical water.

MATERIALS AND METHODS

MATERIALS

As biomass samples, sugi (*Cryptomeria japonica* D. Don) and buna (*Fagus crenata* Blume) woods were selected for supercritical water treatment. In addition, microcrystalline cellulose (avicel), glucomannan extracted from softwood dissolving pulp, larch xylan (Shigma Co., Ltd.), and milled wood lignin (MWL) from buna wood were studied to understand reaction behaviors of wood components during supercritical water treatment.

SUPERCRITICAL WATER BIOMASS CONVERSION SYSTEM

The supercritical water biomass conversion system used in this study was revised to be a batch-type reaction vessel from the semibatch-type employed in a previous work.[10] This system can cover a range of up to 280MPa in pressure and up to 500℃ in temperature.

A schematic diagram of this batch-type reaction system is shown in Fig. 1 in which a 5 ml volume reaction vessel is made of inconel-625. To this reaction vessel, about 5 ml of water was fed with 150 mg of the biomass sample, and then it was attached to the system.

To start a treatment of the sample in supercritical water, the reaction vessel was quickly heated by immersing it in the tin bath preheated at 500℃ and maintained at supercritical temperature between 380 and 400℃ and pressure between 140 and 200MPa for about 5 sec unless otherwise indicated. To quench the reaction, the reaction vessel was moved in the water bath. During this treatment, the temperature of the reaction vessel was monitored by a thermocouple installed into the reaction vessel. At the same time, the pressure was also measured by the pressure gauge attached to reaction vessel.

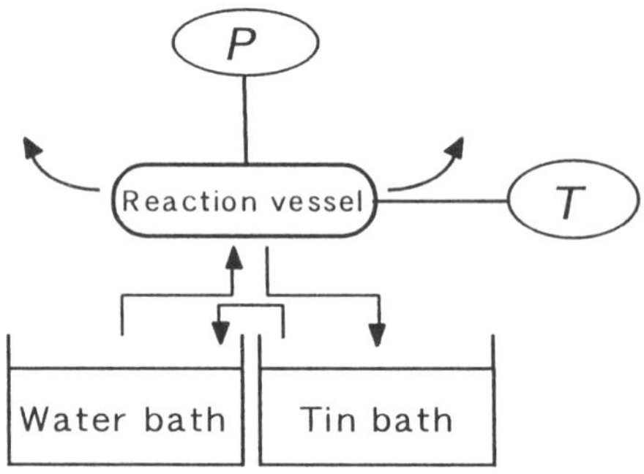

Fig. 1. Biomass conversion system in supercritical water.
P: Pressure monitor, *T*: Temperature monitor

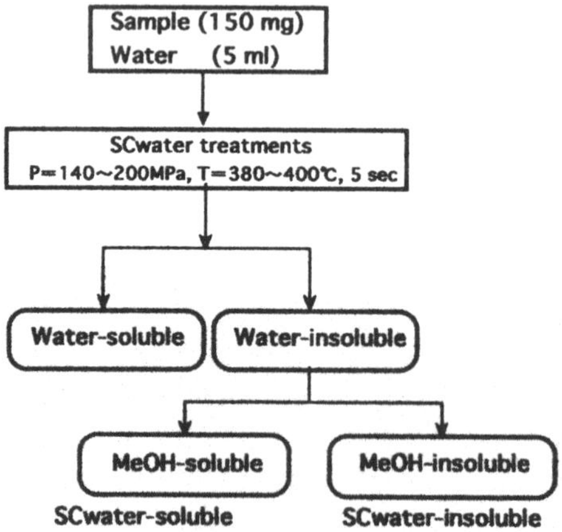

Fig. 2. Schematic flowchart of the experimental procedure for lignocellulosics as treated in supercritical water.

SEPARATION OF THE REACTION MIXTURE AND THEIR ANALYSES

According to the scheme in Fig. 2, the obtained reaction mixture of 150 mg sample in 5 ml water after supercritical treatment (>Tc (critical temperature) = 374℃, >Pc (critical pressure) = 22.1MPa) was filtrated with 0.2 μ m membrane filter (ADVANTEC) to separate water-soluble and water-insoluble portions.

The water-insoluble portion was then washed with 10 mL of MeOH to separate MeOH-soluble and MeOH-insoluble portions. The obtained water-soluble and MeOH-soluble portions, the sum of which nearly corresponds to the supercritical water-soluble portion, were then analyzed by the high performance liquid chromatograph (HPLC) (Shimadzu LC-10A) which consists of a high pressure pump (Shimadzu Co., Model LC-10AT). HPLC analysis conditions were as follows: [Column: STR ODS-II, Column temperature: 40℃, Carrier solvent: CH_3OH/H_2O = 20/80 (0→10 min), 20/80 →100/0 (10→20 min), 100/0 (20→30 min), Flow rate: 0.7 ml/min, Detector: a spectrophotometric detector (SPD) (λ=254 nm) or a refractive index detector (RID)], or [Column: ULTRON PS-80P, Column temperature; 80℃, Carrier solvent: H_2O, Flow rate: 1.0 ml/min, Detector: SPD (λ=254 nm) or RID]. To obtain the yields of monosaccharides and their decomposed products, their standard samples (Nacalai tesque, extra pure reagent) with known concentrations in water were analyzed by HPLC as a standard in a similar manner.

To study the molecular distribution of water-soluble and MeOH-soluble portions, gel permeation chromatography (GPC) (Shimadzu LC-10A) which consisted of a high pressure pump (Shimadzu Co., Model LC-10AT) was performed with a column of KF-803L, operated at 40℃ with flow rate of 1.0 mL/min of tetrahydrofuran. The molecular weight of the fractions was determined from calibration curve obtained with polystyrene as a standard.

Water-insoluble and MeOH-insoluble portions were also studied by light microscopy (Nikon OPTIPHOT) and the latter portion which corresponds to the supercritical water-insoluble portion was weighed to obtain its yield, and studied by X-ray diffractometry with a Rigaku RINT 2000V (Cu-Kα, λ=1.542Å) at 40 kV and 30 mA to examine its crystallographic nature.

RESULTS AND DISCUSSION

SUPERCRITICAL CONDITIONS OF WATER IN BATCY-TYPE REACTION SYSTEM

Figure 3 shows the changes of the temperature and the pressure in the reaction vessel as it was immersed into the tin bath and moved into water bath. From this figure, the water reaches Tc after 15 sec, while for Pc, it takes only 6 sec. Therefore, supercritical condition of water can be achieved after 15 sec. Under such conditions, supercritical water treatment was made for about 5 sec against biomass samples and their cell wall components.

SEPARATION OF SUPERCRITICAL WATER-SOLUBLE AND SUPERCRITICAL WATER-INSOLUBLE PORTIONS OF WOOD

According to the scheme in Fig. 2, the biomass sample was treated with water under supercritical conditions, and the reaction mixtures of sugi and buna woods and their cell wall components were separated into water-soluble and water-insoluble portions by centrifugation. However, the reaction mixture always contains oily substances. Therefore, the latter was further treated with methanol (MeOH) to remove oily substances from water-insoluble residues. Because a dielectric constant of water at an ordinary condition being about 80 is reduced down below 10 in supercritical state,[14] oily substances collected as the MeOH-soluble portion must be solvated with supercritical water during the treatment. Therefore, under supercritical condition of water, water-soluble portion plus MeOH-soluble portion would roughly correspond to a supercritical water-soluble portion. The MeOH-insoluble portion, thus, corresponds to the supercritical water-insoluble portion.

Table 1 shows the yields of these fractions for sugi and buna woods as treated in supercritical water. Since the supercritical water-insoluble residues were in a small quantity in both woods, it is evident that both sugi and buna woods were effectively converted into supercritical water-soluble fraction in about 5 sec treatment. However, the water-soluble portion was more in buna than sugi wood. In turn, the MeOH-soluble portion was more in sugi wood. This would be due to the different chemical compositions of these two woods as described later.

Table 1 The yields of the water-soluble and MeOH-soluble portions and MeOH-insoluble residues.

Species	Lignin* (%)	Supercritical water-soluble(%)		Supercritical water-insoluble(%)
		Water-soluble	MeOH-soluble	MeOH-insoluble
Sugi	32.7	57	39	4
Buna	24.0	76	22	2

* Klason lignin plus acid-soluble lignin

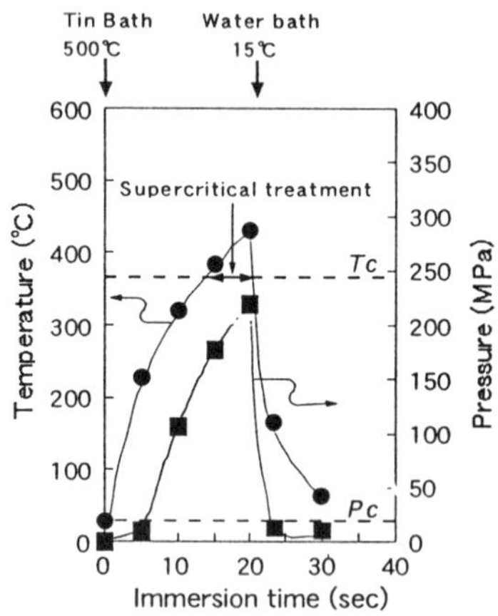

Fig. 3. Changes of the temperature and the pressure in the reaction vessel as it was immersed in the tin bath and moved into water bath. The supercritical treatment of water was made for 5 sec. *Tc*, critical temperature of water=374℃; *Pc*, critical pressure of water=22.1MPa

Water-soluble portion

HPLC analysis has been performed with RID detector on the water-soluble portion of the sample treated in supercritical water for about 5 sec for sugi and buna woods and their cell wall components (Fig.4). For cellulose, it is apparent that glucose appears predominantly as a hydrolysate product. In addition, levoglucosan, 5-hydroxymethyl-furfural (5-HMF) and furfural can be observed. As a minor component, cellobiose and fructose appeared sometimes. On the other hand, for hemicelluloses, glucomannan and xylan, the expected hydrolysate products such as glucose and mannose from glucomannan, and xylose from xylan appear in a very small quantity. Instead, an unknown product appears at 16 min in a retention time in Fig. 4. It seems reasonable that amorphous hemicelluloses are decomposed to a greater extent than crystalline cellulose, under the same supercritical condition of water. Therefore, the supercritical treatment of water for 5 sec is too severe for hemicelluloses. As reported by Mok et al.[15], 90% of hemicelluloses is hydrolyzed to be monomeric sugars, under hot compressed liquid water (2 min, 230℃, 34.5MPa). Thus, more appropriate treatment condition must be found for hemicelluloses.

Fig.5 shows the HPLC chromatograms with RID detector for buna wood as studied in various tin bath temperatures. It is obvious that oligomers and monomeric sugars such as xylose and glucose from hemicelluloses appear at 300 and 350℃ tin bath temperatures and that much higher sugar yields can be obtain in this subcritical state of water. For MWL from buna wood, lignin-derived products sensitive to the ultraviolet light (λ =254nm) could be observed in the water-soluble portion as studied by the HPLC with SPD detector (Fig. 6).

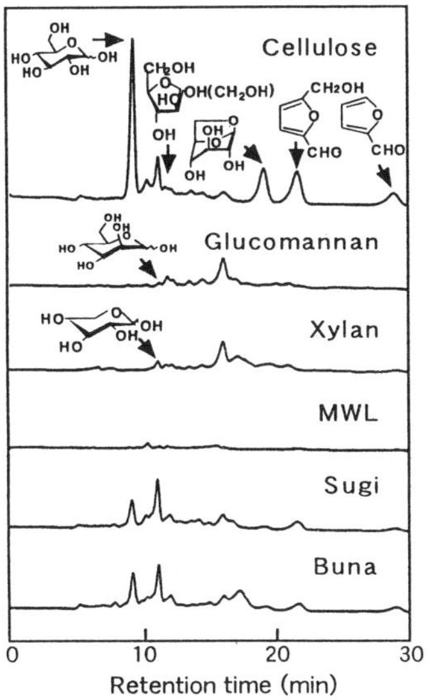

Fig. 4. HPLC chromatograms of the water-soluble portion from woods and their components as treated in supercritical water. (Tin bath 500℃)
Detector : RID, Column : ULTRON PS-80P, Carrier solvent : H_2O

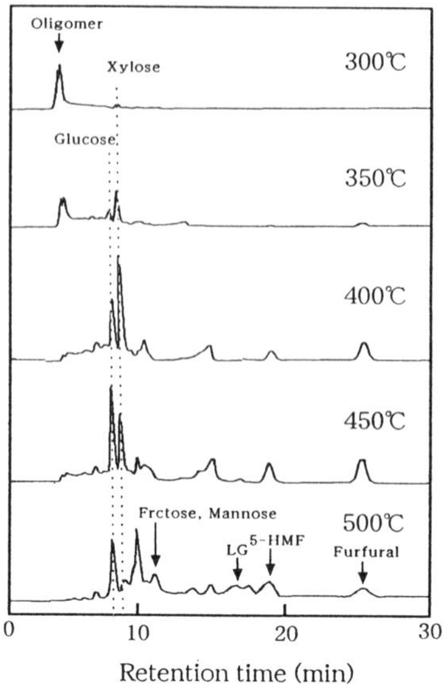

Fig. 5. HPLC chromatograms of the water-soluble portion from buna woods as treated in various temperatures of the tin bath.
Detector : RID, Column : ULTRON PS-80P, Carrier solvent : H_2O

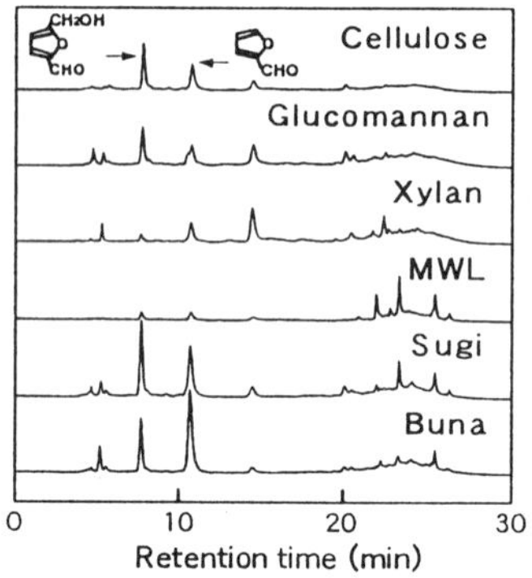

Fig. 6. HPLC chromatograms of the water-soluble portion from woods and their components as treated in supercritical water.
Detector : SPD(λ =254nm), Column : STR ODS-II, Carrier solvent : H_2O/MeOH=20/80(0→10min)→0/100(20min)→0/100(20→30min)

Based on these lines of evidence, compounds appeared in sugi and buna woods in Figs. 4 through 6 seem to be derived mainly from cellulose and hemicelluloses with some from lignin. As shown in Table 1, buna wood contains less lignin than sugi. It is, therefore, reasonable that the water-soluble portion derived mainly from cellulose and hemicelluloses is more in buna than sugi wood.

MeOH-soluble portion

The reaction mixture after supercritical treatment of water consists of the water-soluble portion and water-insoluble residues with oily substances. The oily substances seem to be solvated with water in its supercritical state due to their hydrophobic nature, but separated in ordinary condition of water. Such oily substances shown by the arrows in Fig. 7 are recovered with water-insoluble residues, but can be washed with methanol. Therefore, they may be collected as MeOH-soluble fraction.

On this MeOH-soluble portion HPLC analysis has been performed with SPD detectoor for sugi and buna woods and their cell wall components.It is quite apparent in Fig. 8 that such oily substances collected as MeOH-soluble fraction is only detected in MWL among cell wall components. Therefore, those from sugi and buna woods can be concluded to be lignin-derived products.

Such MeOH-soluble portion derived from lignin is more in sugi wood than buna, as in Table 1. This is consistent with the fact that sugi wood has the higher lignin content than buna as shown in Table 1.

Fig. 9 shows GPC chromatograms of water-soluble and MeOH-soluble portions of sugi wood. It is apparent that, compared with the water-soluble portion, the MeOH-soluble fraction has a higher molecular weight. The former has an average molecular weight in a range between 100 and 200, whereas the latter has on average of 530 .

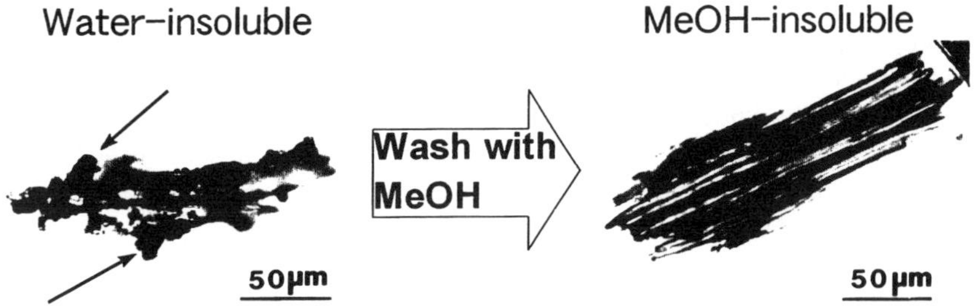

Fig. 7. Light micrographs of water-insoluble residue of sugi wood (left) and MeOH-insoluble residue after washing with MeOH (right). The arrows show the oily substances present with MeOH-insoluble residue.

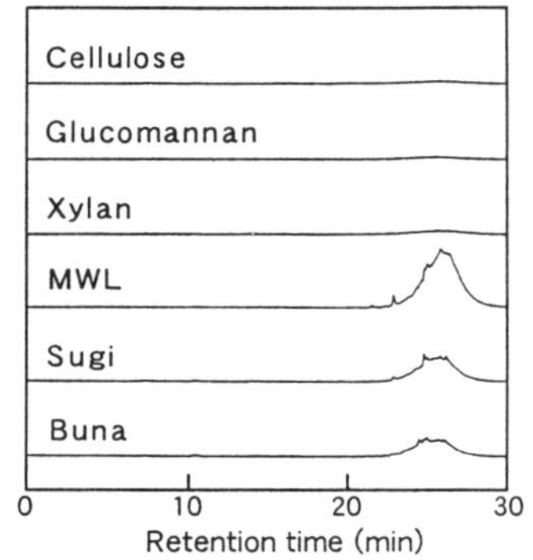

Fig. 8. HPLC chromatograms of MeOH-soluble portions from woods and their components as treated in supercritical water.
Detector : SPD(λ =254nm), Column : STR ODS-II, Carrier solvent : H_2O/MeOH=20/80(0→10min)→0/100(20min)→0/100(20→30min)

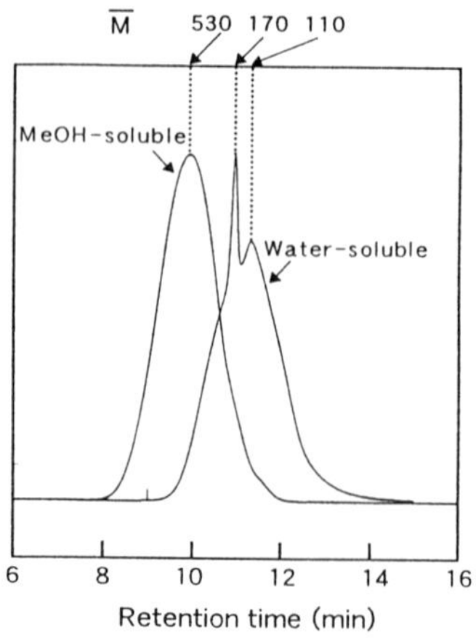

Fig. 9. GPC chromatograms of water-soluble and MeOH-soluble portions of sugi wood as treated in supercritical water.
Detector : SPD(λ =254nm), Column : KF-803L, Carrier solvent : THF

SUPERCRITICAL WATER-INSOLUBLE PORTION OF WOOD

The obtained MeOH-insoluble residues, which are supercritical water-insoluble portion of wood, were only 2 to 4 % as shown in Table 1. Therefore, these can be ignored. However, they were studied by X-ray diffractometry as in Fig. 10. For comparison, untreated wood and cellulose were also studied. It is obvious that the MeOH-insoluble residues are amorphous. Perhaps, these residues would be liquefied in supercritical water if the treatment condition is more appropriate.

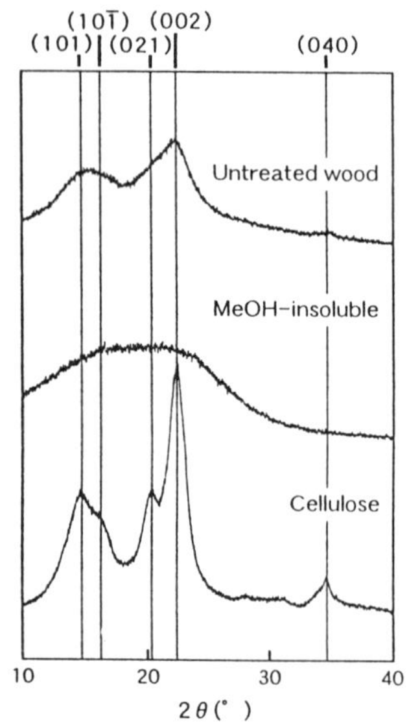

Fig. 10. X-ray diffractograms of supercritical water-insoluble portion (MeOH-insoluble residue) from sugi wood as treated in supercritical.

CONCLUDING REMARKS

Biomass resources will become more important in the future as alternative chemicals and fuel from fossil resources. Such alternatives can be carbon-neutral and renewable. Therefore, the use of biomass is of benefit for our global environments. The present study has clearly demonstrated a potential of woody biomass resources to be converted into useful chemicals in supercritical water. In addition, the products derived from cellulose and hemicelluloses were effectively separated from lignin-derived products. This raises the possibility of converting whole biomass substrates to alternative chemicals, and provides ample opportunities for human beings to produce useful biomass-based products as chemicals and bio-fuels without using fossil resources.

ACKNOWLEGEMENT The authors would like to thank TOYO KOATSU Co. for manufacturing the supercritical water biomass conversion system. This work has been done under the program of the Research for the Future (RFTF) of The Japan Society for the Promotion of Science (JSPS-RFTF 99 P01002). This research was also supported by a Grant-in-Aid for Exploratory Research (11876039, 1999), and a Great-in-Aid for Scientific Research (B) (2) (12460144, 2000) from the Ministry of Education, Science, Sports and Culture, Japan.

REFERENCES

1. Boocock D.G.B. & Sherman K.M. (1985) Further aspects of powdered poplar wood liquefaction by aqueous pyrolysis. *The Can. J. of Chem. Eng.* **63**, 627-633.
2. Meier D. Larimer D.R. & Faix O. (1986) Direct liquefaction of different lignocellulosics and their constituents, 1. Fractionation, elemental composition, *Fuel* **65**, 910-905.
3. Boussaid A. & Saddler N.J. (1999) Adsorption and activity profiles of cellulase during the hydrolysis of two douglas fir pulps. *Enzyme Microbial Technol.* **15**, 138-143.
4. Mok S.W. & Antal J.M. Jr. (1992) Productive and parasitic pathways in dilute acid-catalyzed hydrolysis of cellulose. *Ind. Eng. Res.* **31**, 94-100.
5. Minowa T. Zhen F. & Ogi T. (1998) Cellulose decomposition in hot-compressed water with alkali or nickel catalyst. *J. of Supercritical Fluids* **13**, 253-259.
6. Townsend H.S. Abraham A.M. Huppert L.G. Klein T.M. & Paspek C.S. (1988) Solvent effects during reactions in supercritical water. *Ind. Eng. Chem. Res.* **27**, 143-149.
7. Sakaki T. Shibata M. Miki T. & Hirosue H. (1996) Decomposition characteristics of woody biomass in hot compressed liquid water. *The Second International Conference on Solvothermal Reactions*, December 18-20, pp 180-183.
8. Sakaki T. Shibata M. Miki T. & Hirosue H. (1996) Decomposition of cellulose in near-critical water and fermentability of the products. *Energy and Fuels* **10**, 684-688.
9. Sakaki T. Shibata M. Miki T. & Hirosue H. (1996) Reaction model of cellulose decomposition in near-critical water and fermentation of products. *Bioresource Technol.* **58**, 197-202.
10. Saka S. & Ueno T. (1999) Chemical conversion of various celluloses to

glucose and its derivatives in supercritical water. *Cellulose* **6**, 177-191.

11. Fukuzato R. (1998) Chemical recycling process for waste plastics using supercritical water (in Japanese). *Sigenkankyoutaisaku* **34**, 1165-1171.
12. Ramayya S. Brittain A. DeAlmeida C. Mok W.S.&Antal M.J. Jr. (1987) Acid-catalyzed dehydration of alcohols in supercritical water. *Fuel* **66**, 1364-1371.
13. Sasaki M. Kabyemela B. Malaluan R. Hirose S. Takeda N. Adschiri T.&Arai K. (1998) Cellulose hydrolysis in subcritical and supercritical water. *J. of Supercritical Fluids* **13**, 261-268.
14. Tester J.W. Holgate H.R. Armellini F.J. Webley P.A. Killilea W.R. Hong G.T. & Barnner H.E. (1993) Emerging technologies in hazardous waste management Ⅲ. In : *Supercritical water oxidation technology:Process development and fundamental Reseach.*(Ed. by D.W.Tedder&F.G.Pohland), pp.35 ACS Symposium series **518**, Amer. Chem. Soc., Washington, D.C.
15. Mok W.S. & Antal M.J. Jr. (1992) Uncatalyzed solvolysis of whole biomass hemicellulose by hot compressed liquid water. *Ind. Eng. Chem. Res.* **31**, 1357-1361.

Co-pyrolysis Under Vacuum of Bagasse and Petroleum Residue

A. Chaala, M. Garcia and C. Roy
Department of Chemical Engineering, Université Laval, Sainte-Foy, Québec G1K 7P4, Canada

ABSTRACT: Pyrolysis under vacuum of sugarcane bagasse, petroleum residues (PR) and mixtures thereof was performed. Upgraded liquid and solid products were obtained by thermal decomposition of bagasse combined with petroleum residue. It has been found that vacuum pyrolysis of bagasse alone yielded a large amount of bio-oil (34.2% by wt., bagasse anhydrous basis). Although this bio-oil meets several of the gas turbine fuel specifications, some combustion properties such as the ignition point, the susceptibility to carbonisation and the atomization can be improved by co-pyrolyzing the bagasse with heavy crude oils or petroleum residues. Blending pyrolytic oils separately produced by vacuum pyrolysis of bagasse and PR resulted in an unstable emulsion. In order to make this emulsion more stable, direct incorporation of PR at different proportions in the bagasse feedstock was performed. The oil blends obtained from the 5 wt.%, 15 wt.% and 50 wt.% PR/bagasse mixtures were stable emulsions. However, the oil from the 30 wt.% PR / bagasse mix was unstable. The physico-chemical and rheological characterization of the bio-oils are reported. Ageing tests of bagasse oil were performed at 80°C. The co-pyrolysis process also affected the properties of the pyrolytic charcoal. A surface analysis by ESCA performed on the charcoal samples showed that the bagasse charcoal surface was covered with coke formed during PR decomposition. This increased the ignition point of the bagasse charcoal.

1 INTRODUCTION

Reserves of light crude oils have drastically decreased over the last few decades. If no new crude oil deposits are discovered in the near future, it will be difficult to meet the demand for the energy consumed in fossil fuel power stations. Alternative solutions are to find new extracting and processing technologies for heavy crude oils or to make use of environmentally friendly renewable resources. Many approaches have been investigated which ought to take into account environmental and economical concerns. Thermal decomposition of biomass into transportable, storable and workable fuels is a promising approach to find a solution to the foreseeable shortage of fossil fuels [1-4].

Several studies have been conducted on co-pyrolysis processes in order to elucidate any synergetic effect on the quality of the products generated. Co-pyrolysis of coal and biomass has been studied by thermogravimetry [5]. No strong interactions between the components of the coal and the biomass feed have been observed. McGhee *et al.* [6] investigated the co-pyrolysis of polyvinyl chloride (PVC) and wood mixture with straw to simulate municipal solid waste. They found that the carbon obtained has a reduced reactivity. Co-pyrolysis in packed-bed pyrolyser (PBP) of mixtures of coals and heavy petroleum residues [7] and of Australian oil shale and lignite [8] has revealed the prevalence of synergetic effects. These studies also showed that the initial composition of the mixed feedstock influences the product distribution and properties.

Vacuum pyrolysis of biomass yields a high amount of oils (biofuels) which meet several of the heavy fuel performance specifications [9,10]. Some properties of these bio-oils might be improved by upgrading processes such as co-pyrolysis of biomass with heavy crude oils or petroleum residues.

Co-pyrolysis can also affect the charcoal properties. The carbonaceous material formed by PR thermal decomposition is usually called coke. The solid product obtained by pyrolysis of biomass is called charcoal. The surface chemistry, the bulk composition and the structure of these two materials are different.

2 EXPERIMENTAL

2.1 MATERIALS

The sugarcane bagasse was provided by United States Sugar Corporation, Clewiston, Florida. The feedstock was air-dried to 8 wt.% moisture content and then sieved in order to remove the particles smaller than 0.450 mm. The petroleum residue supplied by Shell Canada, Montréal, was a grade 150-200 penetrability petroleum bitumen.

2.2 SAMPLE PREPARATION

Petroleum residue was dissolved in toluene in order to well mix it with the bagasse fibres. The mixture obtained was then carefully shaken and exposed to evaporation during 12 h at room temperature in order to remove the solvent. Residual toluene was eliminated by heating the sample at a temperature of 106°C for 10 h.

2.3 PYROLYSIS RUN

In the present work, bagasse has been pyrolyzed in a pilot plant and a bench scale retort. PR and bagasse/PR mixtures have been pyrolyzed in the retort.

2.3.1. Bench scale retort

A sample of 80 g was introduced into the batch retort. All the pyrolysis tests were performed at a temperature of 500°C, a total pressure of about 8 kPa and a heating rate of 12°C/min. The holding time of the solid residue after completion of the test was one hour. Vapours formed in the reactor were removed and condensed in three traps connected in series and maintained at –30°C, -78°C and –78°C, respectively. The non condensable gas was removed by the vacuum pump and stored in a vessel previously depressurized. A detailed description of the apparatus is available elsewhere [11]

2.3.2. Pilot plant reactor

A sample of 20 kg of air-dried bagasse (8 wt.% moisture) was introduced into a cylindrical reactor 3 m long and 0.6 m in diameter equipped with two internal heating plates. The test was conducted in a batch mode at a temperature of 530°C and a total pressure of about 12 kPa. The condensing system included four traps connected in series and maintained at 25°C, 0°C, -30°C and –80°C, respectively. The equipment has been described by Roy *et* al. [12]

In order to simulate the condensing system of an industrial vacuum pyrolysis plant which consists of two condensing packed towers continuously operating, the liquids collected in each trap were mixed and then evaporated at 45 °C during half an hour in a rota-vapour (Büchi, RE 111). The heavy fraction which remained in the flask corresponds to the oil from the first condensing tower and is called "bio-oil", while the evaporated fraction which consists of water and light organic compounds corresponds to the aqueous phase of the second tower and is called "aqueous phase".

2.4 CHARACTERISATION OF THE PYROLYSIS PRODUCTS

2.4.1 Bio-oil

The physico-chemical properties of the oil samples were measured according to ASTM methods: density D 369; kinematic viscosity- modified D 445-88; [9], flash point D-93; gross calorific value D4809; water content D-1744; and ash content D 482. The apparent viscosity was measured using a Brookfield viscometer, LVDV III+. The content of the methanol insoluble materials (MIM) was determined according to the method described by Oasmaa *et* al. [13].

The molecular weight distribution was determined by gel permeation chromatography (GPC) using a Waters 510 pump with refractive index detector. The separation was performed using two columns in series: Styragel^R HR 4^E and HR1 with 5µm particle diameter. THF was used as the eluent at a flow rate of 1 ml/min. The bio-oil samples were dissolved in THF at a concentration of 2 g/l. The GPC columns were calibrated using polystyrene standards of 400, 600, 760, 800, 1200, 2000, 2430 and 20700 molecular weights.

For the stability test, 60 g of bio-oil was poured in tightly closed 60 ml volume bottles. The ageing effect was measured in terms of viscosity, MIM content, water content and molecular weight distribution.

2.4.2. Charcoal

Proximate analysis of the charcoal sample was performed in a MAC-400 instrument from LECO, while the C, H and N concentrations were determined in a LECO CHN-600 equipment following the ASTM D 5291-92 method.

A SSC/5200 microbalance from Seiko was used for the thermogravimetric tests. Samples of 3-4 mg were heated from room temperature to 600°C, under air flow of 100 ml/min, at a constant heating rate of 5°C/min.

3 RESULTS AND DISCUSSION

3.1 BAGASSE AND PETROLEUM RESIDUE CHARACTERISATION

The particle size distribution of the bagasse and the proximate analysis of the fractions obtained are shown in Table 1. The fine particles contained in the two last fractions which represented 7 wt.% of the total bagasse, were removed due to their high ash content. All the data reported in this paper refer to the fraction > 0.450 mm.

Table 1. Particle size distribution of the air-dried bagasse and proximate analysis of the fractions.

Sieve Denomination (mm)	Amount (w%)	Volatile	Ash	Fixed Carbon
> 4.75	34	82.3	1.3	16.4
0.85 - 4.75	44	82.1	1.6	16.3
0.45 - 0.85	15	81.8	2.3	15.9
0.25 - 0.45	4	72.5	13.0	14.5
< 0.25	3	60.0	27.7	12.3

Proximate analysis as well as generic and elemental compositions of the sieved bagasse feedstock are presented in Table 2. Some physico-chemical properties of the petroleum residue are also presented in Table 2.

Table 2. Properties of petroleum residue and bagasse feedstocks (anhydrous basis)

Petroleum Residue (PR)		Bagasse	
Penetration (1/10 mm)	163	Volatile matter (wt.%)	82.1
Flash point (° C)	338	Ash (wt.%)	1.6
Viscosity at 135°C (cSt)	185	Fixed carbon (wt.%)	16.3
High heating value* (MJ/kg)	41.2	High heating value* (MJ/kg)	16.7
Carbon (wt.%)	82.8	Carbon (wt.%)	48.8
Hydrogen (wt.%)	10.0	Hydrogen (wt.%)	5.9
Nitrogen (wt.%)	0.7	Nitrogen (wt.%)	0.5
Ash (wt.%)	Trace	Sulphur (wt.%)	Trace
Sulphur + oxygen, by diff. (wt.%)	6.5	Oxygen + ash , by diff. (wt.%)	43.2
Asphaltenes (wt.%)	76.9	Hemicellulose (wt.%)	35.2
Maltenes (wt.%)	23.0	Cellulose (wt.%)	42.4
Toluene insolubles (wt.%)	0.1	Lignin (wt.%) **	20.8

*Calculated

**Ash must be added to reach 100%.

PR is composed of the residues obtained from crude oil vacuum distillation unit, fluid catalytic cracking (FCC) unit, lubricant oil deasphalting unit and other refinery operating units. PR is frequently used as binder for road pavement. PR is rich in high molecular weight aromatic hydrocarbons (asphaltenes). Its carbon content is very high which results in a high heating value. However, it is not a good heating fuel. Co-pyrolysis with bagasse might be a valuable upgrading process for PR.

3.2. *PYROLYSIS AND CO-PYROLYSIS PRODUCT YIELDS*

Vacuum pyrolysis of 20 kg of bagasse carried out in the pilot reactor in a batch mode yielded 30.2 wt.% oils, 25.6 wt.% charcoal, 21.2 wt.% aqueous phase and 22.0 wt.% gas (bagasse anhydrous basis). Meanwhile, vacuum pyrolysis of bagasse carried out at the laboratory scale yielded, on the same basis, 34.2 wt.% oil 19.4 wt.% charcoal, 28.1 wt.% aqueous phase, and 17.6 wt.% gas (Table 3). Differences observed between laboratory and pilot plant tests can be explained by the varying heating rates and the thickness of the feedstock bed inside the two reactors. Heating rate was higher in the retort and the bed thickness was higher in the pilot reactor. In comparison with other biomass feedstocks which have been previously pyrolysed in our laboratory [14], the charcoal yield is low. Table 3 also indicates the variation in the product yields with the addition of PR in the bagasse feedstock

Table 3. Yields on vacuum pyrolysis of bagasse, PR and mixtures thereof (wt.%, anhydrous feed basis).

	Concentration of PR (wt.%)					
Product	0	5	15	30	50	100
Pyrolysis oil	34.2	33.9	31.2	41.3	53.3	85.3
Aqueous phase	28.1	21.1	20.6	10.7	10.3	0.5
Charcoal	19.4	28.5	29.8	27.8	23.0	10.0
Gas	17.6	15.8	17.5	19.4	13.0	3.4
Loss	0.7	0.7	0.9	0.8	0.6	0.8
Total	100	100	100	100	100	100

3.3. *CHARACTERISATION OF THE BAGASSE BIO-OIL*

3.3.1 Physico-chemical properties

The physico-chemical properties of the bagasse–derived bio-oil obtained in the large batch reactor are summarized in Table 4. The bio-oil obtained after evaporation contains 13.8 wt.% water. Like the bio-oils originating from diverse biomasses using various pyrolysis techniques, the oil from vacuum pyrolysis of bagasse is heavier (d_{20} = 1.211 g/ml) than water.

The low viscosity at 20°C (116.5 cSt) means a flowability of the bio-oil with less energy requirement, while the viscosity at 80°C (5.4 cSt) will ensure a good atomization and consequently a complete combustion.

The content in methanol insoluble materials (MIM) which represents not only solid compounds such as charcoal and mineral particles but also waxy-like substances, does not meet the usual gas turbine specifications [15]. The carbon Conradson residue which indicates the susceptibility of the bio-oil to carbonisation is high.

The high heating value (22.4 MJ/kg) makes this bio-oil suitable for combustion in boilers or gas turbines. Its flash point is high enough (> 90°C) from a safety point of view. However, it can be corrosive for certain metals (pH = 2.7). Elemental composition showed that the oil contains a very low sulphur content, a relatively low carbon content and a high oxygen content.

Table 4. Physico-chemical properties of the bagasse bio-oil

Properties	Value
Moisture, wt.%	13.8
Density at 20°C, kg/m^3	1211
Kinematic viscosity at 50°C, cSt	16.4
MIM, wt.%	0.38
CCR, wt.%	18.6
pH	2.7
High heating value (anhydrous basis), MJ/kg	22.4
Flash Point, °C	>90
Carbon (wt.%)	47.0
Hydrogen (wt.%)	5.6
Nitrogen (wt.%)	0.6
Sulphur (wt.%)	traces
Ash, (wt.%)	< 0.1
Oxygen*, (wt.%)	46.8

* Determined by difference

3.3.2 Metal content in bio-oil

In comparison with the bio-oils reported in the literature [2, 3], the investigated bio-oil exhibits relatively low contents of sodium (Na), phosphorous (P), calcium (Ca) and potassium (K) (Table 5). These concentrations are far from the gas turbine normal specifications [15]. One should take into account however that the crude bio-oil has not been submitted to any post-treatment. It is important to note the presence of Mg , Fe and Al. In spite of their low concentrations, these metals have potential to inhibit the corrosion caused by the vanadium contained in PR-derived oil.

Table 5. Metals present in the bagasse bio-oil.

Metals	Na	P	Se	V	Zn	Ca	Fe	K	Mg	Mn	Al
mg/kg	21.5	3.3	0.2	0.02	43.8	33.8	105.9	5.3	2.2	1.1	10.1

3.3.3. Rheological properties

It has been found that oils obtained from pyrolysis of bagasse and PR, on an individual basis, exhibit a non-Newtonian flow behavior. The viscosity decreasing with increasing shear rates (Figures 1, 2) shows that these oils have a pseudo-plastic behaviour. The non-Newtonian character of the bio-oil from bagasse is due to the partial miscibility of the two liquids (organic compounds and water) which compose it, as well as the presence of solid particles. The behaviour of the oil from PR is due to the three-dimensional constituents of the oil (large molecules, colloidal particles and other suspended materials such as asphaltenes and gritty materials). The force required to move these oils is determined by the size, the shape and the cohesiveness of the oil constituents. At any shear rate, the alignment of the constituents is different and more or less force will be required to maintain motion.

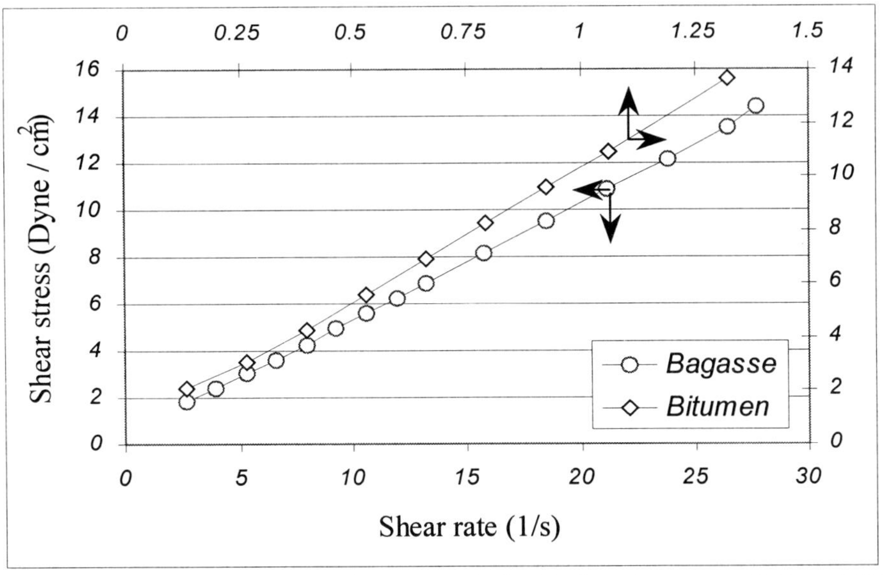

Figure 1.- Shear rate vs shear stress for the pyrolysis oils from bagasse and PR @ 30°C

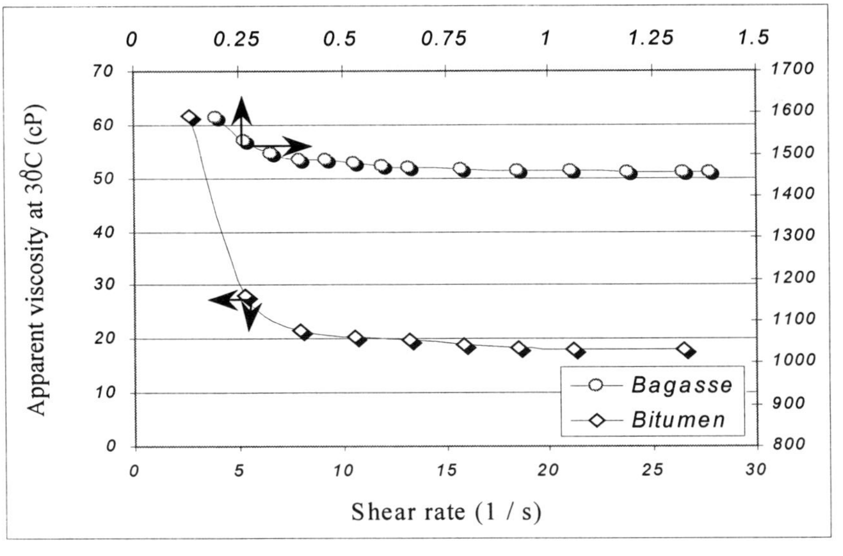

Figure 2.- Apparent viscosity vs. shear rate for the pyrolysis oils from bagasse and PR @ 30°C

3.3.4. *Ageing of the bagasse-derived oil*

A sample of 60 ml of oil was poured into 60 ml tight glass bottles. The bottles were retightened a few times during the heating-up period. The bottles were rapidly cooled under cold water and weighed before the analyses are performed. The possible difference in the weights before and after the test must not exceed 0.1 wt.% of the original weight. The variation in viscosity, water content and methanol insoluble materials (MIM) was determined. The ageing tests were conducted at 80°C during 6 h, 12 h, 24 h, 48 h and 168 h periods.

The variations in the viscosity, water content, MIM content and molecular weight distribution were determined as follows:

$\Delta\gamma$ (%) = $[(\gamma_2 - \gamma_1)/\gamma_1] . 100$ — γ_1, γ_2: viscosity of original and aged samples, respectively [cSt]

ΔI (%)= $[(I_2 - I_1)/I_1].100$ — I_1, I_2: methanol insolubles of original and aged samples, respectively [wt.%]

ΔM (%) = $[(M_2 - M_1)/M_1].100$ — M_1, M_2: molecular weight of original and aged samples, respectively [a.m.u.]

ΔW (%) = $[(W_2 - W_1)/ W_1] . 100$ — W_1, W_2: water content of original and aged samples, respectively [wt.%]

The variation in the viscosity, MIM content and water content shown in Table 6 revealed that the increase in viscosity determined at 40°C is very high, particularly for the oil aged during 168 h (134.4%). The viscosity increase is due to diverse reactions, which can occur during heating, generating high molecular weight compounds which may form three-dimensional colloids and miscelles. The water content variation is a parameter indicating the occurrence of these reactions.

Table 6. Variation in viscosity, water content and methanol insoluble materials (MIM) of the bagasse-derived oil after ageing at 80°C

Heating time (h)	ΔI (%)	$\Delta\gamma$ (%) 40°C	$\Delta\gamma$ (%) 80°C	ΔW (%)
0	0	0	0	0
12	-26.3	29.6	16.9	9.4
24	-26.3	47.2	24.5	10.1
48	-52.6	73.8	49.0	15.2
168	-63.2	134.4	94.3	15.9

The changes in viscosity and water content are confirmed by the molecular weight distribution variation *(Fig. 3)*. It is important to note that the changes are particularly intensive in the queue of the oil fraction (*i.e.* due to the formation of heavy compounds M_w, M_z and M_{z+1}). The changes observed in each fraction of the bio-oil drastically increased from time 0 up to 24 h, then slightly grew up from 24 to 48 h ageing. The increase slowed down during an ageing period longer than 48 h, except for the lighter fraction where the increase had stopped. One can conclude that the ageing process provided more changes in the queue and the middle fractions of the bio-oil than in the head fraction.

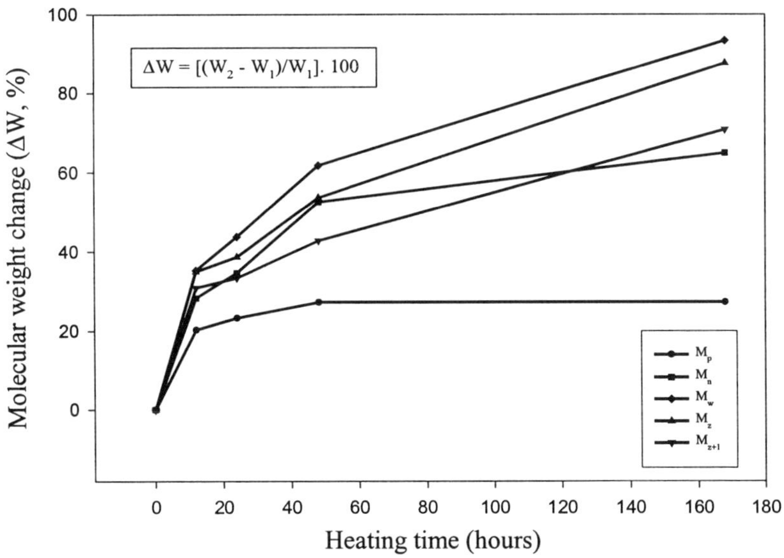

Figure 3 – Molecular weight distribution after ageing at 80°C

3.4 CHARACTERISATION OF THE BAGASSE CHARCOAL

The high content of fixed carbon, and the low volatile matters in the charcoal from bagasse (Table 7) make this material suitable for various uses including production of carbides, reduction of minerals, etc. The heating value (36 MJ/kg) also allows the charcoal to be used as a solid fuel. Due to its relatively high specific area (529 m^2/g), the charcoal might be used directly as an adsorbent material or as a feedstock for activated carbon production.

Table 7. Properties of the bagasse charcoal

Properties	Value
Moisture, wt.%	3.2
High heating value (anhydrous basis), MJ/kg	36
Specific surface area, m^2/g	529
Volatile matters, wt.%	18.9
Ash content, wt.%	6.7
Fixed carbon, wt.%	74.4
Carbon, wt.%	85.6
Hydrogen, wt.%	2.9
Nitrogen, wt.%	1.3
(Oxygen + sulfur)*, wt.%	3.5

* Determined by difference, ash not included

The metals contained in the bagasse charcoal are presented in Table 8. The elements detected in the bio-oils are also present in the charcoal, but at a much higher concentration. In terms of elemental composition, the charcoal obtained meets several specifications which are required for carbon-rich materials used in the production of carbides and for the reduction of minerals.

Table 8. Metals present in the bagasse charcoal.

Metals	Na	P	Se	V	Zn	Ca	Fe	K	Mg	Mn	Al
mg/kg	449	1324	6.5	0.4	27.6	5384	950	5537	2257	73.5	281.5

3.5 CHARACTERISATION OF THE PR/BAGASSE MIXED OILS

The properties of the pyrolysis oils obtained during the pyrolysis of the PR/bagasse mixtures are summarized in Table 9. CCR decreases with the increase of PR concentration in the bagasse feedstock. This means that PR reduces the susceptibility of the bio-oil to carbonisation. The content in pentane soluble compounds is indicative of the amount of PR products present in the oil. The oil calorific value increases almost proportionally with the content of pentane soluble compounds. As petroleum residue and bagasse oils have approximately the same ash content, the ash content of their mixtures was almost constant.

Table 9. Properties of the mixed oils.

	PR Concentration (wt.%)					
	0	5	15	30	50	100
CCR (wt.%)	18.6	19.7	19.4	13.8	8.8	9.3
Ash (wt. %)	0.05	0.05	0.04	0.05	0.06	0.05
HHV (MJ/kg)	22.4	23.0	24.7	33.2	36.5	43.7
Pentane solubles (wt. %)	2.7	4.6	10.0	47.1	62.8	89.9
Water content (wt. %)	13.8	9.5	10.3	6.9	11.7	0.2

The viscosity of the mixed oils is higher than that of the bagasse and the PR individually. This is due to the formation of complex three component emulsions (bio-oil, PR-derived hydrocarbons and water) with dispersed solid particles. As expected, the mixed oils exhibit non-Newtonian flow behaviour (herein not shown). The complex emulsion obtained seems to be more stable than the one obtained by mixing the oils produced separately from bagasse and PR. The oils from bagasse, PR and the mixed oils were also observed by microscopy. The existence of three liquid emulsions was confirmed by microscopic analysis (Figure 4).

The photograph presented in Figure 4 revealed that the mixed oil obtained from the 30% by wt PR mixture is not an homogeneous liquid; it forms emulsion. The mixed oils obtained with the 5 wt.%, 15 wt.% and 50 wt. % PR mixtures seem to be very stable (here not shown). However, the oil from the 30 wt.% PR mixture is not stable. A clear phase separation was observed during the optical test. As an emulsion is generally composed of a dispersed phase and a continuous medium, one can say that at a defined concentration of PR, a phase inversion occurs. Considering the high density of the bagasse oil and its low specific volume, it is postulated that the mixed oils obtained from the mixtures of 5 wt.% PR and 15 wt.% PR are composed of bagasse-derived oil as continuous medium and PR-derived oil as dispersed phase.

At 30 wt.% PR content, no concrete boundaries between continuous medium and dispersed phase were observed; a competition between phases occurred leading to the formation of an unstable oil blend.

At 50 wt.% PR a phase inversion was observed; the PR-derived oil became a continuous medium and the bagasse-derived oil a dispersed phase.

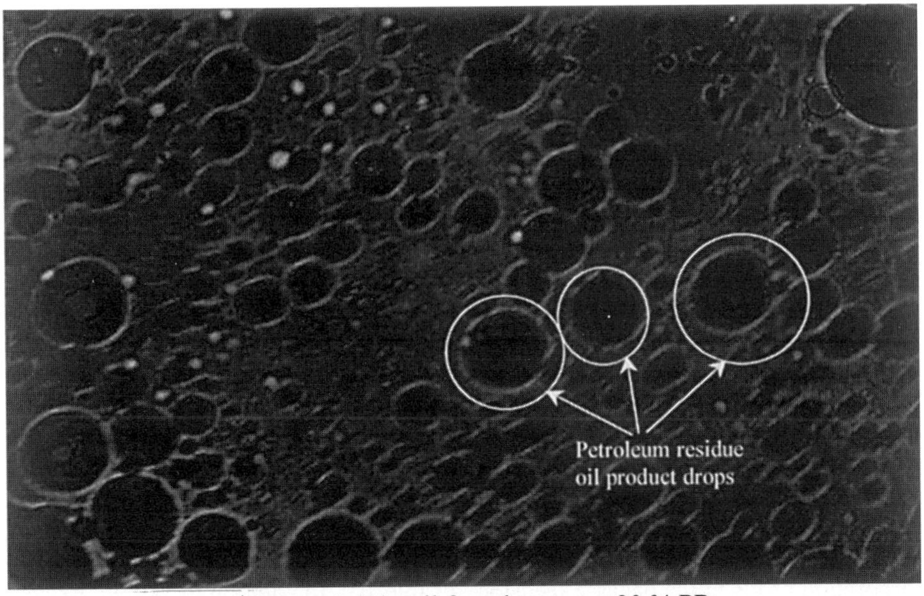

Figure 4.- Bio-oil from bagasse + 30 % PR

Addition of PR in the bagasse feedstock increased the viscosity of the oil. However, the emulsion obtained exhibited a high storage stability. It has been shown that the incorporation of petroleum fuels (5 wt.%) into the bio-oil prior to injecting it into the combustion area of the gas turbine considerably reduced the ignition point of the oil and improved the colour of the flame [16]. Plugging frequency of gas turbine nozzle fed with bio-oils represents a challenge which can be resolved by a special feeding system or by the use of a mixed oil as feedstock. Plugging frequency of the nozzles is due to the high susceptibility of the oil to carbonisation. This property is evaluated by the carbon Conradson residue (CCR) value, which is high for oils very susceptible to carbonisation and low for oils less susceptible. A correlation between the reduction in plugging frequency and the CCR tests has to be investigated in the future (Table 9).

Combustion of the mixed oils may generate less pollutant compounds such as NO_x, PAH and SO_x. The hot corrosion caused by the vanadium contained in the usual gas turbine fuel might also be inhibited by the presence of some metals contained in the biofuels [17]. The sulphur content of the mixed oils is expected to be lower than that of conventional petroleum fuels.

3.6. CHARACTERISATION OF THE PR/BAGASSE MIXED CHARS

Table 10 indicates that the ash content of the chars obtained during the pyrolysis of the PR/bagasse samples decreased as PR concentration increased in the feedstock. As the charcoal from bagasse exhibits a similar elemental composition to the PR-derived coke, no great differences in the elemental composition of the mixed chars were observed.

Table 10. Elemental composition of the mixed chars (anhydrous basis).

Element	PR Concentration (wt.%)					
	0	5	15	30	50	100
Carbon	85.6	84.5	85.4	84.2	87.1	88.2
Hydrogen	2.9	3.5	4.3	3.3	2.9	3.4
Nitrogen	1. 3	1.0	0.9	1.2	1.5	2.1
Ash	6.7	6.2	5.8	5.6	5.5	0.0
(Oxygen + Sulfur)*	3.5	4.8	3.6	5.7	3.0	6.3

*Determined by difference

3.6.1. Surface composition of the mixed chars

The ESCA analysis showed that the elemental surface composition changed with the incorporation of PR in the bagasse feedstock (Table 10). In addition to carbon, oxygen, and nitrogen, potassium and calcium are present on the bagasse charcoal surface. The last two elements disappeared from the surface upon addition of 5 wt.% of PR to the bagasse feedstock. This indicated that the bagasse charcoal particles were covered by products of the PR decomposition. Based on the changes in element contents, one can conclude that the addition of 5 wt. % PR to the bagasse feedstock provides major effects on the surface chemistry of the charcoal particles. An extensive study performed by the authors [18] confirmed that bagasse char surface becomes completely covered when petroleum residue concentration in the feedstock reached 15 wt.%.

Table 11. Surface composition of the mixed chars

[atom %]

Sample	C	O	S	N	Si	Ca	K
Bagasse	87.6	9.8	*	1.1	*	0.5	1.0
Bagasse + 5 % PR	93.1	4.2	2.1	0.6	*	*	*
Bagasse + 15 % PR	93.9	2.3	2.6	1.2	*	*	*
Bagasse + 30 % PR	95.8	1.8	2.4	*	*	*	*
Bagasse + 50 % PR	94.8	1.6	2.9	0.7	*	*	*
PR	90.0	4.4	2.7	1.3	1.6	*	*

* No peak detected

3.6.2 Thermogravimetric analysis

Thermogravimetric tests have been performed in order to determine the influence of the co-pyrolysis on the bagasse charcoal reactivity. The DTG curves under oxidizing conditions of the chars obtained by vacuum pyrolysis of bagasse, PR and mixtures thereof are presented in Figure 4. It is clear that the oxidation process of the bagasse charcoal started earlier than the other investigated chars. Oxidation of the mixed chars is delayed due to the formation of coke on the bagasse charcoal surface. This coke is made of carbonaceous materials formed during the thermal decomposition of the hydrocarbons contained in PR. The coke is more structured and more compact (low porosity) than bagasse charcoal. The presence of coke and intermediate by-products such as carbenes, carboides and other structured compounds is confirmed by ESCA analysis.

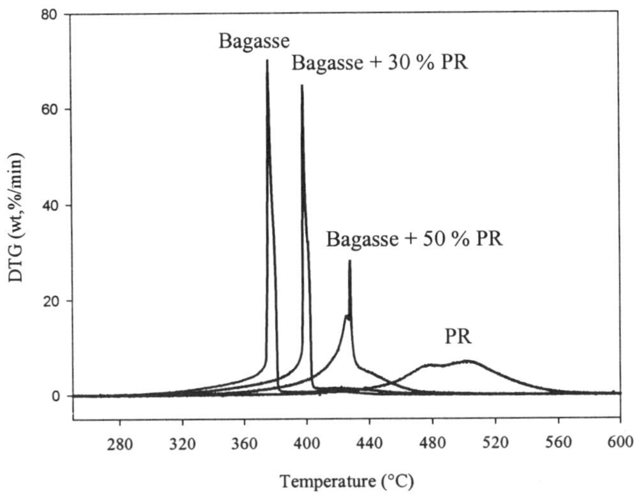

Figure 5.- DTG of bagasse charcoal, coke and composite chars

Covering up the bagasse charcoal with coke is beneficial from the point of view of charcoal storage. In fact, the storage of charcoal particles is hazardous as powderous char can easily catch fire when brought into contact with air. The incorporation of PR in the bagasse feedstock increased the self heating temperature of the charcoal particles. The self heating temperature of the charcoal is the temperature at which the charcoal particles, placed in standardized conditions, warm up and catch fire.

From a combustion point of view, the coke formed will increase the ignition point of the bagasse charcoal, which requires additional oil firing and slower burning in large furnaces to reach complete combustion. This type of chars is used for domestic heating, where heat is transferred directly from the fuel bed. The high reactivity of bagasse charcoal on the other hand is desirable in cyclone burners which carry out rapid, intense combustion to maximise carbon utilisation and minimise smoke emission.

The mixed chars obtained exhibited lower specific area than the bagasse charcoal [18], resulting in a reduction of their adsorption capacity.

4 CONCLUSION

Co-pyrolysis under vacuum of low commercial value products such as sugarcane bagasse and petroleum residue seems to be a promising process. The reactor throughput capacity can be increased by blending petroleum (PR) residue with bagasse feedstock due to the high density of PR. The co-pyrolysis generates two main products: dark mixed oils which have a high heating value, low CCR and relatively high viscosity and mixed chars with high carbon content and low susceptibility to self heating. The mixed oils obtained are complex emulsions consisting of oxygenated organic compounds derived from bagasse, hydrocarbons derived from PR and water. Solid particles are also present in the emulsions. The emulsions obtained are more stable than those prepared by mixing bagasse-derived oil with PR-derived oil. The mixed oils exhibit a pseudo-plastic behaviour.

5 ACKNOWLEDGEMENTS

The authors are thankful to the United States Sugar Corporation (Clewiston, Florida) and Shell Canada (Montréal, Québec) for providing the sugarcane bagasse and petroleum residue (bitumen), respectively. Thanks are also due to Dr. H. Pakdel for the GPC analysis and to Dr. J.Yang for the TGA tests.

6 REFERENCES

1. Moses C., (1994) "Fuel-Specification Considerations for Biomass Liquids". *Proceedings Biomass Pyrolysis Oil Properties and Combustion* Meeting, T.A. Milne, ed., National Renewable Energy Laboratory. Golden, CO, NREL-CP-430-7215, pp. 362-282.
2. Diebold J.P. and Bridgwater A.V. (1999) "Overview of Fast Pyrolysis of Biomass for the Production of Liquid Fuels" *Fast Pyrolysis of Biomass: A Handbook.* Aston University, Bio-Energy Research Group, UK (1999).
3. Diebold J.P.; Oasmaa, A.; Bridgwater A.V.; Piskorz J.; Huffman D.; Cuevas A.; Gust S.; Czernik S. and Milne T.A. (1996) "Proposed Specifications for Various Grades of Pyrolysis Oils" *Bio-Oil Production and Utilisation.* Proceedings of the 2nd EU/Canada Workshop on Thermal Biomass Processing, CPL Press, Newbury, UK., pp. 66-81.
4. Meier D. and Faix O. (1999). *Bioresource Technology*. 68, pp 71-77.
5. Klose W., Stuke V. (1993). Comparison of the Pyrolysis of Different Types of Biomass and Coals. *Fuel Process. Technol.* 21, pp. 283-288
6. McGhee B., Norton F., Snape C. E. and Hall, P.J. (1995). The Copyrolysis of Poly (vinylchloride) With Cellulose Derived Materials as a Model for Municipal Waste. *Fuel* 74, pp. 28 -31
7. Khan M.R., Heshieh, F. Y. and Heaclky L. (1989). *Am. Chem. Soc. Div. Fuel Chem Preprints,* 34, p. 1167.
8. Saxby J.D. and Sato S. (1990). Liquid Products From Pyrolysis of Synthetic and Natural Blends of Australian Low Rank Oil Shales and Lignites. *Fuel* 69, pp. 1109-1112
9. Boucher M.E., Chaala A. and Roy C. (2000) "Bio-Oils Obtained by Vacuum Pyrolysis of Softwood Bark as a Liquid Fuel for Gas Turbines. Part I:

Properties of Bio-Oil and its Blends with Methanol and a Pyrolytic Aqueous Phase. *Biomass & Bioenergy,* In press.

10. Boucher M.E., Chaala A., Pakdel H. and Roy C. (2000). "Bio-Oils Obtained by Vacuum Pyrolysis of Softwood Bark as a Liquid Fuel for Gas Turbines. Part II: Stability and Ageing of Bio-Oil and its Blends with Methanol and a Pyrolytic Aqueous Phase" *Biomass & Bioenergy*, In press.
11. Pakdel H., Couture G. and Roy C. (1994). Vacuum Pyrolysis of Bark Residues and Primary Sludges. *Tappi Journal* 77, (7), p.205-211
12. Roy C., Yang J., Blanchette D. and de Caumia B. (1997). Development of Novel Vacuum Pyrolysis Reactor with Improved Heat Transfer Potential. *In Developments in Thermochemical Biomass Conversion*. Blakie Academic and Professional, London, UK, pp. 351-367.
13. Oasmaa A., Leppämäki E., Koponen P, Levander J., Tapola E. (1997). "Physical Characterisation of Biomass-Based Pyrolysis Liquids : Application of Standard Fuel Oil Analyses". VTT Energy Publication 306.
14. Roy C. (1999). "The Pyrocycling™ Process: New Developments" *4th Biomass Conference of the Americas*, Oakland, CA, U.S.A.
15. Orenda Aerospace Corporation. (1997). Engine Summary Report : Bio-fuel Testing and Optimisation for Gas Turbine Applications. Internal Report.
16. SPE "Mashproekt" (2000). The GT 2500 Gas Turbine Engine Fuel Nozzle and Combustion Liner testing on Pyrvac Biofuel. Technical Report, Kiev, Ukraine.
17. Tiwari, S.N. and Prakash, S. (1998) "Magnesium Oxide as Inhibitor of Hot Oil Ash Corrosion". *The Institute of Materials*, pp. 467-472.
18. Darmstadt, H. ; Garcia-Perez, M. ; Chaala, A. Cao, N.Z. and Roy, C. Co-Pyrolysis Under Vacuum of Sugar Cane Bagasse and Petroleum Residue. Properties of the Char and Activated Char Products. *Carbon.* In press.
19. Anonymous. Transport Canada – Région du Québec. Transport des marchandises dangereuses. In Recommendation on the Transport of Dangerous Goods, *Manual of Tests and Criteria*. United Nations (1993).

Preliminary Results on Wood Waste Pyrolysis

C. Dudouit[1] and Y. Schenkel[2]

[1] *ENSTIB - School of Wood Science and Engineering, 27, Rue du Merle Blanc F - 88051 Epinal Cedex 9 - France*

[2] *CRA, Agricultural Engineering Department, Chaussée de Namur, 146 - B- 5030 Gembloux Belgium*

ABSTRACT A safe and efficient treatment of organic solid waste is becoming more and more important, as many European countries will ban these residues from landfill. Among them, wood waste, such as demolition wood, coated wood, etc., have received little attention as far as their treatment is concerned. Pyrolysis, and more precisely thermolysis, represents a safe and efficient treatment of these contaminated woods. Preliminary experiences have been conducted on the pyrolysis of waste wood, at three pyrolysis temperatures (350, 550 and 750°C), with a heating rate of 20° C/min and a residence time of 60 min.

A particular attention has been given to the behaviour of heavy metals: As, Cd, Cr, Cu, Hg, Ni, Pb, Zn. These heavy metals are mainly concentrated in the residual char with the exception of As, Hg and Cd, which are volatile at low temperatures. A large fraction of these last three metals are recovered in the tars. The same phenomenon is observed with Zn at the temperature of 750° C.

INTRODUCTION

A promising way for the disposal of wood waste (demolition wood or residues of wood processing industries) is pyrolysis. Belgian industries produce some 1 million tons of waste wood [1], France more than 7 millions and Germany 3 to 3.5 millions [2]. In 2004, the landfill of such refuse will not be authorised anymore in Belgium (2002 in France) except for the ultimate waste.

Wehlte S. et al. [2] describe the hazardous potential and characteristics of several wood waste assortments in the former Federal Republic of Germany. They show that these refuse contain various wood preservatives, insecticides or coatings (lacquer, paint, vanish) which may contain heavy metals like copper, chromium, lead, and/or substances like boron, arsenic, fluorine, pentachlorophenol, lindane, formol ... In fact, more than 80 % of demolition wood waste have been treated with compounds of which the toxicity degree could be variable [3]. Some products as pentachlorophenol, lindane have the reputation of being teratogenic, mutagenic, carcinogenic or also neurotoxic. So, it is necessary to identify a suitable treatment respecting environment and health, and creating energy or by-products.

MATERIALS AND METHODS

These preliminary tests of wood waste pyrolysis have been conducted on so called "class B" wood waste; the sample was taken in a waste wood pile of a waste management company located in Belgium. The sample has been characterised and then pyrolysed slowly at low and medium range temperatures in a thermo-balance retort kiln.

WOOD WASTE CHARACTERIZATION

The particle size varies a lot. The finest particles (less than 1 cm long or thick) have been eliminated by sieving in order to not alter the pyrolysis yield results ; they represent about 2 % of the total mass of the feedstock. The bulk density is about 149 kg/m^3 dry basis.

The particles have been classified according to their thickness and nature. The results are presented in figure 1 and 2.

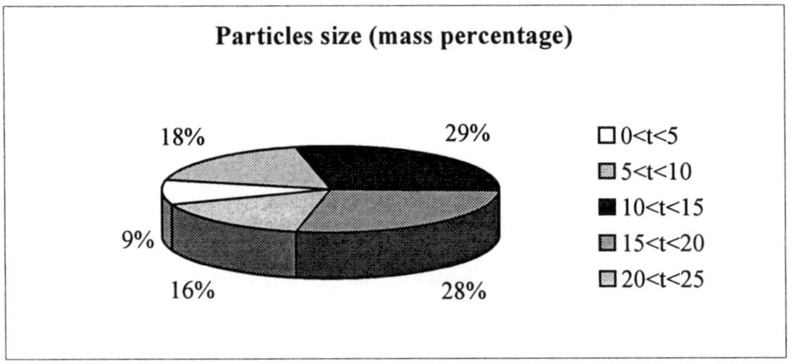

Thickness in mm	*Average (mass %)*	*Standard deviation*
0<t<5	***9,20***	*4,4*
5<t<10	***18,08***	*9,9*
10<t<15	***28,83***	*9,5*
15<t<20	***27,68***	*6,6*
20<t<25	***16,21***	*20,7*

Fig. 1 Classification of the particles according to their thickness (t in mm).

The high standard deviation for a thickness between 20 and 25 mm can be explained as follows: the samples collected for this measure weighed 250 g; the presence of thick particles dramatically change the results and consequently the average. To obtain results with less standard deviation, the samples weight should be higher (that is to say at least 4 kg).

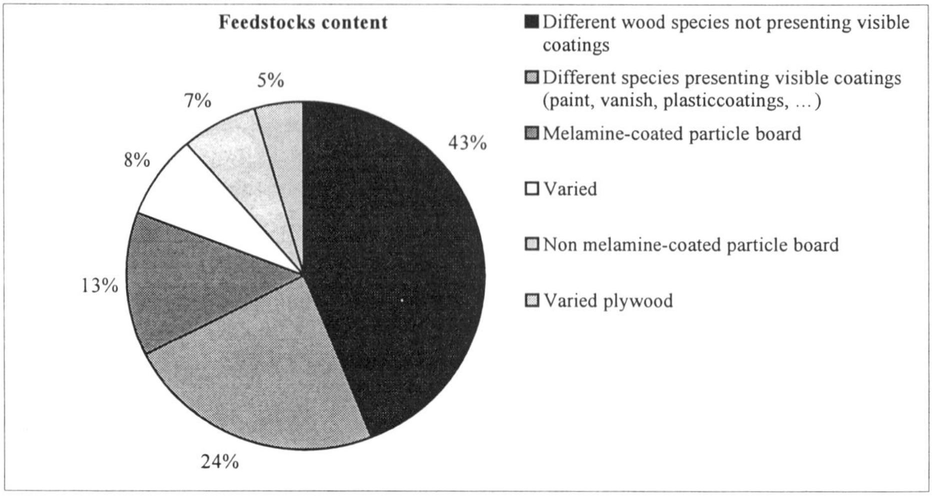

*Varied : Hardboard / Laminated hardboard / Forestry wood (branch, leaf etc) / Wood / board including nails or staples / OSB (Oriented Stand Board) / Varied (linoleum, stone, glass wool, aluminium, cable, hinge, glass, plastics etc) / Fireproof or anti-damp particle board

Fig. 2 Feedstock characterisation from representative samples introduced in the kiln for each pyrolysis experiment

The feedstock composition shows a large variety of products resulting from a great variety of treatments more or less toxic. The heavy metals content of our samples was determined (table 1).

Table 1 Heavy metals content of wood waste (mg/kg of dry matter)

	Cd	Cu	Ni	Pb	Zn	Hg	Cr	As
Test n° 1	0.75	17	74	152	237	0.27	100	0.59
Test n° 2	0.24	7.6	24	81	117	0.13	31	0.08
Test n° 3	0.29	7.4	21	71	106	0.12	28	0.24

The other physico-chemical characteristics of the wood waste samples are given in Table 2.

Table 2 Physico-chemical characteristics of wood waste samples

Proximate composition (% dry mass)		Ultimate composition (% dry mass)	
Volatile content	82.67	Carbon	45.60
Mineral content	2.40	Hydrogen	6.01
"Fixed carbon" content	14.93	Nitrogen	1.77
		Oxygen	44.22
Calorific value (MJ/kg dry mass)		**Moisture content (% mass)**	
Gross	19.63	Dry basis	24.35
Net	18.40	Wet basis	19.58

These pyrolysis experiments have been performed on actual feedstock representing the wood waste that will have to be treated in the future (different particle sizes, wood species, products, treatments etc).

SLOW PYROLYSIS EXPERIMENTATION PLANT

This plant is fully described in [4] and mainly consists of :

(a) *The reactor*: cube-shaped, it has a capacity of 27 litres. The inside walls are made of refractory bricks. The feedstock is put in a metallic basket and the reactor is closed tightly with a cordon of clay, which is replaced at each experiment. Sealed up, the reactor is totally airtight.
(b) *The heating control system*: the electrical resistances are driven by a numeric regulator which sets the temperature profile and gradient as well as the final temperature of pyrolysis. The amplitude variation of the real temperature is more or less 20°C in comparison with the temperature settings.
(c) *The electronic balance*: the reactor is set on an electronic balance to follow continuously the loss of weight (measurement precision: 0.5 % at 10 kg).
(d) *The gas conditioning and analysis system*: at the exhaust of the reactor, the pyrolysis gases are first condensed, filtered and continuously analysed (non condensable fraction). The liquid fraction (condensable gases) is collected and weighed. Thermocouples give the gases temperature at different levels of the plant : in the reactor and along the conditioning line.

The non condensable gases are analysed by the means of a NDIR spectrometer (determination of carbon monoxide (CO), carbon dioxide (CO_2) and methane (CH_4) content), a thermal conductivity analyser (hydrogen (H_2) content) and a magnetomechanics analyser (oxygen (O_2) content).

The pyrolysis oil is collected at the bottom of the condensation columns, weighed to obtain the wet mass before determining its moisture content (Karl Fisher method).

EXPERIMENTAL DESIGN

Four kilograms of anhydrous wood waste have been pyrolysed in the thermobalance retort kiln without air supply. The first set of pyrolysis conditions were:

- Heating rate: 20 °C/min.
- Residence time at final temperature: 60 min.
- Final temperature: 350, 550 and 750 °C.

These conditions were followed for the first run, but some leaks at the oven were observed. As the catch of the facility could not be modified, we decided to decrease the heating rate from 20°C/min to 2°C/min hoping for a lower pressure in the kiln. Unfortunately, the leaks persisted.

RESULTS AND DISCUSSION

PYROLYSIS TESTS

It was demonstrated that the pyrolysis begins earlier and advances faster when the particles (study carried out on cubic particles of 2, 4, 8 cm side) are disposed close by the kiln surfaces. The pyrolysis process is significantly delayed and slowed if the particles are disposed in bulk, in staggered rows and in the centre of the kiln, respectively [4].

With wood waste, the feedstock characteristics (disposition in bulk, thick particles, low bulk density), lead to a very bulky bed. These conditions favour the rate of the pyrolysis process through a enhanced heat transfer and evacuation of the pyrolysis products. The loss of weight has been followed during the experiments. There are not much differences between the loss of weight for pyrolysis temperatures of 550 °C or 750 °C. In the future, it would be interesting to perform such experiments with pyrolysis temperatures from 500 °C to 800 °C, in order to find the best appropriateness between the pyrolysis products quality and their heavy metals contents.

PYROLYSIS PRODUCTS

Gas analysis

At the beginning, the presence of nitrogen and oxygen (inert atmosphere) is constant until the gas flow become so high that air is ejected outside the kiln. From this time, the production of carbon dioxide, carbon monoxide, methane and finally ethane are successively observed. It was impossible to compare carbon dioxide, nitrogen and hydrogen contents in the different runs because of the carbon dioxide sensor breakdown . Only the gas flows and carbon monoxide, methane, oxygen contents could be compared. The different pyrolysis temperatures do not influence the carbon monoxide, methane and oxygen contents: the maximum contents and the general evolution of the composition are the same.

The gas flow increases after about 1h50, which corresponds to a temperature of about 320 °C, until 3h10 (equivalent to a temperature of 480 °C for the final pyrolysis temperature of 550 °C and 750 °C). So, we have a remarkable gas production during 1h20. It is difficult to determine the best gas flow ; more experiments could be carried out for a finer analyse. We could also analyse the calorific value of these gas during the experiments (we have done it in order to not interfere with the pyrolysis process and in the general results).

The installation for the gases conditioning and analysis could be modified. For example, we could imagine:

- To increase the gas exit diameter to favour the gas evacuation and so to diminish the pressure in the kiln.
- To use strainers after the flow meter to capture arsenic [5] or other particles to study
- To install a gas-meter to know the gas volume produced
- The injection of a hot gas (like N_2) in order to have a best control of the gas flow and simulate actual industrial retorts.

Charcoal analysis

Mass yield of waste wood carbonisation are compared to the results of [4]: anhydrous cubes (2 cm) of *Fagus silvatica* were tested in a laboratory kiln in a very controlled way, the heating rate being 2 °C/min and the residence time at final temperature was 15 min. The results obtained on waste wood (Fig. 3) are lower than the mass yields obtained in a laboratory kiln. In other words, more initial matter (solid wood) has been volatilised in the case of waste wood pyrolysis.

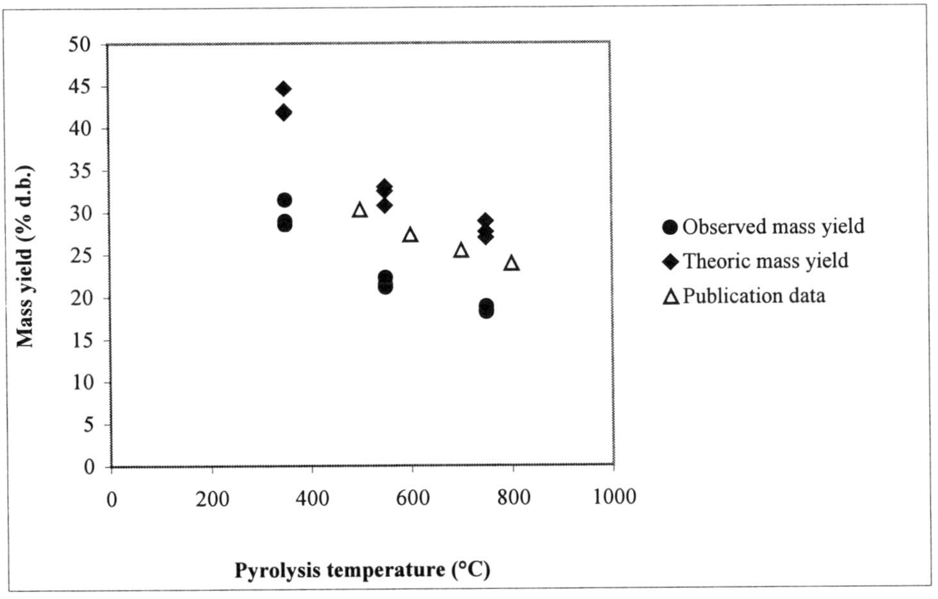

Fig. 3 Comparison of mass efficiencies of waste wood (mass efficiency) and beech wood (publication) charcoal.

The results of proximate composition and calorific value have been also compared to [4] (Figures 4 and 5).

Figures 4 and 5 show that the proximate composition and the calorific value of the waste wood charcoal are equal to the values obtained on beech blocks in a laboratory test. The quality of waste wood charcoal does not seem to be influenced by their heavy metals content.

Pyrolytic liquids analysis

Because of the leaks observed for some tests, the results of the analysis of the pyrolysis oils vary a lot and hence do not gibe any reliable in formation.

Heavy metals contents

The heavy metals contents of the wood waste charcoal and pyrolytic liquids have been compared to the heavy metals contents of the wood waste (Table 3).

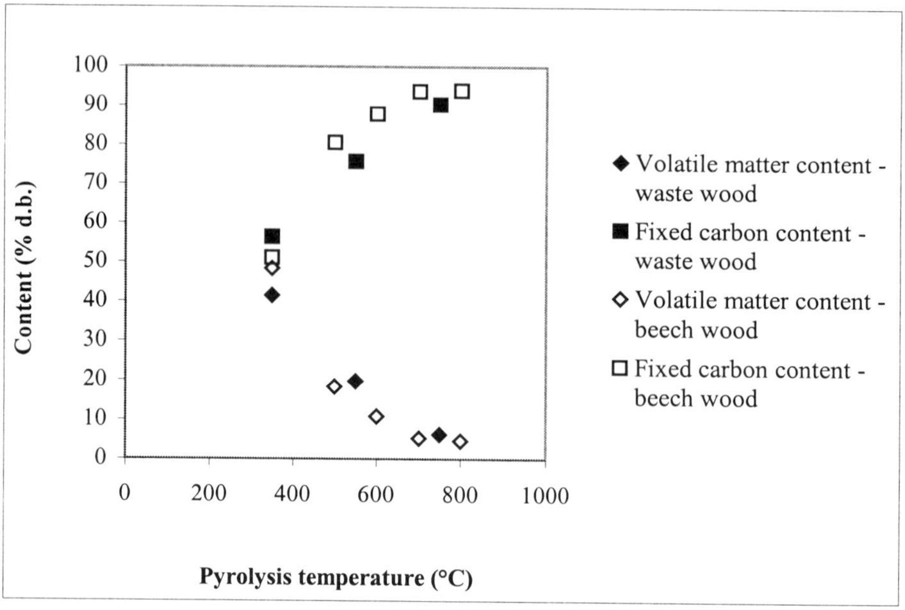

Fig. 4 Proximate composition (volatile and fixed carbon contents) of waste wood and beech charcoals.

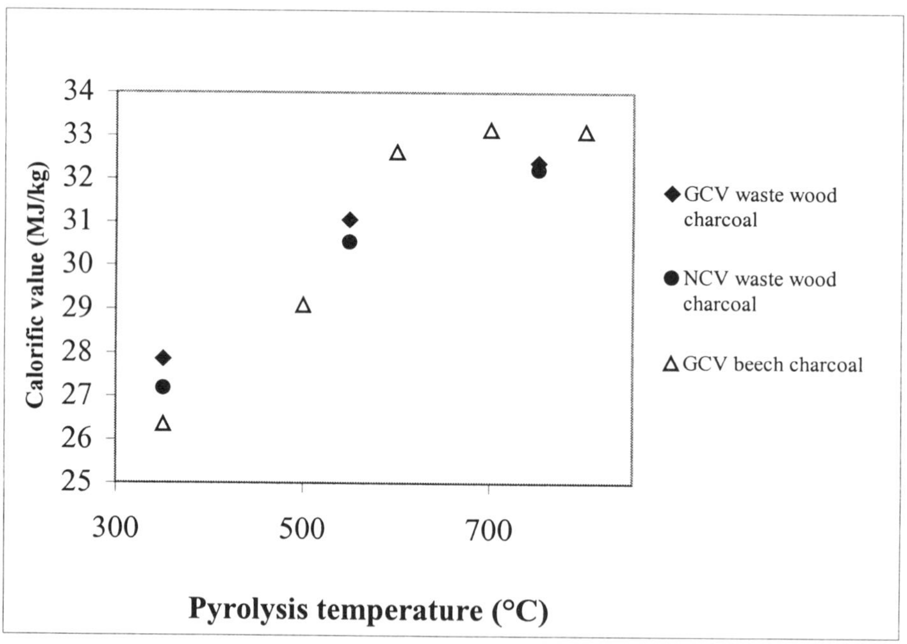

Fig. 5 Comparison of the calorific value (gross – GCV and net – NCV) of waste wood and beech wood charcoals.

Table 3 Average values of heavy metals content of wood waste, wood waste charcoal and pyrolytic liquids (mg/kg).

	Cd	Cu	Ni	Pb	Zn	Hg	Cr	As
Wood waste	0.43	10.67	39.67	101.33	153,33	0.17	53.00	0.30
350 °C								
Charcoal	0.68	6.17	3.27	59.67	153,67	0.02	53.50	0.63
Pyrolytic liquids	0.35	2.47	0.98	1.23	32,60	0.03	0.67	0.46
550 °C								
Charcoal	1.94	12.87	1.33	83.33	188,67	0.07	8.67	1.26
Pyrolytic liquids	0.20	1.13	0.28	0.73	12,27	0.04	<0.25	0.09
750 °C								
Charcoal	<0.05	23.87	62.33	51.00	189,33	0.03	145.33	8.00
Pyrolytic liquids	0.28	0.97	0.15	0.71	77,67	0.02	<0.25	0.69

Mercury, arsenic and cadmium are metals, which are volatile at low temperatures (356.88, 614 and 767 °C respectively). Such metals must be captured at the exit of the gas after the condensation columns. Arsenic appears in treatments like CCA. It was shown that CCA (Copper, Chromium, Arsenic) treated wood could be pyrolysed with a negligible release of copper and chromium and a minimal release of arsenic (76,84 % of arsenic was found in the pyrolysis residues) at a reactor temperature of 350 °C during 20 minutes, using a nitrogen at a flow rate of 5 Nm^3/h. [5]

Figure 6 shows that the increase in zinc and copper content in the charcoal is due to their concentration in this product. The pyrolytic liquids content in these two metals is very variable. But the majority of zinc and copper is found in the charcoal as well as for nickel, lead and chromium. The same conclusion was observed for CCB (Copper, Chromium, Arsenic) treated wood [2].

As the ash content of different biomass could influence the gasification reactivity of biomass chars, a study has been conducted to determine the influence of heavy metals on the gasification process [6]-[7]. They show that the alkali metals increase the reactivity of wood char and that lead, copper and zinc, especially as chlorides, inhibited the gasification of the char (about 2.5 times slower that the char from untreated wood).

The study of the chlorine content of our wood waste could be attractive: the presence of particles board, oriented stand board, hardboard points to the presence of chlorine usually used in glues. The combustion of such products in a boiler generates nitrogen oxide, sulphur dioxide, carbon monoxide, hydrochloric acid and formaldehyde [8].

Finally, it would be necessary to study the heavy metals contents of the waste wood sample of each experiment, in order to increase the accuracy of the mass balance of the heavy metals.

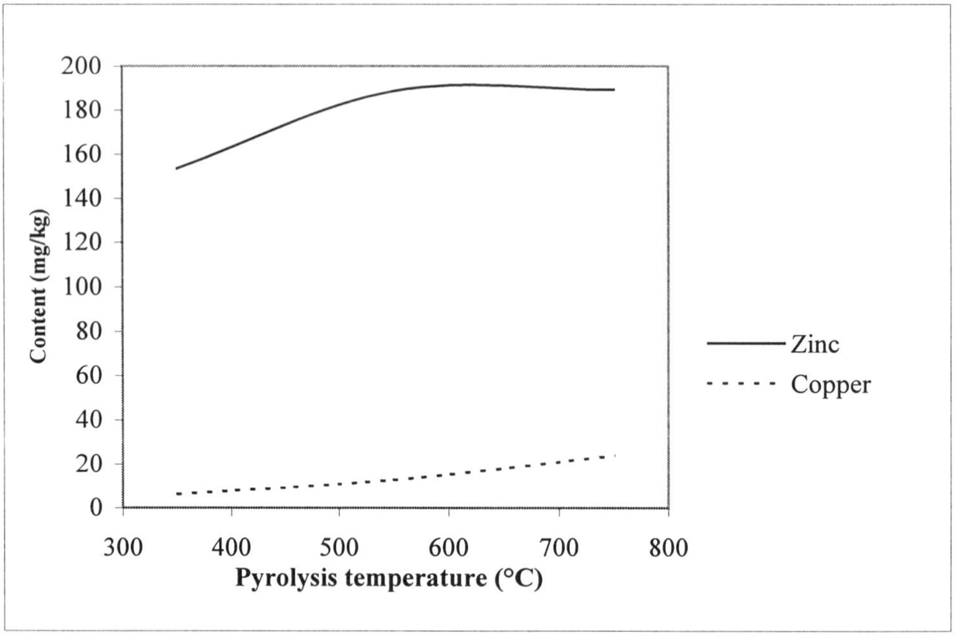

Fig. 6 Zn and Cu content in waste wood charcoal.

CONCLUSION

These orientation tests enabled us to identify the problems encountered in waste wood slow pyrolysis and the modifications to make. In order to gain time and improve the analysis, it is recommended to inject hot gas in the kiln and to inject the cold gas at the completion of the test to decrease the temperature inside the kiln. The pyrolysis experiment required in fact two days to be carried out : one day for the experiment and another to empty and refill the kiln.

With regard to the heavy metals contents of the pyrolysis products, we can say that the heavy metals mainly concentrate in the charcoal except mercury, arsenic and cadmium which are volatile at low temperatures. Significant quantities of these three heavy metals are found in the pyrolytic oil.

In further experiments, tests should be performed in accordance with industrial conditions (high temperatures). Other studies could be conducted on the possible uses of the pyrolysis products, which were contaminated with heavy metals, on the thermochemical aspects of the wood waste pyrolysis (chemical reactions of the wood preservatives during the pyrolysis process).

REFERENCES

[1] Temmerman M., Van Belle J.-F. (1998) *Contribution des filières bois énergie au développement durable en Belgique*. Activity report of the Agricultural Research Centre of Gembloux (Belgium), 195 p.

[2] Wehlte S., Meier D., Moltran J. And Faix O. (1997) The impact of wood preservatives on the flash pyrolysis of biomass. In : *Developments in thermochemical biomass conversion* (Ed. by A.V. Bridgwater & D.G.B. Boocock), pp. 206-219, Blackie Academic & Professional, London.

[3] Guide des déchets de chantiers (1998). Collection : Connaître pour agir. Publié par l'ADEME

[4] Schenkel Y. (1999). *Modelization of mass and energy flows from wood carbonization in retort kilns* (PhD thesis in French). Gembloux (Belgium), Faculté Universitaire des Sciences Agronomiques, 328 p.

[5] Helsen L. And Van Den Bulck E. (1997) Release of metals during the pyrolysis of preservative impregnated wood. In : *Developments in thermochemical biomass conversion* (Ed. by A.V. Bridgwater & D.G.B. Boocock), pp. 220-228, Blackie Academic & Professional, London

[6] Von Scala C., Struis R., Stucki S. (1996) The influence of heavy metals on the gasification of wood. In: *Biomass for energy and the environment* (Ed by P. Chartier, G.L. Ferrero, U.M. Henius, S. Hultberg, J. Sachau, M. Wiinblad), pp 1388-1391, Elsevier Science Limited, Oxford.

[7] Von Scala C., Struis R., Stucki S. (1997) The influence of chlorine on the gasification of wood. In : *Making a business from Biomass in Energy, Environment, Chemicals, Fibres and Materials* (Ed by R.P. Overend & E. Chornet), pp 415-421, Elsevier Science Limited, Oxford.

[8] Falandrin B. (1999). *Wood Board Combustion: the Environment Is Respected.* CTBA report, Bordeaux, France.

Fast Pyrolysis of Industrial Biomass Waste

Ch. Gerdes[1], D. Meier[1], W. Kaminsky[2]
[1]*Federal Research Centre for Forestry and Forest Products, Institute for Wood Chemistry and Chemical Technology of Wood, D-21027 Hamburg, Germany*
[2]*University of Hamburg, Institute for Technical and Macromolecular Chemistry, Bundesstraße 45, D-20146 Hamburg, Germany*

ABSTRACT: In cooperation with several companies different biomass waste such as wood waste, fibre sludge, cocoa shell and panel boards with a high content of phenol-formaldehyde resin have been decomposed by flash pyrolysis into small molecules. The thermal conversion of biomass was performed in a fluidised bed reactor (PDU-scale, throughput 5 kg biomass/h) using the modified and optimised "Hamburg Fluidized Bed Process" ("Hamburger Wirbelschichtverfahren"). The main product was a liquid (bio-oil) with a maximum yield of 50 to 65 wt.%. In addition, pyrolysis gases and solids (charcoal; inorganics) were formed. In many cases pyrolysis of different real biomass waste led to a good yield of bio oil. However, also difficulties with the fluid bed reactor were observed such as large variation in particle size. The oils could be used for production of power and heat, as chemical feedstock or in some cases recycled into the production processes (resin for panel boards).

INTRODUCTION

Industry is not yet able to make complete and rational use of their biomass residue and waste. Intelligent waste disposal and recycling techniques are necessary to meet stronger legislative requirements in future.

Due to the newer legislative requirements in Germany it will not be possible to deposit wood waste by landfill after end of year 2005 [1], [2]. Disposal companies are forced to find the best process to handle waste-wood. Wood waste can be contaminated with inorganic and organic wood preservatives such as CCB, Cu-HDO, Al-HDO, lindane, DDT etc.. Numbers of studies and papers do predict "golden years" for thermo-conversion processes, especially for pyrolysis processes for impregnated-wood [3] [4]. With help of the well known "Hamburg Fluidised Bed Process" [5] investigations were performed to improve this technology for pyrolysis of artificial "self made" waste wood with a homogeneous well defined contamination [6] [7], [8], [9], [10], [11]. Based on these earlier studies real waste wood fractions from a German wood disposal company and residual hardwood from a German/North American charcoal manufacture was pyrolysed. These investigations shall prove the flash pyrolysis technique to be useful in wood waste exploitation.

Looking at the wood processing industries, like furniture, panel boards or charcoal producers, it can be noticed that often a large amount of grinding grit remains from production process. Usually, this grit is burned or deposed by landfill. Going back to the argumentation of the newer legislative requirements concerning waste-wood a new way of disposal should be found for the same reasons. Therefore the pyrolytic behaviour of grinding grit from a German company was studied.

Another possibility for the use of flash pyrolysis is to bring smaller scale plants into the market. Thinkable are plants with a capacity of 1-10 t/d max.. There are a lot of biomass processing companies with a more or less unusual biomass waste fraction. This waste has no value for these companies at the moment, it only entails disposal costs. It will be demonstrated that these biomass can lead to acceptable bio-oil yields. In this study cocoa shell from a German cocoa roasting company with an output of 4-5 t coca shell per day was investigated.

Today, paper recycling is a common practice in paper business. A typical length of fibres for paper production is 0.7-0.8 mm for hardwood and 2.0-2.5 mm for softwood fibres. With each cycle, the paper fibres become shorter. After several cycles the fibres are too short to make paper of acceptable quality (below 0.1-0.2 mm). This is also valid for the production of new paper. In the year 1997 15,953,362 t of new paper, cardboard and carton were produced in Germany. As a result, German paper mills had an output of 239,300 t fibre sludge in 1997 [12]. Additionally 9,457,411 t of used paper were recycled in the same year [13]. Caused by the lower quality as a raw material for paper production, 20-30 wt.% of used paper is removed from the process due to the short fibres. Ultimately, there were at least 1.9 Mio. t. of sludge from paper recycling in 1997. Fibre sludges contain a large amount of organic compounds, inorganic fillings and make-up chemicals. Normally, these sludges are burned or deposed by landfill like grinding grit and other waste. The fibre sludge used for pyrolysis was supplied by a German paper mill. This study wants to demonstrate an alternative way of handling them.

High pressure laminates (HPL) are used as panel boards, tabletop, worktops in kitchen or laboratories, facing for buildings etc.. These panels are produced with an amount of 30 wt.% phenol-formaldehyde resole resin. The costs for the resin ground material are quite high, e.g. the phenol costs are approximately 825 Euro/t [14]. This price and stronger legislative requirements could lead to new ways for the recycling of production waste or recycling of used up laminates from the consumer. One way could be fast pyrolysis of HPL. The aim would be monomer recovery -phenols from resin.

Currently, existing pilot plants in Canada, Netherlands, UK utilize mainly well defined non contaminated biomass fractions such as wood particles, saw dust, and bark. The performed investigations in this work should broaden the knowledge of the pyrolytic behaviour of various industrial biomass waste. This will facilitate the introduction of flash pyrolysis processes into existing industrial processes. Therefore, a new way of biomass exploitation will be demonstrated. In cooperation with several companies different biomass waste such as cocoa shell, wood waste, fibre sludge and panel boards with a high content of phenol-formaldehyde resin were decomposed by flash pyrolysis into smaller molecules to use them for the production of energy and/or chemicals.

EXPERIMENTAL

Thermal conversion of biomass was performed in a Pilot Demonstration Unit (PDU) scale fluidised bed reactor with a throughput of 5 kg biomass/h and in a laboratory fluidised bed reactor 1 g/batch. Therefore the "Hamburg Fluidised Bed Process" ("Hamburger Wirbelschichtverfahren") had to be modified and empirically optimised for biomass feedstock.

PDU-SCALE FLUIDIZED BED PROCESS

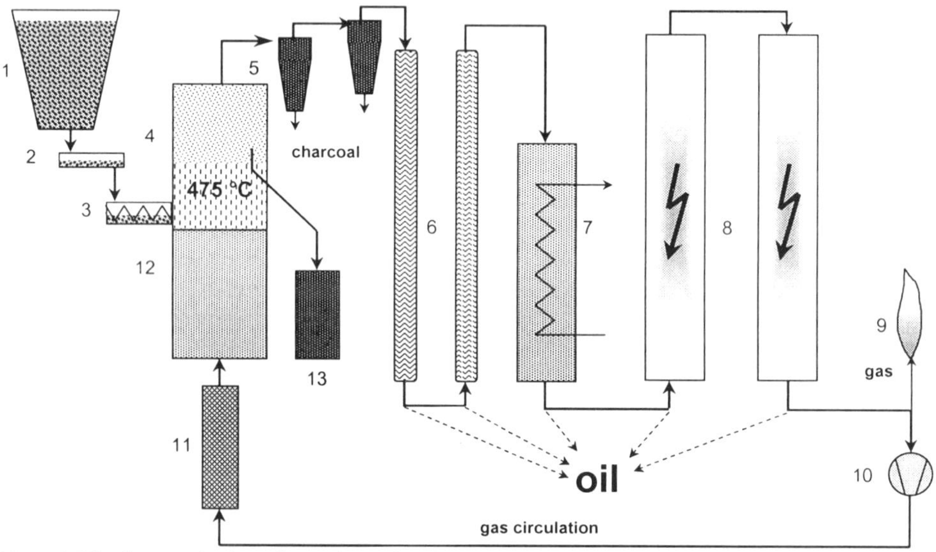

Fig. 1 Flash pyrolysis pilot plant for biomass (PDU-scale); (1 hopper, 2 vibration conveyor, 3 screw feeder, 4 fluidised bed reactor, 5 cyclone, 6 heat exchanger, 7 intensive cooler, 8 electrostatic precipitator, 9 flare, 10 compressor, 11 gas preheater 1, 12 gas preheater 2, 13 overflow container).

The pilot plant (*Fig. 1*) works continuously with a capacity of 5 kg/h (biomass). The feedstock is transported by a vibration conveyor (2) and a screw feeder (3) into the fluidised bed reactor (4). The volatile pyrolysis products and smaller char particles leave the reactor through the reactor head together with the volatiles. Larger charcoal particles leave the reactor directly via an overflow pipe into the collection vessel (13). Micro-char particles are separated from the gas flow with a dual cyclone system (5). The higher boiling components are separated in the water-cooled heat exchangers (6). The remaining condensable components are collected in the intensive cooler (7). Aerosols are separated from the non-condensable gases with electrostatic precipitators (the second is installed for safety reasons) (8). The gas volume necessary for fluidisation is transported with a compressor through two preheaters (11, 12) back into the reactor. The remaining gas volume is burned in a flare (9). The plant runs continuously with recycled pyrolysis gases. The footprint is approximately 4 x 2 m. The height is ca. 2 m.

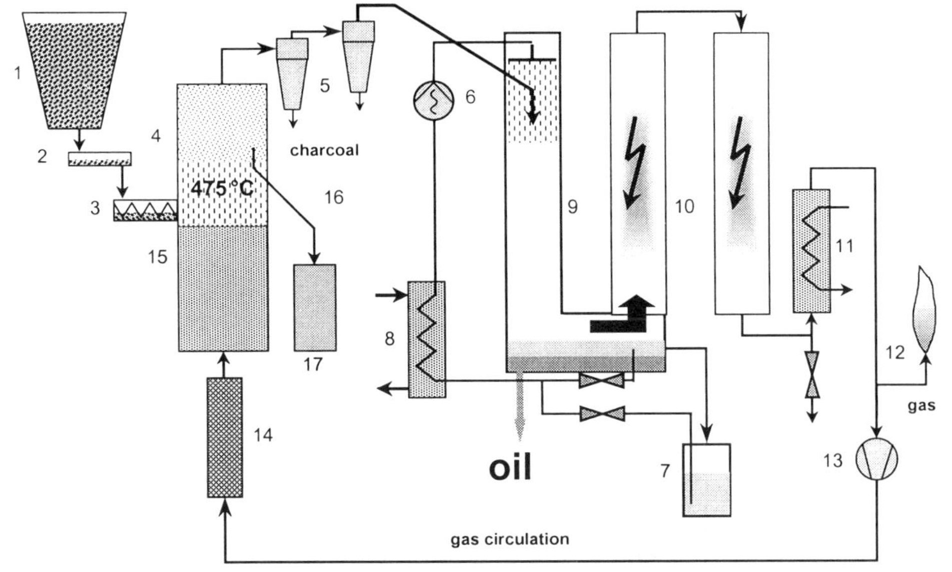

Fig. 2 Updated flash pyrolysis pilot plant for biomass (PDU-Scale); (1 hopper, 2 vibration conveyor , 3 screw feeder, 4 fluidised bed reactor, 5 cyclone system, 6 mono-pump, 7 quench liquid reservoir, 8 heat exchanger, 9 spray tower, 10 electrostatic precipitators, 11 heat exchanger, 12 flare, 13 compressor, 14 gas preheater 1, 15 gas preheater 2, 16 overflow pipe, 17 char collection vessel).

An updated design comprises recirculation of a pre-cooled hydrocarbon liquid with an mono-pump. The quench medium is immiscible with pyrolysis liquids due to the low polarity compared to pyrolysis liquids and can therefore be separated in the lower part of the quenching column and passed into the reservoir (7). Resisting aerosols are collected by electrostatic precipitators. Additionally, a cooler was installed to remove the fluidising gas from traces of lower volatile compounds (water etc.) (11).

LABORATORY FLUIDIZED BED PROCESS

For screening tests of new, unusual feedstock such as cocoa shell or fibre sludge a small fluid bed glass reactor for batch runs of 1 g biomass was designed and built (*Fig. 3*).

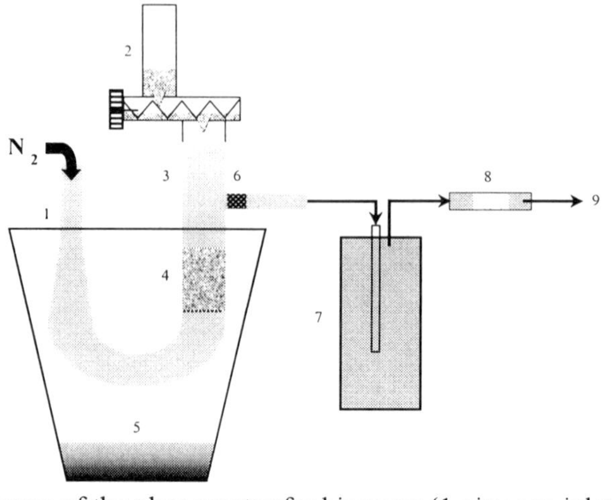

Fig. 3 Flow scheme of the glass reactor for biomass; (1 nitrogen inlet, 2 silo with screw feeder, 3 glass reactor, 4 fluidised bed, 5 heating system, 6 wire screen, 7 heat exchanger, 8 glass tube with cotton filter, 9 gas outlet).

The glass reactor works continuously with a capacity of 1 g (biomass). The feedstock is transported from the reservoir by a screw feeder (2) into the fluidised bed reactor (3). The bed (quartz sand) (4) is fluidised with nitrogen (1) and heated with an electrical heating system. The volatile pyrolysis products are leaving the reactor through a wire screen (6) to prevent the pyrolysis liquids from charcoal contamination. The volatile components are condensed in a cooling trap filled with a mixture of acetone and dry ice (7). The non condensable gases and the aerosols pass through the trap into a cotton wool filter (8). The non condensable gases are leaving the reactor through a gas outlet (9).

Experimental conditions

Pyrolysis temperature has the greatest influence on the quality and quantity of pyrolysis liquids. Earlier studies indicate the optimum temperature for a high oil yield between 475 and 500 °C [6] [15]. *Table 1* shows the experimental conditions for three different feedstock investigated by lab. scale pyrolysis (glass-pyrolysis-reactor). The experimental conditions for the different feedstock pyrolysed with the PDU-scale equipment are presented in *Table 2*.

Table 1 Experimental conditions of laboratory scale pyrolysis.

feedstock		cocoa shell	fibre sludge	grinding grit
feed	[mg]	1155	1308	1185
feed, dry	[mg]	1040	1270	1173
particle size	[mm]	< 1	< 1	< 1
feedstock moisture	[wt.%]	9.9	2.9	1.0
ash content	[wt.%]	0.5	48.0	7.8
pyrolysis temperature	[°C]	475	475	475
nitrogen flow	[l/min]	2.7	2.7	2.7
exp. time	[s]	180	120	129

Table 2 Experimental conditions of the PDU scale pyrolysis.

feedstock		beech waste	fibre sludge	HPL	mixed waste wood
pyrolysis temp.	[°C]	484	486	464	458
particle size	[mm]	1-3	1-3	1-3	1-3
ash content	[wt.%]	4.19	48.00	2.45	5.25
feedstock moisture	[wt.%]	7.47	0.94	5.40	12.42
total feed	[g]	6000	10000	8124	16602
total feed atro*	[g]	5552	9906	7685	14541
org. feed	[wt.%]	88.34	51.06	92.15	85.13
bed material	[-]	quartz sand	quartz sand	quartz sand	quartz sand
size of bed material	[mm]	0,3-0,5	0,3-0,5	0,5-0,6	0,5-0,6
amount of bed material	[g]	6000	6000	5200	5500
experimental time	[h:min]	1:57	2:11	1:55	5:00
throughput	[g/h]	3077	4580	4239	3320
reactor residence time	[s]	2.18	3.16	2.62	2.23
total gas flow/cold	[m³/h]	7.21	4.97	5.99	7.06
total gas flow/hot	[m³/h]	17.84	12.28	14.81	17.46

*: less feedstock moisture

The feedstock was prepared by milling and sieving to the suitable size. Important characteristics of each feedstock such as moisture, ash content and organics are displayed in *Fig. 4* for the lab. scale pyrolysis. The characteristics of the PDU scale pyrolysis can be found in Fig. 5.

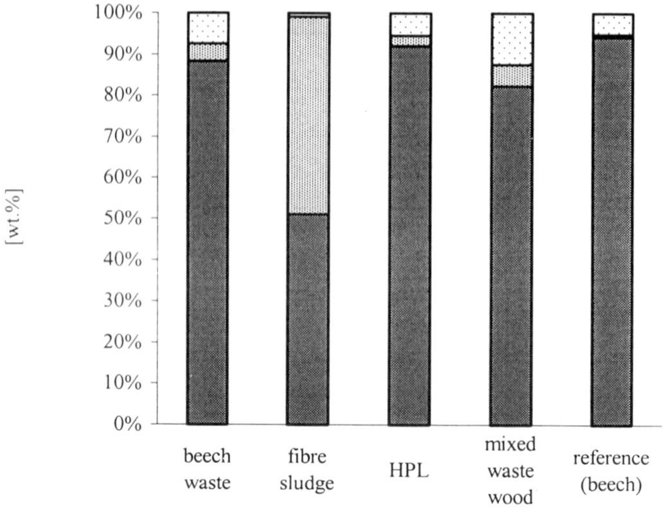

	beech waste	fibre sludge	HPL	mixed waste wood	reference (beech)
water	7.47	0.94	5.40	12.42	5.44
inorganics	4.19	48.00	2.45	5.25	0.58
organics	88.34	51.06	92.15	82.33	99.42

Fig. 4 Characteristics of laboratory scale pyrolysis feedstock.

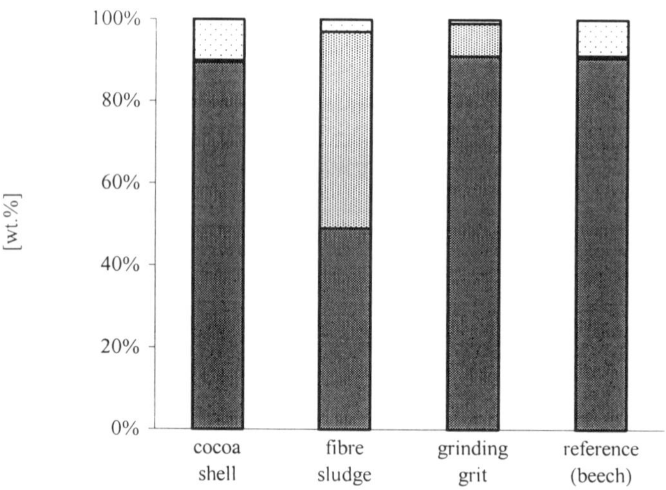

	cocoa shell	fibre sludge	grinding grit	reference (beech)
water	9.89	2.88	1.00	8.67
inorganics	0.50	48.00	7.80	0.58
organics	89.61	49.12	91.20	90.75

Fig. 5 Characteristics of PDU scale pyrolysis feedstock.

Analysis

All pyrolysis products were analysed by chem./phys. methods (GC-MS, GC-FID, GC-WLD a.o.). The feedstock moisture was determined by drying according to DIN 51718. The ash content was measured by annealing according to DIN 51719. Pyrolysis gas was investigated by GC-FID and GC-WLD using definite gas mixtures for calibration. The pyrolysis product with the highest value, the pyrolysis liquid or bio-oil was analysed by GC-MS. Each identified compound in the oil was then quantified by GC-FID. The total amount of the high molecular weight fraction, so-called: pyrolytic lignin was determined by a special precipitation method [16] [17]. The water content of bio-oil was measured by Karl Fischer titration method.

RESULTS AND DISCUSSION

The first step on the way to the decision if fast pyrolysis is a suitable technique for biomass waste exploitation is to take a serious look at the mass balance. In almost every case bio-oil is the pyrolysis product with the highest commercial value no matter weather it is used as bio-fuel for power and heat production or as chemical feedstock.

The by-product charcoal can have the potential to be refined to activated carbon. The pyrolysis gas is the product with the lowest value. It can only be used as a low heating value gas for burners.

Usually the inorganic fraction has no specific value. Looking at fibre sludges from paper mills it consists of cheap filling material (from the paper production process) like $Ca(CO_3)$, kaoline, make-up chemicals etc.. In other cases e.g. grinding grit, HPL or waste wood the inorganic fraction consists of an inhomogeneous undefined mixture of different substances or, in worst cases, of metall and heavy metall mixtures. Existing pilot plants in Canada, Netherlands, UK utilizes mainly well defined non contaminated wood fractions for pyrolysis. For that reason non contaminated beech wood (white wood) was used as reference feedstock.

The organic mass balance of different feedstock pyrolysed with laboratory and PDU scale equipment is presented in *Table 3*. The screening testes of cocoa shell, fibre sludge and grinding grit demonstrate a poor quantity of bio-oil between 24-30 wt.%. Contrary to that, the charcoal yields are quite high. Due to the low oil yields the amounts of calculated pyrolysis gases are nearly 1.5 to 2 times of the usual gas amount achieved in the beech wood reference experiment.

The product yields of the beech wood reference experiment pyrolysed with the laboratory scale equipment are satisfactory compared to the results of the PDU scale pyrolysis.

Table 3 Organic mass balance (bio-oil includes reaction water).

laboratory scale	cocoa shell [wt.%]	fibre sludge [wt.%]	grinding grit [wt.%]	reference (beech) [wt.%]
bio-oil*	27.71	29.85	23.94	66.08
charcoal	43.62	41.50	47.00	15.26
pyrolysis gas**	28.67	28.65	29.06	18.66

* less feedstock moisture;**calculated

PDU scale	beech waste wood [wt.%]	fibre sludge [wt.%]	HPL [wt.%]	mixed waste wood [wt.%]	reference (beech) [wt.%]
bio-oil*	46.11	9.65	44.28	75.79	67.83
charcoal	15.02	31.26	31.62	16.28	15.50
pyrolysis gas	38.87	59.08	24.10	7.93	16.67

* less feedstock moisture

The product yields of the laboratory scale pyrolysis of fibre sludge are differing badly in comparison to the PDU scale results. The oil yield of the PDU scale pyrolysis is 3 times lower than the laboratory scale pyrolysis. In fact the PDU scale data are the more reliable ones. Yields of 8.4 wt.% bio-oil and 25.8 wt.% charcoal formed by fibre sludge pyrolysis can be also found in earlier studies [18].

The organic amount of bio-oil obtained by beech waste and HPL pyrolysis is lower than reference or mixed waste wood pyrolysis but approximately 4.5 times higher than the fibre sludge oil yield. The extremely high oil yield of mixed waste wood is caused by high reaction-water amount included (29.8 wt.%).

These data could lead to the premature conclusion that, except fibre sludge, all investigated feedstocks can easily be pyrolysed. However, this is only half the truth. The total mass balances *Fig. 6* and *Fig. 7* are indicating the importance of the ash content. The complete separation of inorganic solids from the waste is almost impossible if the ash content is very high.

Feedstock such as grinding grit with 7.8 wt.% or beech waste with 4.19 wt.% of inorganic material are difficult to handle in fluid bed reactors. Additionally, they lead to a more or less poor yield of bio oil below 25 wt.% calculated on dry feed. Mixed waste wood is forming an exception in the whole string of investigated feedstock. The inorganic solids with an amount of 5.25 wt.% did not cause any problems during test runs. The oil yield is only 10 wt.% lower than the reference.

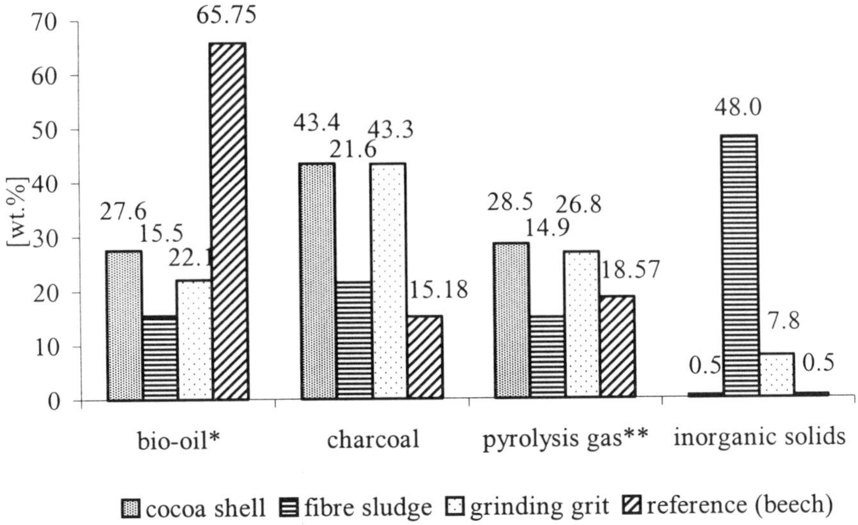

*less feedstock moisture, **calculated

Fig. 6 Mass balance of several feedstocks investigated by laboratory scale pyrolysis.

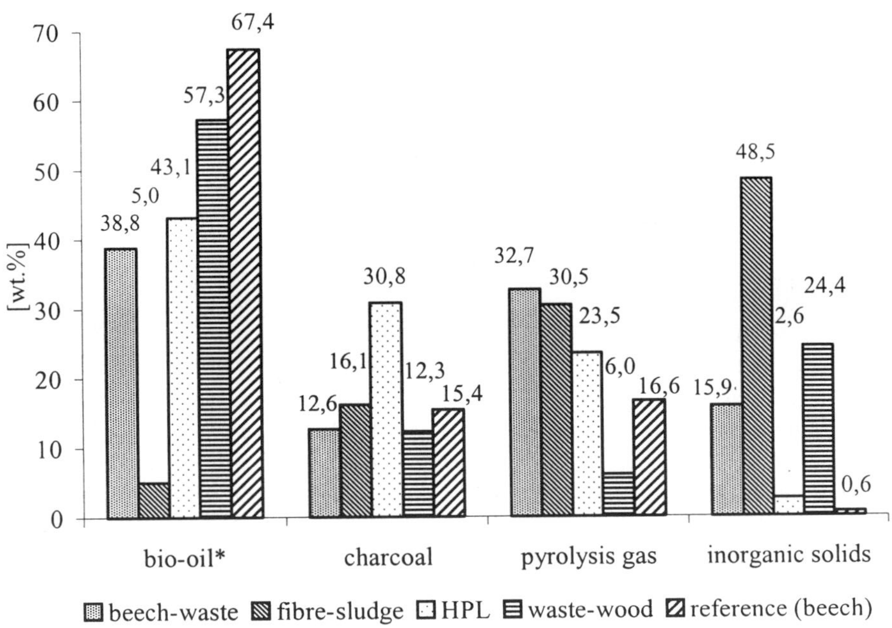

*less feedstock moisture

Fig. 7 Mass balance of several feedstocks investigated by PDU scale pyrolysis.

Pyrolysis plants with an integrated liquid quenching system normally operate the quencher (*Fig. 2*) at temperatures between 20-30 °C. Due to that procedure the water phase (with organic compounds included) can not be collected in the condensation system of pyrolysis plants. Therefore an intensive cooler was installed after the liquid collection system of the pilot plant displayed in *Fig. 2*. This phase is normally burned with the pyrolysis gas and does not enter the mass balances of other existing plants. It consists of volatile organic compounds and water. The balance of organics and water in the water phase is presented in *Table 4*. The organic residue in the water phase varied between 9-20 wt.%. If this water phase is not collected and ignored in the mass balance the mistake will be 10 wt.% related to the total oil-mass and 1-2 wt.% related to the organic oil mass. The total mass balance is then closed only to ca. 90 %.

Table 4 Organic residue in the water-phases obtained by pyrolysis of different feedstock

feedstock	HPL	waste wood	reference (beech)
water content	91.25	86.94	82.78
organic residue	8.75	13.06	17.22

Table 5 Composition (summary) of bio-oils obtained from different feedstocks. quantified by GC-FID

feedstock	waste beech	fibre sludge	HPL	waste wood	reference (beech)
compound	wt.%	wt.%	wt.%	wt.%	wt.%
acids	5.46	2.40	2.20	3.27	5.64
alcohols	0.18	0.04	0.40	0.00	0.18
aldehydes	4.44	0.23	0.13	6.25	2.83
aromatics	0.00	0.00	0.00	0.09	0.00
furans	1.49	1.73	1.14	2.02	1.67
guaiacols	0.83	4.11	0.82	1.94	1.37
ketones	6.00	1.66	4.29	5.63	6.46
phenols	0.20	2.73	11.94	0.39	0.25
pyrans	0.13	0.15	0.01	0.13	0.14
sugars	3.19	0.77	0.75	6.79	2.43
syringols	1.11	2.29	0.78	0.03	4.03
water	54.21	11.38	49.08	40.30	37.35
total organics	23.02	16.10	22.46	26.53	25.01
total identified	77.23	27.48	71.54	66.83	62.36

The mixture of organic compounds in bio oil from pyrolysis of different feedstock is shown in *Table 5*. It can be seen that the amount of phenols is very high in HPL-oil in comparison to others. More than 97 wt.% of these phenols originate from phenol-formaldehyde resins from the HPL production process. Normally the phenol components have an amount below 1 wt.% in the oil. On the other hand *Table 5* demonstrates a large variation in the total amount of each chemical class.

All measured water contents are higher than the reference (except fibre sludge). The water content in the bio-oil is important for its quality and value. The amount of reaction water does not only depend on neat pyrolysis reactions but also on secondary cracking reactions which depend on the design of the pyrolyser. The water/organic ratio should not be lower than 0.7. An increase to a ratio of 1 will entail an oil precipitation into a water soluble and a high molecular weight fraction. The feedstock moisture has no specific influence on the bio-oil formation. Feedstock moisture can be added to the reaction water and causing high process costs and a lower oil quality. A rough calculation shows the minimum reaction water content in bio-oil oil to be nearly 13 wt.%. This was calculated based on dry hard wood assuming that all cellulose (77 wt:%) is decomposed to levoglucosan (*Fig. 8*).

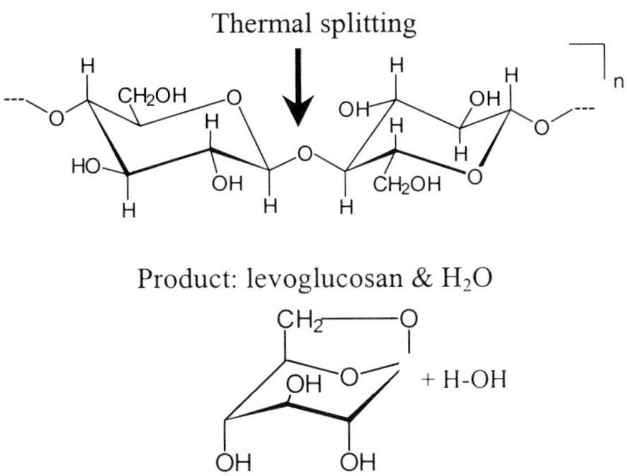

Fig. 8 Thermal splitting of 1-4 linked β-D-glucose with H_2O emission

CONCLUSIONS

Pyrolysis of biomass waste with flash pyrolysis technique has been proved to be possible with all types of investigated feedstock. In comparison with the reference feedstock, beech wood ("white wood") mixed waste wood, HPL and beech-waste are leading to more or less high bio-oil yields between 57.3 to 38.8 wt.% (based on dry feedstock). Feedstock with a high amount of inorganic solids led to a poor oil yield e.g. fibre sludge's (5.0 wt.%), grinding grit (22.1 wt.%). Fibre sludge with 48.5 wt.% of inorganics caused technical problems during bio oil precipitation. The pyrolytic behaviour of the particle is of decisive importance. If the particle is shrinking, crumbling quickly to small sizes, the operation of a fast pyrolysis fluid bed process is extreme difficult, sometimes impossible. If these inorganics are not decomposed to dust, they can be separated with the char. Dusty pyrolysis products can only be handled in fluid bed plants with a "blow-through" design. These dusts are then precipitated from hot pyrolysis gases with multi cyclone systems.

Beside the technical problem with dust separation, the water content and hence the value of pyrolysis products should be looked at seriously. To receive a bio-oil with a high value, the water content should be as low as possible. The feedstock moisture should therefore be minimized. The amount of reaction water does depend on the neat pyrolysis reaction (min. ca. 13 wt.%, see *Fig. 8*). In addition to that, reaction water content does also depend on secondary cracking reactions which depend on the design of the pyrolyser (e.g. vapour residence time). Bio oil from pyrolysis of biomass is a complex mixture of organic compounds (*Table 5*). At the moment there is no possibility to separate individual components economically. The first step to an application of bio oil should be the use of bio-oil as a whole. HPL-oil shows a high amount of phenolic compounds (12 wt.% phenols). The potential of this oil to be used as a basic material for "recycled" phenol resin production should therefore be high.

Pyrolysis of feedstock with a high amount of inorganic solids is useful if the value for the solid product is high enough. The recovery of black liquor from pulping processes could be a practicable way. But, if the value of inorganics is to low, pyrolysis of these feedstocks will not be economic due to the high heating and transportation costs in the process.

The results of the different biomass waste for flash pyrolysis are leading to some recommendations:

(1) Look at the feedstock characteristics (water/organic/inorganic-content and particle size).
(2) Check the pyrolytic behaviour of the particle (shrinking, crumbling to small size?).
(3) Choose the pyrolysis process (e.g. fluidized bed with or without "blow-through" design).
(4) Look seriously at the mass balance of the chosen process and on the phys./chem. analysis of pyrolysis products.
(5) Calculate the production costs and the product values.
(6) Decide if pyrolysis of the feedstock in question is wise or not.

REFERENCES

1. AbfG (Abfallgesetz) (1986) Gesetz zur Vermeidung und Entsorgung von Abfällen v. 27.08.1986 *Bundesgesetzblatt* **Teil I 44**, pp. 1410-1420.
2. TA Abfall (1993) Dritte allgemeine Verwaltungsvorschrift zum Abfallgesetz-Teil 2 v. 14.05.1993. In: *Abfallrecht*, Beck-Texte im dtv, Verlag C. H. Beck , München (1996), pp. 265-315.
3. Willeitner H. & Voß A. (1994) *Gesamtkonzept für die Entsorgung von schutzmittelhaltigen Hölzern*. Umweltbundesamt Abschlußbericht für das Forschungsvorhaben 145 06 76 „Entsorgung".
4. Voß A. (1997) *Aufkommen und Zusammensetzung schutzmittelbehandelter Althölzer und ihre Entsorgung*. PhD thesis, Universität Hamburg.
5. Kaminsky K., Sinn H. & Jannings J. (1979) *Technische Prototypen für die Altreifen-und Kunststoffpyrolyse.* Chem.-Ing.-Tech. **51**, 419.
6. Meier D., Ollesch T., Gerdes Ch. & Kaminsky W. (1998) Herstellung von Bioölen aus Holz in einer Flash-Pyrolyseanlage. In: *DGMK-Fachbereichstagung: Energetische und stoffliche Nutzung von Abfällen und nachwachsenden Rohstoffen*, **9802**, pp. 83-90.
7. Wehlte S. H. C. (1997) *Untersuchungen zur Wirbelschichtpyrolyse von nicht naturbelassenem Holz*. PhD thesis, Universität Hamburg.

8. Meier D., Wehlte S. H. C., Simon C. M. & Ollesch T. (1998) *Stoffliche Verwertung von nicht naturbelassenen Holzresten durch Pyrolyse in der Wirbelschicht*. DBU Abschlußbericht zum Projekt 03 631.
9. Meier D., Ollesch T. & Faix O. (2000) Fast Pyrolysis of Impregnated Waste Wood-the Fate of Hazardous Components. In: *Progress In Thermochemical Biomass Conversion*, (Ed. by T. Bridgwater), submitted for publication, Blackwell Science Ltd, Oxford, UK.
10. Meier D., Wehlte S., Wulzinger P. & Faix O. (1996) Upgrading of bio-oils and flash pyrolysis of CCB-treated wood waste. In: *Bio-oil Production and Utilization-Proceedings of the 2nd EU-Canada Workshop on Thermal Processing*. (Ed. by Bridgwater AV & Boocock DGB) CPL Press, UK, pp. 1402-1412
11. Wehlte S., Meier D. & Faix O. (1995) Wood waste management using flash pyrolysis in a fluidized bed. In: *Proceedings of the Workshop-Frontiers of Pyrolysis*. Breckenridge, CO, June, NREL, USA.
12. Burkhart H. (2000) *Personal communication*. Stora Kabel GmbH, Schwerter Straße 263, 58099 Hagen.
13. VDP (1998) VDP-Statistiken. In: *Papier'98, Ein Leistungsbericht*. (Ed. by Verband Deutscher Papierfabriken e.V.), VDP, Bonn, pp. 34-59.
14. IS-LOR (06/2000) Price list for ground chemicals. ICIS-LOR-Group worldwide
15. Wehlte S. (1997) *Untersuchungen zur Wirbelschichtpyrolyse von nicht naturbelassenem Holz* PhD thesis, Universität Hamburg.
16. Meier D. & Scholze B. (1997) Fast pyrolysis liquid characteristics. In: *Biomass, Gasification and Pyrolysis*, (Ed. by M. Kaltschmitt & A.V. Bridgwater), CLP Press, UK, 431.
17. Scholze B. & Meier D. (2000) Characterization of the water-insoluble fraction from pyrolysis oil (pyrolytic lignin). Part I. Py- GC/MS, FTIR, and functional groups, *J. Anal. Appl. Pyrolysis*, in press.
18. Ying Y. (1991) *Pyrolyse von kommunalem und papierindustriellem Klärschlamm in der Wirbelschicht* PhD thesis, Universität Hamburg.

Co-pyrolysis of wood biomass and plastic wastes of different origin under the pressure of argon and hydrogen

B.N.Kuznetsov[a], V.I. Sharypov[a], N.G.Beregovtsova[a], N.Marin[b] and J.V.Weber[b]

[a] *Institute of Chemistry and Chemical Technology SB RAS, Krasnoyarsk, RUSSIA*

[b] *The Laboratory of Industrial Chemistry, University of Metz, Saint-Avold, FRANCE*

ABSTRACT Influence of wood biomass and plastic wastes origin, iron catalyst and co-processing parameters on the products yield and composition was studied. Experiments were carried out in a rotating autoclave in argon and hydrogen atmosphere at temperatures 340-460 °C and initial pressure 0.3-0.4 MPa. Pine wood, beech wood, cellulose, hydrolytic lignin, polyethylene, isotactic and atactic polypropylene were used as starting materials. Mechanically activated iron-containing ore materials were used as catalysts. Methods of GC, GC-MS, TLC, ^{1}H and ^{13}C NMR and chemical analysis were applied for products investigation.

Some regularities describing the influence of co-pyrolysis process operating parameters, nature of wood biomass and plastics on the yield and composition of liquid products were established and discussed. Obtained data indicate that the optimum temperature of biomass/plastic mixtures conversion which corresponds to the maximum yield of liquids is 390-400 °C.

Hydropyrolysis process gives the higher degree of mixture conversion and higher yield of light liquids as compare to pyrolysis in an inert atmosphere. Observed in some cases non-additive effects indicate that the interaction between wood and plastic derived products takes place during mixture thermal treatment. The more pronounced synergistic effects were detected for hydropyrolysis process. Iron catalysts promote the formation of liquid hydrocarbons from biomass/plastic mixtures and influence on their composition.

INTRODUCTION

Utilization of wood-biomass residues as well as waste polymers is the important direction of recent research activities. It is known that direct catalytic liquefaction of plant biomass can be used to produce liquid fuels and chemicals [1,2]. Co-pyrolysis and co-hydropyrolysis processes have the potential for the environmentally friendly transformation of lignocellulosic and plastic waste to valuable chemicals.

Polymers like polyethylene and polypropylene contain approximately fourteen percent of hydrogen and these materials could provide hydrogen during thermal co-processing with other raw materials. There are some data indicating that the mixture of pine wood with polyethylene can be successfully converted into the distillation liquids [3,4]. The yield of these products depends on the process operating conditions. Liquid products obtained by biomass/plastic mixtures co-processing shall contain both the products of thermal conversion of each separated component of the mixture and the products of their chemical interaction.

Low cost iron-containing ore materials have been intensively studied as catalysts for the hydroprocessing of organic raw materials of different origin. Because the iron ore materials have a moderate level of catalytic activity some methods of their activation have been suggested. Natural minerals can be dispersed into small particles by grinding in tensile-energy apparatus. Such mechanical treatment not only increases the material surface area [5,6] but can change their catalytical properties [7].

In the present work the influence of wood biomas and plastic waste origin, iron catalysts and co-processing parameters on the products yield and their composition was studied.

METHODS

The feedstock materials used in this work included medium density polyethylene (PE), atactic-polypropylene (aPP), isotactic-polypropylene (iPP), beech wood, pine wood, cellulose and hydrolytic lignin. The size of the wood biomass and the plastic particles was less 0,1 mm.

Some characteristics of the wood biomass used are given in *Table 1*.

Table 1 Characteristics of wood biomass.

Wood biomass	Elemental composition (% weight)[a]					Chemical composition (% weight)				
	C	H	N	S	O	cellu-lose	lig-nin	Hemi cellu-lose	Water soluble	Ether soluble
Pine wood	47.4	6.2	0.4	0.2	45.8	48.2	29.4	15.3	6.3	0.8
Beech wood	45.9	6.0	0.2	0.2	47.7	46.4	25.3	22.4	5.1	0.8
Cellu-lose	40.7	6.9	0	0	50.4			-		
Lignin	62.7	6.9	0	0.4	30.0			-		

[a] Dry ash-free basis

Iron ore samples, containing about 40% wt. Fe in the crystal phases haematite, pyrrhotite and pyrite were used as catalysts. In order to increase the ore catalyst activity they were treated in a tensile energy planetary activator mill (PAM) in the presence of water, as described elsewhere [7]. Such treatment results in the increase of catalysts dispersion and changes their structure. Specific surface area of modified catalysts was up to 80 m^2/g.

Liquefaction experiments were carried out in 0,25 l rotating autoclave. A mixture of dry wood biomass (≤1 wt% moisture), plastic and in some cases catalyst (5% wt.) was loaded into autoclave, pressurised with argon up to 0.1 MPa or hydrogen up to 3.0 – 5.0 MPa and then heated up 340-480°C. The duration of thermal treatment was 1 h.. After cooling the autoclave, gas products were collected and analysed by gas chromatography technique. Light liquid products (b. p. < 180°C) and water fraction were distilled *in vacuo*. The other products were extracted with benzene. Than the solvent was evaporated, extracted products were dried to a constant weight in *vacuo* and their yield was evaluated. The degree of the feedstock material conversion was calculated from the difference between the weights of initial mixture and solid residue, insoluble in benzene.

G.c. - m.s. analyses of light liquid products was carried out using 40 m × 0,3 mm silica phased capillary column coated with SE-30 and connected with VG-70HS mass spectrometer. Heavy liquid products (b.p. > 180°C) were analysed by ^{13}C n.m.r. spectroscopy ("Bruker" MSL-400). For characterization of the heavy liquid products, their were separated on a particular families of compounds by a colume chromatography technique. Educed fractions were analysed by g.c. - m.s. technique.

RESULTS AND DISCUSSION

PYROLYSIS OF WOOD BIOMASS AND PLASTIC MIXTURES

The influence of pyrolysis temperature on the products yields was studied using mixtures of wood biomass with plastic (from 2:1 to 2:1 in weight ratio). The conversion degree did not change significantly with temperature variation as it was noted by other researches. *Fig.1* shows the main effect of temperature increase for pine wood / polyethylene co - pyrolysis within the range 360°C - 460°C is the higher amount of gas formed, whilst water fraction yield was decreased. The maximum yield of liquids (50 wt.%) was obtained at 370-400°C.

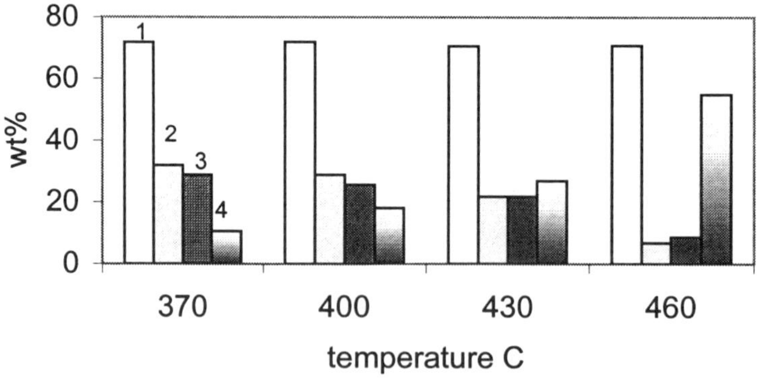

Fig.1 Influence of temperature of co-pyrolysis of pine wood / PE mixture (2:1) in the argon atmosphere on the conversion (1) and the yields of light (2), heavy (3) liquid fractions and gas (4).

The addition of iron ore catalyst did not increase the yield of liquid products and degree of wood / PE mixture conversion. In order to study the influence of the biomass origin on the degree of biomass / plastic mixture conversion and product yields the co-pyrolysis experiments were carried out at 400°C with fixed mixture composition (biomass / plastic ratio 1:1). *Table 2* shows, that mixtures (1:1 weight ratio) of beech and pine woods with atactic-polypropylene (aPP) give the similar yields of liquid and gaseous products of co-pyrolysis process. In the case of cellulose the increased yield of water and light fraction was obtained. In the runs with lignin the higher yield of char and less yields of water and light fraction were observed.

Table 2 Influence of biomass origin on the yield of products of biomass / a-PP (1:1 weight ratio) co-pyrolysis in the argon atmosphere (400°C, 1h).

Biomass origin	Product yields, % wt.				
	Light liquid	Heavy liquid	Water	Gas	Char
Cellulose	25.0	25.6	24.0	8.1	17.3
Beech wood	18.6	30.5	18.8	10.9	21.2
Pine wood	23.4	24.4	17.1	11.5	23.6
Hydrolytic lignin	17.1	27.7	12.5	10.3	32.4

Some experiments have been done to study the influence of wood beach / a-PP and cellulose / a-PP mixtures composition on the degree of their conversion and products yield. In all runs the degree of conversion was increased with the growth of plastic concentration. As shown in *Fig. 2* the maximum yields of light liquids were observed for the mixture consists of 20 wt.% biomass and 80 wt.% a-PP, showing the synergistic effect. The yields of light liquids fractions reached up 43.5 wt.% for the beech wood / a-PP mixture and 37.1 wt.% for the cellulose / a-PP mixture. These yields are by 2.5 and 2.2 times higher respectively, as compared to the expected ones, calculated as a sum of light liquid fractions produced from each separated component of the mixture.

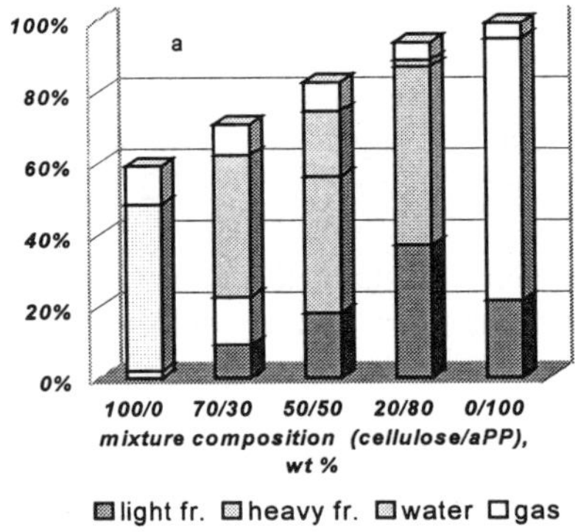

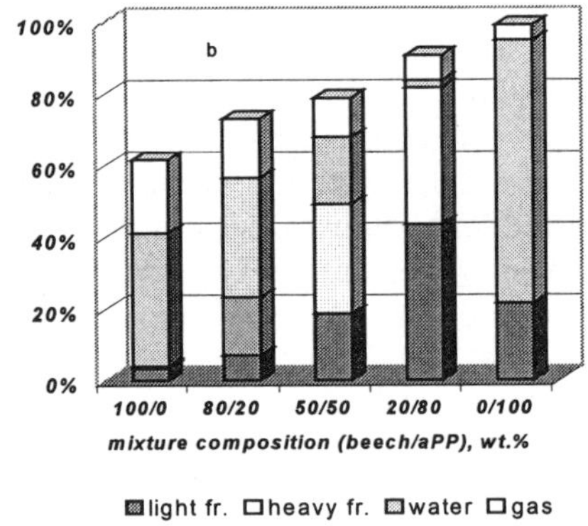

Fig.2 Influence of cellulose / aPP (a) and beech / aPP (b) mixtures composition on the yield of co-pyrolysis products (400°C, 1h).

According to the GC data gaseous products of a-PP pyrolysis contain more than 75% wt. of olefins. In the gaseous products of the beech wood / a-PP mixture (20/80 weight ratio) only approximately 8 % wt.of the unsaturated hydrocarbons were detected. Taking these data into account it is possible to suppose that olefinic products of a-PP thermal conversion react with products of cellulose and lignin depolymerization with liquids formation. This assumption shall explain the synergistic effect, observed in the pyrolysis of the biomass / a-PP mixtures.

The origin of plastic has more influence on the distribution of products of biomass / plastic co-pyrolysis. According to *Table 3* the yield of light liquids from beech-wood / plastic mixtures (1:2 weight ratio) was decreased in the following sequence: isotactic-polypropylene > atactic-polypropylene > polyethylene, whilst the yield of heavy liquids was increased in the same order.

Table 3 Influence of different type of plastics on the yield of products and degree of conversion in the process of beech wood and plastic (1:1) co-pyrolysis (400°C, 1h).

	Gas	Water	Light Liq.	Heavy Liq.	Char	Conversion
i-PP	10,8	17,4	33,6	22,2	16	84
a-PP	10,9	18,7	18,5	30,5	21,4	78,6
PE	10,8	18,9	8,9	39,8	21,6	78,4

HYDROPYROLYSIS OF WOOD BIOMASS AND PLASTICS MIXTURES

In co-hydropyrolysis experiments without catalysts the degree of pine wood/ polyethylene mixture (1:1 weight ratio) conversion was 80% wt. and yield of the light liquid fraction - 23% wt. The addition of iron ore catalyst activated by mechanochemical treatment increased the degree of mixture conversion by 5-13%. This increase was mainly due to light liquid fraction formation. The variation of catalyst nature (pyrite, pyrrhotite, haematite) influences on the product composition. Pyrrhotite catalyst yields the highest amount of the light fraction (about 40% wt.).

The influence of the process temperature on catalytic hydropyrolysis of biomass/plastic mixture was studied in the range 360 – 460°C. *Fig. 3* shows that the highest conversion (91% wt.) of the pine wood / polyethylene mixture (1:1 weight ratio) was observed at 390°C – 430°C in the presence of activated haematite catalyst. Higher temperatures promote increased yields of char and gaseous products. At lower temperatures a reduced yield of distillate fraction was observed. In comparison with pyrolysis in inert atmosphere the increased yields of light hydrocarbon fractions (by 1.6 – 1.8 times) and increased degree of mixture conversion (by 1.2 time) were observed for hydropyrolysis process.

Some hydropyrolysis experiments have been done to study the effect of biomass concentration on the degree of a biomass / plastic mixture conversion and yields of products at 390°C. In all runs with different types of biomass the degree of conversion was increased with the growth of plastic concentration. Dotted line on *Fig. 4* represents the expected degree of pine wood / polyethylene mixture conversion assuming that no interaction between wood and polyethylene derived products takes place during hydropyrolysis. But the experimental data are differed from expected ones, showing the synergistic effect. In all range of a lignite / polyethylene mixture composition the degree of conversion was higher as compared to theoretical one. Maximum synergistic effect

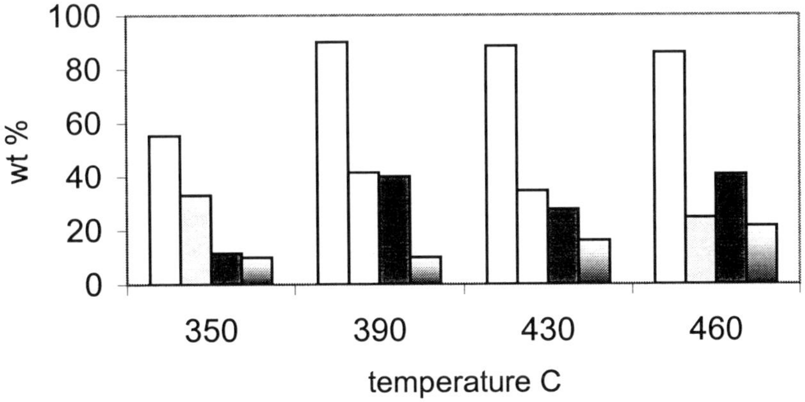

Fig. 3. Influence of temperature of pine wood / PE mixtures (1:1) hydropyrolysis in the presence of activated haematite catalyst on the mixture conversion (1) and the yields of light (2), heavy (3) liquid fractions and gas (4).

was observed for the mixture, containing 30-70 wt.% of pine. In this range of mixture composition the observed degree of conversion in the experiment with 3.0 MPa initial H_2 pressure was higher by 10–15 wt% as compared to theoretical one.

The increasing of pressure up 5.0 MPa results in higher conversion degree (*Fig.4*). Under these experimental conditions the mixture conversion riches up to 100% at concentration of pine-wood more then 40% wt.

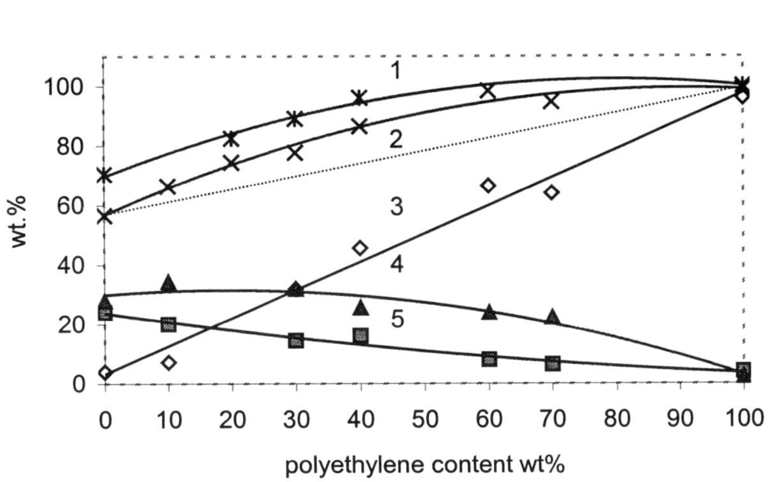

Fig.4 Influence of polyethylene / pine-wood ratio on the degree of mixture conversion at 390°C and initial pressure H_2 5 MPa (1), 3 MPa (2) and on yields of light (3), heavy (4) liquid fractions and gaseous products (5) at hydropyrolysis time 1 h.

The hydropyrolysis of pine wood at 390°C in the presence of activated haematite catalyst gives the wood conversion degree 54 wt.%. Under the same hydropyrolysis process conditions about 99.5% wt. of polyethylene was converted to liquid and gaseous products.

Brown coal, pine wood with the content of lignin 29.4% wt., pine wood after extraction with dimethyl sulfoxide (with the content of lignin 17.2 % wt.), as well as industrial cellulose were tested in co-hydropyrolysis with polyethylene at the same process conditions.

Fig. 5 shows that the degree of pine wood conversion was permanently increased with the increase of polyethylene / wood ratio in the feedstock materials. The addition of 85 wt.% polyethylene to pine wood results in of wood conversion degree 80 wt.%.

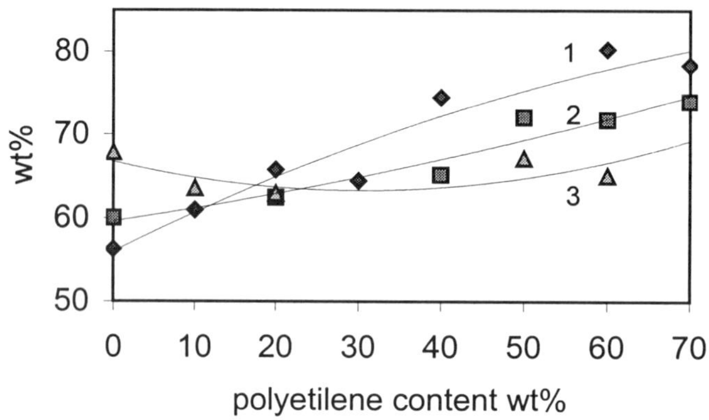

Fig.5 Influence of polyethylene concentration on the conversion of brown coal (1), pine wood (2), cellulose (3) and pine wood after lignin extraction with dimethyl sulfoxide (4) in the co-hydropyrolysis process (initial H_2 pressure 3 MPa, 390°C, 1h).

As was noted early, under the action of polyethylene the conversion of pine wood was increased by 1.4 times. After removing 11% wt. of lignin from pine wood by extraction with dimethyl sulfoxide this effect was less (1.2 times). In the runs with cellulose / polyethylene mixtures no positive influence of polyethylene on pine wood conversion was defected. The maximum increase of the conversion degree (1.5 times) was observed for coal / polyethylene mixture.

According to GC data the gaseous products of polyethylene catalytic hydropyrolysis contain only alkanes. In the runs with polyethylene / biomass mixture at similar conditions carbons oxides were detected in gaseous products and the lower yields of C_2 - C_4 alkanes was observed.

GC-MC data show that the light liquid fractions of plastic / biomass co - hydropyrolysis contain mainly normal paraffins C_7-C_{12}. Their content was 75% for pine wood / i-PP mixtures. Alkyl derivatives of benzene were also detected in the light liquid fractions. The content of unidentified substances was 15%, alkylbenzenes and alkylfuranes compounds - approximately 10% relative.

The difference in the composition of gaseous and liquid products of hydropyrolysis of plastic and biomass / plastic mixtures indicates the interaction between intermediate products of natural and synthetic polymers thermal conversion. The iron ore catalysts increase the yields of light liquid products and olefinic hydrocarbons. Producing of light liquid products is promoted by pressurised hydrogen which suppresses the char formation and facilitates the thermal cracking of heavy liquid products.

CONCLUSION

Obtained data show that, the mixtures of the different types of the natural and synthetic organic polymers can be successfully converted with a high yield to light distillate fraction by pyrolysis under inert atmosphere and catalytic hydropyrolysis in the autoclave conditions. The optimum temperature of biomass / plastic mixtures conversion which corresponds to the maximum yield of liquids is 390 – 400°C. In the co - liquefaction processes the interaction between products of natural and synthetic polymers thermal decomposition takes place.

Iron ore catalysts, modified by mechanical treatment were found to show a catalytic activity in the process of hydropyrolysis of biomas / plastic mixtures. Some synergistic effects were observed in this process resulting in the increase of conversion degree and yield of light liquid fraction and in the decrease of hydrocarbons content in gaseous products.

REFERENCES

1. Meier D., Rupp M. (1990) *Biomass pyrolysis liquids upgrading and utilisation.* (Ed.by A.V.Bridgwater and G.Grassi), pp. 155-76. Elsevier applied science, London.
2. Kuznetsov B.N. (1990) *Catalysis of chemical conversions of coal and biomass.* Novosibirsk, Nauka.
3. Gulyurtlu F.P., Gongalves M., Cabrita I. (1994) Co-pyrolysis of plastics with biomass. In: *Proc. 8th Europ. Conf. on Energy, Environment, Agriculture and Industry*, p. 318. Vienna.
4. Doroginskaya A.N., Sharypov V.I., Beregovtsova N.G., Parmon V.N., Kuznetsov B.N. (1997) Combined thermocatalytic conversion of plant biomass and polyethylene. In: *Proc. 3th Int. Symp. on Catalysis in Coal Conversion*, pp. 319-24. Novosibirsk, Russia (rus).
5. Pavlyukhin Y.T., Medikov Ya.Ya., Boldyrev V.V. (1984) On the consequences of mechanical activation of zink and nicel ferrites. *Solid State Chem*, **53**, 155 –60.
6. Popov B.I., Shkuratova L.N., Kuznetsov V.I., Pavlyukhin Y.T. (1998) Effect of mechanical treatment on some properties of Fe_2O_3-MoO_3 catalysts. *React. Kinet. and Catal. Lett.*, **25**, No 3-4, 255-9.
7. Sharypov V.I., Kuznetsov B.N., Beregovtsova N.G., Reshetnikov O.L., Baryshnikov S.V. (1996) Modification of iron ore catalysts for lignite hydrogenation and hydrocracking of coal-derived liquids. *Fuel*, **75**, No 1, 39-42.

Fate of Arsenic After Fast Pyrolysis of Chromium-Copper-Arsenate (CCA) Treated Wood

T. Hata[*], D. Meier[#], T. Kajimoto[+], H. Kikuchi[%], Y. Imamura[*]
[*]*Wood Research Institute, Kyoto University, Uji, Kyoto 611-0011, Japan*
[#]*Institute of Wood Chemistry and Chemical Technology of Wood, Federal Research Centre for Forestry and Forest Products, Leuschnerstr, 91, D-21031 Hamburg, Federal Republic of Germany*
[+]*Industrial Technology Center of Wakayama Prefecture, Ogura 60, Wakayama 649-62, Japan*
[%]*S S Alloy, Ltd., Techno Plaza 308, Kagamiyama 3-13-26, Higashi Hiroshima, Hiroshima, Japan*

***ABSTRACT*:** The purpose of this study is to find pyrolysis conditions for chromium-copper- arsenate (CCA) treated wood in order to maximize the retention of arsenic in wood charcoal along with a high oil yield. An experimental facility was built to examine the influence of process parameters such as pyrolysis temperature and total pyrolysis time. Milled powder was prepared from CCA treated wood. The powder was pyrolized on top of the fluidized bed at temperatures between 300 to 500 °C and during total pyrolysis times of 80 to 3600 s.

The concentration of arsenic in each product was measured by AAS and ICP and complete yield balance were established. The yield of gas at low temperature increases drastically at longer pyrolysis time. At higher temperatures more arsenic was volatized. When one type of chemical is used for catching arsenic compounds, there exist optimum pyrolysis conditions for the maximum removal of As.

The concentration of arsenic in wood charcoal increases linearly and reaches a plateau level at 400 °C. The concentration of arsenic in pyrolysis oil drastically increases with low temperatures and longer pyrolysis times. The optimum condition in this study for the pyrolysis temperature and total pyrolysis time is 450 °C and 80 s. There will be good possibility to apply fast pyrolysis techniques for the treatment of CCA treated wood because of the minimum secondary reaction and maximum mass reduction.

INTRODUCTION

Waterborne salts such as chromium-copper-arsenate (CCA) have been used to protect wood from attack by insects or fungi for some time. Substantial amounts of CCA remain in the wood for long periods, which make CCA the preferred agent for wood treatment. Most of the arsenic is present in the treated wood as $CrAsO_4$, which is precipitated on cellulose or complexed with lignin [1]. The disposal of scrap wood is a growing problem not only in Europe and America but in Japan as well. CCA-treated wood waste includes old telephone poles, wooden playground equipment, and timber from the landscape and basement of wooden houses. Most of this waste is disposed of at non-hazardous sites, burned in conventional incinerators, or recycled to wood-based materials.

Pyrolyzing the CCA-treated wood may be one solution to the problem. Pyrolysis results in three products, wood charcoal, pyrolysis oil and gas [2]. Since the first report into the volatilization of arsenic during the combustion of arsenic-treated wood in the 1950's, many studies have been carried out on the burning of contaminated wood [3]. The percentage of volatilized arsenic increases with temperature and oxygen partial pressure [4]. The losses of arsenic versus range of temperatures and times are 22 % at 400 °C but rose to 77 % at 800 to 1000 °C for 6 hours. Pyrolyzing CCA-treated wood at lower temperature without any oxidizing agent, resulted in a lower loss of arsenic, and thus is a promising approach for disposal. The percentages of volatilized Cu and Cr are thought to be low enough not to be an environmental problem [5]. Although the possible production of bio-oil from CCA treated wood was already mentioned [2], no study has so far focused on making bio-oil by pyrolysis of CCA-treated wood. Such a fuel could be used as an important source of energy.

The purpose of this study is to maximize the fraction of arsenic in wood charcoal while minimizing mass reduction of the bio-oil. Lab-scale pyrolysis was conducted in order to determine mass balances of yield and percentage of arsenic over the total system. The experimental set-up was built to examine the influence of process parameters such as pyrolysis temperature and total pyrolysis time. The optimum combination of temperature and total pyrolysis time, at which the amount of arsenic retained in the wood charcoal is maximized and that in oil is minimized, was tried to be found.

MATERIALS AND METHODS

DESCRIPTION OF THE SYSTEM

An experimental facility for the pyrolysis of CCA-treated wood powder was built. A general schematic diagram of the pyrolysis system with a fluidized bed was shown in Fig.1. The middle part of the glass reactor held the fluidized bed forming a cylindrical passage that kept both sand and feedstock and carried the volatile products to the exit. The glass reactor had a diameter of 30 mm and a distance of 100 mm separates the bottom of the fluidized bed and the exit. The major part of the reactor was placed inside the heating system. One end of the reactor lead to a silo through a screw feeder. The other end extended upward above the heating system and was connected with the inlet tube from a N_2 gas cylinder. The side arm of the glass reactor, in which a wire screen was inserted to filter dust, was connected to a cold trap and to a glass washing-bottle through a cotton filter in series. Inside the cooling trap was liquid CO_2 acetone solution. The washing bottle had 30 ml of acetone base tetra-butyl-ammonium hydroxide (TBAH) with a volume ratio of 2 to 1, which collects arsenic (II) oxides [4].

The reaction temperature was monitored with a thermocouple connected to a temperature controller that regulates the temperature of the heating system. The total pyrolysis time was defined here as the length of time. The glass reactor is inside the

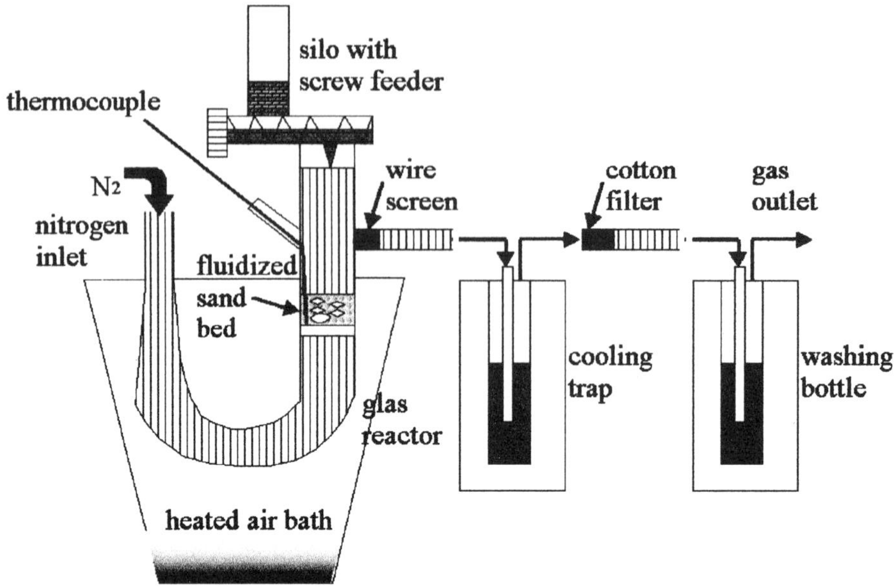

Fig. 1 Schematic diagram of fluidized bed reactor.

heating system before the target temperature was attained. The complete yield balance was determined with this set up. The calculation of a complete arsenic balance in the system is possible.

PREPARATION OF WOOD SAMPLES

The feedstock used in the experiment, supplied by Koshii Preserving Ltd., was western hemlock (*Tsuga heterophylla* Sarg, 100 x 100 mm cross section) treated with type III CCA salt. Type III was a CCA formulation containing 47$\pm$3 parts by weight of hexavalent chromium as CrO_3, 19$\pm$2 parts by weight divalent copper as CuO, and 34 $\pm$4 parts by weight of pentavalent arsenic as As_2O_5, of which the concentration was measured to be 7.73, 6.42, and 7.18 kg/m^3, respectively.

The CCA-treated pieces were broken down into chips and passed through a hammer mill. The resulting particles were screened through a 2 mm sieve and used for the pyrolysis experiment.

PYROLYSIS

One gram of the CCA treated wood was heated in N_2 atmosphere. Wood powder was fed with the screw feeder. The wood powder entered the glass reactor above the fluidized bed and fell on top of it. Total pyrolysis time is the period from the start of feeding up to moment when no further smoke was observed.

A constant rate of N_2 flow (2.25 l/min) was maintained throughout the pyrolysis reaction. N_2 passed through the accumulated wood charcoal residue and swept the volatiles to the reactor exit. The residence time of the vapors was calculated as 1.9 sec. The gas stream was passed through a wire screen in order to separate fine char particles to separate the solids and light compounds. The smoke caused by the pyrolysis can be characterized as a combination of oil vapor, micron-sized droplets and polar molecules bonded with water vapor. The gas went through a condenser to separate fine char particles with an acetone/dry ice mixture kept at –45 °C and a cotton filter to collect aerosols. The pyrolysis condensates were collected and the non-condensable gases went through a washing bottle filled with an acetone-based TBAH solution.

VARIABLES AND ANALYTICAL METHOD

Important variables in the pyrolysis of wood are temperature, heating rate and duration of pyrolysis. The reactor temperature was varied from 300 to 500 °C, and the duration of pyrolysis was 80, 200, 1800 and 3600 s. After the experiment, the glassware, fluidized bed including sand, wire screen and cotton filter were washed with 20 ml of distilled acetone.

The yields of char, oil, and gas were calculated from the mass differences of the glassware before and after the pyrolysis. The TBAH solution was analyzed for arsenic using an inductively coupled plasma (ICP) or atomic adsorption spectroscopy (AAS).

RESULTS AND DISCUSSION

MASS BALANCE AND YIELD OF PYROLYSIS PRODUCTS

Fig.2 shows the mass balance of pyrolysis products. The error bars show the standard deviation. The gas yield was determined by difference. Between 300 and 400 °C, the yield of char was drastically reduced in the pyrolysis of CCA-treated wood, whereas there was a reverse trend for pyrolysis oil. The increase in yield of pyrolysis oil at higher temperatures can be related to the higher heating rates in the wood particles.

More pyrolysis oil was produced from treated than untreated wood at 400 °C at 200 s. of pyrolysis time. CCA showed a significant influence on the thermal behavior of wood. The metals may have a catalytic effect on the thermal decomposition of the wood, resulting in more pyrolysis oil [6]. According to the thermal analysis study, thermal decomposition of hemicellulose is shifted to lower temperatures in the presence of CCA resulting in a higher release of volatile compounds [7].

Pyrolysis for 1800 and 3600 s at 350 °C resulted in less oil yield and almost the same amount of wood charcoal. At 450 °C, pyrolysis for 200 s produced more oil than at 80 s, but little difference in char yield. This is probably due to microstructural changes (small droplets) in charcoal occurring as the result of the thermal degradation process [5].

When the total pyrolysis time was less than 200 s, the yield of gas gradually increased with temperature, except for 450 °C. Arsenic had little influence on the yield. The gas yield at 350 °C and 1800 and 3600 s pyrolysis time increased significantly.

MASS BALANCE AND PERCENTAGE OF ARSENIC IN PYROLYSIS PRODUCTS

Fig.3 shows the mass balance and relative arsenic distribution in pyrolysis products.

The percentage was based on the initial amount of arsenic calculated for each specimen. Any reduction in the percentage of arsenic in wood charcoal was strongly dependent on the temperature. The percentage was increased in pyrolysis oil with increasing temperature. The percentage of As at 450 °C and 80 s. was less than at higher temperatures or longer pyrolysis times. The losses of arsenic increased with higher temperatures.

About 20 % of the arsenic escaped from the TBAH solution at 500 °C. Increasing the temperature may cause secondary reactions, which make it difficult to trap all the arsenic compounds using only the TBAH solution. Approximately 80 to 90 % of the initial arsenic in wood was evolved during pyrolysis. When using the TBAH solution, a reaction temperature of 450 °C or less is preferred to minimize losses of gas including arsenic compounds.

The arsenic content in the gas after passing cotton filter increased with temperature up to 400 °C before leveling off at 2 to 3 % at pyrolysis times of 80 or 200 s. The content of arsenic for 1800 and 3600 s was larger at 350 °C. A wider variation reflects some instability in the amount of volatile products. However, the As content in the gas is generally low. At 450 °C it is higher for 80 than for 200 s (approx. 0.05 %). The difference may be influenced by the gas yield, which is higher at 450 °C for 80 s.

The percentage in pyrolysis oil increased with temperature up to 400 °C, then leveled off at 200 s. The shorter pyrolysis gave a higher percentage at 450 °C by around 5 %. An increase in temperature produced no significant increase in the arsenic content. Pyrolysis for 3600 s gave higher arsenic content in the pyrolysis oil at 350 °C. A shorter pyrolysis time resulted in a lower concentration in the oil.

The relative portion of arsenic in wood charcoal was over 75 %, and independent of temperature up to 450 °C. It decreased markedly to 60 % at 500 °C. The total pyrolysis time caused no significant difference at temperatures between 350 and 450 °C. It is believed that arsenic is strongly bound to cellulose and lignin [1]. The decomposition of both chemical components occurs as temperature or the total pyrolysis time increase, resulting in more arsenic compounds being released. The combination of a temperature between 350 and 450 °C and total pyrolysis time of 200 s and lower is optimal for maintaining arsenic in the wood charcoal under these experimental conditions.

CONCENTRATION OF ARSENIC IN WOOD CHARCOAL AND PYROLYSIS OIL

The concentration of arsenic, which was calculated based on the initial weight of arsenic in wood charcoal, increased linearly up to 400 °C and then leveled off. A longer pyrolysis gave a higher concentration of arsenic at 350 °C. On the other hand, there was no difference at 450 °C between 80 and 200 s. The lower the arsenic content in the bio-oil, the lower the yield of bio-oil. These are thought to be the optimal conditions for the application of the bio-oil. This can be obtained between pyrolysis temperature of 450 and 500 °C (Fig.4).

Once the volatile products escaped from wood, the N_2 flow swept them from the hot zone to the exit in less than 1 s. In this study, 80 s was the minimum total pyrolysis time at 450 to 500 °C and 200 s at 500 °C. The combination of 80 or 200 s and 450 °C led to the desired result that is high retention of As in char along with a high oil yield. At longer pyrolysis times such as 1800 and 3600 s the products in the hot-zone undergo

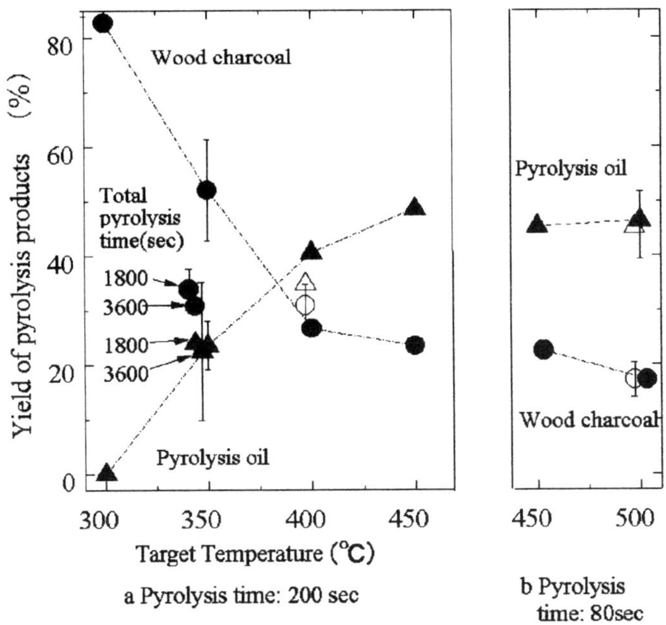

Fig.2. Yield of pyrolysis oil and wood charcoal as a function of target temperature.

Note: ●;wood charcoal from CCA-treated wood, ○;wood charcoal from original wood, ▲;pyrolysis oil from CCA-treated wood, △;pyrolysis oil from original wood

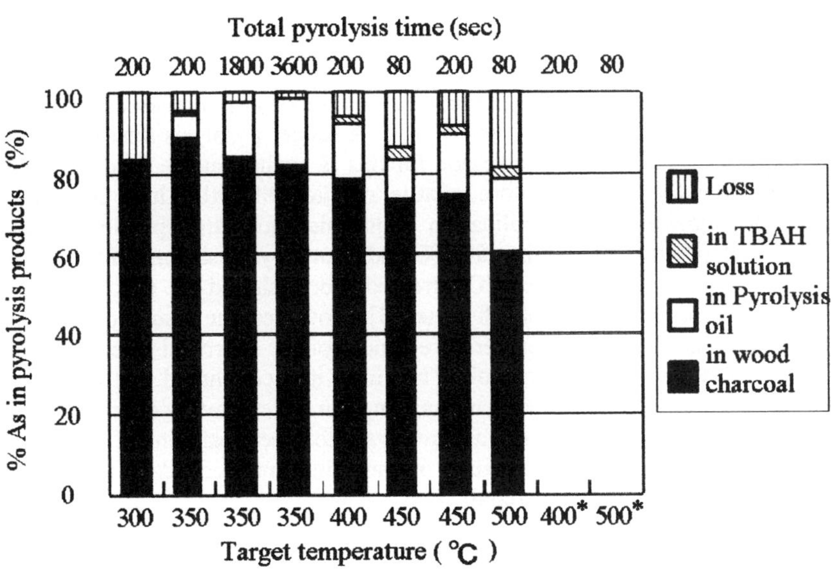

Fig. 3 Mass balance of As in pyrolysis products. (* : Original wood)

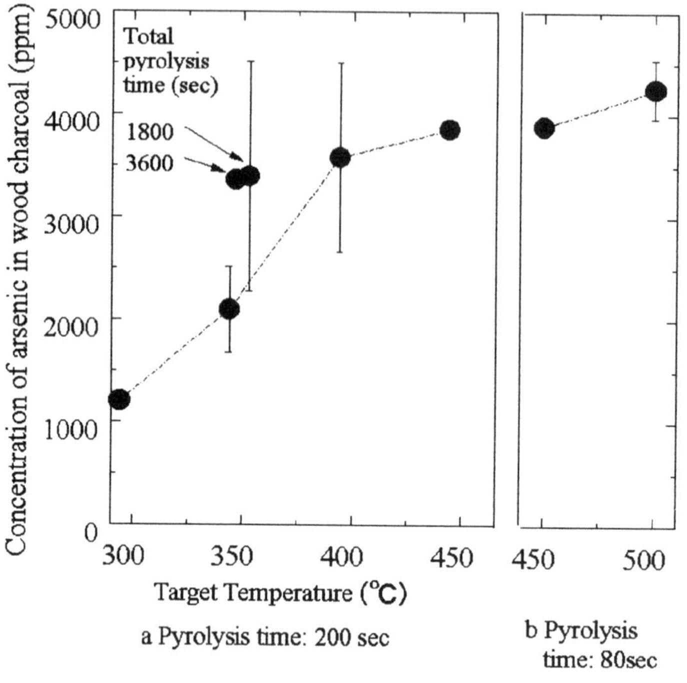

Fig.4. Relation between concentration of arsenic in wood charcoal and target temperature.
Note: ●;wood charcoal from CCA-treated wood

secondary reactions, leading to a higher concentration of arsenic in the wood charcoal.

Fig.5 shows the relation between the concentration of arsenic in pyrolysis oil and target temperature. The concentration of arsenic in pyrolysis oil increased up to 350 °C, levels off and increased again at 450 and 500 °C. A longer pyrolysis time gave a much higher concentration at 350 °C. The reason for this is again that the products undergo secondary reactions that generate more volatile arsenic. Oil with a high concentration of arsenic is not suitable for any application. From this figure, the optimum conditions appear to be a temperature of less than 500 °C and a total pyrolysis time of 80 to 200 s.

Based on data from literature, CCA-treated wood should be pyrolyzed at low temperature from the point of safety disposal [8]. However, the results of the present study indicate that a lower temperature and longer pyrolysis led to higher concentrations of arsenic in pyrolysis oil. Therefore, it is concluded that the optimum conditions for pyrolysis with respect to minimizing the concentration of arsenic in oil is 450 °C and 80 s. The application of fast pyrolysis for the treatment of CCA-treated wood seems to be advantageous because of the minimum secondary reactions and maximum retention of arsenic in charcoal.

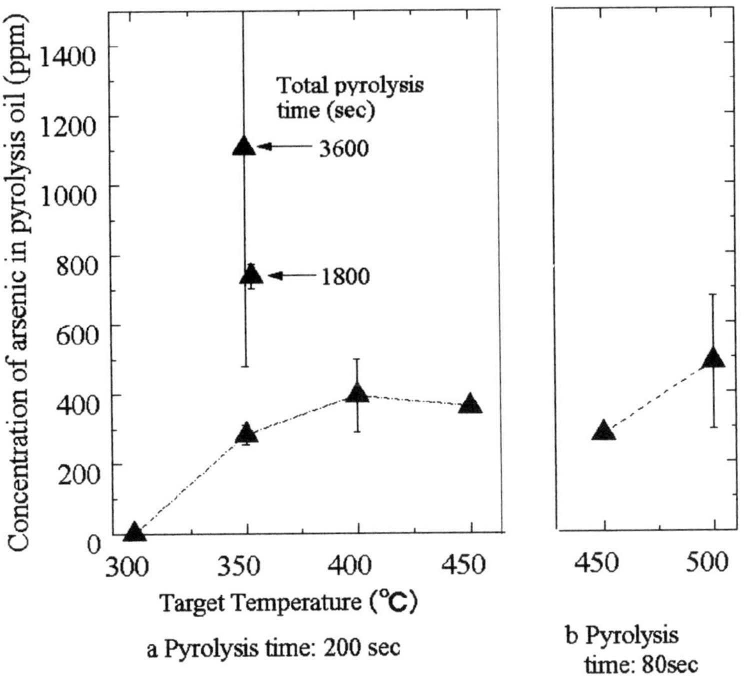

Fig. 5. Relation between concentration of arsenic in pyrolysis oil and target temperature.
Note: ▲;pyrolysis oil from CCA-treated wood

CONCLUSIONS

The pyrolysis system at laboratory scale described in this work proved to be suitable for producing bio-oil with a low (>0.04 %) arsenic content from CCA-treated wood samples. The complete yield balance was determined using the experimental facility with a fluidized bed reactor

In principle, treated wood behaves similar to the untreated with respect to the yields of pyrolysis products. However, more pyrolysis oil was produced from treated than from untreated wood at 400 °C potentially caused by the catalytic effect of CCA. The gas yield at low temperature is increased markedly by prolonging the pyrolysis time. A lower content of arsenic in oil was obtained at higher temperature. Although TBAH was used to trap arsenic compounds from the gases losses of almost 20 % were measured.

The concentration of arsenic in charcoal increased linearly with temperature before leveling off at 400 °C. The concentration of arsenic in pyrolysis oil increased markedly at low temperature and at long pyrolysis times. It is therefore concluded that the optimal temperature and duration of pyrolysis is 450 °C and 80 s, respectively. The fast pyrolysis technique has potential for the treatment of CCA-impregnated wood because of minimized secondary reactions and maximum arsenic retention are expected.

ACKNOWLEDGEMENTS

This research was supported by a Grant-in-Aid for Scientific Research (No. 10876038) from the Ministry of Education, Science, and Culture of Japan.

REFERENCES

1. Pizzi A.(1982) The chemistry and kinetic behavior of Cu-Cr-As/B wood preservatives. IV fixation of CCA to wood. *J. Polym. Sci.*, **20** 739-764.
2. Broek K.V.D'., Helsen L., Vandecasteele C.& Bulck E.V.D. (1997) Determination and characterization of copper, chromium and arsenic in chromated copper arsenate (CCA) treated wood and its pyrolysis residues by inductively coupled plasma mass spectrometry. *Analyst,* **122**, 695-700.
3. Dobbs A.J., Phil D., Grant C. (1978) The volatilization of arsenic on burning copper-chrome-arsenic (CCA) treated wood. *Holzforschung* **32**, 32-35.
4. Helsen L.& Bulck E.V.D. (1997) Release of metals during the pyrolysis of preservative impregnated wood. In: *Developments of Thermochemical Biomass Conversion*, (Eds. A.V. Bridgewater and D.G.B. Boocock) Vol. 1, pp. 220-228. Blackie Academic & Professional.
5. Helsen L. & Bulck E.V.D. (1998) The microdistribution of copper, chromium and arsenic in CCA treated wood and its pyrolysis residue using energy dispersive X-ray analysis in scanning electron microscopy. *Holzforschung* **52** 607-614.
6. Wehle S., Meier D., Moltran J. & Faix O. (1997) The impact of wood preservatives on the flash pyrolysis of biomass. In: *Developments of Thermochemical Biomass Conversion*, (Eds. A.V. Bridgewater and D.G.B. Boocock) Vol. 1, pp. 206-219. Blackie Academic & Professional.
7. Wilkins E. & Marray F. (1980) Toxicity of emissions from combustion and pyrolysis of wood. *Wood Sci. and Technol.*, **14**, 281-288.
8. Hirata T., Inoue M. & Fukui Y. (1993) Pyrolysis and combustion toxicity of wood treated with CCA. *Wood Sci. and Technol.*, **27**, 35-47.

Fast Pyrolysis of Impregnated Waste Wood – The Fate of Hazardous Components

D. Meier, T. Ollesch, O. Faix
Federal Research Centre for Forestry and Forest Products, Institute for Wood Chemistry and Chemical Technology of Wood, D-21027, Hamburg, Germany

ABSTRACT: The behaviour of contaminated wood wastes during fast pyrolysis in fluidized bed reactors has been investigated. The waste samples comprised painted wood, demolition wood, cable drums, fences, window frames, railway sleepers, and wood impregnated with chlorinated organic compounds. The study was aimed at establishing mass balances for inorganic and organic contaminants.

A continuously operating small pilot plant with a capacity of 5 kg/h and a bubbling fluidized bed reactor was used. All experiments were performed at the same pyrolysis temperature of 475 °C. Wood samples were impregnated with known amounts of both inorganic preservatives containing chromium, copper, boron salts (CCB), complexes of Cu-HDO and Al-HDO, and 10 typical organic compounds such as lindane, pentachlorophenol, dichlofluanide, toluylfluanide, DDT, propioconazole, tebuconazole, permethrine, cypermethrine, and deltamethrine. Additionally, real wood samples from a commercial waste collection site were pyrolyzed.

The heavy metals are collected with the char and the liquid bio-oil contains only traces of metals. The real samples show the same behaviour as those prepared in the laboratory. Fast pyrolysis of wood impregnated with Al and Cu complexes gave similar results: aluminum and copper are predominantly found in the char fraction. The oil contains only traces of these metals. The stability of the organic compounds during pyrolysis was different. DDT was degraded completely, the most stable compounds were propioconazole and tebuconazole. Highly substituted PCDD/F were degraded to less substituted, but more toxic PCDD/F. Almost 99 % of PCDD/F can be found in the oil.

INTRODUCTION

In Germany there are increasing disposal problems in managing the amount of 8-10 million tons of waste wood which are accumulated each year. More restrictive environmental laws will not allow landfill in the near term. Therefore, the application of new technologies is under investigation. On the other hand, there already exists an increasing market for waste wood. Actual average prices for different assortments are listed in Table 1 (1). As can be seen, prices are slightly decreasing. But still, the price credits might have positive effects on the economics of thermal processing.

Table 1 Average prices in Germany at gate of the wood collector (EURO/t).

		April 1999	October 1999	January 2000	May 2000
H1 (untreated)	unshredded	0 - 25	0 - 25	0-30	0-30
	shredded	10 - 30	10 - 30	20-40	15-45
H2 (treated)	not chopped	10 - 50	20 - 50	10 - 45	10 - 45
	chopped	0 - -35	10 - 35	0 - 30	0 - 30
H3 (contaminated)	not chopped	100 -250	100 - 250	80 - 160	60 - 140

Table 2 shows the materials streams for German waste wood. Most of the input material comes from demolition and furniture. About one third is dumped on landfills and more than a quarter exported.

Table 2 Materials streams for German waste wood.

INPUT (%)		OUTPUT (%)	
demolition	41	landfill	31
furniture	35	export (to Sweden for energy, to Italy for particle boards)	27
packaging	12	incineration & combustion	22
sleepers & poles	12	particle boards	20

As German regulations favor materials recycling over simple heat production, fast pyrolysis, which gives a liquid (bio oil) as main product, is getting more and more attention, because bio oil can also be used as a source for chemicals (2). On the other hand, fast pyrolysis has to compete with existing alternatives such as combustion and gasification. Therefore, the quality of the oil from pyrolysis of contaminated waste wood is of paramount importance, if this process is to be considered an attractive alternative approach for costumers such as disposal site operators or waste wood collection sites.

Hence, this study aimed at pyrolyzing various waste wood samples with different contaminants, both organic and inorganic components. The fate of the contaminants is presented through mass balances.

EXPERIMENTAL

MATERIALS AND IMPREGNATION

Beech wood (*Fagus sylvatica L.)* particles of 1.5-3 mm and a moisture content of 10 % were generally used for the experiments. Impregnation was performed by firstly soaking biomass with the dissolved compounds and secondly applying vacuum for penetration of the compounds into the wood matrix.

Where metal-organic complexes were used for impregnation, 10 g of each compound (see Table 3) were used to impregnate 3 kg of wood.

Table 3 Metal-organic complexes used for wood impregnation.

copper HDO	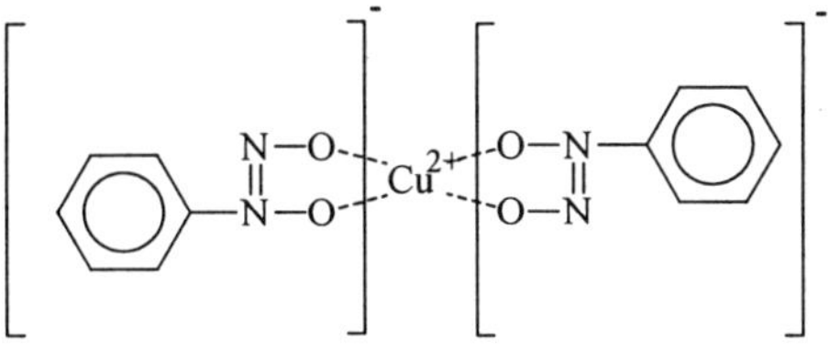
aluminum-HDO	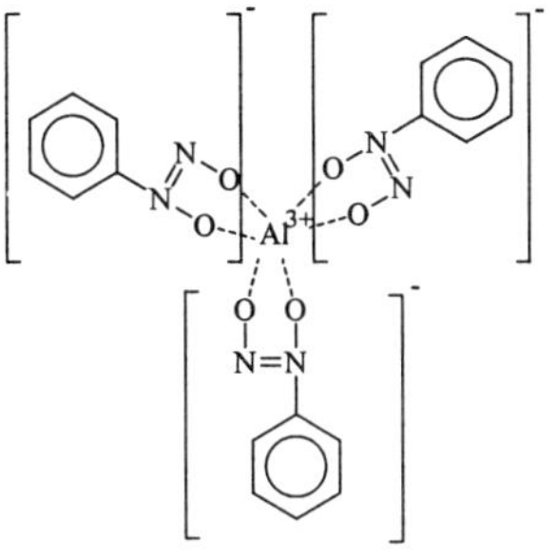

Three kilograms of biomass were impregnated with 3 g of the corresponding single organic contaminant. The list of contaminants tested appears in Table 4.

Table 4 List of organic compounds used for wood impregnation.

No.	Chemical Name	Chemical Formula
1	Hexachloro-cyclohexane (lindane)	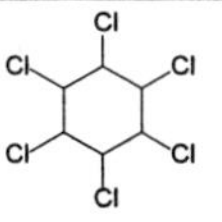
2	Pentachlorophenol (PCP)	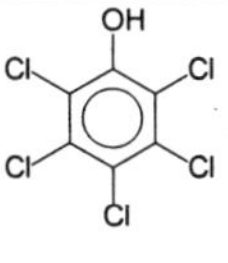
3	Dichlofluanide (DCFN)	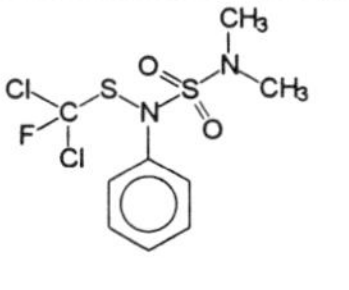
4	Toluylfluanide (TFN)	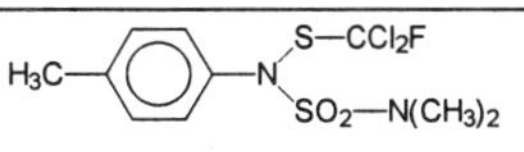

5	(1,1,1-Trichloro-2,2-bis-(p-chlorophenyl)-ethane (DDT)	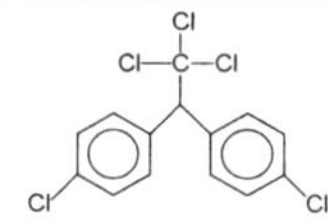
6	Propioconazole	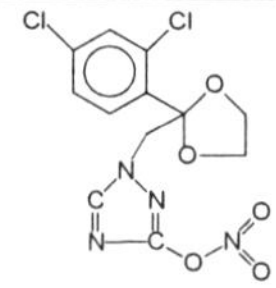
7	Tebuconazole	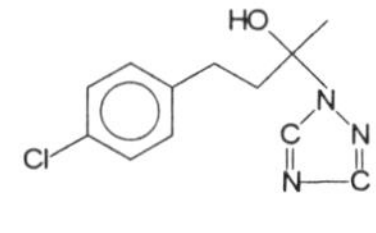
8	Permethrine	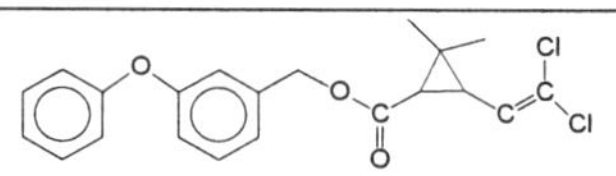
9	Cypermethrine	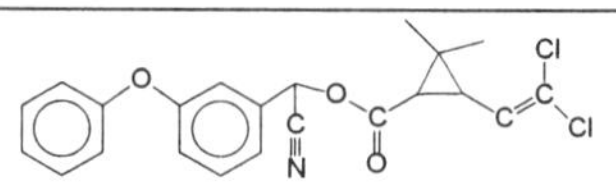
10	Deltamethrine	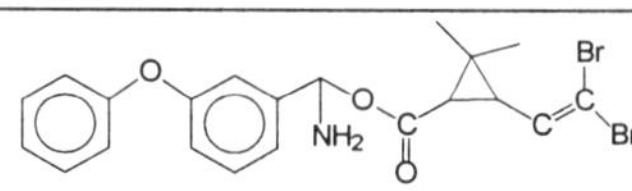

Biomass was also impregnated with an inorganic salt solution (CCB, 20 % concentration) consisting of 38 % $K_2Cr_2O_7*2H_2O$, 34% $CuSO_4*5H_2O$, 25 % HBO_3 and 3 % $NaHSO_4$. 300 ml of the solution were used to impregnate 3 kg of wood.

Furthermore, representative samples from a waste wood collection site (Otto Dörner, Hamburg) were subjected to fast pyrolysis and analysis. Two classes, H2 and H3, were selected. H2 contains typically particle boards and window frames, H3 consists of railway sleepers, fences, cable drums, and other materials impregnated with organic chemicals.

ANALYTICAL METHODS

Organic contaminants

Organic contaminants were analyzed by gas chromatography using a CHROMPACK CP 9000 model and a DB-1 column (30 m x 0.25 mm, 0.32 µm film thickness) with the following temperature programme: initial: 120 °C, heating rate: 5 °C, final 320 °C, hold for 10 minutes. For calibration solutions with two concentration levels (10 and 100 µg/ml), following DIN 38402, were used. Fluoroanthene was taken as internal standard.

In order to reduce the influence of the bio oil on the results, a simple clean-up method was applied, using one-way 500 mg silica cartridges. They were conditioned with 2 ml of ethyl acetate. Then 200 µg of whole bio oil was placed on top of the silica and all the relevant toxic compounds eluted with 2 ml of ethyl acetate. To this solution the internal standard was added and the mixture subjected to GC.

Metallic contaminants

Metals were analyzed using dispersive X-ray fluorescence spectroscopy (XRF) on a SPECTRACE 6000. 2 g of sample were used. For metals in char multi-level calibrations within a range of 10-20,000 ppm were carried out. For the liquid samples the range was set between 1 and 10,000 ppm.

Dioxines (PCDD/PCDF)

Analysis of PCDD/PCDF was carried out at a certified laboratory (ERGO, Hamburg) using high resolution GC/MS methods. Identification and quantification with a dissolution method was performed following VDI procedure 3499. Samples of bio oil and gas prior to combustion were sent to the laboratory. Gaseous contaminants were collected on a special polyurethane cartridge.

PYROLYSIS PROCEDURE

Fast pyrolysis of waste wood has been studied in a small pilot plant with a nominal capacity of 5 kg/h. A schematic diagram is presented in Figure 3.

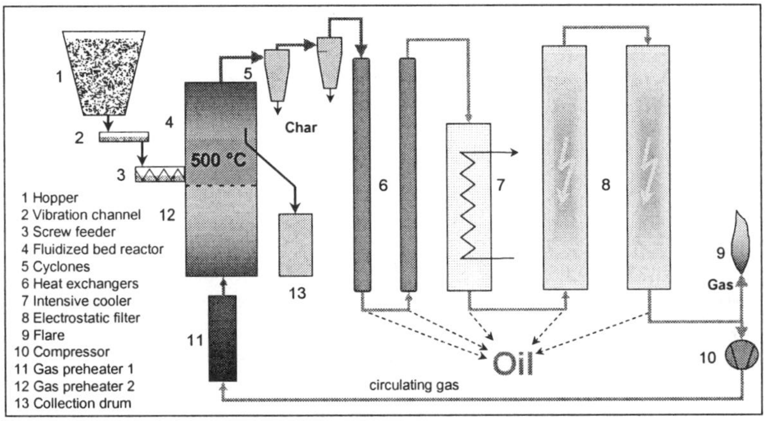

Fig. 1 Schematic diagram of pilot plant for fast pyrolysis of waste wood.

Before starting the experiment the plant was checked for leaks. A pressure of 150 mbar nitrogen was applied and the losses measured over a time of 30 min. The plant was declared sealed when the losses were <20 mbar. After the leak test the plant was flushed with 1.5 m^3 of nitrogen and the system heated to the desired temperature within approximately 2 hours. The on-line micro-GC, attached to the system for gas analysis, was calibrated with standards. Wood feeding was from the hopper via a vibration channel to a water-cooled screw feeder running at constant velocity. Throughput was controlled by the vibration frequency. Wood particles entered the reactor just above the distribution plate of the fluidized bed reactor. A weir inside the reactor was used to remove larger charcoal particles. Because of their lower density they sit on top of the sand bed and fall through the pipe into a collection drum. Vapors and smaller char particles were entrained with the fluidizing gas. Two cyclones separated the fine char particles and the residual vapors were condensed in a condensation train consisting of two tubular water-cooled heat exchangers (operated at 200 and 80° °C, respectively), an intensive cooler operated at –15 °C, and two electrostatic precipitators working at approximately –15 kV.

The non condensable gases were circulated as fluidizing gas. Surplus gas was burnt in a flare.

RESULTS AND DISCUSSION

No significant differences in the yield of the main products liquid, gas, and char from untreated and treated wood were observed (see Fig. 2).

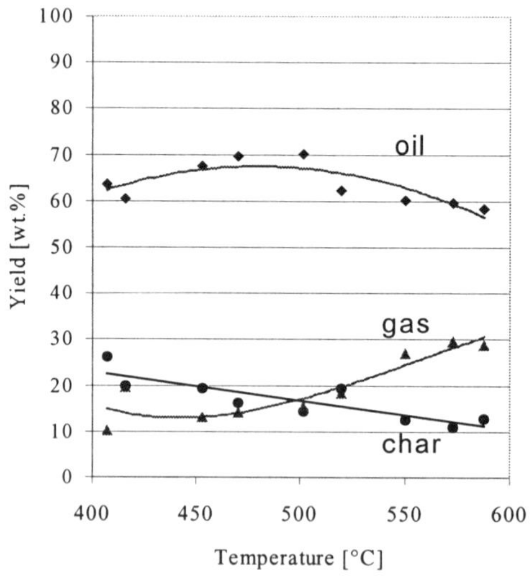

Fig. 2 Typical yields of fast pyrolysis experiments with the pilot plant.

Oil yields reached almost 70 % at 475 °C reactor temperature. With increasing temperature the gas yields increase and the char yields decrease correspondingly.

CCB impregnated wood was pyrolyzed at 500 and 600 °C and the heavy metals copper and chromium were determined in the char and liquid by XRF (see Table 5). The data clearly show that typical heavy metals from wood preservatives are trapped together with the char fraction. If the char is burnt for energy recovery, the heavy metals would remain in the ash. From here they might be recovered or dumped. The oil is practically uncontaminated. These results are consistent with those found already in a laboratory fluidized bed reactor (3).

Table 5 XRF analysis of Cu and Cr in pyrolysis oil and char.

contaminant	500 °C CCB alone		600 °C CCB with organic preservative	
	char	oil	char	oil
copper (ppm)	22560	12	8257	16
chromium (ppm)	31561	4	19768	3

XRF analysis was also performed on pyrolysis oil and char from fast pyrolysis of real wood waste obtained from a commercial collection and separation site. The XRF data show similar results compared to the samples impregnated in the laboratory. Various mono-fractions from the collection site were pyrolyzed such as cable drums, fences, railway sleepers, particle boards and window frames. The results are presented in Table 6 and are based on the whole oil whereas the data presented in (4) only refer to the cleaning oil fraction. Nothing could be detected in the oils collected from the condensation train.

Table 6 XRF analysis (in ppm) of Cu and Cr from authentic waste wood samples

	cable drums		fences		railway sleepers		particle boards		window frames	
	char	oil	char	oil	char	oil	char	oil	char	oil
Cu	6921	0.42	1628	0.48	932	0,49	270	1.64	681	0.74
Cr	14103	1.12	4748	0.14	267	0.11	527	0.36	465	0.21

The metal-organic preservatives copper HDO and aluminum HDO are commonly used for timber used in landscape improvement. Therefore, samples impregnated with these preservatives were also pyrolyzed at 500 °C. Aluminum could not be measured in the oil because the chamber had to be evacuated which would have contaminated the inner walls with pyrolysis liquids. Therefore, only the char was analyzed for Al resulting in a recovery of 82.7 %. Hence, 17.3 % of the input aluminum is expected to remain in the oil. This is much more compared to Cu and Cr due to the lower melting and boiling point of Al.

In the past, various chloro-organic compounds were used in preservative formulations in Germany. They are highly toxic not only towards bacteria, fungi and insects but also for humans. Therefore, it was of special interest to study their fate with fast pyrolysis. Each beech wood sample was impregnated with one compound from Table 4. In order to obtain measurable amounts after pyrolysis it was decided to use a high concentration of 1 g/kg which was thought to represent the worst case. In practice this concentration is unrealistic as no discrimination between the different organic preservatives can be made at the wood collection site.

Table 7 Recovery rate (%) of chloro-organic compounds based on initial concentration

	Temperature °C			
	450	500	550	580
Lindane	33.06	29.84	16.33	2.02
PCP	42.18	25.13	25.03	13.04
DCFN	10.65	7.45	3.13	2.40
Toluylfluanide	18.72	17.00	13.10	7.92
Propioconazole	40.47	26.92	23.80	16.66
DDT	0	0	0	0
Tebuconazole	40.16	28.11	26.19	15.93
Permethrine	32.51	22.76	13.51	6.83
Cypermethrine	4.26	0.68	0.47	0.32
Deltamethrine	3.00	2.10	1.61	0

Pyrolysis experiments were conducted at different temperatures: 450, 500, 550, and 580 °C. Quantitative analyses of each compound were performed by GC using the internal standard method. A special clean-up procedure with silica cartridges was used to enrich the compounds in the sample solution and to reduce matrix effects from the pyrolysis liquid.

Generally speaking, the concentrations of the contaminants decrease with higher pyrolysis temperatures. PCP, HCH, propioconazole, tebuconazole, and permethrine are destroyed by 60-70 % already at the lowest pyrolysis temperature of 450 °C. At 580 °C the reduction reaches 80-95 °% based on the initial concentration (see Fig. 3).

The pyrethroids deltamethrine and cypermethrine as well as toluylfluanide and dichlofluanide are even further destroyed. At 450 °C they lost 80-98 % of their initial concentration and at 580 °C destruction rate is between 90 and 99 % (see Fig. 4). DDT is already completely destroyed at 450 °C.

In general, the presence of chloro-organic preservatives in thermal processing makes the formation of polychlorinated dibenzodioxins and polychlorinated dibenzofurans (PCDD/F) highly probable. Their formation rate is highest at around 300 °C 5. In contrast to polyaromatic hydrocarbons (PAH), PCDD/F are formed especially at temperatures below 600 °C (deNovo synthesis). A prerequisite for reaction is the presence of char particulates, chlorine particulates and gaseous oxygen (6). As fast pyrolysis occurs in an inert atmosphere the formation of PCDD/F might be reduced. To test this thesis, two experiments were carried out: one blank run with beech wood and another with beech wood spiked with

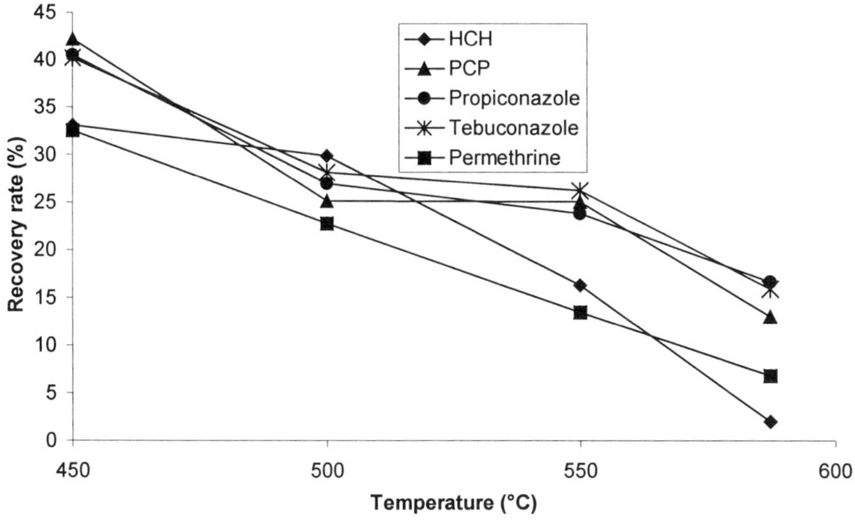

Fig. 3 Decrease of recovery rate with increasing temperature of selected chloro-organic compounds

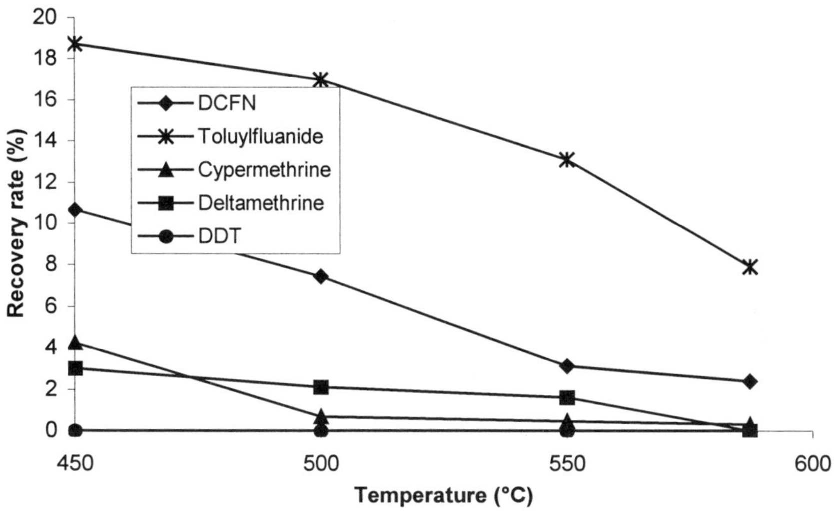

Fig. 4 Decrease of recovery rate with increasing temperature of selected chloro-organic compounds

CCB salt and technical grade PCP containing high concentrations of substituted hepta- and octa-PCDD/F (see Table 8). The highly substituted compounds are degraded in favor of less substituted, but more toxic compounds. Therefore, the toxicity of product bio-oil increased by a factor of 2.57 as demonstrated by the increase of the International Toxicity Equivalent (I-TE).

However, it is not clear whether a de novo synthesis took place or just recombination reactions. From the total PCDD/F content the oil contains 98.65 %, the char has 1.34 and the gas has only 0.01 %. If the oil is used as fuel one has to carefully select the combustion conditions in order to destroy the PCDD/F completely.

The results presented here show the general trend if chloro-organic compounds are present in the feedstock. The amount of chloro-organics in the feedstock used here were somewhat unrealistic but selected to better understand and measure PCDD/F formation and concentration in the products of fast pyrolysis. In reality, there are no mono-fractions of wood contaminated only with chloro-organic compounds. Generally, pyrolysis of waste wood mixtures results in a much less contaminated oil as presented in this study. However, further investigations are necessary to prove if de novo synthesis takes place during fast pyrolysis of contaminated waste wood.

CONCLUSIONS

The results of this study show that fast pyrolysis of contaminated waste wood is technically feasible without any restrictions. Heavy metal compounds introduced by wood preservatives are almost completely mechanically separated by cyclones together with the charcoal.

The destruction rate of chlorinated organic compounds depends on the temperature, the higher it is the higher the extent of degradation. At 580 °C the recovery was between 0 and 16.7 %. Polychlorinated dioxins or furans present in preservatives are partly destroyed. However, more toxic compounds with a lower degree of chlorine substitution are formed. Thus, high temperature combustion of these liquids is necessary to meet the German limit of 0.1 ng/m^3. It should be noted that for the sake of analysis worst case samples were pyrolyzed in order to get measurable amounts of substances. In reality much less contamination is expected. No organic preservatives (except from railway sleepers) could be detected in the oil from fast pyrolysis of realistic waste wood mixtures.

ACKNOWLEDGEMENTS

We acknowledge the financial support of Deutsche Bundesstiftung Umwelt, Osnabrück, project no. 03 631

Table 8 Concentration of PCDD/F in feedstock sample and pyrolysis products

	reference		spiked sample		
	bio oil (ppb)	***preservative*** (ppb)	**bio oil** (ppb)	**char** (ppb)	**gas** ng/m^3
PCDD					
Tetra	n.d.	n.d.	51.0	0.45	0.0043
Penta	n.d.	n.d.	272.4	1.50	0.0237
Hexa-1	n.d.	n.d.	166.6	0.80	0.0212
Hexa-2	0.0195	55	375.6	2.23	0.0363
Hexa-3	n.d.	5	285.3	2.63	0.0257
Hepta	1.896	4194	2163.0	26.83	0.1870
Octa	32.12	47007	4719.8	76.24	0.3190
total PCDD	**34.04**	**51261**	**8033.7**	**110.68**	**0.6173**
PCDF					
Tetra	0.0019	n.d.	3.3	0.04	0.0004
Penta-1	n.d.	n.d.	6.1	0.06	0.0005
Penta-2	0.0013	n.d.	4.8	0.04	0.0003
Hexa-1	0.0111	29	14.9	0.12	0.0012
Hexa-2	0.0035	n.d.	16.5	0.16	0.0010
Hexa-3	n.d.	n.d.	1.30	0.02	0.0002
Hexa-4	n.d.	n.d.	4.40	0.05	0.0003
Hepta-1	0.2748	709	33.2	0.32	0.0026
Hepta-2	0.0288	158	11.5	0.08	0.0017
Octa	6.1900	11598	99.6	0.91	0.0010
total PCDF	**6.51**	**12494**	**195.6**	**1.8**	**0.01**
total PCDD/PCDF	**40**	**63755**	**8229**	**112**	**0.63**
I-TE	**0.064**	**118**	**303**	**2**	**27**

n.d. = not detected

REFERENCES

1. Speckels, L., BFH, Institute for Wood Physics, personal communication (2000)
2. Bridgwater, AV, Peacocke, GVC. (2000) Fast pyrolysis processes for biomass. *Renewable and Sustainable Energy Reviews*, **4**, 1-73.
3. Wehlte S., Meier D., & Faix O. (1997) The impact of wood preservatives on the flash pyrolysis of biomass. In: *Developments in Thermochemical Biomass Conversion,* (Ed by A.V. Bridgwater & D.G.B. Boocock), pp.206-219. Blackie Academic & Professional, London.
4. Bridgwater AV, Meier D, Radlein D. (1999) An overview of fast pyrolysis of biomass. *Org. Chem.*, **30**, 1479-1493.

5 Kolenda J., Gass H., Wilken M., Jager J., Zeschmar-Lahl B.(1994) Determination and reduction of PCDD/PCDF emissions from wood burning facilities, *Chemosphere* **29**, 1927 –1938.

6 Mohr K, Wilken M, Nonn Ch, Jäger J. (1995) Behaviour of PCDD under pyrolysis conditions, *Organohalogen Componds*, **23**, 293-297.

Low-temperature pyrolysis as a possible technique for the disposal of CCA treated wood waste: metal behaviour

L. Helsen and E. Van den Bulck
Katholieke Universiteit Leuven, Department of Mechanical Engineering, Celestijnenlaan 300 A, B-3001 Heverlee, Belgium

***ABSTRACT*:** Low-temperature pyrolysis is evaluated as a possible technique for the disposal of CCA treated wood waste. A theoretical and experimental study of the low-temperature pyrolysis of CCA treated wood waste is performed in order to gain more insight in: (1) the metal (Cr, Cu, As) behaviour during the pyrolysis process and (2) the influence of CCA on the pyrolysis process. The experimental study focuses on the determination and characterisation of Cr, Cu and As in CCA treated wood and its pyrolysis residue and the study of the pyrolysis process. Based on the experimental observations some important conclusions are drawn with respect to the metal behaviour during the pyrolysis process. Furthermore, kinetic models are derived for the low-temperature pyrolysis of CCA treated wood and the As release during the process. CCA is identified as a promoter of charcoal production and pyrolysis temperature and residence time are the most important parameters with respect to metal release during the pyrolysis of CCA treated wood waste. At the low temperatures considered, Cu and Cr do not pose any problems, As on the other hand is already released at temperatures between 300 °C and 400 °C. The mechanism responsible for As release is identified as the reduction of As(V) to As(III), which occurs sharply at 327 °C. Arsenic is probably released as As_4O_6, which is very difficult to capture and toxic. Due to the high leachability of As and Cu in the pyrolysis residue, this residue can not be landfilled without a pre-treatment. The presence of metal agglomerates in the pyrolysis residue, however, may allow a dry separation of the metals and the charcoal.

INTRODUCTION

Waterborne salts have been used to preserve wood from insects, fungi and water damage for many years. One of the more common formulations contains copper, chromium and arsenic salts and is known as chromated copper arsenate or CCA. After impregnation of the wood with a CCA solution the metal compounds will be fixed to the cell walls of the wood matrix. Substantial amounts of CCA remain in the wood for many years and the disposal of scrap wood is a growing problem in Europe and in the United States. Telephone poles, railway sleepers, timber from landscape and cooling towers, wooden silos, hop-poles, cable drums and wooden playground equipment generate wood waste for which environmental benign disposal technologies need to be developed.

Pyrolysing the CCA treated wood at low temperature is a promising solution to the growing disposal problem since low temperatures and no oxidising agents are used. In turn, this leads to a smaller loss of metals than in combustion or even a total recuperation of the metals. However, the current literature clearly shows that the mechanism of metal volatilisation during the thermal decomposition is not yet completely understood. Fast pyrolysis of impregnated waste wood is studied by Meier [1, 2]. In the present study, a low-temperature pyrolysis facility for CCA treated wood waste is optimised with respect to a minimal metal release upon a maximal mass reduction. During this study several experiments have been carried out with the aim to gain more insight in the behaviour of the metals (Cr, Cu, As) during the pyrolysis process. Among these experiments are:

(1) Determination of metal mass balances over the total system using inductively coupled plasma mass spectrometry (ICP-MS).
(2) Sequential extraction of the pyrolysis residue to study the leaching behaviour of the residual solid matrix.
(3) Speciation study of As in the pyrolysis residue using hydride generation coupled to ICP-MS (HG-ICP-MS).
(4) Study of the metal microdistribution in CCA treated wood and its pyrolysis residue by energy dispersive X-ray analysis coupled to scanning electron microscopy (SEM-EDXA).
(5) Labscale pyrolysis study: examination of the influence of temperature and residence time on the release of metals and resulting mass reduction.
(6) Thermogravimetric analysis (TGA) of CCA treated wood and the major arsenic compound in CCA treated wood (chromium arsenate ($CrAsO_4$)).

In this paper, the most important experimental observations with respect to metal behaviour during pyrolysis of CCA treated wood are outlined and discussed. Based on the experimental observations and the new insights resulting therefrom, a reaction scheme for the pyrolysis of CCA treated wood and a kinetic model for the As release during pyrolysis have been developed. A critical evaluation of these kinetic models results in a hypothesis concerning the behaviour of metals, especially As, during pyrolysis of CCA treated wood.

EXPERIMENTAL STUDIES

DETERMINATION OF METALS THROUGH ANALYTICAL TECHNIQUES

Fixation of CCA to wood

The mechanism of interaction of the CCA preservative with the wood structure was initially elucidated by Dahlgren [3, 4] and Dahlgren and Hartford [5]. Subsequently, this scheme was modified by Pizzi [6, 7, 8, 9], who studied the kinetics of reactions between CCA or CCA component elements and cellulose or lignin model compounds. The kinetic data combined with results from analysis of reaction precipitates were used to determine the distribution of preservative elements between the major compounds in the wood structure. A summary of the products formed when CCA reacts with wood is presented in Figure 1.

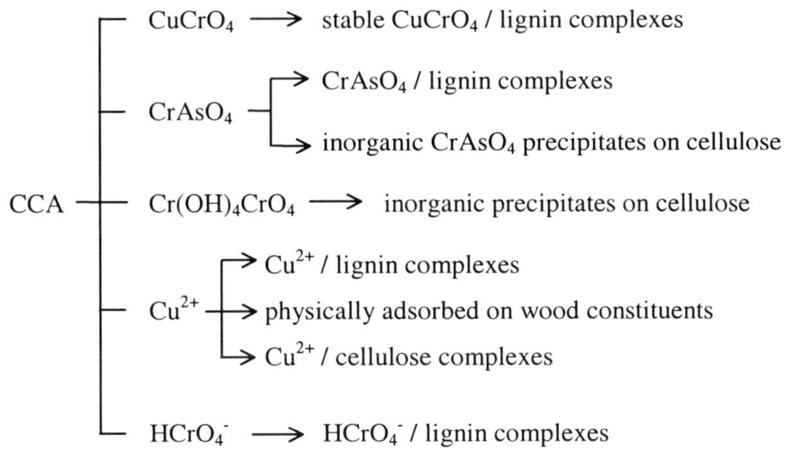

Fig. 1 Scheme of the interaction of CCA with wood.

As can be seen, the CCA interacts mainly with the lignin and cellulose components of the wood structure, resulting in complexation, precipitation or adsorption. The fundamental reaction during the fixation process is the oxidation of wood components by Cr(VI) resulting in the reduction of Cr(VI) to Cr(III). This can be described as an interaction of CCA with the functional groups (carbonyl, carboxyl, methoxy and phenolhydroxyl groups) of the wood components [10], during which Cr is partly reduced, As and Cu are being fixed and some wood components are oxidised. The products formed when CCA reacts with wood are composed mainly of the $CuCrO_4$-lignin complexes, $CrAsO_4$-lignin complexes and $CrAsO_4$ precipitates on cellulose. These compounds make up 95 % of the system CCA-wood (excluding Cu^{2+} unreacted with chromium). Copper arsenates are not involved as final fixation products unless a preservative of high As content is used. The actual proportions of the various chemical species depend on the relative proportions of Cu, Cr and As in the preservative formulation.

Determination of the metal content

To determine the total amount of Cr, Cu and As in CCA treated wood and its pyrolysis residue by inductively coupled plasma mass spectrometry (ICP-MS), the metals have to be brought in solution. Several leaching and dissolution methods were tested and compared to find the optimal procedure [11, 12]. To determine the total amount of Cr, Cu and As in the dried CCA treated wood, a leaching procedure using H_2O_2 and H_2SO_4 (BSI method) seems to be the most suitable method: reproducible, easy to handle and it guarantees that all metals are brought into solution. Since Cr is not completely dissolved when applying the BSI method to the pyrolysis residue, a more aggressive leaching procedure or a dissolution method (for example the dissolution in concentrated HNO_3) is necessary. This indicates that the Cr in the pyrolysis residue is even more strongly bound or less easily leached compared to Cr in the CCA treated wood.

Sequential extraction

A sequential extraction procedure, consisting of five steps, was applied to the pyrolysis residue [11, 12]. The different extraction steps correspond to different fractions, sequentially liberated in more aggressive environments. This extraction procedure gives information about the mobility of the metals if the pyrolysis residue has to be disposed on a landfill. About 30 % of both active compounds (As and Cu) retained in the pyrolysis residue is relatively easy leached in a neutral environment. The leachability of As and Cu will thus pose problems upon landfilling. The fixing agent (Cr) on the other hand, is only leached in a strongly oxidising or strongly reducing environment and will pose no problems upon disposal on a landfill.

Speciation study

A hydride generator was coupled on-line with the ICP-MS system to determine the oxidation state of As in the first two extraction steps (neutral environments) [11]. Identification of As(III) in the reducing leach in the third step and the oxidising leach in the fourth step does not provide evidence for the valence state of As in the pyrolysis residue. Therefore, the leachates of these steps were not subjected to hydride generation. This speciation study shows that the major part of As in the pyrolysis residue is present in the trivalent state. The presence of As(III) in the pyrolysis residue indicates that the As, present as As(V) compound ($CrAsO_4$) in the CCA treated wood, is partly reduced to As(III) during the pyrolysis process. If the pyrolysis residue would be landfilled, mainly As(III), which is the more toxic form, will be liberated into the environment. Hence, the pyrolysis residue of CCA treated wood waste can not be landfilled without pretreatment.

Micro-analytical study

Scanning electron microscopy coupled with energy dispersive X-ray analysis (SEM-EDXA) was used to study the preservative distribution and the anatomical structure at the micro-level in CCA treated wood and its pyrolysis residue [13]. Comparison of the micro-distribution between CCA treated wood and its pyrolysis residue provides information about what happens to the wood metal bonding during the pyrolysis process. SEM-EDXA shows that the micro-distribution of metals is not necessarily the same before and after pyrolysis. The metal compounds may have migrated towards the ray cells during the pyrolysis process, the radial way through the ray cells being the main pathway of preservative penetration in softwoods. Furthermore, the metals appear as agglomerates rather than isolated elements (see Figure 2). These observations suggest the possibility of the use of a pneumatic centrifugal grinder-separator system, which aims at recovering clean charcoal, based on the density difference between the carbon and the metal particles [14].

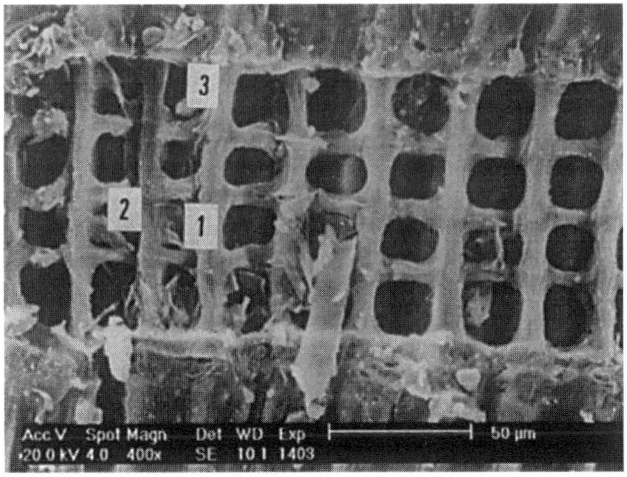

Fig. 2 Longitudinal radial sectional view of ray cells in the pyrolysis residue of CCA treated *Pinus sylvestris* sapwood (400×) with metal agglomerates at the locations 1, 2 and 3 in the ray parenchyma cells.

EXPERIMENTAL PYROLYSIS STUDY

Labscale pyrolysis experiments

An experimental facility for the low-temperature (T < 500 °C) pyrolysis of CCA treated wood particles (between 2 and 135 mm long, between 2 and 17 mm wide and between 0.5 and 2 mm thick) was built with the aim of maximising the fraction of metals that is contained in the ash upon a maximal burnout of the wood. Two different configurations were tried: updraft [15] and downdraft [11, 12, 14] fixed bed.

In the updraft system the mixture of pyrolysis gases and nitrogen is cooled in a highly compact coiled water-cooled heat exchanger before passing through a filter. Different types of filters were tried in order to capture the As and tar compounds as much as possible. The influence of pyrolysis temperature and residence time of the wood chips on the metal content of the resulting pyrolysis residue and the mass reduction was determined. Therefore, the reactor temperature was varied between 300 °C and 450 °C and the duration of the pyrolysis process between 20 and 60 minutes. More details of these pyrolysis experiments are given elsewhere [15]. Since the liquid tar droplets and heavy gaseous compounds condensed on the spiral heat exchanger and formed there a dense sticky layer that could not be removed, nor analysed, the calculation of a complete metal mass balance over the system was impossible. To overcome these problems a downdraft system was built.

The gas flow, leaving the downdraft pyrolysis reactor, is quenched in a water scrubber. The resulting aerosol is then forced through a glass fibre filter. This gas cleaning system has later been optimised to release as little metals as possible: the aerosol at the outlet of the water scrubber is forced through a plate-column, a liquid-gas separation vessel, an extra ice-water condenser and finally through a tube filled with cotton wool that acts as a filter. A metal mass balance was calculated over this system for the three metals (Cu, Cr and As). This downdraft pyrolysis system is described in more detail by Helsen et al. [11, 12, 14].

The influence of reactor temperature and duration of the pyrolysis process on the weight percentage of As in the pyrolysis residue and on the mass reduction of the wood sample is illustrated in Figure 3. Since the same trends are observed for the two other metals (Cr and Cu), the retainment of these metals in the pyrolysis residue is not presented here.

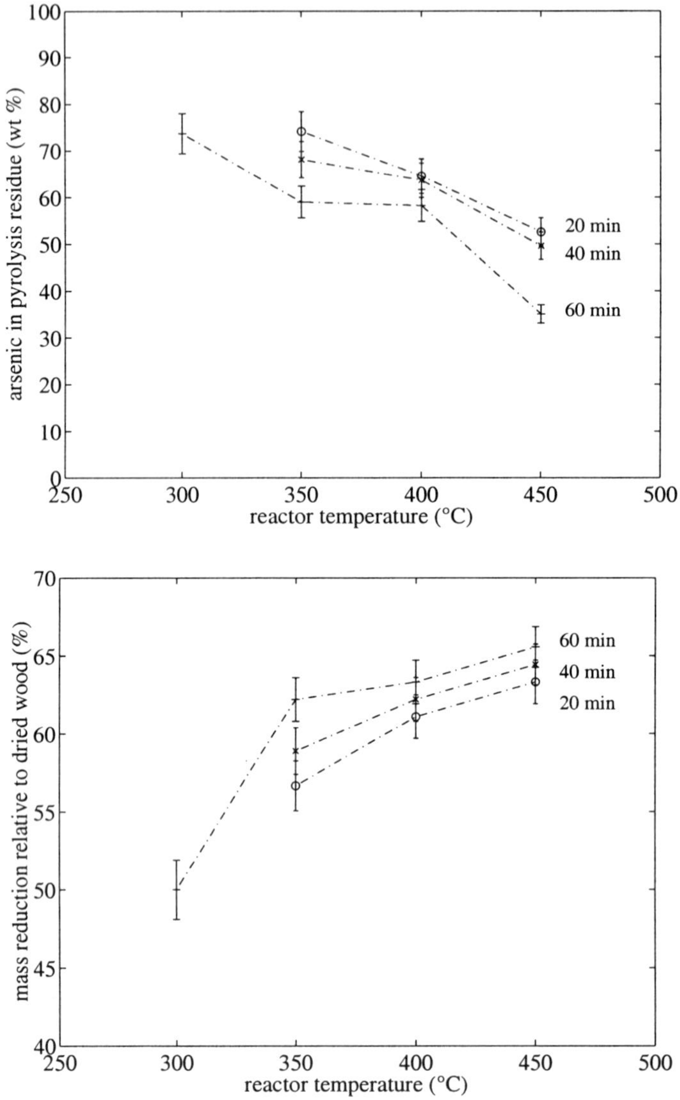

Fig. 3 Percentage As retained in the pyrolysis residue and percentage mass reduction relative to dried wood as a function of the mean reactor temperature for different residence times (o for 20 minutes, × for 40 minutes, + for 60 minutes).

Figure 3 shows that the As content in the pyrolysis residue decreases as the temperature or duration of the process increases. The mass reduction is influenced in the opposite way. The influence of both parameters on the release of Cu and Cr is the same, but the amounts retained in the residue are higher. The arsenic compounds are thus more volatile than the copper and chromium compounds. The optimal combination of pyrolysis temperature and duration with respect to minimal metal release and maximum mass reduction, resulting from these experiments, is 350 °C and 20 minutes. At these conditions, the release of Cu and Cr is negligible, while the As release is about 25 %, using nitrogen at a flow rate of 5 Nm^3/h. The corresponding reduction in mass is 56.7 % relative to dried wood.

The percentages of Cr, Cu and As found in the CCA treated wood, the pyrolysis residue, the scrubber liquid, the glass wool filter and the gas stream obtained in a typical pyrolysis experiment using the downdraft fixed bed reactor are given in Table 1.

Table 1 Metal content (wt %) of the different products used as input or obtained as output in the pyrolysis process.

product	Cr (wt %)	Cu (wt %)	As (wt %)
dried wood	100	100	100
pyrolysis residue	96.9 ± 2.5	94.1 ± 1.8	85.4 ± 2.0
scrubber liquid	0.051 ± 0.001	3.6 ± 0.08	3.3 ± 0.06
filters	0.0092 ± 0.0002	0.021 ± 0.0005	4.0 ± 0.08
gas (losses) *	3.0 ± 2.5	2.3 ± 1.9	7.3 ± 2.1

* calculated by subtraction

The dried wood was heated according to the following reactor temperature profile: hold at 200 °C for 10 minutes, ramped to 350 °C in 20 minutes and hold at 370 °C for 10 minutes. The resulting mass reduction was 57 % relative to the dried CCA treated wood. The metal content of the gaseous product (losses) was calculated by difference, while all other products were analysed by ICP-MS.

From Table 1 it can be seen that low-temperature pyrolysis of CCA treated wood (with a precisely controlled temperature and residence time of the wood particles) may be a promising technique for this type of wood waste. The percentages Cu and Cr volatilised are low enough to conclude that they will pose no problems under the process conditions used. The amount of As volatilised, on the other hand, can not be neglected: it is partly captured in the water scrubber, partly by the filters, but a non-negligible amount (about 7 %) still escapes with the gas flow. Arsenic can thus be identified as the problematic compound with respect to metal release.

Thermogravimetric analysis

A thermo-analytical study of untreated and CCA treated wood samples (more or less cylindrically shaped with diameter less than 2 mm) was performed in order to examine the influence of the presence of CCA on the pyrolytic behaviour of wood samples (TG study), as well as the release of metals (Cu, Cr and As) during the pyrolysis process in the kinetically controlled regime. Since arsenic is the most problematic compound during pyrolysis of CCA treated wood and almost all arsenic is present in the treated wood as $CrAsO_4$ (precipitated on cellulose or complexed with lignin) [9], a thermogravimetric analysis was also carried out on the compound chromium(III) arsenate. TG analysis was performed using the apparatus and samples described in [16, 17].

Comparison of the TG and DTG curves obtained for untreated and CCA treated wood (see Figure 4, corresponding to slow heating experiments) shows that CCA has a significant influence on the thermal behaviour of wood.

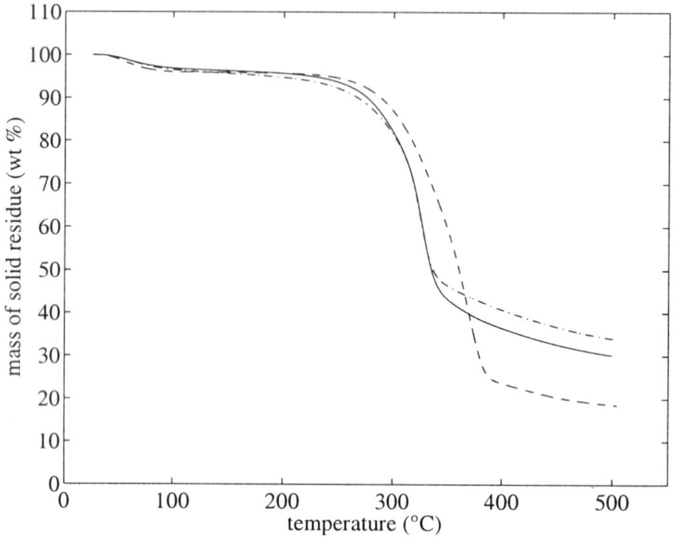

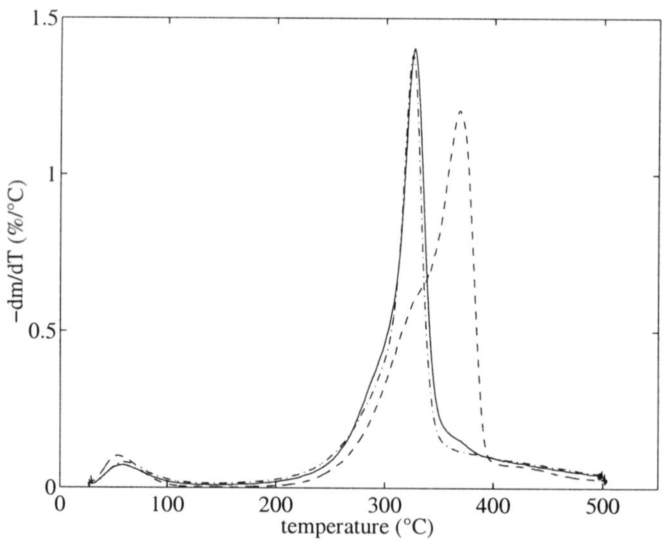

Fig. 4 TG and DTG curves for untreated wood (dashed line) and CCA treated wood (solid line for CCA1 and dash-dotted line for CCA2 with CCA2 retention > CCA1 retention): slow heating (10 °C/min) experiments.

The temperature at the onset of pyrolysis as well as the temperature where the maximum rate of decomposition occurs are lowered, the rate of weight loss is much more peaked (the shoulder at the low-temperature side of the DTG curve has disappeared) and the final char yield is higher for CCA treated samples. A promoting action of the CCA compounds upon the pyrolysis reactions which favours the formation of char can be postulated.

Based on the TG study of hydrated chromium(III) arsenate, which is the major arsenic compound in CCA treated wood, and on the results of equilibrium calculations carried out by other researchers [18, 19, 20], it is concluded that the decomposition of hydrated chromium arsenate results in the formation of solid Cr_2O_3 and gaseous H_2O, O_2 and As_4O_6, according to the scheme:

$$4\ CrAsO_4.6H_2O\ (s) \rightarrow 24\ H_2O\ (g) + 2\ Cr_2O_3\ (s) + 2\ As_2O_5\ (s)$$

$$2\ As_2O_5\ (s) \rightarrow 2\ As_2O_3\ (l) + 2\ O_2\ (g)$$

$$2\ As_2O_3\ (l) \rightarrow As_4O_6\ (g)$$

This scheme suggests that arsenic is released in trivalent state, as As_4O_6.

From the study of the metal (Cr, Cu and As) release during pyrolysis, it can be concluded that the metal volatilisation is strongly dependent on both temperature and residence time of the wood sample at a given temperature. A critical point (10 minutes at 400 °C) is identified, below which the release of Cr and Cu is negligible and the release of As is lower than 10 %. Above this critical point (longer times at 400 °C), there is a dramatic increase in metal release for all three metals. Arsenic is again recognised as the most problematic metal compound, being released (or volatilised) at temperatures between 300 °C and 400 °C. These observations suggest that the release of As is not solely made up by processes that are governed by thermodynamical equilibrium principles, such as sublimation and evaporation. The time dependence suggests that processes that involve chemical kinetics are substantial to the As release mechanism as well.

THEORETICAL STUDIES: KINETIC MODELS

LOW-TEMPERATURE PYROLYSIS OF CCA TREATED WOOD WASTE

A least square evaluation of the DTG curves has led to an acceptable mathematical description of the thermal decomposition of untreated and CCA treated wood [17]. The low-temperature pyrolysis of untreated wood is described by a scheme of three independent parallel reactions, while a combination of two independent parallel reactions and one subsequent reaction is needed to give a good presentation of the thermal decomposition of CCA treated wood [17]. Comparison between the reaction schemes for untreated and CCA treated wood, each having its own kinetic constants, reveals the following observations.

The first peak for CCA treated wood is characterised by lower activation energy, pre-exponential factor and peak temperature, resulting in a slightly higher peak with approximately the same shape but shifted towards lower temperatures. The second peak for CCA treated wood, on the other hand, is characterised by a higher activation energy and pre-exponential factor, resulting in a narrower peak located at lower temperatures. The third peak is better represented by a subsequent reaction, rather than a third independent reaction. The kinetic parameters obtained for this last peak have very low values, resulting in an ill-defined peak.

Based on all previous experiments and the kinetic scheme derived for the pyrolysis of CCA treated wood, the following scenario with respect to the influence of the metals on the pyrolysis behaviour is suggested:

(1) Originally, the metals Cu, Cr and As are preferentially bound to the cellulose and lignin compounds in the CCA treated wood, which means that no metals are bound to the hemicellulose compound [9].
(2) The thermal decomposition of hemicellulose (first DTG peak) is shifted to lower temperatures in the presence of CCA. The metals may have a catalytic effect on the thermal decomposition of hemicellulose, resulting in a lower activation energy.
(3) The cellulose compound is originally loaded with $CrAsO_4$ precipitates and small amounts of other Cu and Cr precipitates or complexes [9], which may hinder the thermal decomposition at first, resulting in a high activation energy for the second DTG peak. However, once the decomposition of cellulose starts, it continues very fast (very high value for pre-exponential factor), resulting in a very narrow peak. As temperature increases, the $CrAsO_4$ precipitate decomposes into Cr_2O_3 and As_2O_5, which further dissociates in As_2O_3 and O_2 at 327.2 °C. The other metal (copper and chromium) compounds, precipitated on or complexed with cellulose, are not volatile. The DTG peak temperature for the second independent reaction (which may be attributed to cellulose decomposition) is 327.9 °C, which is remarkably close to the dissociation temperature of As_2O_5. There may be a correlation between the decomposition of both compounds. An explanation could be that once the major part of the $CrAsO_4$ precipitate on cellulose has decomposed, the decomposition of cellulose continues very fast and may be catalysed by the metals that remained in the solid matrix after the decomposition of $CrAsO_4$ or accelerated by the oxidation by O_2 that comes from the dissociation of As_2O_5, giving rise to the lower DTG peak temperature compared to untreated wood (371 °C).
(4) The intermediate products formed by the thermal decomposition of hemicellulose and cellulose may decompose further at higher temperatures, together with lignin. The small values of pre-exponential factor and activation energy give rise to a very wide DTG peak and thus a very slow charring process. Lignin was originally complexed with $CuCrO_4$ and $CrAsO_4$ and small amounts of other Cu and Cr compounds [9], delaying the decomposition process.
(5) At the end of the pyrolysis process, and maybe even earlier, the metals have been combined to agglomerates and are no more accessible for catalysis, which may be another reason for the slow charring process at higher temperatures.

ARSENIC RELEASE DURING THE LOW-TEMPERATURE PYROLYSIS OF CCA TREATED WOOD WASTE

Arsenic may be assumed to be released according to its own kinetic scheme, independent of the release of other volatile compounds. A simple first order reaction is used to model the arsenic release. The derivation of this kinetic scheme is based on the arsenic releases that were experimentally determined for different combinations of temperature and residence time during the pyrolysis studies (labscale and TGA) [21, 22]. According to the following equations, with m_0 the mass of As in the pyrolysis residue at time t_0 and m at time t (t_0 is defined as the time that the reactor temperature is 20 °C lower than the pyrolysis temperature, such that in the period $[t_0,t]$ the pyrolysis process can be classified as nearly isothermal):

$$-\ln\left(\frac{m}{m_0}\right) = \int_{t_0}^{t} K(T)\,dt \overset{\text{isothermal}}{=} K(T)(t - t_0)$$

$$K(T) = A\exp\left(\frac{-E}{RT}\right)$$

linear regression of $-\ln(m/m_0)$ as a function of $(t-t_0)$ for a constant temperature T delivers the reaction rate constant K(T) at that particular temperature, and linear regression of ln(K) as a function of 1/T for the different temperatures allows one to calculate the pre-exponential factor A and the activation energy E from intercept ln(A) and slope (-E/R). The resulting pre-exponential factor and activation energy are:

$A = 0.3875\ \text{min}^{-1} = 6.4586\ 10^{-3}\ \text{s}^{-1}$
$E = 20.37$ kJ/mol.

To reversely check the kinetic model, the integral rate equation (for non-isothermal conditions) describing the As release during pyrolysis of CCA treated wood is used in combination with the measured temperature profiles T(t) in order to calculate the corresponding As content of the pyrolysis residues. The calculated arsenic content of the pyrolysis residues is compared with the experimental values (labscale and TGA experiments) in the parity plot, presented in Figure 5.

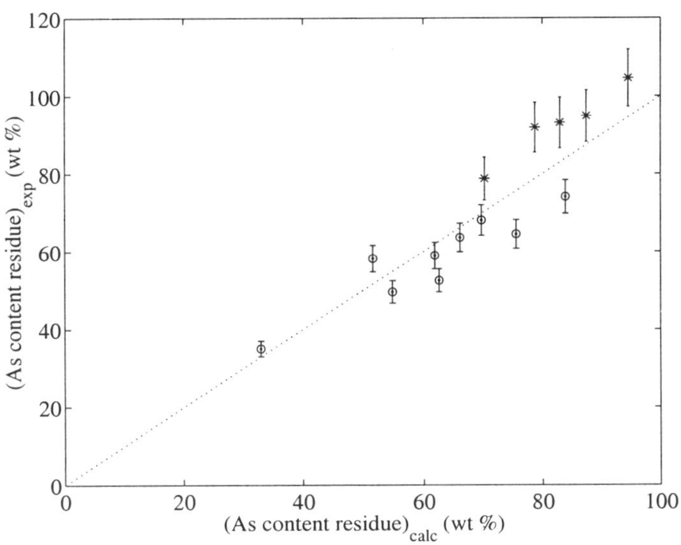

Fig. 5 Parity plot: experimental As contents of the pyrolysis residues (o for labscale, * for TG experiments) are compared with the calculated values, using the first order single reaction kinetic scheme. The errorbars represent 95 % confidence intervals.

Figure 5 shows that the agreement between calculated and experimental values is fairly good, which means that this simple first order single reaction kinetic scheme succeeds in describing the release of As during the low-temperature pyrolysis of CCA treated wood. It should be noted that the experimental As contents are higher than the calculated values for all TGA results, while for most of the labscale experiments the opposite is true. This may be explained by the fact that the labscale experiments do not consider one particle at one particular temperature, but a reactor filled with particles at a mean pyrolysis temperature. Consequently, the collection of these labscale results combined with the TGA results, obtained in the kinetically controlled regime, results in the description of an "averaged" As behaviour. Furthermore, the treated wood samples used in the TG experiments have a higher CCA content compared to the samples used in the labscale experiments. TG analysis performed on CCA treated wood samples with different CCA content [16] shows that the relative concentration of metals (Cr, Cu and As) in the pyrolysis residue increases with the CCA concentration of the original sample. Consequently, the relative amount of metals that is found in the pyrolysis residues resulting from the TG experiments should be higher, which is in agreement with the experimental observations. The kinetic scheme derived here does not account for the influence of the original CCA content.

The kinetic constants (A, E) fall within the range of chemically meaningful values. Rate constants of the same order of magnitude were found for the reduction reaction of Cr(VI) to Cr(III) by Pizzi [6, 9], who studied the fixation of CCA in wood. Because of this similarity the reaction responsible for the As release may be identified as a reduction reaction. The presence of As(III) in the pyrolysis residue (see speciation study), while As is originally present as As(V) in CCA treated wood (as $CrAsO_4$), confirms this hypothesis. As(V) is thus reduced to As(III), which is more mobile and more toxic, during the pyrolysis process and the resulting As(III) is probably released as As_4O_6, which is a volatile compound. This scheme agrees with the thermal decomposition behaviour of $CrAsO_4$ (see TGA).

The volatilisation reaction is not instantaneous, since As(III) was found in the pyrolysis residue. Consequently, the release of As during the pyrolysis of CCA treated wood may be better described by two consecutive reactions (reduction followed by volatilisation) instead of one single reaction. The examination of this hypothesis may be a subject for further research. The good agreement between experimental and calculated values (see Figure 5), using the simple scheme derived here, already shows that the single reaction scheme with the corresponding rate constants is of high value as a first approximate to describe the As release during pyrolysis. It should be noted that the validity of this kinetic scheme is only tested for temperatures in the range from 350 °C to 450 °C, which means that good predictions are only guaranteed when low-temperature pyrolysis processes at temperatures in this range are simulated. A more advanced model should account for the influence of the initial CCA content and incorporate the volatilisation reaction that follows the reduction reaction. Both extensions need more experimental work: (1) pyrolysis experiments using CCA treated samples with different CCA content and (2) speciation studies of the pyrolysis residues. Furthermore, extrapolations to temperatures lower than 300 °C and/or higher than 450 °C should be compared with experimental results.

DISCUSSION

Originally, to prepare the CCA solution, As and Cr are mixed in the form of As_2O_5 and CrO_3, respectively. They are thus both in the high oxidation state, being As(V) and Cr(VI). During the fixation process Cr(VI) is partly reduced to Cr(III) in order to fix the As as $CrAsO_4$. The decomposition of $CrAsO_4$ results in Cr_2O_3 and As_2O_5, which further dissociates into As_2O_3 and O_2. Both metals appear thus in the lower oxidation state, being As(III) and Cr(III), after pyrolysis. The resulting chromium(III) compound (Cr_2O_3) is insoluble in water, acids, alkali and alcohols, which could be the explanation for the fact that Cr is more strongly bound in the pyrolysis residue compared to the CCA treated wood and for the lower mobility. Only at temperatures higher than 1000 °C, the chromium(VI) oxide (CrO_3), which has a higher mobility and toxicity, is formed.

All experimental observations (from this study and the studies of other researchers [23, 24, 25]) confirm the hypothesis that As is released as the volatile arsenic trioxide As_2O_3 or As_4O_6, which is very difficult to capture and therefore rarely detected. Arsine (AsH_3) is not present in the gas exhaust since it decomposes at 300 °C. The only way to release arsenic as As(V) compounds is in the particulate or aerosol phase, as condensed arsenates. These can be captured by scrubbing and filtering. The decomposition of wood components may give rise to methyl groups which can methylate arsine (at temperatures below 300 °C), resulting in volatile organic As(III) compounds, which are less toxic than arsenic trioxide.

According to the kinetic model that describes As release during pyrolysis, As is already released at temperatures lower than 300 °C, since the kinetic parameters A and E give rise to a very broad and low DTG peak. This is in contradiction with the reaction scheme derived for the thermal decomposition of chromium arsenate. According to the latter, As is only released for temperatures higher than 327 °C. The volatilisation of arsenic is preceded by a reduction of As(V) to As(III). This reasoning suggests that the kinetic scheme for As release during pyrolysis, that was derived based on experimental results in the temperature range of 350 °C to 450 °C, may not give accurate predictions for temperatures outside this range. In the temperature range of 350 °C to 450 °C, the kinetic scheme for As release predicts a little rise in the rate of As volatilisation. This rise in volatilisation rate with temperature may also result from liquid-vapour equilibrium controlled by the temperature dependent vapour pressure of As_2O_3 (l). Vapour pressures as high as 100 mmHg at 332.5 °C and 200 mmHg at 370 °C, which give rise to non-negligible volatilisation, are reported for As_2O_3 [26]. Due to the continuous outflow of pyrolysis vapours, As_2O_3 (g) will be removed once it has been formed. Consequently, a rise in temperature results in a rise in volatilisation rate due to the increasing vapour pressure.

According to this hypothesis As(V) is reduced to As(III) before volatilisation occurs. For temperatures lower than 327 °C As_2O_5 is the most stable compound [21], but a sharp reduction reaction takes place at 327 °C. Since the melting point of As_2O_3 (l) is 313 °C [26], the As_2O_3 is formed as a liquid. For temperatures higher than 327 °C, the combination of phase equilibrium between As_2O_3 (l) and As_2O_3 (g) and the carrying off of As_2O_3 (g) by the outflowing pyrolysis vapours, determines the amount of As_2O_3 (l) volatilised. To check the validity of this hypothesis, additional pyrolysis experiments at temperatures lower than 350 °C and different residence times are needed.

CONCLUSIONS

The most important conclusions with respect to metal behaviour during low-temperature pyrolysis of CCA treated wood waste are summarised in Figure 6.

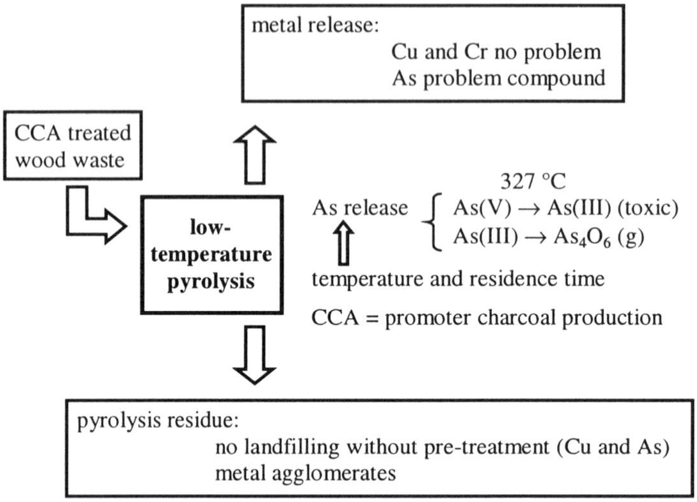

Fig. 6 Schematic presentation of the most important conclusions with respect to the metal behaviour during low-temperature pyrolysis of CCA treated wood waste.

Low-temperature pyrolysis is suggested as a possible technique to dispose CCA treated wood waste. The presence of CCA influences the pyrolysis process significantly: CCA can be identified as a promoter of charcoal production. Pyrolysis temperature and residence time of the wood particles are the most important process parameters with respect to metal release during the pyrolysis of CCA treated wood waste. At the low temperatures considered, Cu and Cr do not pose any problems. Arsenic, on the other hand, is the problematic compound, being released already at temperatures between 300 °C and 400 °C. The mechanism responsible for the release of As is identified as the reduction of As(V) to As(III), which occurs sharply at 327 °C. Limiting the As release implies thus controlling (avoiding) this reduction reaction, which means that the As has to remain in pentavalent form and thus, the temperature has to be kept low enough. Arsenic is probably released as arsenic trioxide As_4O_6, which is very difficult to capture and toxic.

The pyrolysis residue (solid product) can not be landfilled as such due to the high leachability of Cu and As. The presence of As(III), which is more toxic than As(V), in the pyrolysis residue makes this leaching behaviour even more problematic. The recombination and agglomeration of metals during the pyrolysis process suggest the possibility of using a dry separation (pneumatic centrifugal separation based on density differences) of the metal-containing char residue, which makes the metal recycling process complete and therefore environmentally friendly.

ACKNOWLEDGEMENTS

The authors would like to thank the Department of Chemical Engineering of the K.U.Leuven for their valuable help in conducting the quantitative analysis of the samples. We are especially grateful to Prof. C. Vandecasteele from the Chemical Department for his comments through the course of the study. Also the assistance of Rudy de Vos in the SEM-EDX analysis at the Department of Metallurgy and Materials Engineering of the K.U.Leuven and of Steven Mullens and prof. Jules Mullens in the TG analysis at the Laboratory of Inorganic and Physical Chemistry of the "Limburgs Universitair Centrum" are acknowledged. Furthermore, we are grateful to Beaumartin S.A. and Mr. J.S. Hery in particular for the financial support, the helpful discussions and the wood samples.

REFERENCES

1. Meier D., Wehlte S. and Faix O. (1997) Flash pyrolysis – a possibility to utilise contaminated wood. *Plant Research and Development*, **46**, 46-55.
2. Wehlte S., Meier D., Moltran J. and Faix O. (1997) The impact of wood preservatives on the flash pyrolysis of biomass. In: *Developments in Thermochemical Biomass Conversion*, (Ed. by A.V. Bridgwater and D.G.B. Boocock), pp. 206-219. Chapman & Hall, London.
3. Dahlgren S.E. (1974) Kinetics and mechanism of fixation of Cu-Cr-As wood preservatives IV. Conversion reactions during storage. *Holzforschung,* **28** (2), 58-61.
4. Dahlgren S.E. (1975) Kinetics and mechanism of fixation of Cu-Cr-As wood preservatives VI. The length of the primary precipitation fixation period. *Holzforschung*, **29** (4), 130-133.
5. Dahlgren S.E. and Hartford W.H. (1972) Kinetics and mechanism of fixation of Cu-Cr-As wood preservatives I. pH behaviour and general aspects on fixation. *Holzforschung*, **26** (2), 62-69.
6. Pizzi A. (1981) The chemistry and kinetic behaviour of Cu-Cr-As/B wood preservatives. I. Fixation of chromium on wood. *J. Polym. Sci.*, **19**, 3093-3121.
7. Pizzi A. (1982) The chemistry and kinetic behaviour of Cu-Cr-As/B wood preservatives. II. Fixation of the Cu/Cr system on wood. *J. Polym. Sci.*, **20**, 707-724.
8. Pizzi A. (1982) The chemistry and kinetic behaviour of Cu-Cr-As/B wood preservatives. III. Fixation of a Cr/As system on wood. *J. Polym. Sci.*, **20**, 725-738.
9. Pizzi A. (1982) The chemistry and kinetic behaviour of Cu-Cr-As/B wood preservatives. IV. Fixation of CCA to wood. *J. Polym. Sci.*, **20**, 739-764.
10. Chou C.K, Chandler J.A. and Preston R.D. (1973) Microdistribution of metal elements in wood impregnated with a copper-chrome-arsenic preservative as determined by analytical electron microscopy. *Wood Sci. Technol.*, **7**, 151-160.
11. Van den Broeck K., Helsen L., Vandecasteele C. and Van den Bulck E. (1997) Determination and Characterisation of Copper, Chromium and Arsenic in CCA treated Wood and its Pyrolysis Residues by Inductively Coupled Plasma Mass Spectrometry. *The Analyst*, **122**, 695-700.
12. Helsen L., Van den Bulck E., Van den Broeck K. and Vandecasteele C. (1997) Low-temperature pyrolysis of CCA treated wood waste: chemical determination

and statistical analysis of metal in- and output; mass balances. *Waste Manage.*, **17** (1), 79-86.
13. Helsen L. and Van den Bulck E. (1998) The Microdistribution of Copper, Chromium and Arsenic in CCA Treated Wood and Its Pyrolysis Residue Using Energy Dispersive X-Ray Analysis in Scanning Electron Microscopy. *Holzforschung*, **52** (6), 607-614.
14. Helsen L., Van den Bulck E. and Hery J.S. (1998) Total recycling of CCA treated wood waste by low-temperature pyrolysis. *Waste Manage.*, **18**, 571-578.
15. Helsen L. and Van den Bulck E. (1997) Release of metals during the pyrolysis of preservative impregnated wood. In: *Developments in Thermochemical Biomass Conversion*, (Ed. by A.V. Bridgwater and D.G.B. Boocock), pp. 220-228. Chapman & Hall, London.
16. Helsen L., Van den Bulck E., Mullens S. and Mullens J. (1999) Low-temperature pyrolysis of CCA treated wood: thermogravimetric analysis. *J. Anal. Appl. Pyrolysis*, **52**, 65-86.
17. Helsen L. and Van den Bulck E. (2000) Kinetics of the low-temperature pyrolysis of chromated copper arsenate-treated wood. *J. Anal. Appl. Pyrolysis*, **53**, 51-79.
18. Ebbinghaus B.B. (1993) Thermodynamics of gas phase chromium species: the chromium oxides, the chromium oxyhydroxides, and volatility calculations in waste incineration processes. *Combust. Flame*, **93**, 119-137.
19. Frandsen F.J., Helble J.J., Erickson T.A. and Kuhnel V. (1997) Prediction of trace element partitioning in utility boilers, *presented at the EPA/DOE/EPRI Symposium*.
20. Wu C.Y. and Biswas P. (1993) An equilibrium analysis to determine the speciation of metals in an incinerator. *Combust. Flame*, **93**, 31-40.
21. Helsen L. (2000) *Low-temperature pyrolysis of CCA treated wood waste.* Ph.D. Thesis, Katholieke Universiteit Leuven, Belgium.
22. Helsen L. and Van den Bulck E. (2000) Metal Behaviour During the Low-Temperature Pyrolysis of Chromated Copper Arsenate Treated Wood Waste. accepted for publication in *Environ. Sci. Technol.*.
23. McMahon C.K., Bush P.B. and Woolson E.A. (1986) How much arsenic is released when CCA treated wood is burned. *For. Products J.*, **36**, 45-50.
24. Hirata T., Inoue M. and Fukui Y. (1993) Pyrolysis and combustion toxicity of wood treated with CCA. *Wood Sci. Technol.*, **27**, 35-47.
25. Cornfield J.A., Vollam S. and Fardell P. (1993) Recycling and disposal of timber treated with waterborne copper based preservatives. *24th annual meeting*, Orlando, USA, IRG/WP 93-50008.
26. Perry R.H. and Chilton C.H. (1973) *Chemical Engineers' Handbook*, 5th edn. McGraw-Hill, New York.

Pyrolysis of Biomass as Pre-Treatment for Use as Reburn Fuel in Coal-Fired Boilers

C. Storm, S. Unterberger, K. R. G. Hein
Institut für Verfahrenstechnik und Dampfkesselwesen (IVD), University of Stuttgart, Pfaffenwaldring 23, D-70550 Stuttgart, Germany

ABSTRACT Biofuels attract increasing interest in power plant technology as sources of carbon dioxide neutral fuels. Besides using solid pulverised biomass as an additional fuel in coal-fired boilers a further possibility to run a combined coal and biomass process is to pre-pyrolyse or pre-gasify biomass and to inject the gas as reburn fuel into the coal-fired boiler. Within the pretreatment process the solid feedstock is separated into a high energy gas and a solid residue. The gas can be used as a reburn fuel in the coal-fired boiler with excellent NO_x reduction properties. The separation of coal and biofuel ash within this process enables a specialised use of both residuals.

At the Institute for Process Engineering and Power Plant Technology the pyrolysis and gasification of different biomass has been investigated with special emphasis on the gas, tar and char composition and on the NO_x reduction efficiency of the produced gas. The gasification experiments have been carried out in an electrically heated lab-scale entrained flow reactor in understoichiometric atmosphere, for the reburn experiments a pulverised fuel combustion reactor was available. The gas, tar and composition as well as the NO_x reduction efficiency have been analysed at temperatures of the entrained flow reactor between 400°C and 1300°C and of the fluidised bed reactor between 600°C and 900°C and air ratios of the pyrolysis/gasification process between 0 and 0,4.

Detailed results of the investigations using pyrolysis gas from coal [7, 8], sewage sludge [17], or biomass [18] for NO_x reduction are published elsewhere.

INTRODUCTION

Various forms of biofuels represent interesting feedstocks for combustion, gasification and pyrolysis processes in power plant technology as a source of CO_2 neutral fuels. Biomass has an estimated potential in Germany of about 660 PJ (1 Petajoule = 10^{15} Joule). This is equivalent to about 5 % of the annual primary energy consumption in Germany. [10]

The aim of the investigations is the optimisation of a co-combustion process using biofuel pyrolysis/gasification gas as a reburn fuel with special view on parameters for NO_x reduction, parameters for alkali, chlorine and heavy metal release.

The advantage of ash separation within this combined combustion of coal and biomass offers a separate ash utilisation. While the ash utilisation of pure coal ash is technically approved and economically feasible already, different kinds of utilisation of biomass ash have to be proved. The utilisation of mixed ashes in building industry, which is the main market for ashes from coal combustion, is not foreseen in the legislation of the European Union. The investigated pre-treatment process is a suitable alternative for biomass co-combustion.

FUNDAMENTALS

BIOMASS

Biomass contains more linear and weaker connected macromolecules than the highly cross-linked coal with its aromatic components. Photosynthesis is the basic step in the formation of biomass. An endothermic reaction creates glucose from carbon dioxide and water. In a second step, the main biomass components cellulose, hemi-cellulose and lignin are created. The composition of cellulose (% wt.) is 44,4 % carbon, 6,2 % hydrogen and 49,4 % oxygen. Due to the chemical composition it is evident, that biomass is an oxygen rich feedstock. Hemi-cellulose are like cellulose chains from glucose molecules, but their length is much shorter and they can have subchains. Unlike cellulose and hemi-cellulose lignin consists of aromatic elements, which are arranged three-dimensional. The share of oxygen in lignin is much below the oxygen content of celluloses.

NITROGEN, CHLORINE, ALKALINES, HEAVY METALS

Inorganic components differ between the different types of biofuels. For energetic use of biofuels the most interesting inorganic elements are chlorine and alkalis as they cause problems in conventional combustion systems. Nitrogen is another element, which is important when burning biofuels, because it can either cause NO_x emissions or it can help minimising them, when the nitrogen is bound in certain compounds i.e. like HCN or ammonia. Furthermore heavy metals and other trace elements are of interest as they can cause emission problems. Those elements usually occur only in anthropogenic made residues like sewage sludge, waste wood etc.

PYROLYSIS OF BIOFUELS

Pyrolysis (or devolatilisation) is a thermal decomposition of organic matter in an inert atmosphere. In the gasification or partial gasification oxidation medium is added to the process. This oxidation medium usually is air or oxygen. Pyrolysis is also the first step in combustion and gasification.

Besides solid residues carbon-, hydrogen-, and oxygen containing gases as well as condensable aromatic hydrocarbons, which are designated as tars, are formed during pyrolysis. [4]

Homogeneous and heterogeneous reactions in the pyrolysis process cause a changing product composition, depending on the reaction temperature. The main reactions are shown in *Figure 1*.

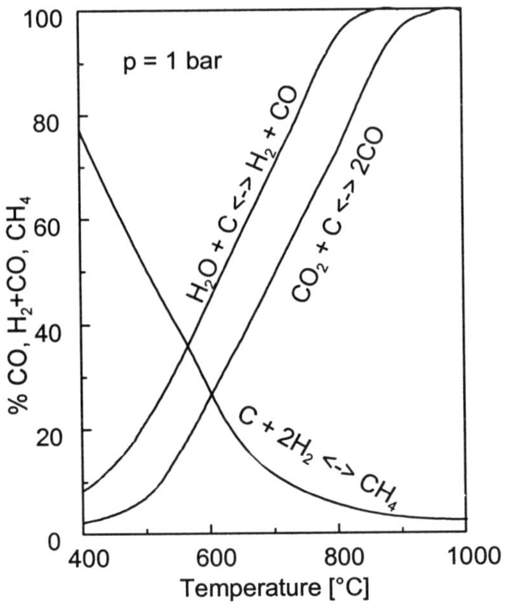

Figure 1 Equilibrium of different gasification reactions depending on the reaction temperature

NO_X REDUCTION WITH PYROLYSIS GAS

Nitrogen oxides (NO_x) are formed during combustion at high temperatures by the oxidation of air nitrogen and of nitrogen bound in the fuel [1, 5, 6, 20]. The sustainingly increasing demand of energy, which is mainly met by combustion of fossil fuels, makes the use of technologies for the minimising of NO_x emissions essential.

NO_x Formation

There are known three different kinds of NO_x formation in coal combustion: thermal NO_x formation; prompt NO_x formation; NO_x formation from fuel nitrogen.

All three kinds of formation result from different reaction zones in the flame and from two principle sources of nitrogen, the fuel nitrogen and nitrogen from air. NO_x formation from fuel nitrogen plays the most important role for NO_x emissions from coal combustion [3, 14]. After the primary pyrolysis of the coal the nitrogen occurs in the solid phase bound into the char and in the gaseous phase. A small part is directly converted to NO, released as N_2 or reduced to N_2. From the fuel nitrogen in the gaseous phase mainly ammonia and HCN are formed. The char nitrogen is either oxidised heterogeneously or decomposed to N_2 or further gaseous components like HCN or radicals are formed.

NO_x Reduction

NO_x emissions can either be influenced by primary measures (less NO_x formation) or by secondary measures (reduction of NO_x) [2, 3, 7, 11, 19]. The chemical reactions of

NO_x deformation during secondary measures are much more complex than those of the NO_x formation. Not all ways of reactions are known by now. The reactions are differentiated into two types, the homogeneous and the heterogeneous, which characterise the reactions in the gas phase and the reactions between the gas phase and the particle surfaces.

Homogeneous reduction: Nitrogen oxides can be reduced during the combustion process either by intermediate products or by added reducing agents. Reducing agents are divided into selective and non-selective agents. Non-selective reducing agents, like carbon monoxide, hydrogen, and hydrocarbons, reduce by catching the oxygen, which is still in the flue gas due to reasons of the combustion technique. This results in a high demand of reducing agent. In contrast to the non-selective agents are the selective reducing agents, which only reduce nitrogen oxides independent on the oxygen content in the flue gas. Both reaction mechanisms are supported exclusively by radicals. An increase of temperature leads also to an increase of radicals. There is also the possibility to influence the formation of radicals by decreasing the O_2 partial pressure.

Heterogeneous reduction: The heterogeneous reduction of NO directly at the surface of char particles is extremely dependant on the temperature [15]. At low temperatures no reduction takes place at the particle surfaces in contrast to high temperatures, where the rate of reduction is about 95 %. This behaviour is due to the much faster oxidising reaction of nitrogen than the reduction of NO. The heterogeneous reduction mainly takes place in fluidised bed and grate firing systems. In pulverised fuel combustion systems the influence of this reduction mechanism is much lower, because a high particle concentration and a high share of combustible solids is not given.

Reburning – In-Furnace NO_x Reduction

Methods of minimising NO_x emissions already in the furnace by staged burner or air staging minimise the often very complex reduction systems after the furnace like DeNOx systems, injection of ammonia, which are often used for NO_x reduction in flue gases. Air staging is already a technically feasible option of minimising NO_x emissions and it is already integrated in many lignite combustion systems [9] and hard coal combustion systems [12].

Reburning, also called in-furnace reduction or fuel-staging, is a three step combustion. In the first step the main fuel (usually coal) is burnt completely at an air ratio above one, within the second step the reburn fuel is injected into the furnace without adding any oxygen. In this reduction zone the air ratio is below one and creates conditions to reduce the NO_x, which has been formed in the main combustion zone. To ensure a complete burnout air is injected into the third zone.

The potential of NO_x reduction by reburning depends on the air ratio in the reduction zone, the temperature in the reduction zone, the mixing conditions of the reburn fuel in the reduction zone, the residence time in the reduction zone, and the fuel and reburn fuel composition.

Figure 2 shows the process of the thermal pre-treatment of biomass fuels in combination with the injection of biomass derived gas into a pulverised fuel combustion as reburn fuel.

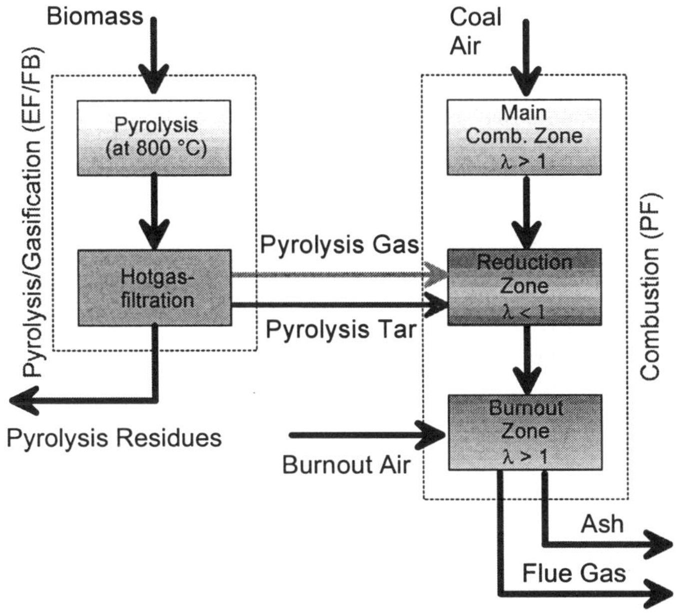

Figure 2 Process scheme: thermal pre-treatment of biomass and reburning of the biomass derived gas in a coal-fired boiler

The thermal pre-treatment process consists of a pyrolysis or gasification process, where the biomass is fed into.

After the pre-treatment process the solid pyrolysis residues are separated from the energy-rich pyrolysis gas within a hot gas filtration unit. The gas is then used as reburn fuel.

TEST FACILITY

The BTS (Brennstofftrennstufung – fuel splitting and staging) test facility was built within the scope of the project 'Combined minimizing of NO_x-production and reduction of formed NO_x during combustion of coal dust' in co-operation with a German coal mining utility [16]. *Figure 3* shows the scheme of the combined test facility.

The test facility consists of following two main components:

- entrained flow reactor with fuel feeding unit, carrier gas pre-heating, product gas analysis and hot gas filtration with precipitator, cyclone and ceramic candle filter.
- combustion reactor with fuel feeding unit, air pre-heating, pyrolysis gas and burnout air supply, gas analysis and flue gas filtration.

The two facilities can be operated either together for i.e. investigation of NO_x reduction by the means of pyrolysis gas as reburn fuel or separate for investigations concerning pyrolysis, gasification and combustion of various solid fuels.

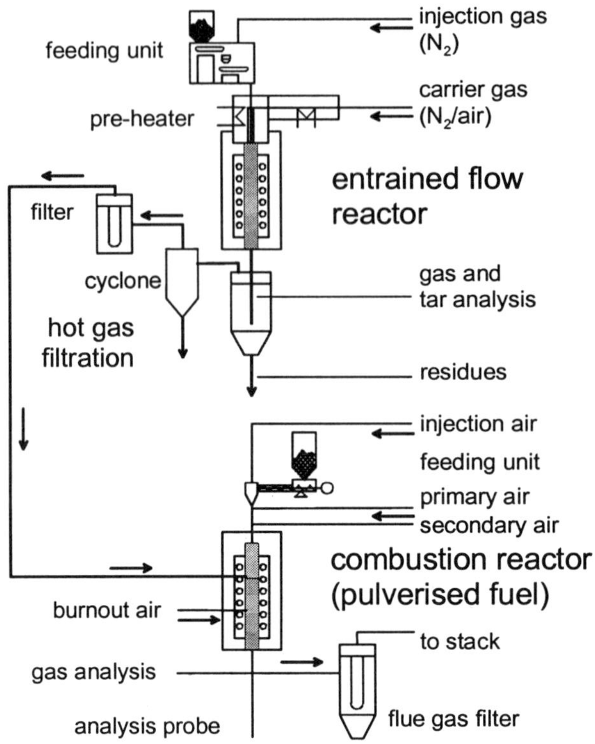

Figure 3 Test facility for the research of pyrolysis and reburning

The entrained flow pyrolysis reactor (20 kW_{th}) can be connected to the entrained flow combustion reactor (50 kW_{th}). The pyrolysis residues are separated from the pyrolysis gas in the cyclone and in the hot gas filter. Afterwards this gas is led as reburn fuel into the reduction zone of the combustion reactor. All pipes are trace heated up to 350 °C to avoid tar condensing.

The results of pyrolysis of different solid fuels discussed here were performed at the entrained flow reactor. The results of NO_x reduction experiments have been gained using the pyrolysis gas as reburn fuel in the pulverised fuel combustion reactor.

ENTRAINED FLOW REACTOR

The dosing of the solid fuel into the entrained flow reactor takes place gravimetrically. The dosing rate of the fuel is usually between 0,5 and 2,5 kg/h. The reactor itself consists of an electrically heated ceramic pipe of the material „Pythagoras“ (ceramic with 60 % Al_2O_3 heatable up to 1300 °C) with a total length of 2,7 m, an outer diameter of 100 mm, an inner diameter of 80 mm and a heated length of about 2 m. At the outlet of the reaction pipe the hot gas filtration unit with the connection for gas analysis is located. A heated pipe leads the gases from the hot gas filtration to the combustion reactor or to the flare, respectively.

COMBUSTION REACTOR (PULVERISED FUEL)

The pulverised fuel combustion reactor was used for reburn test runs using pyrolysis gas generated in the entrained flow reactor as reburn fuel. The main part of the reactor is the electrically heated furnace with a total electric output of 57 kW_{el}, devided into five heating zones. A ceramic pipe with a length of 2,5 m and an inner diameter of 200 mm is installed inside the furnace as reaction pipe. The feeding unit, the reactor itself and the flue gas filter are designed for a coal feeding rate between 1 and 5 kg/h. The fuel and the combustion air, split in primary and secondary air, are injected through the movable block burner. The maximum wall temperature of the combustion reactor is 1500 °C. At a position of 0,9 m distance below the burner of the combustion reactor the pyrolysis gas is injected through a nozzle. To ensure a complete burnout air can be injected through a height adjustable burnout air probe.

The reburn test runs have been performed under constant conditions in the combustion reactor. Besides the wall temperature (1300 °C) the fuel, the fuel feeding rate, and the probe positions stayed unchanged. The hard coal Göttelborn was burnt in the main combustion zone at an air ratio of 1,15. The fuel feeding rate amounts to 1 kg/h. The position of the burnout air probe remained unchanged. It was adjusted to a calculated residence time of 1 m_n^3/h pyrolysis gas of 2 s in the reduction zone. Those settings (residence time, wall temperature) have been proved to be optimal for reburning in earlier projects [7, 8]. Only the burnout air was varied in order to get a constant O_2 concentration of 3 % in the flue gas.

FUEL ANALYSIS

Table 1 shows the composition of the fuels used. Besides natural like biomass also anthropogenic treated biofuels like sawdust and sewage sludge have been used for experiments. The wheat straw and beech have been milled for the experiments in a cutting mill with sieve sizes of 1,5 mm, 1 mm and 0,5 mm, respectively. The sawdust or sanding dust comes from a chip board manufacturer and was delivered with the used particle size. The sawdust contains melamine from glue constituents. Melamine is nitrogen rich compared to the nature like biomass. This kind of biomass was chosen for NO_x reduction experiments, because nitrogen plays a major role for the reburn efficiency of the pyrolysis gas. Several sewage sludges have been pyrolysed and the pyrolysis gases have been tested for their reburn and NO_x reduction efficiency [17]. The hard coal Göttelborn was used for combustion in the pulverised fuel combustion reactor as main fuel.

RESULTS

PYROLYSIS GAS, TAR AND CHAR

Main components of the pyrolysis gas are the gaseous components CO, CO_2, H_2, and aliphatic hydrocarbons as well as the condensable aromatic hydrocarbons, usually designated as tars.

Table 1 *Composition of the raw fuels*

fuel	straw	beech	saw-dust	sew. sludge	sew. sludge	hard coal
abbr.	ST	BU	SM	DD	HG	GB
proximate analysis [$\%_{dry}$]						
fixed C	17,0	15,4	21,2	5,6	4,7	57,0
volatile	77,8	84,0	77,5	54,5	46,9	33,9
ash	5,2	0,6	1,3	39,9	48,4	9,1
ultimate analysis [$\%_{daf}$]						
C	49,7	49,8	49,5	53,0	56,5	81,4
H	6,1	6,3	6,3	7,3	7,2	5,6
N	0,5	0,1	5,1	7,7	5,4	1,5
S	0,2	0,1	0,1	2,8	1,5	1,0
Cl	1,1	*	*	0,2	0,1	0,5
O (diff)	42,4	43,7	39,0	29,1	29,3	10,6
mois %	8,3	5,9	5,3	5,7	5,1	5,3
particle size [mm] sieve						
	1,5	0,5	1,0	90 µm	30 µm	60 µm
calor. value [kJ/kg_{dry}]						
	16473	*	*	13694	11287	30723
ash [$\%_{ash}$]			*			
SiO_2	36,6	20,0		39,8	41,6	42,3
Al_2O_3	0,8	7,0		11,7	11,3	24,3
Fe_2O_3	0,6	1,4		14,2	9,1	11,3
CaO	10,2	26,1		6,9	8,6	6,5
MgO	2,5	9,2		1,1	1,1	4,1
K_2O	36,7	23,5		0,8	0,8	3,3
Na_2O	2,3	1,8		0,6	0,5	0,7
SO_3	*	*		*	*	6,6

• not determined

Basic investigations on the pyrolysis of different biofuels under different process and fuel parameters have been carried out in previous investigations. Especially gaseous and condensable pyrolysis products have been the focus of those investigations [13, 17]. For the sake of completeness main results of straw pyrolysis are presented here.

Straw has been chosen as reference fuel. As shown in the previous investigations other biofuels behave similar regarding the main components. Even sewage sludge in spite of its high ash content produces almost the same gaseous and tar components like straw compared on a water and ash free basis (daf).

Gaseous Components

Figure 4 shows the yield of the main components of straw pyrolysis. The composition of the pyrolysis gas changes with increasing pyrolysis temperature to the favour of light gases like H_2 and CO. Carbon monoxide and hydrogen are generated in a higher share with increasing reaction temperatures. Carbon monoxide production starts already at 400 °C; hydrogen is significantly generated at temperatures higher than 800 °C. Aliphatic hydrocarbons are also produced at low temperatures and reach a maximum at about 800 °C. The aliphatic hydrocarbons are decomposed to less complex compounds with rising temperature. Carbon dioxide is not strongly influenced by the reaction temperature. The high oxygen content in the raw fuel leads to an increased formation of CO and CO_2.

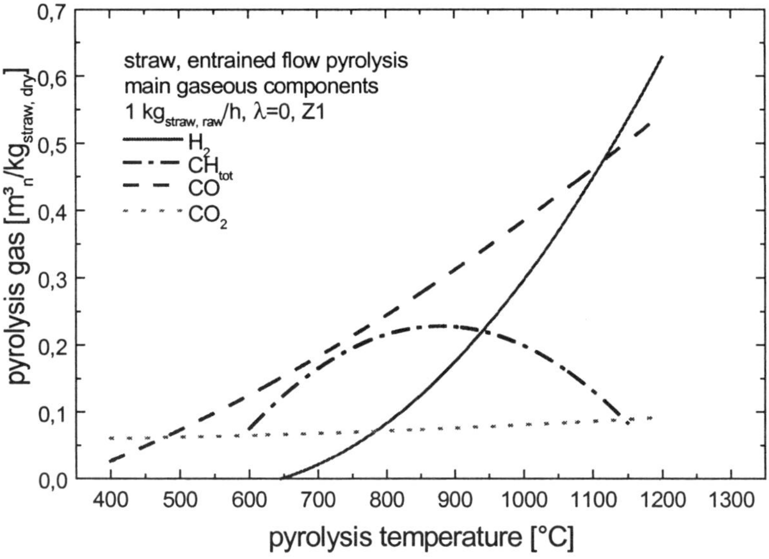

Figure 4 *Main gaseous components of straw pyrolysis gas*

Tar Components

The tar species are divided into two groups. The term heavy tars describes the sum of all species with a boiling point above about 200°C. These are all aromatic hydrocarbons with more than two benzene rings (naphthalene and above). The light tars are mainly benzene, toluene and xylene. The boiling point of the compounds is below 200°C and the number of benzene rings does not exceed two. Figure 5 gives information about the tar production of straw. At low temperatures, more heavy components are produced. Light tars are formed by cracking out of heavy components with higher temperatures and reach a maximum between 800 °C and 900 °C comparable to the aliphatic hydrocarbons. With further increasing temperatures, both tars, heavy and light, are decomposed rapidly to less complex compounds like the light gases H_2 and CO.

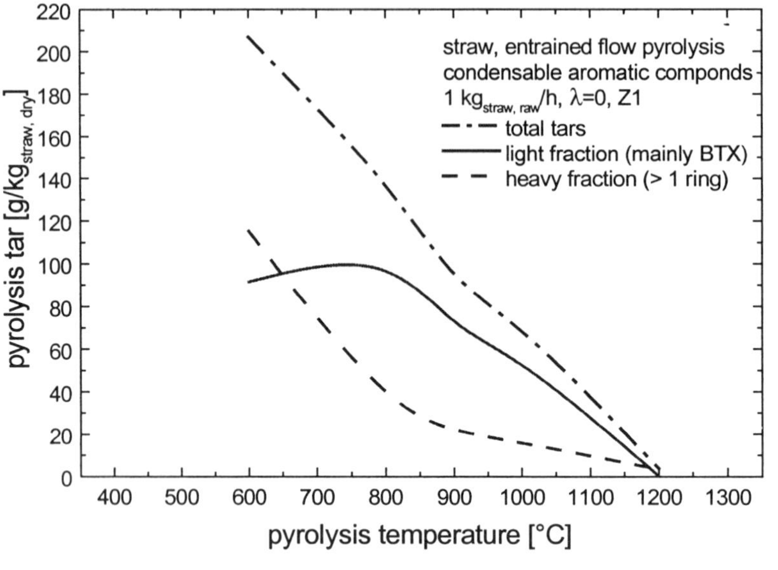

Figure 5 Production of tars during straw pyrolysis in dependence on the pyrolysis temperature

Pyrolysis Residues

Char samples were taken from the pyrolysis experiments at the entrained flow reactor. Figure 6 shows the devolatilisation and the composition of the char from straw pyrolysis at different temperatures and air ratios and of the raw straw. The results of the thermogravimetric measurements are related to the ash content of the raw fuel. Straw shows a high devolatilisation already at low temperatures. The devolatilisation rate is almost not dependant on the air ratio in the entrained flow reactor. Deviations between the values are within the tolerance of the sampling and measurement system.

Other biofuels (beech, sewage sludge) behave similarly to straw during the pyrolysis.

The release of nitrogen during the pyrolysis is nearly independent from the pyrolysis temperature and from the fuel. Around 20 % of the nitrogen stays in the char independent of the fuel. Even the share of nitrogen release during pyrolysis of a nitrogen rich fuel like sewage sludge lays in the same order as the nitrogen release out of N poor fuels, like straw and beech. The independency of the temperature on the nitrogen release is only true for temperatures above 600 °C. Due to the low degasification rate of straw pyrolysis at low temperatures there is only a minor nitrogen release below 600 °C.

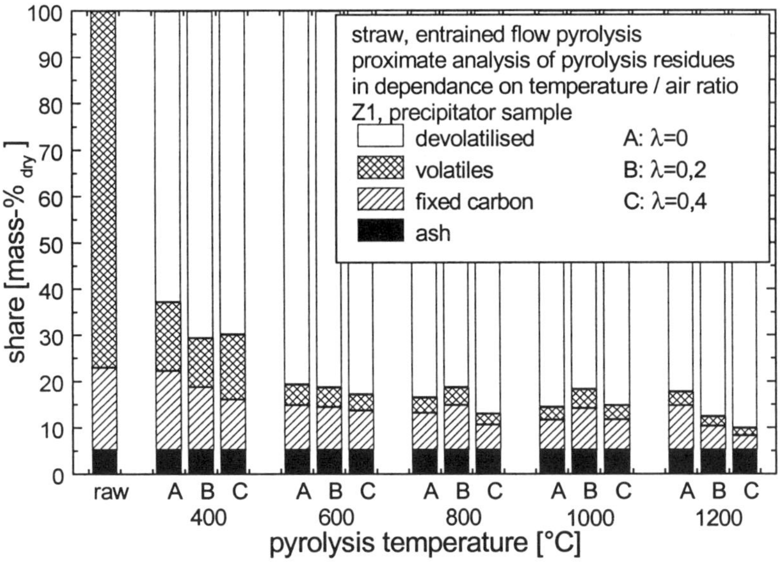

Figure 6 Proximate analysis of straw (raw material and residual char), values related to ash content of raw fuel

NO_X REDUCTION WITH PYROLYSIS GASES

Using pyrolysis gas from biofuels as reburn fuel with good NO_x reduction efficiency enables a way of contributing to the net CO_2 reduction. It opens up a way of co-combustion without affecting the coal ash and causing operational problems in the boiler, and with concurrent minimising of emissions.

The results are usually outlined in the diagrams as NO_x emissions of the combustion reactor related to 6 % O_2 in the flue gas over the fuel mass flow rate of the pyrolysis process. The higher the fuel mass flow rate of the pyrolysis process the more pyrolysis gas and thus reburn fuel is injected into the combustion. Directly related to the flow rate of the pyrolysis gas into the combustion is the air ratio in the reduction zone of the combustion reactor. The outline of the NO_x emissions over the air ratio describes the reburn efficiency of the injected gas, whereas the outline over the fuel mass flow rate gives information on the biomass ratio on the overall process.

Influence of Pyrolysis Temperature

Figure 7 shows the NO_x emissions using pyrolysis gas from straw in dependance on the straw mass flow and the pyrolysis temperature. The pyrolysis gases were generated in the entrained flow reactor with the ceramic reaction pipe. With increasing straw mass flow into the pyrolysis process and thus with increasing pyrolysis gas flow into the reduction zone of the combustion reactor the NO_x emissions of the combustion decrease until a minimum is reached at a certain straw mass flow rate. A distinct influence of the pyrolysis temperature on the reburn efficiency can be determined. Pyrolysis gas generated at 800 °C showed the best NO_x reduction efficiency with a NO_x minimum of 110 mg/m³ at a flow rate of 1,3 kg_{dry}/h. Gases generated at lower or higher pyrolysis temperature resulted in worse NO_x emissions. More straw must be pyrolysed to reach NO_x emissions below 200 mg/m³. When increasing the pyrolysis gas flow after reaching the NO_x emission minimum the NO_x emissions rise again due to an excessive supply of nitrogen and combustible compounds in the pyrolysis gas. Those compounds lead to higher NO formation in the burnout zone of the reactor.

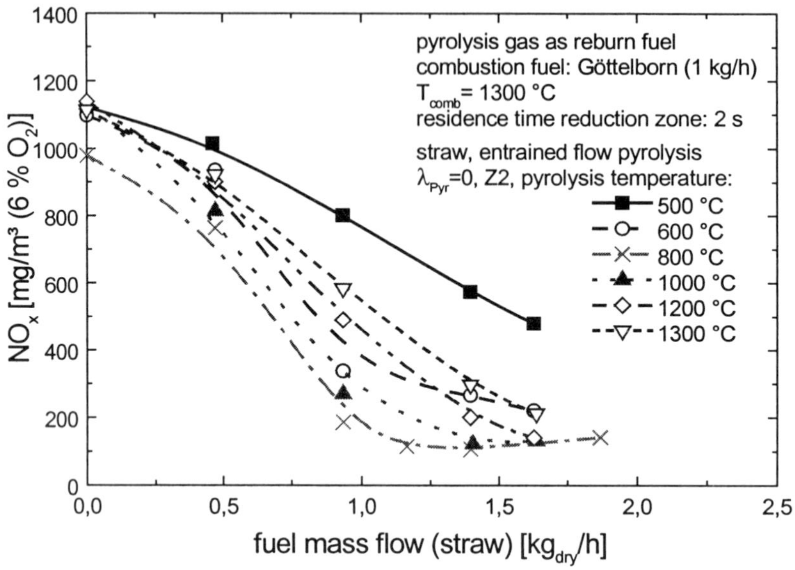

Figure 7 *NO_x emissions using pyrolysis gas from straw in dependance on straw mass flow and pyrolysis temperature*

The optimum settings for the pyrolysis process regarding the NO_x reduction efficiency is 800 °C pyrolysis temperature and a straw mass flow of about 1,3 kg_{dry}/h. This is a mass ratio for the complete co-combustion process of 1,3 straw to coal mass flow. This is equivalent with a thermal power input of 21415 kJ/h straw or an overall share on the total thermal power input of about 40 % of straw.

Outlining the air ratio in the reduction zone over the fuel mass flow rate of the pyrolysis process for different pyrolysis temperature (*Figure 8*) the air requirement of the pyrolysis gases is shown. There is a linear relation between the fuel mass flow and the air ratio. There is no big difference in air consumption between the gases generated at different pyrolysis temperatures. That means that the composition of the different gases is nearly the same regarding the oxygen demand. Oxygen compounds like CO and CO_2, which lower the oxygen demand of a gas, exist in all gases. The low air demand of the pyrolysis gas generated at 500 °C is due to the insufficient devolatilisation of straw in the pyrolysis process.

The data of *Figure 7* are outlined related to the air ratio in the reduction zone in *Figure 9*. As the air demand of all pyrolysis gases are nearly the same influence of the gases on the NO_x emissions can be seen in *Figure 9*. Taking the information of the two figures the result is a difference between reburn efficiency and NO_x reduction efficiency. Reburn efficiency means in that case the capability of producing a certain air ratio in the reduction zone. This is directly related with the air demand of the gas mixture. The reduction efficiency takes into consideration additionally the capability of reducing NO_x. Therefore a sufficient amount of radicals have to be present.

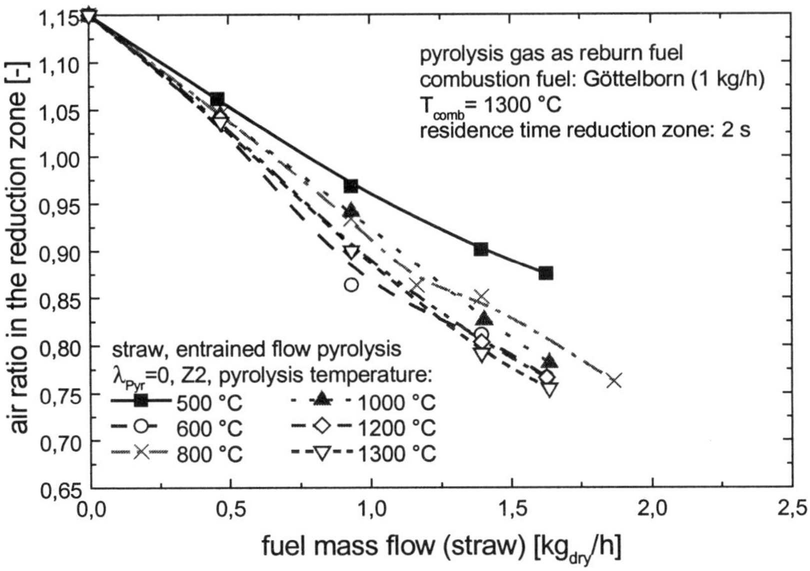

Figure 8 Relation between the air ratio in the reduction zone and fuel mass flow rate of the straw pyrolysis process

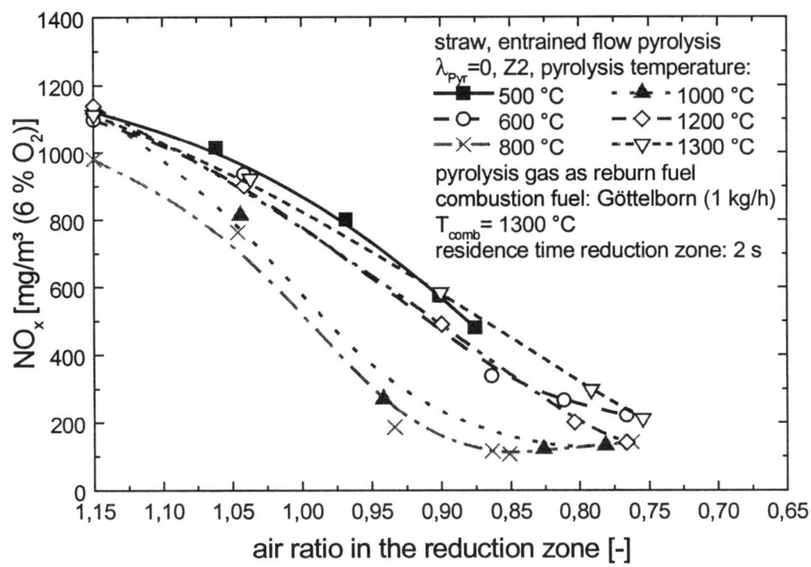

Figure 9 NO_x emissions using pyrolysis gas from straw in dependance on air ratio in the reduction zone and pyrolysis temperature

The air demand of pyrolysis gases produced at high temperatures comes from light gases like CO and H_2, which catch the oxygen very fast, but do not form radicals. The NO_x reduction efficiency of a gas generated at 800 °C comes from the radicals, which are produced by cracking hydrocarbons (aliphatic or aromatic). The air demand of the gas is equal to the 'lighter' gas, but the oxygen consumption carries radicals with it. The content of hydrocarbons in the gas have their maximum in a pyrolysis gas generated at 800 °C. Because all investigated fuels have similar hydrocarbon maxima at around 800 °C in further comparisons of influences on the reduction the evaluation focuses on those settings.

Influence of Air Ratio in the Pyrolysis Process

Having a technical feasible process in mind an addition of air to the pyrolysis process is practically. Experiments with air ratios of 0,15 and 0,3 in the entrained flow reactor have been performed with several fuels. Results gained with all fuels are similar therefore only straw results are presented.

In *Figure 10* a distinct influence of the air ratio in the pyrolysis process on the reburn efficiency is determinable. The NO_x emission minimum of below 200 mg/m³ are achieved with a very high mass flow of straw. For the experiment with an air ratio of 0,3 the minimum could not be reached due to the limited feeding rate of the reactor.

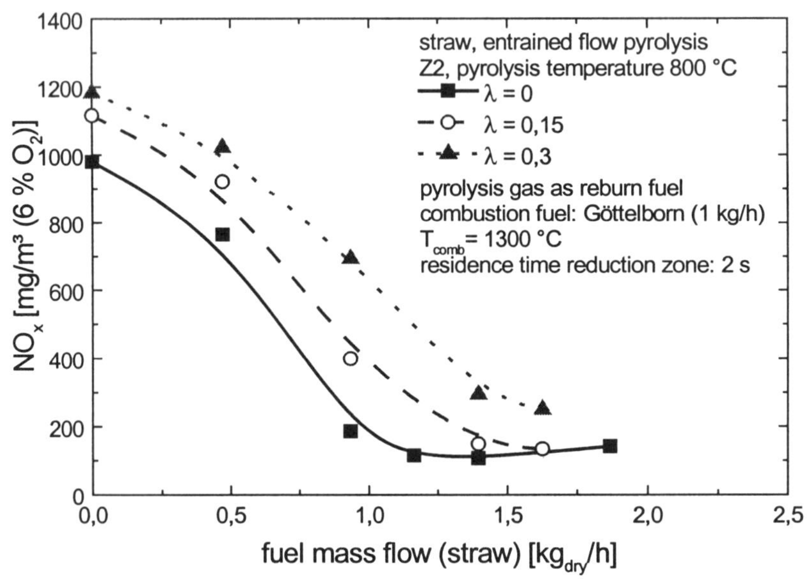

Figure 10 NO_x emissions using pyrolysis gas from straw in dependance on straw mass flow and air ratio in the pyrolysis

The air demand of the gas shows a dependance on the air ratio in the pyrolysis reactor (*Figure 11*). Because the addition of air into the pyrolysis process already serves a part of the demand the air consumption in the combustion process is less.

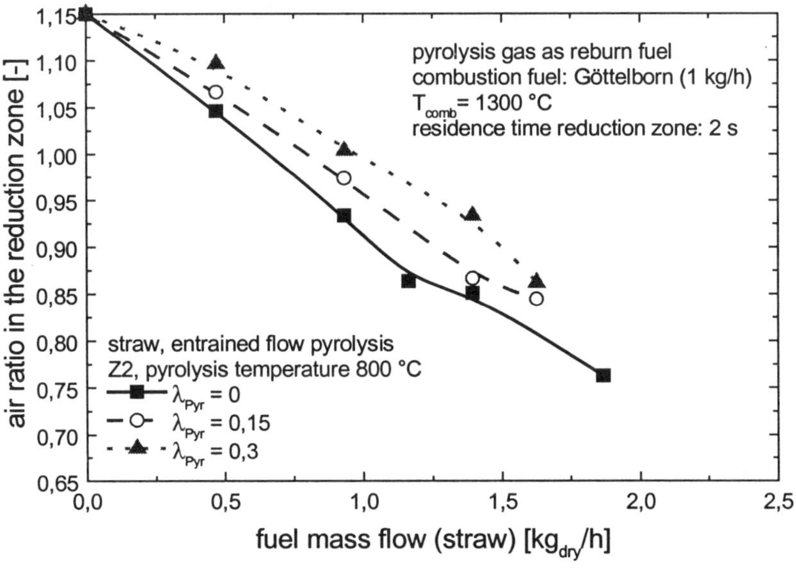

Figure 11 Relation between the air ratio in the reduction zone and fuel mass flow rate of the straw pyrolysis process (λ test runs)

The reduction efficiency of the gases is not as much influenced by the air ratio in the pyrolysis process like when varying the temperature (*Figure 12*). The minimum NO_x emission is shifted to lower air ratios in the reduction zone with increasing air ratio in the pyrolysis. Combining the information from Figure 10 to *Figure 12* the result is, that during the pyrolysis process mainly gas components are oxidised, which do not form the necessary radicals.

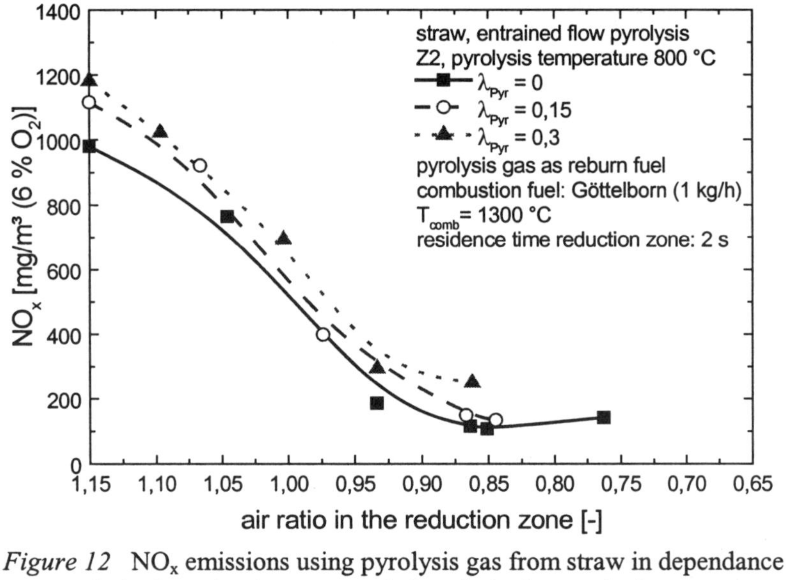

Figure 12 NO_x emissions using pyrolysis gas from straw in dependance on air ratio in the reduction zone and air ratio in the pyrolysis process

Influence of Fuel

Several fuels have been tested on their NO_x reduction efficiency when pyrolysing them under different conditions. For further comparison the setting of 700 °C pyrolysis temperature and inert atmosphere in the entrained flow reactor has been chosen. Detailed results of experiments with coal [7, 8], biomass [18] and sewage sludge [17] pyrolysis gas have been published elsewhere. The results are integrated into Figure 13 for comparison.

The used synthetic gas mixture (26,5 % CO, 5,35 % CO_2, 7,75 % H_2, 53,1 % CH_4, 7,3 % C_2H_x) characterises the composition of a pyrolysis gas from coal generated at a reactor temperature of 700 °C without any N and tar compounds. The NO_x minimum of 200 mg/m³ is reached at an air ratio of 0,75 in the reduction zone. Changes of the composition of the gas mixture (different CO, CO_2, H_2, CH_4 contents) did not have any effect on the NO_x emissions and on the air ratio. Even 100 % of CH_4 results in same NO_x emissions like reburning with synthetic gases. The better NO_x reduction behaviour of pyrolysis gas compared with the synthetic gas mixtures examined is due to the nitrogen compounds in the raw fuel and thus in the pyrolysis gas and due to the higher hydrocarbon content (especially aromatic hydrocarbons, tars). This can be seen from the results achieved with several reburn experiments with pyrolysis gas from coal and different biomass generated in the entrained flow reactor shown in Figure 13.

The coal pyrolysis gas corresponds with the synthetic gas. But additional tar is contained in the real pyrolysis gas. A NO_x emission minimum of 200 mg/m³ is already achieved at an air ratio of 0,9 in the reduction zone.

Biomass with a rather low nitrogen content in the raw fuel (straw, beech) and biomass with a very high N content (sawdust, sewage sludge) have been chosen to show the occurring effects more clearly. The sawdust is a residue from wood-processing industry and it is contaminated with melamine ($C_3H_6N_6$), which is a constituent of the adhesives used.

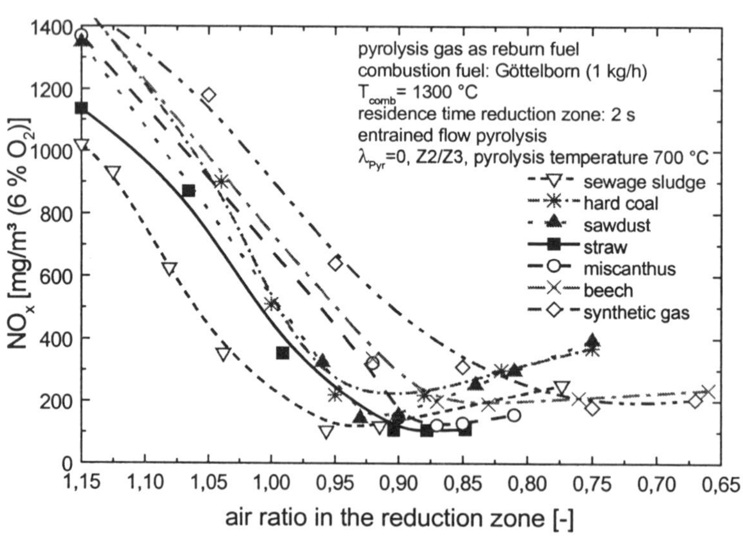

Figure 13 NOx emissions using pyrolysis gas from different fuels in dependence on air ratio in the reduction zone and pyrolysis fuel

Straw and beech produce comparable gases because of their similar composition. The minimum NO_x emission under same combustion conditions of about 200 mg/m³ can be found at air ratios between 0,84 and 0,87 in the reduction zone.

Unlike the pyrolysis gases from beech the generated gases from sawdust and sewage sludge are very rich concerning nitrogen compounds. Due to the high N content in the pyrolysis gas and tars the distinct minimum of NO_x emissions (about 160 mg/m³, 110 mg/m³, respectively) is shifted to higher air ratios ($\lambda = 0{,}92$ and $\lambda = 0{,}95$) in the reduction zone as shown in Figure 13. The increasing NO_x emissions at low air ratios is caused by unreacted N-species of pyrolysis gas and tars, which are oxidised in the burnout zone of the combustion reactor.

CONCLUSION

The results of the reburning test runs have shown that pyrolysis gases from biofuels are very well suitable for using them as a reburn fuel with excellent NO_x reduction quality. Trials with pyrolysis gas of different biomass types resulted in NO_x emissions of a coal furnace below 200 mg/m³. Especially with nitrogen rich pyrolysis gases low emissions were achieved. Nitrogen compounds bound in pyrolysis tars and gaseous compounds, like ammonia and hydrogen cyanid, have a big influence on the reduction capacity of gasification/pyrolysis gas, which has been proven by using a gas from a high melamine contaminated sawdust as feedstock for the entrained flow reactor. The reburn efficiency increases with a higher share of fuel nitrogen. But besides nitrogen compounds also radical forming hydrocarbons are important for NO_x reduction in the reduction zone of the combustion reactor. Therefore pyrolysis gases generated at 800 °C show the best reduction efficiency due to their high content of aliphatic and aromatic hydrocarbons. Excellent NO_x reduction can be achieved at pyrolysis temperatures around 800 °C with an overall straw to coal ratio of 1,3. This means that the share on the thermal input of 40 % is met by straw.

Investigations on the behaviour of volatile and ash components have been carried out. The solid pyrolysis residues of several fuels have been analysed for their main ash composition and their devolatilisation. Increasing reaction temperatures result in a higher devolatilisation for all fuels. E.g. straw shows a devolatilisation of approx. 80 $\%_{dry}$ above 600 °C pyrolysis temperature. But already at low pyrolysis temperatures of 400 °C a rather high devolatilisation of about 70 % can be reached. Other biofuel show a similar behaviour on ash and water free basis. Focusing the devolatilisation efficiency high temperatures in the pyrolysis process give best results, but also pyrolysis at low temperatures yields in sufficient devolatilisation rates.

Gaseous hydrocarbons have a production maximum at about 800 °C reaction temperature for all feedstocks. Carbon monoxide and hydrogen are increasingly formed at high pyrolysis temperatures due to gasification reactions. Tar components are decreasing with increasing temperatures. A distinct release of about 70 % of the fuel nitrogen begins above 600 °C, but with rising temperatures the level of release is not increasing anymore. About 20 % of the fuel nitrogen stays in the char and is separated from the gas within the filtration.

ACKNOWLEDGEMENT

The research is funded in part by the Commission of the European Community, Joule III Research Programme (JOR3-CT95-0057), and the Bundesministerium für Wirtschaft (Federal German Ministry of Economic Affairs) via the AiF Arbeitsgemeinschaft industrielle Forschung 'Otto von Guericke' e.V. (industrial research foundation).

REFERENCES

1 Albrecht W.: NO_x-Emissionen aus Kohlestaubflammen. VGB-Kraftwerkstechnik 72, Heft 7, 1992

2 Beer J.: NO-Formation and Reaction in Fluidized Bed Combustion of Coal. J. Inst. Energy 54, 1981

3 Dürselen H.J.: Feuerungsseitige NO_x-Minderung bei Braunkohlenstaubfeuerungen, Einfluß von brennstoff- und betriebsseitigen Parametern. Dissertation, Ruhr-Universität Bochum, 1992

4 Evans R.J., Milne T.A.: Molecular Characterization of the Pyrolysis of Biomass. Energy & Fuels, Vol. 1, No. 2, 1987

5 Fenimore C.P.: Formation of Nitric Oxide in Premixed Hydrocarbon Flames. XIII Comb. Symp., New York, 1971

6 Fenimore C.P.: Reactions of Fuel-Nitrogen in Rich Flame Gases. Combustion and Flame 26, 1976

7 Greul U.: Experimentelle Untersuchung feuerungstechnischer NO_x-Minderungsverfahren bei der Kohlenstaubverbrennung. VDI Fortschrittberichte, Reihe 6, Nr. 388, VDI Verlag, Düsseldorf, 1998

8 Greul U., Magel C., Moersch O., Rüdiger H., Storm C., Schnell U., Spliethoff H., Hein K.R.G.: Einsatz von kohlestämmigem Pyrolysegas als Reduktionsbrennstoff. Final Report, BMBF-Research Project No. 0326766A; IVD Berichte, Bd. 1, Institut für Verfahrenstechnik und Dampfkesselwesen, Universität Stuttgart, Stuttgart, 1997

9 Hein K.R.G.: NO_x-Minderung bei Feuerungen für Rheinische Braunkohlen. Konferenz Einzelbericht, VGB Seminar Rauchgasreinigung und Reststoffentsorgung in Kraftwerken, Industrie- und Heizkraftwerken, Technische Vereinigung der Großkraftwerksbetreiber VGB, No. 7, Pp. 1-24, Cottbus, 1990

10 Kicherer A., Gerhardt Th., Spliethoff H., Hein K.R.G.: Co-Combustion of Biomass / Sewage Sludge with Hard Coal in a Pulverized Fuel Semi-Industrial Test Rig. Final Report, EC-Research Programme, APAS-Contract COAL-CT92-0002, Institute for Process Engineering and Power Plant Technology, University of Stuttgart, Stuttgart, 1995

11 Pershing D.W., Wendt J.O.L.: Pulverized Coal Combustion: The Influence of Flame Temperature and Coal Composition on Thermal and Fuel NO_x. XVI Symposium on Combustion, 1976

12 Petzel H.-K., Scholl G., Tigges K.-D.: Modernste Verbrennungstechnologie zur Primärredizierung non NO_x. VGB Kraftwerkstechnik, Jg. 73, No. 3, Pp. 231-237, 1993

13 Rüdiger H.: Pyrolysegas von festen biogenen und fossilen Brennstoffen zur Erzeugung eines Zusatzbrennstoffes für Feuerungsanlagen. Berichte aus der Energietechnik, Shaker Verlag, Aachen, 1997

14 Schu G.: Experimentelle Untersuchungen zur selektiven nichtkatalytischen Reduktion von Stickoxiden an einem Flammrohrkessel. Dissertation, Technical University of Munich, Munich, 1989

15 Schuler J.: Laboruntersuchungen zur NO-Bildung bei der Kohleverbrennung sowie NO-Reduktion an Pyrolysekoksen. Dissertation, University of Essen, Essen, 1988

16 Spliethoff H., Rüdiger H., Greul U., Spliethoff H., Hein K.R.G.: Kombinierte Minderung

der NO_x-Bildung und Reduzierung von gebildetem NO_x bei der Verbrennung von Steinkohle, Phase 3. Final Report, Research Project No. 0326535C, Bundesministerium für Forschung und Technologie BMFT, Bonn, 1993
17 Storm C., Harter M., Spliethoff H., Hein K.R.G.: Untersuchungen von NO_x-Minderungsmaßnahmen mittels Reduktionsgasen aus Klärschlämmen in Kohlenstaubfeuerungen. Final Report, AiF Research Project No. 10640 N; IVD Berichte, Bd. 3, Institut für Verfahrenstechnik und Dampfkesselwesen, Universität Stuttgart, Stuttgart, 1998
18 Storm C., Harter M., Spliethoff H., Hein K.R.G.: Generation of a Gaseous Fuel by Pyrolysis of Biomass for Additional Use in Coal-Fired Boilers. Final Report, EC Research Project No. JOR3-CT95-0057; Institut für Verfahrenstechnik und Dampfkesselwesen, Universität Stuttgart, Stuttgart, 1999
19 Suuberg E.M., Teng H., Calo J.M.: Studies on Kinetics and Mechanism of the Reaction of NO with Carbon. XXIII Comb. Symp., 1990
20 Zeldovic Y.B.: The Oxidation of Nitrogen in Combustions and Explosions. Acta Physicochimica URSS 21, 1946

Combustion of Bio-oil in a Gas Turbine

R. Strenziok, U. Hansen, H. Künstner
Institute of Energy and Environmental Technology
University of Rostock, Germany

ABSTRACT: Based on a series of test runs in a combustion test facility and a gas turbine it could be shown that bio-oil from pyrolysis of wood can be employed for energy purposes. The fuel has to be clean without solids and of a homogenous consistency. The gas turbine combustion behaviour was examined in numerous tests during 1999 and 2000. The optimisation of the operating parameters was performed with adjustment of the fuel viscosity and fuel pressure in different load regions. The results set the priority for further investigations. The adaptation of the industrial gas turbine T216 to the wood based bio fuel was restricted by the design to part load. The emissions were measured for both bio-oil and diesel fuel operation. When compared to diesel fuel, characteristically, the emissions of CO and HC are higher for bio-oil at part load operation. The results indicate slightly lower NOx emissions for bio-oil.

INTRODUCTION

A study of the future energy supply in Germany has shown that, five percent of the primary energy demand and ten percent of the electricity production in the year 2010 could be supplied from renewable energy. Also bio-mass as a renewable resource and an environmental friendly energy carrier is assigned an increasing importance for future energy supply. As a clean fuel bio-oil has a number of competitive advantages over diesel fuel, no sulphur dioxide and lower nitrogen oxide emissions. Furthermore, it has a favourable carbon balance contributing little to greenhouse gas emissions.

Bio-oils have so far mainly been in the focus as an alternative fuel for internal combustion engines. The fuel obtained from fast pyrolysis, however, displays properties which do not lend themselves easily to this application. Intermittent combustion under high pressure in an reciprocating machine appears more demanding than continuous combustion under lower pressure in the combustion chamber of a gas turbine. This is the working assumption for the studies reported below.

At the Institute of Energy and Environmental Technology the combustion behaviour of wood based bio-oil has been investigated in a combustion test facility of 300 kW-thermal capacity as well as in a gas turbine of 75 kW-electric capacity. The investigations were carried out within the framework of the EU- demonstration project

"Scaling up and operation of a flash pyrolysis system for bio-oil production and application on the basis of rotating cone technology" (1997-2000).
The bio-oil used for the test programme was produced by BTG in Enschede (NL) with the Rotating Cone Technology from poplar as the bio-mass material. As alternative raw materials for pyrolysis wheat straw, Spanish thistle and other wood wastes have been investigated. The main objective of this paper is to demonstrate the power generation from bio-oil in a gas turbine, the optimisation of the combustion behaviour with respect to emissions and operation characteristics and the comparison of emission and operation data from bio-oil and conventional diesel fuel. For technical reasons the gas turbine had to be adapted to the operation in dual fuel mode with the possibility to switch between diesel and bio-oil.

BIO-OIL CHARACTERISTICS

The bio-oil used in the tests was delivered from the project partner BTG in April 2000. For the chemical and physical analysis of bio-oil we used modified standard fuel oil methods. Table 1 shows the chemical and physical properties of the oil. The analysis was carried out in the Institute's chemical laboratory.

Table 1 Properties of the bio-oil used in the gas turbine

Date of analysis 23.5.2000					
Sample number	2232				
Short name	**K3**				
Raw Material	Poplar				
	units				
Density @ 15 °C	kg/m³	1175	Calorific value	[J/g]	14067
Viscosity @ 40 °C	mm²/s	13.5	Neutralization number	[KOH/g]	93
Coke tendency	[Mass.-%]	13.4	pH-value	[-]	2.65
Carbon	[Mass.-%]	36.9	Pour point	[°C]	-18
Hydrogen	[Mass.-%]	7.6	Flash point	[°C]	56
Nitrogen	[Mass.-%]	0	Sodium	[mg/kg]	8
Water content	[Mass.-%]	26.1	Aluminium	[mg/kg]	16
Oxide ash	[Mass.-%]	0.018	Silicon	[mg/kg]	10
Calcium	[mg/kg]	38	Calcium	[mg/kg]	38
Zinc	[mg/kg]	5.7	Lead	[mg/kg]	1.5
Nickel	[mg/kg]	4	Sulphur	[mg/kg]	0.03
Copper	[mg/kg]	700	Phosporus	[mg/kg]	30

The oil was produced from non-contaminated soft wood from Spain. Due to the low pH-value, acid proof materials had to be used in the fuel supply system. The pure bio-oil need to be preheated up to 50 °C to achieve a sufficient viscosity. Bio-oil from pyrolysis contain considerable amount of water between 20 to 30 per cent. However, combustion in the gas turbine was not hampered by the presence of water as the ignition of the bio-oil was initiated by the pilot flame burning diesel fuel.

Bio-oil from pyrolysis is immiscible with petroleum based fuels and therefore a stable mixture of bio-oil and diesel fuel could not be obtained and tested in the gas turbine.

EXPERIMENTS

The experimental work was performed with a small commercial gas turbine type T 216 with a rated electric power output of 75 kW. It has a single shaft, single stage radial turbine and a single stage radial flow compressor with a pressure ratio of approx. 2,5. The turbine shaft speed ranges from 50 000 to 30 000 min^{-1} and with a two stage reduction gear the turbine is coupled to a synchronous generator.
For test purposes it is important that the combustion chamber is of a silo type arranged tangentially. A picture of the gas turbine is shown in Fig. 1.

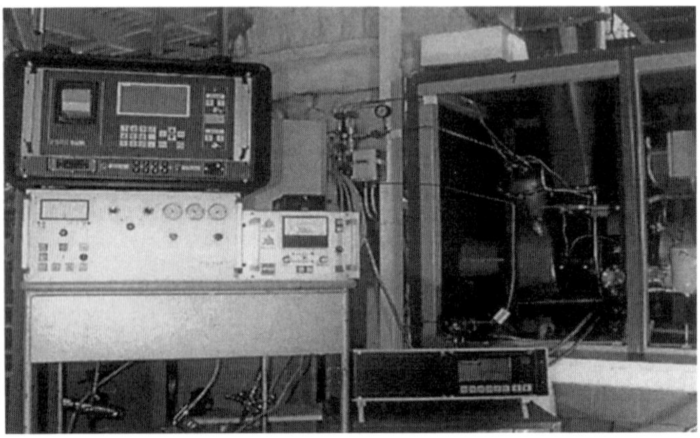

Fig. 1 Gas turbine T 216 with exhaust measurement device

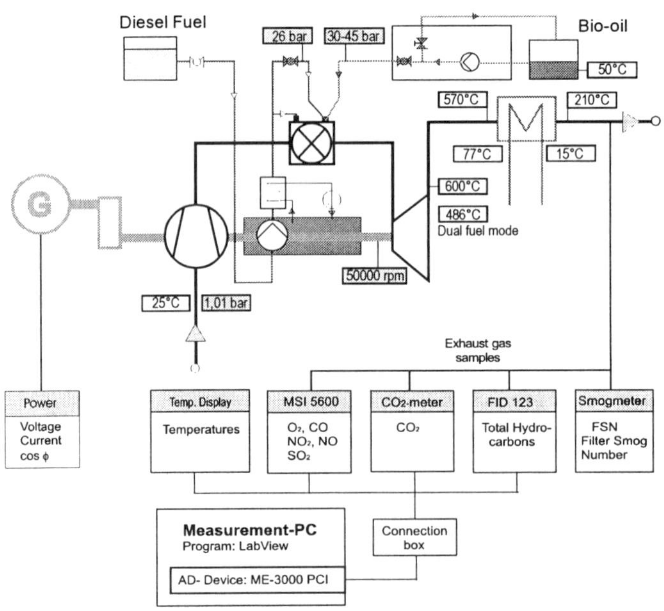

Fig. 2 Gas turbine test bed for Dual Fuel operation

Extensive alterations to the fuel system was necessary in order to burn bio-oil from pyrolysis. The flow diagram in Fig. 2 shows the layout of the gas turbine system after modifications. The temperatures and pressures refer to diesel fuel operation.

The operation on bio-oil alone is not possible because the speed safety governor and the lubricating oil cooler is designed for diesel oil operation. Therefore some proportion of fuel must be diesel. The combustion chamber of the gas turbine was fitted with two inline fuel nozzles, a main nozzle and an ignition nozzle (Fig. 2). The ignition nozzle was operated with diesel oil. The supply of bio-oil to the main nozzle was made through a separately driven, external fuel pump (Dual Fuel Mode). It is designed for a maximum flow of 455 l/h at a pressure up to 80 bar and the effective flow controlled in a bypass. The flow through the main nozzle can be switched between diesel and bio-oil. The proportion of fuel mass flow of the main nozzle and ignition nozzle is given in Fig. 3.

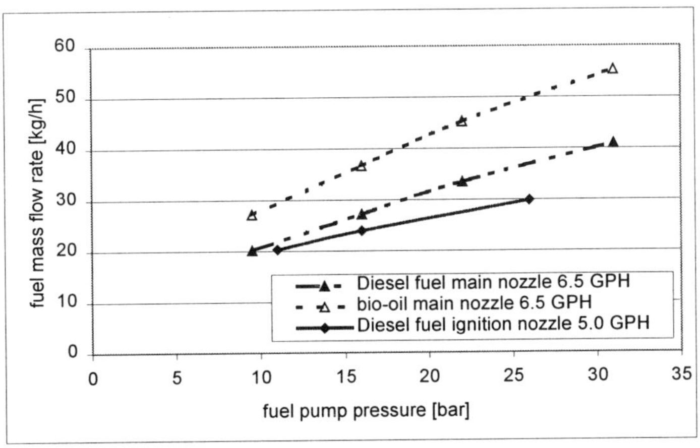

Fig. 3 Mass flow rate of ignition nozzle and main nozzle

Fig. 4 External bio-oil pump

The bio-oil from pyrolysis may contain solid residues and is highly viscous. Filtering and pre-heating are necessary before injection in the main nozzle. For the gas turbine tests the following properties were achieved:

- particle size < 20 Micrometer
- injection viscosity about 12 cSt.

The gas turbine was started with diesel oil. Then followed a change over of the fuel through the main nozzle from diesel oil to bio-oil. Due to the lower heating value of the bio-oil it was only possible to operate the gas turbine at part load in the dual-fuel-mode. With the mass flows in Fig. 3 the following power levels are calculated and shown in Table 2.

Table 2 Comparison of turbine power in diesel fuel and dual fuel operation

Turbine power	Diesel fuel Operation	Dual fuel operation
Ignition nozzle 5.0 GPH	P = 354 kWth (p = 26 bar)	P = 354 kWth (p = 26 bar)
Main nozzle 6.5 GPH	P = 437 kWth (p = 26 bar)	P = 226 kWth (p ~ 30 bar)
Total	P = 791 kWth	P = 580 kWth

The mass flow of bio-oil in relation to diesel fuel increases in the main nozzle with 6.5 GPH (gallons per hour) by up to 1.33. The power in the dual fuel mode is reduced to 73 % of the full power in the diesel fuel mode. The reduction is due to the specific limitations of the gas turbine used for the experiments and not inherent in bio-fuel operation. However, for later commercial use further alterations to the fuel system and the nozzle size would be required.

The high viscosity makes pre-heating necessary and with high pre-heating temperatures the danger of coking in the main nozzle arises. In order to prevent overheating in the nozzle it is equipped with a cooling adapter and a thermocouple.

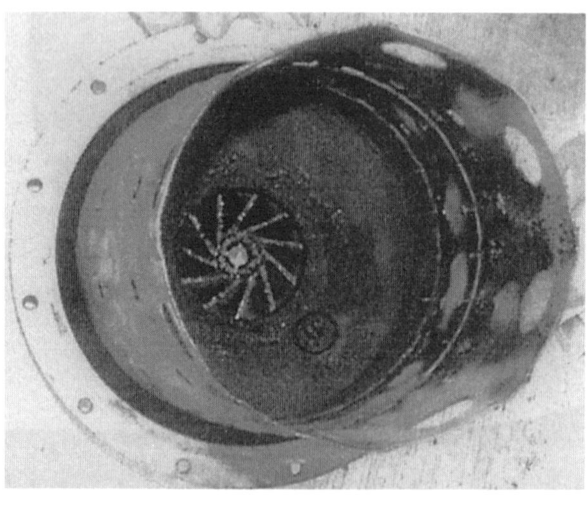

Fig. 5 Deposits in the gas turbine combustion chamber after bio-oil operation

After short combustion tests in the gas turbine deposits were found in the combustion chamber as shown in Fig. 5. The deposits are of a lacquer like nature and can only be removed mechanically. At present, the deposits in the combustion chamber do not constitute a problem for the operation of the gas turbine. However, there are also slight deposits on the turbine blades.

If the deposits are to be removed, the turbine needs to be taken apart. Due to the high speed of the turbine larger deposits cannot be tolerated. Means to prevent fouling of the turbine is a priority requirement. The extent to which fouling occurs may depend on the quality of the bio-oil and can only be determined from duration tests.

EXHAUST EMISSIONS

The equipment for measuring and analysing emissions is shown in Fig. 1 and 2. The oxygen content "as measured" is 17.5 %. The excess air ratio is in the diesel fuel operation 3.3 and in the dual fuel operation about 6.

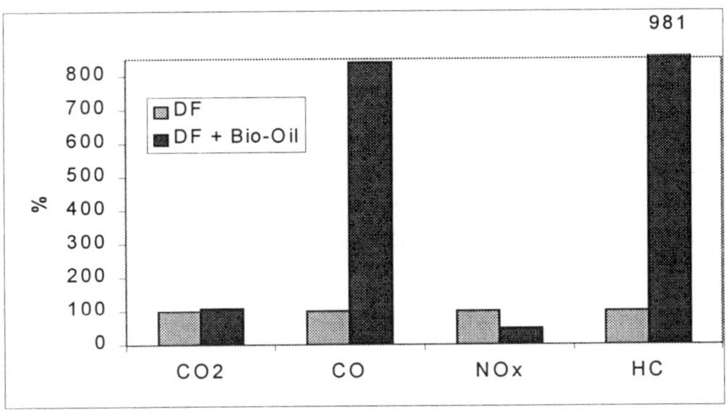

Fig. 6 Bio-oil emissions in the gas turbine in relation to diesel operation (100 %)

When compared to diesel fuel the emissions of CO and HC are higher for bio-oil. The results indicate slightly lower NOx emissions for bio-oil. All measurements were taken at part load operation due to the limitation of the gas turbine. The initial tests in dual-fuel-mode were performed with a relatively "cold" combustion at part load which may be the reason for the incomplete combustion of the hydro carbons. Also the combustion air fuel ratio was not adjusted for bio-oil operation at part load. The combustion of bio-oil is clearly recognised by the exhaust gas odour and a light exhaust gas cloudiness.

The tests performed at the Institute for Energy and Environmental Technology are among the first involving exact measurements of emissions and are comparable to emissions data found in the technical literature (Andrews and Fuleki, 1997). Due to limitations of the particular gas turbine and restricting the test series to one type of bio-oil the results cannot claim to be generally valid for the comparison of bio-oil from pyrolysis to a standard fuel like diesel oil but should give some indication of the qualitative outcome.

The studies have demonstrated the possibility to burn bio-oil in a gas turbine for the production of electricity and heat.

The principal modifications involve the fuel handling and the fuel supply system. Basic results and insights of bio-oil combustion were obtained in the combustion test facility.

REFERENCES

1. Bridgwater A. & Hogan E. (1996) *Bio-oil Production and Utilisation.* Proceedings of the 2nd EU- Canada Workshop on Thermal Biomass Processing. Cpl press , ISBN 187269151 X
2. Andrews R. G. & Fuleki D. (1997) *Results of industrial gas turbine tests using a biomass derived fuel.* Proceedings of the Third Biomass Conference of the Americas.
3. Oasmaa A. & Leppämäki P. (1997) *Physical characterisation of biomass-based pyrolysis liquids.* VTT Publications No. 306, Espoo
4. Bridgwater A. & Czernik S. (1999) *Fast pyrolysis of biomass: A Handbook.* Cpl press, ISBN 1872691072
5. Hansen U. (1999) *Einsatz von Pyrolyseöl als regenerativer Energieträger zur dezentralen Versorgung mit Strom und Wärme.* Freiberg, Germany.
6. Strenziok R. (1999) *Methods and problems of bio-oil analysis.* PyNe-Workshop Montpellier, France.
7. Morris K. & Piskorz J. (1999) *Bio ThermTM: A System for Continuous-Quality, Fast Pyrolysis BioOil.* 4th Biomass Conference of the Americas.
8. Meier D. & Faix O. (1999) *State of the art of applied fast pyrolysis of lignocellulosic materials- a review.* Bioresource Technology 68

Stirling Engine with Flox® Burner Fuelled With Fast Pyrolysis Liquid

Andreas Bandi and Frank Baumgart
Center for Solar Energy and Hydrogen Research Baden-Wuerttemberg, (ZSW),Hessbruehlstr. 21C, 70565 Stuttgart, Germany

ABSTRACT Fast pyrolysis liquid was combusted with a modified propane gas FLOX® burner (WS) mounted to a Stirling CHP unit (SOLO 25 kW_{th}). The propane burner was modified by an air pressure atomiser designed for pyrolysis liquids. Nearly 110 hours of operation with different fuel loads have been achieved, engine performance and emissions were recorded. The Stirling tests proved that pyrolysis liquids can be burned efficiently and with low emissions in a FLOX® mode operating burner attached to a Stirling CHP unit.

INTRODUCTION

In the frame of the EU-project "Small-Scale Combined Heat And Power Production (Chp) From Bio-Crude Oil (Bco) Fuelled To A Stirling Engine" (JOR3-CT98-0310, co-ordinator CRES, Greece) ZSW is investigating in close collaboration with industrial partners, WS, SOLO, Germany, the use of BCO in a Stirling CHP unit with a FLOX® burner. "FLOX" is acronym for "**FL**ame-less **OX**idation". The special feature of FLOX® burners, developed by WS, Germany, is that at temperatures as high as 850-900°C thermal NO_x is drastically lowered by internally mixing combustion air and exhaust gas, avoiding temperature peaks of the flame. This is even true at high air preheating temperatures. FLOX® has an additional advantage for burning of pyrolysis liquids. As the residence time of the fuel in the burning chamber, is higher than in a normal burning, a more efficient carbon burnout becomes possible. At present FLOX® burners fuelled by propane or natural gas are used for high temperature process heat production. A future, promising application of these burners is the use in Stirling CHP units (SOLO, Germany).

Within this project ZSW designed and realised an injection-atomisation system for BCO. A WS propane burner [1, 2] was modified with a ZSW atomiser and attached to a SOLO Stirling engine 161 (4-9 kW_e, 10-25 kW_{th}) [3]. This paper presents experimental results on the burning of BCO in a FLOX® burner and in a Stirling CHP unit with a FLOX® burner.

The combustion of bio crude oils has been tested so far in stationary Diesel engines, boilers and gas turbines [4]. There are several atomisation systems available for combustion of mineral oils. Based on physical properties, BCO can be classified between light fuel oil and low sulphur heavy oil. However BCO has a number of unique properties, such as high acidity (pH $\approx$ 2–3), low heating value (Lower Heating Value, LHV = 15-19 MJ/kg), high viscosity (30-2000 cSt at ambient temperature) [5] and high instability, that makes it difficult to use as a combustion fuel. The wide range of organic components with different boiling temperatures and the char content have a negative effect on the emissions (CO, HCs and soot) by burning BCO. In addition, the high water content (generally > 20%) and the high oxygenated molecule concentration impair the ignition and increase the coking tendency.

ATOMISATION EXPERIMENTS

Viscosity, solid particle content, particle size and heating value of the liquids are determining factors in the selection of the proper atomisation system. The atomisation technologies available include pressure atomisers, air or steam pressure atomisers and rotary atomisers. Pressure atomisation is mostly used for the combustion of oils with low viscosity (2.5-5 cSt), e.g. light oils. In this system a pump raises the pressure up to 10 and more bar and forces the oil through a narrow nozzle hole producing a fine spray of oil with droplet sizes ranging between 30 and 100 μm. Due to the generally narrow sized nozzle holes, the pressure atomisers tend to get blocked. These atomisers are less suitable for liquids with high viscosity, relative high solid particle content and low fuel charge.

The rotary atomisation is suitable only for large liquid flows. For the atomisation of BCO a pressurised air atomiser with external mixing chamber from (Schlick-Düsen GmbH, Germany) was selected. Using this type of atomiser the liquid mass flow can be varied independently from the air flow, providing a constant atomisation quality at different BCO mass flow. The selected nozzle hole was 0.8 mm and the spray cone angle 30°. In order to provide the proper atomisation temperature for BCO, the nozzle has been equipped with a temperature control (cooling/heating) system. The atomiser was designed to operate with different BCO mass flow up to 8-10 l/h, by a liquid pressure of about 0.2-1.0 bar. For an air flow of 15 m^3/h (2 bar), the droplet size was estimated from the propellant air/BCO mass ratio to about 10-50 μm (Sauter mean diameter). In order to study the atomisation properties of BCO, an experimental set-up has been constructed, including the atomiser, storage tanks for BCO and methanol (for cleaning), and control systems for the BCO temperature in the nozzle, liquid/air pressure and atomising air pressure. A schematic presentation of the experimental set-up is given in Fig. 1. A picture of the equipment is shown in Fig. 2.

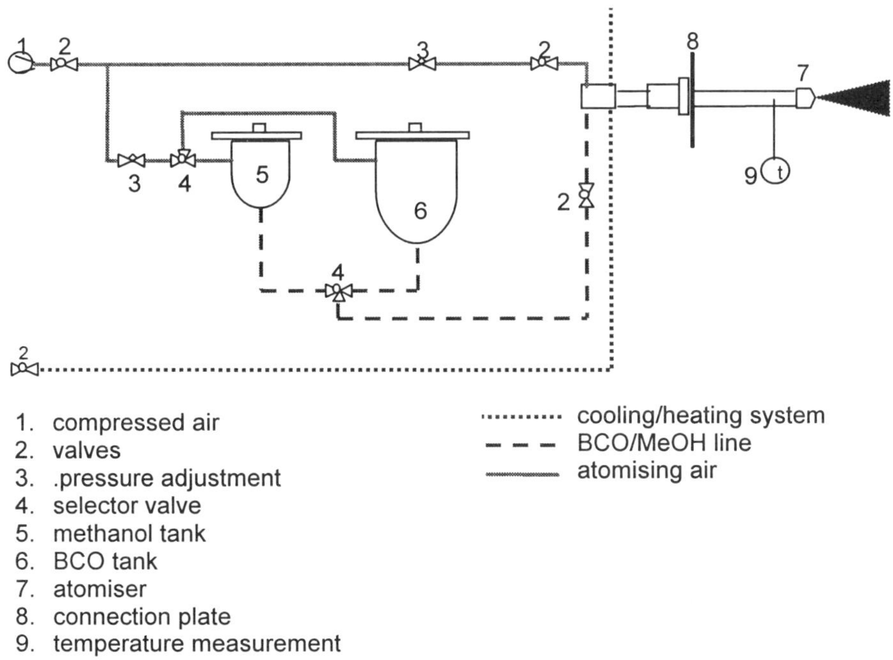

Fig. 1: Scheme of the experimental installation

Fig. 2: Experimental set-up for BCO atomisation tests

For the atomisation and burning tests BCO supplied by Ormrod Diesels was used. Some of the characteristic properties of this BCO are presented in Table 1.

Table 1: Characteristic properties of the BCO used in experiments

Feedstock	Pine
Density, g/cm^3	1.188 – 1.191
Ash, %	0.02
Carbon, %	46.24 – 47.07
Hydrogen, %	6.81 – 7.09
Nitrogen, %	< 0.1
Sulphur, %	< 0.1
Oxygen, %	46.95 – 45.84
Flash point, °C	44 - 55
Pour point, °C	-27 ...-24
pH	2.2 – 2.4
H_2O, %	24.1 – 24.4
Char content, %	0.31 – 0.37
Particle size of char, μm	8.3 – 8.4
Absolute viscosity, centipoise (cP)	< 100
Kinematic viscosity, centistokes (cSt): 20°C 50°C	 53.3 – 61.6 11.6 – 18.0
Higher heating value, MJ/kg	16.5 – 16.8
Lower heating value, MJ/kg	15.0 – 15.3

Mass flow experiments at different liquid pressures were performed at constant temperature. For 30°C and 50°C the results are shown in Fig. 3. A typical spray cone at 2 bar and a BCO mass flow of 8 l/h is shown in Fig. 4. The mass flow and atomising experiments have been carried out without any difficulties related to BCO properties.

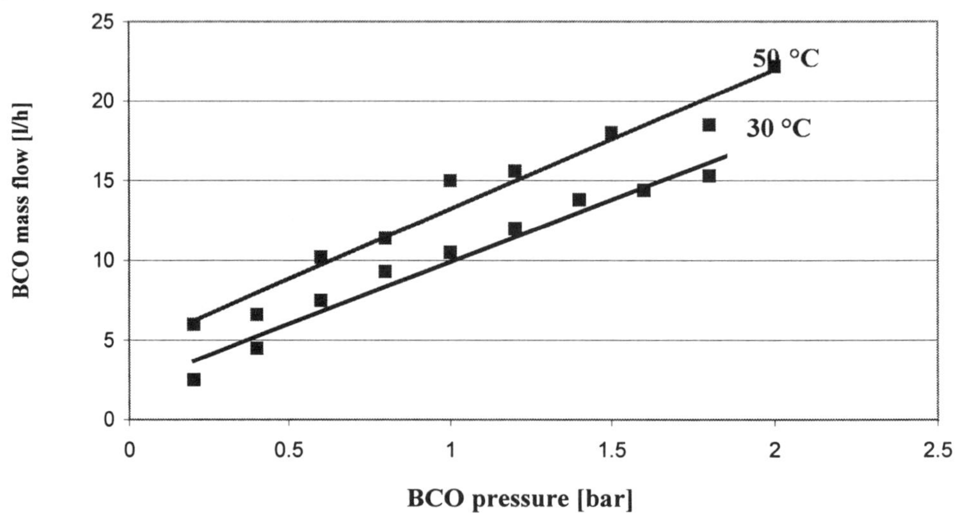

Fig. 3: BCO mass flow dependence on pressure (nozzle hole diameter: 0.8 mm)

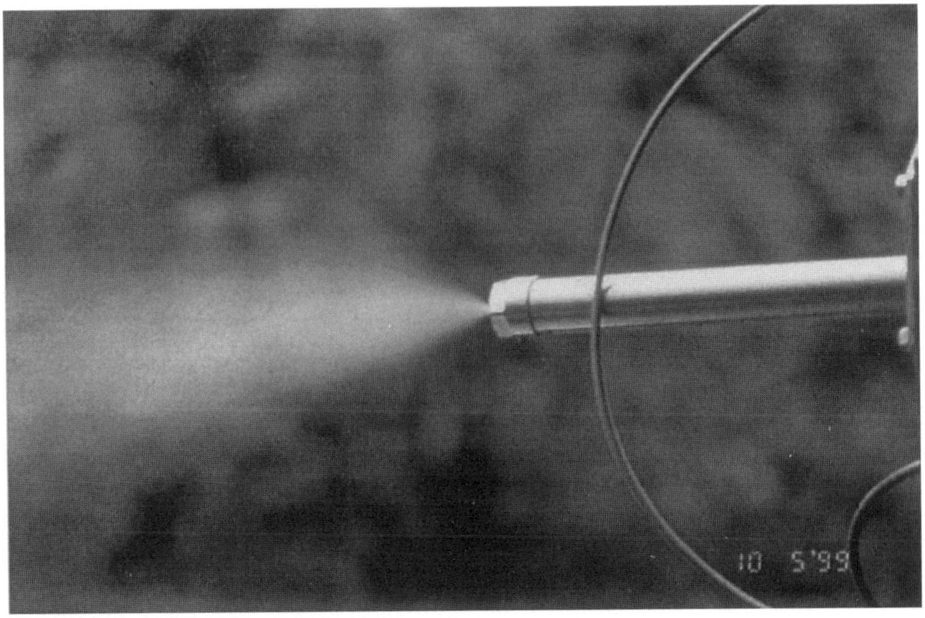

Fig. 4: Atomisation cone with BCO (nozzle hole: 0.8 mm; BCO flow: 8 l/h; air pressure: 2 bar

COMBUSTION EXPERIMENTS WITH A FLOX® BURNER

A traditional propane burner of WS was modified for BCO combustion. The burning chamber with the modified burner was pre heated up to 900°C with an auxiliary burner placed opposite to the BCO atomiser (Fig. 5). The nozzle was cooled down to 30°C. Once 900°C was reached in the burning chamber, the BCO injection was started. After the ignition of BCO, which occurred immediately with supply of the first amount of BCO, the auxiliary burner was turned off. The supply of BCO was started with about 2 l/h and than increased to about 9-10 l/h (liquid pressure 0.9 bar). The air supply was adjusted to about 3% O_2 in the flue gases. The burning of BCO in FLOX® mode continued stationary without any difficulties for several hours. Table 1 summarises the main experimental results.

The experiments proved that BCO can be burned without any noticeable residues or soot formation and with practically no CO emission in stationary FLOX® mode. Improvement in NO_x emission can be expected by varying excess air and exhaust gas recirculation ratio. However, the nitrogen content of BCO is itself a source of NO_x which can not be reduced below a certain limit. The low CO concentration provides evidence for a high quality atomisation which allows a total carbon burn out. Fig. 6 shows the combustion chamber fuelled with BCO (FLOX® mode). The temperatures in the burning chamber (on different places) and in the nozzle were recorded. (see Fig. 5). In order to avoid the nozzle plugging, the BCO temperature in the nozzle was controlled closed to the nozzle tip and was kept at 25-30°C, therefore a temperature caused blocking (coke formation) can be excluded.

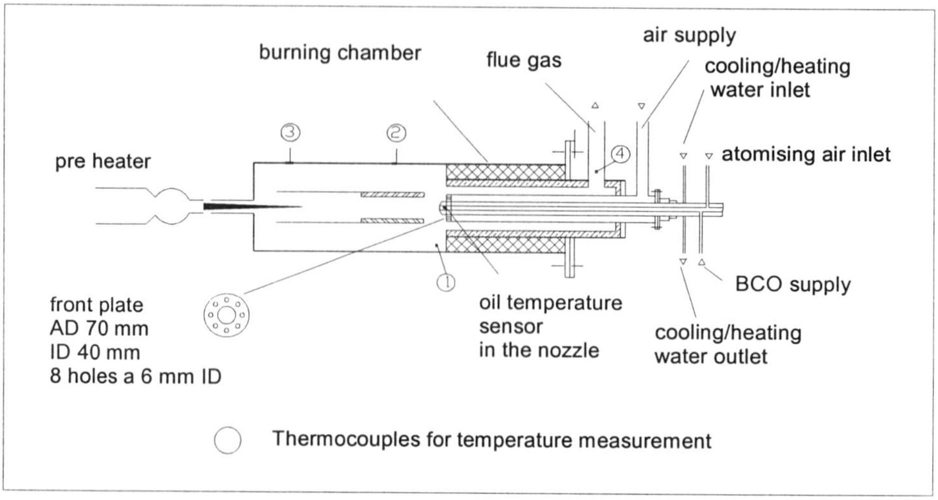

Fig 5: Modified WS propane FLOX® burner for BCO combustion

Table. 1: Experimental results to BCO FLOX® burning in a modified WS propane burner

BCO supply	l/h	9-10
Oven temperature	°C	970-1060
Liquid pressure	bar	0.9
Thermal output	kW_{th}	40
Oxygen conc. in flue gas	% Vol	3±0.5
NO_x conc. in flue gas	ppm	100-120
CO conc. in flue gas	ppm	<6
Atomising air pressure	bar	2
Air supply (excluding atomising air)	m^3/h	45
Heat density	kW/m^2	50
Burning chamber charge	MW/m^3	2
Exhaust gas temperature	°C	970

COMBUSTION EXPERIMENTS WITH A STIRLING CHP UNIT EQUIPPED WITH A FLOX® BURNER

The BCO-Stirling experiments were conducted with different pyrolysis liquid loads and air pressures. In order to prevent the blocking of the nozzle during heating-up the engine, the atomising air supply was started simultaneously with the start the heating. After reaching the operation temperature, BCO supply was started and the auxiliary burner was turned off. The burning of BCO in FLOX® mode was stable without any difficulties related to the engine and the burning of BCO. The CHP test rig is shown in Fig. 7.

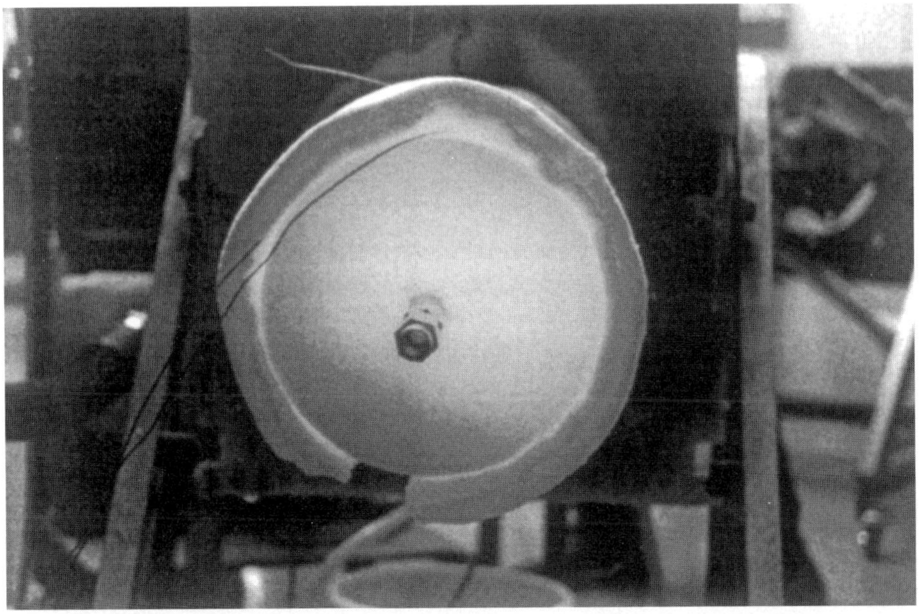

Fig. 6: Combustion chamber fuelled with BCO (FLOX® mode, WS)

In order to avoid the blocking of the nozzle due to the solid particles and agglomerates in the BCO, a mechanical cleaning system was integrated in the atomisation system. The cleaner was activated only when due to break of BCO supply a temperature decrease at the heater was observed. During the experiments engine performance and emission data were recorded. A summary of the records is presented in Table 2.

The Stirling tests proved that fast pyrolysis oils can be efficiently burned in FLOX® mode with low emissions. Several trial runs were carried out in order to improve the engine performance and emission characteristics. BCO flow rate, air/fuel ratio have been varied and emissions have been measured. Ash deposition on the burning chamber walls and on the heat exchanger was not observed. Electrical and thermal efficiencies of the CHP unit were not satisfactory (overall efficiencies 50-60 %). It is considered that the main reasons for low efficiencies are the introduction into the engine of not preheated atomiser air, and the burning chamber geometry. As indicated by the temperature measurements, at high fuel loads the burning takes place behind the heat exchanger, causing heat loss for the engine.

Table 2: Emission and engine performance data for Stirling CHP unit operation

BCO flow, l/h	Atom. air press., bar	Emissions				Q_{th}, kW_{th}	P_{el}, kW_e
		Soot number	HC, mg/m^3	NO_x, mg/m^3	CO, mg/m^3		
4.40	2.5	0.07-0.09	20-40	58	100	8	3.1
6.46	2.0			95	125	13	5.4
7.85	3.0			94	35	15	6.1
7.97	5.0			20	37	14	6.2

German Emission Standards for diesel engine < 5 MW_{th} (O_2 concentration 5%):
NO_x: 500 mg/m^3; CO: 650 mg/m^3; HC: 100 mg/m^3; Soot number: 2
O_2 concentration in flue gas: 6-10%

Fig. 7: Stirling CHP unit test system

CONCLUDING REMARCS

(1) Experiments carried out with modified propane burner proved that BCO can be combusted in FLOX® mode without noticeable residues with low emissions.

(2) Due to its special feature, the FLOX® mode is highly suitable for BCO combustion.

(3) Combustion experiments of BCO in a Stirling CHP unit with FLOX® burner were conducted for nearly 110 hours with different loads.

(4) After preheating the burning chamber up to 800-850 °C with an auxiliary burner, the BCO ignition and combustion were carried out without any difficulties.

(5) Blocking of the atomiser nozzle due to the agglomerates and solid particles contained in the pyrolysis liquid created difficulties in the fuel supply.

(6) The measured emission figures with Stirling CHP unit were below the German Standards for diesel engines in similar power range

REFERENCES

1. Wünning, J., "*GASWÄRME International*", Vol. 47, Vulkan-Verlag, Essen, 1998, pg. 322
2. Wünning, J., Regemat Burner, WS report, 1999
3. Baumüller, A., G. Lundholm, L. Lundström and W. Schiel, Development History of the V 160 and SOLO 161 Engines, International Stirling Engine Conference, South Africa, 1999
4. Shaddix, R. C., G. R. Hardesty and S. Gust, *Combustion Properties of Biomass Flash Pyrolysis Oils: Final Report*, Sandia National Laboratories, Albuquerque, 1999
5. D. Meier, A. Oasmaa, G.V.C. Peacocke, in "Developments in thermochemical biomass conversion", ed. A. V. Bridgwater and D. G. B. Boocock, Blackie Academic & Professional, London, 1998, pg. 394

Pyrolysis oil combustion tests in an industrial boiler

Anja Oasmaa*, Matti Kytö**, Kai Sipilä*
* *VTT Energy, P.O.Box 1601, 02044 VTT, Finland*
** *Oilon Oy, P.O.Box 5, 15801 Lahti, Finland*

ABSTRACT Combustion properties of various pyrolysis oils were studied at Oilon (Lahti, Finland) in a test boiler on 4 MW output level. Burner settings and boiler dimensions as well as oil properties were tested. A special emphasis was on oil specifications for boiler application. It was concluded that pyrolysis oils can be combusted in slightly modified conventional oil furnaces and boilers fairly well. Some existing oil burners are also capable to combust pyrolysis oils, but they might need slight modification. However, oil specifications must be defined, including at least water and solids. Special care should be taken in handling and pumping of oils. Support fuel is needed at the burner to start the combustion and recommended for supporting pyrolysis oil in case of poor-quality oil. Emissions depend on oil quality. They are generally equal to those of light fuel oil or light bunker oil, but the content of particles in exhaust gases is higher. The results are applicable with small modifications in the boiler size class down to 1 MW_{th}. Research into replacing light fuel oil in boilers below 1 MW_{th} continues.

INTRODUCTION

The industrial organisations supporting VTT in the biomass pyrolysis R&D are Vapo Oy, Fortum Oil & Gas, and Oilon Oy. Vapo, the biggest domestic biofuel producer in Finland, is interested in the production of fast pyrolysis liquid from domestic fuels. Fortum is an international energy company, which was formed from the Neste and IVO Groups. Fortum's Oil & Gas Division explores and produces oil and gas, refines, sells and markets petroleum products, and transmits, distributes and sells natural gas. In pyrolysis area, Fortum is focusing on the use of pyrolysis liquid in medium-size boilers [1]. Oilon is the biggest burner manufacturer in Finland and is interested in boiler applications in burner size class of 350 kW-45 MW. Additionally, Wärtsilä NSD Oy has carried out a study on using pyrolysis liquid as diesel power plant fuel together with VTT in the 1990s [2].

At VTT, the following issues are addressed in the present study: pyrolysis of solid biomass [3, 4], hot vapour filtration, pyrolysis oil quality [5] and fuel oil specifications [6], storage and handling properties of pyrolysis oil, boiler applications [7], and techno-economic assessment of pyrolysis systems [8]. A considerable amount of work has been carried out on testing and further developing standard fuel oil analyses and test methods for pyrolysis oils [9, 10, 11).

In the EU Project JOR3-CT95-0025, co-ordinated by VTT Energy, i.a. the quality of pyrolysis oil of pilot plants was controlled, using Swedish pine as raw material [12]. The oils produced were used in combustion tests at Oilon Oy.

Two main targets were set for the combustion tests at Oilon:

- Optimisation of the boiler unit and combustion conditions in 4 - 5 MW size class. With small modifications, the research results will be applicable up to the burner output of 45 MW.
- Comparison of combustion properties of various qualities of pyrolysis oil.

Other aims included:

- Identification of burner types, in which pyrolysis oil can be burned, and definition of critical issues in successful combustion
- Homogenisation of phase-separated inferior-grade pyrolysis oil, and determination of its combustion characteristics
- Development of a feed system for pyrolysis oil
- Study of material corrosion.

MATERIALS AND METHODS

TEST ARRANGEMENT

The combustion tests were carried out in the combustion test laboratory of the R&D Center of Oilon Oy employing a test furnace of 8 MW nominal capacity manufactured by TKH (Finland). The test furnace was cylindric, the inner diameter of the furnace being 2.4 m and length 5.2 m. A number of changes were made in the furnace during the combustion tests with pyrolysis oils.

The flue gases escape upwards from the rear of the furnace. The furnace is water-cooled. During the run, the temperature of incoming circulation water ranged 60 - 85 °C and that of return water 80 - 105 °C (1.27 bar). Lots of the heat released in combustion escaped along with flue gases, and hence, the flue gases were rather hot (300 - 500 °C). There were several viewports in the furnace for visual monitoring of the flame. The main components and apparatuses of the combustion test equipment were

- Feed tanks of pyrolysis oil: two 1.5 m^3 steel tanks with propeller mixers
- Feed pump
- Preheater of pyrolysis oil: electrical mass preheater of Oilon, oil flowing in pipes inside it
- Dual fuel oil lance and nozzle suitable for dual fuel combustion
- Modified Oilon Lenox GRT-5L dual fuel burner
- Forced-draught fan controlled by an inverter
- Oilon test furnace: nominal power output 8 MW, test run output about 4 MW
- Measuring assembly for flue gas emissions.

The burner of type Oilon Lenox GRT-5L is a mixing burner for oil and gas combustion in boilers. The burner was equipped with the dual fuel lance, which enables the use of two different liquid fuels either separately or simultaneously. Heavy fuel oil (PORL 180) was used as support and start up fuel. A KM-nozzle type

2388H02 (Y-jet principle) of Oilon with atomisation medium was used in the dual fuel oil lance. Compressed air of about 6 bar was mainly used for atomisation in the tests.

The combustion tests were divided into three main tasks:

1. Testing and optimisation of combustion conditions with one pyrolysis oil
   - The aim was to test the combustion of pyrolysis oil and to adjust the combustion equipment and slightly modify the used burner
2. Combustion tests with typical pyrolysis oils
   - The aim was to run comparable combustion tests with different "typical" pyrolysis oils
3. Combustion tests with pyrolysis oil high in water
   - The aim was to apply, if needed, support combustion for difficult pyrolysis oils (high water content, inhomogeneity) and run comparable combustion tests.

MEASUREMENTS AND SAMPLES

The following systematic emission measurements were carried out during the tests:

- Gasmet gas analysis for emission gases: CO, CO_2, NO, NO_2, N_2O, CH_4, C_2H_6, H_2O as momentary measurements (IR-principle)
- Continuous backup measurement for emission gases: CO, NO and NO_2
- Continuous oxygen measurement with two separate measuring devices
- Particles content measurement according to standard SFS 3866, isokinetic sampling for the filter
- Soot measurement according to standard EN 267 with a Bacharac standard pump, and numerical value with an optic tone value gauge.

NOx concentration in the exhaust gases is calculated on an anhydrous oil basis from dry exhaust gases according to SFS 5624.

In this article, "particles" mean dust, particulates and solid particles contained in flue gases and measured as weight increase of the filter. Particles differ from soot due to the measuring method. Soot means the darkness level of the spot taken with the Bacharac pump and soot paper,and this sample is compared between absolute black and white. The cleanliness of the boiler was determined by visual observations inside the boiler after the test runs. Comparisons to the heavy fuel oil runs were done. The cleanliness of the nozzles was determined by checking each hole of the nozzle by a steel wire after the test run. The blockage of the nozzle holes was also observed visually through the small window on the back wall.

Samples were taken as follows:

- Pyrolysis oil samples after the preheater for each measuring point
- Possible coke or impurity samples from the nozzle after the test cycle
- Possible coke samples from the boiler after the test cycle.

The samples were analysed according to fuel standard modified for pyrolysis oils [9, 11, 10] .

PYROLYSIS OILS

The pyrolysis oils chosen for the Oilon combustion tests were produced from various hardwoods and softwood using either a bubbling fluidised bed or circulating fluidised bed. The capacity of production ranged up to 1 t/h. Since the oils are not yet commercial and producers still try to improve them, the oils are marked as in Table 1.

Table 1 Pyrolysis oils in Oilon combustion tests.

Oilon No.	Class	Feedstock	Process	Production date
1 - 6.6	Testing	Hardwood 1	CFB*	11/1996
9	Combustion tests	Hardwood 2	CFB*	2/1999
10 + 11	Combustion tests	Hardwood 3	CFB*	2/1999
12 + 13	Combustion tests	Pine 1	CFB*	3/1999
15	Combustion tests	Pine 2	BFB**	3/1999
7 - 8	High-water content	Pine 3	BFB**	6/1998

* Circulating fluidised bed **Bubbling fluidised bed

Hardwood pyrolysis oil 1 (Oilon 1-6.6) was used as "test oil", its amount being largest. The oil was inhomogeneous due to a long storage time (2.5 years) outdoors. The oil was homogenised with 20 vol% (abt. 10 wt%) methanol addition. Simultaneously, the viscosity of oil reduced from about 200 cSt to 35 cSt (@ 50 °C). Oil was stirred with a propeller mixer slowly continuously after methanol addition to keep it homogeneous.

Hardwood oils 2 and 3 and Pine oils 1 and 2 were classified as "typical oils" for the combustion tests.

The most important fuel characteristics are heating value, viscosity, and water and solids content. These characteristics of all oils used in the measurements are presented in Table 2. Nitrogen content was below 0.1 wt% for each oil. Figure 1 shows the viscosity curves as a function of temperature for all pyrolysis oils used for combustion.

Table 2 Fuel oil analyses of pyrolysis oils used for combustion.

OILON No.		Feedstock	LHV MJ/kg	Viscosity 50 °C cSt	Water wt-%	Solids wt-%
6.2	(Testing) *	Hardwood 1	17.4	42	20.4	0.40
6.3	(Testing) *	Hardwood 1	17.3	38	21.0	0.41
6.4	(Testing)*	Hardwood 1	17.4	40	19.3	0.38
6.5	(Testing) *	Hardwood 1	17.2	37	21.1	0.40
6.6	(Testing)*	Hardwood 1	17.5	42	19.9	0.40
9	(Typical)	Hardwood 2	17.6	41	20.7	1.20
10+11	(Typical)	Hardwood 3	16.4	33	24.8	1.86
12+13	(Typical)	Pine 1	15.3	27	22.6	0.17
15	(Typical) *	Pine 2	15.3	6	24.5	0.25
7+8	(High-water content)	Pine 3	11.2	6	35.7	0.03

* Contains about 10wt% methanol.

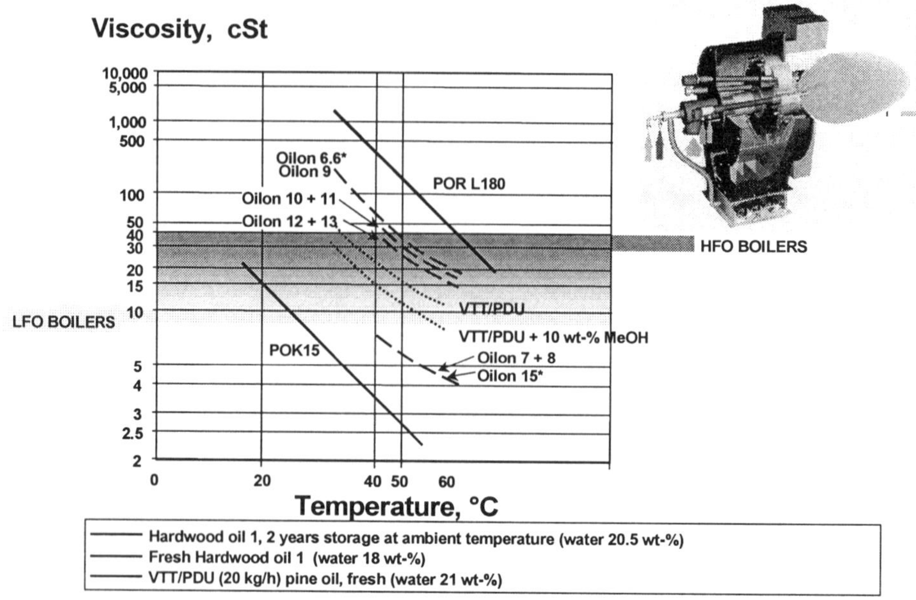

Fig. 1 Viscosity curves of pyrolysis oils used in the Oilon combustion tests. Testing oils: light oil POK15 used by big real estates, marine purposes and industry, and heavy fuel oil PORL 180, most generally used in size class of 1 - 10 MW. * Contains about 10 wt% methanol.

RESULTS

ADJUSTMENT OF COMBUSTION SET-UP

Several adjustment combinations to the feeding and boiler set-up were tested. During previous tests at Birka Energi [13] it was concluded that a special type of front head was needed inside the boiler for preventing the heat loss. This was used also at Oilon. The use of an extra cylinder inside the furnace yielded a more "dense" flame, higher temperature and hence faster volatilisation.

The burner head was changed from a diverging type to a converging one which yielded a narrower and more intense flame. A swirler of 60° yielded a stable flame although a large amount of primary air was used.

ADJUSTMENT OF COMBUSTION CONDITIONS

Combustion conditions were optimised and basic regulations of the combustion system were adjusted by burning "test oils". Simultaneously, an operation model for the basic measurement was finished and the operation of measuring assembly was tested. In total 11 m^3 of hardwood 1 pyrolysis oil was burned in the test runs.

10 wt% of MeOH was added to the test oils to increase the homogeneity of oil, to improve the combustibility and ignition and to reduce particles emissions. In part for the same reason the highest NO_x contents were measured for these oils. The highest NO_x level, 121 mg/MJ was measured for the run, in which the pyrolysis oil had been heated to 80 °C prior to atomisation. However, the difference in NO_x emissions was

not great when compared to the test run, in which the temperature of oil was 54 °C (117 mg/MJ). On the contrary, the amount of incombustibles increased clearly when the temperature of the oil was raised. The mean combustion results (oils at about 50 °C prior to atomisation) for the oils are given in Table 3.

Table 3 Mean combustion results for test oils.

O_2 vol%	NO_x mg/MJ	CO mg/MJ	HC mg/MJ	Soot Bac.	Particles mg/MJ
3.45	109	1.2	0.3	2.1	19

In general, the test oils burned relatively well and the emission values were fairly good. No significant problems appeared in the combustion assembly, in particular after determining the basic regulations. The flame was usually unbroken and stable, and no significant splash occurred. The boiler did not foul due to pyrolysis oil after the initial stage.

COMBUSTION TESTS WITH TYPICAL PYROLYSIS OIL GRADES

The lowest NO_x contents were achieved for "typical pyrolysis oils" (Table 2). The reason for this was apparently the lack of MeOH and perhaps also the higher water content. It was typical of these oils that the particle emissions were on average 4.5-fold, but respectively, the NO_x emissions were about 20% lower, compared to the values measured for the test oils. The soot number (average 2.4) was not much higher than that measured for the test oils (average 2.1).

The combustibility of the oils was relatively good, with the exception of Pine 2 pyrolysis oil. The problems with this oil appeared primarily in the nozzle, which clogged quickly despite regulation measures and other efforts. Hardened, floury, dry and crumbly particles were find in the holes of the clogged nozzle. The measurement run could be carried out only by adding methanol. This phenomenon was not found for the other oils of the group. Methanol might have dissolved crust and/or prevented its formation for a time. The oil was otherwise of relatively "high grade". The average combustion results for typical pyrolysis oils are presented in Table 4. The emission values for heavy and light fuel oil and natural gas using same boiler with same adjustments are also presented as Testing. The emission values using a low-NOx system are in brackets. In Finland the current emission standards for 50 - 150 MW boilers are 120 mg/MJ NO_x. In smaller boilers NO_x is not specified and the limits for particles are: 90 mg/MJ in 1-3 MW, 60 mg/MJ in 3-5 MW and 40 mg/MJ in 5 - 50 MW boilers.

Table 4 Average combustion results for typical pyrolysis oils. Emission results for heavy fuel oil (HFO), light fuel oil (LFO), and natural gas (GAS) are included.

	O_2 vol%	NO_x mg/MJ	CO mg/MJ	Soot Bac.	Particles mg/MJ
Pyro	3.5	88	4.6	2.4	86
HFO	3.5	193 (75)	3 (10)	2 (3)	23 (28)
LFO	3.5	70 (38)	1 (5)	0.5 (1)	2 (2)
GAS	3.5	55 (27)	1 (6)	0 (0)	0 (0)

The values presented in Table 4 are average emission values obtained for all typical pyrolysis oils in combustion tests.

The combustion tests with typical pyrolysis oils usually proceeded relatively well. The expected difficulties and possible problems were related to the stability of flame, combustion characteristics of oil, simultaneous achievement of low emission levels, fouling of furnace, and cleanness of nozzle devices. Changes also occurred in these during the tests. For example, the difficult pine 2 oil would not have burned during the measuring cycle (30 min) without special monitoring and "manipulation". On the other hand, the combustion of the best oils could have been continued a long time without any changes or problems in combustion or fouling. Figures 2 - 4 show photos of a typical stable flame of clean pyrolysis oil, a flame of a difficult pyrolysis oil, and a flame of pure support fuel.

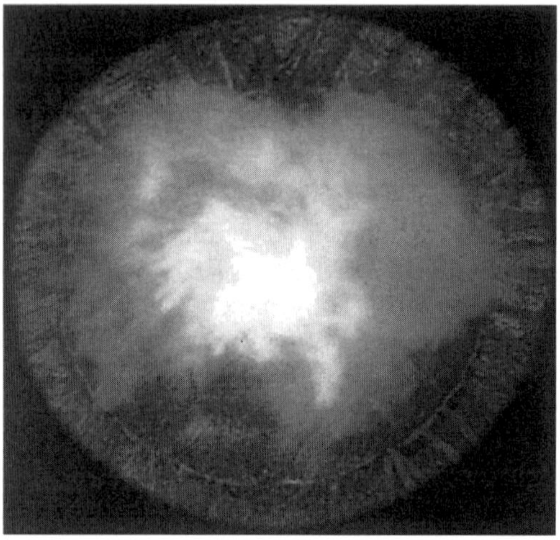

Fig. 2 Typical stable pyrolysis oil flame without problems.

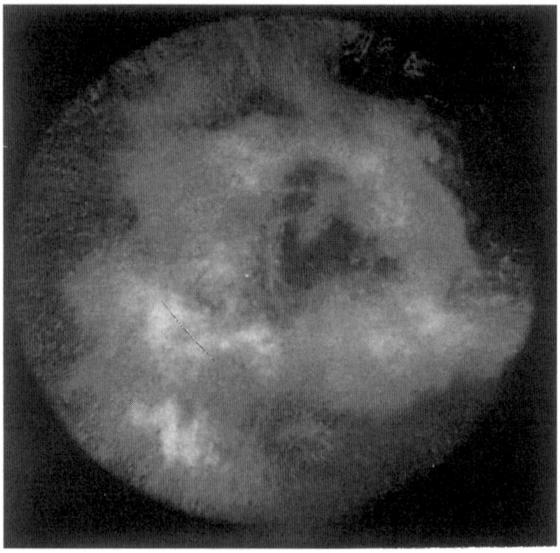

Fig. 3 Difficult unstable and broken pyrolysis oil flame.

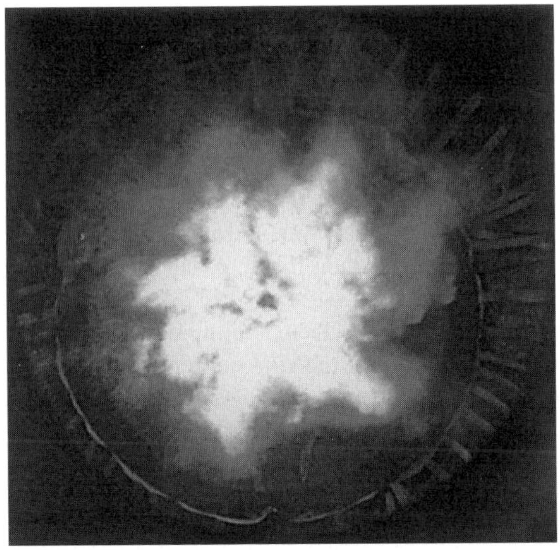

Fig. 4 Flame of support fuel.

A particle sample collected from the flue gases of the No. 12+13 pine pyrolysis oil was studied by light and electron microscopy. Single particles contained a lot of inorganic substances, mainly calcium, other substances being iron, silicon, magnesium, aluminium, potassium, chrome and vanadium. The amount of organic substances (e.g. tars) or carbon is likely small.

COMBUSTION TESTS WITH "WATER OILS"

The combustion tests with pyrolysis oils containing an abundance of water (Table 2) proceeded in general relatively well. The expected difficulties and possible problems were related to the stability of flame, combustion characteristics of oil, simultaneous achievement of low emission levels, fouling of furnace, and cleanness of nozzle devices. Changes also occurred in these during the tests. The use of support flame immediately improved the stability and combustion in general, as i.a. the particles emission immediately reduced by about 15%, but respectively the nitrogen oxides increased by about 50%. Support fuel power of about 1 MW (25%) is sufficient for securing good combustion. On the other hand, combustion proceeds reasonably well without any support fuel, if the other parameters are correct. Combustion of water-rich oil could have been continued much longer than the measurement required. No dramatic changes or problems were monitored in combustion or fouling during the measuring cycle. The average results for water-rich pyrolysis oils in combustion with pure pyrolysis oil and with support fuel are presented in Table 5.

Table 5 Average combustion results for water-containing pyrolysis oils.

Support fuel	O_2 vol%	NO_x mg/MJ	CO mg/MJ	HC mg/MJ	Soot Bac.	Particles mg/MJ
No	3.50	75	0.6	0.1	6.0	98
Yes	3.50	114	0.3	0.1	6.4	83

MATERIAL CORROSION

Acidity of pyrolysis oil (pH 2 - 3) and the combined effect of specially acids and high water content when raising the temperature set special requirements for materials. Acid-proof steel, certain special steels [1], brass [14], and plastics stand well pyrolysis oil [9].

In these tests, usual carbon steel was wittingly used, as the aim was to test combustion of pyrolysis oil in a common fossil oil burner system. The pipeline and the heat exchanger of pyrolysis oil feed were of St 37.4 DIN 2391 precision steel pipe (size 10 x 1), so-called Ermeto-pipe. The pipeline clearly corroded during the tests, which was also seen in an increase in the iron content of oil. It is self-evident that the durability of materials, both steel and seals, has to be ensured in the design of a special pyrolysis oil burner and fuel line to it.

OIL HANDLING AND SPECIFICATIONS

The homogeneity of pyrolysis oil is one of the main issues to be notified. A fairly slow (max. 1,200 rpm) propeller type mixing after homogenisation of the oil ensures a steady feed and combustion of the oils. Even an old phase-separated pyrolysis oil can be kept homogenous by slow mixing after adequate methanol addition. An Eccenter screw pump ('mono'-pump) was used successively for oil feeding. A typical coarse filtering (45 μm) was used through the whole test period without any problems.

The best preheating temperature for pyrolysis oils varied depending on the original viscosity of the oil. A good atomisation was obtained for this type of burner and boiler system in viscosity range of 15 - 20 cSt.

The most important fuel oil properties for pyrolysis oils besides the homogeneity were water, solids in the oil, viscosity, and lower heating value. High water content decreased the heating value, and the viscosity and in combustion decreased NO_x and increased solid emission in the exhaust gas. A high content of solids in the oil yielded high particle emissions in exhaust gas. The optimum viscosity range yielded good atomisation as mentioned above. The lower heating value has to be defined for adjusting steady power output.

DISCUSSION

A summary of typical emission values for "test oils", "typical oils" and "water oils" at the fuel power of about 4 MW under as identical conditions as possible is presented in Table 6. There were no principal problems in the feed or combustion of the pyrolysis oils except for pine oil 2 (Oilon 15), which caused blockages in the nozzle. Despite these problems one basic measurement was successfully carried out for this oil by adding MeOH to the oil.

Table 6 Typical emission values for different oil grades at nominal fuel power of 4 MW. Combustion conditions: total output 3.4 - 3.8 MW, compressed air used as atomisation medium of pyrolysis oil, oil temperature 48 - 58 °C, regulation of primary air 50%, regulation of secondary swirl S′=0.7, primary air swirler 60°. Particles had been measured during a longer-term test cycle.

	Test oils	Typical oils			Water oils
OILON oil	6.6	10+11	12+13	15	7+8
MeOH, wt%	10	0	0	10	0
Raw material	Hardwood 1	Hardwood 3	Pine 1	Pine 2	Pine 3
O_2, vol%	3.4 - 3.5	3.3 - 3.6	3.3 - 3.4	3.5	3.4 - 3.7
CO, ppm	2 - 5	1 - 2	10 - 25	2 - 6	2 - 4
NO_x, ppm	193 - 205	159 - 164	108	110	88 - 105
NO_2, ppm	0	0	0	5	1
Soot, Bac.	1.3	2.0	2.8	3.5	5.9
PARTICLES, g/MJ	15	100	92	53	98

The reason for the nozzle blockage is unclear. There may have been some operational problems during the production of pine oil 2, which could have contributed to this kind of peculiarity. However, the GC-MS analyses showed that pine oil 2 contained significantly more methoxy-substituted phenols than the other oils tested. These monosubstituted fragments are particularly reactive (due to free position on aromatic ring) and prone to recombination/condensation reactions. In pine oil 2, flaky sticky precipitation increased in the course of time, which may consist of polymerisation products of these fragments and be the reason for the nozzle blockage. This precipitation was left on the bottom of the container and the rest of the oil was used in the tests. This kind of precipitation did not appear in other pine oils. It is quite clear that the chemical composition of pyrolysis oil affects significantly the atomisation behaviour of the oil. However, more data is needed on the reactivity of various commercial pyrolysis oils before final conclusions can be drawn.

The test oils were relatively easy to burn and emission values were fairly good. The combustability was significantly improved by methanol addition. After adjusting the burner, no greater problems appeared and the boiler remained clean. Neither did the nozzle clog easily.

The combustion of pyrolysis oils includes two main phases: 1) vaporisation and combustion of light fraction, vaporisation of water and gasification of heavier fractions, and 2) combustion of the heaviest lignine fractions. As the initial stage requires a long residence time at a sufficiently high temperature, the front chamber improves the total combustion process. It is essential in the dimensioning of the front chamber that the flame shall not touch its surfaces in any stage, as coke is rapidly deposited on the surface and the particle content increases. If the shape of flame is well suitable in the front chamber, the particle emission is sharply reduced. The heat release rate per volume (kW/m^3) of the furnace should not be as high as in mineral oil combustion, as the pyrolysis oils burn slower and the flame of respective power is larger. Hence, the power output of existing mineral oil boilers reduces when pyrolysis oil is used. It might be possible to apply co-combustion in large boilers, e.g. by running, for instance, the highest burner level or some of the burners with mineral oil. The possible tar-like particles, soot and PAH compounds of pyrolysis oil flames burning on the lower level are mixed in oil flames and burn prior to convection surfaces.

The atomisation viscosity should be adjusted suitably so that the size and size distribution of oil droplets are correct and their penetration into the air pattern of the burner is as desired. A burner dimensioned for oil combustion can be used or it should at least be slighly modified. However, if the best result is wanted a special burner should be designed. A too high temperature (> 60 °C) increases the risk of nozzle blockage. Particle emissions, soot and usually also nitrogen oxide emissions increase compared to a situation, where the oil temperature is lower and the viscosity is suitable.

As regards emission values, a general conclusion is that the nitrogen oxide level was rather high for the test oils, while the particles emissions were small. The NO_x level increased due to the reduction of water content and the heat of initial flame due to rapid combustion of added methanol. Also the burner had to be adjusted so that the flame was suitable, stable and the level of particle emissions reasonable. On the other hand, this contributed to oil vaporisation and particles gasification and hence to small particles emission.

The typical pyrolysis oils were nearly invariably relatively easy to burn. The emission values were fairly good, but not as good as those for the test oils. A general conclusion from the emission values is that the nitrogen oxide level of typical pyrolysis oils was lower than that of the test oils, but the particles emissions respectively clearly higher. The NO_x level was reduced by the higher water content and the disuse of methanol, which on the other hand resulted in higher particles emissions.

The "water oils" were surprisingly easy to burn in dual fuel mode and the emission values were relatively good. The soot number rose to a higher level than that of typical oils. One reason for the high soot number may be the unproper (too low) viscosity level of these oils. As soon as the suitable burner adjustments were found, no great problems in combustion occurred and the boiler remained clean. The nozzle also remained fairly clean in all test runs.

Mean NO_x and particles emissions for different pyrolysis oil groups are shown in Figure 5.

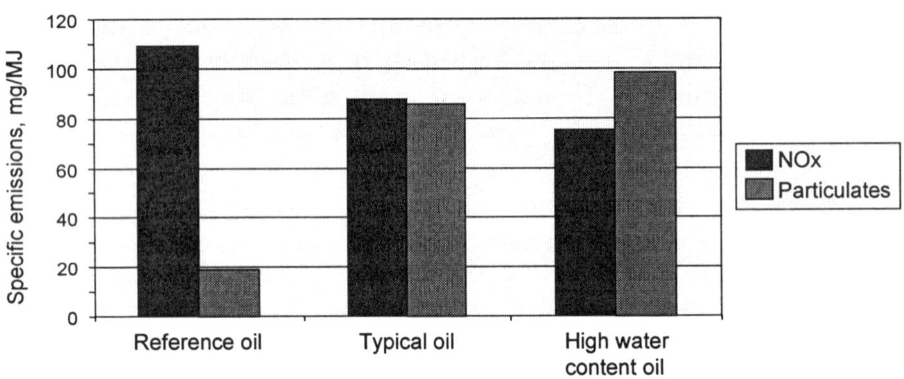

Fig. 5 Mean NO_x and particles emissions for different pyrolysis oil groups.

It can be concluded from the emission values, that the nitrogen oxide level of water oils was fairly low without support oil, but grew when increasing support combustion. Support fuel increased NOx emissions by two ways: the average nitrogen content of the fuel mixture increased and the flame temperature increased yielding to an increase in thermal NOx. The particle emissions were relative high. The NO_x level was reduced by the high water content and the disuse of methanol. Due to these factors, the effect on particles emissions was contrary. On the other hand, the particle content of the water-rich oils was very low, and hence, the final results was not poor at all.

It should be borne in mind, that the composition of soot and particles of pyrolysis oils typically differs from that of conventional oil combustion. As there is no separate soot classification for pyrolysis oils, the Bacharac soot method of oil combustion was used for describing the tone value. The particles consisted mainly of very fine brownish material on the filter.

CONCLUSIONS

Different grades of pyrolysis oil were burned at 4 MW nominal power output in a test furnace of Oilon Oy's R&D Centre. Operation, dimensions, adjustment parameters, and characteristics of pyrolysis oils were tested and compared, and emissions in different cases were measured.

A general conclusion is that the results were relatively clear and the main issues were retrieved well. There were clear differences in combustibility and in particular in emissions for different oil grades. The most important parametres of pyrolysis oil are viscosity, water and particles content, methanol addition, oil raw material, and oil age. Good and poor oils or at least difficult oils were distinguished in combustion. The burner and boiler modifications improve the combustion result but cannot help much if the oil quality is poor.

When optimising combustion conditions for the test oils, i.a., the following effects of oil grade were found: feedstock and/or pyrolysis process yields various reactivities of oil components, which may cause blockages in the feed line, oil age/unhomogenity yields uneven combustion, methanol addition homogenises pyrolysis oil and improves combustion, solids content affects mainly the amount of incombustibles, and an increase in water content reduces NO_x emissions and increases particle and soot emissions.

As regards adjustments, the following factors improved combustion and flame: clean nozzle, strong swirl, intense symmetrical flame, pressure air atomisation (compared to steam), increase of air coefficient and combustion power (having enough residence time though), suitable atomisation viscosity (abt 15 - 20 cSt). At the optimum adjustments of this combustion system, the mean combustion results and emission values of typical pyrolysis oils were as follows: O_2 3.5 vol%, NO_x 88 mg/MJ, CO 4.6 mg/MJ, hydrocarbons 0.1 mg/MJ, soot 2.4 Bac., and particles 86 mg/MJ.

In addition to the characteristics of pyrolysis oils and burner settings, different furnace constructions were compared in the test runs. By insulating the forepart of the furnace the mean temperature levels of flame were increased and hence combustion improved.

The main results of the combustion tests are:

- Pyrolysis oils can be burned relatively well in conventional furnaces and boilers. Boilers and oil burners may require small modifications or additions. The flame is larger and combustion takes a longer time than with mineral oils.
- Handling and pumping of pyrolysis oil should be performed according to exact recommendations.
- Support fuel is required at the start of combustion and possibly in the combustion of pyrolysis oil of poor quality to maintain good and stable combustion.
- The nozzles should be kept clean and in good conditions. Extra cooling air for the nozzle could be useful during combustion.
- Emissions from combustion are in general between those from light fuel oil and the lightest heavy oil, but the particle content is higher. There are no SO_x and net CO_2 emissions.

- Quality specifications should be defined for pyrolysis oil, including especially water and solids contents. Viscosity range is significant for good atomisation. The quality of pyrolysis oil has a strong impact on emissions. High solids content in pyrolysis oil yields high particulate emissions. Hence, solids removal from pyrolysis vapours or oil is highly recommended. High (above 30 wt-) water content also yields high particulate emissions. These emissions can be decreased to a certain extent by using a support fuel and optimizing the atomization viscosity. Methanol addition (max 10 wt%) homogenizes the inferior-grade oil and decreases particulate emissions. The costs for methanol addition and oil combustion in a commercial boiler are most probably lower than those of incinerating poor-quality oil in a special incineration plant for hazardous wastes.
- Further research is required on combustion properties of various commercial pyrolysis oils in order to identify the reasons for emission behaviour, nozzle blockages and related phenomena.

ACKNOWLEDGEMENTS

This work was funded by the VTT Tekes PROGAS Programme, VTT Energy, Birka Energi (Sweden) and Oilon Oy. Thanks are due to Eero Pekkola, Managing Director, and Seppo Hotti, Product Group Manager, for the leadership of the project, and to Mikko Mikkola, Timo Nironen, Hannu Sipiläinen and Kauko Tuovinen for the technical assistance at Oilon. From Birka Energy, thanks are due to Björn Hallgren, Lars-Erik Hägerstedt, and Harri Vuoluterä, and from Fortum, to Steven Gust. At VTT Energy thanks are due to Jarmo Kleemola, Ilkka Isoksela, Eija Tapola, Johanna Levander, Eeva Kuoppala, Jaana Korhonen, Kaisa Lanttola, Raili Silvasti, and Paula Käyhkö for their technical assisstance, and to Antero Moilanen.

REFERENCES

1. Gust, S. (1997). Combustion of pyrolysis liquids. In: Kaltschmitt, M. & Bridgwater, T. (eds.). Biomass gasification & pyrolysis. State of the art and future prospects. Newbury: CPL Scientific Ltd. Pp. 498 - 503.
2. Jay, D., Rantanen, O., Sipilä, K. & Nylund, N.-O. (1995). Wood pyrolysis oil for diesel engines. Proc. 1995 Fall Technical Conference, Milwaukee, Wisconsin, 24 - 27 Sept. 1995. New York: ASME.
3. Sipilä, K., Oasmaa, A., Solantausta, Y., Arpiainen, V. & Nyrönen, T. (1999). Perspectives for pyrolysis oil production and market in Scandinavia. In: Sipilä, K. & Korhonen, M. (eds.). Power production from biomass III. Espoo: VTT. Pp. 277 - 292. (VTT Symposium 192).
4. Sipilä, K., Oasmaa, A., Solantausta, Y., Arpiainen, V. & Nyrönen, T. (1999). Pyrolysis oil - a new product for heat and power. To be published.
5. Oasmaa, A. & Czernik, S. (1999). Fuel oil quality of biomass pyrolysis oils - state of the art for the end users. Energy & Fuels, vol. 13, no. 4, pp. 914 - 921.
6. Diebold, J., Milne, T., Czernik, S., Oasmaa, A., Bridgwater, A., Cuevas, A., Gust, S., Huffman, D. & Piskorz, J. (1997). Proposed specifications for various grades of pyrolysis oils. In: Bridgwater, A. & Boocock, D. (eds.). Developments in thermochemical biomass conversion. Vol. 1. London: Blackie Academic & Professional. Pp. 433 - 447.

7. Sipilä, K., Oasmaa, A., Arpiainen, V., Westerholm, M., Solantausta, Y., Angher, A., Gros, S., Nyrönen, T. & Gust, S. (1996). Pyrolysis oils for power plants and boilers. In: Chartier, P. et al. (eds.). Biomass for energy and the environment. Vol. 1. Kiddlington: Elsevier. Pp. 302 - 307.
8. Solantausta, Y., Podesser, E., Beckman, D., Östman, A., Overend, R. P. (2000). IEA Bioenergy Task 22: Techno-economic assessments for bioenergy applications 1998-1999. Final report. Espoo: VTT Energy. 241 p. (VTT Research Notes 2024).
9. Oasmaa, A., Leppämäki, E., Koponen, P., Levander, J. & Tapola, E. (1997). Physical characterisation of biomass-based pyrolysis liquids Application of standard fuel oil analyses. Espoo: VTT. 46 p. + app. 30 p. VTT Publications 306).
10. Oasmaa, A. & Meier, D. (1999). Analysis, characterization, and test methods of fast pyrolysis liquids. In: Biomass - a growth opportunity in green energy and value-added products. Vol. 2. Kidlington: Elsevier Science. Pp. 1229 - 1234.
11. Bridgwater, A. , Czernik, S., Diebold, J., Meier, D., Oasmaa, A., Peacocke, G., Piskorz, J. & Radlein, D. (1999). Fast pyrolysis of biomass: A handbook, Newbury: .CPL Scientific Ltd. 188 p.
12. Bio fuel oil for power plants and boilers. Confidential Final Report. Contract JOR3-CT95-0025. Espoo: VTT Energy, 1999.
13. Hallgren, B. (1996). Test report of Metlab Miljö AB. Skelleftehamn: Metlab Miljö AB. 17 p. (Reg. no. ALL-1668, 1996 02 08-09.)
14. Fuleki, D. (1999). Bio-fuel system materials testing. PyNe Newsletter, March 1999, Issue 7, pp. 5 - 6.

Transport, Handling and Storage of Biomass Derived Fast Pyrolysis Liquid

G.V.C. Peacocke[1] and A.V. Bridgwater[2]

[1] *Conversion And Resource Evaluation Ltd. 9 Myrtle House, 5 Cassowary Road, Birmingham, B20 1NE, UK*

[2] *Bio-Energy Research Group, Chemical Engineering and Applied Chemistry, Aston University, Birmingham, B4 7ET, UK*

ABSTRACT Biomass fast pyrolysis liquid is being developed for fuel and chemical applications. As these developments proceed, the liquid product is increasingly being transported by air, water, rail and road to satisfy user demands for products. This paper addresses the legislative requirements and regulations for the safe transport of this liquid. As biomass derived fast pyrolysis liquid is not on the UN approved carriage lists; its own classification has been determined from the UN manual as:

UN 1993 Flammable Liquid [Fast Pyrolysis Liquid], n.o.s., 3, 1°(a), 2°(a), 1

This classification should be used on all packages containing biomass fast pyrolysis liquid. Protocols for the labelling of packages and containers of all sizes are given with the aim of compliance with transport regulations in the EU, Canada and the USA. In conjunction with the requirements for packaging and labelling, guidance on the details to be enclosed on the transportation documents are given, with appropriate MSDS for the liquids. Guidance on the handling of fast pyrolysis liquid and its storage are given with procedures for treatment of spills.

INTRODUCTION

Biomass derived fast pyrolysis liquid is now being actively produced for research, testing and evaluation purposes, for use as a chemical feedstock and/or as an alternative fuel for use in boilers, engines and turbines.

As fast pyrolysis technologies develop, and utilisation of the liquid increases, there will be a greater demand for the transportation of the liquid by all possible routes. To ensure that the liquid is transported in a safe and environmentally secure manner, due care and attention must be taken to ensure that the appropriate national and international regulations are met. To this end, it is likely that fast pyrolysis liquid will be classed by the regulatory authorities as a "dangerous" or a "hazardous" substance for transportation purposes.

The UK Department of Transport Dangerous Goods Division and the UK Health and Safety Executive were consulted to discuss the classification of this liquid for transport. (1). Based on discussions with these organisations, their opinion was that fast pyrolysis liquid would be classed as a "dangerous good" or hazardous material, due to the chemical complexity and composition.

The pyrolysis community may feel that the classification of fast pyrolysis liquid as "dangerous" or "hazardous" material does not reflect the true nature of the liquid. However, due to the highly variable chemical and physical properties, a wide spectrum of liquids must be considered. This paper addresses legislative requirements, transport, storage and handling of the liquid and the mitigation of spills.

This paper and its recommendations applies only to biomass derived fast pyrolysis liquid, although it may well be equally applicable to other liquids such as carbonisation tars, gasification tars and other liquids produced by thermal processing of biomass or other solid materials or wastes.

INTERNATIONAL REGULATIONS ON THE TRANSPORTATION OF DANGEROUS GOODS

The scope of this legislative review pertains to the transportation of fast pyrolysis liquids in the EU, USA and Canada for all modes of transport. This paper cannot cover all the national regulations in force, however, most national transportation regulation are based on, or use the UN Regulations, as described below.

UN REGULATIONS

At the United Nations level, all work related to the transport of dangerous goods is co-ordinated by the Economic and Social Council [ECOSOC] Committee of Experts on the Transport of Dangerous Goods, which produces the "Recommendations on the Transport of Dangerous Goods", also called the "Orange Book" (2). These Recommendations and Regulations are addressed to all Governments for the development of their national requirements for the domestic transport of dangerous goods, and also to international organisations such as:

- The International Maritime Organisation [IMO];
- The International Civil Aviation Organisation [ICAO] and;
- Regional commissions such as the Economic Commission for Europe [ECE];

These are included for regulations and international/regional agreements or conventions governing the international transport of dangerous goods by sea, air, road, rail and inland waterways. The UN Recommendations addresses the following areas:

1. Dangerous goods most commonly carried, their identification and classification,
2. Consignment procedures: labelling, marking, and transport documents,
3. Standards for packaging, test procedures, and certification,
4. Standards for multi-modal tank-containers, test procedures and certification.

These recommendations contain all basic provisions for the safe carriage of dangerous goods, but they may have to be supplemented by additional requirements that are applicable at national level or for international transport depending on the mode or modes of transport envisaged. These recommendations are presented in the new form of Model regulations so that they can be more easily transposed into national or international legislation (2).

INTERNATIONAL CARRIAGE OF DANGEROUS GOODS BY AIR

There is no EU Agreement concerning the international carriage of dangerous goods by air. The ICAO issues technical instructions for the transport of dangerous goods by air (3). Regulations are also produced by the International Air Transport Association [IATA] (4) and these regulations are typically followed by cargo carriers for international transport. The IATA Dangerous Goods Regulations manual is based on the ICAO Technical Instructions. For example, the Civil Aviation Authority [CAA] in the UK has produced a guide for the transport of dangerous goods based on the IATA regulations (5).

Dangerous goods can be transported safely by air transport provided certain principles are strictly followed. Air transport incorporates additional operational requirements that provide a harmonised system for airlines to accept and transport dangerous goods safely and efficiently. Users of the IATA Dangerous Goods Regulations are assured that they are meeting all legal requirements for shipping dangerous goods by air internationally and its use is strongly recommended. The IATA Regulations include a detailed list of individual articles and substances specifying the United Nations classification of each article or substance and their acceptability for air transport as well as the conditions for their transport.

INTERNATIONAL CARRIAGE OF DANGEROUS GOODS BY WATER

As noted above the IMO is responsible for the provision of guidance on the safe transport of dangerous goods on water. The International Maritime Dangerous Goods [IMDG] Code (6) is accepted as an international guide to the transport of dangerous goods by sea and is recommended to governments for adoption or for use as the basis for national regulations. It is intended for use not only by the mariner but also by all those involved in industries and services connected with shipping and contains advice on terminology, packaging, labelling, stowage, segregation, handling and emergency response.

INTERNATIONAL CARRIAGE OF DANGEROUS GOODS BY RAIL

The transport of dangerous goods by rail is serviced by the Intergovernmental Organisation for International Carriage by Rail [OTIF]; it is responsible for ensuring harmonisation between ADR [Regulations concerning the international carriage of dangerous goods by road - see next section], RID [Regulations concerning the international carriage of dangerous goods by rail – see next section] and ADN [Regulations concerning the international carriage of dangerous goods by water – see next section].

EU TRANSPORT REGULATIONS

EC subsidiary bodies deal with the transport of dangerous goods. These bodies are subsidiary bodies of the Inland Transport Committee, and therefore they are concerned only with inland transport, i.e. road, rail and inland waterway. These bodies are part of The Working Party on the Transport of Dangerous Goods [known as WP.15], which is responsible for:

- The European Agreement concerning the International Carriage of Dangerous Goods by Road [ADR] and;
- The European Provisions concerning the International Carriage of Dangerous Goods by Inland Waterways [ADN] and;
- The Joint Meeting of the Working Party on the Transport of Dangerous Goods and the RID Safety Committee also called the RID/ADR/ADN Joint Meeting.

The RID/ADR/ADN Joint Meeting is serviced jointly by the ECE secretariat and the secretariat of the Intergovernmental Organisation for International Carriage by Rail [OTIF]; responsible for ensuring harmonisation between ADR, RID [Regulations concerning the international carriage of dangerous goods by rail) and ADN.

European Agreement concerning the International Carriage of Dangerous Goods by Road [ADR] and Rail

ADR is based on the UN Recommendations as regards the listing and classification of dangerous goods, their marking and labelling and packaging standards (7), but it also contains much more detailed provisions as regards:

1. The types of packaging which may be used,
2. The consignment procedures,
3. Transport equipment [vehicle to be used, vehicle construction and equipment],
4. Transport operation [training of drivers, supervision, emergency procedures, loading and unloading, placarding of vehicles].

The ADR is intended primarily to increase the safety of international transport by road, but it is also an important trade facilitation instrument. Except for dangerous goods which are totally prohibited for carriage, and except when carriage is regulated or prohibited for reasons other than safety, the international carriage of dangerous goods by road is authorised by ADR on the territory of Contracting Parties provided that the conditions laid down in annexes A and B are complied with. There are at present 34 Contracting Parties to ADR and RID, including all of EU, USA and Canada. The guidance within RID provides similar guidance to ADR (8).

European Agreement concerning the International Carriage of Dangerous Goods by Inland Waterways [ADN] (9)

The status of the European Provisions for the International Carriage of Dangerous Goods by Inland Waterways is different from that of ADR as ADN is only a recommendation directed to Governments for their national regulations and to river commissions for regulating the international carriage of dangerous goods on specific inland waterways under their responsibility. One well-known example of such regulations is the "Regulations for the Carriage of Dangerous Substances on the Rhine [ADNR]" developed by the Central Commission for the Navigation of the Rhine

[CCNR]. A draft European agreement concerning the international carriage of dangerous goods by inland waterways ["ADN" agreement] has recently been completed under the joint auspices of the UN/ECE and the CCNR (9).

USA AND CANADA REGULATIONS

USA Regulations on the Transport of Hazardous Materials

The United States Department of Transportation [USDOT] regulates the transportation of Hazardous Materials to, from and through the United States. Title 49, Code of Federal Regulations, Subtitle B "Other Regulations Relating to Transportation" Parts 100 - 199 set forth the standards for Hazardous Materials transportation, commonly known as "49 CFR" and these are readily available (10). There are significant differences between 49 CFR and other international regulations such as ICAO, IATA and IMO, despite efforts to reduce these differences and in addition, the US Regulations change regularly. All shipments of Hazardous Materials to, from and through the United States must comply with 49CFR and other regulations.

These regulations are administered by the Research and Special Programs Administration of the Department of Transportation [RSPA] and enforced on a federal level by the Federal Aviation Administration [FAA] for air transport; the Federal Highway Administration [FHWA] for ground transportation and the United States Coast Guard [USCG] for water transportation. Enforcement on a local level is usually the State Police or Highway Patrol. Exporters, importers, shipping lines, airlines, forwarders and couriers have to comply with 49 CFR, and failure can lead to delays in transport or in the case of non-compliance to potentially massive fines.

49 CFR Part 171.11 allows the use of ICAO Technical Instructions to be used for air shipments as long as certain additional requirements set forth in 49 CFR are also complied with. 49 CFR Part 171.12 allows the use of the International Maritime Organization's International Maritime Dangerous Goods Code for import and export shipments by vessel as long as certain additional requirements set forth in 49 CFR are also complied with.

Canadian Regulations on the Transport of Hazardous Materials

Transport Canada is responsible for the transportation of dangerous goods in Canada legislated by the Transportation of Dangerous Goods [TDG] Regulations (11). These regulations are based on the UN Regulations and are very similar in content. A detailed manifest of all shipments is required to allow easy tracking of shipments until they arrive at their destination or exit the country. An emergency response plan must also be submitted to Transport Canada and a Summary of Emergency Response Plan; the reference number of the Emergency Response Plan filed with Transport Canada and a 24 hour telephone number must be included on the transportation documents with the consignment.

PROPERTIES OF FAST PYROLYSIS LIQUID

The nature of fast pyrolysis liquid means that there is no "generic" analysis to cover the wide spectrum of liquids producible from biomass. In the UN or EU regulations, there is no classification for fast pyrolysis liquid or its derivatives, fractions or by-products.

As noted in the introduction, a submission to the EC to have fast pyrolysis liquid listed in ADR, RID and ADN could take 2 to 3 years, therefore a self assessment of the substance was made, using the methods described in the UN Manual (2). It is also likely that the outcome of a submission would be that fast pyrolysis liquid would be classed as a "dangerous substance". The assessment of a substance as a dangerous good considers the chemical and physical properties of fast pyrolysis liquid and these are summarised below.

PHYSICAL PROPERTIES OF FAST PYROLYSIS LIQUID

Specific physical properties, if not known, can be determined to UN test methods (12). As fast pyrolysis liquid is not listed as a substance in its own right in ADR or the UN Regulations, it can be categorised with a not otherwise specified [n.o.s.] classification. The relevant physical properties of the fast pyrolysis liquid used are given in Table 1, based upon average values in the literature.

Table 1. *Applicable hysical properties of fast pyrolysis liquid (13)*

Physical property	Fast pyrolysis
Moisture content	~20%
pH ~2	
Specific gravity	~1.2
Viscosity [cp @ 40°C]	~50
Kinematic viscosity [cSt]	20-1000 @ 25°C
Flash point [°C]	50-61
Pour point [°C]	-23

From these properties, fast pyrolysis liquid is a Class 3 substance - Flammable Liquid. The exact specification is then related to its chemical composition to determine its level of hazard, as described below.

CHEMICAL COMPOSITION OF BIOMASS FAST PYROLYSIS LIQUID

There are numerous references that include chemical analyses of fast pyrolysis liquids from a variety of sources, including slow pyrolysis tar, fast pyrolysis liquid and fractions thereof. As fast pyrolysis liquid may be raw, treated, filtered, catalytically derived, upgraded and dependent on the process parameters, a "worst case" must be taken for the liquid composition, i.e. high variability. As noted, fast pyrolysis liquid falls within Class 3, based on its physical properties, however, this is further complicated by the presence of other chemicals that are in Class 6.1 – Toxic substances [phenols, etc.] which are in concentrations above 0.1 wt%. A typical analysis of several fast pyrolysis liquids from different laboratories is given in Table 2.

The Aston and IWC liquids were produced in a fluid bed and the BTG liquid was produced in the rotating cone reactor, all using wood feedstocks.

Table 2 *GC-MS Analysis of Fast Pyrolysis Liquids (14)*

Chemical	Aston	BTG	IWC
(5H)-Furan-2-one	0.63	0.34	0.61
2,4- and 2,5-Dimethyl phenol	0.04	0.07	0.37
2.5-Dimethoxytetrahydrofuran (cis)	0.12		0.47
2-Furaldehyde	-	0.35	-
2-Furfuryl alcohol	-	0.01	-
2-Hydroxy-1-methyl-1-cyclopentene-3-one	0.22	0.13	0.15
4-Ally- and 4-Propyl syringol			0.19
4-Ethyl guaiacol	0.12	0.08	0.07
4-Methyl guaiacol	0.73	0.30	0.13
4-Methyl syringol	-	-	0.27
4-Vinyl guaiacol	0.05	0.03	0.05
5-Hydroxymethyl-2-furaldehyde	0.37	0.28	0.00
Acetic acid	2.65	2.56	4.23
Acetoguaiacone	0.16	0.17	0.07
Acetol	5.78	3.57	3.54
Eugenol	0.18	0.13	0.05
Guaiacol	0.44	0.20	0.14
Homovanillin	0.16	0.13	0.08
Hydroxyacetaldehyde	10.40	10.89	7.07
Isoeugenol (cis)	0.21	0.13	0.06
Isoeugenol (trans)	0.55	0.20	0.27
Levoglucosan	4.47	4.46	3.20
m-Cresol	0.26	0.08	-
o-Cresol	0.04	0.08	-
p-Cresol	0.02	0.08	-
Phenol	0.05	0.13	0.03
Syringol	-	-	0.29
Vanillin	0.21	0.29	0.09
Water	21.4	18.6	10.0

UN CODE FOR FAST PYROLYSIS LIQUID

To this end, the following generic classification is proposed based on the guidance in the UN Manual and the ADR for basic labelling purposes. The nearest UN n.o.s. classification that can be used [with the addition of "Fast Pyrolysis Liquid"] is:

UN 1993 Flammable liquid, [Fast Pyrolysis Liquid], n.o.s., 3, 1°(a), 2°(a), 1

Only two chemical groupings need to be indicated for the components comprising the most significant risk. This classification is an interim classification until an application is made to the European Union for the inclusion of fast pyrolysis liquid on the dangerous goods list. This classification should be used on all labels and for all sizes of shipments, in particular on the transportation documents and the MSDS. For tank containers, and bulk shipments, additional placarding is required and this is discussed below.

PACKAGING OF FAST PYROLYSIS LIQUID

The most critical aspect of transport of dangerous goods is packaging. Appendix II gives the fuller details of packaging codes and weight restrictions for specific UN approved packaging types. Fast pyrolysis liquid can be shipped from small samples of the order of a few mg to tonne quantities, in a variety of receptacles [single package or combination packaging] and for different purposes.

RECEPTACLE REQUIREMENTS TO COMPLY WITH UN REGULATIONS

Receptacles that are acceptable for international transport must meet the specifications of packaging described in the UN regulations (2). Packaging relates to samples of liquids of all sizes. However, depending on the classification of the dangerous goods, there are limitations to the quantities, which may be shipped per package. Packages may also be single, e.g. drums, or combination packages, e.g. plastic bottles inside a cardboard box. In summary, the physical requirements for packaging are in Table 3. There are also particular limits for Class 3 substances, depending on the package, materials of construction and type. The wide ranges of combinations are not discussed here, but a summary of the limits for inner packages is given in Table 4.

Table 3. *Minimum Package requirements for Fast pyrolysis liquid*

Packing group	1 or "X"
Receptacle required minimum test pressure	250kPa g
Degree of filling of receptacle [at 15°C]:	90%
Hazard symbols	see below

Table 4. Limitations on inner packages

Type of inner packaging	Max. permissible capacity [l]
Glass, porcelain or stoneware packaging	5
Plastic packaging	30
Metal packaging	40
Other types of small packaging, e.g. tubes	1

PACKAGING REQUIREMENTS

Fast pyrolysis liquid can be transported in varying quantities, from grams to tonnes. There is no stated maximum for the shipment of Class 3 liquids; however, the labelling and packaging requirements will vary, depending on the size of package. Guidance is given on the package specification for a range of shipment sizes.

Very small quantities [up to 1 litre]

For small samples, it is recommended that polypropylene [e.g. Nalgene™] bottles be used with a cap insert inside the neck. This type of plastic packaging is extremely resilient to compression and damage. Glass bottles should be avoided where possible. All very small quantities should be shipped as a combination package - an inner package in a UN approved box containng a suitable fabric adsorbent.

Small quantities [1-10 litres]

For larger quantities, individual 1 litre containers should be used either with a cap insert, or small plastic drums with non-removable heads. A supplier of such packaging can provide a suitable receptacle and it is recommended that for quantities of less than 10 litres, an outer package is used, e.g. a cardboard box filled with adsorbent or a wooden box. If a 10 litre plastic drum were used, it would be preferable to place it for shipment in an outer package, e.g. a steel drum or wooden box filled with adsorbent. Drums can also be shipped as a single package.

Moderate quantities [10-450 litres [maximum 400 kg]]

For moderate quantities, the UN limitations mean that maximum volumes are only 30 l for plastic packages [drums] or 40 l for metal drums [see Table 4]. Metal drums should be stainless steel, or a PTFE lined mild steel drum if used as an inner package. For single packages, e.g. drums, the maximum weight is 400 kg; therefore, standard drums can be used, provided they are stainless steel or polypropylene barrels [or lined steel drums] and are filled to less than 90% of maximum capacity.

In addition to the packaging types noted above, the other important packaging, which is occasionally used for fast pyrolysis liquid, is the Intermediate Bulk Container [IBC]. An IBC is a rigid, or flexible portable packaging, other than those specified in Appendix A.5 of the UN guide (2). According to the requirements of ADR, IBCs are not to be used for Packing Group 1 liquids.

Large quantities [more than 400 kg]

Large samples need to be transported in larger containers or tanks. Containers are specifically defined in the ADR regulations (2). The additional requirement for tank/containers is the use of a placard on road containers, displayed the appropriate UN code of 1993 on the bottom and 33X on the top [see next section]. Containers and tanks will be the preferred method with time for larger quantities for land transport. Again, acid resistant containers and tanks are required.

LABELLING OF PACKAGES

Marking

Each package shall be clearly marked with the substance identification number of the goods to be entered in the transport document, preceded by the letters "UN". For fast pyrolysis liquid, this classification is 1993.

Danger labels

Packages containing substances or articles of this class shall bear a label conforming to model No. 3 [Class 3 – Flammable liquids] and shall in addition bear a label conforming to model No. 6.1 [Class 6 – Toxic substances] as shown in Figure 1. Packages containing receptacles, the closures of which are not visible from the outside, and packages containing vented receptacles with or without outer packaging shall in addition bear on two opposite sides a label conforming to model 11 – see Figure 1.

There is a requirement for this label, due to the presence of phenols in the liquid. Labels No. 3 and No. 6.1 shall be diamond shaped and measure at least 100 x 100 mm. They have a line of the same colour as the symbol appearing on the label 5 mm inside the edge and running parallel to it. If the size of the package so requires, the dimensions of the label may be reduced, if they remain clearly visible [see Table 5]. Labels to be affixed to vehicles, to tanks of more than 3 m^3 or to large containers shall measure not less than 250 x 250 mm.

Class 3 Flammable Liquid
Label 3

Class 6 Toxic substances
Label 6.1

Store this way up
Model 11

Figure 1. *Labels for fast pyrolysis liquid receptacles*

Any label required to be carried on a package shall be securely fixed to the package with its entire surface in contact with it and the label shall be clearly and indelibly printed. The colour and nature of the marking shall be such that the symbol [if any] and wording stand out from the background to be readily noticeable and the wording shall be of such a size and spacing as to be easily read. The package shall be so labelled that the particulars can be read horizontally when the package is set down normally. The dimensions of the labels required for the package are given in Table 5. For tank containers, as defined above, a placard is required. The top code for the placard is: 33X, the bottom code is 1993. Large tanks are fitted with replaceable codes. The placard dimensions are typically a minimum of 30 cm high by 40 cm wide, numerals to be a minimum of 10 cm high, black on an orange background.

Table 5. *Label Requirements for Packages*

Capacity of Package	Dimensions of label
Not exceeding 3 litres if possible	at least 52 x 74 mm
Exceeding 3 litres but not exceeding 50 litres	at least 74 x 105 m
Exceeding 50 litres but not 500 litres	at least 105 x 148 mm
Exceeding 500 litres	at least 148 x 210 mm

MIXED AND EMPTY PACKAGING

Mixed packaging

Fast pyrolysis liquid should not be shipped with other dangerous goods. Up to 5 litres [inner packaging] may be shipped with goods not subject to the provisions of ADR, provided they do not react dangerously with each other. Fast pyrolysis liquid shall not be packed together with substances and articles of Classes 1 and 5.2 [explosives substances and articles and organic peroxides respectively, other than hardeners and compound systems] and material of class 7 [radioactive material].

Only 0.5 litres of fast pyrolysis liquid per inner packaging and 1 litre per package, which are classed under (a – highly hazardous), may be shipped with up to 5 litres of Class 3 substance not exceeding 5 litres if they classified under (b- moderately hazardous) or (c- low hazard). This applies if mixed packaging is also permitted for substances or articles of these classes, and/or with goods that are not subject to the provisions of ADR, provided they do not react dangerously with each other. Dangerous reactions are classed as:

1. Combustion and/or giving off considerable heat,
2. Emission of flammable and/or toxic gases,
3. Formation of corrosive liquids,
4. Formation of unstable compounds.

If wooden or fibreboard boxes are used, a package shall not weigh more than 100 kg relating to the inclusion of adsorbent packaging and the provision of a leak proof outer package and that all individual packages are clearly singly labelled (2)].

Empty packaging

In the case of empty tank vehicles, empty demountable tanks and empty tank containers, uncleaned, this description shall be completed by adding the words, "Last load", together with the name and item number of the goods last loaded, e.g.:

Last load 1993 Flammable liquid [Fast Pyrolysis Liquid], 1°(a)

HANDLING AND STORAGE OF FAST PYROLYSIS LIQUID

HANDLING OF FAST PYROLYSIS LIQUID

Generally, handling and storage activities are a common feature of the majority of workplaces. They are also one of the principal causes of death and injury due to this interface between people and the wide range of materials handled. This section sets out the measures necessary on the part of employers, employees, manufacturers, designers, importers and suppliers of fast pyrolysis liquid used at work to ensure safety and the minimisation of risks to health in connection with the use, handling, storage and transport of fast pyrolysis liquid. The guidance in this section is based on the UK legislation (15) and Guidance from the Environment Agency, UK on the storage of hazardous materials. Handling and storage covers a broad range of areas, including:

1. The use of fixed and mobile handling equipment;
2. Manual handling operations;
3. Design of the workplace;
4. The provision of a suitable working environment;
5. Specific requirements for the handling and storage of identified hazardous substances;
6. Controls on the use of hazardous materials;
7. Specific requirements for labelling of hazardous substances; and
8. The selection, provision and use of personal protective equipment.

The law on handling and storage is diverse. For example in the UK these range from the more general requirements such as the Health and Safety at Work Act [HSWA] to the specific requirements of the Control of Substances Hazardous to Health [COSHH] Regulations 1994 and the Highly Flammable Liquids and Liquefied Petroleum Gases Regulations 1972. Each country has its own legislation and national guidance for the handling of goods and these should be used where appropriate.

The handling and storage of materials has, in many cases, great potential for pollution incidents, particularly in the case of hazardous materials that may be discharged by natural seepage to water and land resulting in groundwater pollution in particular. Within the EU, this issue is tasked within the new Integrated Pollution Prevention and Control [IPPC] regulations that came into force on 1st August 2000. Guidance from the Management of Health and Safety at Work Regulations [UK], 1992, proposes that the following activities should be instigated in the case of the handling and storage of hazardous materials:

1. Risk assessments;
2. Implementation of management systems for the effective planning, organising, controlling, monitoring and review of any preventive and protective measures arising from a risk assessment;
3. Appointment of competent persons;
4. Establishment of emergency procedures to be followed in the event of serious or imminent danger;
5. Provision of comprehensible and relevant information;
6. Consideration of human capability;
7. Provision of health and safety training; and;
8. Consultation with safety representatives.

STORAGE OF HAZARDOUS SUBSTANCES

Prior to storing and handling fast pyrolysis liquid, it is essential to consult sources of hazard data, typically the MSDS or other available sources (16, 17). The chemical compatibility of hazardous materials must be given particular consideration. Potentially reactive material must be stored separately [mixing may occur due to spillage, leakage or accident e.g. during a fire]. The following precautions are necessary to ensure the safe handling and storage of dangerous goods and/or chemical substances with fast pyrolysis liquid:

1. Meticulous standards of housekeeping should be maintained at all times;
2. Smoking and the consumption of food or drink should be prohibited in any area in which substances are used or stored;
3. Staff must be reminded regularly of the need for good personal hygiene, in particular washing of hands after handling chemical substances;
4. The minimum quantities only should be stored in the working area; extra bulk storage may be required separately and well away from the work area;
5. Containers and transfer containers should be clearly and accurately marked;
6. Chemical substances should always be handled with care and carriers used for Winchester and other large containers;
7. Fume cupboards should operate with a minimum face velocity of approximately 0.4 m/sec when measured with the sash opening set at 300 mm maximum, and performance should be checked frequently in accordance with the COSHH Regulations;
8. Staff should always wear personal protective clothing and equipment e.g. eye protection, face protection, aprons, gloves, protective boots, whenever handling fast pyrolysis liquid, in particular when handling large quantities;
9. Responsibility for safe working should be identified at senior management level, and written procedures published and used in the training of staff.

Bulk chemical storage [drums, barrels, tanks and similar containers]

In the design and use of bulk storage facilities, the following aspects need attention:

1. The range and quantities of substances to be stored;
2. Dependent upon (1) above, the degree of segregation by distance of:
    a. The store from any other building; and
    b. Certain chemical substances within the store from other chemical substances stored.

Purpose-built chemical stores should be of the detached single-storey brick built type or constructed in other suitable materials, such as concrete panels, with a sloping roof of weatherproof construction. The structure should have a notional period of fire resistance of at least one hour. Other features include [based on advice from the Environment Agency, UK]:

1. Permanent ventilation by high and low level airbricks set in all elevations, except in those forming a boundary wall; low-level airbricks should be sited above doorsill level;

2. Access doors constructed from material with at least one hour notional period of fire resistance; doorways should be large enough to provide access for fork lift trucks, with ramps on each side of the door sill [also to contain any internal spillage]; separate pedestrian access, which also serves as a secondary means of escape, should be provided;
3. An impervious chemical-resistant finish to walls, floors and other surfaces;
4. Artificial lighting by sealed bulkhead or fluorescent fittings, to provide an overall luminance level of 300 lux;
5. Provision of adequate space, with physical separation and containment for incompatible substances, each area to be marked with the permitted contents, the hazards and the necessary precautions, and incorporating an area for the storage of empty containers;
6. Fire separation of individual areas sufficient to prevent fire spreading;
7. Provision of the following equipment in a protected area outside the store:
   - Fire appliances (dry powder and/or foam extinguishers);
   - Fixed hose reel appliance;
   - Emergency shower and eyewash station with water heating facility to prevent freezing;
   - Personal protective equipment i.e. safety helmet with visor, impervious gloves, disposable chemical-resistant overall, with storage facilities for same; and
   - Respirator and breathing apparatus in a marked enclosure;
8. A total prohibition on the use of naked flames and smoking, appropriate warning signs should be displayed;
9. A prohibition on the use of the store for storage of other items or for any other purpose; and
10. Provision of racking or pallets to enable goods to be stored clear of the floor.

External drum storage

Drums, barrels, carboys and other similar containers for fast pyrolysis liquid should be stored in the external air on an impervious and durable surface, which is in excess of 4 m to any risk area, bund or open boundary. The area should be protected by a bund wall, dished or ramped to contain spillages, with the walls and floor impervious to the materials stored. The bunded area should contain no drains or valves. Vehicular access to such areas should be protected by a ramp or a channel ensuring that the ramp itself does not cause regular spillages.

Generally, no container should be stored within 2 m of any window, escape route or door. Much will depend on the nature of the substances stored and the design of the storage area. Where a storage area is constructed with fire resistant walls, these distances can be reduced.

Ensure that overflow pipes on all tanks discharge within the bunded area. Any tank situated on a roof may drain to the surface water system via the guttering therefore roof storage should be avoided. Flammable liquids should be stored in a purpose-built external flammable materials store and not in a warehouse. Much will depend upon the quantities to be stored. Small quantities should be stored in a lockable metal cupboard, suitably marked. Drums should be stored within a bunded area to contain any spillages. Vehicular access to such areas should be protected by a ramp or a channel ensuring that the ramp itself does not cause regular spillages. Automatic cut-offs should be provided on all delivery pipes to prevent spillage due to overfilling.

TREATMENT OF SPILLS

Spills can range in size from grams to thousands of litres. The prevention of spills and the problems associated with spillages can be minimised by good handling, storage practices, and other preventative measures. Spillages of hazardous substances can arise because of poor storage systems, the use of unsuitable or defective containers, during refilling of tanks and other containers, or because of human error. Procedures for dealing with both small and large spillages should be outlined in the Material Safety Data Sheet [MSDS].

In most cases, small spillages can be dealt with immediately by absorption in sand, sawdust or proprietary absorbent granules for subsequent disposal to a waste container. Large spillages will need a considerable degree of attention to ensure, in particular, that substances do not enter a drainage system or natural watercourse in concentrated form. This may entail containment of the spillage using drain cones, sand bags, polythene sheeting and a range of other materials, so that it can eventually be pumped into a disposal container. Spillage of hazardous materials should never be washed to surface water drains. An absorbent material used to clean up a spillage of a pyrolysis liquid may have to be disposed of as special waste. The potential for water and ground pollution because of spillages very much depends upon the handling systems operated on site. The following recommendations with regard to materials handling are made:

1. All loading and discharge points should be designated, marked and isolated from the surface water drainage system.
2. Routes of transfer for all materials should be identified and the complete route should be protected against spillages to the surface water system.
3. Underground pipework should be avoided where possible, as faults are more difficult to detect and can lead to groundwater contamination.
4. Manual handling should be avoided where possible to reduce the risk of human error and accidents.
5. Yard areas used for materials handling or materials handling processes must be isolated from the surface water drainage system by bunding. Roofing over such an area is an advantage to prevent the accumulation of rainwater, fire regulations permitting.
6. Appropriate containers should always be used for different materials. They should be sturdy, in good condition, clearly labelled and not liable to leak.
7. The necessity for materials handling and transfer should be minimised to reduce risks.

To handle spills, the following actions are proposed:

<u>Small Quantities [up to 1 litre]</u>

1. Wear rubber gloves and suitable eye and face protection.
2. Cover contaminated area with sawdust, or other suitable inert adsorbent, e.g. vermiculite or montmorillonite,
3. Collect contaminated absorbent and place in closed container,

4. Transport to approved landfill or incinerator for disposal.

Large Quantities [above 1 litre]

1. Evacuate area.
2. Wear rubber boots, rubber gloves, suitable eye/face protection and NIOSH/MSHA approved respirator.
3. Cover contaminated area with sawdust or vermiculite. Take up sawdust or vermiculite and place in closed container. Transport to approved landfill or incinerator.

For large spills, a spill kit is advised to contain the spill and prevent its incursion to local watercourses. The recovery of fast pyrolysis liquid as an adsorbed waste may mean that it must be treated as "special waste" for disposal. Where possible, contaminated adsorbent should be recovered and stored in sealed containers for subsequent disposal.

CONCLUSIONS

Fast pyrolysis liquid may be classed as a Class 3 Flammable liquid for the purposes of transport by any mode. The use of the UN manual for the assessment of fast pyrolysis liquid has allowed a generic classification to be made to ensure that liquid that is shipped in UN approved packaging, or packaging conforming to UN requirements is meeting the relevant international regulations. The recommended coding is:

UN 1993 Flammable Liquid [Fast Pyrolysis Liquid], n.o.s., 3, 1°(a), 2°(a), 1

The use of the UN code will allow samples to be shipped in a manner complying with the regulations currently in force in the EU, Canada and the USA. The use of an approved supplier of UN packaging will ensure that samples are shipped in the correct packaging and facilitate transport, which is more rapid, with the required information for the transportation document and the provision of MSDS.

RECOMMENDATIONS

The following recommendations are made:

1. Data on fast pyrolysis liquid should be provided to the UK DETR – Dangerous Goods Branch to permit an application to the EU for the listing of fast pyrolysis liquid. This automatically provides access to the analogous approvals bodies in 34 countries.
2. Data is required on the toxicity of fast pyrolysis liquid to ensure that MSDS can provide sufficient information for handlers and transporters of the liquids,
3. The relative quantities of adsorbents required to deal with spills should be assessed,
4. Work should be carried out to understand how to remediate spill areas and assess environmental consequences.

ACKNOWLEDGEMENTS

The authors wish to thank the EC and IEA Bio-energy Funded Pyrolysis Network (PyNe) for providing the funding to enable this work to be carried out. The full report from which this paper is taken will be published by the PyNe Network.

GLOSSARY

49 CFR	United States Regulations Relating to standards for Hazardous Materials transportation
ADN	The European Provisions concerning the International Carriage of Dangerous Goods by Inland Waterways
ADNR	Regs. for the Carriage of Dangerous Substances on the Rhine
ADR	EU Agreement on the Carriage of Dangerous Goods by Road
CAA	Civil Aviation Authority
CCNR	Central Commission for the Navigation of the Rhine
COSHH (UK)	Control of Substances Hazardous to Health Regulations 1994
DETR (UK)	Department of the Environment, Transport and Regions, UK
ECE	Economic Commission for Europe
ECOSOC	Economic and Social Council Committee of Experts
FAA (USA)	Federal Aviation Administration
FHWA (USA)	Federal Highway Administration
HSWA (UK)	Health and Safety at Work Act
IATA	International Air Transport Association
IBC	Intermediate Bulk Container
ICAO	International Civil Aviation Organisation
IMDG	International Maritime Dangerous Goods
IMO	International Maritime Organisation
MSDS	Material Safety Data Sheets
MSHA	Mine Safety and Health Administration, USA
NIOSH	National Institute of Occupational and Safety Hazards, USA
OTIF	Organisation for International Carriage by Rail
PTFE	Polytetrafluoroethene
RID	EU Agreement on the Carriage of Dangerous Goods by Rail
RID/ADR/ADN	The Joint Meeting of the Working Party on the Transport of Dangerous Goods and the RID Safety Committee
RSPA (USA)	Research and Special Programs Administration of the Department of Transportation, USA
TDG (Canada)	Transportation of Dangerous Goods Regulations, Canada
USCG	United States Coast Guard
USDOT	United States Department of Transportation

REFERENCES

1. Approved Supply List [4th Edition] - Information Approved for the Classification and Labelling of Substances and Preparations Dangerous for Supply, HM Stationery Office, 1998.
2. UN Recommendations on the Transport of Dangerous Goods Model Regulations, 11th Edition, United Nations, New York, January 2000.
3. *Technical Instructions for the Safe Transport of Dangerous Goods by Air, 1999-2000 Edition*. Doc 9284-AN/905, ICAO.
4. Dangerous Goods Regulations CD-ROM (ref#9515-41) IATA Dangerous Goods Regulations manual, IATA, Montreal, Quebec, Canada.
5. Technical Instructions for the Safe Transport of Dangerous Goods by Air [Doc 9284-AN/905 and supplement] from CAA, Printing and Publications Service, Cheltenham, UK.
6. International Maritime Dangerous Goods Code (IMDG Code) - 1994 [including Amendment 29-98], IMO-213E, 1999, IMO.
7. European Agreement concerning the international carriage of dangerous goods by road [ADR] and protocol of signature, United Nations, New York and Geneva, ECE/TRANS/130 (Vols 1 and 2), 1998, ISBN 92-1-139062-1.
8. Regulations concerning the international carriage of dangerous goods by rail [RID) 1999 edition [Annex 1 to Appendix B to the Convention concerning international carriage by rail [COTIF)), HM Stationery Office 1998.
9. European Provisions concerning the International Carriage of Dangerous Goods by Inland Waterway [ADN], ECE/TRANS/WP.15/148, EU, 2000.
10. See Internet site http://hazmat.gov.com/
11. The Export and Import of Hazardous Wastes Regulations and the Transportation of Dangerous Goods [TDG] Regulations, 1992, Transport Canada, Ottawa, 1992.
12. Recommendations on the Transport of Dangerous Goods - Manual of Tests and Criteria - Third revised edition, United Nations, New York, **2**, January 2000.
13. Bridgwater A.V. (1996) Production of high-grade fuels and chemicals from catalytic pyrolysis of biomass. *Catalysis Today,* **29**, 1-4, pp. 285-295.
14. Meier D., Oasmaa A. and Peacocke G.V.C., (1996) "Properties of Fast Pyrolysis Liquids: Status of Test Methods", In: Developments in Thermochemical Biomass Conversion, (Ed. by A.V. Bridgwater and D.G.B. Boocock), volume 1, pp 391-408, Blackie Academic and Professional.
15. Health and Safety At Work Act [UK], 1974 and Management of Health and Safety at Work Regulations [UK], 1992
16. Diebold J.P. (1999) "A Review of the toxicity of biomass pyrolysis liquids formed at low temperatures", In: *Fast pyrolysis of biomass: a handbook*, (Ed. by A.V. Bridgwater, S. Czernik, J. Diebold, D. Meier, A. Oasmaa, C. Peacocke, J. Piskorz and D. Radlein), pp. 135-163. CPL Press Ltd, UK.
17. Chemicals (Hazard Information and Packaging for Supply) (CHIP 2) Regulations 1994, UK.

Levoglucosenone – a Product of Catalytic Fast Pyrolysis of Cellulose

G. Dobele, G. Rossinskaja, G. Telysheva
Latvian State Institute of Wood Chemistry, 27 Dzerbenes St., LV-1006, Riga, Latvia
D. Meier, S. Radtke, O. Faix
Federal Research Centre for Forestry and Forest Products, Institute for Wood Chemistry and Chemical Technology of Wood, D-21027, Hamburg, Germany

***ABSTRACT*:** Five celluloses, differing by polymerization degree and crystallinity index, impregnated with phosphoric acid and thermally pretreated have been subjected to fast pyrolysis in order to investigate the formation of levoglucosenone, a dehydrated 1,6-anhydrosaccharide. It was established that the yield of levoglucosenone depends on both properties of cellulosic raw material and conditions of thermal pretreatment. Thermal pretreatment allows to change the degree of polymerization of cellulose. When the pretreatment temperature exceeds 100 °C, monophosphates of cellulose are formed, the development of intermolecular cross-linking is promoted and depolymerization of cellulose is hindered.

The highest yield of levoglucosenone (29 % b.o. cellulose) was obtained by pyrolysis of microcrystalline Munktell cellulose impregnated with 3.5 % of phosphoric acid and pretreated at 100 °C. Realization of high yields of levoglucosenone from celluloses with a higher degree of polymerization requires larger amounts of phosphoric acid.

INTRODUCTION

It is known that the qualitative and quantitative composition of the thermal degradation products from polysaccharides can be altered by use of different catalysts [1]. Inorganic acids are not well selective catalysts as they affect both the degradation and condensation reactions in the pyrolysis process. A relationship between dehydration, degradation and condensation reactions is determined by the individual properties of the acid, the characteristics of the cellulose structure and by the pyrolysis conditions [2].

In the presence of acid additives thermal degradation of cellulose is intensified at lower temperatures, due to the occurrence of the dehydration reactions [3]. Condensation reactions result in the decrease of volatile products at 450°C combined with an increase of their water portion [4]. Slow heating rates and atmospheric pressure promote the dehydration and subsequent condensation reactions [5].

Depending on the properties of the acid used as an additive, the cellulose dehydration process proceeds according to two mechanisms, i.e. *via* formation of intermediate carbon ions or esters. In the case of phosphoric acid dehydration is accomplished predominantly *via* a repeated formation and splitting of esters [5]. The cellulose phosphate esters are sufficiently thermostable [6], therefore cellulose condensation reactions in the low-temperature range (up to 300 °C) do not develop considerably. The phosphoric acid dehydrating action as well as its hydrolytic activity leads to formation of dehydrated bicyclic 1,6-anhydrosaccharide - levoglucosenone (more than 20% b.o. cellulose) during cellulose pyrolysis at 350 °C [7, 8]. The unique chemical structure of levoglucosenone offers a variety of synthesis of biologically active products on this basis [9].

Latest reports demonstrated that levoglucosenone can be obtained from slow and fast pyrolysis of various cellulose containing raw materials impregnated with phosphoric acid [10-12]. Slow pyrolysis of impregnated cellulose (3-7 % of phosphoric acid) promotes preferentially levoglucosenone formation. Fast pyrolysis experiments at analytical scale revealed that apart from levoglucosenone also levoglucosan is formed which is the main depolymerization product of untreated cellulose. The yields of levoglucosan and levoglucosenone in the composition of pyrolysis volatile products amount to 75 to 85%, regardless of the cellulose type, amount of phosphoric acid (3.5-7%) used for impregnation and pretreatment temperature (100-160°C) [12].

The goal of the present study was to investigate the effect of structural peculiarities of different types of cellulose on levoglucosenone formation under catalytic fast pyrolysis conditions. Moreover, amount of phosphoric acid and pretreatment temperature should be determined to provide maximum possible yields of levoglucosenone.

EXPERIMENTAL

MATERIALS

Five cellulose samples were used: microcrystalline celluloses (Munktell, Avicell and Thermocell), cellulose from sulphate pulping process (Taircell), and recycled Kraft-pulp (IFAB UKP, Sweden). The method for preparation of microcrystalline Thermocell cellulose was developed at IWCh, Latvia [13].

Crystallinity index was calculated based on data of X-ray diffractometry [14]. The degree of polymerization was determined by the viscosity measurements from cellulose cadoxen solutions (the relative error of the method was 5 %) [15].

IMPREGNATION AND PRETREATMENT

87 % phosphoric acid (puriss., Fluka) was used. An aqueous solution of phosphoric acid (3.5, 5 and 7 wt.% based on dry cellulose) was mixed with cellulose (cellulose/acid solution = 1:5). Water was evaporated under vacuum, and the impregnated material was dried at 40 °C under the reduced pressure. The dried samples were heated at 100, 160 and 200 °C for 1 hr.

DETERMINATION OF PHOSPHORUS

The amount of phosphorus present in the sample was measured through the combustion of impregnated cellulose and the subsequent determination of phosphorus as a triple phosphorus-molybdate-vanadate complex by photometry on a Specord UV/VIS at a wave length of 400 nm [16].

For the determination of phosphorus chemically bound to cellulose, impregnated samples were washed with water with a modulus of 1:200, then dried under reduced pressure at 40 °C. The washed sample (20-100 mg) was suspended in a potassium chloride solution (0.75 g of KCl in 25 ml of water), and was titrated potentiometrically with 0.025 N NaOH. The degree of substitution for phosphoric acid esters (mono-, di- and tri-esters) was determined by titration curve differentiation [17].

ANALYTICAL PYROLYSIS (Py-GC)

A CDS Pyroprobe 100 combined with a gas chromatograph (CP 9000) was used to pyrolyze and analyze a 70 mg sample. Pyrolysis temperature was set to 500 °C. Heating rate was 600 $°C^{s-1}$ and pyrolysis time was 10 s. A DB 1701 column (60 m×0.25 mm, 0.25 mm film thickness) was used for the separation of monomeric volatile compounds. Further details of the procedure can be found elsewhere [18]. External calibration was used for quantification of levoglucosenone.

RESULTS AND DISCUSSION

CELLULOSE CHANGES DURING IMPREGNATION PRETREATMENT

Effect of interaction of acid and cellulose at the stages of impregnation and thermal pretreatment depends on the cellulose properties. The celluloses under study have different ratios of ordered and amorphous regions. They differ also by their degree of polymerization (Table 1) and hydrophilic properties. The presence of phosphoric acid affects the system of the cellulose hydrogen bonds, the crystallinity index and leads to the formation of esters [11,12, 19].

Table 1 Characteristics of cellulose supramolecular structure and degree of polymerization (DP).

Cellulose	Crystallinity index		DP
	untreated	7 % H_3PO_4, 160 °C	
Munktell	0.86	0.85	200
Avicell	0.81	0.78	220
Thermocell	0.72	0.71	300
Taircell	0.78	0.76	1040
Kraft-pulp	0.72	0.59	2450

To evaluate the cellulose reactivity with respect to the esterification reactions, the amount of chemically bound phosphorus in impregnated materials was determined and related to the amount of acid and pretreatment temperature. Two celluloses with a different crystallinity index and DP were selected for comparison: microcrystalline Munktell cellulose and Taircell cellulose from sulphate pulping process.

The data presented in Table 2 demonstrate that pretreatment of impregnated celluloses at 100 °C does not lead to chemically bound phosphorous and formation of esters, i.e. phosphoric acid was completely washed off from cellulose. However, phosphoric acid esters as monophosphates could be detected in the impregnated cellulose samples after heating them at 160 and 200 °C.

Table 2 Changes in the amount of chemically bound phosphorus and cellulose DP depending on the quantity of acid and pretreatment temperature.

H_3PO_4, (%)#	P (%)#	Pretreatment temperature, (°C)	Amount of bound P (%)#	Amount of bound P (%) b.o. P	DP
Taircell					
3.5	1.2	100	-	-	310
3.5	1.2	160	0.35	29.2	265 (9*)
3.5	1.2	200	1.16	92.5	n.d.
5	1.7	100	-	-	n.d.
5	1.7	160	0.61	35.8	n.d.
5	1.7	200	1.50	95.9	n.d.
7	2.5	100	-	-	380
7	2.5	160	1.02	42.9	230 (41*)
7	2.5	200	2.44	97.2	n.d.
Munktell					
3.5	1.1	100	-	-	190
3.5	1.1	160	0.18	16.1	380 (10*)
3.5	1.1	200	0.84	76.8	n.d.
5	1.6	100	-	-	n.d.
5	1.6	160	0.48	30.0	n.d.
5	1.6	200	1.36	81.2	n.d.
7	2.3	100	-	-	195
7	2.3	160	0.78	33.9	320 (43*)
7	2.3	200	1.82	81.5	n.d.

*amount of cellulose insoluble in cadoxene, b.o. cellulose, %
based on cellulose

By increasing the pretreatment temperature and the amount of the acid the content of chemically bound phosphorus increased for both celluloses. At 200 °C the less ordered cellulose Taircell binds practically all phosphorous (92.5 to 97.2%). For the more ordered microcrystalline Munktell cellulose the amount of chemically bound phosphorus was lower compared to Taircell and reached after thermal treatment at 200 °C 81 % of the amount introduced.

Besides ester formation another effect of phosphoric acid is hydrolytic splitting of glycosidic bonds. This effect is more pronounced for celluloses with a higher DP and depends also on the amount of acid used for impregnation and the pretreatment temperature.

After thermal pretreatment at 100 °C of Taircell cellulose, which has the highest DP amongst other celluloses under study, its DP decreased from 1040 to 310 and 380 after introduction of 3.5 and 7 % phosphoric acid, respectively (Tables 1 and 2). An increase of DP after introduction of larger amounts of acid (7 % vs. 3.5 %) could be explained by intermolecular linking reactions, which compete with the hydrolysis of glycosidic bonds. At similar impregnation and pretreatment conditions the DP of Munktell cellulose, having the lowest DP (200), was practically left unchanged.

Increase of pretreatment temperature promotes intermolecular cross-linking reactions, which are additionally stimulated by increasing amount of phosphoric acid. Pretreatment at 160 °C resulted in a decrease of cellulose solubility in cadoxene. At 3.5 % of phosphoric acid concentration, the amount of the cadoxene-insoluble fraction for both celluloses was approximately the same, i.e. 9-10 %. Increasing the acid addition from 3.5 to 7 % gave four times higher values. (Table 2). At the same time the DP of the soluble fraction of Taircell cellulose, impregnated with 3.5 and 7% of phosphoric acid and pretreated at 160°C, decreased to 265 and 230, respectively, indicating the ongoing hydrolysis of glycoside bonds. Unlike Taircell cellulose, the DP of the soluble fraction of microcrystalline Munktell cellulose increased, to 380 and 320, respectively, i.e. in this case the intermolecular cross-linking reactions are prevailing.

PYROLYSIS OF IMPREGNATED CELLULOSE, YIELD OF 1,6-ANHYDROSACCHARIDES

Cellulose changes taking place during impregnation and thermal pretreatment stages find their reflection in the alteration of the composition of volatile pyrolysis products (Table 1, 2). As a result of dehydration and depolymerization reactions the yield of levoglucosan (Lg) decreased, while levoglucosenone (Lg-none) became the main pyrolysis product of cellulose impregnated with phosphoric acid (Table 3, Fig. 1). Thermal pretreatment of non-impregnated cellulose also decreased the Lg yield during pyrolysis (Table 3).

The changes in the yields of 1,6-anhydrosaccarides, under the conditions of analytical pyrolysis which model the fast pyrolysis conditions, were different for each cellulose (Fig.1.). The celluloses studied can be subdivided into two groups with respect to the Lg-none yield: Avicell and Kraft pulp in one group, and Munktell, Thermocell and Taircell in another.

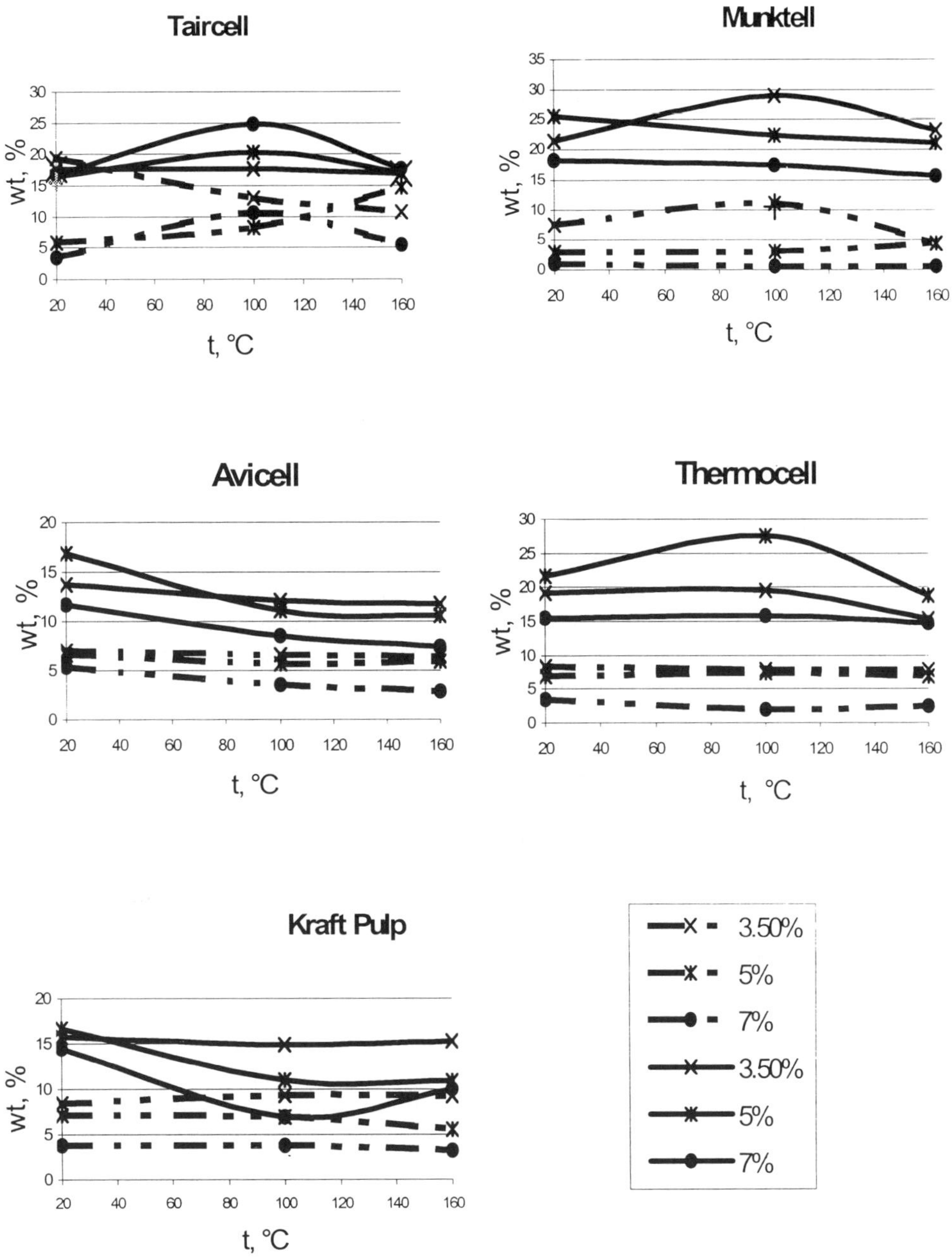

Fig 1. Dependence of levoglucosan (dashed line) and levoglucosenone (solid line) yields on the amount of phosphoric acid and pretreatment temperature

Table 3 Yield of levoglucosan obtained by analytical pyrolysis of non-impregnated cellulose.

Cellulose	Levoglucosan yield, b.o. cellulose (%)	
	blank	after pretreatment @ 160 °C
Munktell	79.3	46.2
Avicell	36.6	28.2
Thermocell	13.6	12.4
Taircell	19.6	17.0
Kraft	8.2	7.3

Despite the fact that celluloses of the first group (Avicell and Kraft pulp) differ considerably by their structure, DP and chemical composition (Table 1), the yields of 1,6-anhydrosaccharides during pyrolysis alter in a similar way depending on the impregnation and pretreatment conditions (Fig. 1). The highest levoglucosenone yield was 16% by introducing 5% phosphoric acid without thermal treatment. Pretreatment negatively affects the formation of levoglucosenone in this case. The yield of levoglucosan slightly depends on impregnation and pretreatment conditions and changes in the range of 3 to 9 %.

For the second group of celluloses (microcrystalline celluloses Munktell, Thermocell and cellulose Taircell) the highest yields of levoglucosenone were 29, 28 and 25 %, respectively. They were obtained of samples impregnated with phosphoric acid and pretreated at 100°C (Fig.1). However, the amount of the acid to obtain the highest yield of Lg-none from celluloses of this group differs for the row of above mentioned celluloses and is equal to 3.5, 5 and 7 %, respectively. The least amount of phosphoric acid is required for cellulose Munktell which is characterized by a homogeneously ordered supramolecular structure and low polymerization degree.

The results obtained show that the formation of 1,6-anhydrosaccharides was affected by both the properties of the cellulose and its structural changes obtained by impregnation and thermal pretreatment. The highest yield of levoglucosenone was achieved from pyrolysis of impregnated Munktell cellulose having the highest index of crystallinity and levelling-off DP. Pyrolysis of Munktell also gave higher levoglucosan yields compared to the other celluloses. However, for the tested series of celluloses no correlation was observed concerning the yields of levoglucosan and levoglucosenone obtained from pyrolysis of initial and impregnated celluloses.

Formation of cellulose phosphates during pretreatment at temperatures higher than 100 °C inhibited levoglucosenone formation. Obviously, the main positive effect of pretreatment at 100 °C was the change of the hydrogen bonds system under the action of phosphoric acid and the decrease in DP. Celluloses with a low initial DP needed less acid to give a maximum yield of Lg-none (Table 1, Fig.1). So, in the order of celluloses Taircell (DP 1040), Thermocell (DP 300) and Munktell (DP 200) produced the highest yield of Lg-none at 7, 5, and 3 % of phosphoric acid, respectively. The yield of levoglucosan decreased with an increase of phosphoric acid. For lower acid amounts (3.5 %) the yield of LG obtained from Munktell and Taircell varied within 11 and 7 %.

Taircell impregnated with 3.5% acid and without pretreatment gave levoglucosenone yields of up to 17 %. The amount of levoglucosan is the same as for initial cellulose, namely 19 %, which is a proof for a catalytic effect of phosphoric acid on cellulose depolymerisation. Thermocell and Kraft pulp showed an increase in the summary yield of 1,6-anhydrosaccharides after impregnation with of 3.5 % acid compared to initial samples. This indicates that phosphoric acid catalyzed thermal depolymerization of celluloses having a more disordered supramolecular structure. Probably, under fast pyrolysis conditions dehydration and depolymerization processes with levoglucosenone formation proceed first of all in more disordered regions of the cellulosic structure where phosphoric acid is concentrated during impregnation. Thermal depolymerization takes place simultaneously in more ordered regions of cellulose where the dehydrating effect of phosphoric acid is less pronounced resulting in the formation of the non-dehydrated 1,6-anhnydrosaccharide, levoglucosan.

CONCLUSIONS

From the results obtained, the studied celluloses could be arranged in the following order with the respect to the yield of levoglucosenone obtained by catalytic flash pyrolysis:

Kraft = Avicell < Taircell < Thermocell < Munktell

The highest yield of levoglucosenone (29 %) has been obtained by pyrolysis of microcrystalline Munktell cellulose, which is characterized by the highest index of crystallinity, levelling-off DP and the lowest hydrophility.

In general, a high yield of levoglucosenone by catalytic pyrolysis of cellulosic raw materials can be achieved when:

- the amount of phosphoric acid needed for the process of cellulose hydrolytic depolymerization correlates with DP of cellulose, i.e. the higher the DP, the higher the amount of acid needed;
- thermal pretreatment should be realized in the temperature range of 40-100 °C, because higher temperatures promote formation of monophosphates, which slow down the depolymerization processes.

REFERENCES

1. Einsele U., Meier P., Herlinger H.(1979) Zusammenhänge zwischen Dehydratisierungsreaktion und Flammschutzeffekt bei Cellulosefasern. *Cell. Chem. And Technol.*, **13**, 57-75.
2. Hendrix I.E., Rostic I.E., Olsen E.S. (1970) Pyrolysis and combustion of cellulose. 1. Effect of triphenylphosphate in the presence of nitrogenous bases. *J. Appl. Polymer Sci.*, **14**, 1701-1723.
3. Lewin M., Basch A. (1978) Structure, pyrolysis and flammability of cellulose. In: *Flame Retardant Polymeric Materials*, **2**, pp. 1-41. New York- London.
4. Dobele G., Dizhbite T., Rossinskaja G., Telysheva G. (1995) Thermocatalytic destruction of cellulose. In: *Cellulose and Cellulose Derivatives: Physico-chemical Aspects and Industrial Application (*Ed. by. J.F.Kennedy, G.O.Phillips, & P.A.Williams) pp. 125-130, Woodhead, Cambridge, UK.
5. Jermolenko I.N., Ljubliner I.P., Gulko N.V. (1982) *Element-Containing Carbon Fibrous Materials*. Nauka I Technika, Minsk.
6. Jacevskaja M., Komarov V. (1979) The properties of activated carbon obtained from metalsubstituted phosphorylated wood. K*imija Drev.*, **2**, 91-95.
7. Rossinskaja G., Dobele G., Domburg G. (1986) Thermocatalytic transformation of cellulose and lignin in the presence of phosphoric acid. 3. Characteristics of the higher - boiling oil fraction of carbohydrates pyrolysis. *Khimija Drev.*, **6**, 72-76.
8. Shafizadeh F., Furneaux R.H., Stevenson T.T. (1979) Some reactions of

levoglucosenone. *Carbohydr. Res.*, **71**, 169-191.
9. Witczak Z.I. Levoglucosenone; past, present and further application (1994) In: *Levoglucosenone and Levolucosans, Chemistry and Application*, pp. 3-16, ATL Press, Mount Prospect, IL
10. Dobele G., Rossinskaja G., Telysheva G., Meier D., Faix O. (1999) Cellulose dehydration and depolymerisation reactions during pyrolysis in the presence of phosphoric acid. *J. Anal. Appl. Pyrolysis*, **49**, 307-317.
11. Dobele G., Rossinskaja G., Dizhbite T., Telysheva G., Meier D., Faix O. (1999) Cellulose as a raw material for levoglucosenone production by catalytic pyrolysis. In: *Recent Advances in Environmentally Compatible Polymers* (Ed. by I. Kennedy, G.Phillips, P.Williams), Woodhead publ., Cambridge, UK (in press).
12. Dobele G., Meier D., Faix O., Radtke S., Rossinskaja G., Telysheva G. (2000) Volatile products of catalytic fast pyrolysis of cellulose. *J. Anal. Appl. Pyrolysis* (in press).
13. Maskavs M., Kalninsh M., Reihmane S., Laka M., Chernyavskaya S. (1999) Effect of water sorption on some mechanical parameters of composite systems based on low-density polyethylene and microcrystalline cellulose. *Mechanics of Composite Materials*, **35**, 1, 79-90.
14. Ioelovich M., Tupureine A., Veveris G. (1989) Investigation of the crystalline structure of cellulose in plant materials. *Khimija Drev.*, **5**, 3-9.
15. Bolotnikova L., Danilov S., Samsonova T. (1966) The method for the measuring of cellulose viscosity and degree of polymerization. *Zhurn. Prikl. Khim.*, **39**, 176-180.
16. Umland F., Jansen A., Tyring D., Wunsh P. (1975) *Complex Compounds in Analytical Chemistry*, Mir, Moskaw.
17. Koch H., Bommer H.D., Koppers I. (1982) Analytische Untersuchungen von Phosphatvernetzten Stärken. *Starch*, **34**, 1, 16-21.
18. Meier D. and Faix O. (1992) Pyrolysis - gas chromatography - mass - spectrometry. In: *Method in Lignin Chemistry* (Ed. by S.Y.Lin & C.W.Dence), pp.177-199, Springer, Berlin.
19. Dobele G., Rossinskaja G., Rone B. (1996) Thermodestruction of cellulose and levoglucosenone obtaining. In: The Chemistry and Processing of Wood and Plant Fibrous Materials (Ed. by J.Kennedy, G.Phillips, & P.Williams) pp. 345-350, Woodhead, Cambridge.

Microporous sorbents produced by pyrolysis and gasification of hydrolytic lignin

G.V.Plaksin, O.N.Baklanova, V.K.Duplyakin, V.A. Drozdov
Institute of Catalysis of the Siberian Branch of the Russian Academy of Sciences, Omsk Department, Russia
644040, 54, Neftezavodskaya st., Omsk, Russia
Tel.: +7 (8) 381-2-664411, Fax: +7 (8) 381-2-646156

ABSTRACT: Plant raw processing via acid hydrolysis is widely spread in Russia. Hydrolysis lignin is a large scale waste of this processing. In Russia hydrolysis lignin deposits mount to more than 1 million tons. Hydrolysis lignin contains 60-80% carbon and is a promising raw for production of adsorbents.

We have studied the process of pyrolysis and gasification of hydrolysis lignin from the Krasnoyarsk Biochemical Plant. It has been found that as pyrolysis temperature increases from 400 to 1300 C, specific adsorption surface of pyrolysis product decreases from 297 m^2/g до 19.8 m^2/g, and its micropores volume decreases from 0.17 cm^3/g до 0.01 cm^3/g. The width of pores in carbonized lignin 2X, passes through an extreme depending on pyrolysis temperature. We have also studied the effect of burn-off on the texture of lignin adsorbent. As burn-off grow, adsorbent specific surface, micropores volume and average pores size increases.

Chromatography was used to estimate the separation capacity of lignin adsorbent as applied to gas mixtures He-CH_4. Maximum separation coefficient is found to correspond to adsorbent with pores 0.8-0.9 nm wide.

INTRODUCTION

Microporous carbon materials are widely used in adsorption processes for separating gaseous and liquid components [1]. Fossil peat and coal [2], polymers and resins [3], wood pulp and other plant raw materials [4] are widely used as raw to produce microporous carbon-containing materials.

There is rather a broad assortment of plant raw materials used as precursors in carbon sorbent production: wood pulp, nuts shell, rice husks, corn cobs, wood bark, and that is not a complete list of carbon-containing raw materials.

In Russia hydrolytic lignin (a waste of large scale wood processing) is a quite promising raw material to produce microporous carbons. Note that hydrolytic lignin differs from natural lignin by polymerization degree and polymer structure. Lignin utilization is stimulated by the following factors:

(1) At present Russian enterprises, working with wood pulp hydrolysis, annually produce more than 1 million tons of lignin as secondary waste [5]. A large quantity of lignin is concentrated in Siberia, but not more than 30% of lignin is efficiently used, mainly as a fuel.
(2) Hydrolysis lignin has some particular features in comparison to other lignin obtained by various methods. It has a well developed porous capillary structure, and contains more carbon. Experiments show hydrolysis lignin to be a rather promising raw for carbon adsorbents production. Adsorbents of lignin origin are characterized by large pore volume (1.0-1.5 cm^3/g) and bimodal pore size distribution with the main contribution of micro- and macro-pores [5].

Among other industrially evolved lignins hydrolysis lignin is the most condensed polymer owing to its production conditions. Acidic treatment at high temperature and pressure (160-190^{0}C, 0.8-1.0 MПа) produces a strong three-dimensional polymer net. It contains many carbon-carbon bonds between aromatic nuclei. This spatial structure makes lignin hardly soluble in water and organic solvents. Structural forms in hydrolysis lignin may be destroyed only at pyrolysis or by very strong oxidizers [6].

In Russia many researchers focused their efforts on the development of technology for the synthesis of porous materials from hydrolytic lignin, and certain results have been achieved in the field. According to some publications, not only common sorbents production was tried to solve environmental problems [7,8], but also the synthesis of sorbents for special purposes, in particular for gas mixtures separation [9].

Separation of mixture H_2(He) – hydrocarbons is a well known chemical problem. This process is known to base on equilibrium sorption [10] due to a different thermodynamic nature of adsorbate-adsorbent interactions. At present the most efficient separation of this particular gas mixture is achieved with uniform microporous adsorbents. Macro-pores are needed to provide a free access of gas to material bulk.

This study focuses on the processes producing a microporous structure of carbon sorbents during a thermal treatment of hydrolytic lignin in inert and oxidizing media, and verifies adsorbents capability of separating He-hydrocarbon mixtures.

EXPERIMENTAL

Pyrolysis and gasification of lignin.

Hydrolysis lignin, produced at the Krasnoyarsk Biochemical Plant, was used as raw to produce micro-porous sorbents. Hydrolytic lignin consists of lignin (61.3%), cellulose (13.3%), substances extracted by alcohol-benzene mixture (15%), ash _(3.3%), sulfur (0.4%). The summed pore volume of raw lignin was 1.1$cм^3/g$, its specific surface was A_{BET}=12 m^2/g [8].

Prior to laboratory studies lignin was dispersed to a fraction of 0.1-0.25 mm. This powder and liquid medium were for paste preparation. Thus prepared paste was extruded through a die to produce granules 6 mm in size. Then lignin granules were dried at 105°C .

Dried lignin granules were exposed to carbonization and graphitization in rotatable reactor providing reaction zone volume of 0.5 dm^3. Reactor was equipped with electric heater, stabilizing reaction temperature with an accuracy of ±2°C. Lignin granules were loaded into reactor at 20-25^{0}C. Reactor heating to desired temperature was done with a constant rate of 1-4 ^{0}C/min. Lignin pyrolysis occurred in the inert gas atmosphere at 400-1300^{0}C. Pyrolysis time at adjusted temperature was constant for all samples - 120 min. Carbonized product gasification was performed in steam at 800°C. Gasification time was 5-120 min. After gasification carbon material burn-off was estimated.

Analysis of texture and separation capability

Thus obtained carbon material was studies regarding its texture and adsorption properties using Sorptomatic-1900 and Sorpty-1750 ("Fisons", Italy). Sorptomatic-1900 is an automated static vacuum device to measure full adsorption isotherms by volumetric method. Sorpty-1750 – is a static device for the express analysis of specific surface by a single point of nitrogen adsorption isotherm (in our case measurements were done at a relative pressure of nitrogen vapors P/P_0= 0.178). The total volume of sample pores was calculated from nitrogen adsorption isotherms (77.4K) at P/Po=0.996, assuming a molar volume of nitrogen adsorption layer to equal 34.68 cm^3/mole. Specific adsorption surface (A_{BET}) for carbons was estimated at relative equilibrium pressures ranging within 0.05-0.33, assuming an adsorbed nitrogen molecule to occupy an area of 0.162 nm^2 in the filled monolayer.

The Dubinin-Radushkevitch equation (TOZM theory) was applied to calculate the micropores volume. Affinity coefficient for nitrogen was taken as 0.33, and adsorption phase density at adsorption temperatures was taken as 0.808 g/cm^3 [11]. Micropores size was estimated from adsorption energy E_o, defined from the Dubinin-Radushkevitch equation using correlation: $X=10/E_o$, (nm), where X is a semi-width of a slit micro-pore [12].

Efficiency of H_2(He)-CH_4 separation was estimated by chromatography. Adsorbent separation capacity was verified with a model mixture H_2(He)-CH_4, helium content being 60-80%. Mixture was separated on a column 3mm in diameter and 1 m long. Lignin adsorbent with 0.1-0.5 mm particles was used for the purpose. Argon is gas carrier. Flow rate of gas carrier was 40 ml/min. Argon/mixture ratio was 1:40. As a criteria of separate sorbent capability was used the coefficient K_p calculated from chromatography pattern. Coefficient K_p was calculated by equation:

$$K_p = \frac{b}{m_1 + m_2},$$

where b is a distance between the He and CH_4 peaks, and m_1 and m_2 are the peak widths.

RESULTS AND DISCUSSION

Effect of pyrolysis parameters

The porous structure and specific surface of activated carbons are determined by precursor type [13] and pyrolysis parameters, i.e. temperature [14] and heating rate [15,16]. Many papers are dedicated to the synthesis of active carbons based on lignin-cellulose materials of various types [17]. There are empirical dependencies of texture on thermal treatment parameters for carbon materials from various precursors of plant nature [13,14,18]. Models for cellulose fibers pyrolysis are suggested [15].

Hydrolysis lignin is less studied. Major studies in this field were performed by the Russian researchers [19,20].

It is known [21], that hydrolysis lignin differs from both native lignin and cellulose by its chemical composition and structure. In hydrolysis lignin there are many carbon-

carbon bonds and secondary multi-nucleus aromatic structures [22]. Lignin samples, obtained at wood pulp hydrolysis contain aromatic nuclei linked to each other directly or through side chains. To some extent they may be considered as polymers with conjugated bonds. Hydrolysis lignin pyrolysis produces carbon with a yield of 40-50%, as referred to dry lignin mass, and liquid phase containing 50-55 % of phenols. Therefore, it is not correct to transfer regulations found for cellulose pyrolysis onto that of hydrolysis lignin. .

With this regards we have tried to elucidate the effect of main carbonation parameters (temperature and heating rate) on micro-pores volume (V_{mi}) and size (2X), specific adsorption surface (A_{BET}). Figure 1a,b gives parameters dependence on carbonation temperature (heating rate 4 °/min). As carbonization temperature increases from 400 up to 700°C at other pyrolysis parameters being constant samples specific adsorption surface and micro-pores volume reduce significantly Fig. 1(a). Micro-pores size varies considerably from 1.48 to 1.04 nm, revealing minimum values 2X=1.04-1.24 nm (Fig. 1b). At 700-900°C a sharp decrease of A_{BET} and V_{mi} is observed, which becomes even more dramatic at a further increase of carbonation temperature up to 1300°C. A_{BET} and V_{mi} decrease by several times. Under these conditions pores width increases to 2.74 nm.

Similar dependencies for pores volume and diameter were obtained by other researchers, studying the pyrolysis of polymer and lignocellulose materials [23,24]. However, there is no good explanation to this phenomenon yet.

As a working hypothesis we may suggest the following. At 400-700°C two main processes occur simultaneously. One of them is a thermal destruction of three-dimensional polymer structures formed at wood hydrolysis. Another process is a poly-condensation of polymer chain residues provided by the labile side propane chains and phenol rings with reactive groups.

These processes produce carbon-containing layers formed chaotically. They bring no essential changes to specific surface and micro-pores volume, but cause a drastic change of micropores size. A further increase of carbonization temperature reconstructs the carbon-containing substance to more regular structures, decreasing the micro-porosity. In this case micropores size increases to 2X=1.8-2.7 nm with the growing carbonization temperature. Thus, for hydrolysis lignin carbonization temperature range 600-700°C is optimum regarding the smallest size of pores.

However, we need further studies to confirmed the above said hypothesis and to reveal more details related to the mechanism of hydrolysis lignin pyrolysis.

Since the processes of lignin carbonization have different rates, texture characteristics of materials produced depend on the rate of temperature growth (Table 1).

Table 1 Effect of the heating rate (V_t^o) in pyrolysis of hydrolysis lignin (T=700°C).

V_t^0, °C/min	A_{BET}, m^2/g	V_{mi}, cm^3/g	2X, nm
1.0	410	0.24	0.76
2.2	371	0.18	0.80
3.3	319	0.17	0.94
4.0	304	0.15	1.24

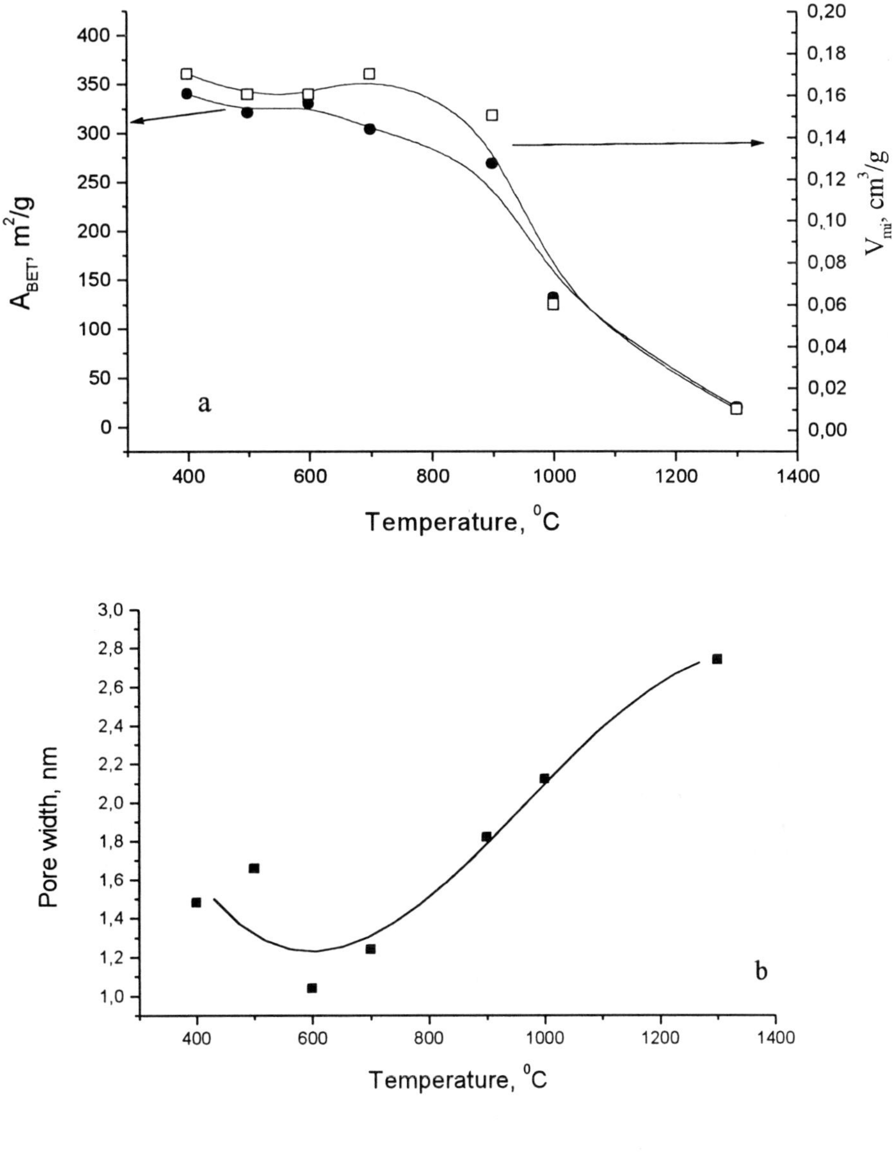

Fig.1(a,b) Carbonization temperature influence on specific adsorption surface, micropores volume (a) and micropores width (b).

Specific surface and pore volume decrease with the growing heating rate results, and micro-pores width increases from 0.76 up to 1.24 nm.

Therefore, most likely high heating rates prevail de-polymerization of hydrolytic lignin accompanied by removal of low molecular fragments. At low heating rates important become the slow processes of poly-condensation and thickening of low molecular aromatic compounds required to form microstructures with the smallest pores.

Effect of burn-off

Gasification (activation) of lignin carbonized products, causing burn-off of various degree, was performed for a further development of microporous structure of lignin carbons. Table 2 shows specific adsorption surface (A_{BET}), micropores volume (V_{mi}) and micropores width (2X) of lignin carbon versus burn-off.

Table 2 Effect of burn-off on the lignin adsorbent texture.

Burn-off,%	V_{mi}, cm^3/g	A_{BET}, m^2/g	2X, nm
-	0.15	304	1.24
18	0.22	444	0.80
39	0.25	526	0.86
48	0.26	526	0.90
57	0.30	653	0.90
82	0.36	749	1.26

Experimental data demonstrate a monotonous increase of micropores volume and specific adsorption surface with growing degree of burn-off. Average pore size at a loss of 18% decreases as compared with the non-activated material, and then grows attaining 2X=1.26 nm. Probably, a higher content of small pores content in the material at a small degree of burn-off is caused by a partial removal of volatile and resinous substances providing pores opening. Then average pore size increases, when carbonaceous material is burnt away. Activation of lignin sorbents was compared with the results obtained by M.T.Gonzalez, et al. [16] for olive stones carbon activation. Independently of a raw material used (lignin or olive stones), at activation porous structure proceeds in a similar manner in both cases.

Separation of model gas mixture He-CH_4

Thus the texture of produced lignin sorbents indeed depends on thermal treatment parameters. Sorbent separation capacity on its turn depends on the texture parameters. Therefore, we tried to reveal correlation between the coefficient of He-CH_4 separation and micropores width in then lignin adsorbent by experimental data simulation (see Fig 2).

Apparently, lignin sorbents with 0.76-1.0 nm micropores show a rather high ability of He-CH_4 (K_p) separation, which corresponds to an outlet helium content of not less than 99.9%.

Thus, regarding the effect of thermal treatment of hydrolytic lignin on the size of lignin carbon pores, the optimum regimes of thermal treatment produce sorbents with a pore size of 0.7-1.0 nm. A pilot batch of such microporous carbon was produced, its texture

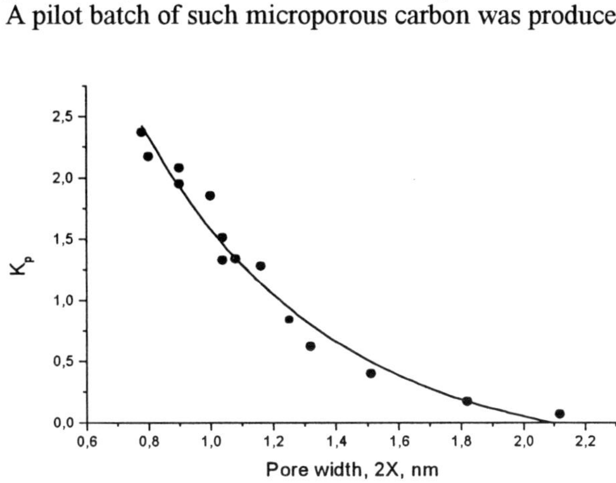

Fig. 2 Correlation between the coefficient of He-CH_4 separation and width of micropores

parameters were estimated, and its capacity in the separation of He-CH_4 mixtures was tested. Test results show that this carbon has micropore size 2X = 0.78 nm, and it provides separation coefficient Kp=2.37.

CONCLUSIONS

Hydrolysis lignin is a large scale waste of wood processing and also a promising raw for the synthesis of various carbon adsorbent. At present synthesis of adsorbents from hydrolysis lignin is not yet studied well enough.

In the present paper we report about experimental results and discuss the revealed regulations of carbon adsorbents synthesis. The width of pores in adsorbents as a function of carbonization temperatures is found to be of extreme character. We suggest a working hypothesis explaining the pore width dependence. However, we need more experimental results to confirm our hypothesis.

We have synthesized carbon materials from hydrolysis lignin and tested them in the separation of gaseous mixtures He-CH_4. The adsorbents show a helium concentration capacity of up to He до 95-99 %vol. They appear proved to be promising for application in the pressure swing adsorption (PSA) plants for H_2 и He concentration.

REFERENCES

1. Foley H.C. (1995) Carbogenic molecular sieves: Synthesis, properties and application. *Microporous materials*, **4**, 407.
2. Chagger H.K., Ndaji F.N., Sykes M.L. & Thomas K.M. (1995) Kinetics of adsorption and diffusional characteristics of carbon molecular sieves. *Carbon*, **33**, 1405.
3. Hatori H., Yamada Y., Shiraishi M., Nakata H. & Yoshitomi S. (1992) Carbon molecular sieve films from polimide. Carbon, **30**, pp. I-III.

4. Braymer T.A., Coe C.G., Farris T.S., Gaffney T.R., Schorc J.M. & Armor J.N. (1994) Granular carbon molecular sieves. *Carbon*, **32**, 445.
5. Chudakov M.I. (1983) *Industrial application of lignin*. Moscow, Lesnaya Promishlennost.
6. Babkin V.A., Levanova V.R. & Isayeva L.V. (1994) Medicine preparation from waste hydrolysis industry. *Chimiya v interesach ustoichivogo razvitia,* **2**, 559.
7. Plachenov T.G., Akhmina E.I., Boikova G.I., Vorozhbitova L.N. & Sokolovskaya L.I. (1983) Carbon adsorbents from hydrolytic lignin. *Journal pricladnoi chimii,* **56**, 1296.
8. Schipko M.L., Baklanova O.N., Duplyakin V.K. & Kuznetsov B.N. (1996) Powdery and molded sorbents from lignite and hydrolytic lignin. *Chimiya v interesach ustoichivogo razvitia,* **4**, 467.
9. Timofeev A.F., Kolosentsev S.D., Gavrilov D.N. & Plachenov T.G. (1986) Molecular-sieve properties of carbon adsorbents on the basis of hydrolytic lignin. *Journal pricladnoi chimii*, **59**, 1210.
10. Juntgen H., Knoblauch K. & Harder K. (1981) Carbon molecular sieves: production from coal and application in gas separation. *Fuel,* **60**, 817.
11. Dubinin M.M. (1972) *Adsorption and porosity*, Moscow, VAKHZ.
12. Carrsco-Marin F., Lopez-Ramon M.V. & Moreno-Castilla C. (1993) Applicability of the Dubinin-Radushkevich equcation to CO_2 adsorption on activated carbons. *Langmuir*, **9**, 2758.
13. Gonzalez J.C.,.Gonzalez M.T, Molina-Sabio M. & Rodriguez-Reinoso F. (1995) Porosity of activated carbons prepared from different lignocellulosic materials. *Carbon,* **33,** 1175.
14. Mackay D.M.&.Roberts P.V (1982) The influence of pyrolysis conditions on yield and microporosity of lignocellulosic chars. *Carbon,* **20**, 95.
15. Brunner P.H.&.Roberts P.V (1980) The significance of heating rate on char yield and char properties in the pyrolysis of cellulose. *Carbon,* **18,** 217.
16. Gonzalez M.T., Rodriguez-Reinoso F., Garcia A.N.& Marcilla A. (1997) CO_2 activation of olive stones carbonized under different experimental conditions. *Carbon,* **35**, 159.
17. Wigmans T. (1989) Industrial aspects of production and use of activted carbons. *Carbon*, **27**, 13.
18. Mackay D.M &.Roberts P.V (1982) The dependence of char and carbon yield lignocellulosic precursor composition. *Carbon,* **20,**87.
19. Stalugin A.B, Kondratenok B.M., Dudkin B.N., Lyubitova S.G., Mordvanyuk S.A. (1993) Some problems of carbonization of hydrolytic lignin and production of active carbons on its basis. *Trudi Komi nauchnogo centra Ur O RAN*, **129**, 113.
20. Ciganov E.A, Akhmina E.I. &.Galaudina V.V (1978) Formation of porous structure chars of hydrolitic lignin by pyrolysis. *Chimiya drevesini*, **5** , 97.
21. Gogotov A.F.&.Babkin V.A (1994) Lignin is a potential source of valuable low-molecular compounds. *Chimiya v interesach ustoichivogo razvitia,* **2**, 507.
22. Obolenskaya A.V. (1993) *Chemistry of lignin*. Sanct-Peterburg, Lesotecknicheskaya Academy.
23. Verma S.K & Walker P.L. (1990) Alternation of molecular sieving properties of microporous carbons by heat treatment and carbon gasification. *Carbon,* **28**, 175.
24. Hayashi J. (1999) Preparing molecular sieve carbon from palm oil shell *Carbon,* **37**, 524.

The Formation of Petrodiesel by the Pyrolysis of Fatty Acid Methyl Esters over Activated Alumina

Boocock, D.G.B., Konar, S.K. and Glaser, G.
Department of Chemical Engineering and Applied Chemistry, University of Toronto, Toronto, Ontario, Canada M5S 3E5

ABSTRACT Fatty acid methyl esters are currently being manufactured and sold in many countries as biodiesel, a renewable petrodiesel substitute. The use of biodiesel is limited by the lowest temperature at which the particular product will pass from the fuel tank through the injectors to the combustion chambers. This temperature is variously represented by the cloud point (CP), the pour point (PP) and the cold filter plugging point (CFPP). Very recently in Europe, North America and elsewhere there is an interest in using waste fats and oils as the source of the fatty acids and triglycerides required to make the methyl esters. Unfortunately, many of these feedstocks contain saturated fatty acids, the methyl esters of which have higher CFPP's than those of unsaturated acids. One strategy to address this problem is to cold-filter[1] the ester, thereby removing the more solid components. This presents the problem of what to do with the 'solid' methyl ester fraction.

It has been shown that when fatty acid methyl esters are pyrolysed over activated alumina at 400°C at a weight-hourly space velocity of 0.45, they are completely deoxygenated to form linear hydrocarbons[2]. These are completely compatible with petrodiesel with which they can be blended in all proportions. Because the blending of biodiesel with petrodiesel is one strategy for addressing the CFPP, the pyrolytic deoxygenation offers several options for the use of waste fats and oils as biodiesel precursors.

INTRODUCTION

Liquid fuels play a key role in modern lifestyles. Their liquid nature offers convenience in being transportable and easy to use. Diesel, spark-ignition and jet engines are the result of this convenience. Unfortunately, the hydrocarbons, which play the major role as liquid fuel are finite, and upon combustion contribute to the accumulation of carbon dioxide in the atmosphere. It is not surprising that alternate renewable liquid fuels are receiving attention. These include ethanol, methanol, fatty acid methyl esters and fast pyrolysis oils. All can be produced from renewable resources. Ethanol may be formed by fermentation of sugars and thereby indirectly from cellulose and starches. Methanol can be produced from carbon monoxide and hydrogen, formed from the gasification of biomass. Fast pyrolysis oils result from the fast pyrolysis of biomass and they are the least homogeneous and stable of the four liquids listed. Methyl esters are produced from vegetable oils and fats. They are the only liquids of the four groups, which can currently be substituted for a conventional fuel without any engine modification. In Europe, they are at this time made mainly from rape oil obtained from dedicated seed crops. In the USA, soybean would likely be the oil of choice. In Canada, either soy or canola could be used. There has been some discussion of using canola containing salt resistant genes. This could be grown on the high salt acreage currently unused in Alberta.

More recently, waste fats and oils have become attractive substrates. They are not only cheaper, but are becoming even more so as their use in cattle feed diminishes. More and more of this material is becoming "waste". In addition, more jurisdictions are banning such wastes from landfills in which case it can only be incinerated. These wastes are mostly fatty acids and triglycerides from which methyl esters can be formed under appropriate conditions. Unfortunately many of these wastes differ from vegetable oils in that they contain higher percentages of saturated fatty acids and triglycerides. As a result, the methyl esters produced from them have a higher cloud point (CP) and pour point temperatures as well as high cold filter plugging points (CFPP). These parameters all relate to the usability of the material as fuel at low temperatures. The cloud point and CFPP relate to the appearance of solid material in the fuel, whereas the pour point is the temperature at which the liquid just flows. In the methyl esters, solid usually appears before the pour point is reached. A typical methyl ester made from a waste fat such as tallow would have a CP of 10-17°C, whereas the methyl esters of rapeseed have a value of approximately -2. This may be contrasted with the cloud point of 8°C for palm oil, which contains close to 50% saturated fatty acids. Therefore, the cloud points of methyl esters are potential problems for all biodiesel methyl esters. In fact once the technology is established for producing biodiesel 'standard' material, cloud point is the most important concern.

At the present time the problems associated with cloud points which are too high are addressed in three ways. These are (a) cold filtration, (b) the use of additives, and (c) the blending with conventional diesel fuel. The recent inclusion of B20, a 20% blend of methyl esters with diesel fuel, as an allowable fuel under the US EPACT program, is not arbitrary. At the 20% level, significant reductions in CFPP of the esters can be achieved. This reduction obviously is related to the individual cloud points of the diesel fuel and methyl esters separately. As an example, a beef tallow methyl ester (CFPP = +20), when blended at the 20% level with a petrodiesel having a CP of -3, had a resulting CP of 11[3]. The reduction effect is not linear over the whole range and has the maximum benefit below 20%.

The "winterisation" of methyl esters describes a process whereby the esters are cooled, and the solid material which results is filtered off from the liquid fraction[1]. This raises the obvious question of what to do with the solid portion.

Several years ago we studied the pyrolysis of vegetable oils and fats in a trickle bed reactor over selected catalysts[4]. Surprisingly we found that the pyrolysis over a certain type of activated alumina at 400°C and a weight-hourly space velocity of 0.45 gave "mechanistic" yields of close to 100% with virtually complete de-oxygenation. The products were almost exclusively linear alkanes and alkenes, which is to be contrasted with the results of other workers who, under similar conditions, obtained mixtures of oxygenates and aromatic compounds. The linear alkanes and alkenes we obtained were completely compatible with conventional diesel fuels in terms of miscibility, but as expected exhibited a wider boiling point range.

In a later study we identified the two major routes by which triglycerides were converted to hydrocarbons[5] (see Fig. 1). The first starts with cleavage of the side chain, which is believed to involve a γ-hydrogen transfer to the carbonyl group. This mechanism accounts for the formation of the thermodynamically unfavourable terminal 1-alkenes. The second mechanism involves β-cleavage to form fatty acids[6]. Two fatty acids then combine to form a ketone which in turn undergoes the γ-hydrogen transfer mechanism to form a linear alkene and the enol of a methyl ketone. This enol rearranges to the terminal aldehyde which then disproportionates to an alcohol and a fatty acid. The fatty acids then go through the ketone route while the alcohol dehydrates to an alkene. Unfortunately, the disproportionation step leads to deactivation of the catalyst. Because this step is part of the route from fatty acids, it is the latter which are indirectly responsible for the deactivation of the catalyst. This was confirmed by using fatty acids as the substrate. It is clear that fatty acid methyl esters should not be capable of producing fatty acids.

This study describes the pyrolysis of soybean methyl ester under similar conditions as were previously used for vegetable oils.

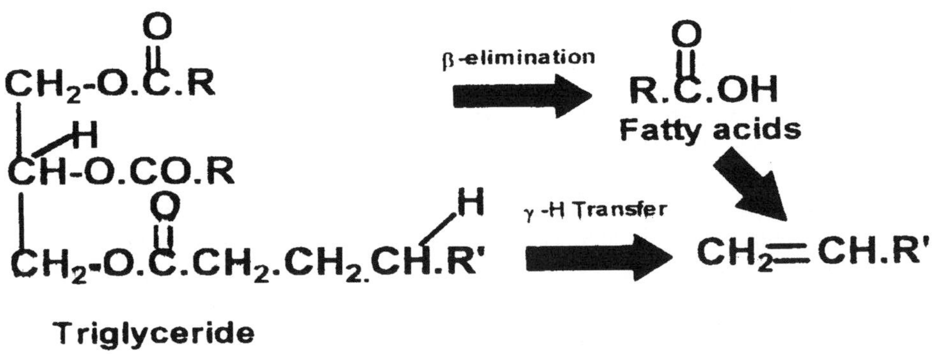

Figure 1 β-elimination and γ-hydrogen transfer on a triglyceride molecule.

EXPERIMENTAL SECTION

MATERIALS AND ANALYSIS PROCEDURES

Soybean oil was supplied by Procter and Gamble Inc., Toronto, Ontario, Canada. Methyl ester was made by the one-phase method employing a methanol/oil molar ratio of 27:1, tetrahydrofuran as cosolvent and 1.0 wt.% sodium hydroxide based on the oil. The methyl ester accounted for 99.7% of the total ester present. The Alcan AA 200 activated alumina used was supplied by Alcan Chemicals, Brockville, Ontario, Canada. The catalyst's BET surface area is 270-290^2/g, and its pore volume is 40 cm^3/g. The pore distribution is binodal, with 66% of pore volume in pores less than 30 radius and the rest in larger pores. Its bulk density is 0.75 g/cm^3.

Infrared spectra were recorded on a Nicolet model 598 IR spectrophotometer using the thin film method. Liquid samples were analysed using a Hewlett Packard 5880A series gas chromatograph (GC) equipped with a flame ionisation detector and a DB17 (30 m x 0.53 mm i.d.) fused silica capillary column. The operating parameters were as follows: detector temperature 225°C; injector temperature 225°C; temperature program, 5 min at 50°C; heated at a rate of 5°C/min to 210°C; held for 23 min. The gas flow rate was 59.6 was 59.6 mL/min, which included both the carrier gas He and the makeup gas. Gas analyses were performed on a similar GC equipped with a flame ionisation detector and a 316 stainless steel column (2.44 m x 3.18 mm i.d.) with a Poropak type QS packing material. The operating conditions were as follows: detector temperature, 120°C; injector temperature, 120°C; temperature program, 1 min at 50°C; heated at a rate of 30°C/min to 150°C; hold for 20 min. Carrier gas (Ar) flow rate was 40.05 mL/min. Gas chromatography-mass spectrometric (GC-MS) analyses were done on a Hewlett Packard 5890 GC equipped with a J&W Scientific DB-5 capillary column (30 m x 0.250 mm i.d.) and using helium carrier gas at a volumetric flow rate of 1 mL/min. The linear flow velocity was 32 cm/s. GC-MS was performed by the Department of Chemistry, University of Toronto, courtesy of Professor Tom Tidwell and Dr. Alex Young. Analytical hydrocarbon standards were supplied by Polyscience Corp., IL. Carbon and hydrogen elemental analyses were performed by Guelph Chemical Laboratories, Guelph, Ontario, Canada.

PYROLYSIS APPARATUS AND EXPERIMENTAL PROCEDURE

The pyrolysis unit consisted of an insulated 316 stainless steel preheater tube (1.3 cm i.d. x 50 cm length) which extended 1 in. into a 316 stainless steel fixed bed tubular reactor (2.5 cm i.d. x 46 cm length), which was heated by a cylindrical block heater. Two type J (iron-constantan) thermocouple probes were used to both monitor the internal catalyst bed temperature and maintain a consistent reactor wall temperature in combination with a temperature controller. A syringe pump, condenser, vacuum adapter, receiving flask, nitrogen cylinder, and gas collection system were connected as shown in Figure 2. The reactor midsection was packed with 40 g of activated alumina, which was held in place by a circular stainless steel screen. The preheater and reactor were operated at 180-190 and 450°C, respectively. The entire process remained at normal atmospheric pressure throughout the run.

Prior to a run, fresh catalyst was demoisturised at 450°C for a 3 h period by intermittently passing nitrogen over the catalyst bed. High demoisturisation was necessary in order to activate the catalyst. One hour of treatment was inadequate, and two hours was only barely adequate to activate the catalyst. Immediately before commencing liquefaction, a nitrogen environment was established in the pyrolytic system.

A syringe pump (Sage Instruments (subsidiary of Orion Research Inc.), Model 355, Sage Instruments Inc., White Plains, New York) was used to inject liquid substrate (ranging between 30 and 100 mL volumes) from a 100 mL syringe into the preheater. From previous studies,[3,11] the optimum feed rate which gave the highest organic liquid yield was 0.34 mL/min, which is equivalent to a 0.46 h^{-1} weight hourly space velocity (WHSV) for crude canola oil. Consequently, this was also the rate employed for these studies. The preheater was angled so the injected substrate proceeded by gravity flow into the reactor. Substrates which were solids at room temperature (i.e. dodecanol) were melted and maintained as liquids by wrapping both the syringe and the preheater entrance area with heating tapes during injection.

The pyrolysed product exiting the reactor was cooled with a water condenser and collected in one or more 100 mL receiving flasks immersed in an ice bath. Noncondensable products passed through the gas trap into the brine solution and were measured by brine displacement into a 500 mL graduated cylinder. After all substrate was injected, an addition 8-15 min period was allowed for the residual substrate to exit the reactor. The liquid product and syringe were then weighed to determine product yield, and the spent catalyst was also weighed to obtain coke-deposition quantities.

Hydrogenation of the pyrolysis product was carried out using a palladium on carbon catalyst. Viscosity was measured according to ASTM D445. The boiling point curve was measured according to ASTM D86. Heat of combustion was measured in a 1241 oxygen bomb calorimeter. Pour point was determined according to the ASTM D97 method.

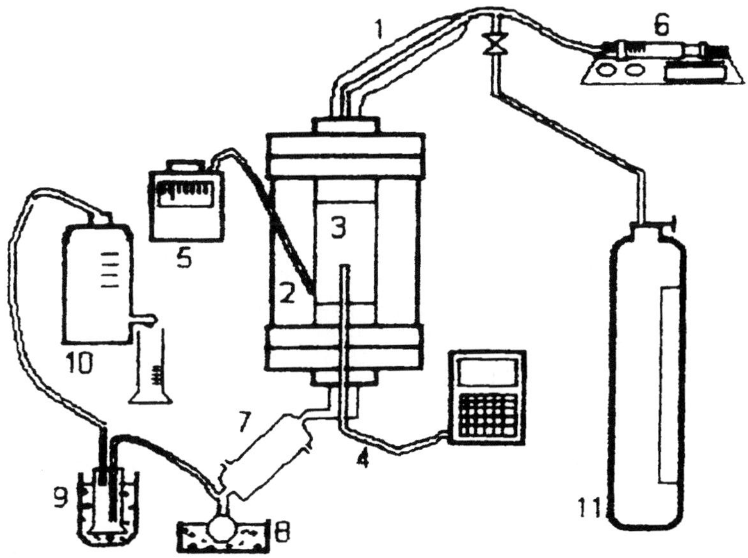

Figure 2 Experimental set up: (1) 316 stainless steel feed preheater tube (1.3 cm i.d. x 50 cm length); (2) block heater containing a 316 stainless steel fixed bed reactor tube (2.5 cm x 46 cm length); (3) catalyst bed; (4) Type J (iron/constantan) thermocouple probe; (5) Type J (iron/constantan) thermocouple with temperature controller; (6) syringe pump; (7) condenser; (8) receiving flask; (9) gas trap; (10) gas collection vessel; and (11) nitrogen cylinder.

RESULTS

YIELD AND DEOXYGENATION

A total of 310.3 g of methyl ester were pyrolysed to yield 204.4 g of liquid product which corresponds to a 65.9% actual liquid yield. The theoretical yield based on a typical C_{16}, C_{18} composition of soybean oil, and the γ-hydrogen transfer mechanism is approximately 76%. Therefore, the yield based on this limitation is approximately 87%. This shows that random cracking of the ester chain does not play a significant role in the reaction. The infrared spectrum shows no carbonyl peaks at 1720 cm^{-1}. Peaks appearing at 907 cm^{-1} and 965 cm^{-1} are probably absorptions due to terminal alkenes and trans alkenes, otherwise the spectrum looks like that of a typical alkane. Consistent with these results, elemental analysis gave 83.7% carbon, 14.56%

hydrogen with the balance (1.74%) presumably being oxygen. A typical diesel fuel (D2) analysed in the same way contained 86.7% carbon, 12.8% hydrogen, 0.22% sulphur and 0.05% nitrogen, the balance (0.86%) presumably being oxygen.

UNSATURATION AND CHAIN LENGTH

The infrared spectrum indicated the presence of double bonds. The hydrogenation experiment showed one double bond for every 19.8 carbon atoms. The gas chromatograph of the pyrolysed product is shown in Fig. 3.

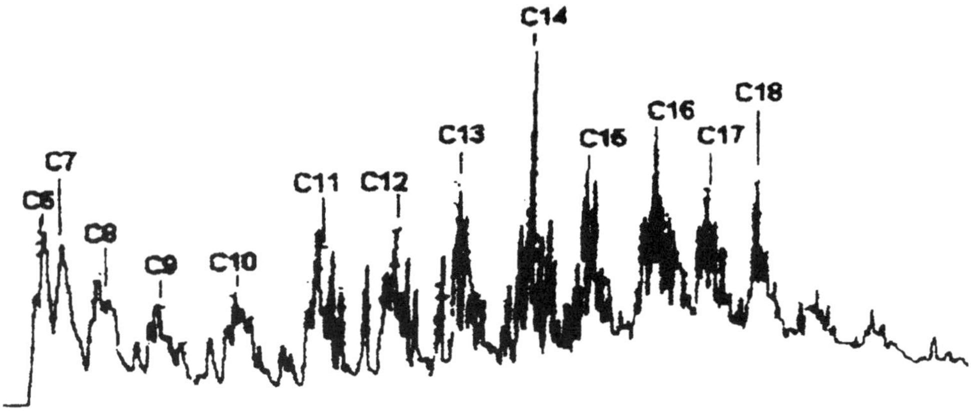

Figure 3 Gas chromatograph of product from methyl ester pyrolysis.

Approximately 50% of all carbon atoms are in chains of C_{12} or less (down to C_6). In addition very few carbon chains are above C_{18}. Therefore many of the individual components must be alkanes with the balance being mostly alkenes. Figure 4 shows the gas chromatograph of the hydrogenated material. The dominance of the n-alkene peaks confirms that the majority of the pyrolysis product is linear material, either alkanes or alkenes. However, other minor peaks show that some non-linear material is also present.

SPECIFIC GRAVITY, VISCOSITY AND BOILNG RANGE

The specific gravity of the product was 0.815 at 23°C resulting in an API gravity of 42.2. The specific gravity was slightly lower than a number 2 diesel fuel (0.851). This was due to the significant quantities of C_6, C_7 and C_9 material in the product. However, the specific gravity does fall in the range of typical diesel fuel specific gravities (0.813-0.852). The kinematic viscosity of the pyrolysis product (1.803 cSt at 39.6°) was again lower than that of a typical number diesel fuel (2.6 cSt at 39.6°C). The prescribed range for number 2 diesel fuels is 1.3-4.1 cSt at 40°C. The boiling point curves for the pyrolysis product and number 2 diesel fuel are shown in Figure 5.

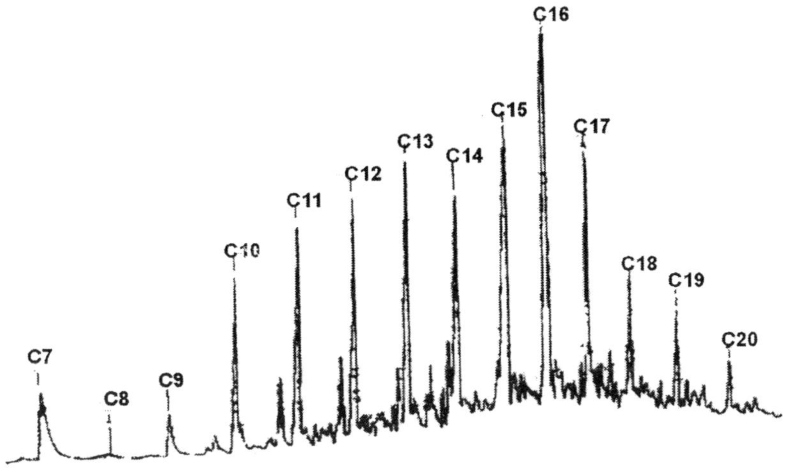

Figure 4 Gas chromatograph of hydrogenated pyrolysed methyl ester product.

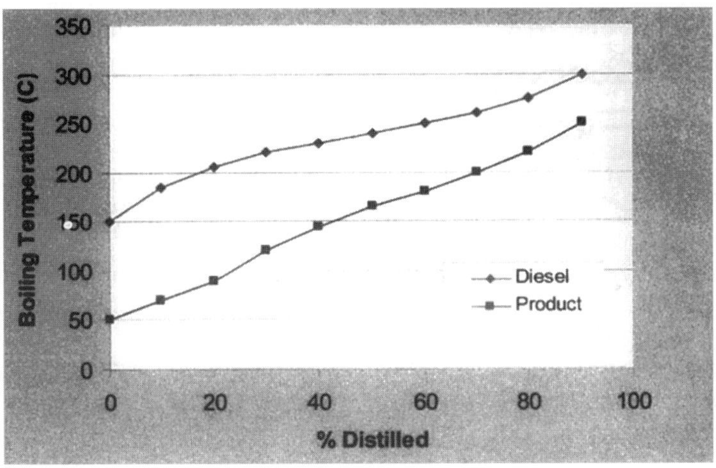

Figure 5 Distillation curves for a number 2 diesel fuel and pyrolysed methyl ester product.

Approximately 50% of the pyrolysis product boils below the initial boiling point of the diesel fuel, which is entirely consistent with the specific gravity and viscosity results. The pour point (-39°C) of the product was also considerably lower than a number 2 diesel fuel (-27°C). It is lower than the average value (-21°C) of diesel fuel in the Great Lakes and Eastern Region of Canada and similar to the average value (-39°C) in the Western Region of Canada.

One purpose of the pyrolysis was to produce a liquid, which could be blended with methyl esters, without raising the cloud point of the blend, and preferably lowering it. At a blend of 30% of soybean methyl ester with the pyrolysis product, precipitation of solids did not occur at 5°C. Soybean methyl ester normally throws down considerable solid over a period of time at this temperature.

CONCLUSION

The pyrolysis of soybean methyl ester at 400°C over activated alumina (WHSV = 0.45 h^{-1}) produces a high yield of hydrocarbon liquid which is comprised mostly of linear alkanes and alkenes. This material can be blended with methyl esters, thereby lowering the cloud point of the latter. This discovery presents one option for dealing with the solid methyl esters produced in the formation of "winterised" methyl ester biodiesel.

REFERENCES

1. Rathbauer, J. (1994) Fatty acid methyl esters (FAME) as special winter fuel. In: *Biomass for Energy, Environment, Agriculture and Industry*, vol. 2, 8th E.C. Conference (Ed. by Ph. Chartier, A.A.C.M. Beenackers & G. Grassi. Pergamon Press.
2. Badal, C. & Woodward J. (1997) ACS Symposium Series 666, *Fuel and Chemicals from Biomass*, Chapter 10, pp. 173-208.
3. Lohrlein, H-P., Anggraini-Suss, A. & Krause, R. (2000) *Possibilities and limits of the re-use of cooking oils with animal fat content as motor and heating fuel*, proceedings of the 1st World Conference and Exhibition on Biomass for Energy and Industry, Seville, Spain, June 5-9.
4. Boocock, D.G.B., Konar, S.K. , Mackay, A., Cheung, P.T.C. & Liu, J. (1992) The production of alkanes and alkenes by the pyrolysis of triglycerides over activated alumina. *Fuel*, **71**, 1291-1297.
5. Vonghia, E., Boocock, D.G.B., Konar, S.K. & Leung, A. (1995) Pathways for the deoxygenation of triglycerides to aliphatic hydrocarbons over activated alumina. *Energy and Fuels*, **9**, 1090-1096.
6. Leung, A., Boocock, D.G.B. and Konar, S.K. (1995) Pathways for the catalytic conversion of carboxylic acid to hydrocarbons over activated alumina. *Energy and Fuels*, **9**, 913-920.

Bio-Crude-Oil/Diesel oil emulsification: main achievements of the emulsification process and preliminary results of tests on Diesel engine

P. Baglioni, D. Chiaramonti, M. Bonini
Consorzio Interuniversitario Sviluppo Sistemi a Grande Interfase, University of Florence, Via G. Capponi 9, Florence, Italy
I. Soldaini
Pasquali Macchine Agricole, Via Nuova 30, Calenzano, Florence, Italy
G. Tondi
Department of Energetics "S. Stecco", University of Florence, Via S. Marta 3, Florence, Italy

ABSTRACT: In the context of an increasing use of Renewable Energy Sources, biomass derived pyrolysis oil is a very attractive solution: Bio-Crude-Oil (BCO) has potential to replace fuel oil or Diesel in many applications such as boilers, turbines and alternative engines for electricity production. However, technological development for BCO upgrading in order to improve its utilisation is still a problem to be solved. This paper describes the main results of a research project (supported by the European Commission, DG XII, JOULE Programme) aimed at the development of a low-cost physical-chemical and mechanical process for facilitating BCO utilisation in small Diesel engine units. This process is based on the preparation of an emulsion between BCO and Diesel oil. BCO and Diesel oil are not miscible, therefore a third component has to be added to obtain a stable emulsion. This third component is called emulsifier (or surfactant). It changes the interfacial properties of the system avoiding (or delaying) the emulsion's breaking. The developed emulsions - based on different BCOs in terms of feedstock and production facility - have been tested in a conventional small size Diesel unit (6.25 kW) at Pasquali Macchine Agricole (Italy). These tests were aimed at assessing combustion quality, operating performances and emission levels of the Diesel engine fuelled with the BCO/Diesel oil emulsion: some not structural modifications were made on the engine, mainly adding of components for insulating and cooling.

INTRODUCTION AND BACKGROUND

Renewable Energy (RE) sources are currently insufficiently exploited in the European Union: the energetic potential is considerable, but currently they provide a small contribution of less than 6% to the EU's overall energy consumption. RE sources are expected, however, to grow considerably in the future, especially to comply with the commitments at both European and international level on environmental protection.

They can significantly contribute to the **Kyoto targets** and EC "**White Paper**" (reduction of 8% of the GHG emissions between 2008 and 2012 and doubling of the RE share in the EU energy balance from 6% to 12% in 2010). REs exploitation is important not only from the environmental point of view: REs are indigenous sources and therefore can contribute to reducing dependency on energy imports (the EU energy imports are currently 50%, a figure that is expected to rise over the coming years if no action is taken, reaching 70% by 2020) and increasing security of supply; furthermore, the RE development can actively contribute to job creation and can be a key point in local (for instance rural) development. In this scenario, a fundamental role will be certainly played by **Bioenergy**.
Technically, bioenergy could be developed in the EU and many other countries on a very large scale given the significant potential of the agro-forestry sector to quickly supply a huge amount of biomass resources: large areas of cultivated land could be dedicated in the near future to biomass crops for energy production.

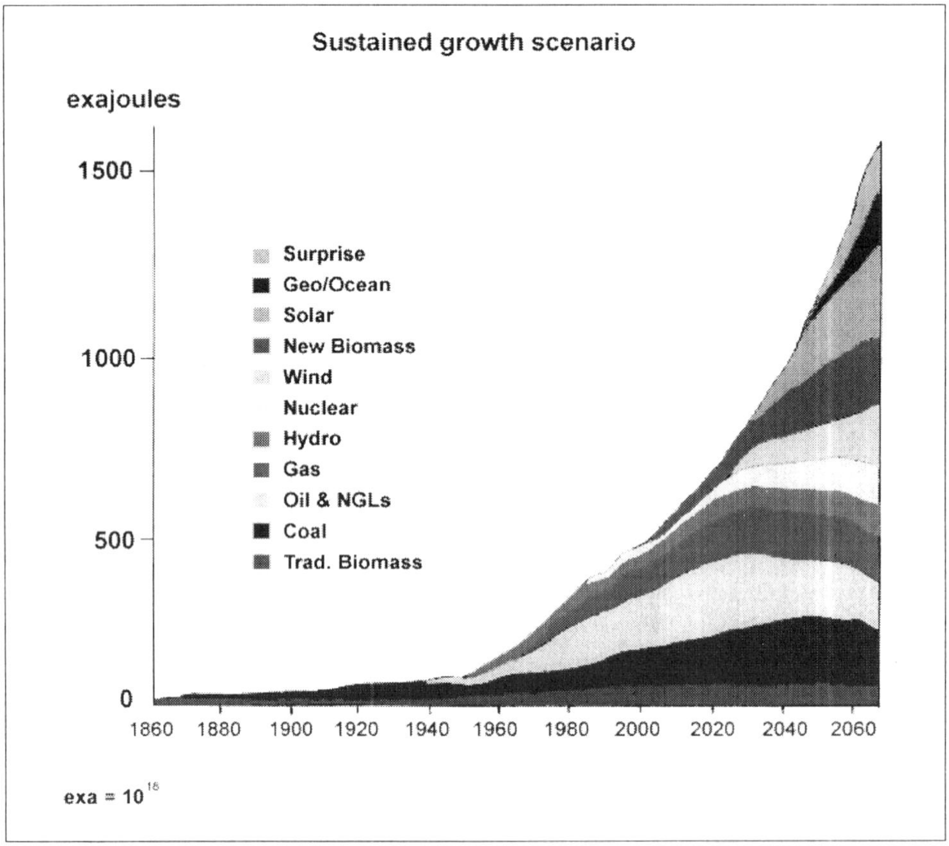

Fig. 1 "Sustained Growth Scenario" foreseen by Shell Oil Company.

Taking into account that biomass resources can be produced (in a sustainable way) by good practice with yields in the range of 5-10 Toe/ha per year, bioenergy contribution to the primary energy needs may be substantial: by 2010 - according to **"White Paper"** recommendations - bioenergy should provide in the EU 135 Mtoe/year.

This means that an energy production three times the 1995 amount of 45 Mtoe is considered to be a possible target for the year 2010. This additional capacity - 90 Mtoe/year - will require an estimated total investment of 84 billion Euro, with a CO_2 reduction of 255 MTon/year in 2010. The potential offered by biomass in fact is widely recognised; for instance, Fig. 1 shows the "Sustained Growth Scenario" foreseen by Shell Oil Company (1 ExaJoule = 23.8 Mtoe).
For biomass conversion technologies, one of the most promising route seems to be the production of biomass derived oil (Bio Crude Oil, BCO) through **pyrolysis**, given its economic viability and simple integration into conventional energy systems. Pyrolysis is attractive because converting solid biomass and wastes into liquid products presents many advantages in transport (energy density is increased about four-fold), storage, handling, retrofitting, combustion and flexibility in production and marketing.
This liquid is of moderate heating value, easily transportable, has potential to be used in thermal power plants, in modified Diesel engines or gas turbines. However, further technological development aimed at facilitating BCO utilisation is still necessary: one promising route for BCO upgrading is the production of an emulsion between BCO and conventional Diesel oil.

THE BIOEMULSION PROJECT

The aim of the BIOEMULSION project is to develop a low-cost physical-chemical and mechanical process for improving operational properties and performances of pyrolysis oil in small and medium Diesel engine units. The project partners are:

- **CSGI**, **University of Florence** (IT)
- **PMA**, Pasquali Macchine Agricole (IT)
- **IEE**, Institute für Elektrische Energietechnik, **Kassel University** (DE)
- **Ormrod Diesels** (UK)

and, as project subcontractors, DEF, Department of Energetics University of Florence (IT), Bio-energy Research Group, Aston University (UK) and WIP (DE).
The principal activities of the project are the preparation of the BCO/Diesel oil emulsion and the consequent experimental campaign aimed at assessing the operation of three Diesel engines fuelled by these emulsions. The use of a mixture of BCO and conventional Diesel oil in fact is relevant in both socio-economic (e.g. fuel cost, new job creation, rural development) and environmental terms (mainly CO_2 and sulphur emissions reduction). The final objective of this research project is therefore to achieve a low-cost upgrading process of biomass derived oil for facilitating its utilisation in Diesel units for power/heat generation. The project is supported by the European Commission, DG XII, in the framework of the JOULE Programme. The present paper describes the activities accomplished up to now by the Italian partners (CSGI, PMA and DEF) within the project. These activities can be summarised as follows:

- BCO upgrading
- Development and optimisation of the BCO/Diesel oil emulsification process
- Emulsions production and characterisation
- Experimental campaign in a small Diesel engine at PMA factory

The first task has been the provision of a sufficient amount of BCO, to be emulsified with the Diesel oil and hence tested in the engine. Four (4) types of BCO have been selected; these BCOs are different in terms of production facility and feedstock:

- ENEL1: produced by **ENEL** (Bastardo plant, IT), from Canadian oak
- ENEL2: produced by **ENEL** (Bastardo plant, IT), from beech wood
- Dyna1001: produced by **Dynamotive** (Vancouver, Canada), from California pine
- Dyna1009: produced by **Dynamotive** (Vancouver, Canada), from California pine

The BCOs produced by Dynamotive Technologies - types 1001 and 1009 - are both from California pine and have very similar characteristics. Table 1 shows the main physical-chemical properties of the selected BCOs.

BCO UPGRADING

The BCO upgrading has been useful to reduce noxious fractions, to increase pH and to make the selected BCOs more similar to each other. This last point is important as it permits to adopt few classes of surfactants and/or additives and to develop and optimise only one BCO/Diesel emulsification process. However the developed emulsification method does not require BCO upgrading, reducing considerably the operational costs.
On the basis of BCO characterisation (both raw and upgraded), finally BCOs referred as ENEL2 and Dyna1009 have been selected for the emulsification process: they have been chosen since they are more manageable and easier to be emulsified in comparison with ENEL1, which is characterised by high viscosity and ash content, thus causing a low quality emulsion in terms of combustion.

EMULSIFICATION PROCESS DEVELOPMENT AND OPTIMISATION

Diesel oil and BCO is a two phases system since Diesel is insoluble in BCO and vice versa. They are not miscible. If the Diesel oil/BCO system must be used as fuel, a stable emulsion is necessary. In the simplest emulsion a phase (oil or water) is dispersed in the continuous medium (water or oil) in the form of droplets. In this case Diesel oil has been considered as the "oil" phase and BCO as the "water" phase because of its consistent water percentage. Three kinds of emulsions can be prepared according to the value of the BCO/Diesel oil ratio:

- Water in oil emulsions (W/O) are obtained when up to about 45% by weight of BCO is added to the Diesel oil phase.
- Oil in water emulsions (O/W) are obtained when up to about 45% by weight of the Diesel oil is added to the BCO phase.
- Bicontinuous emulsions are obtained when the percentage by weight of the two phases is close to 50%.

In the first case, the BCO's droplets dispersed in the Diesel oil (continuous medium) form the emulsion. In the second case Diesel oil's droplets are dispersed in BCO (continuous medium). The third case's description is more complex: in fact "theoretically" there are no droplets and Diesel oil and water phases are continuous and form a "bicontinuous emulsion".

Table 1 Physical-chemical properties of the BCOs.

Physical-chemical properties	ENEL1	ENEL2	Dyna1009
Viscosity (cPoise): 25 °C	701	9.8	46.2
Viscosity (cPoise): 35 °C	361	6.7	27.1
Viscosity (cPoise): 45 °C	150	4.1	14.1
Elemental analysis (% w/w):			
H	5.79	5.94	6.31
C	44.06	34.96	43.52
O	48.68	58.70	50.07
N	0.08	0.11	0.07
S	<0.01	<0.01	<0.01
Alkali/BCO (ppm):			
Na	103	179	6
K	339	694	8
Ca	1096	1561	123
Ash weight (% w/w)	1.38	0.28	0.03
Alkali/Ash (mg/g):			
Na	7.5	64	20.5
K	24.6	248	26.9
Ca	79.4	558	410
Water weight (% w/w)	22.1[23.1][1]	43.6[45.1]	30.5[31.1]
Density @ 20 °C (g/cm^3):	1.230	1.165	1.175
pH	3.2	3.0	2.6
LHV (kJ/kg)	16942	14577	16779
HHV (kJ/kg)	18205	15873	18155

Underlining that BCO and Diesel oil are not miscible[2] a third component must be added to obtain a stable emulsion. This third component is called emulsifier (or surfactant). It changes the interfacial properties (namely interaction potential between droplets) of the system avoiding (or delaying) the emulsion's breaking.

The choice of an emulsifier is a step not completely random but it is driven by several empirical rules. These rules are based on the primary condition for the emulsifier to be efficient in a two phase, two-liquid emulsion. It has to be localised at the oil/water interface to a maximal extent. This criterion is exemplified in the HLB (Hydrophilic Lipophilic Balance) number. The HLB number for a certain molecule denotes the balance between its hydrophilic/lipophilic properties only; the number itself does not give any information about the stabilising efficacy of the emulsifier.

[1] Numbers in square brackets indicate water content values measured adopting the Karl Fisher method.

[2] Shaking a BCO/Diesel oil system, an emulsion is obtained but its stability is approximately only one minute at 25°C.

The HLB number method is based on the fact that the emulsifier is optimal in a water-oil system in which the properties of the oil matches the surfactant. Hence, each water-oil combination is characterised by an HLB number. More practically, emulsifiers characterised by HLB number:

- from 4 up to about 8 stabilise W/O emulsions;
- from 8 up to about 10 stabilise bicontinuous emulsions;
- from 10 up to about 18 stabilise O/W emulsions.

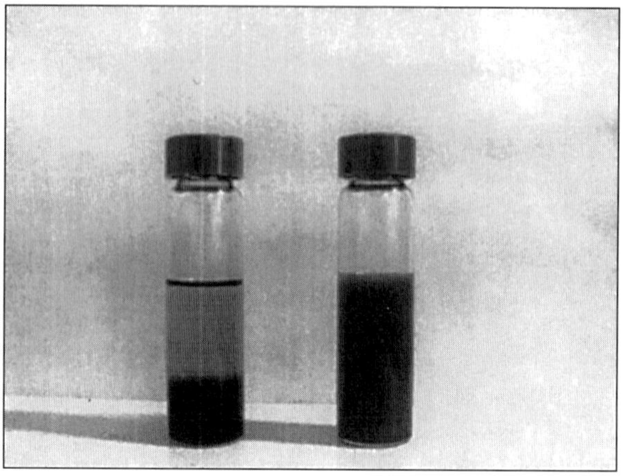

Fig. 3 BCO/Diesel oil mixture (left) and emulsion (right)

The Phase Inversion Temperature (PIT) or HLB temperature concept relates the emulsifier selection to the temperature at which an emulsion stabilised by a non-ionic emulsifier of the polyethylene glycol type changed from oil-in-water to water–in-oil with rising temperature. The tested surfactants were a large number (approximately one hundred), both of commercial type and chemically pure composition. Among them, cationic, anionic, zwitterionic and non-ionic stabilisers were tested. Moreover, for each of these classes of additives, polymeric and non-polymeric surfactants were tested.
Additives presenting sulphur or nitrogen in their composition have been neglected, to avoid an increase in NO_x and SO_x gaseous emissions during the emulsion combustion.

Preparation of the emulsions

The water-in-oil emulsions have been prepared by adding the surfactant to Diesel oil and thereafter adding the BCO to the resulting mixture. Bicontinuous emulsions have been formulated by adding the surfactant to Diesel oil, and then mixing to BCO the resulting mixture. Oil-in-water emulsions have been formulated by adding the surfactant to BCO and thereafter adding to the resulting mixture the Diesel oil during emulsification.
The emulsification process has been carried out using a homogenisation unit. The temperature during mixing has been preferably maintained between 60-65 °C and the emulsification process continued until a homogeneous single phase has been obtained. During the emulsifier choice phase, emulsions were prepared using a discontinuous process on a little amount of mixture (from 20 to 500 g).

Having identified the emulsifier, a continuous process has been employed on greater amount of material (up to 10 kg). In this configuration, the tank was a heating case controlling the desired temperature in the continuous homogenisation unit. Such emulsions present a stability up to some months; emulsions stable even for one year have been produced. It is important to underline that emulsions can be prepared also at room temperature; in this case the emulsion stability decreases (about 20 days). However, the destabilisation of the emulsions, prepared at both high and room temperature, is not an irreversible process, since it is sufficient to gently shake the fluid to obtain again a homogenous system.

The phase diagram of the system BCO\Diesel oil\emulsifier has been investigated in order to identify the best w/w ratio in terms of homogeneity, stability, engine performance, and operational costs.

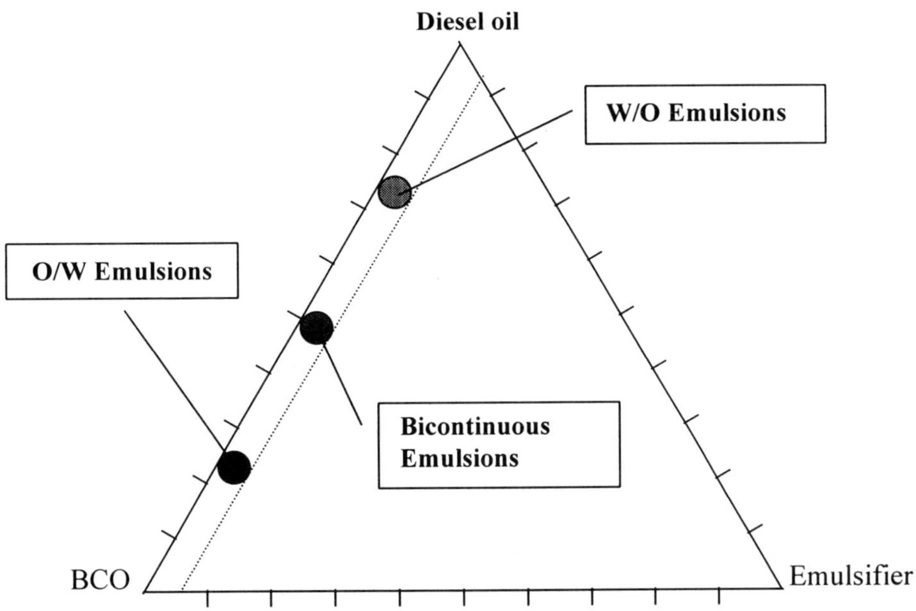

Fig. 4 Phase diagram for the BCO\Diesel oil\emulsifier ternary system.

Fig. 4 shows the phase diagram of the analysed system. The interesting region as far as regards the present work is located on the left of the dashed line (any concentration of BCO and Diesel oil but emulsifier <5% by weight). The circles are the regions that correspond to the 25% (W/O), 50% (bicontinuous) and 75% (O/W) of BCO by weight. These three regions have been studied more accurately.

It has been noted that:

- The emulsions are more stable than correspondent BCO.
- The emulsions were stable and opaque with no evidence of phase separation.
- Higher the emulsifier content, higher the viscosity and stability of emulsions. The optimal range to have an acceptable viscosity is between 0.5 and 2%. Emulsifier up to 4% has been added, in this case an additive (like n-octanol) must be employed to decrease emulsion viscosity.

- The evaluated emulsions stability versus time at high temperature (approximately 70 °C) is about 3 days.
- The high BCO content emulsions had very high viscosity which would make them difficult to use in most applications.
- Recycling cycles of the emulsion in the feeding engine system through the injectors and pipes have not changed the emulsions properties.

The destabilisation of an emulsion goes through several consecutive and parallel steps before the final stage of separated layers is reached. As a first step, the droplets move due to diffusion (or stirring), and if the repulsion potential is too weak, they become aggregated to each other; flocculation has taken place. The single droplets are now replaced by twins (or multiplets) separated by a thin film. The thickness of the thin film is reduced due to the Van der Waals attraction, and when a critical value of its dimension is reached, the film bursts and the two droplets form a single droplet. Coalescence has occurred. In parallel with these phenomena, the droplets rise through the medium (creaming) or sink to the bottom (sedimentation) due to differences in density of the dispersed and continuos phase. The final result is a highly concentrated emulsion at the top or bottom of the container, and the increased number of droplets per unit of volume increases the flocculation rate in a most decisive manner. The flocculation and coalescence process lead to larger and larger droplets until, finally, a phase separation has occurred. Prepared emulsions tend to flocculate and finally (after one-two months) the flocks sediment and coalescence.
It seems that coalescence process is very delayed by the emulsifier. In order to investigate the flocculation process, a study through an optical microscopy of the droplets and flocks has been employed. Unfortunately, a little difference in the contrast (brown against pale yellow) between droplets and continuous phase has not permitted a good determination of mean droplets dimensions. Probably, mean diameter should be few microns (1-5).

EMULSIONS PRODUCTION AND CHARACTERISATION

Six (6) different emulsions with different weight/weight ratio (from 25 to 75%) and based on two different BCOs (ENEL2 and Dyna1009) have been produced for utilisation in the experimental campaign.
The following tables report the principal chemical-physical properties of the **emulsions** derived from **Dyna1009** and **ENEL2**. It can be noticed that the physical-chemical properties of the emulsions are strictly related to the corresponding properties of the initial fuels (that is Diesel and Bio Crude Oil). Other additives (conventional lubricating and amidic or aminic co-surfactants) have been added (up to 0.5% by weight) to the emulsions to improve the lubricating properties and decrease the corrosive behaviour due to BCO.
In order to have large amounts of BCO/Diesel oil emulsions, an apposite unit has been designed and manufactured. The constructed emulsifier system is able to produce ≅ 50 l/h of emulsion, since the existing laboratory facility was insufficient to the production of the amounts (several kilograms) necessary for the experimental campaign.

Table 2 Physical-chemical properties of Dyna1009 emulsions.

Physical-chemical properties	BCO25 D1009	BCO50 D1009	BCO75 D1009
Viscosity (cPoise): 25 °C	9.79	25.41	n.a.
Elemental analysis (% w/w):			
H	12.27	10.50	8.98
C	74.63	61.82	51.03
O	13.04	27.6	39.88
N	0.03	0.05	0.07
S	<0.01	<0.01	<0.01
Alkali/emulsion (ppm):			
Na	2	4	6
K	3	5	8
Ca	41.1	83.2	125.1
Ash weight (% w/w)	0.01	0.02	0.03
Alkali/Ash (mg/g):			
Na	20.1	20.3	20.4
K	26.0	26.4	26.7
Ca	411	415	417
Water weight (% w/w)	7.3 [7.7]	14.9 [15.3]	22.5 [23.1]
Density @ 20 °C (g/cm^3):	0.923	1.008	1.094
pH	3.0	2.9	2.8
LHV (kJ/kg)	35810	28347	21558
HHV (kJ/kg)	38486	30637	23517

EXPERIMENTAL CAMPAIGN

The experimental campaign is aimed at assessing combustion characteristics, operating performances (e.g. efficiency, flexibility and hours of operation) and noxious emission levels of a small size Diesel fuelled with the BCO/Diesel oil emulsions.

The engine used to perform the test on the emulsions is a mono-cylinder direct injection Diesel engine manufactured by the Italian company Lombardini. The engine is part of a generator set of 5.4 kW and it has been only slightly modified, because the aim of the project was to succeed in running an unmodified engine. However, some minor modifications have been necessary, mainly an adding of components (for insulating and cooling) aimed at optimising the operation of the injection system when the engine is run with the emulsion.

Table 3 Physical-chemical properties of ENEL2 emulsions.

Physical-chemical properties	BCO25 ENEL2	BCO50 ENEL2	BCO75 ENEL2
Viscosity (cPoise): 25 °C	9.43	28.3	n.a.
Elemental analysis (% w/w):			
H	12.11	10.09	8.83
C	71.59	56.64	44.98
O	16.13	33.05	45.88
N	0.04	0.07	0.09
S	<0.01	<0.01	<0.01
Alkali/emulsion (ppm):			
Na	45	91	134
K	174	347	520
Ca	390	781	1171
Ash weight (% w/w)	0.07	0.14	0.21
Alkali/Ash (mg/g):			
Na	64.3	65.0	63.8
K	248.6	247.9	247.6
Ca	557.1	557.9	557.6
Water weight (% w/w)	10.7 [11.1]	21.6 [22.0]	32.4 [32.9]
Density @ 20 °C (g/cm^3):	0.921	1.003	1.084
pH	3.2	3.1	3.0
LHV (kJ/kg)	35706	27134	19259
HHV (kJ/kg)	38348	29335	21185

The main characteristics of the Lombardini engine (see figure 4) are reported in the following table.

Table 4 Characteristics of the Lombardini engine.

Model	Lombardini 6LD400
Cylinders	1
Bore x stroke	86 x 68 mm
Displacements	395 cm^3
Max rotational speed	3600 rpm
Max power	6.25 kW
Max power for electrical generation	5.4 kW (a 3000 rpm)
Max torque	19.6 Nm (a 2200 rpm)
Dry weight	45 kg
Injection hole's diameter	0.2 mm

Fig. 4 Lombardini 6LD400 engine

A large number of tests have been performed, using the emulsions prepared during the previous activities. The most used percentage was the 25% BCO. However, also emulsions with higher BCO content have been tested. Above all one test with 50% BCO Dyna 1009 gave promising results: the engine was able to run correctly for over 30 minutes; when the engine was switched to diesel oil its performance drastically decreased.

Among all the tests executed at PMA, the most significant in terms of running time and performances are the four reported in table 5.

Table 5 Main data of the engine tests

Test number and fuel type		Warm up Diesel oil	Running with emulsion	Diesel oil after-test	Developed Power	Emulsion consumption	Engine modification
		min.	**min.**	**min.**	**kW**	**litres**	
1	**BCO25 D1009**	20	78	8	3.1	1.1	none
2	**BCO5 D1009**	25	53	12	3.2	1.5	water spray
3	**BCO25 D1009**	32	116	25	3.2-3.3	4.5	water spray
4	**BCO25 D1009**	30	81	15	3.0-3.1	2.6	water jacket

The **first test** was performed to evaluate the maximum time the engine could operate with 25% emulsions (used BCO: Dyna1009). During the initial 4-6 minutes the engine ran correctly, without significant increase in smoke and emissions, only a little higher than Diesel oil. Then the engine operation suddenly started to get worse and in about 20 minutes the smoke became heavy black and the emissions were so high that it was better to detach the analysis probe. The engine continued running for 78 minutes, but it must be stated that during the last 15 minutes the engine worked very erratically.

The engine was then serviced. The injector was completely stuck in its clearance and a notable enlargement of the holes occurred: a final diameter of 0.45 mm was measured, the starting value being equal to 0.2 mm. All the injector items in contact with the emulsion were covered with a black sticky deposit.

The **second test** was performed using a low content BCO emulsion (5%). In this test it was used a water atomiser spraying the water on the visible part of the injector to keep the fuel as cool as possible. Even the recirculation pipe has been cooled. The test was positive: the emissions deteriorate slightly with time and the engine ran for more than 50 min. very plain and without smoke increasing. The injector showed less sign of wear: black deposit were concentrated on the tip of the needle only, no blockage of the needle, holes only a little larger (0.25 mm).
After this good result the 25% BCO emulsion has been tested with the cooling system. The **third test** lasted for approximately 2 hours, the longest one: it was stopped because the emissions became very bad after two hours of operation. However, during an initial period of about 30 minutes, the engine performances were quite constant. Afterwards, the engine operation started to get worse, first slowly and then more rapidly: the reason is the progressive enlargement of the injection hole, probably due to the erosion caused by polymerisation/carbonisation of the BCO particles occurring on the tip of the injector and of the needle. The final effect was an incorrect fuel injection in the combustion chamber and the consequent worsening of the engine performance. However, as in test n° 2, the arranged cooling system avoided the blockage of the needle and delayed the enlargement of the injection holes.
Further modifications on the engine were mainly in the direction of keeping the fuel as cool as possible, preventing BCO polymerisation and formation of particles not only sticky but also eroding. In fact, these particles, although they are soft, when accelerated to ≅100 m/s on the tip of injector probably erode the metal of the injector itself.
Therefore, for the **fourth test** a water jacket has been developed to be installed around the tip of the injector, in order to cool as much as possible this zone, where the temperature is higher. This modification brought to a further slight improvement of the engine performance in comparison with previous tests: a smoother behaviour of the engine and a more regular decreasing of performances. The injector's needle presented less sticky deposits than previous tests.

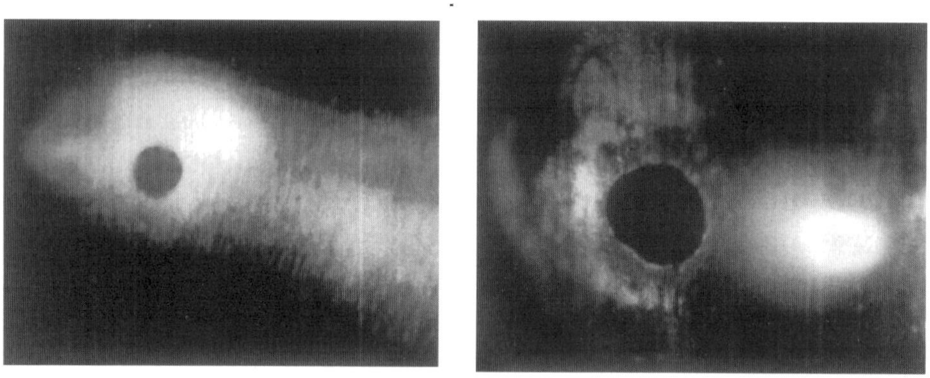

Fig. 5 Injector's hole before and after emulsion running (test n° 4)

Fig. 5 and 6 shows the injection hole and the needle tip as new (hole's diameter: 0.2 mm) and after 80 minutes of emulsion running (diameter: 0.45 mm). The pictures have been taken by a camera with a microscope adapter. Finally, the next charts show CO and NO_x emissions as well as exhaust gas temperature and engine efficiency vs. running time, in order to make an analysis comparing the different performances of the engine with the corresponding modification on the engine.

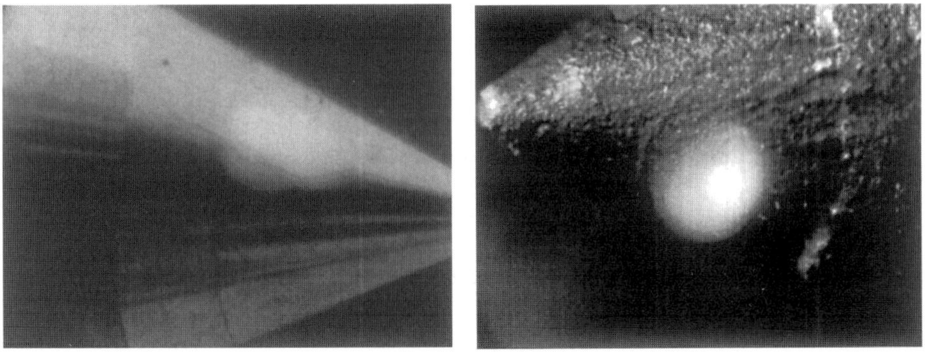

Fig. 6 Needle tip as new and after emulsion running (test n° 4)

The behaviour of the emissions can give an idea of what happens in the engine. The increase in CO emissions is mainly due to the poor injection (large fuel droplets) and consequently a bad combustion. The NO_x decrease versus running time is due to low temperature in combustion chamber caused by bad combustion. This also cause the higher temperatures measured in the exhaust pipe, since the combustion reactions, not completed in the combustion chamber, continue outside in the pipe.

From the chart it is possible to note that the engine performance increased when the cooling system was adopted: for a long period of time the emissions remained constant until the engine operation got worse due to the deterioration of the injector's components.

Moreover, the good operation of the engine when fuelled by an emulsion with a low BCO content (5%) should be recognized: the measured parameters have been constantly positive and the deterioration of the injection system is limited in comparison to the other tests executed in similar conditions.

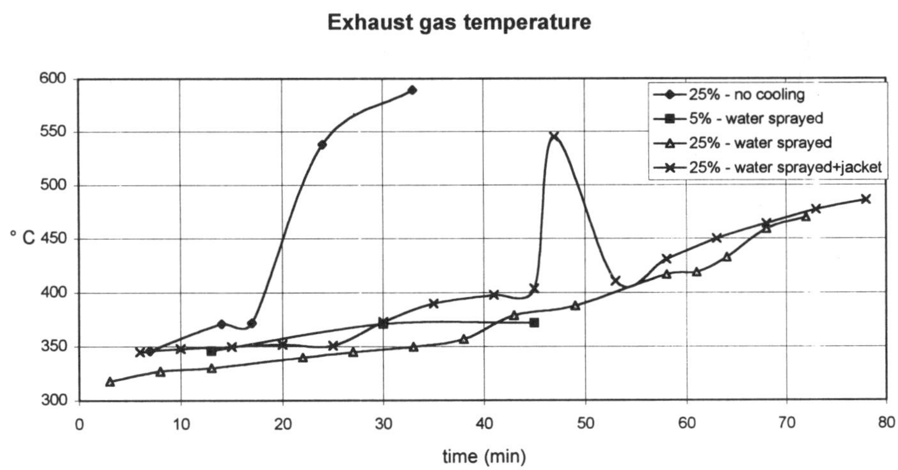

Fig. 7 Exhaust gas temperature versus running time

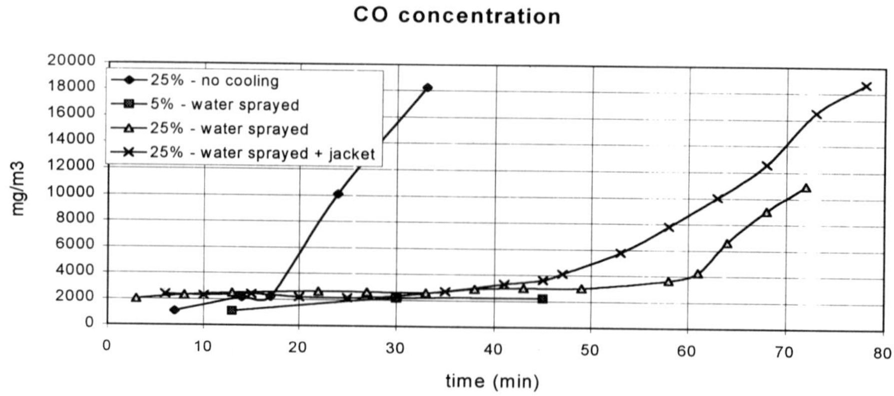

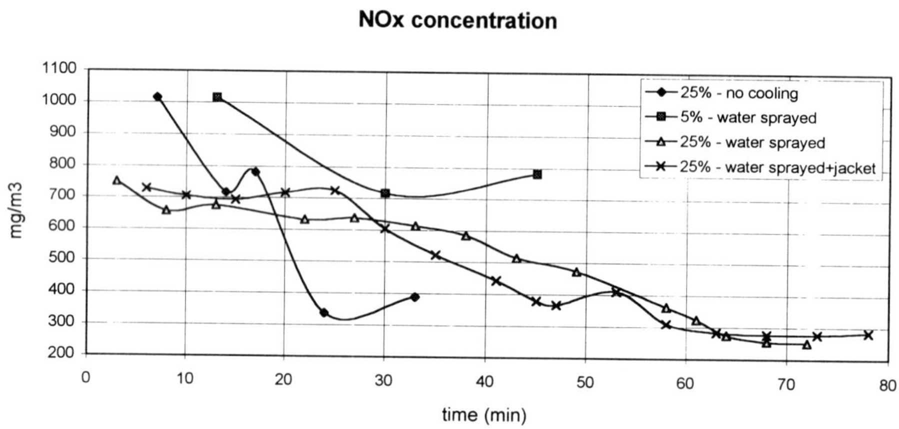

Fig. 8 CO and NO_x emissions versus running time

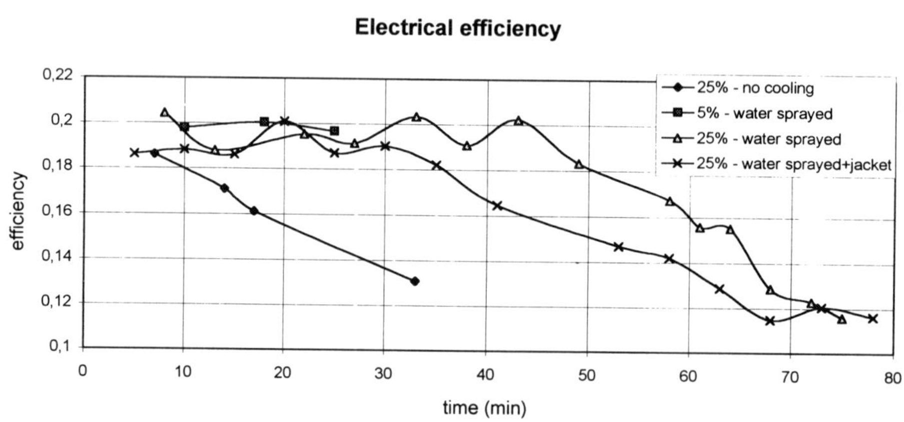

Fig. 9 Engine efficiency versus running time

CONCLUSION

The BCO/Diesel oil emulsification process has been developed and optimised, to obtain a fuel easier to be handled, stored and used in Diesel engine units. On the basis of this process, stable emulsions have been prepared: in some cases the stability reached even one year, thus representing a notable improvement in comparison with whole BCO, in general presenting a limited stability over the time.
The produced emulsions have been tested on a 6.25 kW Diesel unit: the most interesting results have been achieved with emulsions having a BCO content not greater than 50% by weight. The performed tests pointed out the following issues

- The deterioration of the injection system is probably due to erosion phenomena;
- The modifications on the engine aimed at cooling the fuel at the tip of the injector caused a considerable improvement of the engine performance;
- The problems related to the utilisation of emulsions as fuel are limited to the injection system: further tests with the adoption of special coating materials are scheduled in the future.

In conclusion, the tests performed at PMA facility showed that the engine is able to burn this kind of fuel, however some problems have been encountered. Some of them have been successfully solved by insulation and cooling. The remaining problems, especially the erosion phenomena on the tip of the injector, are expected to be solved by means of surface treatments of the injector itself.

ACKNOWLEDGEMENTS

Tha authors wish to thank the persons and Institutions that worked and/or are working at this project, in particular Prof. Bridgwater from Aston University, Graziano Crocetti from PMA, Klaus Gartner from IEE, Archie J. Webster from Ormrod Diesels.
The research was funded in part by the European Commission in the framework of the Non Nuclear Energy Programme JOULE III. A particular mention to Mario Frias, responsible on behalf of the European Commision for the Bioemulsion Project.

REFERENCES

1. Contract JOR3-CT98-0307. Midterm (July 1999), Progress reports (January 2000)
2. Leech J. (1997) Running a dual fuel Diesel engine on crude pyrolysis oil. In: *Biomass gasification and pyrolysis. State of the art and future prospects* (Edited by M. Kaltschmitt and A.V. Bridgwater), pp. 495-497. CPL Press, Newbury, UK
3. PyNE Newsletter, Bio-Energy Research Group, Aston University, UK.
4. A.V. Bridgwater *et al.* (May 1999) *Fast Pyrolysis of Biomass: a handbook* (Edited by IEA Bioenergy). CPL Press, Newbury, UK
5. Proceedings of the "10th European Conference & Technology Exhibition Biomass for Energy and Industry" (Edited by H. Kopetz *et al.* June 1998).

Production Of Diesel Fuel Additives From The Rosin Acid Fraction Of Crude Tall Oil

R. Coll[1], S. Udas[2], W.A. Jacoby[2]

[1]*Servei de Tecnologia Química, Rovira i Virgili University, Av. dels Països Catalans, s/n, 43007 Tarragona (Spain)*

[2]*Department of Chemical Engineering, University of Missouri-Columbia, W2033 Engineering Bldg. East, Columbia, MO 65211 (USA)*

ABSTRACT: This work aims to develop processes for converting the rosin acid fraction of crude tall oil into diesel fuel additives. Fractionation of crude tall oil produces an excess of rosin acids. This material is an abundant, inexpensive and chemically acceptable feedstock. Two components, abietic acid and dehydroabietic acid, comprise about 70% of the rosin acid fraction of tall oil. Hydrotreatment of these compounds allows obtaining saturated hydrocarbons that can be used as diesel fuel additives. In this paper, we report the first results obtained in a laboratory-scale batch reactor. Three different factors (temperature, hydrogen partial pressure and space velocity) were studied using two commercial NiMo catalysts. Operating conditions for each experiment were selected making use of factorial design techniques. The resulting liquid products, mainly saturated tricyclic ring hydrocarbons, were analyzed to determine their suitability to be used as diesel fuel additives. The results are promising enough to encourage further research on this topic, which could lead to the future use of tall oil rosin acid as a source of fuels.

INTRODUCTION

The increasing use of biomass-derived fuels is mainly due to their environmental advantages in comparison with fossil fuels. Besides these advantages, it is widely accepted that biomass-derived fuels do not show any particular technical problem when they are used in standard engines. Ethanol, for example, has been successfully employed as a gasoline additive or gasoline surrogate. For diesel engines, the use of bio-diesel derived from esterification of vegetable oils has also achieved a considerable success. In this case, however, the process is still in an incipient stage. The main problem associated with all these attempts is the question of economic feasibility. One of the ways to overcome this difficulty would be the use of an almost-free raw material that could be obtained in large quantities from a single source.

Tall oil fulfils all of these requirements. Tall oil is a natural product of pine trees that is isolated by means of the Kraft pulping process. It is composed of the ether extractable, non-lignin, non-cellulosic portion of the pine tree, and it must be fractionated via steam-vacuum distillation for commercial use. Fractionation of one metric ton of crude tall oil produces about 350 kg of rosin acids, 300 kg of fatty acids and 350 kg of distillated tall oil, head and pitch [1]. The economic value of tall oil is primarily derived from its fatty acid fraction.

Tall oil rosin production has recently been driven by demand for its co-product, tall oil fatty acid. United States production of tall oil rosin in 1994 was 265,000 metric tons, exceeding consumption by nearly 50,000 metric tons. Therefore, the rosin acid fraction of crude tall oil is an under-used byproduct in abundant supply, with attractive characteristics as a precursor to high value chemicals. Two components, abietic acid and dehydroabietic acid, comprise about 70% of the rosin acid fraction of tall oil. Therefore, it is a relatively simple mixture, which is also a beneficial characteristic for its use as a feedstock.

Despite its low cost and abundance, tall oil has only recently been considered as a potential fuel source. Initial studies have focused on the unfractionated material [2-8]. Direct use of tall oil is not possible since it produces excessive corrosion on certain parts of the diesel engines; it has deficient rheological properties at low temperatures and produces unacceptable contamination of the lubricating oil and coking of the diesel engine [6]. Therefore, some kind of previous processing becomes necessary. Esterification and hydrotreatment are two interesting possibilities. Hydrotreatment has one important advantage: the process is already employed in refineries so it can be applied without substantial modifications in producer plants. Moreover, the product has the same chemical characteristics as the regular present fuels, so no changes would be required in present-day engines.

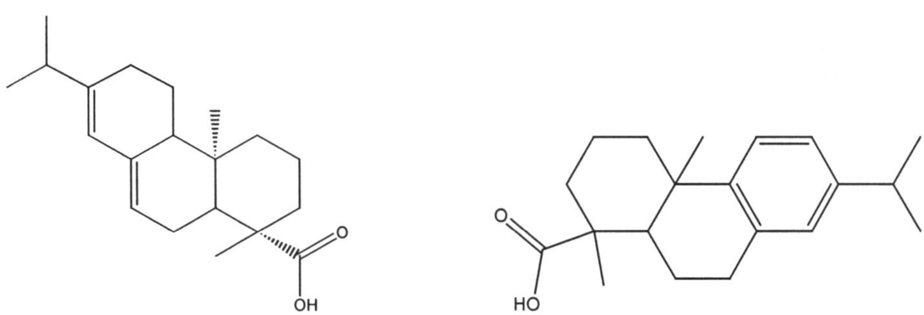

Abietic acid Dehydroabietic acid

Fig. 1 Chemical structures of abietic acid and dehydroabietic acid.

This work, however, does not deal with crude tall oil but with the under-used portion of fractionated tall oil, the rosin acids, which, as we mentioned before, show some advantages over crude tall oil. To our knowledge, this is the first time that the focus is on this fraction, although the work of Schuller's group provides a relevant basis for exploration [9-11].

TECHNICAL APPROACH

The chemical structure of all rosin acids is similar to those of abietic or dehydroabietic acids (*'Fig. 1'*). To obtain fuel additives, carboxylic acid functionality must be removed and double bonds saturated. A two-stage process can achieve both objectives. The first step would consist of the cleavage of carboxylic groups (either thermally or catalytically) producing a hydrocarbon and CO_2. Then, hydrogenation of the unsaturated bonds with an appropriate catalyst (Ni, NiW, Pd or Pt, for example) should be performed. An alternate process involves the simultaneous hydrogenation of carboxylic groups and unsaturated bonds. Partial hydrogenation of carboxylic groups would yield oxygenated compounds (aldehydes or alcohols), as shown in Paths 1 and 2. These products can be interesting as additives with specific properties. Complete hydrogenation would produce a hydrocarbon structure (Path 3).

$$R\text{-}COOH + H_2 \rightarrow R\text{-}CHO \qquad \text{(Path 1)}$$
$$\rightarrow R\text{-}CH_2OH \qquad \text{(Path 2)}$$
$$\rightarrow R\text{-}CH_3 \qquad \text{(Path 3)}$$

At the same time as the carboxylic functionality is removed, double bonds and aromatic rings would probably be hydrogenated. This second process would require significant amounts of hydrogen, which should be taken into account when the economic feasibility of the process is evaluated. The characteristics of this process suggest the use of hydrotreating catalysts, most of them containing molybdenum and either cobalt (CoMo) or nickel (NiMo) on an alumina carrier. This process is simpler than the first one, although the extent of double-bond saturation would probably be lower. The higher simplicity of this second alternative, however, makes it more interesting to be studied.

Gasoline-additive production would require one more step: thermal cracking or hydrocracking to reduce the molecular weight of the material. In this work, however, we have just focused on the first part of the process, which intends to obtain diesel fuel additives.

EXPERIMENTAL

Hydroprocessing reactions were carried out in a 500-mL batch reactor. The reactor is charged with 4 g of a commercial sulfided NiMo catalyst, supplied by Haldor Topsoe, Inc. and Criterion Catalysts, and 40 g of tall oil rosin (Unitol® NCY), supplied by Arizona Chemical. The reactor is first purged with Nitrogen and then filled with hydrogen to 3 bar. After reaching 150°C, the hydrogen partial pressure is increased to the run value. The last part of the heating period is carried out at high pressure. No additional hydrogen is added during the experiment, so its partial pressure is decreasing as the reaction goes on. A stirrer is used to improve the reactive-catalyst contact. When the reaction is over, the reactor is cooled and then depressurized.

Three operating variables have been analyzed: hydrogen partial pressure, temperature and reaction time. A factorial design of experiments was used to choose the conditions of each run. The levels of the three factors used during the runs are shown in '*Table 1*'. In the same table we can see the expressions employed to normalize the factors.

Table 1 Operating conditions employed.

Factor	Level -	Level +	X_i
Temperature	350°C	400°C	(T (°C) - 375) / 25
H_2 partial pressure	100 bar	150 bar	(P_{H2} (bar) - 125) / 25
Time	2 hr	4 hr	(time (hr) - 3) / 1

Product analysis

The amount of liquid product was measured to find out the reaction yield for each experiment. Density and boiling-point distribution were analyzed by ASTM D97 and ASTM D2887 (simulated distillation) standard methods. Simulated distillations were converted to ASTM D86 distillations to allow calculating cetane indexes according to ASTM D4737. Liquid products were also analyzed by MS-GC to obtain their qualitative composition, and by elemental analysis (C, H, S and O) to check the extent of hydrogenation.

RESULTS

Liquid Yield

Liquid yields, calculated as mass of product recovered divided by mass of rosin acid introduced in the reactor, range from 57 to 85 % ('*Table 2*'). There are significant differences between the catalysts employed. Catalyst A yields amounts of liquid product markedly lower than those obtained with catalyst B. This suggests that the first catalyst favors cracking reactions whereas the second one does not, at least at the operating conditions explored.

For liquid yield, as well as for the other variables, the effects of temperature, hydrogen partial pressure and duration of the experiment have been calculated.

Quantification of these effects was achieved by adjusting the response variables versus the normalized factors by the least-square method. The general form of these equations is:

$$\textit{Variable measured} = A0 + A(T)\, X_T + A(P_{H2})\, X_{PH2} + A(time)\, X_{time} \qquad (Eq.\ 1)$$

where $A0$, $A(T)$, $A(P_{H2})$ and $A(time)$ are the parameters adjusted by the least-square method and X_i are the normalized factors. Observing the values of these effects ('*Table 3*'), the differences between the two catalysts become clearer. For catalyst A, liquid yields are strongly affected by temperature (in a negative way) and by hydrogen partial pressure (in a positive way). Duration of the experiment does not seem to have a significant effect, at least for the times employed. For catalyst B, on the other hand, the results are just the opposite. Neither temperature nor hydrogen partial pressure affect significantly to liquid yield, whereas longer experiments produce lower amounts of liquid recovered, probably due to an increase in the extent of cracking. Correlation coefficients between these variables confirm this conclusion. These results are consistent with the idea that catalyst A favors cracking more than catalyst B does.

Table 2 Liquid yields for the two catalysts.

Exper.	Catalyst	T (°C)	P (bar)	time (h)	X_T	$X_{P\,H2}$	X_{time}	Yield (%)
EXP01	A	350	100	2.0	-	-	-	66
EXP02	A	400	100	4.0	+	-	+	57
EXP03	A	350	150	4.0	-	+	+	78
EXP04	A	400	150	2.0	+	+	-	73
EXP05	B	350	100	2.0	-	-	-	85
EXP06	B	400	100	4.0	+	-	+	75
EXP07	B	350	130	4.0	-	+	+	77
EXP08	B	400	150	2.0	+	+	-	84

Product Composition

We have performed two kinds of analyses in order to determine the chemical composition of the liquid products. First, we analyzed the elemental composition of the product. The raw material is composed of 76.2% of C, 9.6% of H and 14.2% of O (percentages in mass). Sulfur was also analyzed, but it was not detected (detection limit = 0.05%). These percentages suppose an H-to-C ratio of 1.52. For the hydrogenated products, no oxygen was detected, except for experiment EXP04 (1.25% in mass). Therefore, we can affirm that carboxylic groups are either completely reduced to $-CH_3$ or removed as CO_2.

Hydrogen consumptions were about 200 NL/kg rosin acid. The extent of double bond and aromatic hydrogenation in each experiment can be measured by comparing the product H-to-C ratios. Supposing that no cracking is produced and an abietic or pimaric acid structure, the maximum H-to-C ratio achievable would be 1.80. Results for catalyst A range from 1.40 (EXP02) to 1.72 (EXP03), with 1.59 as average. EXP02, as we saw before, yielded the smallest amount of liquid, surely due to a high extent of cracking. This is also the reason for the low extent of hydrogenation in that experiment. Experiment EXP03, on the other hand, which yielded the highest amount of product with this catalyst, also gave the highest extent of saturation. For catalyst B, conclusions are very similar. Hydrogenation is maximum at 350°C, 150 bar and 4 hr, and minimum at 400°C and 100 bar. Liquid yield and extent of hydrogenation, hence, seem to be proportional.

Table 3 Effects of temperature, hydrogen partial pressure and reaction time.

	Catalyst	A0	A(T)	A(P_{H2})	A(time)
Yield (%) :	A	68.4	-7.3	14.3	-2.3
	B	80.3	-1.8	0.8	-8.5
Density (°API):	A	28.0	0.9	4.8	-1.4
	B	29.9	4.9	3.1	1.0
BP 10 % (°C) :	A	179	-78	-46	-12
	B	164	-81	-19	-21
BP 50 % (°C) :	A	296	-47	-1	2
	B	284	-80	3	-9
BP 90 % (°C) :	A	346	-15	-3	-7
	B	351	-24	-7	-6
C.I. :	A	35.1	-5.8	6.8	-4.8
	B	36.7	0.0	4.4	0.6
H / C :	A	1.59	-0.18	0.14	-0.05
	B	1.58	-0.10	0.10	0.00

The effects of the three operating variables (temperature, hydrogen partial pressure and reaction time) on extent of hydrogenation have also been quantified following the same procedure as for liquid yield (as well as for the other response variables studied). Results, shown in '*Table 3*', reveal that the three operating conditions also affect in the same way to both liquid yield and extent of saturation. Temperature and duration of the experiment decrease hydrogenation and yield, whereas a high hydrogen partial pressure improves them.

We have also analyzed the products by GC-MS. Identifying all the compounds present in the liquid product is not possible because a perfect separation is not possible by GC. However, these analyses are still useful to:

- Observe the amount of C20 and C19 compounds. C20 are produced by reduction of the carboxylic group to $-CH_3$. C19 are produced by cleavage of the carboxylic group (producing CO_2).
- Observe the presence of aromatics.
- Observe the characteristics of the cracked compounds, that is observe whether the tricyclic compounds are present or have been cracked, or what the amount of linear compounds present in the product is.

The MS-GC analyses reveal that most of the product is composed of C19 compounds. Therefore, we can affirm that the main mechanism for carboxylic group removal is by producing CO_2. Tricyclic structures of the raw material are still preserved in the final product, especially in the experiments performed at 350°C. The products also show significant amounts of aromatic compounds. This fact points out that aromatic hydrogenation is not good enough. Therefore, future efforts should be carried out to improve this aspect of the reaction and produce better quality fuels. The use of more specific hydrogenation catalysts should help to achieve this purpose.

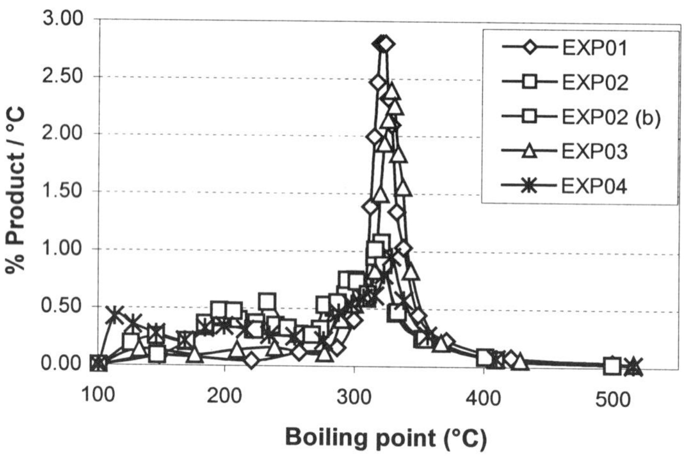

Fig. 3 Boiling point distribution for products with catalyst A.

Boiling Point Distribution

In '*Fig. 3-4*', boiling point distributions for all the products are shown. The extent of cracking in each experiment is also easily deduced from these graphics. As we can see, there exist important differences between the various operating conditions employed. Experiments at 400°C produce significant higher amounts of cracked compounds, which confirm the conclusions obtained from liquid yields. However, the characteristics of the products are not so different depending on the catalyst employed. The effects of operating conditions on boiling point distribution were calculated ('*Table 3*'). Temperature appears to be the most important factor in all the cases. Hydrogen partial pressure and, especially for catalyst B, reaction time also show significant effects on BP 10% (10% recovery temperature, in °C). All these variables affect cracking reactions in a negative way (decreasing the boiling points). The trends are the same for both catalysts.

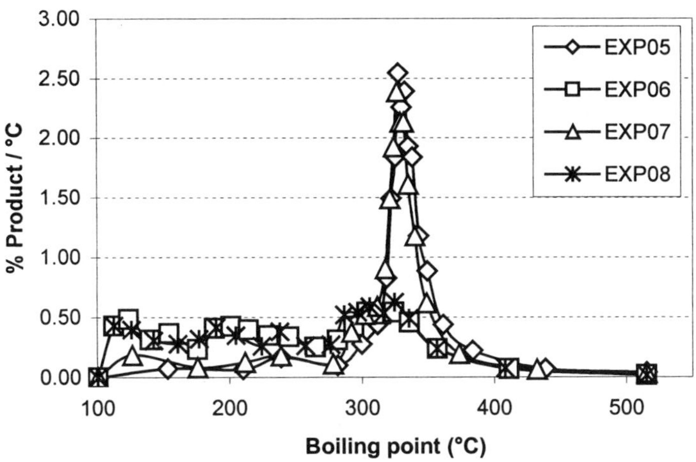

Fig. 4 Boiling point distribution for products with catalyst B.

Density

Density is another variable that must be taken into account to study the applicability of our product as a diesel fuel. In general, diesel fuels usually have densities between 30 and 40°API (diesel fuel No. 2) or between 37 and 45°API (diesel fuel No. 1). Our product densities range between 25 and 35°API depending on operating conditions. Therefore, we can conclude that the product we obtain can fall into the specifications for diesel fuels.

Cetane Index

Cetane index is an estimation of the cetane number, which measures the anti-knock properties of the fuel. This index is more used than the cetane number because it can be calculated by a four-variable equation (density, BP 10%, BP 50% and BP 90%) according to ASTM D4737 standard method, instead of the test engine needed to measure cetane number. In our case, the relatively low amount of product obtained in these initial experiments does not allow determination of the cetane number, so we have used the cetane index to evaluate and compare our products, an approximation perfectly valid at this initial stage of investigation.

Results show that rosin acid produces fuels with low cetane indexes, between 28 and 42 for catalyst A and between 34 and 43 for catalyst B. According to ASTM D975 (Standard Specifications for Diesel Fuel Oils), diesel fuels must have a minimum cetane number of 40 (diesel no. 2). Therefore, some of our products seem to fall into specification for diesel fuels. However, it must be taken into account that cetane index is just an estimation of cetane number, and this estimation may be inaccurate. According to ASTM D4737 standard method, this equation possesses some limitations:

- It is not applicable to fuels containing additives for raising the cetane number.
- It is not applicable to pure hydrocarbons, or to non-petroleum fuels derived from coal.
- It can be inaccurate if it is applied to residual fuels or crude oils

A study about different diesel fuels [12], for example, showed that cetane index underestimates cetane number when it is lower than 43, sometimes more than 5 units (C.N.=40, C.I. = 35.1). Therefore, the actual cetane numbers for our products can be higher than the measured ones. In the future, producing higher amounts of product to measure actual cetane numbers appears to be necessary to confirm their suitability as diesel fuel.

Regarding the effects of the operating conditions, we can affirm that cetane indexes improve when hydrogen partial pressure is high. For catalyst A, they decrease at high temperatures and reaction times. For catalyst B, these last two effects are not significant.

CONCLUSIONS

The research on the use of hydrotreated rosin acids as fuel additives is in an incipient phase of development. However, the first set of experiments reported in this work gives promising results that encourage continuing the work on this topic. Liquid yields are high, especially when hydrotreatment is performed at 350°C and high hydrogen partial pressures. No coking or deactivation was observed. Further research is necessary to optimize operating conditions in order to maximize hydrogenation and to extend the applicability of the products for specific functions in the fuel. However, tall oil rosin appears to be a suitable source of diesel fuel and/or diesel fuel additives.

ACKNOWLEDGEMENTS

We would like to thank the University of Missouri Research Board for the economical support provided, Criterion Catalysts and Haldor Topsoe for catalyst samples, Arizona Chemical, NREL and Dr. Esteban Chornet for his comments and suggestions.

REFERENCES

1. 1997 North American Pulp and Paper Fact Book
2. Sharma, R.K. and Bakhshi, N.N. Catalytic Conversion of Crude Tall Oil to Fuels and Chemicals over HZSM-5: Effect of Co-feeding Steam. *Fuel Process. Technol.* 27, 113-130, 1991.
3. Sharma, R.K. and Bakhshi, N.N. Upgrading of Wood-derived Bio-oil Over HZSM-5. *Bioresource. Technol.* 35, 57-66, 1991.
4. Sharma, R.K. and Bakhshi, N.N. Upgrading of Tall Oil to Fuels and Chemicals Over HZSM-5 Catalyst Using Various Diluents. *Can. J. Chem. Eng.* 69, 1991.
5. Sharma, R.K. and Bakhshi, N.N. Catalytic Upgrading of Biomass-derived Oils to Transportation Fuels and Chemicals. *Can. J. Chem. Eng.* 69, 1991.
6. Wong, A. Tall Oil-based Cetane Enhancer for Diesel Fuel. *Pulp & Paper Canada*, 96 (11), 37-40, 1995.
7. Stumborg, M., Wong, A., Hogan, E. Hydroprocessed Vegetable Oils for Diesel Fuel Improvement. *Bioresource Technol.* 56, 13-18, 1996.
8. Liu, D.D.S., Monnier, J., Tourigny, G., Kriz, J., Hogan, E., Wong, A. Production of High Quality Cetane Enhancer from Depitched Tall Oil. *Petr. Sci. & Tech.* 16 (5&6), 597-609, 1998.
9. Severson, R.F., Schuller, W.H., Lawrence, R.V. Pyrolysis of Certain Resin Acids at 800°C. *Journal of Chemical & Engineering Data* 17 (2), 1972.
10. Severson, R.F., Schuller, W.H. The Thermal Behavior of Some Resin Acids at 400-500°C. Canadian *Journal of Chemistry* 50, 2224, 1972.
11. Takeda, H., Schuller, W.H., Lawrence, R.V. Thermal Behavior of Some Resin Acid Esters. *Journal of Chemical & Engineering Data* 13 (2), 1968.
12. Diesel Fuel Quality and Effects of Fuel Additives

Preliminary Study on Fungicide and Sorption Effects of Fast Pyrolysis Liquids Used as Wood Preservative

D. Meier[1], B. Andersons[2], I. Irbe[2], J. Chirkova[2], O. Faix[1]
[1]*Federal Research Centre for Forestry and Forest Products, Institute for Wood Chemistry and Chemical Technology of Wood, D-21027 Hamburg, Germany*
[2]*Latvian State Institute of Wood Chemistry, 27 Dzerbenes St., LV-1006 Riga, Latvia*

Abstract: Pine wood samples were impregnated with fast pyrolysis liquid to study the fungicide and sorption effects. Three fungi were selected: *Coniophora puteana, Poria placenta*, and *Lentinus lepideus*. Impregnation was performed according to EN 113, and ageing procedures such as leaching and evaporation according to EN 84 and EN 73, respectively.

The treated specimens without the ageing procedure did not suffer from fungal decay. After leaching, the samples lost approx. 5% of their initial weight. However, after evaporative ageing of the samples, the mass loss was below 1%.

The alteration in the porous structure of the wood cell wall as a result of the impregnation and the ageing procedures, as well as after the action of fungi (*Coniophora puteana* and *Lentinus lepideus*) was investigated by the water vapours sorption method.

The impregnation of wood with oil leads to the intensification of hydrophobic properties, which remain practicaly the same during evaporation (EN 73). As a result of leaching, the hydrophilic properties are restored incompletely, and the structure of the sample changes.

Distinctions in the destructive action of *Coniophora puteana* and *Lentinus lepideus* have been established: from the viewpoint of structure, the activity of *Lentinus lepideus* relative to impregnated wood is more than twice that in the case of *Coniophora puteana*. After the leaching procedure (EN 84), the destructive action of *Coniophora puteana* increases dramatically, while the structural characteristics of leached samples upon contacting with *Lentinus lepideus* are not practically changed.

Pyrolysis oil has shown a protective effectiveness against wood destroying basidiomycetes, so that it could be used as a wood preservative. However, some properties, such as stability in wood after ageing procedures, reduction of hydrophilicity and retention in wood, call for improvement.

INTRODUCTION

Creosote (tar oil) preservatives are used in industrial applications to treat ties, poles, posts, pilings etc. Creosotes and other tar oils are produced when a naturally occurring carbon-rich substance (coal, lignite or wood) is heated without air (1). The nature of creosote depends upon the type of coal used and the nature of the distillation processes. Typically, creosote is a complex mixture of over 200 compounds consisting mainly of polyatomic aromatic hydrocarbons (PAH) and phenolics. It may contain more than 30 different PAHs amounting up to 85 % (2).

The wood-preservative properties of creosote depend upon many factors and, whilst the tar acids are good fungicides, they are also the components, which are most susceptible to loss from wood by volatilisation and leaching (3). Due to the relatively high toxicity of creosote based predominantly on the presence of benzo[a]pyrene (b[a]p), there exist special regulations and restrictions within the EU and in some member states such as Sweden, Denmark, Austria, Germany, and the Netherlands (4).

In this study, the use of pyrolysis oil from wood was tested as an alternative wood preservative. In the initial laboratory evaluation, the toxic effects of the oil against wood destroying basidiomycetes were determined. The fungus *Lentus lepideus* was included as an especially resistant organism to tar oils.

MATERIALS AND METHODS

Pyrolysis oil was obtained from pine wood fast pyrolysis in a fluidised bed reactor operated at Union Fenosa, Spain. Pine (*Pinus sylvestris* L.) sapwood blocks (20 x 20 x 5 mm) were vacuum impregnated with pyrolysis oil according to the modified European Standard EN 113 method (5), based on the agar wood block decay test.

Additionally, ageing procedures such as the leaching (modified EN 84) (6) and evaporation (modified EN 73) (7) of impregnated wood blocks were carried out. The modifications of the Standard methods were based on the shortening of the test procedures. In all cases, the test duration was 6 weeks, the leaching of treated blocks lasted 2 weeks and evaporation 6 weeks before exposure to the fungal attack.

A series of treated and control specimens was exposed to the brown rot basidiomycetes *Coniophora puteana* BAM Ebw. 15, *Poria placenta* FPRL 280 and *Lentinus lepideus* Ebw. 20 at 22 C and 70% RH. Treated check test specimens were placed in uninoculated culture vessels for calculation of the correction value. At the end of the test, the specimens were assessed and mass losses were calculated to determine any protective effectiveness of the given preparation.

Changes in the structure and hydrophilic properties of the wood cell wall in specimens after the impregnation and ageing procedures samples were studied by the water vapours sorption method. The sorption-desorption isotherms measured on a vacuum sorption balance at 295 K were analysed by the comparative method in combination with the BET method (8). The accessible specific area **A** (m^2/g), reflecting swelling capacity, the mass hydrophilicity **a** (mM/g) and the surface concentration of the hydrophilic centres α (groups/nm^2) was determined.

If the structure of the samples is identical, the comparative graphs are linear. The deviations from linearity indicate that the cell wall of the sample under study contains pores which are absent in the control sample. The sizes of such pores are calculated from the theory of capillary condensation (8), based on the location of the bend region corresponding to a definite P/Po value on the isotherm. The bends on the comparative desorption-adsorption graphs unequivocally indicate the presence of pores of definite size in the sample.

RESULTS AND DISCUSSION

EFFECTS ON FUNGI

For each fungus and screening method, four replicates were taken and the mean values of results are shown in Table 1 - 3. The blocks were treated both with distilled water (0 concentration) and pyrolysis liquid in concentrated form.

As can be seen from the Tables 1 - 3, uncorrected and corrected mass losses for water-impregnated wood had a positive value and exceeded 30 % for each test fungus. On the contrary, uncorrected mass losses for oil-treated specimens showed negative values indicating an increase in mass. Negative values were obtained also for check (correction) test specimens. Since they were not placed on fungi, the mass variations of these specimens allowed determination of the changes in mass resulting from other factors than fungal attack.

The corrected mass losses of the treated specimens were obtained in mathematical calculations and might show irrelevant theoretical values. These theoretical values are attributed to the results of the modified EN 113 method where, in fact, no mass losses of the treated blocks were observed. According to EN 113, especially for creosotes, the increase in the mass of the treated check test specimens may be very high. Therefore, the use of the correction values may produce erroneous theoretical losses in the mass of the treated specimens (5). To avoid these misassessments, the treated specimens should also be carefully investigated for any visible attack by the fungus. Our investigations confirmed that the treated specimens without ageing procedure did not suffer from the fungal decay.

The screening test with a leaching procedure revealed some loss of mass after the fungal attack. Oil-treated and leached specimens after attack by *C. puteana, L. lepideus* and *P. placenta* showed mass losses of 5.5%, 5.1% and 25.9%, respectively. Regarding the first two fungi, mass losses of about 5% seemed to be believable, because some visible signs were noticed after splitting of the specimens. The highest mass loss of *P. placenta*-attacked blocks evidently had an erroneous value. Some small traces of decay were noticed but they could not be as high as 25.9% and exceed the mass loss of the corresponding control specimens (24.2%). These results indicate that pyrolysis oil, is leachable to some extent, hence, it could promote further fungal decay.

The screening test with an evaporative ageing of wood after the fungal attack indicated a negligible mass loss (less than 1.0%) i.e. 0.0%, 0.1% and 0.8% for *C. puteana*, *L. lepideus* and *P. placenta*, respectively. A visual assessment of the treated blocks confirmed the calculated results. Wood blocks were intact and showed no signs of decay. Mass losses of the corresponding untreated control specimens reached acceptably high values (Tables 1 - 3).

The virulence of fungal strains was controlled using untreated specimens in separate culture vessels. Mass losses of these specimens were 51.2%, 56.2% and 45.5% for *C. puteana, L. lepideus* and *P. placenta*, respectively. This is in good agreement with the EN 113 demands where the minimum mass loss for virulence specimens should be 20%.

Moisture contents of the treated and control specimens are shown in Tables 1 - 3 as the lowest and highest values of the specimen batch. Changes in moisture content were determined on a basis of the final dry mass. In general, specimens treated with water

had a moisture content comparable with that of the control. On the contrary, wood treated with pyrolysis liquid had a comparatively high moisture content after-modified EN 113 and EN 113/ EN 73 tests, i.e. ca 80-120% depending on the fungus. These elevated hygroscopic properties resulted in an extra moistening of wood and the accumulation of liquid in the culture medium. The liquid around the specimens could affect the fungal activity. The screening tests with a leaching procedure presented different results. In this case, when a part of oil components was leached out, wood moisture did not exceed 60% depending on the fungus. The specimens were covered with a sparse mycelium and some loss in mass was observed.
Retention of the oil in wood comprised approximately 315 kg/m^3. This is a higher amount in comparison with creosote preservatives used at present, which are introduced into wood up to 200 kg/m^3 for ground contact use or in a marine environment (9).
Further investigations will be carried out to improve the properties of pyrolysis products for their potential use in wood preservation.

Table 1. Summary of results for *Scots pine* sapwood blocks treated with pyrolysis oil and attacked by the brown-rot fungus *Coniophora puteana*.

Procedure	Concentration, % (m/m)	Uncorrected mass loss, %	Correction value, %	Corrected mass loss, %	Mass loss of control, %	Moisture % treated.	Moisture % control
EN 113	0	36.9	-0.7	37.6	35.4	61.7-78.8	64.1-73.9
	preparation	-36.5	-46.2	10.2* (0.0)	44.5	87.6-107.9	62.0-74.2
EN 113/ EN 84	0	37.1	0.5	36.6	36.7	64.4-80.6	68.5-82.6
	preparation	-10.8	-16.3	5.5	44.6	36.1-59.8	75.9-90.8
EN 113/ EN 73	0	35.1	-0.1	35.2	35.0	74.6-94.7	82.5-87.7
	preparation	-34.0	-31.9	0.0	43.5	86.0-110.9	65.5-73.0

*Irrelevant theoretical value due to the calculation of the correction value

Table 2. Summary of results for *Scots pine* sapwood blocks treated with pyrolysis oil and attacked by the brown-rot fungus *Poria placenta*.

Procedure	Concentration, % (m/m)	Uncorrected mass loss, %	Correction value, %	Corrected mass loss. %	Mass loss of control, %	Moisture % treated.	Moisture % control
EN 113	0	32.4	-0.7	33.1	24.1	68.8-97.1	71.8-84.9
	preparation	-25.3	-46.2	20.9* (0.0)	33.4	79.9-102.9	65.3-78.9
EN 113/ EN 84	0	34.9	0.5	34.4	24.8	91.0-122.0	61.9-103.8
	preparation	6.3	-16,3	25.9*	24.2	46.7-54.8	51.7-67.4
EN 113/ EN 73	0	35.2	-0.1	35.3	31.8	72.1-116.4	65.4-81.2
	preparation	-32.6	-31.9	0.8	36.4	70.6-103.2	63.4-69.4

*Irrelevant theoretical value due to the calculation of the correction value

Table 3. Summary of results for *Scots pine* sapwood blocks treated with pyrolysis oil and attacked by the brown-rot fungus *Lentinus lepideus*.

Procedure	Concentration, % (m/m)	Uncorrected mass loss, %	Correction value, %	Corrected mass loss, %	Mass loss of control, %	Moisture %	
						treated.	control
EN 113	0	40.6	-0.7	41.3	40.3	58.1-72.0	53.7-68.1
	preparation	-32.7	-46.2	13.5* (0.0)	38.3	81.1-126.1	55.1-75.6
EN 113/ EN 84	0	33.1	0.5	32.6	36.4	59.8-69.1	55.3-73.2
	preparation	-11.2	-16.3	5.1	29.5	24.4-35.4	49.7-56.2
EN 113/ EN 73	0	39.7	-0.1	39.8	38.6	60.0-75.6	58.7-69.5
	preparation	-33.8	-31.9	0.1	42.6	85.7-109.3	58.0-65.5

*Irrelevant theoretical value due to the calculation of the correction value

EFFECT ON SORPTION

Some isotherms of water vapours sorption-desorption are presented in Fig. 1. In the first cycle of sorption, the isotherms have a pronounced stepwise character, and the values of water vapour sorption are unusually low for wood specimens. This is connected with the strict drying conditions of the specimens (378K at a rapid rise in temperature) after the exposure procedure, which resulted in the fixation of the non-equilibrium structure. After saturation with water vapours and the subsequent vacuum treatment at 295K during the procedure of the sorption experiment, the second cycle of isotherms was carried out, and this state was regarded as an equilibrium one. Thus, the analysis of the specimens structure was performed on the basis of the sorption-desorption isotherms of the second cycle.

In pine wood impregnated with water according to EN 113, pores with $r > 2.1$ nm are practically absent, and the filling of the pores proceeds without the capillary condensation, which is testified by the linearity of the comparative desorption-adsorption plots.

As a result of impregnation with oil, in comparison with water impregnation, a dramatic (by 37%) decrease in the accessible surface and the mass concentration of the hydrophilic groups occurs (Table 4). Although the total volume of the pores tends to decline (from 0.15 to 0.12 cm^3/g), a minor volume of pores in the range $r = 2.1 - 4.8$ nm appears (Fig. 2,A).

Table 4. Structural characteristics of pine wood specimens after impregnation with water and pyrolysis oil (EN 113), leaching (EN 84), and evaporation (EN 73)

Procedure	2nd cycle		
	A, m^2/g	**a**, mM/g	α, groups/nm^2
Water EN 113	252	2.06	4.92
Oil EN 113	159	1.28	4.85
Oil EN 113/EN 84	206	1.64	4.80
Oil EN 113/EN 73	161	1.32	4.94
Oil EN 113 – *C. puteana*	170	1.40	4.96
Oil EN 113/EN 84 – *C. puteana*	285	2.33	4.92
Oil EN 113/EN 73 – *C. puteana*	180	1.46	4.88
Oil EN 113 – *L. lepideus*	184	1.52	4.97
Oil EN 113/EN 84 – *L. lepideus*	210	1.71	4.90
Oil EN 113/EN 73 – *L. lepideus*	185	1.52	4.95

Of interest is the comparison of specimens leached according to EN 84 after impregnation with oil with those impregnated with water according to EN 113. Such a comparison makes it possible to evaluate the degree of leaching of oil components. As can be seen from Table 4, an increase in **A** (by 30%) and **a** (by 28%) values of the oil-impregnated samples after leaching (EN 84) occurs. Therefore, a considerable part of the more light oil components is leached from the specimen with water. However, the structural characteristics do not reach the corresponding values for the blank specimens washed only with water, and the **A** and **a** values remain lower by about 20 %. In this case, a considerable change in the cell wall structure occurs, i.e. very narrow pores (radius about 0.5 nm) and rather a large pore volume in the radial range 1.0-1.3 nm tends to appear. This is unequivocally indicated by the bends in the comparative plots (Fig. 2,B).

As a result of the blowing (EN 73) of pine wood impregnated with oil, no dramatic changes in the sorption properties and structure are observed (Table 4): only a minor increase in **A** occurs, and the surface becomes more hydrophilic as a result of the weathering of more volatile hydrocarbon components of oil.

After the exposure of the impregnated wood specimens to *C. puteana* and *L. lepideus*, the hydrophilic properties of wood increase, i.e. **A** and **a** values increase by 6.9% and 9.4% in the case of *C. puteana*, respectively, and by 15.7% and 18.8% in the case of *L. lepideus*, respectively. The increase in α values in both cases indicates the appearance of new hydrophilic centres in wood. Obviously, it occurs owing to a partial destruction of wood substance by the microorganisms.

From the above mentioned data it follows that the destructing action of microorganisms on wood impregnated with pyrolysis oils in the case of *L. lepideus* is almost twice as intensive as in the case of *C. puteana*. According to comparative plots, both cultures cause some decrease in the volume of the wider pores (Fig. 2,A). However, in the case of *L. lepideus*, in contrast to *C. puteana*, a minor volume of pores in the radial range 1.5-2.1 nm appears.

After exposure of impregnated and leached samples to *C. puteana* and *L. lepideus*, the picture is being changed: in the case of *C. puteana*, they exhibit a dramatic increase in **A** (by 38.5%) and **a** (by 42%), respectively. Moreover, some increase in α is observed. A considerable alteration in the structure, i.e. a new pore volume in the radial range 1.8-2.5 nm (mesopores) appears (Fig. 2,B, 3,A). Obviously, the increase in the accessible surface of the specimen impregnated with oil after leaching facilitates the destruction process upon the contact with this microorganism.

On the contrary, in the case of *L. lepideus*, only minor changes occur in the **A** and **a** values (a decrease by 1.9% and 4.3%, respectively). The analysis of the structure indicates a decrease in the volume of wider pores and an increase in the pores volume in the radial range 1.3-1.5 nm (Fig. 2,B, 3,A).

Hence, the oil components removed in the EN 84 procedure, inhibiting the considerable activity of *L. lepideus* relative to impregnated wood, play some role in their vital activity. The dramatic increase in the activity of *C. puteana* after leaching may be explained by facilitating the diffusion of the fungicide in wood cell wall.

The specimens exposed to *L. lepideus* after the weathering procedure (by EN 73) do not practically differ in terms of structural characteristics from the specimens not subjected to this procedure (Table 4). In the case of *C. puteana*, some growth in the hydrophilic properties is observed, mainly owing to the appearance of a new surface (the **A** value increases almost twice), which is important for this culture from the viewpoint of facilitating diffusion, as has been suggested earlier.

As follows from Fig. 3,B, the specimen exposed to *C. puteana* is practically identical to the specimen before exposure, while, in the comparative plot of the specimen exposed to *L. lepideus*, there are bends indicative of a porous structure distinct from that of the non-exposed specimen.

CONCLUSIONS

The impregnation of wood with pyrolysis oils (by EN 113) decreases its hydrophilic properties mainly owing to the decrease in its swelling ability. The surface concentration of hydrophilic centres decreases negligibly.
In the evaporation procedure (by EN 73), in which the light components of oil are removed, there is practically no the change of cell wall structure, moreover, the concentration of the surface hydrophilic groups somewhat increases.
In the leaching process (EN 84), water-soluble oil components are removed, and the specific surface accessible to water as well as mass hydrophilicity increase.
L. lepidius has a destructive action upon oil-impregnated wood, which is twice as high as that in the case of *C. puteana*, i.e. in particular, specific surface is increased by 7% and 16%, respectively. In this case, the hydrophilic properties of the surface are changed not only at the expense of the loosening of the structure, but also owing to the increase in the surface concentration of the hydrophilic centres.
The activity of *C. puteana* relative to the sample after the leaching procedure tends to increase dramatically, i.e. specific surface and mass hydrophility tend to increase by 40% at some simultaneous increase in the concentration of the surface hydrophilic centres. At the same time, the activity of *L. lepideus* relative to this sample is negligible, i.e. specific surface is increased only by 2%.
The activity of both the cultures relative to the sample subjected to the evaporation procedure (EN 73) are close, i.e. the destructive action of *L. lepideus* is only by 10% stronger, in comparison with the case of *C. puteana*.
Pyrolysis oil has shown a protective effectiveness against wood-destroying basidiomycetes, so that it has a potential for use as a wood preservative. However, some properties of the oil need to be improved, particularly, the stability in wood after ageing procedures, the reduction of a high hydrophilicity and a high retention of the oil in wood.

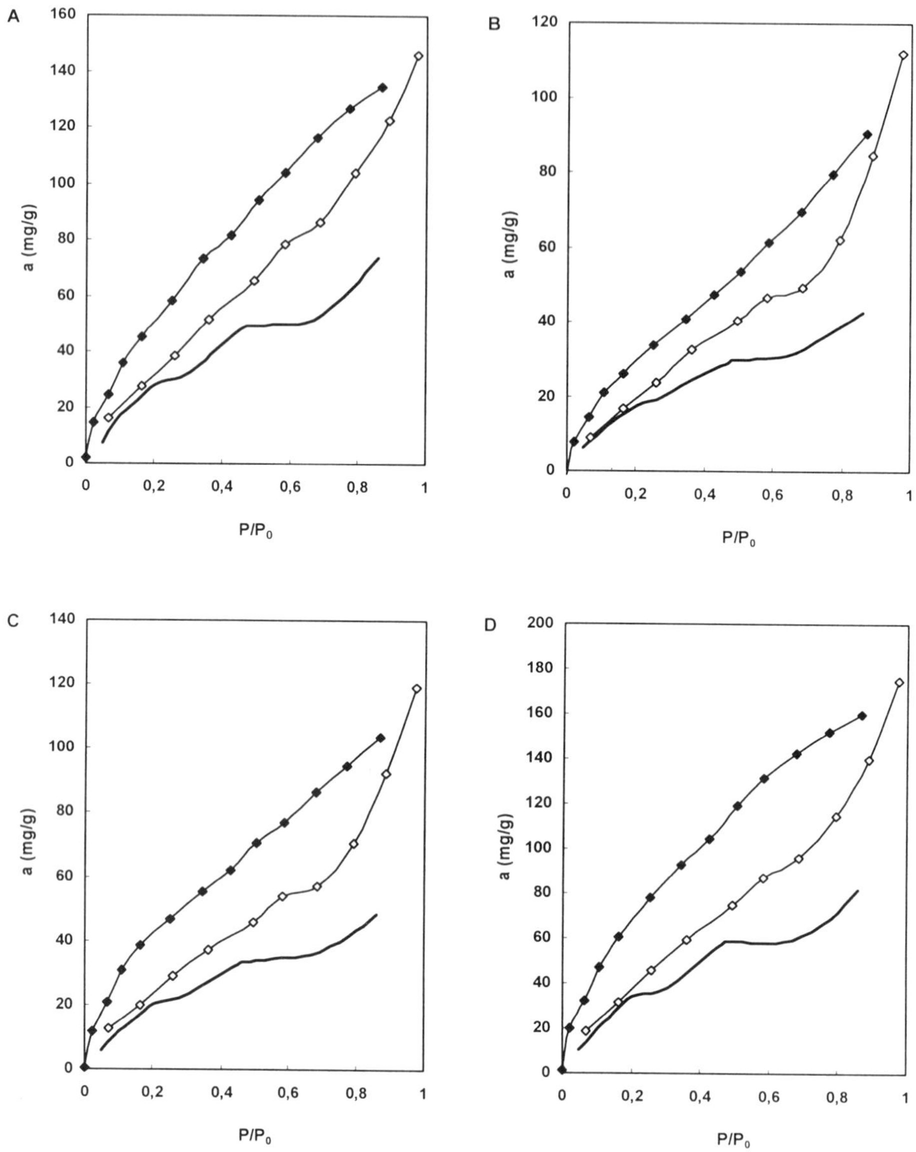

Fig. 1. Isotherms of water vapours sorption-desorption at 295K on pine wood after the procedure: A – water (EN 113); B – oil (EN 113); C – oil (EN 113) / *C. puteana*; D – oil (EN 113/EN 84) / *C. puteana* (bold line – 1st cycle; labels - 2nd cycle: empty labels - sorption; filled labels - desorption)

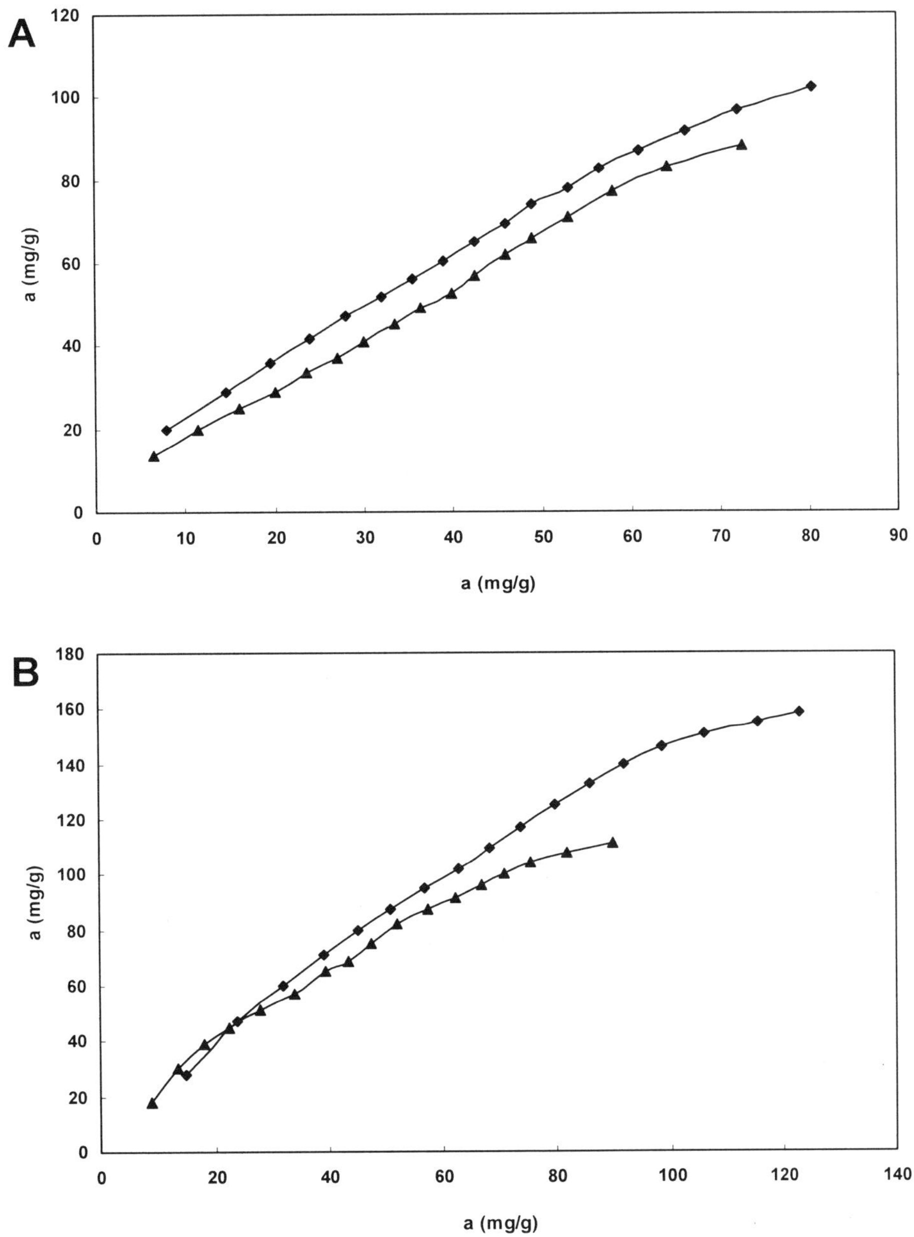

Fig. 2. Comparative plots of the desorption vs. sorption on the specimen:
"oil (EN 113)" (◆) and "oil (EN 113) - *C. puteana*" (▲) (A);
"oil (EN 113/EN 84)" (◆) and "oil (EN 113/EN 84) – *C. puteana*" (▲) (B)

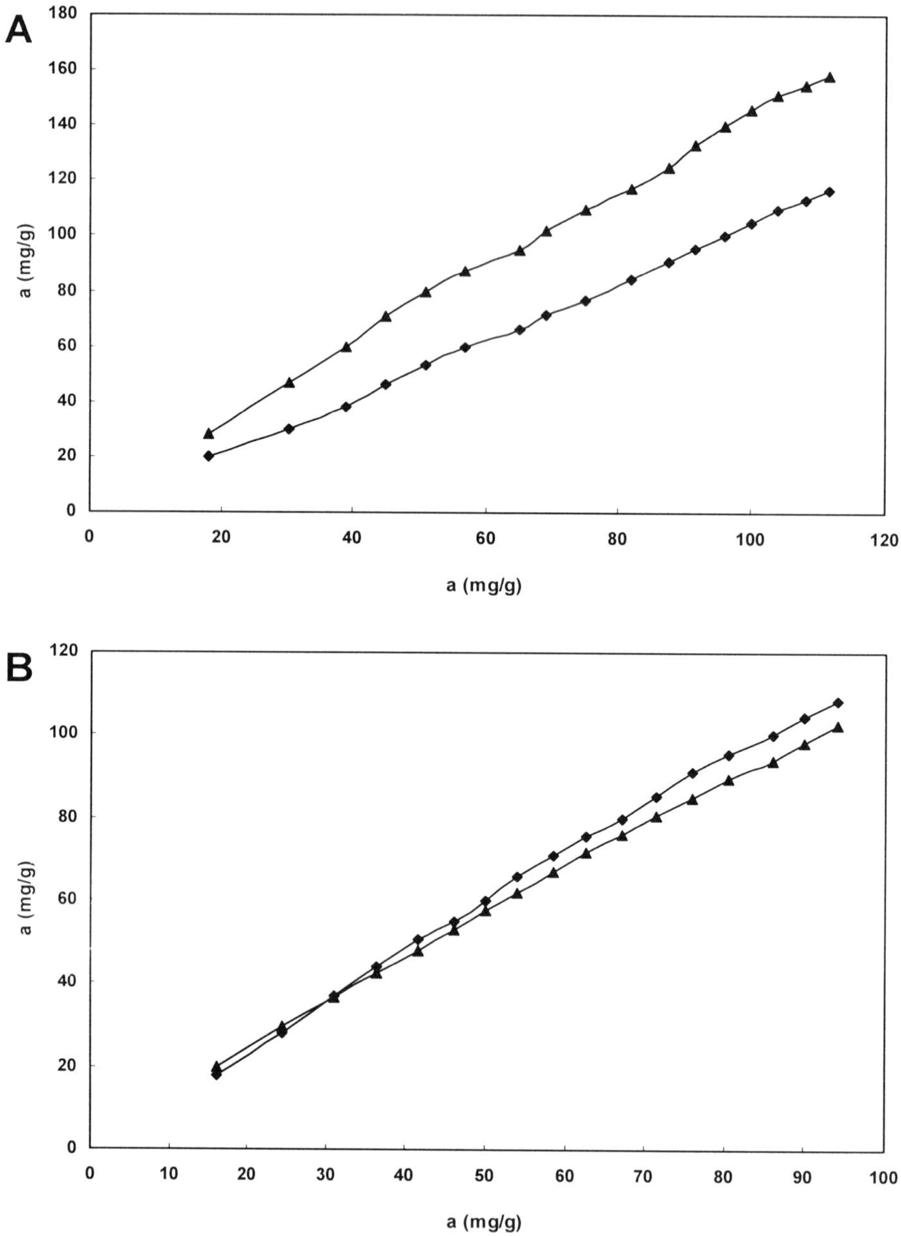

Fig. 3. Comparative plots of the water vapours desorption on the wood specimens: "oil (EN 113/EN 84 – *C. puteana*)" (▲) and "oil (EN 113/EN 84) – *L. lepideus*." (◆) to "oil (EN 113/EN 84) (A); "oil (EN 113/EN 73) –*C. puteana*" (▲) and "oil (EN 113/EN 73) - *L. lepideus*." (◆) to "oil (EN 113/EN 73) (B)

REFERENCES

1. Milton F.T. (1995) The Preservation of Wood. Minnesota Extension Service.
2. Official Journal of the European Communities, October 26,1999.
3. Richardson B.A. (1978) Wood Preservation. The Construction Press Ltd, Lancaster, England.
4. EU Directive 94/60/EG.
5. European Standard EN 113: 1996. Wood Preservatives. Method of Test for Determining the Protective Effectiveness against Wood Destroying Basidiomycetes. CEN, Brussels.
6. European Standard EN 84: 1990. Wood Preservatives. Accelerated Ageing of Treated Wood Prior to Biological Testing: Leaching Procedure. CEN, Brussels.
7. European Standard prEN 73: 1987. Wood Preservatives. Accelerated Ageing of Treated Wood Prior to Biological Testing: Evaporative Ageing Procedure. CEN, Brussels.
8. Gregg, S.J. & Sing K.S.W. (1982) Adsorption, Surface Area and Porosity, Academic Press, London.
9. Willeitner H. & Liese W. (1992) Wood Protection in Tropical Countries: a Manual on the Know-How. Deutsche GTZ GmbH.

Fractional Vacuum Pyrolysis of Biomass for High Yields of Phenolic Compounds

H. Pakdel, J. N. Murwanashyaka and C. Roy
Department of Chemical Engineering, Université Laval, Sainte - Foy, Québec Canada G1K 7P4

ABSTRACT: Phenolic compounds with potential industrial applications are major pyrolysis products of biomass lignin. Fractional pyrolysis allowed the evolution of chemical compounds such as phenols to be monitored as a function of temperature. Due to the less restrained pyrolysis conditions, phenolic compounds were produced in high yields by fractional pyrolysis in five steps. The pyrolysis oils were analysed and the phenolic compounds were quantified. The feedstock moisture and volatile components evolved during the first step below 200 °C. The pyrolysis oil obtained in the second step (200-275 °C) was composed of oxygenated compounds which were mainly produced by the degradation of cellulose and hemicelluloses. Low molecular weight carboxylic acids and aldehyde derivatives were abundant in the second pyrolysis step; they tended to polymerize with phenolic compounds if they were not quickly removed from the reactor hot zone. The pyrolysis oil concentrated with guaiacol and syringol-derivatives was obtained by the rupturing of β-O-4 (guaiacylglycerol-β-arylether) bonds during the third step in the temperature range of 275-350 °C. As the pyrolysis temperature increased, ArO-CH_3 bonds cleaved to form alkyl phenols and diols in the 350-450 °C temperature range in the fourth step. Polycyclic aromatic hydrocarbons were produced at temperatures above 450 °C by rupturing of the Ar-O bonds in the final step. Methanol detected in the final step was formed by demethoxylation reactions.

INTRODUCTION

In wood pyrolysis, it is known that several parameters influence the yield of pyrolytic oil and its composition. Among these parameters, wood composition, heating rate, pressure, moisture content, presence of catalyst, particle size and combined effects of these variables are known to be important[1]. The thermal degradation of wood starts with free water evaporation. This endothermic process takes place at 120 to 150°C, followed by several exothermic reactions at 200 to 250°C, 280 to 320°C, and around 400°C, corresponding to the thermal degradation of hemicelluloses, cellulose, and lignin respectively[2]. In addition to the extractives, the biomass pyrolytic liquid product represents a proportional combination of pyrolysates from cellulose, hemicelluloses, and lignin when these compounds are separately pyrolysed[3]. Pyrolysis of different wood components produces different classes of compounds. Cellulose for example, produces acetic acid, gases, water, sugars and occasionally a small amount of furans and phenols[4]. Hemicelluloses on the other hand principally produce acids, furans and sugars. Lignin is biosynthetically constructed by copolymerization of phenylpropanoid monomers namely coumaryl-, coniferyl- and sinapyl alcohols[5] which is principally decomposed to phenols and aromatic hydrocarbons during pyrolysis[6]. In addition to pyrolysates, free organic compounds or extractives present in wood are also released during pyrolysis, making pyrolytic oils a very complex product.

Pyrolysis under vacuum of aspen poplar wood chips exhibited an active zone of decomposition occurring between 240 and 300°C[7]. A 58 % by wt. yield of pyrolytic oil was obtained at 350°C which contained miscellaneous compounds such as phenols, carboxylic acids, alcohols and esters[7]. The vacuum process removes the reaction products before they are further decomposed[8]. The yield of relatively volatile phenols (monolignols in this paper) depends on the pyrolysis materials. The yield of phenols varied from 1 to 3.7 % when different biomass samples were subjected to vacuum pyrolysis[1].

Phenolic compounds typically possess anti-diarrheal and anti-motility properties[9], germicidal activity[10], herbicidal effect and antiseptic properties[11]. They have also been used in the tanning of leather[12], as dyes[13], as a thermal insulating material[14], as food aroma and liquid smoke[15]. In some cases, the smoke flavour has been attributed to the presence of phenols and sometimes to a single component. The phenols forming the "liquid smoke" which is used to smoke foods are well studied but their role and extent of flavouring have been clarified recently[16].

The present study was undertaken to determine the evolution of the main monolignols at different temperatures by fractional vacuum pyrolysis of birch wood.

MATERIAL AND METHODS

Birch (*Betula papyrifera*) wood chips used in this work were obtained from Scierie John Lewis, a woodmill in the province of Quebec, Canada. The sample was composed of 53.5 % of sapwood and 46.5 % of inner and outer bark. The moisture content was 8.5 % as determined by placing a known quantity of sample in an oven at 102 ± 3°C until a constant weight was reached. The elemental composition of the sample was 51.3 % C, 6.2 % H, 0.7 % N and 41.8 % O (by difference). Its ash content was 2.0 %.

FRACTIONAL VACUUM PYROLYSIS

A 800 g sample of the milled wood was pyrolysed in a batch reactor under vacuum (run # G72). A detailed description of the batch pyrolysis reactor system used in this work has been described elsewhere[17]. Two vacuum pumps in series were used to achieve a total pressure of 0.7 kPa and three dry-ice-in-limonene condensers (-72°C) were used to trap the pyrolysis vapours. When the maximum pyrolysis temperature was reached, it was held for 1 hour prior to cooling to room temperature. After the pyrolysis in each step, the system was kept under nitrogen until the next pyrolysis step was started. Each pyrolysis step was carried out by using the solid residue from the previous step. Table 1 shows the different pyrolysis steps and product yields. The pyrolysis oils which were recovered in the different traps were stored in sealed vials in the refrigerator (4°C) for further chemical analysis.

Table 1 Fractional pyrolysis conditions and yields (run # G72).

Pyrolysis step #	Temperature range (°C)	Oil yield (wt. %, anhydrous wood basis)
1	30-200	6.02
2	200-275	18.14
3	275-350	27.28
4	350-450	9.47
5	450-550	1.48
Total		62.39
One step	25-550	63.43

PYROLYSIS OIL CHEMICAL CHARACTERIZATION

The identification and quantification of the phenols were carried out using gas chromatograph coupled to a mass spectrometer (GC/MS) and using standard phenols. The pyrolysis oil was acetylated before analysis. A 200 to 300 mg sample of well mixed pyrolytic oil, together with 5 mL of acetic anhydride and two drops of pyridine, were heated in a sealed vial in a water bath at 60°C for 90 minutes. The derivatized solution was then eluted over approximately 2 g of silica gel with 100 ml of 80 % CH_2Cl_2 in hexane solution. The eluate was concentrated under vacuum. A 100 μl of a solution of 25 mg anthracene dissolved in 10 ml of ethylacetate was added to the pre-concentrated solution as an internal standard. Three sets of derivatized standard phenols diluted in 1:1, 1:2 and 1:3 volumes of ethylacetate were used for the determination of the absolute response factors of acetylated phenols with respect to anthracene for quantitative analysis. The GC/MS analysis was performed on a HP-5890 gas chromatograph with split injection at 290°C. The column was a 30 m x 0.25 mm i.d. HP5-MS fused silica capillary with 0.25 μm film thickness from Hewlett Packard. Helium was the carrier gas with a flow rate of about 1 ml min^{-1}. The GC initial oven temperature was held at 50°C for 2 min, then programmed to increase to 210°C at 5°C min^{-1} and then to 250°C at 10°C min^{-1}. The oven temperature was held at 250°C for 10 min. The end of the column was introduced directly into the ion source of a HP-5970 series quadruple mass selective detector. The transfer line was set at 270°C and the mass spectrometer ion source was set at 250°C with 70 eV ionization potential. A volume of 1 μl of sample was injected into the GC using a HP-7673 automatic sampler. Data acquisition was done with a PC base G1034C Chemstation software and a NBS library data base. The mass range of m/z = 30-350 Dalton was scanned every second. The identification of targeted compounds were based on NBS library mach values typically over 50 %. Furthermore, the identifications were confirmed by matching their mass spectra and retention times with the standard authentic compounds.

RESULTS AND DISCUSSION

FRACTIONAL PYROLYSIS

Compositional analysis of the pyrolysis oils at various steps allowed five distinct pyrolysis steps to be established in the temperature range of 200-550 °C as illustrated in Figure1. All five oil fractions were analysed by GC/MS. In the first step, about 6.02 % volatile compounds, which contained 99.9 % moisture, evolved at a temperature lower than 200 °C. That step involved mainly the sample drying process and partially the evaporation of wood natural extractives. Volatiles, hydrophilic and lipophylic compounds like terpenes and carboxylic acids have been detected earlier during the drying of fresh wood at about 200°C[18]. Table 2 lists the main components which were produced in the first step.

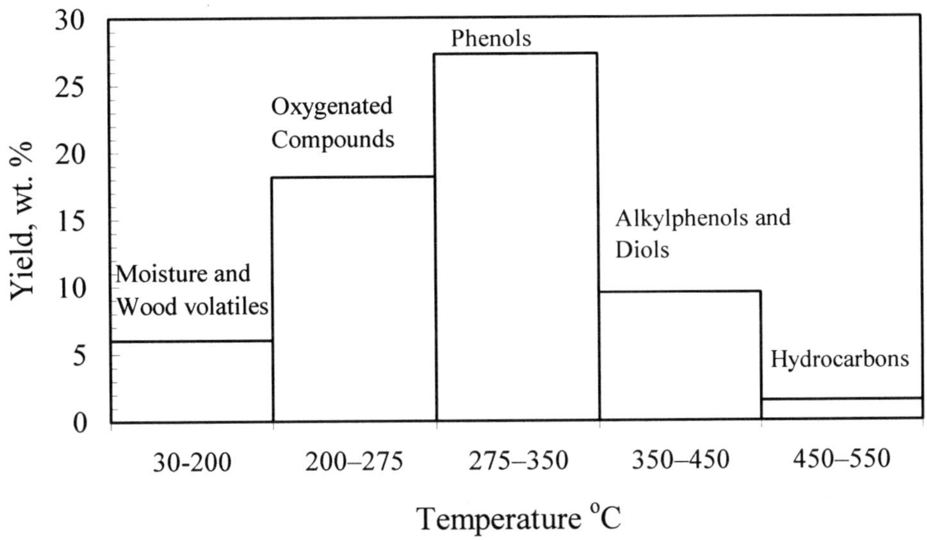

Fig. 1 Product distribution following fractional pyrolysis (wt. %, anhydrous wood basis).

The pyrolysis oil collected between 200 to 275°C in the second step was 18.14 % by wt. on an anhydrous wood basis and corresponded to the hemicelluloses degradation[19]. In general, hemicelluloses decompose to yield furans and its derivatives as well as a series of aliphatic carboxylic acids. Hardwoods are composed of pentosan-based hemicelluloses, while hexosan-based hemicelluloses are found in softwoods. Hardwood produces more low molecular weight carboxylic acids and furans than softwood upon decomposition[20]. In contrast, hardwood contains a lower quantity of fatty acids than softwood. C1 – C6 carboxylic acids, fatty acids (*e.g.* C_{18}), furan and derivatives were identified in this fraction. In the second pyrolysis step, a little rearrangement of the lignin polymer occurred which was based on the formation of small amount of phenolic compounds (0.395 % by wt., anhydrous wood basis). The main components of this fraction are listed in Table 2. Betulin and lupeol were abundant in this fraction. Furthermore, levoglucosan which has been reported earlier as a major compound in the cellulose-derived pyrolysis oils[8], was identified in the second and third fractions. Upon further heating, the fragmentation is followed by dehydration, disproportionation, decarboxylation and decarbonylation reactions. Several phenolic compounds as well as aromatic and cyclic hydrocarbons, aliphatic and cyclic alcohols, ketones, aldehydes, acids, esters and furans associated with the recondensation reaction of cellulose degradation products have been reported earlier[4].

A pyrolysis liquid yield of 27.28% was obtained in the third step in the temperature range of 275-350°C. This step corresponds mainly to the lignin decomposition where C-C bonds cleaved to form radicals, followed by recombination reactions. A 2.176 wt. % mixture of volatile phenols on an anhydrous initial feed basis was identified in this fraction. Guaiacyl and syringyl compounds like guaiacol, 4-methylguaicol, syringol, syringol, 4-methylsyringol, isoeugenol, allyl- and propenylsyringol were the main phenolic compounds. The guaiacol and syringol concentration leveled off at about 370°C and 350 °C respectively. Catechol and its derivatives (catechol, 3-methyl- and 4-methylcatechol, 3-methoxycatechol and resolcinol) were also identified in this fraction. The main components of this fraction are listed in Table 2.

The oil yield in the fourth step represented 9.47 wt. % of the initial anhydrous sample. In the fourth step, 1.845 wt % of volatile phenols on an anhydrous initial feed basis were produced. The content of catechol and its isomers were high, being the most abundant compounds, whereas the guaicyl and syringyl content were low in the fourth step. The main components of this fraction are listed in Table 3.

C-O bond cleavage of the remaining lignin residue in the temperature range of 450-550 °C, mainly resulted in the formation of some polycyclic aromatic hydrocarbons (PAHs) as listed in Table 3. The total yield of hydrocarbons produced in the final step was 1.48 wt. % on an initial anhydrous sample basis. The residual phenolic compounds represented < 0.1 % by weight in the final step. The principal components of this fraction are listed in Table 3. Interestingly, the contamination of pyrolysis oil with PAHs can be reduced without losing a significant mass of the pyrolysis oil if the pyrolysis reaction is stopped at approximately 450 °C.

Table 2 Pyrolysis oil composition as a function of pyrolysis temperature.

25- 200 °C	200-275 °C	275 – 350 °C
Acetic acid	Hydroxytetrahydrofuran	Hydroxytetrahydrofuran
Hydroxypropanone	Hydroxypropanone	Propanol
Pentanol	Hexanal	3-methylpropanol
Hexanal	Furaldehyde	Pentanoic acid
3-furaldehyde	Acetic acid	4-methylcyclopentanone
Hexan-1-ol	Furaldehyde	3-furaldehyde
Pentanoic acid	Dihydromethylpyran	2-furaldehyde
Hexan-3-ol	Propanone	2-furanmethanol
Ethylfuraldehyde	Pentanoic acid	5-methylfuranone
Hexanoic acid	Hexanoic acid	Tetrahydro-2,5-
Furanone	(*5H*)-furan-2-one	dimethoxyfuran
Methyldihydrofuran	2-hydroxycyclopent-2-en-1-one	Cyclopentanedione
Heptanoic acid	Methylfuraldehyde	2-cyclopentenone
trans-linaloloxide	Methylmethoxypropenoate	Furanone
Methylheptadienone	Hexanoic acid	2-hydroxycyclopenten-1-one
Heptanol	Dimethylfuranone	Methylfuranone

25- 200 °C	200-275 °C	275 – 350 °C
Octanol	3-methylcyclopentanedione	5-Methylfuraldehyde
1,4-dimethoxybenzene	Dimethylhydroxyfuran-3-one	Phenol
Heptan-3-one	Guaiacol	1,2-cyclohexanedione
Hexacosane	Maltol	3-methylcyclopentanedione
Heptacosane	Octan-3-one	Dimethylcyclopentenone
Methylbetulinate	4-hydroxybenzeneethanol	*o*-cresol
Methyloleanolate	4-methylguaiacol	*m*-cresol
	Catechol	2,5-dimethylhydroxyfuranone
	1,6-anhydro-α-D-glucopyranose	Guaiacol
	4-(2-propenyl)phenol	2-2-ethylcyclohexanone
	3-methoxycatechol	Maltol
	4-ethylguaiacol	3-ethylhydroxycyclopentenone
	4-methylcatechol	2,5-dimethylhydroxyfuranone
	5-acetoxymethylfuraldehyde	4-ethylphenol
	Syringol	Octanoic acid
	Eugenol	4-methylguaiacol
	Vanillin	Catechol
	Cis-isoeugenol	3-methoxycatechol
	4-methylsyringol	Naphthalene
	trans-isoeugenol	4-ethylguaiacol
	4-propylguaiacol	Hydroquinone
	Acetovanillone	4-methylcatechol
	Levoglucosan	1,2,4-benzenetriol
	2-propiovanillone	1-(3-methoxyphenyl)ethanone
	Hydroxyphenylbutanone	Syringol
	4-allylsyringol	Eugenol
	cis-4-propenylsyringol	3,4-dimethoxyphenol
	Syringaldehyde	4-propylguaiacol
	Trans-4-propenylsyringol	4-ethylcatechol
	1,2-dimethoxybenzene	1,2,3-benzenetriol
	Trihydroxyphenylpentanone	Vanillin
	Hexadecanoic acid	4-ethylsyringol
	Heneicosane	Levoglucosan
	9,12-octadecadienoic acid	1-(2,6-dihydroxy-4-methoxyphenyl)ethanone
	Octadecanoic acid	4-allylsyringol
	Docosane	*cis*-4-propenylsyringol
	Tricosane	Syringaldehyde
	Tetracosane	*trans*-4-propenylsyringol
	Pentacosane	1-(4-hydroxy-3,5-dimethoxyphenyl)ethanone
	Hexacosane	(2,4,6-trihydroxy-3-methylphenyl)butanone
	Heptacosane	Hexadecanoic acid
	Octacosane	
	Nonacosane	
	Lupeol	
	Betulin	

25- 200 °C	200-275 °C	275 – 350 °C
		Octadecadienoic acid
		Methyloctadece-9-noate

Table 3 Pyrolysis oil composition as a function of pyrolysis temperature.

350 – 450 °C	350 – 450 °C	450 – 550 °C
Toluene	Dihydroindenone	Dimethylnaphthalene
Octene	Tridecene	Pentadecene
Octane	4-methylcatechol	2-methylbiphenyle
2-methylpropanol	1-methylnaphthalene	Acenaphthene
Ethylbenzene	Tridecane	Pentadecane
1,2-dimethylbenzene	2-methylnaphthalene	4-methylbiphenyl
Styrene	1-(2-hydroxy-5-	Dibenzofuran
Nonene	methylphenyl)ethanone	1,4,6-trimethylnaphthalene
Nonane	1,3-dimethoxybenzene	1,6,7-trimethylnaphthalene
Methylcyclopentenone	2,3-dihydroindenol	Hexadecene
Ethylmethylbenzene	Syringol	Fluorene
Phenol	4-methylcatechol	Dimethylbiphenyl
Decene	3,4-dimethoxyphenol	Fluorenol
Trimethylbenzene	Decanoic acid	2-methylfluorene
Benzofuran	4-ethylcatechol	*o*-hydroxybiphenyl
Decane	Tetradecane	Azulene
4-methylguaiacol	1,4-dimethylnaphthalene	Ethyldimethylmethylfluorene
Trimethylbenzene	7-dimethylnaphthalene	Dimethylethylazulene
Indene	4-methylsyringol	Phenanthrene
o-cresol	1,2-dimethylnaphthalene	Anthracene
m-cresol	1,6-dimethylnaphthalene	2-phenylnaphthalene
Hexanoic acid	Pentadecene	2-methylanthracene
2-ethyl-1,3-	Pentadecane	1-phenylindene
Dimethylbenzene	Trimethylnaphthalene	2-phenylnaphthalene
Guaiacol	Hexadecene, Hexadecane	Eicosane
Undecane	2-methylnaphthalenol	3,6-dimethylphenanthrene
2-methylbenzofuran	Heptadecene	Pyrene
2-ethylphenol	Heptadecane	Heneicosene
Dimethylcyclohexanedi	1-methyl-7-(1-	Heneicosane
one	methylethyl)naphthalene	Fluoranthene
2,4-dimethylphenol	Otadecene	Benzo[*a*]fluorene
1-methylindene	Fluorenol	Benzo[*c*]fluorene
2,5-dimethylphenol	Octadecene	Methylpyrene
4-ethylphenol	Octadecane	*o*-terphenyle
Octanoic acid	Dihydromethylphenylbenz-	Triphenylene
1,4-dimethoxybenzene	ofurane	Benz[*a*]anthracene
Dodecene	Nonadecene	Dihydromethylbenz[*a*]anthracene
Methylguaiacol	Nonadecane	
Dodecane	Hexadecanoic acid	
Catechol	Eicosene	

350 – 450 °C	350 – 450 °C	450 – 550 °C
Trimethylphenol	Eicosane	
Dimethylbenzofuran	Heneicosene	
4-ethyl-2-methylphenol	Heneicosane	
4-propylphenol	9,12-octadecadienoic acid	
3-methylcatechol	Tetradecane	
2,3,6-trimethylphenol	2-methylanthracene	
Octanoic acid	2-phenylnaphthalene	
4-ethylguaiacol		

EVOLUTION OF THE PHENOLIC COMPOUNDS

At a pyrolysis temperature lower than 200 °C no phenolic compounds were detected. During the second pyrolysis step, 0.395 % phenolic compounds (anhydrous feedstock basis) including phenol and cresol (low quantities), allylphenol, guaiacol, 4-ethylguaiacol, eugenol, isoeugenol, vanillin, ethylvanillin, syringol, 4-methyl, 4-allyl and propenylsyringol were detected (Table 4). The presence of these phenols indicated the initiation of lignin degradation and disproportionation reactions following the drying of the wood in the first step. Similar compounds have been reported earlier during pyrolysis of a lignin model compound in the 230-260°C temperature range[21]. These compounds indicate the rupturing of bridge C-C bonds in aromatic compounds or β-ether or β-arylether bonds (β-O-4). These two former bonds possess a low energy of dissociation in the range of 217-235 kJ/mol[22]. The dissociation of β-ether bonds was accelerated as the pyrolysis temperature exceeded 300 °C and guaiacyl compounds were formed. The formation of guaiacol and derivatives ended at *ca.* 400°C or below. The evolution of phenols reached a maximum in the third step in the temperature range of 275-350 °C (see Table 4). Normally, the natural lignin contains no catechol unit, which leads to the conclusion that catechols are the secondary and guaiacol is the intermediate pyrolysis products which were formed at high temperatures. Petrocelli[23] emphasized that the secondary decomposition reactions of guaiacol to catechol at high temperature were favoured at a long holding time. The cleavage of the O-C bond in methoxy groups of guaiacol at high temperature reduced the guaiacol and increased the dihydroxybenzene content. In this work, catechol was detected at about 300°C and its concentration was increased considerably and ended at 500°C, the temperature at which the pyrolysis was almost over.

Pyrolysis of 4-ethylguaiacol yielded 4-ethylphenol by cleavage of the O-C (alkyl) and O-C (aryl) bonds[24]. Similarly, methyl-, dimethyl- and vinylphenols originated from guaiacol intermediates. It has also been reported that phenol is a secondary pyrolysis product of guaiacol and derivatives[23]. Phenol, cresol, ethylphenol, dimethylphenol, propylphenol and methylcatechol were the main components of the oil fraction in the fourth step. Traces of hydrocarbons such as toluene and styrene were produced by the dehydroxylation of phenols in this fraction.

The maximum phenolic compounds which could be produced and identified was about 2.5 wt. % on an anhydrous feedstock basis during the one-step pyrolysis approach in the temperature range of 25-550 °C. Interestingly, 4.4 wt. % phenols were identified following the fractional pyrolysis approach in a lower temperature range of 25-450 °C. One can conclude that the removal of volatile phenols during the fractional approach prevents some secondary thermal decomposition reactions which occur during the one-step pyrolysis process. These results are in agreement with the data reported earlier by Vohler and Schweers[25]. According to Avni et al.[26], at a low pyrolysis temperature, the decomposition of lignin leads to the release of CO_2, water and hydrocarbon gases from aliphatic and methoxy side chain groups. Aldehyde functional groups tend to decompose to CO. At high temperatures, thermal cracking reactions and rearrangements of lignin sub-units lead to the evolution of H_2 from aromatic hydrogen as well as additional CO from the tightly bonded oxygen functionalities such as diaryl ether and phenols in lignin. This information reveals that lignin is very sensitive to temperature. The lignin functional groups are easily decomposed or modified and C-C and C-O bonds were cleaved within the wide temperature range of 200 to 400°C. However, it is believed that a pyrolysis process can still be designed to efficiently degrade the lignin polymeric structure into simple phenols without modifying its original structure by preventing dehydroxylation, decarbonylation, demethoxylation and dehydrogenation reactions which occur during the thermal degradation process.

Catechol and its derivatives and alkylphenols were very abundant in the fourth step due to the decomposition of the primary phenols by demethylation reactions which require a higher dissociation energy (about 356-414 kJ/mole) than the β-arylether bonds. Ceylan and Bredenberg[27] have shown that 29-40% of guaiacol could be transformed into catechol after 2 h of pyrolysis at 305-345 °C. Similarly, 4-ethylguaiacol has been shown to transform to 4-ethylguaiacol by pyrolysis[28].

In rather restrained pyrolysis conditions, phenolic compounds produced at low temperatures were decomposed as they were exposed to high temperatures in the one-step pyrolysis (Table 4). Moreover, phenolic compounds in the acidic medium polymerize with aldehydes to form resins. The result was a progressive decline in the phenolic content. Under both fractional and one-step pyrolysis conditions, the oil yield was about the same (Table 1).

Figure 2 illustrates the evolution of both total phenols and pyrolysis water as a function of temperature. The evolution of water at temperatures above 200 °C was indicative of the dehydration reaction of oxygenated compounds during pyrolysis. The abundance of phenolic compounds declined during the dehydration reaction at temperatures above 350 °C. A similar result has been reported earlier in the literature[29].

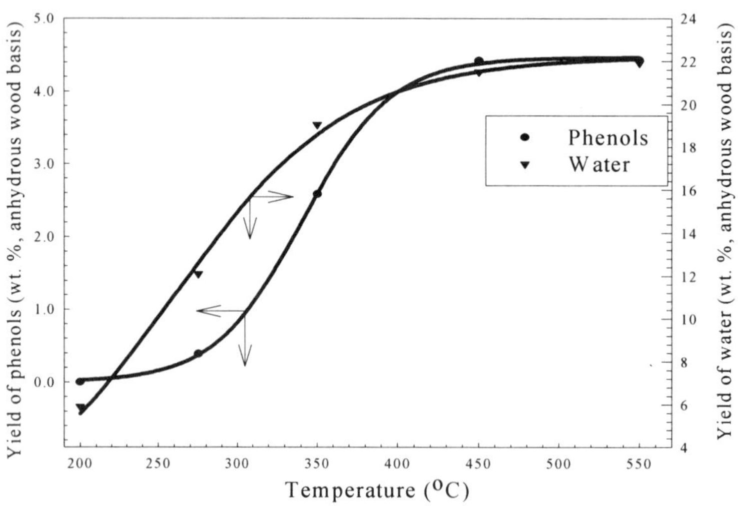

Fig. 2 Evolution of phenols and water as a function of pyrolysis temperature.

Table 4 Evolution of phenols as a function of pyrolysis temperature (°C).

Compound	Yield on an anhydrous feed basis (wt. %)					
	200-275	275-350	350-450	450-550	Total	25-550
Phenol	0.004	0.023	0.082	0.004	0.113	0.151
o-cresol	nd.	nd.	0.080	0.002	0.082	0.089
m-cresol	nd.	nd.	0.052	0.001	0.053	0.064
p-cresol	0.002	0.026	0.072	0.001	0.102	0.144
2,5-xylenol	nd.	nd.	0.011	0.000	0.012	0.014
2,3-xylenol	nd.	nd.	0.028	0.001	0.029	0.050
Xylenol	nd.	nd.	0.017	nd.	0.017	Nd.
4-ethylphenol	nd.	nd.	0.035	nd.	0.035	0.054
Guaiacol	0.017	0.251	0.041	nd.	0.309	0.182
3,5-xylenol	nd.	nd.	0.010	nd.	0.010	Nd.
4-allylphenol	0.010	nd.	nd.	nd.	0.010	Nd.
4-methylguaiacol	0.011	0.097	0.001	nd.	0.108	0.107
Catechol	0.011	0.197	0.331	0.001	0.539	0.246
Hydroquinone	nd.	nd.	0.064	nd.	0.064	Nd.
Resorcinol	0.005	0.024	0.048	nd.	0.078	0.034
4-ethylguaiacol	0.027	0.089	nd.	nd.	0.116	0.090
4-methylcatechol	0.005	0.014	0.178	nd.	0.197	0.103
Syringol	0.031	0.370	0.025	nd.	0.425	0.113
4-methylcatechol	0.005	0.066	0.200	nd.	0.270	0.108
Vanillin	0.012	nd.	nd.	nd.	0.012	Nd.
Methylresorcinol	nd.	nd.	0.058	nd.	0.058	Nd.
Dimethylcatechol	nd.	nd.	0.075	nd.	0.075	0.027
Dimethylcatechol	nd.	0.017	0.088	nd.	0.105	0.069
4-methylsyringol	0.014	0.143	0.028	nd.	0.184	0.089
Methoxycatechol	nd.	0.110	0.075	nd.	0.185	Nd.
Dimethylresorcinol	nd.	nd.	0.063	nd.	0.063	Nd.
Dimethylresorcinol	nd.	nd.	0.019	nd.	0.019	Nd.
Isoeugenol	0.069	0.020	0.012	nd.	0.101	0.069
4-ethylsyringol	0.011	0.057	0.017	nd.	0.085	0.029
Propylguaiacol	0.067	0.031	nd.	nd.	0.098	0.067
1,2,4-benzenetriol	nd.	0.128	0.067	nd.	0.195	0.034
Allylsyringol	0.012	0.119	nd.	nd.	0.132	0.093
Syringaldehyde	0.017	0.100	nd.	nd.	0.117	0.048
1,3,5-benzenetriol	nd.	0.110	0.061	nd.	0.171	0.273
Propenylsyringol	0.065	0.130	nd.	nd.	0.195	0.161
Methoxyresorcinol	nd.	0.056	0.007	nd.	0.063	nd.
Total	0.395	2.176	1.845	0.010	4.426	2.504

*nd. : not detected.

CONCLUSION

The thermal decomposition of wood under vacuum in a fractional pyrolysis approach can simplify the pyrolysis oil composition and facilitate the separation and purification of phenols produced during the pyrolysis. The evolution of pyrolysis water and phenols by fractional vacuum pyrolysis of birch wood bark exhibited a similar trend. The major volatile phenolic compounds were obtained in the temperature range of 200 to 350 °C following free radical C-O and C-C bond cleavage. A pyrolysis temperature higher than 350 °C was detrimental for the recovery of syringyl compounds which tended to decompose to catechol and its derivatives. Guaiacol and its derivatives were produced at a low temperature in the range of 250-350 °C followed by syringol and its derivatives. The transformation of the methoxy group to catechol was accelerated at temperatures higher than about 350 °C. Finally, the aromatic hydrocarbons were produced by dehydroxylation and demethoxylation reactions at temperatures higher than 450 °C. Under vacuum, the maximum yield of volatile phenolic compounds with 4.4 wt. % was achieved in the temperature range of 270-400°C following a fractional pyrolysis approach.

REFERENCES

1. Pakdel H., Roy C. & Lu X. (1996). Effect of Various Pyrolysis Parameters on the Production of Phenols from Biomass. In : *Developments in Thermochemical Biomass Conversion.* (Ed. by A. V. Bridgwater & D.G .B. Boocock), pp.509-522. Blackie Academic & Professional, London.
2. Fengel D. & Wegener G. (Eds) (1984) *Wood: Chemistry, Ultrastructure and Reactions.* Walter de Gruyter, Berlin.
3. Elder T. J . (1991) Pyrolysis of Wood. In : *Wood and Cellulosic Chemistry.* (Ed. by D. N. S. Hon & N. M. Shiraishi), pp. 665-702. Dekker Inc. New York.
4. Russell J.M., Miller R.K & Molton P. (1983) Formation of Aromatic Compounds from Condensation Reactions of Cellulose Degradation Products. *Biomass*, **3**, 43-57.
5. Monties B. (1989) Lignins. In : *Methods in Plant Biochemistry.* Vol.1 (Ed. by J. B. Harborne and M.P. Dey), pp. 113-157. Academic Press.
6. Alen R., Kuoppala E. & Oesch P. (1996) Formation of the Main Degradation Compound Groups from Wood and Its Composition During Pyrolysis. *J. Anal. Appl. Pyrolysis,* **36**, 137-148.
7. Roy C., de Caumia B., Brouillard D. & Menard, H. (1985) The Pyrolysis of Aspen Poplar. In : *Fundamentals of Thermochemical Conversion.* (Ed. by R. P. Overand, T. A. Milne & L. K. Mudge), pp. 237-255. Elsevier Applied Science Publication, New York.
8. Shafizadeh F. (1984) The Chemistry of Pyrolysis and Combustion. In: *The Chemistry of Solid Wood,.* (Ed. by R. Rowell), chapter 13. ACS, Washington D.C.
9. Ogata N. B. T. & Shibata T. (1993) Demonstration of Antidiarrheal and Antimotility Effects of Wood Creosote. *Phalmacol.* **46**, 173-180.
10. Guha G., D. Das P.D. Grover, & Guha, B.K. (1987) Germicidal Activity of Tar Distillate Obtained from Pyrolysis of Rice Husk. *Biol. Wastes.* **21**, 93-100.
11. Azhar L.P., Levin E.D. & Sokolova, N.A. (1972) Antiseptic Properties of Oil Obtained from Lignin Sedimentation Tar. *Izv.VUZ, Lesnoi Zh.* **15** , 118-20.

12. Ratner M.E., Smetanina S.S., Kovalev V.E. & Korotova, O.A. (1979) Preparation of Syntans Based Phenols of Soluble Resin and Wood-resin Pyrolysate. *Izv. Vyssh. Uchebn. Zaved., Lesn Zh*, 72-76.
13. Rijkuris A., Biseniece S. & Sergeeva, V.N. (1978) Composition and Properties of Tar Formed During Thermolysis of Lignocellulose. Preparation of Azo Dyes from Thermolysis Tar. *Khim. Drev*, **1:68**, CA. 90: 153480.
14. Garalevicius R., Medzevicius V., Roshchupkin V.I., Faintsimer R.Z., Yakolvev, D.A. & Romanauskas, A. (1978) Composition for Preparing a Heat-Insulating Material. USSR Patent SU 339173.
15. Guillen M.D. & Ibargoitia M.L. (1998) New Components with Potential Antioxidant and Organoleptic Properties, Detected for the First Time in Liquid Smoke Flavoring Preparations. *J. Agric.and Food Chem.* **46**, 1276 - 1285.
16. Wittkowski R., Ruther J., Drinda, H. & Rafiei-Taghanaki F. (1992) Formation of Smoke Flavor Compounds by Thermal Lignin Degradation. In: Flavor Precursors-Thermal and Enzymatic Conversions. (Ed. R. Teranishi, G. R. Takeoka & M. Guntert), pp. 232-243. ACS symposium series 490, American Chemical Society, Washington DC.
17. Pakdel H., Couture G. & Roy, C. (1994) Vacuum Pyrolysis of Bark Residues and Primary Sludges. *Tappi J.*, 77, 205-211.
18. Fagernas L. (1993) Formation and Behaviour of Organic Compounds in Biomass Dryers. *Bioresource Technology*, **46**, 71-76.
19. Goldstein I.S . (1980) Organic Chemicals from Cellulose. In: *Organic Chemicals from Biomass*. (Ed. I. S. Goldstein), pp. 101-124. CRC Press, Boca Raton.
20. Sarkanen K.V., Connors W.J., Johanson L.N. & Winslow, P. (1980) Thermal Degradation of Kraft Lignin in Tetralin. *Holzforschung,* **34**, 29-31.
21. Fiddler W., Doer R. C., Wasserman A. E. & Parker W.E. (1967) Thermal Decomposition of Ferulic Acid. *J. Agric. Food Chem.*, **15**, 757-761.
22. Benson S.W. (Ed) (1976) *Thermochemical Kinetics*, 2nd edn. Wiley, New York.
23. Petrocelli P. F. & Klein T. M. (1985) Simulation of Kraft Lignin Pyrolysis. In : *Fundamantals of Thermochemical Biomass Conversion*. (Eds R.P. Overend, T.A. Milne & L.K. Mudge), pp. 257-273. Elsevier Applied Science Publisher, London.
24. Connors W.J., Johanson L.N., Sarkanen K.V. & Winslow P. (1980) Thermal Degradation of Kraft Lignin in Tetralin. *Holzforschung*. **34**, 29-37.
25. Vohler W. & Schweers H.M. W. (1975) Utilization of Phenol Lignin. *Appl. Polym. Symp.*, **28**, 277-274.
26. Avni E., Suib L.S. & Coughlin W.R. (1985) Free Radical Formation in Lignin During Pyrolysis. *Holzforschung*, **39**, 33-40.
27. Ceylan R. & Bredenberg J.B. (1982) Hydrogenolysis and Hydrocraking of the Carbon-Carbon Bond. 2. Thermal Cleavage of the Carbon-Oxygen Bond in Guaiacol. *Fuel*, **61**, 377-382.
28. Masuku, S.P. (1991). Thermal Reactions of Bonds in Lignin. IV. Thermolysis of Dimethoxyphenols. *Holzforschung*, 45, 181-190.
29. Samolada M.C., Stoïkos T. & Vasalos I.A. (1990) An Investigation of the Factors Controlling the Pyrolysis Product Yield of Greek Wood Biomass in a Fluidised Bed. *J. Anal. Appl. Pyrolysis*, **18**, 127-141.

ACKNOWLEGEMENTS

The authors would like to thank Dr. Annette Schwerdtfeger for her technical assistance in the preparation of this manuscript.

Production of Hydrogen from Biomass-Derived Liquids

S. Czernik, R. French, C. Feik, and E. Chornet
Chemistry for BioEnergy Systems Center, National Renewable Energy Laboratory, 1617 Cole Boulevard, Golden, CO 80401

ABSTRACT Renewable biomass can become an attractive feedstock for producing hydrogen especially because of its essentially zero net CO_2 impact. Unfortunately, hydrogen content in lignocellulosic biomass is only 6-6.5% compared to almost 25% in natural gas. Therefore, on a cost basis, producing hydrogen by the biomass gasification/water-gas shift process cannot compete with the well-developed technology for steam reforming of natural gas. However, an integrated process, in which biomass is partly used to generate hydrogen and partly to produce more valuable materials or chemicals can be an economically viable option.

In the presented concept hydrogen is produced from lignocellulosic biomass by pyrolysis/steam reforming or by steam-aqueous fractionation/steam reforming processes that also generate valuable co-products. Effluents from other biomass processing technologies such as transesterification of vegetable oils can also be attractive feedstocks for the production of hydrogen. This work focused on the second stage of the process, catalytic steam reforming of various biomass-derived liquids. We employed a fluidized bed reactor configuration with commercial nickel catalysts developed for processing natural gas and naphtha. The hydrogen yields obtained approached or exceeded 90% of the values possible for stoichiometric conversion.

INTRODUCTION

Hydrogen is the most environmentally friendly fuel that can be efficiently used for power generation. At present, hydrogen is produced almost entirely from fossil fuels such as natural gas, naphtha, and inexpensive coal. In such case, the same amount of CO_2 as that formed from combustion of those fuels is released during the hydrogen production stage. Renewable biomass is an attractive alternative to fossil feedstocks because of essentially zero net CO_2 impact. However, the hydrogen content of lignocellulosic biomass is only 6-6.5%, compared to almost 25% in natural gas, and on a cost basis, producing hydrogen by a direct conversion process such as gasification/water-gas shift cannot compete with the well-developed technology for steam reforming of natural gas. Vegetable oils have a better potential for producing hydrogen than lignocellulosic materials but their high costs make the process economics not favorable either. Only an integrated process, in which biomass is partly used to produce more valuable materials or chemicals with the residual fractions utilized for generation of hydrogen, can be an economically viable option. The concept of our approach to producing hydrogen from biomass is shown in Figure 1.

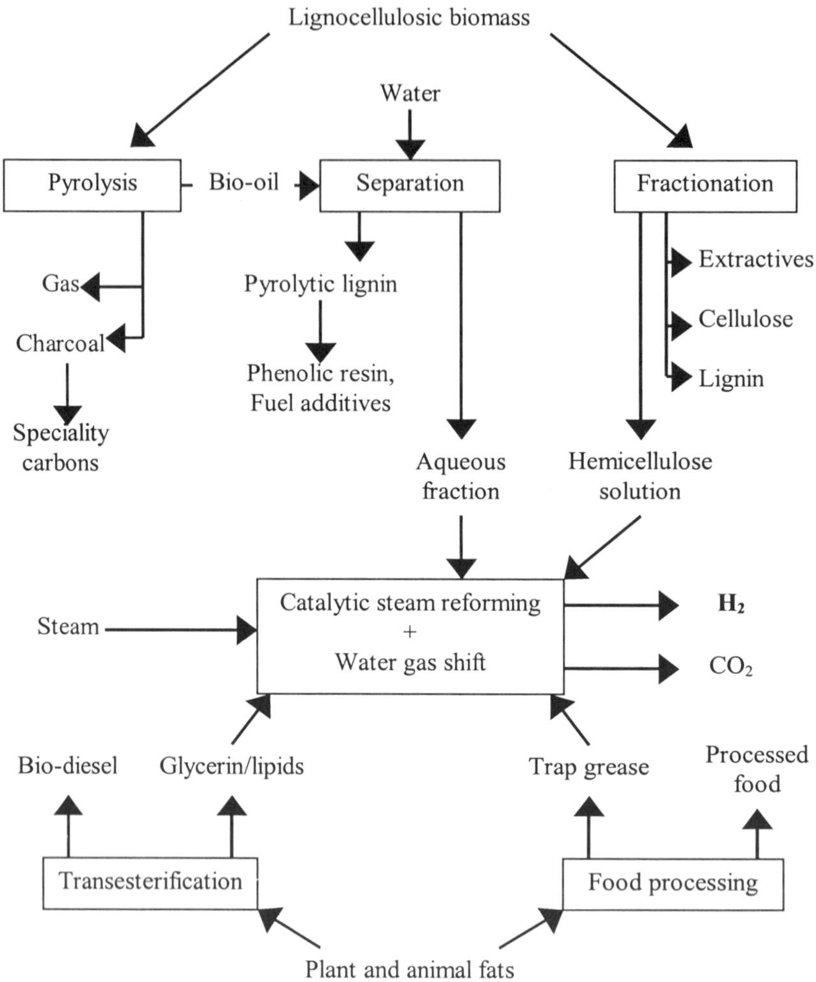

Figure 1. Biomass to hydrogen – Process concept

In earlier papers[1-3] we proposed a method, which combines two stages: fast pyrolysis of biomass to generate bio-oil and catalytic steam reforming of the bio-oil to hydrogen and carbon dioxide. This concept has several advantages over the traditional gasification technology. First, bio-oil is much easier to transport than solid biomass and therefore, pyrolysis and reforming can be carried out at different locations to improve the economics. For instance, a series of small-size pyrolysis units could be constructed at the sites where low cost feedstock is available then the oil transported to a central reforming plant located at a site with existing hydrogen storage and distribution infrastructure. The second advantage is the potential for production and recovery of higher-value co-products from bio-oil that could significantly impact the economics of the entire process. The lignin-derived fraction can be separated from bio-oil and used as a phenol substitute in phenol-formaldehyde adhesives[4] or converted to cyclohexyl ethers (fuel additives[5]) while the carbohydrate-derived fraction is catalytically steam reformed to produce hydrogen.

An alternative concept for producing hydrogen that we are developing combines steam-aqueous fractionation of biomass with catalytic steam reforming of the lower-value hemicellulose-rich liquid by-product while cellulose and lignin components would be used for other applications. The concept can be extended to steam reform other biomass-derived liquids that are by-products from existing technologies. In this work we studied "crude glycerin", a by-product from a bio-diesel plant based on transesterification of vegetable oils.

In previous years we demonstrated, initially through micro-scale tests[1] then in bench-scale fixed-bed reactor experiments[2,3] that bio-oil model compounds as well as bio-oil carbohydrate-derived fraction can be efficiently converted to hydrogen. Using commercial nickel catalysts the hydrogen yields obtained approached or exceeded 90% of those possible for stoichiometric conversion. The carbohydrate-derived bio-oil fraction contains substantial amounts of non-volatile compounds (sugars, oligomers) which tend to decompose thermally and carbonize before contacting the steam reforming catalyst. Even with the large excess of steam used, the carbonaceous deposits on the catalyst and in the reactor freeboard limited the fixed-bed reforming time to 3-4 hours. The limitations of the fixed-bed reactor were even more obvious for processing the whole bio-oil or the hemicellulose-rich solution from steam-aqueous fractionation. The hydrogen yield obtained from the whole oil was only 41% of that stoichiometrically possible and the reforming duration was less than 45 minutes. For this reason we decided to employ a fluidized bed reactor configuration that can overcome some limitations of the fixed-bed unit for this application. Though carbonization cannot be completely avoided, the bulk of the fluidizing catalyst remains in contact with the liquid droplets fed to the reactor.

EXPERIMENTAL

MATERIALS

Bio-oil used for this study was generated from pine sawdust using the NREL fast pyrolysis vortex reactor system[6]. The oil comprised 47.7% carbon, 7.4% hydrogen, and 44.8% oxygen with water content of 26.7%. It was separated into aqueous (carbohydrate-derived) and organic (lignin-derived) fractions by adding water to the oil in a weight ratio of 2:1. The aqueous fraction (75% of the whole oil) contained 21.8% organics ($CH_{1.25}O_{0.55}$) and 78.2% water.

Steam-aqueous fractionation of poplar wood was performed at the University of Sherbrooke, Canada using a continuous Stake II unit[7]. This treatment led to solubilization, after washing, of 30% of the biomass into a hemicellulose-rich aqueous solution, which contained 32.1% of solutes of the elemental formula $CH_{1.36}O_{0.67}$. These solutes were mostly oligomeric pentosan and a small amount of dissolved lignin.

"Crude glycerin" samples were obtained from West Central Co-op bio-diesel plant in Ralston, Iowa. Transesterification of vegetable oils with methanol produces a mixture of bio-diesel (methyl esters of fatty acids) and glycerin. Glycerin settles down at the bottom of a separation tank while bio-diesel forms the top layer. "Crude glycerin" is a very viscous liquid, only partially miscible with water. Its elemental composition includes 54.7% carbon, 9.9% hydrogen, and 35.5% oxygen, which suggests that the phase separation in the tank was not very clean and the liquid is a mixture of glycerin (55%) with methyl esters of fatty acids (45%).

C11-NK, a commercial nickel-based catalyst used for steam reforming of natural gas and naphtha, was obtained from United Catalysts and ground to the particle size of 300-500μ.

FLUIDIZED BED REFORMER

The bench-scale fluidized bed reactor is shown in Figure 2. The two-inch-diameter inconel reactor supplied with a porous metal distribution plate was placed inside a three-zone electric furnace. The reactor contained 150-200g of commercial nickel-based catalyst ground to a particle size of 300-500μ. Superheated steam was used as fluidizing gas as well as a reactant in the reforming process. Steam was generated in a boiler and superheated to 750°C before entering the reactor at a flow rate of 2-4 g/min. Liquids were fed at a rate of 4-5 g/min using a diaphragm pump. A specially designed injection nozzle supplied with a cooling jacket was used to spray liquids into the catalyst bed. The temperature in the injector was controlled by coolant flow and maintained below the feed boiling point to prevent evaporation of volatile and deposition of nonvolatile components.

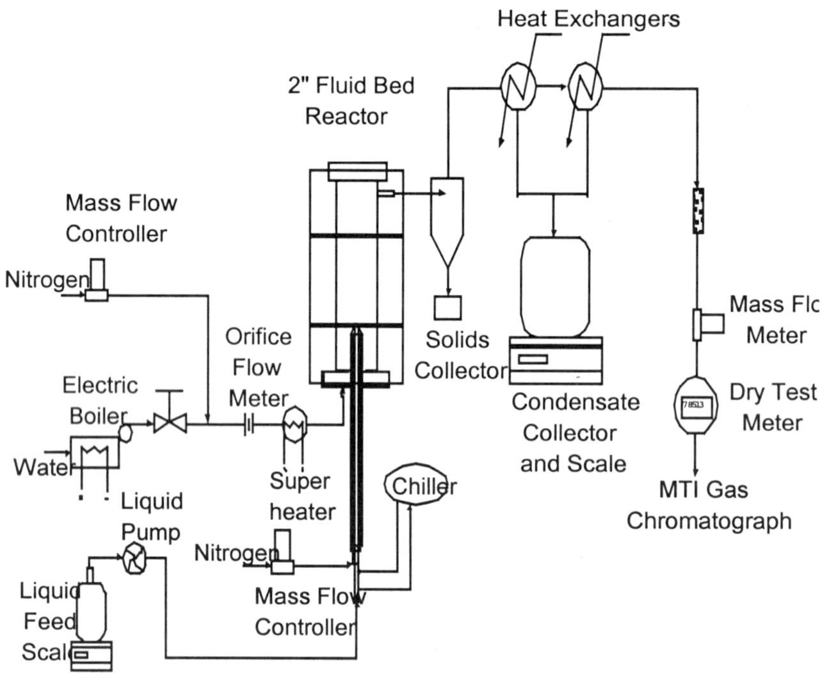

Figure 2. Fluidized bed reformer system

The product gas passed through a cyclone that captured fine catalyst particles and any char generated in the reactor, then two heat exchangers to remove excess steam. The condensate was collected in a vessel whose weight was continuously monitored. The outlet gas flow rate was measured by a mass flow meter and by a dry test meter. The concentrations of CO_2, CO, and CH_4 in the reforming gas composition were monitored by a non-dispersive infra-red analyzer (NDIR Model 300 from California Analytical

Instruments) and that of hydrogen by a thermal conductivity monitor TCM4 manufactured by Gerhard Wagner. In addition, the gas was analyzed every 5 minutes by an on-line MTI gas chromatograph, which provided concentrations of hydrogen, carbon monoxide, carbon dioxide, methane, ethylene, and nitrogen as a function of time of the test. The temperatures in the system as well as the flows were recorded and controlled by the OPTO data acquisition and control system. Total and elemental balances were calculated as well as the yield of hydrogen generated from the feed.

RESULTS AND DISCUSSION

BIOMASS PYROLYSIS OIL – AQUEOUS FRACTION

The steam reforming experiments on aqueous extract of the pine bio-oil were carried out at the temperature of 800°C and 850°C. The steam to carbon ratio was in the range of 7-9 while methane-equivalent gas hourly space velocity $G_{C1}HSV$ was 1200-1500 h^{-1}. During the experiments at 800°C a slow decrease in the concentration of hydrogen and carbon dioxide and an increase of carbon monoxide and methane in the gas generated by steam reforming of the carbohydrate-derived oil fraction was observed. These changes resulted from a gradual loss of the catalyst activity, probably due to coke deposits. As a consequence of that, the yield of hydrogen produced from the oil fraction decreased from the initial value of 85% of stoichiometric (3.24 g of hydrogen from 100 g of feed) to 77% after 12 hours on stream. If a water-gas shift reactor followed the reformer the hydrogen yields would be 94% and 84% respectively.

At 850°C the formation of char and coke was much lower or their gasification by steam was more efficient than that at 800°C. During over 90 hours of uninterrupted reforming of the bio-oil carbohydrate-derived fraction, the product gas composition remained almost constant and only a small decrease in the concentration of hydrogen was observed, as presented in Figure 3.

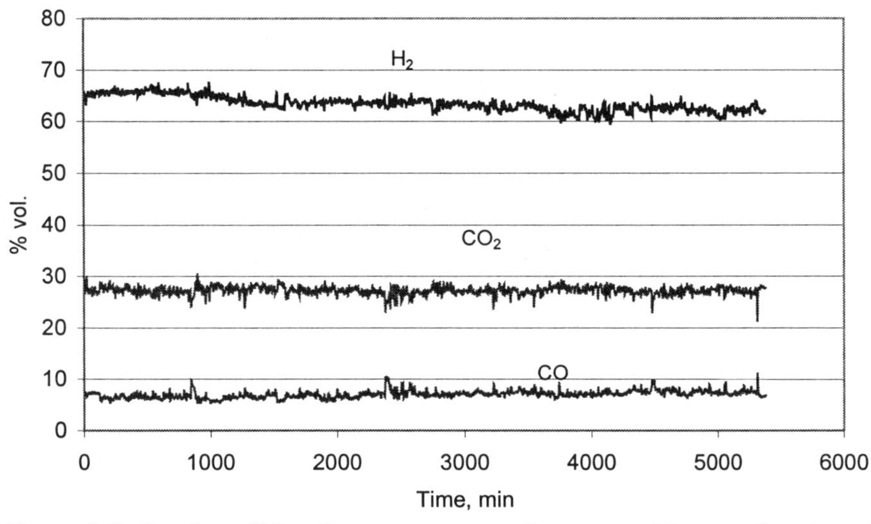

Figure 3. Reforming of bio-oil aqueous extract. Gas composition as a function of process time.

HEMICELLULOSE-RICH AQUEOUS SOLUTION

The steam reforming experiments were carried out at the temperature of 800-850°C with the hemicellulose solution feed rate of 240-300 g/h and steam flow of 140-180 g/h. This corresponds to the methane equivalent space velocity $G_{C1}VHSV$ of 1000 h^{-1} and the molar steam to carbon ratio S/C of 9.5-14. The product gas composition is shown in Figure 4. For over five hours the reforming gas composition was almost constant. However, after 2.5 hours on stream at 800°C the amount of gas generated started to decline. Consequently, the hydrogen yield, which at the beginning reached 85% of the stoichiometric potential decreased to 55% (Figure 5). Increasing the temperature to 850°C and the steam flow (S/C=14.2) resulted in the improvement of the catalyst performance.

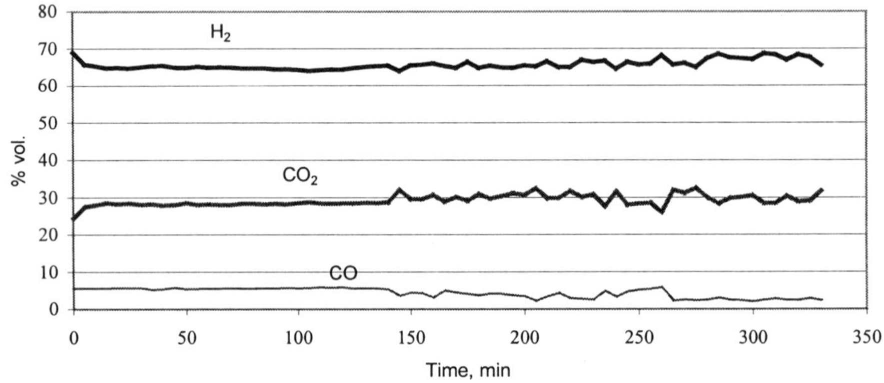

Figure 4. Reforming of hemicellulose aqueous solution. Gas composition as a function of process time.

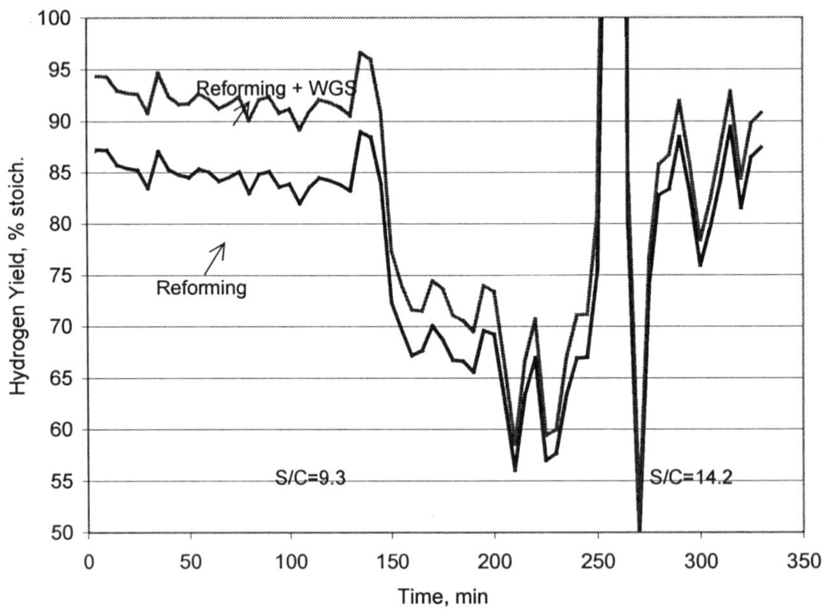

Figure 5. Yield of hydrogen from steam reforming of hemicellulose solution.

Mass balances indicated that 85% of carbon from hemicellulose was converted to CO_2 and CO in the first phase of the experiments. The remaining 15% could thus form char entrained from the system and coke deposits on the catalyst surface, which would explain the loss of its activity. The activity of the catalyst used for reforming was restored by steam gasification of the deposits and the catalyst was reused in the next experiments showing almost the same efficiency. Though the carbon to gas conversion reached again 85%, the hydrogen yield was in the range of 75-80% (85% initially) of the stoichiometric value and more methane was observed in the produced gas than during the tests using fresh catalyst. This indicates some permanent loss of the catalyst activity.

In general, hemicellulose is more difficult to reform than bio-oil aqueous fraction due to a higher content of oligomeric material that tends to carbonize during the process. Therefore, it requires a higher steam to carbon ratio to reduce carbon deposits on the catalyst.

CRUDE GLYCERIN FROM BIODIESEL PRODUCTION

"Crude glycerin" is a high viscosity liquid and, therefore, it had to be preheated to facilitate pumping and atomizing (the whole feeding line was maintained at 60-80°C). The liquid was fed at a rate of 78 g/h (G_{C1}VHSV = 1600 h^{-1}) and steam at a rate of 145 g/h, which corresponds to the molar steam to carbon ratio of 2.3. The experiments proceeded very smoothly with only occasional fluctuations in the liquid feed rate resulting from a non-complete homogeneity of the feed. The concentration of the major gas products was constant during the run time but a gradual increase in methane production was noticed (Figure 6). The process performance measured as the yield of hydrogen did not decrease significantly during several hours on stream. The overall mass balance closure was close to 100% at the beginning of the tests then decreased to 95-96% after four hours on stream. Similar closure was also observed for elemental balances of carbon, hydrogen and oxygen.

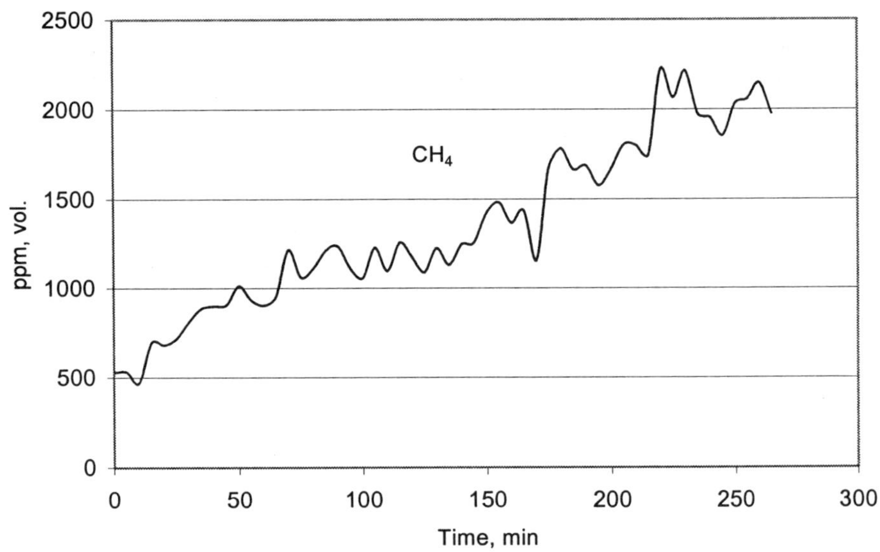

Figure 6. Methane concentration during steam reforming of "crude glycerin"

The hydrogen yield oscillated around 77% of the stoichiometric potential, which was 23.6 g per 100 g of feed. It could be significantly higher if more steam were used in the reaction. Conversion of CO in the gas through water-gas shift to CO_2 and H_2 would increase the hydrogen yield to 95% of that theoretically possible. These promising results suggest that a low-value by-product from bio-diesel production could become a viable renewable raw material for producing hydrogen. An integration of these two technologies could significantly improve the economics of both processes.

SUMMARY AND CONCLUSIONS

1. Biomass can be a valuable resource for producing hydrogen if utilized in an integrated process that also generates higher value co-products. Following this strategy we have considered three process options: fast pyrolysis/steam reforming, steam-aqueous fractionation/steam reforming, and transesterification of vegetable oils/steam reforming.
2. Bio-oil from pyrolysis or its aqueous, carbohydrate-derived fraction, hemicellulose-rich solution from steam-aqueous fractionation, and glycerin from bio-diesel production can be catalytically steam reformed to generate hydrogen using commercial nickel-based catalysts.
3. The hydrogen yield obtained in a fluidized bed reactor from the aqueous fraction of bio-oil was about 85% of the stoichiometric value, which corresponds to almost 6 kg of hydrogen from 100 kg of wood.
4. Hydrogen yield from the hemicellulose solution was about 70% of the stoichiometric potential. Lower performance was due to the higher content of oligomeric material, which is more difficult to reform.
5. The hydrogen yield from "crude glycerin" was 18 g per 100 g of the feedstock, which corresponds to 76% of the stoichiometric potential. If the steam reforming were followed by a water-gas shift process or a higher amount of steam were used in the reforming stage, the hydrogen yield could increase above 90%.
6. Catalysts are readily regenerated by steam or CO_2 gasification of carbonaceous deposits.
7. The process needs to be optimized to determine conditions that allow for maximum yields of hydrogen and minimum coke formation.

REFERENCES

1. Wang, D., Czernik, S., Montané, D., Mann, M., and Chornet, E. (1997) Biomass to hydrogen via pyrolysis and catalytic steam reforming of the pyrolysis oil and its fractions. *I&EC Research*, **36**, 1507-1518.
2. Wang, D.; Czernik, S., and Chornet, E. (1998) Production of hydrogen from biomass by catalytic steam reforming of fast pyrolysis oils. *Energy & Fuels*, **12**, 19-24.
3. Czernik, S., French, R., Feik, C, and Chornet, E. (1999) Fluidized bed catalytic reforming of pyrolysis oils for production of hydrogen. In: *Proceedings of the Fourth Biomass Conference of the Americas,* (Ed. by R.P. Overend and E. Chornet), pp. 827-832. Elsevier Science Ltd., Oxford.
4. Kelley, S. S., Wang, X.-M., Myers, M. D., Johnson, D. K., Scahill, J. W. (1997) Use of biomass pyrolysis oils for preparation of modified phenol formaldehyde resins. In: *Development in Thermochemical Biomass Conversion*, (Ed. By

Bridgwater, A.V. and Boocock, D.G.B.), pp. 557-572. Blackie Academic & Professional, London.
5. Shabtai, J.S., Zmierczak, W., and Chornet, E. (1997) Conversion of lignin to reformulated gasoline compositions. In: *Proceedings of the Third Biomass Conference of the Americas,* (Ed. by Overend, R.P. and Chornet, E.), pp. 1037-1040. Elsevier Science Ltd., Oxford.
6. Diebold, J.; Scahill, J. (1988) Production of primary pyrolysis oils in a vortex reactor. In: *Pyrolysis Oils from Biomass: Producing, Analyzing and Upgrading*; (Ed. by Soltes, E. J. and Milne, T. A.), pp. 31-40. American Chemical Society, Washington, D.C.
7. Heitz, M., Capeck-Ménard, E., Koeberle, P.G., Gagné, J., Chornet, E., Overend, R.P., Taylor, J.D., and Yu, E. (1991) *Bioresource Technology*, **35**, 23-32.

Co-Firing of Bio-Oil with Simultaneous SO_x and NO_x Reduction

R.H. Venderbosch[1], B.M. Wagenaar[1], E. Gansekoele[1], S. Sotirchos[2], H.D.T. Moss[3]

[1] *BTG Biomass Technology Group BV, P.O. Box 217, 7500 AE Enschede, The Netherlands*

[2] *Foundation for Research and Technology – Hellas, Institute of chemical engineering and high temperature chemical processes, P.O. Box 1414, GR - 265 00 Patras, Greece*

[3] *DynaMotive Europe Ltd., 33 Waterloo Road, Bedfort, UKGB - MK40 3PQ, United Kingdom*

ABSTRACT Calcium-enriched bio-oil (CEB) can be used for flue gas desulfurisation in coal and waste combustion chambers. It is produced by mixing biomass derived fast pyrolysis oil with calcium oxide. The aim of the proposed project is to develop a technology i) to produce calcium-enriched bio-oil with a calcium content of 13 wt.%, and ii) to test the CEB in a combustion chamber by co-firing it with a sufur-containing fuel. In this paper the production method of CEB will be elucidated, and small-scale experiments related to CEB spraying will be presented. Finally, co-combustion experiments of a sulfur-containing fuel with CEB in a small flame tunnel (20 kW_{th}) will be reported.

INTRODUCTION

Industrialised nations impose strict limits on emissions of pollutants such as SO_x and NO_x. Guidelines referred to by the World Bank Pollution Prevention and Abatement Handbook for fossil fuel based thermal plants of 50 MW_e or greater, aim at a 40% emissions reduction in NO_x. In the USA, for example, the 1990 Clean Air Act Amendments require a 2-stage drop in the level of pollution. It aims at a reduction in SO_2 emissions for coal and oil fired boilers by 10 million tons and for NO_x by 2 million tons by the year 2000 (vs 1980 levels). National standards for SO_2 became increasingly stringent throughout the 1980's and 1990's: a range of national emission standards for sulfur emissions in almost 40 countries is given by Soud (3).

Technologies to control emissions of SO_2 and NO_x include flue gas desulphurisation (FGD) for SO_2 control and NO_x emissions abatement and control by primary measures, or by flue gas treatment. Flue gas desulfurization can be classified into the following six main categories:

- wet scrubbers;
- spray dry scrubbers;
- sorbent injection processes;
- dry scrubbers;
- regenerable processes; and
- combined SO_2 / NO_x removal processes

An overview of these technologies can be found on the website of the IEA Coal Research reviews (1) on FGD technologies (including low cost retrofit options) and from the reviews provided by Soud (2, 3). In the present paper more details will be presented about sorbent injection techniques.

SORBENT INJECTION

Depending on the location (and temperature) where the sorbent is injected, there are four types of injection techniques:

- furnace sorbent injection (750-1250°C);
- economiser sorbent injection (300-650°C);
- duct sorbent injection (approx. 150°C); and
- hybrid sorbent injection (a combination of two or more of the above).

The simplest flue gas desulfurisation technology is furnace injection, where a dry sorbent is injected into the upper part of the furnace to react with the SO_2 in the flue gas. The finely grained sorbent is distributed quickly and evenly over the entire cross section in the upper part of the furnace in a location where the temperature is in the range of 750-1250°C. Commercially available and cheap limestone ($CaCO_3$) or hydrated lime ($Ca(OH)_2$) is used as sorbent. The sorbent reacts with SO_2 and O_2 to form $CaSO_4$. Below 750°C the reaction rate is too low. At temperatures over 1250°C the surface of the sorbent will be sintered, and the structure of the pores will be destructed, reducing the active surface area. The major part of SO_2-removal takes place within 1 to 2 seconds.

For $Ca(OH)_2$, removal efficiency of up to 50% are reported at a Ca:S ratio of 2. For $CaCO_3$, the removal efficiency will be even lower. Fine sorbent particle size (<5 μm) and an even distribution of the sorbent over the cross-section of a boiler significantly improves the process performance. Bjerle and others (1993), for example, reported results of laboratory scale experiments showing that SO_2-removal efficiencies > 95% can be achieved with furnace sorbent injection with particle size <3 μm.

At lower temperatures hydrated lime, $Ca(OH)_2$, can be injected into the flue gas stream near the economiser zone (300-650°C). In this temperature range, $CaCO_3$ can be formed, which is undesirable because it not only consumes sorbent but pore closure also blocks the access of SO_2 to the active sorbent surface. Carbonation significantly increases with reaction temperature, and therefore, the flue gas duct process where the temperature is about 150°C, may be more effective. This process yields SO_2-removal efficiencies of approx. 80% in actual commercial installations if small particles with an open pore structure are applied.

The aim of the present work is the development of a Calcium Enriched Bio-oil (CEB) that it is suitable to be co-fired with a sulfur containing fuel in a combustor to simultaneously reduce SO_2 and NO_x. Where the capture of NO_x is important, CEB has greater flexibility. Additionally, CEB will reduce the higher N_2O-concentrations associated with this technology. The data below have been abstracted from the IEA Coal Research publication entitled "Air Pollution Control Costs for Coal-Fired Power Stations" (1995).

Table 1 Investment and operating costs for various de-SO_x and de-NO_x processes

Technology	Capital cost US$/kWe	Avg 15yr cost US$/t	SO_2 % Red.	NO_x % Red.
Wet FGD	180 - 260	460 - 580	90	
Dry FGD	140 - 170	430 – 480	90	
Selective Cat. Red.	100 - 150	1800 – 2000		80
Selective Non Cat. Red.	10 - 25	926 – 1352		30 - 50
Gas reburn	14 - 18	1200 - 2400		60
Low NO_x burner (LNB)	20 - 30	629 – 889		30 - 60
Overfire air (OFA)	10 - 20	439		10 - 25
LNB - OFA	30 - 45	592 - 1065		40 - 60
CEB (7.3% Ca)	2 - 5	726	90	40
CEB (11.0% Ca)	2 - 5	562	90	40

Table 1 demonstrates that the use of CEB can be of commercial interest. It may provide NO_x/SO_2 control at costs well below those of conventional emission control systems and imply significant savings on up-front capital expenditure. Additionally, it has the advantage of reducing the production of greenhouse gases by substitution of fossil fuels.

CALCIUM ENRICHED BIO-OIL

CEB is produced by reacting biomass fast pyrolysis oil with CaO. The desulphurization technology is applicable both for coal and for heavy fuel oil fired combustion chambers. The installed coal-fired power station capacity in Europe approximates 75,000 MW_{el} but this is decreasing. The sulphur content of the coal ranges from 0.5 to 3.0 wt.% depending on the mining site. On the other hand, waste and orimulsion firing is increasing in capacity.

PREVIOUS WORK

First experiments using CEB were carried out in a 90 kW_{th} Down-Fired Coal Combustor at Penn State (4, 5, 6, 7) These experiments showed that it can be sprayed with slightly modified spray nozzles. Atomisation of a CEB-like material yielded droplets < 40 μm for 70 to 75 wt.% of the material, with all droplets < 170 μm. A 5 wt.% Ca-containing material was tested in the combustion chamber, showing a significant SO_2 and NO_x reduction potential. The SO_2 reduction appeared to be a strong function of the temperature of the combustor, and values of up to 94 % were reported at a Ca:S ratio of 1. The optimum wall temperature for SO_2 capture was observed to be around 1150°C, above which the performance dropped. The NO_x reduction appeared to be a strong function of the CO concentration, which in turn depends on the oxygen concentration. The location of the introduction of additional air for complete combustion was an important factor to balance NO_x reduction and to minimise CO emissions. A reduction in NO_x levels of up to 44 % was observed at Ca:S ≈1.

PRESENT WORK

In the present work, CEB with a calcium content of 13 wt.% (higher than in the previous work referred to above) will be tested as a flue gas desulfurisation agent in a small (25 kW_{th}) and a large combustion (7 MW_{th}) facility. The work discussed in this paper focuses on the demonstration of CEB, and includes:

- Production of bio-oil and optimisation of the oil;
- Production of CEB and its characterisation, and
- Demonstration of CEB to remove SO_2 in a small scale 25 kW_{th} flame tunnel.

SMALL SCALE PRODUCTION OF BIO-OIL AND OPTIMISATION

Upon heating biomass to about 650°C in the absence of oxygen, bio-oil is produced with a possible yield of > 75 wt.% of the feedstock material. This oil contains high concentrations of aldehyde- and carboxyl-groups, which can react with lime to produce the CEB. The calcium uptake of the bio-oil is proportional with the aldehyde and carboxyl concentration of the bio-oil. Both are dependent on the pyrolysis process conditions. The first specific objective was to maximise the amount of aldehyde and carboxyl groups in the bio-oil, and to produce sufficient and reproducible amounts of bio-oil under the best optimal process conditions.

The optimisation of the bio-oil was aimed at adjusting the operating conditions of the fast pyrolysis process to maximize the concentration of these reactive groups in the bio-oil, while maintaining a high overall oil yield. Figure 1 shows the main part of the small-scale production facility for small amounts of the oil. It consists of the biomass feeder, the reactor and a bio-oil collection system. Before each experiment, a batch of sand was preheated inside an electrical furnace to about 600°C, where after it is mixed with cold sawdust in the bottom of the cone. The produced vapours could be immediately removed from the hot reactor, and collected in several water-cooled vessels.

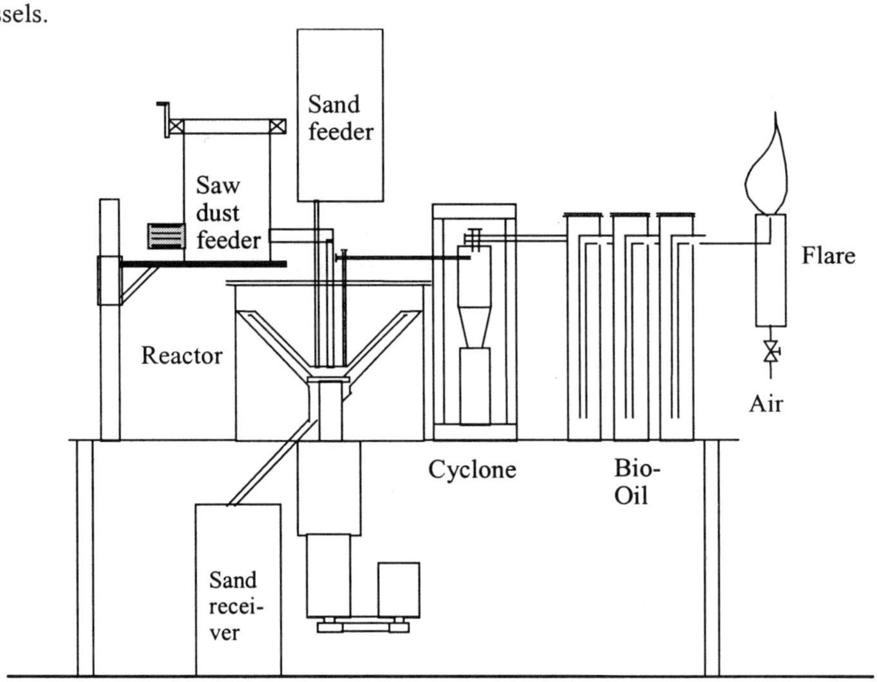

Figure 1 A schematic representation of the small-scale pyrolysis plant.

Bio-oil can react with lime to Calcium Enriched Bio-oil (CEB) according to:

- carboxylic acids + $Ca(OH)_2$ → Ca-carboxylates
- phenols + $Ca(OH)_2$ → Ca- (hydroxy)phenoxides
- aldehydes + $Ca(OH)_2$ → Ca-carboxylates + alcohols
- aldehydes + $Ca(OH)_2$ → Ca-polyol (aldol) complexes
- ketones + $Ca(OH)_2$ → Ca-polyol (ketal) complexes
- esters + $Ca(OH)_2$ → Ca-carboxylates + alcohol
- carbohydrates + $Ca(OH)_2$ → Ca-polyol complexes

The optimum conditions for these reactions is depending on the type of reaction taking place, and an experimental procedure much be undertaken to investigate the 'best' CEB preparation conditions.

It seems reasonable to define the oil quality in terms of the functional groups present, mainly acids, ketones and aldehydes, and phenols. The keton/aldehyde and acid groups can be easily measured by techniques derived for liquid smoke. Details have been presented elsewhere (12). The results showed that the keton content increased with temperatures, while the acid content only slightly decreased. This indicates that higher pyrolysis temperatures are beneficial for the bio-oil to be used in the CEB. Fortunately, in the temperature interval of 450 to 600°C the oil yield is almost independent of the temperature. The operating temperature at which the bio-oil should be produced for CEB production is up to 600°C. At 600°C and in this small-scale set-up, oil yields were up to 70 wt.% (on wood as received). For straw, the oil yields were significant lower in comparison (approx. up to 60%), most probably due to the higher ash-content of straw in comparison with wood, catalyzing secondary cracking reactions.

SMALL-SCALE CEB PRODUCTION

A stirred tank reactor enabling continuous measuring of pH, temperature and torque (at pre-set stirrer speeds), was used to produce the first small samples of CEB. Oil was slowly added to slaked lime at various operating conditions, viz. the reaction temperature, the stirrer speed, the total reaction time, the bio-oil addition rate and the bio-oil/slaked lime ratio. In a typical CEB production experiment it was shown that the point of addition of bio-oil was clearly marked by a significant rise in temperature and drop in pH. The experiments demonstrated that slaked lime had a high capacity for bio-oil, and that calciumhydroxide could be added far in excess of the theoretical acid content without the pH increasing above 12. This indicates that 'excess' calciumhydroxide reacts further with the other functional groups in the bio-oil.

The CEB product was assessed visually (by colour), pH and water content, and settled for several days.. Several batches of 20 kg, comprising a mixture of both the upper and lower fraction, were produced to be used later in the small-scale combustion trials. Characteristics of the CEB are presented in Table 2.

BENCH-SCALE INVESTIGATION OF CEB DECOMPOSITION / SULFATION

SO_2 removal experiments with CEB were performed in a thermogravimetric analysis (TGA) set-up to simulate the process during its combustion. The sample was first heated in a mixture of 70% CO_2 in N_2 to produce $CaCO_3$, followed by the decomposition of $CaCO_3$ to CaO under N_2 and O_2. This calcined material was sulfated using a mixture containing 3000 ppm SO_2, 12% O_2, and the balance N_2.

Table 2 CEB characteristics

Parameter		value
Bio-oil fraction	wt.%	up to 50
Water content	wt.%	up to 40
pH	-	11-12
Calcium content	wt.%	13
Viscosity	cP	20°C - 3430 50°C - 1900 100°C- 1290
Density	kg/m^3	1210
Caloric value	MJ/kg	up to 10

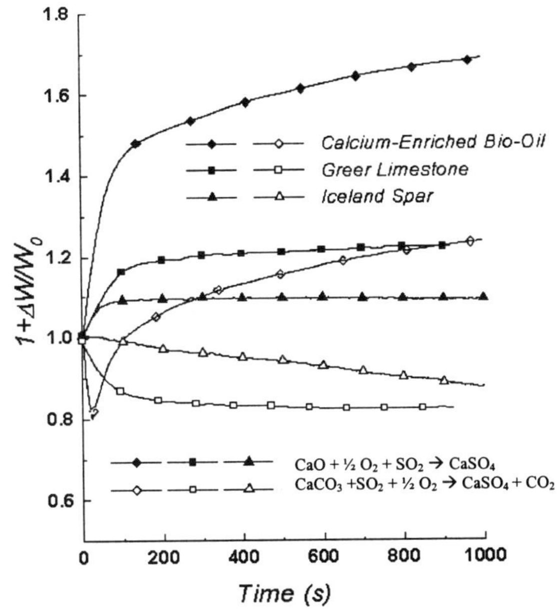

Figure 2 Comparison of the behavior of CEB with 60 μm Greer Limestone and Iceland Spar at 750°C

Although not shown in this paper, at 750 °C almost complete sulfation of the CEB material was observed, whereas conversions at 850°C were lower. Obviously, the temperature during the decomposition has a very strong influence on the performance of the sorbent. It may be that exposure to high temperatures increases the rate of sintering, which reduces the porosity and the surface area of the sorbent. The sulfation at 750°C is compared to that of similarly treated limestone particles (see Figure 2: open symbols). A strong decrease in weight at the start of the experiments with $CaCO_3$ can be clearly noted, which is due to the calination of $CaCO_3$ to CaO. For the limestone material, the decrease in weight due to the calcination reaction is not compensated anymore by sulfation. The CEB material exhibited a higher sulfation conversion than calcium carbonate solids particles. Shown in Figure 2 by the closed symbols is the increase in weight of the samples after sulfation of the calcined material. For complete conversion of $Ca(OH)_2$ to $CaSO_4$, $(1+\Delta W/W_0) = 1.84$, and for $CaCO_3$, $(1+\Delta W/W_0) = 1.36$.

To establish if the effect of the temperature on the sulfation is due to changes in the pore structure, the CEB was analysed with mercury porosimetry and nitrogen sorption. Results indicated that pores in the material treated at 750°C had a smaller surface area and volume than in the materials treated at 850°C. The latter has more pore volume for pores close to 100 nm. However, on the basis of the sorption results, one would expect the material treated at 850°C to show a higher overall reactivity. This is characterized not only by a higher porosity but also by a larger fraction of small pores and a larger surface area. One explanation for the lower conversion for the material treated at 850°C can be that the large pores produced at these conditions are accessible only through pores of smaller size. This is demonstrated also by the fact that a large fraction of mercury is trapped within the pore structure, and relatively broad hysteresis loops are encountered in the nitrogen sorption results. At 110°C, CEB forms a structure with a very high porosity of approx. 90-95%. This material already exhibited pores in the size order of a few hundreds microns, resulting in the so-called 'popcorn-effect' yielding a larger effective surface area. Because plugging of the pores is the main cause for the incomplete utilization of CaO-based materials, the high porosity of CEB decomposed material is a clear indication of the potential of CEB as a desulfurization sorbent.

PRODUCTION OF LARGE QUANTITIES OF BIO-OIL

Several tonnes of wood derived bio-oils are required for the production of sufficient quantities of CEB for the combustion experiments in a small (25 kW_{th}) and in a large (7MW_{th}) combustion facility. These oils are produced in a 200 kg/hr pilot plant, shown in Figure 3. Details of the set-up are presented elsewhere (8).

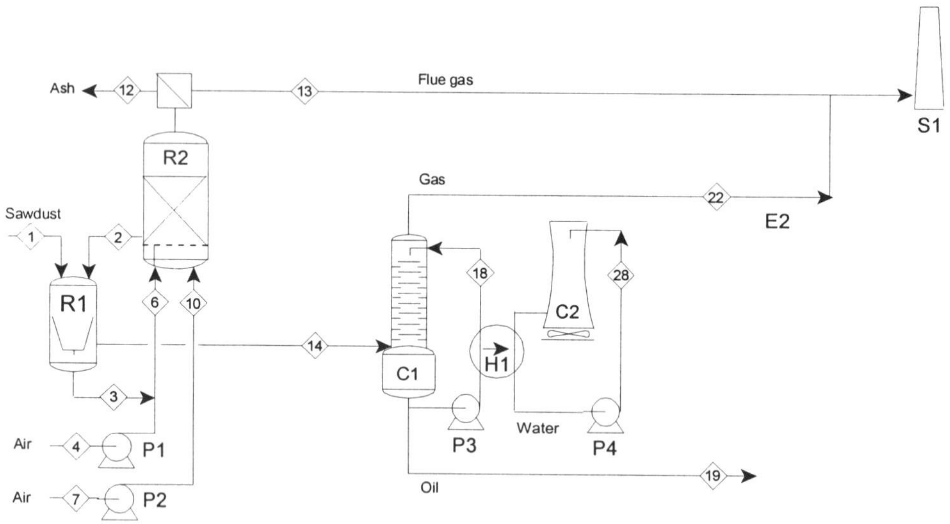

Figure 3 A schematic flow diagram of the proposed 2 t/hr fast pyrolysis system

The pyrolyser is a patented rotating cone reactor. Biomass particles (#1) are fed near the bottom of the pyrolysis reactor (R1) together with an excess flow of hot heat carrier material such as sand, where it is being pyrolysed at 500 to 600°C. The produced vapours (# 14) pass through several cyclones (not shown) before entering the condenser (C1), in which the vapours are quenched by re-circulated oil (# 18). The pyrolysis reactor is integrated in a circulating sand system composed of a riser (the sand and char, stream 3, is transported by air, # 4), a fluidized bed char combustor (R2), the pyrolysis reactor (R1), and a down-comer. In this concept, char is burned with air (# 10) to provide the heat required for the pyrolysis process. Oil is the only product (# 19), as non-condensable pyrolysis gases (# 22) are combusted and flared (S1).

Oil yields up to 75%, based on the feedtsock as received, could be obtained (see Figure 4: the line only represents a trendline). In the early oil samples, water contents of 25 to 35 wt.% were produced, while currently the water content has a consistent value of 24 to 26 wt.% with an ash content below 0.1 wt.%. As of date, almost 20 tons of oil has been produced, while approx. 3 tons oil has been made to produce the CEB material.

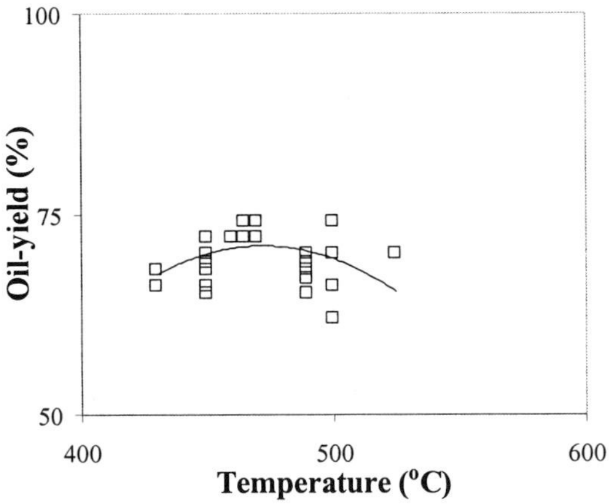

Figure 4 The oil yield as a function of the pyrolysis temperature (gas phase residence time < 1 s) for several biomass feedstock types. The line represents a trend.

CEB PREPARATION FOR LARGER-SCALE COMBUSTION TRIALS

The experimental facility to produce larger amounts of CEB is shown in Figure 5. Batch-wise, approx. 1 ton of CEB can be produced in one week. The recipe to produce CEB is to mix $Ca(OH)_2$ with water to produce so-called slaked lime at the optimised conditions referred to above. This slaked lime is pumped into a stirred stainless steel reactor, and bio-oil is added over a period of about 1.5 hours. The mixture was slowly heated to 70°C over 30 minutes and kept at this temperature for a further 2 hours. The CEB to used for the large-scale testing was stored at ambient conditions. The mixture will be thoroughly mixed before spraying and co-combustion.

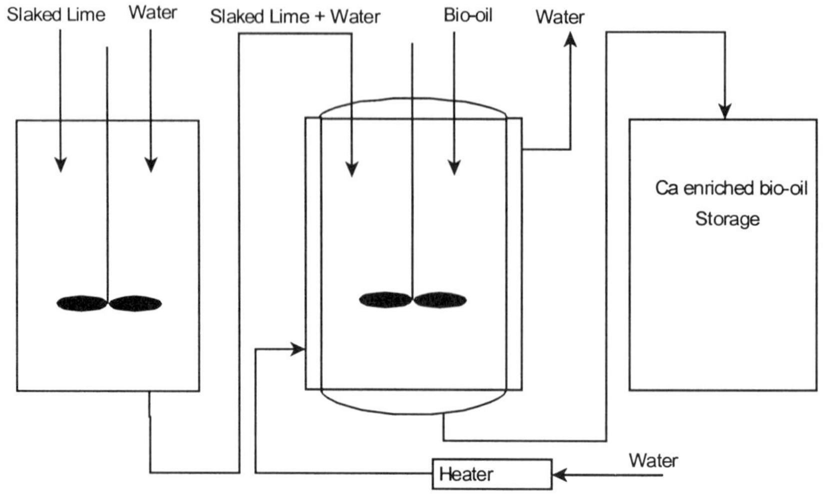

Figure 5 The CEB production facility

DEMONSTRATION OF CEB IN A SMALL EXPERIMENTAL SET-UP

Due to the abrasive character of the CEB together with its high viscosity, spraying of CEB is possible only if a suitable atomiser is applied in combination with an appropriate CEB pump. Various atomisers and oil pumps were tested for this purpose, after a first selection on basis of the physical properties of the feed (viscosity, density, solids content, moisture, surface tension, etc.), and the required spray characteristics (droplet size and its size distribution, spray angle, spatial distribution, power consumption, etc.).

The atomisation system finally selected was tested with respect to the droplet diameter. A simple technique was used, in which the atomiser was positioned approx. 25 cm above a liquid collection vessel. The produced droplets were collected in a specific type of oil, yielding a dispersion of CEB droplets as shown in Figure 6. By means of a microscope the droplet diameter could then be easily determined.

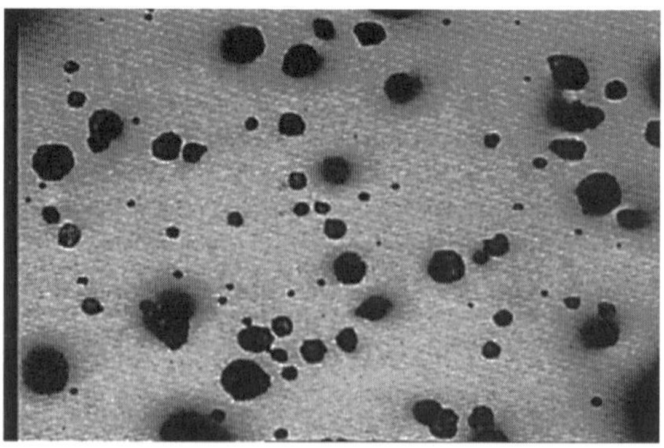

Figure 6 Droplets observed in the oil-CEB mixture after spraying.

The droplet size distribution could be measured with high accuracy and reproducibility by collecting the droplets in a microscope glass in oil. A droplet size distribution (Figure 6) could be easily constructed from the photographs.

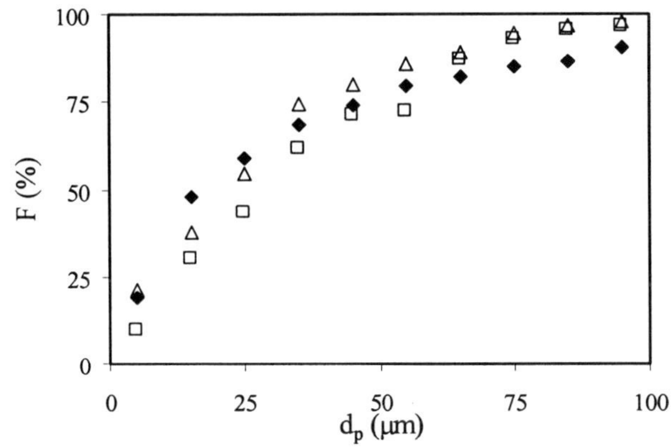

Figure 7 Droplet size distribution upon spraying CEB at a flow rate of 0.37 kg/hr, $\Phi_{air, CEB}$=2 Nm^3/hr for three different spray cap sizes: small (~), medium (□) and large (Δ).

Results can be presented as the fraction of particles smaller than a certain value versus the measured particle diameter in a cumulative F-curve (Figure 7). From these figures, an average droplet size of approx. 22 μm can be calculated for all nozzles and at the conditions presented. Apparently, larger nozzles do not influence the droplet size distribution. For two CEB flow rates, Φ_{CEB}, and two gas flow rates for CEB atomisation, $\Phi_{air,CEB}$, similar plots are presented in Figure 8. Representative values for the (number) average droplet diameter for CEB range from 30 to 50 μm, the lower values being observed at the higher air and CEB flow rates.

Contaminant removal with CEB

CEB will be used for coal and waste combustion, and not primarily for heavy fuel oil. However, the projected large-scale test 7 MW_{th} facility can only use heavy fuel oil. Therefore, this oil was used to test it on a small scale. A small flame tunnel with separate air and liquid fuel injectors was constructed to demonstrate the potential of CEB to recover SO_2. Figure 10 shows a photograph of the set-up, with the control panel at the right hand, and the flame tunnel in the middle. The three injection ports for CEB material are visible at the front side. Through the opening in the middle (the nozzle has been removed), the flame from diesel combustion is clearly visible. In all experiments, CEB is sprayed perpendicular to the fuel stream. A portable Mass Spectrometer has been purchased to analyse the exit gases from the flame tunnel experiments on SO_2.

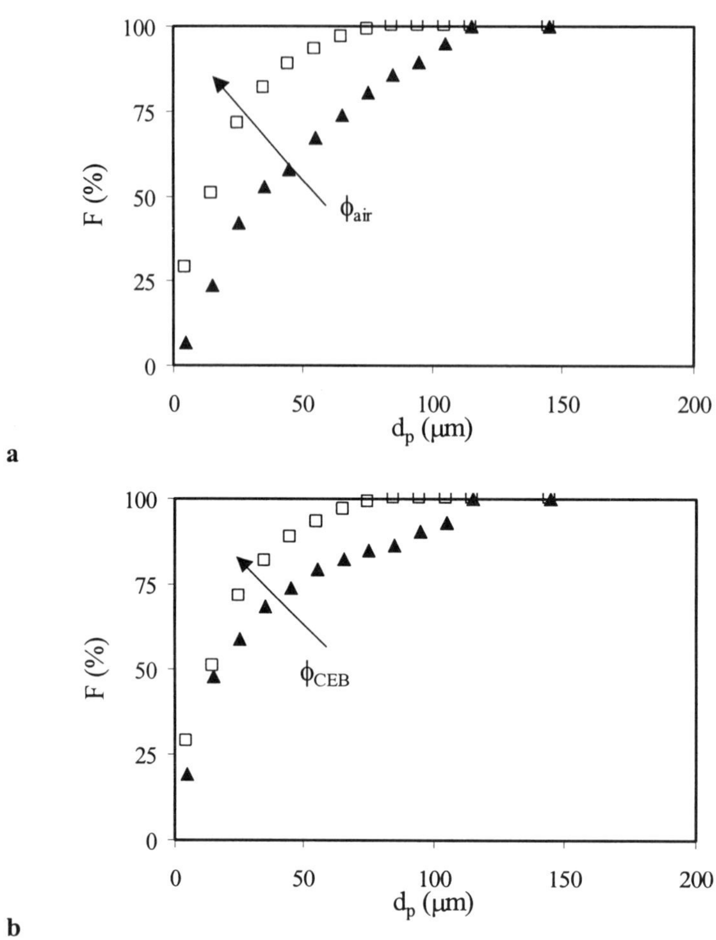

Figure 8 Droplet size distribution after spraying CEB for a) two gas flow rates (1 and 2 Nm^3/hr) and Φ_{CEB}=0.37 kg/hr and b) two CEB flow rates (0.18 and 0.37 kg/hr) and $\Phi_{air,\ CEB}$=2 Nm^3/hr.

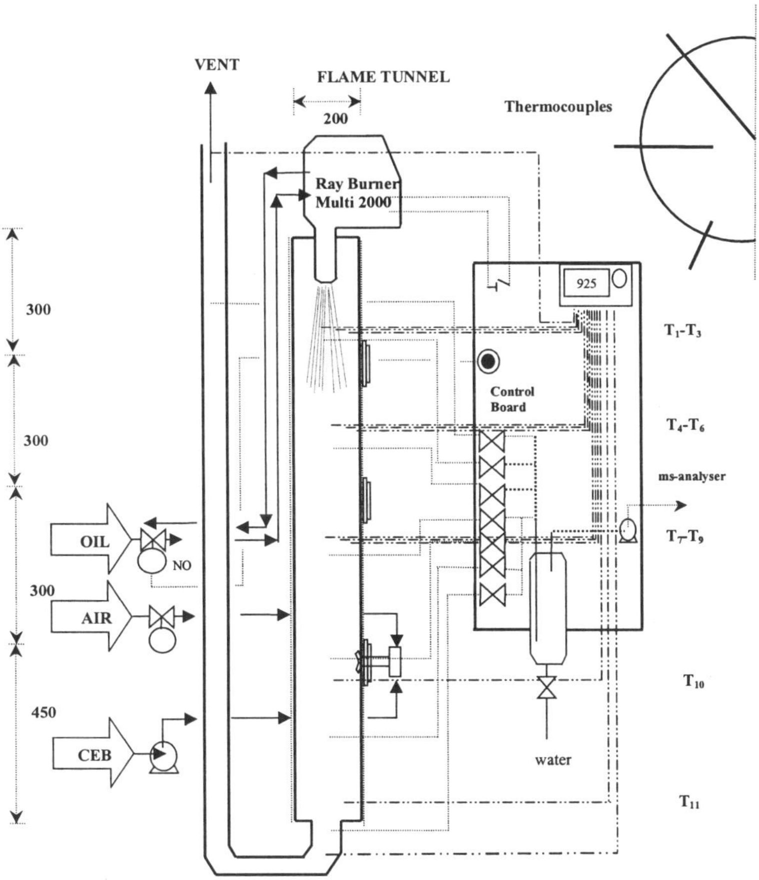

Figure 9 A schematic presentation of the 20 kW_{th} flame tunnel. Indicated at the right side of the flame tunnel are the CEB injection ports.

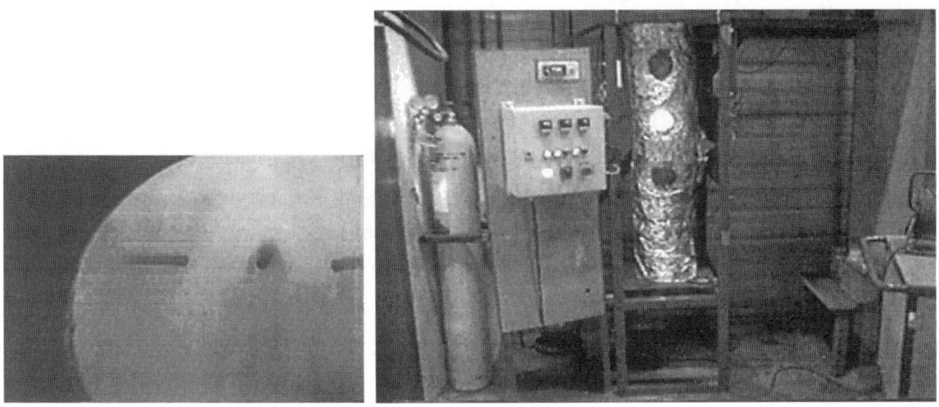

Figure 10 Photographs of the flame tunnel, with the control panel at the right hand side, and the flame tunnel in the middle. The burner is located at the top (not shown). At the front side, the three injection nozzles for CEB material are visible. Through the opening in the middle, the clear flame from diesel is visible (right).

Results SO_2 removal

For the first preliminary experiments, a 'standard' fuel oil burner system was purchased. Due to several reasons, however, a new pump and atomisation system was required, viz. a dedicated fuel pump and a different spray nozzle. In this new set-up, primary air for a proper fuel atomisation and secondary air for the combustion could be controlled separately. Furthermore, pumping and combustion of heavy fuel oil (sulphur content of 3.35 wt.% and a viscosity of 180 cP) was possible only, when mixed with diesel in a ratio of 60:40 (diesel:heavy fuel oil). Besides, the flame tunnel had to be preheated to approx. 750°C using ordinary diesel or ethanol as a start-up fuel. When a steady state operation was observed (after approximately 10 minutes), the diesel flow to the flame tunnel could be replaced by a mixture of heavy fuel oil and diesel (40:60), and the emissions of SO_2 are recorded (approx. 400 ppm). Then CEB is injected and subsequently the SO_2 levels are recorded every 30 s. Usually, almost instantly a sharp decrease in the SO_2 concentration was observed, due to the reaction of the Ca with SO_2. The experiments were carried out at various fuel flow rates, ranging from 0.5 to 3 kg/hr, and at a CEB flow rate corresponding to Ca:S ratios from 0 to 3.

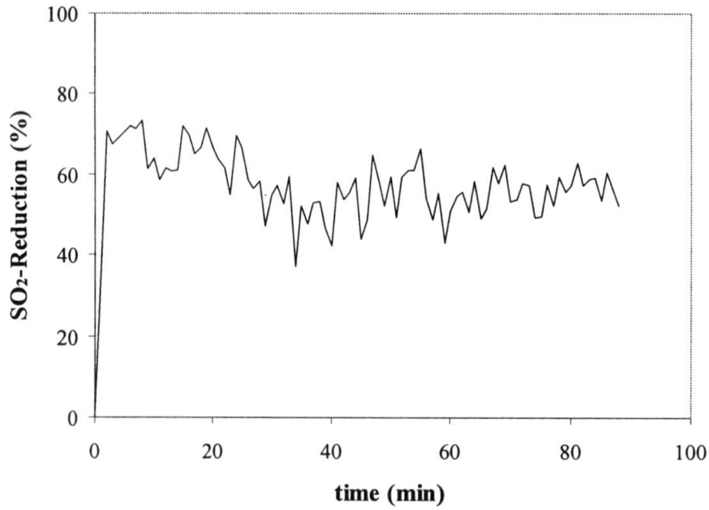

Figure 11 The SO_2-reduction versus the time while combusting the 60:40 diesel/fuel oil with injection of CEB. Operating conditions: Φ_{oil}=1.5 kg/hr, $\Phi_{air,p}$=2.5 Nm^3/hr, $\Phi_{air,s}$=20 Nm^3/hr, $\Phi_{air,CEB}$=2.2 Nm^3/hr and Ca:S=1. Sulfur content oil approx. 1.36 wt.

Figure 11 shows a typical result for the SO_2-reduction when CEB (corresponding with Ca:S = 1) is injected in the flame of the diesel/fuel oil mixture. On the average, the sulphur reduction was 55 to 60%. This relatively low value is due to the high fuel and air flows (leading to a short residence time of approx. 1.35 s), and the low Ca:S ratio. In real practice, however, residence times of several seconds are more realistic. The fluctuation in the results shown is due to the inevitable pulsation of the CEB flow rate.

Results for different Ca:S ratios have been plotted in Figure 12. CEB apparently gives a SO_2-reduction of nearly 100% at Ca:S=2. Compared to the injection of $Ca(OH)_2$, an increase in the SO_2-reduction is obtained. In Figure 12 the theoretical line for complete SO_2-removal is given while assuming that only the following reactions are taking place:

$$Ca(OH)_2 \quad \rightarrow CaO + H_2O$$

$CaO + SO_2 + \frac{1}{2} O_2 \rightarrow CaSO_4$

One of the explanations for this difference between the sulfation results for CEB and $Ca(OH)_2$ may be that the effective particle size of the active components for the CEB material is smaller than those for the $Ca(OH)_2$ material. This effect is already demonstrated in Figure 8 by varying the air for atomisation of CEB. Apparently, and due to a finer atomisation of the CEB material, the effective particle size decreases, resulting in an improved SO_2-reduction.

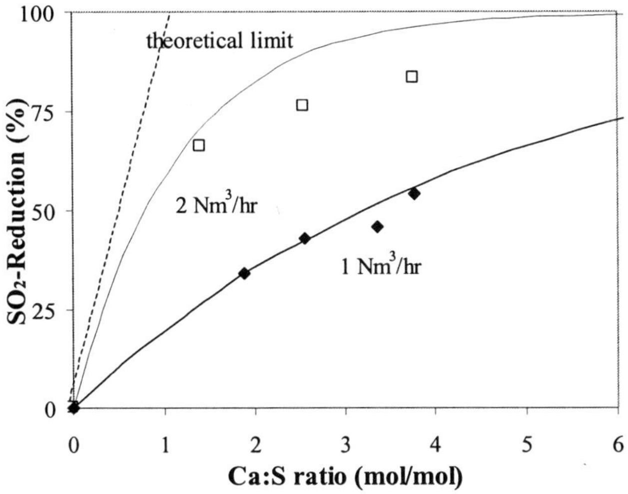

*Figure 12*a SO_2 removal as a function of the ratio Ca:S for CEB (~) and $Ca(OH)_2$ (□). Operating conditions: Φ_{fuel}=0.75 kg/hr fuel; $\Phi_{air,p}$=2.75 Nm^3/hr, $\Phi_{air,s}$=14 Nm^3/hr, and $\Phi_{air,CEB}$=2 Nm^3/hr. The lines presented are theoretical lines calculated for mass transfer controlled conditions for effective particle diameters of 2.5 and 3 μm respectively, whereas the dashed line represents the theoretical limit.

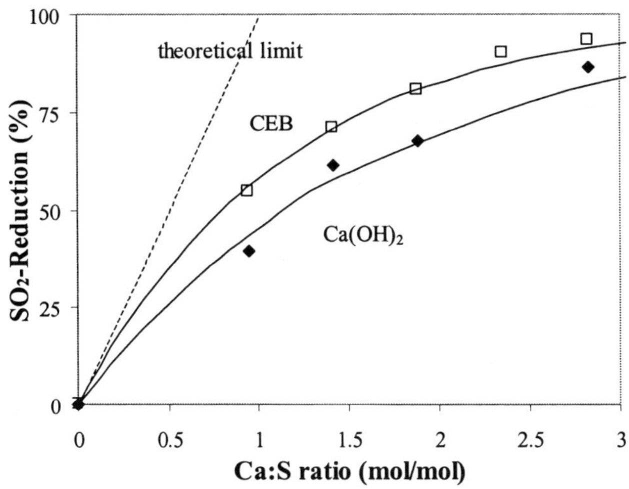

*Figure 12*b SO_2 removal versus Ca:S for $\Phi_{air,CEB}$=1 Nm^3/hr (~) and $\Phi_{air,CEB}$=2 Nm^3/hr (□). Operating conditions: Φ_{fuel}=1.5 kg/hr fuel; $\Phi_{air,p}$=2.5 Nm^3/hr, and $\Phi_{air,s}$=13 Nm^3/hr. The lines presented are theoretical lines calculated for mass transfer controlled conditions for effective diameters of 2.5 and 5 μm respectively, whereas the dashed line represents the theoretical limit.

Results NO_x removal

The NO_x concentrations were measured using a combination of the MS analysis system and a conventional fluw gas analyser. With CEB NOx-reduction levels of 18 and 38 % could be realised at a Ca:S ratio of 1 and 2 respectively. In comparison, a reduction of only approx. 12% was realised when spraying water in the flame, at a similar water-to-oil ratio comparable with the CEB experiments.

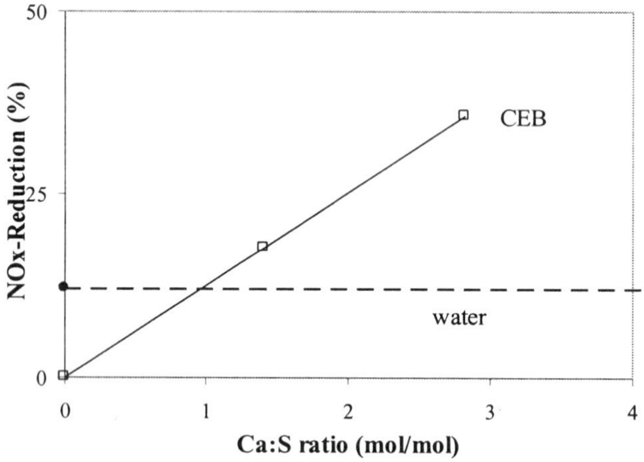

Figure 13 NO_x removal as a function of the ratio Ca:S for CEB (~) and water (dashed line). Operating conditions: Φ_{fuel}=0.75 kg/hr fuel; $\Phi_{air,p}$=2.75 Nm^3/hr, $\Phi_{air,s}$=14 Nm^3/hr, and $\Phi_{air,CEB}$=2 Nm^3/hr. The lines presented are trendlines.

Theoretical calculations

Only a small amount of Ca-particles and a hort residence time are required for the complete removal of SO_2. In combination with the high temperatures in the combustion chamber (> 800°C), it seems reasonable to state that the removal of the sulfur is a mass transfer controlled reaction. For simplicity, it is assumed that the reaction of CaO with SO_2 is first order in SO_2, and zero-order in Ca and oxygen:

$$CaO + SO_2 + \tfrac{1}{2} O_2 \rightarrow CaSO_4$$

Assuming that the reaction rate, represented by k_r ($m^3_g/(m^3_s s)$) is limited by the external gas-to-solid interface mass transfer to spherical particles, $k_r = k_g \cdot a = k_g \cdot 6/d_p$ yielding:

$$\frac{\partial c}{\partial z} = -\frac{k_r \varepsilon_s V_r}{\phi_v} = -\frac{k_g a_s \varepsilon_s V_r}{\phi_v} = -\frac{6 k_g \varepsilon_s V_r}{\phi_v d_p}$$

The unknown parameters are the characteristic particle diameter, d_p, the gas-to-particle mass transfer coefficient, k_g, and the solid hold-up ε_s.

For the very small particles and the very dilute two-phase system considered, the mass transfer number can be estimated by $Sh = k_g d_p / D = 2$ (Ranz and Marshall, 1952). The diffusion coefficient, D, can be estimated by Fuller et al. (9) The solids hold-up equals $G_s/\rho_s u_s$ (10), with G_s as the solids flux in kg/(m^2_rs), ρ_s the particle density with u_s approximating the gas velocity u_g. Values for these parameters are listed in Table 3.

Table 3 Values and relation used for the calculations

Parameter			Value	Ref.
D	Diffusion coefficient	m^2/s	$2\cdot10^{-6}$	9
L	Length Flame Tunnel	m	1.35	
d_p	Effective particle diameter	μm	1-10	
G_s	Solids flux	kg/m^2s	$\Phi_{CEB}/(1/4*\pi*Dt^{\wedge 2})$	11
Sh	Sherwood number	-	2	
ρ_s	Density solids	kg/m^3	1000	
u_s	Solids velocity	m/s	u_g	

The conversion data suggest that the initial SO_2-absorption rate is controlled by mass transfer of SO_2 from the gas phase to the active surface of the Ca-particles, but later by diffusion rates inside the particle. The effective diameter of the calcium particles range from 2 to 5 μm, depending on the spray conditions and on the Ca-containing material (CEB or $Ca(OH)_2$). Most likely, for CEB the CaO-efficiency is 1. In comparison, using lime instead of CEB yields CaO efficiencies < 1.

FUTURE WORK

Combustion test with SO_2 removal are ongoing, together with tests to demonstrate the NO_x-reduction potential. Future tests will be carried out to reveal if the Ca in the CEB is bonded to organic groups or present in the form of suspended calciumhydroxide. Tests will also be carried out in a 7 MW_{th} heay fuel oil combustion facility.

CONCLUSIONS

Wood derived fast pyrolysis oil and slaked lime produce a Calcium-Enriched Bio-oil (CEB) capable to remove SO_2 while combusting sulphur-containing fuels. The bio-oil required is produced in a 200 kg/hr fast pyrolysis plant, while CEB is made in batch-wise operation by slowly mixing slaked lime with bio-oil at controlled conditions.

The prepared CEB material has pores in the size order of a few hundreds μm, and this may result in the so-called 'popcorn-effect' yielding a larger effective surface area. As plugging of the pores is probably the main cause for the incomplete utilization of CaO-based materials, the high porosity of CEB is an indication of the potential of CEB as a desulfurization sorbent, as has been shown in TGA experiments.

Co-combustion experiments are carried out in a small 25 kW_{th} flame tunnel, combusting a mixture of diesel and heavy fuel oil (60:40 resulting in 1.34 wt.% S). An essential parameter in SO_2-removal is shown to be the atomisation of the CEB material. Higher gas and or CEB flow rates improved the atomisation characteristics, yielding high SO_2-removal > 60 % at Ca:S=1.

Compared to $Ca(OH)_2$, injection of CEB during simultaneous heavy fuel oil combustion gave higher SO_2 removal efficiencies. Further experiments should reveal whether this is due to the higher reactivity of the CEB, or to better atomisation.

The conversion data suggest that the SO_2-absorption rates are controlled by mass transfer of SO_2 from the gas phase to the active surface of the Ca-particles. The effective diameters are 2 to 5 μm., depending on the spray conditions and on the Ca-containing material (CEB or $Ca(OH)_2$). Conversion rates are controlled by external mass transfer rate at the start of sulfidation, but then by diffusion inside the particle.

ACKNOWLEDGMENT

The financial support of the EU under contract JOR3-CT97-0179 and of NOVEM under the EWAB program is gratefully acknowledged.

REFERENCES

1. http://www.iea-coal.org.uk
2. Soud H.N., 1994, FGD installations on coal fired plants, IEA Coal Research, ISBN 92-9029-239-3
3. Soud H.N., 2000, Developments in FGD, IEA Coal Research, ISBN 92-9029-339-X
4. Pisupati S.V., Clark D.A., Hill M.A., 1996, Evaluation of Simultaneous SO_2 and NO_x Reduction potential of BioLimeTM, Pennstate Energy and Fuels Research Center, 1996
5. Pisupati S.V., Clark D.A., Hill M.A., 1997, A study on the Efect of Atomization Characteristics of BioLimeTM Simultaneous SO_2 and NO_x Reduction Pennstate Energy and Fuels Research Center
6. Oehr K., Zhou J., Simons G., Wójtowicz M., 1997, Simultaneous SO_2 and NO_x control with BioLimeTM derived from biomass pyrolysis oil, in Developments in Thermochemical Biomass Conversion (ed. by A.V. Bridgwater and D.G.B. Boocock), 1477-1481
7. Zhou J., Oehr K., Simons G., Barras G., Put B., 1997, Simultaneous SO_2 and NO_x control using BioLimeTM, in Biomass Gasification and Pyrolysis: State of the Art and Future Prospects (ed. by M. Kaltschmitt and A.V. Bridgwater), 490-494
8. Wagenaar B.M., Venderbosch R.H., Carrasco J., Strenziok R., van der Aa B., 2000, Scaling-up of the Rotating Cone Technology for Biomass Fast Pyrolysis, in: 1st World Conference and exhibition on Biomass for Energy and Industry.
9. Fuller E.N., Schettler P.D., Giddings J.C., 1966, A new method for predicting of binar gas phase diffucion coefficients, Ind. Engng. Chem., 58, 18
10. Kunii D., Levenspiel O., 1990, Fluidization Engineering, J. Wiley & Sons, New York
11. Ranz W.E., Marshall Jr. W.R., 1952, Evaporation from drop: part 1 and part II, Chem . Engngn. Prog., 48, p.141, p.173
12. Meier D., 1999, New Methods for Chemical and Physical Characterization and Round Robin Testing, in: Fast pyrolysis of Biomass: A Handbook (edited by A.V. Bridgwater *et al.*), CPL Press, Berkshire, UK, ISBN 1 872691 07 2, p. 91-101

Improving Charcoal Kiln Performance - Do Fundamental Studies Have a Role?

M.A. Connor
Department of Chemical Engineering, University of Melbourne, Vic. 3010, Australia

ABSTRACT: Though not the subject of much research at present, wood carbonisation or charcoal making is still one of the most widely used thermochemical biomass conversion processes. Existing charcoal kilns are comparatively inefficient and many researchers have sought to find ways of improving kiln efficiency. Prospects for using fundamental research to enhance kiln efficiency and improve kiln productivity and profitability are explored. Processes within kilns can be divided into microscale processes and macroscale processes. The former category encompasses processes that take place in and around individual wood pieces. A good deal is known about these processes but it is concluded that practical considerations limit the economic benefits that further research into these processes can bring. Much less is known about the latter category, which includes large-scale heat transport, mass transport and reaction processes within the wood stacked in the kiln. A better understanding of the fundamentals of these processes has a much better chance of contributing to improved kiln designs and operating procedures.

INTRODUCTION

Recent research into thermochemical biomass conversion processes seems to have been concerned mainly with areas such as gasification and rapid pyrolysis. As a result, research inputs into other areas such as carbonisation or charcoal production have diminished greatly. Yet charcoal making is still one of the most important of the thermochemical biomass conversion processes, affecting the lives and livelihoods of millions of people. The global significance of charcoal can be gauged from the large amount consumed worldwide - in 1996 this was estimated to be around 100 million tonnes annually [1]. What is more, despite the widely held view that charcoal making is a declining industry, global demand for charcoal continues to grow [1][2].

The biggest single user of charcoal remains the Brazilian iron and steel industry. In 1990 this industry was estimated to have an annual charcoal consumption rate of around 5 million toe (tonnes of oil equivalent) [3]. However, the amounts used can vary markedly from year to year. This is reflected in the figures for Brazilian industry as a whole, which reportedly used over 11 million tonnes of charcoal in 1991 but only 7.3 million tonnes in 1992 [1].

Although industry is a significant user of charcoal, the major users are people living in Third World towns and cities. For households and small-scale commercial enterprises, charcoal has distinct advantages over wood for both cooking and heating [4]. These advantages have made it a preferred fuel in many urban centres across the developing world, including the Middle East, Southeast Asia, East and West Africa, South and Central America and the West Indies [4].

Virtually all the charcoal made in these regions - other than that used for specialised industrial purposes - is made in kilns. A wide range of kiln types is in use. Some are quite primitive, like the simple earth and vegetation covered kilns used by itinerant charcoal burners. Others, such as the beehive kilns used by Brazil's iron and steel industry, are comparatively sophisticated [5][6]. Perhaps most widespread, and typical of the kilns used by smaller commercial enterprises in the developing world, are traditional earth or earthmound kilns like those to be found in Zambia [7] and Sudan [8].

The nature of the carbonisation process is such that the charcoal removed from a kiln weighs a lot less than the original wood charge. For kilns of the earthmound type, charcoal yields are around 23% to 27% (based on the original mass of oven-dry wood), or as low as 13% when based on the original air-dried wood mass [7][8]. These values are well below the theoretically attainable maximum of 44-55% [9].

The inefficiencies of kilns, and the implications these have as far as deforestation rates are concerned, have prompted many researchers to seek ways of increasing charcoal yields. One approach has been to use larger or more sophisticated kilns. The benefits of this approach are well illustrated in a study [8] comparing a traditional Sudanese earthmound kiln with a metal kiln of the type developed by the Tropical Products Institute [10]. In this study, the metal kiln gave a 33% yield, as compared to 27% for the earthmound type. Further improvements can be achieved by going to large, top of the range kilns such as the beehive kiln or the Missouri kiln but reported yields rarely exceed 35% for charcoals with at least a 70% fixed carbon content.

Yield values, such as those given above, are given much prominence in the scientific literature on charcoal. In fact, scientific papers often give the impression that charcoal yield is the key parameter of concern to the charcoal industry. In practice, this is not the case. Researchers may be preoccupied with charcoal yield but commercial charcoal producers are much more concerned about profitability. A good yield can certainly contribute to a high profitability but other factors are of equal or greater importance. Some of these factors may be largely unrelated to the processes that take place within the kiln; examples include the capital cost of the kiln and its ancillary equipment, and local labour costs. Such costs can be the determining factor when it comes to choosing what type of kiln to use. In a study carried out in Montserrat, the economics of four quite different small to medium sized kilns were analysed [11]. By far the most profitable of these proved to be the comparatively primitive, local 'coal pit' type of kiln. The low capital and labour costs of this kiln, together with its ten year lifetime, made it far more economically attractive than more short-lived, costly and sophisticated kilns offering only a marginal improvement in charcoal yield.

However, other factors affecting profitability are much more closely linked to what happens inside the kiln. One of these is the marketability of the charcoal. If charcoal is to command a high price, its properties have to conform to users' requirements. Iron and steel plants, for instance, may require a charcoal with a higher fixed carbon content and a higher strength than is needed by domestic consumers. However, these plants may also tolerate a fines content greater than that acceptable to domestic users. The latter generally prefer a lumpy, dense, slow-burning charcoal with a good resistance to breakage during transport and handling. Domestic users also dislike charcoals that emit a pungent smoke or give off showers of sparks while burning.

Some of the above characteristics are determined in part by the tree species from which the wood charge comes. Charcoal makers are well aware of this and are selective in their choice of woods. However, other charcoal characteristics are more dependent on what takes place inside the kiln. For example, prolonging the carbonisation period is known to enhance the charcoal yield as well as reducing the incidence of cracking (and hence the percentage of fines) [12].

It should be evident from the above discussion that maximising the profitability of a charcoal making enterprise is not a straightforward process. Charcoal quality has to be maintained, yet at the same time the carbonisation period must not be too long or productivity will suffer. Heating rates must be high but not high enough to give too many fines. Capital, operating and labour costs also have to be taken into account, as do disparities in selling price for charcoals of differing quality.

In the case of large-scale plants using wood of consistent quality, experience alone may enable a good compromise between all these factors to be achieved. Such a situation exists in Brazil where the large producers source much of their wood from large *Eucalyptus* plantations. Wood harvested from these plantations arrives at the kilns as pieces of uniform length, reasonably similar in diameter, and with consistent physical and chemical characteristics. This uniformity makes it possible, over time, to refine kiln designs and operating practices and come up with a set of procedures that gives a satisfactory product in an acceptable time and at an affordable cost.

However, kiln operators elsewhere generally lack access to a uniform wood supply or else see advantage in being able to switch between a variety of different wood sources as prices and availability alter. Such a policy may enable savings in wood costs, but it also makes it harder to establish appropriate operating procedures. What is needed is some way of predicting how kiln operating procedures should be modified to allow for changes in wood characteristics. To do this requires a better understanding of the fundamentals of kiln processes than we have at present. Some of these fundamentals are fairly well understood but others are poorly known. It is the aim of this paper briefly to review these fundamentals and discuss the extent to which, through better understanding, producers using established kiln types might be helped to achieve better yields and greater profitability when using a variety of wood types and sizes.

KILN PROCESSES

Charcoal kiln processes can be divided into two groups, referred to here as microscale processes and macroscale processes. The former group encompasses those processes occurring close to or within individual wood particles. The latter group includes larger scale processes such as the convective transport of heat through the wood stacked in the kiln.

When reviewing these processes it is important to remember that the production of charcoal in kilns is itself a batch process rather than a continuous one. For each batch there are four stages [4]:

(1) The loading stage, during which as much wood as possible is stacked in a predetermined pattern inside the kiln.
(2) The carbonisation stage, during which a portion of the charge is burnt; the heat released is used to drive off moisture and raise the remainder of the wood charge to a temperature high enough for charcoal to form.
(3) The cooling stage, during which the charcoal is allowed to cool to a temperature low enough to prevent it catching fire when exposed to the air.
(4) The unloading stage, when the charcoal product is removed from the kiln.

If profitability is to be maximised, all four of these stages need to be properly managed. The first and fourth stages are purely mechanical and it might seem unnecessary to cover them in a review of process fundamentals. This is true for the fourth stage but not for the first; this is because the structure of the woodpile formed when loading the kiln has a strong influence on gas and heat transport during subsequent stages. The stage of greatest importance is obviously the second, since it is during this stage that all the processes accompanying charcoal formation occur. As its name implies, the third or cooling stage does involve some heat transfer but the processes occurring are much more straightforward than those in the second stage.

MICROSCALE PROCESSES

Heat transfer, mass transfer and reaction processes all play a role in the transformation of a piece of wood into charcoal. Past research has shown that the carbon laid down within a carbonising particle is a product of both primary and secondary decomposition reactions. The larger part of the carbon is formed during the initial (primary) pyrolytic decomposition reactions that occur when the wood is heated. During these reactions the organic components of the wood break down to yield a mixture of gases, condensable vapours, and a solid carbonaceous residue known as char. Significant further amounts of char can also be produced during secondary reactions. These are reactions undergone by volatile products of the primary reactions as they move from the interior of the carbonising particle to its surface [12].

Both the primary and the secondary decomposition reactions produce char and hence have the potential to affect charcoal yield and hence kiln profitability. Factors known to influence the nature, extent or rates of these reactions are reviewed below.

Wood origin

It was mentioned earlier that charcoal makers are well aware that woods from different tree species yield charcoals of differing quality. These differences can be linked to differences in wood anatomy and composition. There are well known and well-characterised anatomical and chemical differences between softwoods and hardwoods, for example [13], and softwood charcoal is widely regarded as inferior to that from hardwoods. Significant differences can also exist between the charcoals and charcoal yields from superficially similar species of the same genus. This has been shown in the case of *Eucalyptus* species [14][15]. There can even be differences in charcoals made from trees of the same species when these come from regions with dissimilar climates or soil types.

Informed and careful charcoal makers no doubt make some allowance for most of the above differences by segregating their wood stocks on the basis of species or source. A more difficult problem is how to allow for the differences that exist between woods from different parts of the same tree. The heartwood and sapwood of many tree species can be quite different in structure [15], as can the so-called earlywood and latewood layers. In addition, wood from areas subjected to non-uniformly distributed tensile or compressive stresses (as happens where branches intersect the main trunk) can differ significantly from wood that has developed in a more evenly stressed environment. Segregating wood pieces from the same tree would seem quite impracticable. The only option open to charcoal makers appears to be to choose a set of kiln operating conditions that gives an acceptable result for all the structural types present.

Wood composition

Given the diversity of tree species from which the wood used in charcoal kilns could be drawn, chemical composition might be expected to be a factor of some importance in charcoal making. Heightening this expectation is the fact that the major organic components of wood contribute unequally to char formation, with the bulk of the char coming from the lignin fraction [16][17]. However, because the relative proportions of the major organic components vary little between species - see Table 1 below - variations in these components appear of only minor consequence to charcoal makers.

Table 1 Composition of softwoods and hardwoods [13].

Component	Mass % in Softwoods	Mass % in Hardwoods
Cellulose	42 ± 2	45 ± 2
Hemicelluloses	27 ± 2	30 ± 5
Lignin	28 ± 3	20 ± 4

One wood component that could be expected to influence thermal decomposition patterns is its mineral matter content. Introduction of various inorganic compounds into wood is known to modify wood behaviour during heating. This change in behaviour has been exploited by makers of fire retardants. These chemicals, largely inorganic in nature, suppress the formation of volatiles in wood during heating and promote char formation. Although these introduced compounds have a well-characterised impact on wood decomposition, it has yet to be demonstrated that the mineral matter naturally present in wood has any effect on levels of char formed during pyrolysis. This apparent lack of impact is rather surprising since it has been shown that leaching of various salts from agricultural residues does lead to changes in decomposition behaviour [18][19]. Nevertheless, on the basis of the evidence available, it would seem that ash content is another property that can be disregarded by charcoal makers - except, of course, as it affects the calorific value, ash levels and burning characteristics of the final charcoal product.

The wood component likely to be of greatest interest to charcoal producers is water. Freshly cut wood can have a moisture content in excess of 50% and even air-dried wood has an equilibrium moisture content of 10 to 15%. In kilns, the heat required to drive off this moisture comes from the burning of a fraction of the wood charge. It follows that the greater the wood moisture content, the more wood has to be burnt and the lower is the overall charcoal yield. The longer drying time will also increase the length of the carbonisation cycle, which could be expected to further reduce kiln productivity. Despite these apparent disadvantages, it is reported that a charcoal producer in the Marianas found it better to use green wood (which entailed firing its metal kiln for a longer time than normal) since this gave a better quality charcoal that was easier to market [20].

This practical observation ties in with theory, which suggests that using wood with a higher moisture content could lead to an increased charcoal yield (based on wood dry weight). This is because an extended drying period will prolong the time wood spends at lower temperatures during pyrolysis, and this can favour decomposition pathways promoting char formation [21]. However, whether or not an increased char yield always eventuates is not clear. A fairly recent paper did review published work on the impact of moisture content on biomass behaviour during pyrolysis but the information available was found to be conflicting and contradictory [22]. Based on the evidence presently available, and despite the findings of the charcoal producer in the Marianas, the safest and most profitable approach would seem to be to do what most charcoal producers normally do and keep the wood moisture content as low as is practicable.

Wood structure and physical properties

As discussed earlier, of the char formed during thermal decomposition, a significant fraction can be laid down as a result of secondary reactions. These reactions occur as the volatile products of primary decomposition migrate out of the decomposing wood. Evidence is accumulating that suggests that the extent of secondary char formation depends on how long volatiles remain within the wood/char matrix before escaping into the gas phase. It follows that any property that affects migration rates through the wood/char matrix could influence the overall charcoal yield.

Density is one such property. It might seem logical that when the same wood constituents are more closely packed, as must be the case in woods of higher densities, the rate at which volatiles can diffuse to the wood surface should fall. Woods of a higher density might therefore be expected to give greater charcoal yields. Support for this hypothesis comes from a study showing that higher density woods from species such as oak and hickory gave higher yields of charcoal than lower density woods from species like cottonwood [23]. However, a study of ten *Eucalyptus* species in Brazil showed no correlation at all between charcoal yield and density [14]. Similar conclusions were reached in a study of various Australian and South-East Asian hardwoods [15]. However, the latter study did show that higher charcoal yields were obtained from higher density sections of a single wood sample of reasonably uniform structure. This implies that density may well influence charcoal yield slightly but that its influence is often overshadowed by other factors.

In practice, these findings are probably of little relevance to charcoal producers. Producers are well aware that their customers generally prefer a dense charcoal and that a dense wood is needed to make such a product [11]. This means that denser woods are in any case being targeted by charcoal makers.

In the above rather simplistic discussions on density, the arguments put forward implicitly assume that wood is a homogeneous, isotropic material. Of course, in reality wood has a non-uniform, anisotropic, vascular structure. A low permeability wood could be expected to delay the escape of volatile pyrolysis products for a longer time than more permeable woods, increasing the opportunities for carbon-depositing secondary reactions to occur. This suggests that permeability could well be a better predictor of charcoal yield than density - provided, of course, that the permeability of the wood is representative of that of the carbonising wood/char matrix through which the volatiles make their escape. Whether or not this is the case is not known at present and further research into permeabilities could provide some useful insights.

Permeabilities have been measured for a wide range of tree species and span six orders of magnitude [24]. This suggests that if there does exist a correlation between permeability and charcoal yield, this should be easy to confirm. In practice it is not that simple. Because of local structural differences, samples of wood from the same tree can show permeabilities that vary by an order of magnitude or more [25]. Moreover, permeabilities are markedly anisotropic; in a study of beechwood, for example, longitudinal, radial and tangential permeabilities were found to be in the ratio of 13000: 0.03: 0.02 [24]. From a practical viewpoint, the longitudinal permeability is likely to be the determining value. A recent study of this parameter for a number of *Eucalyptus* species showed that, for the group of species studied, longitudinal permeability was a much better predictor of charcoal yield than density [15]. However, when the study was broadened to take in a wider range of tree species, a much poorer level of correlation between permeability and charcoal yield was found [15].

The usefulness of the above findings to charcoal makers is uncertain. Few would have the equipment, the expertise or the time to make accurate measurements of permeability. The need to make such measurements could be partly overcome if a reference set of permeability (and density) values for commonly used woods were available. Even then, predicting charcoal yield from such values remains a chancy business. There is evidence that woods with a high density and a low permeability will generally give a high charcoal yield, and vice versa, but exceptions do occur. It is also probable that related woods with similar properties will give similar yields, but again, nothing is certain.

One set of properties over which charcoal makers do have some control is the size and shape of the wood pieces loaded into the kiln. Many practical considerations affect the way wood pieces are cut and shaped before being put in the kiln. The first of these is the wood source. If waste wood from wood-processing industries is being used as a feedstock, then the size and shape of the wood pieces are largely predetermined. However, if the wood comes from plantations grown specifically to supply the kilns, there is greater scope for optimising size and shape characteristics. Desirable features include:

(1) A diameter that is not too large - a maximum of 200 mm has been suggested [10]. If diameters are too large, the wood interior will not be fully carbonised - unless the heating period is prolonged (and overall kiln productivity decreased).
(2) A length that is suited to the kiln design - 450 to 600 mm has been recommended for transportable metal kilns [10] but much greater lengths are usual in beehive kilns. Transport and handling considerations will also affect the choice of wood length.
(3) Straightness and a lack of protrusions - these features make it easier to create a low voidage, closely packed wood stack and to maximise the mass of charcoal made in each batch.

In practice, compromises have to be made. Where a wood is very hard, for example, it may be more cost effective not to split large wood pieces but to recycle them through the kiln until they are fully carbonised. Separation of "brands" (wood pieces that are not fully carbonised) for reprocessing is in fact a frequent occurrence on kilns; these brands can result from poor distribution of heat within the kiln as well as from the use of oversized wood pieces.

Charcoal producers undoubtedly learn a lot from experience about how to adapt kiln operating conditions to suit woods with different properties. Nevertheless, a better understanding of how size and shape properties affect wood decomposition processes should help them do this more efficiently.

For instance, research has shown that the larger the wood particle, the higher is the charcoal yield, providing that the particle does not crack [12]. Cracking provides volatiles from the wood interior with a rapid escape route out of the wood. This is disadvantageous as it reduces the extent of secondary carbon-forming reactions. Cracking also decreases the average size of the charcoal product. The extent of cracking can be reduced by decreasing the rate of heating. However, if taken too far, the gains in the area of charcoal yield and quality will be offset by the fall in annual output.

In the case where particles are non-isometric, the orientation of the longest wood dimension relative to the wood grain can also have a small effect on charcoal yield [12]. This finding would seem to be of most relevance to those producing charcoal in retorts, where the wood is often cut into uniformly sized pieces before being carbonised. If kiln operators tried to exploit this finding, more extensive preprocessing of the wood charge would be needed; given that the likely gains in yield are small, the cost effectiveness of doing this is questionable.

It should be evident from the above discussion that, in theory at least, for any given wood charge there is a set of operating procedures that will maximise the benefits to the charcoal producer. Altering procedures to suit each wood batch may be possible on small-scale owner-operated kilns but it seems unlikely that larger kiln batteries could show the same flexibility. Here there is a need to coordinate the activities of employees responsible for loading, running and unloading a group of kilns. Making the best use of their time requires close adherence to a preset schedule, and scope for modifying operating procedures is limited. In such cases, it seems that theory has relevance only insofar as setting the original cycle times is concerned or in helping producers to know how best to blend wood stocks from different sources or to work out what extent of further preprocessing of wood stocks is worthwhile.

MACROSCALE PROCESSES

During a kiln's carbonisation stage, part of the wood charge is burnt, and the heat released provides the driving force for pyrolytic decomposition of the remaining wood. This involves a variety of heat transport, mass transport and reaction processes. These are dealt with individually below.

Heat transport

A key tactic in maximising kiln profitability is to minimise the fraction of the wood charge that is burnt. This means making the best possible use of the heat generated during combustion of this (sacrificial) material. One obvious action is to limit heat losses to the surroundings. Various approaches would appear feasible, for example decreasing the thermal conductivity of the kiln walls or surrounding the kiln with insulation. However, reducing heat losses from the kiln exterior will also lengthen the kiln cooling stage, with fewer charcoal batches being produced annually. This loss of productivity will soon offset any savings made by reducing heat losses during carbonisation. It also helps explain why measures to reduce external heat losses are missing from many kiln designs.

One way round this problem is to find another means of cooling the charcoal. Circulating cool gases through a kiln is feasible in theory, but use of air would cause the charcoal to catch fire, and inert gases are most unlikely to be available at kiln sites. Quenching with water has been used [1] but the thermal shock when the cold water strikes the charcoal surface can cause additional fracturing and an unwanted reduction in average particle size. A promising alternative, observed under test at a kiln site in Brazil, is to introduce a fine spray of water into the top of the kiln. By the start of the cooling stage, the shrinkage that occurs during carbonisation has created a sizeable space in the top half of the kiln. The injected water droplets evaporate in this space before reaching the charcoal surface. This cools the gas and accelerates the cooling of the kiln contents without affecting charcoal quality.

There is another approach that may not reduce the amount of wood burnt but that could still improve profitability. This is to stack the wood in such a way that those pieces least likely to yield good quality charcoal are the pieces most likely to be burnt. A logical extension of this approach is to design kilns so that the heat needed for carbonisation is obtained by burning wood unsuitable for charcoal production. This would be easiest to achieve if the wood combustion zone were separated from the carbonisation chamber. It was reported fairly recently that the Brazilian company, Acesita, had built a test kiln with such an external combustion chamber. This was done to see if better use could be made of forest residues unsuitable for conversion to charcoal [1]. It was reportedly found that the cost of harvesting these residues had made the proposed scheme uneconomic and that the external chamber was now being used only as an ignition point and to control air flows into the main chamber [1].

The most energy efficient way of providing the heat needs of a kiln would be to burn the vapours given off as the wood inside the kiln decomposes. If the hot gases formed in this combustion step could be diverted back into the kiln, the fraction of the wood charge sacrificed to provide heat could be substantially reduced. (This is the approach adopted in continuously operated charcoal retorts, which lie outside the scope of this paper.) What makes this idea impractical for a single kiln is that much of the heat needed has to be available before evolution of vapours from the wood commences. However, it is tempting to ask whether better use couldn't be made of this heat at kiln sites where a number of kilns operate simultaneously; if kilns were operated on a staggered schedule, heat recovered from the vapours of one kiln could be used in the next kiln, and so on.

Even if it is impractical to direct the energy content of these vapours back into the kiln, capturing some of this energy nevertheless remains an attractive idea for kiln owners. What has been done at some kiln sites in Brazil is to pass kiln off-gases through condensers [1]. These serve mainly to alleviate air pollution problems but they also recover the heavier volatile components in the kiln off-gases. The recovered condensate has been used as a supplementary fuel in nearby blast furnaces. Whilst this may not change the energy efficiency of the charcoal production process itself, it does mean that more effective use is being made of the energy content of the original wood.

Within the kiln, the most important heat transport problem is ensuring that the heat generated during partial combustion of the wood charge is distributed quickly to where it is needed. In my opinion, this is the aspect of kiln operation where fundamentals can contribute most to efficient charcoal making.

In an ideal kiln, the hot gases would distribute themselves through the wood stack in such a way that all wood pieces would be exposed to the same pattern of surface temperature changes. In practice this is impossible to achieve. What normally happens is that some readily combustible waste or fuel is ignited at one or more points within the kiln. Enough air is admitted to sustain limited combustion at each of these points and heat is carried away into the rest of the woodpile by the expanding volume of hot combustion gases. As these gases pass over the surfaces of wood pieces surrounding the ignition point they lose heat progressively to these wood pieces, which in turn heat up and start to decompose.

Initially, when the surrounding wood is cold, the gases lose their heat within a comparatively short distance of the ignition point. However, as the surrounding wood becomes hotter, heat is carried to wood pieces further from the point of ignition. In effect a heat front moves outwards through the woodpile from each of the ignition points. Behind this heat front follows a so-called carbonisation front; this defines the boundary of a zone within which the wood has been exposed, for a sufficient length of time, to temperatures high enough for charcoal to form. Within the kiln, carbonisation can only be said to be complete when the carbonisation fronts from the various ignition points have converged and extended throughout the woodpile.

In effect, the time taken for the carbonisation front to reach the extremities of the woodpile defines the length of the carbonisation stage in the overall charcoal production cycle. Factors affecting the movement of carbonisation fronts should therefore be of great interest to charcoal producers, since any shortening of the carbonisation stage should improve kiln productivity. Some experimental information on carbonisation fronts is available. For example, Briane & Doat [6] have illustrated how the carbonisation front develops and moves over time in an earthmound kiln with its ignition point situated centrally and near the top of the woodpile inside. Also Shah *et al.* have monitored the temperature changes occurring during charcoal production in a 2 m^3 pilot scale Magnien kiln fitted with six thermocouples [26]. However, I am unaware of any comprehensive study of factors influencing the rate of movement of carbonisation fronts.

Various strategies are available to producers wishing to minimise the length of the carbonisation stage. One approach is to segregate the wood on the basis of size so that the larger wood pieces are closer to the point of ignition and the smaller pieces further away. Such an arrangement, for a modified Casamanca kiln, is illustrated in Briane & Doat [6]. This approach takes account of the fact that the time taken for heat to penetrate to the centre of a log of wood decreases as the log's diameter decreases. The time needed for the wood to carbonise fully decreases correspondingly. The delays before the carbonisation front reaches the kiln extremities are therefore less serious if the wood at these points is of a smaller size.

Another strategy is to increase the number and spread of ignition points. With a properly designed arrangement of ignition points it should, in theory, be possible to reduce the time needed for the carbonisation front to envelop the entire woodpile. However, to do this most effectively calls for careful coordination of the rates of advance of the temperature fronts moving outwards from each ignition point. Coordination of these fronts on an existing kiln would necessitate measuring temperatures continuously at a large number of points through the kiln. Not only is this practically inconvenient for the charcoal producer but it is made very difficult by the marked shrinkage of the woodpile that accompanies carbonisation. It would also add significantly to operating costs.

Using measuring instruments for control purposes may well be practically unrealistic during normal kiln operations. However, specific research investigations using a range of instruments can be rewarding. This is well illustrated by an investigation carried out in South Africa on some Armco-Robson kilns [27]. These kilns were made from corrugated iron sheets bent and joined to form a roughly semi-circular tunnel 12 metres long, 2.5 metres high and 3.8 metres across the base. Each end of the tunnel was closed off by a masonry wall. The intention was to use these kilns to produce a consistent high quality charcoal product with a fixed carbon content in excess of 90%.

The original kiln design had a limited number of air inlet points distributed along the base of the kiln. With this design, unacceptably large variations in charcoal quality occurred across the kiln. It was felt that this could be due to the air inlet points being too widely spaced. This was investigated using oxygen probes to map oxygen concentrations around each inlet (ignition) point. In this way, the sphere of influence of each inlet point was delineated. From this information it was determined how many more air inlet points would be needed to obtain more uniform conditions in the kiln and hence a more consistent product [27].

It would be even more helpful to kiln operators and designers if, for a given set of conditions, the progress of temperature and carbonisation fronts through the woodpile in a kiln could be predicted theoretically. To do this would require a comprehensive model that describes reaction, heat transfer and mass transfer processes in the kiln for wood pieces of a known size and moisture content stacked in a particular way. Development of such a model is probably still a long way off but some information is being gained on the fundamental aspects of this challenging problem.

One crucial factor in determining how fast heat moves outwards from the ignition/combustion points is the permeability of the woodpile. This has been shown by work carried out on a 3 ton/day pilot-scale wood residue pyrolysis plant in Papua New Guinea [28]. Air was admitted to the base of this unit through a rotating 'airgitator'. An experiment undertaken using a fine sawdust/shavings fraction of less than 4 mm in size showed that the heat produced in the partial combustion zone penetrated only a short distance upwards, creating a narrow intensely hot zone immediately above the combustion zone. Temperature gradients at the top edge of this intensely hot zone were extremely high, with wood particles only a short distance above this zone remaining at temperatures close to ambient. This pattern of behaviour can be explained in terms both of the resistance to flow posed by the packed bed of small particles together with the large surface to volume ratios of the fine particles. Together these created a situation where both the heat transfer coefficient and the heat transfer surface were large, and hence the combustion gases lost their heat content very rapidly.

A markedly different pattern of behaviour was observed when a coarse fraction of wood shavings greater than 4 mm in size was used. In this case the hot combustion gases could pass so easily through the packed bed of particles that they were still quite hot when they reached the top of the bed of particles. Temperature gradients through the unit were gradual, and outside the combustion zone temperatures were low and well below those needed to initiate pyrolysis [28]. Not only was there no pyrolysis but some of the heat produced was escaping unused from the unit - a waste of wood resources.

It is evident that what was needed in the pyrolysis unit was a pattern of behaviour somewhere in between the above. This was shown to be achievable when the fine and coarse fractions of the feed were combined, though maintenance of stable operating conditions was difficult [28].

Given the comparatively high permeability of a woodpile composed of logs up to 200 mm in diameter, and containing very few small wood pieces, temperature profiles in the kiln might be expected to be gentle, as in the above-mentioned pyrolysis unit when using a feed of coarse shavings. Partial confirmation of this comes from a recent study of temperature profiles in a packed bed of comparatively large wood particles (80 mm long and 60 mm in diameter) heated by hot inert gases [29]. Temperature gradients were gradual across the upper part of this bed but a lot steeper close to the gas inlet. The latter finding was attributed to the fact that the wood used was green and that a high rate of moisture removal was occurring in the inlet region. It was also noted that temperatures at different levels in the bed were not particularly uniform, suggesting (not unexpectedly) that some channelling of the gases was occurring.

What is not clear from the above discussion is why the pyrolysis unit using a coarse feed of wood shavings was unable to initiate pyrolysis whereas in kilns, where the permeability of the wood particles is even greater, pyrolysis is readily achieved. This is attributed to the fact that the combustion zones in kilns are localised at a few inlet points whereas in the pyrolysis unit the combustion zone covered the unit's entire base. The effect of limiting the number of air inlet points is to increase both the volumetric heat release rate and the gas flow rate. This will increase both the temperature difference between the gas and the wood particles surrounding the inlet point and also the velocity and turbulence levels in the gas flows (and hence the heat transfer coefficient). So a greater rate of heat transfer and a correspondingly steeper temperature gradient can be expected in the region around each air inlet point.

One implication of the above discussions is that for any given kiln there will exist an optimum number and arrangement of air inlet points. If there are too few points, it could be expected that the carbonisation fronts will take longer than is desirable to reach the kiln extremities. If there are too many points, it may be necessary to increase rates of air addition in order to achieve desired temperature levels, and this may bring about a reduction in charcoal yield.

Mass transport

A number of the mass transport/mass flow problems of relevance to kiln operation have been alluded to above. It is evident that modelling of kiln behaviour will require a better understanding than we have at present of the gas flow paths within wood stacks of various patterns. Also needed is a knowledge of how the channels through such wood stacks alter as wood pieces carbonise and shrink - in a non-uniform way - and the wood/charcoal pile partly collapses. A technique that could be helpful in characterising gas flows in such piles has been developed [30]. It involves measuring the air permeability of woodchips and appears readily extendable to heaps of charcoal and other wood/charcoal mixtures where the permeability is not too high.

Another mass transport process about which more needs to be known is the migration of condensable vapours that occurs ahead of the carbonisation front. Both water and other volatile substances are known to vapourise as temperatures rise and to be carried by gas flows into cooler regions where they condense. As temperatures at their new position increase these compounds may again evaporate and move on to cooler regions. This cycle of vapourisation and condensation is important since, as is widely appreciated in briquette-making, it can enhance heat transfer rates quite significantly.

Reaction processes

Many of the reaction processes of relevance in a charcoal kiln are microscale processes in the sense that they are associated with the decomposition processes in individual particles. Reactions not falling into this category are largely those associated with the combustion of wood pieces close to the air inlet points. Given the global importance of wood combustion, it was initially assumed when preparing this paper that there is already sufficient information available on wood combustion reactions for this not to be a limiting factor in developing a charcoal kiln model. However, in a recent paper it was stated that 'no generally applicable model exists to describe the thermal conversion of a packed bed of solid fuel particles' [31]. This paper goes some way towards remedying this lack but only for a wood combustor operating with an adequate air supply. Further work is needed to determine how this model needs to be modified before it can be applied to the more oxygen-limited combustion environments in kilns.

CONCLUDING COMMENTS

It is evident that the contribution fundamental research can make to improving charcoal kiln efficiency is constrained by practical and economic considerations. Further research into the processes occurring within individual wood pieces could in theory help in the fine tuning of kiln operating procedures. However, especially on sites where a number of kilns are being operated simultaneously, kiln schedules are strongly influenced by the availability of manpower and this can severely limit operating flexibility. It is nonetheless likely that adding a good understanding of microscale processes to wisdom gained through practical experience could still be of help to producers when wood characteristics alter and operating procedures have to be modified.

A better understanding of the fundamentals of macroscale processes seems to have far more potential to assist charcoal producers improve kiln designs and operating methods. However, there is much still to be learnt about these processes and the development of a comprehensive and accurate model of kiln behaviour appears a long way off. The benefits of developing such a model would nevertheless be considerable and certainly sufficient to warrant further research on the heat transport, mass transport and reaction processes taking place in existing kilns.

ACKNOWLEDGEMENTS

Ms Melanie Viljoen's help in obtaining much of the information used in preparing this paper is much appreciated.

REFERENCES

1. Rosillo-Calle F., de Rezende M. A. A., Furtado P. & Hall D. O. (1996) The charcoal dilemma. Intermediate Technology Publications, London.
2. D'Apote S. L. (1998) IEA biomass energy analysis and projections. In: *Biomass Energy:.Data, Analysis and Trends*, pp. 151-79. IEA/OECD, Paris.

3. Ackerman F. & de Almeida P. E. F. (1990) The industrial fuelwood crisis in Minas Gerais. *Energy Policy*, **18** (7), 661-8.
4. Foley G. (1986) Charcoal making in developing countries. *Technical report No. 5*, Earthscan - International Institute for Environment and Development, London.
5. Emrich W. (1985) Handbook of charcoal making. D. Reidel Publishing, Dordrecht.
6. Briane D. & Doat J. (1985) Guide technique de la carbonisation. EDISUD, Aix-en-Provence.
7. Chidumayo E. N. (1991) Woody biomass structure and utilisation for charcoal production in a Zambian miombo woodland. *Bioresource Technology*, **37**, 43-52.
8. Khristova P. & Khalifa A. W. (1993) Carbonization of some fast-growing species in Sudan. *Applied Energy*, **45**, 347-54.
9. Mok W. S-L., Antal M. J., Szabo P., Varhegyi G. & Zelei B. (1992) Formation of charcoal from biomass in a sealed reactor. *Ind. Eng. Chem. Res.*, **31**, 1162-1166.
10. Paddon A. R. & Harker A. P. (1980) Charcoal production using a transportable metal kiln. *Rural Technology Guide 12*, Tropical Products Institute, London.
11. Wartluft J. L. & White S. (1984) Comparing simple charcoal production technologies for the Caribbean. Volunteers in Technical Assistance, Arlington, USA.
12. Connor M. A. & Salazar C. M. (1988) Factors affecting the decomposition processes in wood particles during low temperature pyrolysis. In: *Research in Thermochemical Biomass Conversion* (Ed. by A. V. Bridgwater & J. L. Kuester), pp. 164-78. Elsevier, London.
13. Thomas R J (1977) Wood structure and composition. In: *Wood Technology: Chemical Aspects* (Ed. by I. S. Goldstein), pp. 1-23. American Chemical Society, Washington, DC.
14. Brito J. O. & Barrichelo L. E. G. (1977) Correlações entre características fisícas e químicas da madeira e a produção de carvão vegetal: I. Densidade e teor de lignina da madeira de eucalypto. *IPEF, Piracicaba*, **14**, 9-20.
15. Connor M. A., Viljoen, M. H. & Ilic, J. (1996) Relationships between wood density, wood permeability and charcoal yield. In: Developments in Thermochemical Biomass Conversion (Ed. by A. V. Bridgwater & D. G. B. Boocock), pp. 82-96. Blackie, London.
16. Browne F. L. & Tang W. K. (1962) Thermogravimetric and differential thermal analysis of wood and of wood treated with inorganic salts during pyrolysis. *Fire Research Abstracts and Review*, **4**, 76-91.
17. Hirata T., Kawamoto S. & Nishimoto T. (1991) Thermogravimetry of wood treated with water-insoluble retardants and a proposal for development of fire-retardant wood materials. *Fire and Materials*, **15**, 27-36.
18. Jenkins B. M., Bakker R. R., Baxter L. L., Gilmer J. H. & Wei J. B. (1997) Combustion characteristics of leached biomass. In: *Developments in Thermochemical Biomass Conversion* (Ed. by A. V. Bridgwater & D. G. B. Boocock), pp. 1316-30. Blackie, London.
19. Connor M. A., Kisler J. P., Alesich N. I., Kane M. G., Watkins R. N. & Shallcross D. C. (1997) The pyrolytic decomposition and subsequent combustion of rice husks. In: *Developments in Thermochemical Biomass Conversion* (Ed. by A. V. Bridgwater & D. G. B. Boocock), pp. 67-81. Blackie, London.
20. Paeniu B. (1988) Charcoal Kiln Project. An Innovative Rural Technology of the Energy Office of the Commonwealth of the Northern Marianas. *Case Studies of*

Rural Development Experiences in the Pacific: Case Study No. 2, South Pacific Commission, Noumea, New Caledonia.

21. Kilzer R. J. & Broido A. (1965) Speculations on the nature of cellulose pyrolysis, *Pyrodynamics*, **2**, 151-63.
22. Antal M. J., Croiset E., Dai X., DeAlmeida C., Mok W. S-L., Norberg N., Richard J-R. & Al Majthoub M. (1996) High yield biomass charcoal. *Energy and Fuels*, **10** (3), 652-8.
23. Cutter B. E. & McGinnes E. A. (1981) A note on density change patterns in charred wood. *Wood and Fiber*, **13** (1), 39-44.
24. Smith D. N. & Lee E. (1958) The longitudinal permeability of some hardwoods and softwoods. *Department of Scientific and Industrial Research, Forest products Research: Special report No. 13*, H.M.S.O., London.
25. Comstock G. L. (1965) Longitudinal permeability of green Eastern Hemlock. *Forest Products Journal*, **15**, 441-9.
26. Shah N., Girard P., Mezerette C. & Vergnet A. M. (1992) Wood-to-charcoal conversion in a partial-combustion kiln: an experimental study to understand and upgrade the process. *Fuel*, **71**, 955-962.
27. Johnstone D. A. & Gore W. T. (1985) The manufacture of industrial grade charcoal in Armco-Robson kilns. Paper presented at a Conference: *Forest Products Research International - Achievements and the Future*, Pretoria, 19pp.
28. Connor M. A. (1983) Heat and mass transfer considerations in fuel production from wood wastes by pyrolysis. *Regional Journal of Energy, Heat and Mass Transfer*, **5**, 179-94.
29. Aganda A. A., Murray P. W. & Kionga-Kamau S. (1997) Temperature profiles in a wood packed bed heated by hot inert gases. *Trans IChemE*, **75**, Pt A, 677-84.
30. Ernstson M-L. & Rasmuson A. (1992) Field and laboratory measurements of the air permeability of chipped forest fuel materials. *Fuel*, **71**, 963-70.
31. Bruch C., Peters B. & Nussbaumer T. (2000) A general model for the investigation of packed bed combustion with respect to wood. Paper presented at a Conference: Progress in Thermochemical Biomass Conversion, Tyrol, Austria, 17-22 Sept., 13pp.

Effect of Four Physical Characteristics of Wood on Mass and Energy Flows from Slow Pyrolysis in Retorts

Y. Schenkel
CRA, Agricultural Engineering Department
Chaussée de Namur, 146 - B 5030 Gembloux Belgium

ABSTRACT : The optimisation of charcoal production in a retort kiln calls for control of the carbonization mass and energy flows. These depend essentially on three types of factors : the physico-chemical characteristics of the raw material, the operational parameters and the reactor parameters. Carbonization experiments have been conducted to assess and to model the effect of four physical characteristics : moisture content (two levels : 0 %, 37 % dry basis), density (two levels : beech wood (650 - 740 kg/m^3 anhydrous basis) and poplar wood (398 - 426 kg/m^3 anhydrous basis)), dimension and shape (two levels : cubes of 4 cm side and blocks of 4x4x16 cm, length parallel to the fibres orientation). The carbonization final temperature was 500° C, the residence time at this temperature was 100 min. The heating rates were 2 and 20° C/min. These experiments show the considerable importance of moisture content regarding mass and energy flows ; the other characteristics are of secondary importance. The results lead to the development of a phenomenological modelling of mass and energy flows of the total volatile matter produced through wood carbonization in retort kilns, based on the logistic symmetric function.

INTRODUCTION

The physical characteristics of wood can exert a considerable effect over the carbonization process and products. Water present in wood solids influences heat transfer : the evaporation of the water requires a significant part of the heat and limits the increase of temperature of the solid material. This results in a longer process of carbonization [1-3], lower mass and energy yields [2, 4, 8], and the quality of charcoal is modified [1, 4].

Heat transfer inside the wood so!id depend on its thermal conductivity. This property is anisotropic and increases with the density, the moisture content and the temperature [10-13]. The thermal conductivity is higher in the longitudinal direction than in the radial and tangential directions : from 1.8 [12] to 2.5 [14, 15].

The specific heat capacity of wood depends on its temperature and moisture content but not on its density or on the wood species [12, 16, 17]. Several authors have proposed a formula to determine the specific heat capacity of wood [13, 16, 18, 19].

The permeability of wood is a characteristic of the relative resistance it poses to internal material (water, steam, gases) transfers when heated. Permeability is also an anisotropic property. It is higher in the longitudinal direction (parallel to the wood fibres) in comparison with the transverse direction [20 - 22]. Permeability varies considerably from species to species and increases with the temperature of the wood solid matrix [22 - 23].

Because of the anisotropy of wood, the dimension and the shape of wood solids are determining factors of the process of thermal decomposition. Heat transfer is more rapid in the longitudinal direction, the limiting factors are related to the transversal direction. Consequently the thickness of the wood solid and the ratio transversal surface/tangential surface are critical [7, 24, 29]. An increasing dimension of wood particles results in a longer carbonization time, an increased production of char and non condensable gases to the detriment of pyrolysis oil, although it is very difficult to predict the effect of the dimension and shape of wood particles on the composition and the quantity of each product of pyrolysis. The dimension determines also the type of thermal decomposition of the wood particle. If the particles are small ("thermally thin"), pyrolysis is controlled by the chemical kinetics ; if the particles are thick ("thermally thick"), pyrolysis is controlled by the heat transfers inside the wood solid. The limit in thickness is 5 mm [30], 0.2 mm [31] or 1 mm [32]. But the dimension and the shape of wood particles also determine several characteristics of the bed of particles, such as the porosity, the bulk volume and the ratio between the surface of reaction and the volume of the particles. These parameters have an effect on the pyrolysis process, especially in fixed or moving beds. But many quantitative relations are still largely unknown.

The density and the porosity of wood seem also to play a role in the pyrolysis process. But the effects observed are rather qualitative than quantitative. Wood with a high density and a low porosity tends to produce more charcoal [23, 29, 33]. The density influences also the productivity of pyrolysis reactors through the bulk volume : the same volume of a reactor will produce more charcoal if a high density wood is pyrolysed compared to a light wood.

The effect of the physical characteristics of wood on slow pyrolysis or carbonization has thus been quite extensively studied. However, these studies focus mainly on the qualitative effects of the physical parameters of the material. The optimization of the carbonization process as operated by charcoal producers around the world requires quantitative relations, particularly regarding mass and energy flows as determining operational parameters of a dynamic process such as carbonization.

The carbonization process aims at the production of charcoal, preferably in large pieces (> 2 cm). Consequently, the carbonization process is mainly the thermal decomposition of thermally thick particles.

A considerable amount of work has been done on the characterization and modelling of the pyrolysis of thermally thick particles of wood, [13, 17, 25, 28, 29, 32, 34 - 47]. These experiments have been conducted on single particles of wood (cylinder, sphere, parallelepiped). However, few experiments have been carried out on a bed of thermally thick particles of wood [48-51], and we have not identified in the literature any work on the modelling of the pyrolysis of a bed of thermally thick particles.

We have therefore conducted an experimental work aiming at :

1° Identifying the physical characteristics of wood that have a significant effect on the dynamics of carbonization of beds of thermally thick particles.

2° Building a model linking the mass and energy flows of this carbonization process to the physical characteristics of wood.

MATERIALS AND METHODS

The carbonization tests were conducted in a 27 l thermobalance kiln. A short description of the installation is given below. A complete description is presented in [52].

(a) *The reactor* : cube-shaped, it has a capacity of 27 litres. The inside walls are made of refractory bricks. The feedstock is put in a metallic basket and the reactor is closed tightly with a cordon of clay which is replaced at each experiment. Sealed up, the reactor is totally airtight.
(b) *The heating control system* : the electrical resistances are driven by a numeric regulator which sets the temperature profile and gradient as well as the final temperature of pyrolysis. The amplitude variation of the real temperature is more or less 20°C in comparison with the temperature settings.
(c) *The electronic balance* : the reactor is set on an electronic balance to follow continuously the loss of weight (measurement precision : 0.5 % at 10 kg).
(d) *The gas conditioning and analysis system* : at the exhaust of the reactor, the pyrolysis gases are first condensed, filtered and continuously analysed (non condensable fraction). The liquid fraction (condensable gases) is collected and weighed. Thermocouples give the gases' temperature at different levels of the plant : in the reactor and along the conditioning line.

The non condensable gases are analysed by means of a NDIR spectrometer (determination of carbon monoxide (CO), carbon dioxide (CO_2) and methane (CH_4) content), a thermal conductivity analyser (hydrogen (H_2) content) and a magnetomechanics analyser (oxygen (O_2) content).

The pyrolysis oil is collected at the bottom of the condensation columns, weighed to obtain the wet mass before determining its moisture content (Karl Fisher method).

The main experimental plan is based on 4 factors and 6 fixed parameters :

1. Species : *Fagus sylvatica* (beech), *Populus x canadensis* (poplar).
2. Dimension/shape : cubes 4 cm side, blocks 4 x 4 x 16 cm (length parallel to fibres).
3. Moisture content : anhydrous (H0) and wet (moisture content ranging between 23 and 55 % d.b., average 37 % d.b. - H37).
4. Heating rate : 2 and 20 °C/min.

1. Final temperature : 500 °C.
2. Initial temperature : 20 °C (at time 0).
3. Residence time at final temperature : 100 min.
4. Mass of samples : 4.5 kg (anhydrous), bulk arrangement.
5. Replications : 3.
6. Cooling : natural in the kiln, at least 24 hours.

The scheme is a full factorial plan. Other factors have been tested in orientation experimentation : cubes of 2 and 8 cm side, medium moisture content (11 and 21 %

d.b.) and wood blocks arrangements (peripheral and central piles, alternate rows). The two chosen wood species present very different physical characteristics : density, thermal conductivity, permeability. Table 1 gives the mean values of the major physical characteristics ; the detailed values are available in [52].

Table 1 Mean values of moisture content, density, thermal conductance, specific heat capacity, void volume, ratio surface of reaction/volume of the particles of the wood samples (*Fagus sylvatica* - beech, *Populus x canadensis* - poplar).

		Beech		Poplar	
		Cubes 4	Blocks 16	Cubes 4	Blocks 16
Moisture content (% d.b.)					
	H0	0.00	0.00	0.00	0.00
	H37	38.25	36.10	36.48	38.88
Density (kg/m^3 d.b.)		650 to 740		398 to 415	
Thermal conductance (W/K at 293 K)					
// fibres	H0	17.434	4.153	11.028	2.767
	H37	24.299	5.698	14.843	3.774
⊥ fibres	H0	7.748	7.382	4.902	4.919
	H37	10.799	10.130	6.597	6.709
Specific heat capacity (kJ/kg K at 293 K)					
	H0	1.235			
	H37	2.034		2.038	
Void volume (%)		0.56	0.64	0.56	0.64
Surface/volume		140	112.5	150	112.5

The mass and energy flows of carbonization have been analysed through a phenomenological approach, as regularly developed in chemical engineering. To determine the parameters of the mass and energy flows, we have considered the production of non condensable gases and pyrolytic liquids as well as the volatilization of the solid, as a growth function : the masses and energy produced by the carbonization process, measured according to the time of carbonization, are cumulated. In order to analyse and compare the results obtained, a mathematical function has been searched and the symmetric logistic function identified. The equation of this function is :

$$y = M / \left(1 + e^{-[(x-a)/b]}\right)$$

where y = mass of product (g).
x = time of carbonization (min).
M = maximal mass corresponding to a infinite time (g).
a = time corresponding to 50 % of the produced mass (abciss of the inflexion point (min).
b = indicator of the reaction speed at the point of 50 % of produced mass.

The symmetric logistic function is shown in figure 1. For the solid volatilization function, a transformation of variables has been applied to convert the decreasing function (loss of weight) into a growth function (quantity of volatile matter produced).

After adjustment, a statistical analysis of variance is applied to the three parameters of the symmetric logistic function, M, a, b.

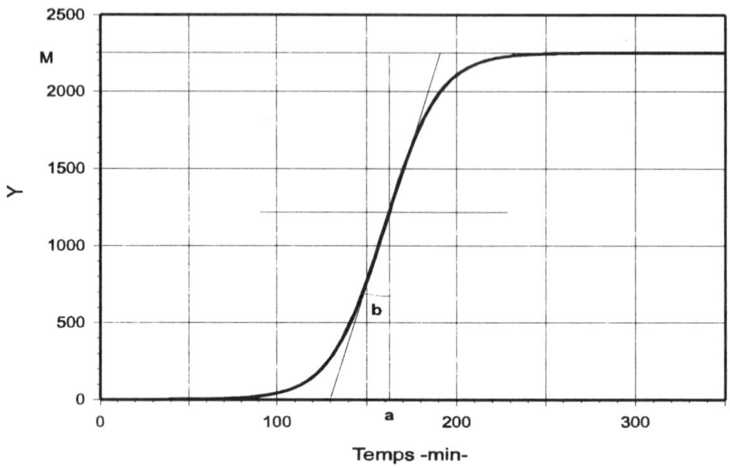

Figure 1 The symmetric logistic function and its parameters, M, a, b.

The energy flows of carbonization have been determined for the pyrolytic liquids and the non condensable gases. In the case of pyrolytic liquids, the energy content is determined in two steps. Firstly, the actual net calorific value of pyrolytic liquids is measured on a wet basis. Multiplying the net calorific value by the corresponding mass of pyrolytic liquids, we obtain an energy flow. In a second step, we must take into account the fact that, in an industrial process, the two fractions of gaseous effluents – condensable and non condensable phases – are generally burned in the gaseous phase. This means that the heat required by the vaporisation of the two fractions of condensable gases (water and pyrolytic liquids) is not necessary. Knowing the proportion of each phase for each sample of condensable gases, we have added to the actual net calorific value of the wet condensed matter, the product of the mass of each phase and its latent heat of vaporisation (water : 2500 kJ/kg, pyrolytic liquid : 450 kJ/kg).

The calorific value of non condensable gases is calculated on the basis of their chemical composition in terms of CO, H_2, CH_4, C_2H_4. Then, the calculation is identical to the one used for the condensable fraction : multiplication of the net calorific value by the corresponding mass of non condensable gases and obtention of an energy flow per unit of time. A total energy flow is obtained by adding the energy flows of the condensable and non condensable fractions of the gaseous effluents of carbonization

Like the mass flows, the energy flows have been transformed into a growth function, to which the symmetric logistic function is adjusted. The parameters of this function, M, a, b, are analysed by means of an analysis of variance. These parameters do not correspond to a production of mass but to a production of energy (kJ).

RESULTS

In this paper, the results are analysed for the two total functions, the total volatile matter production and the total energy production. The results related to each fraction of the gaseous effluents, condensable and non condensable phases, are presented in detail in [52].

MASS FLOWS

Moisture content

A very highly significant effect of the moisture content of wood on the symmetric logistic function is observed as would be expected (table 2 and figure 2).

Table 2 Effect of the moisture content on the parameters of the symmetric logistic function adjusted to the total volatile matter mass flows.

	M	a	b
H0	3 200	101	8.13
H37	5 162	159	27.76
Standard error	*97*	*1.4*	*1.21*

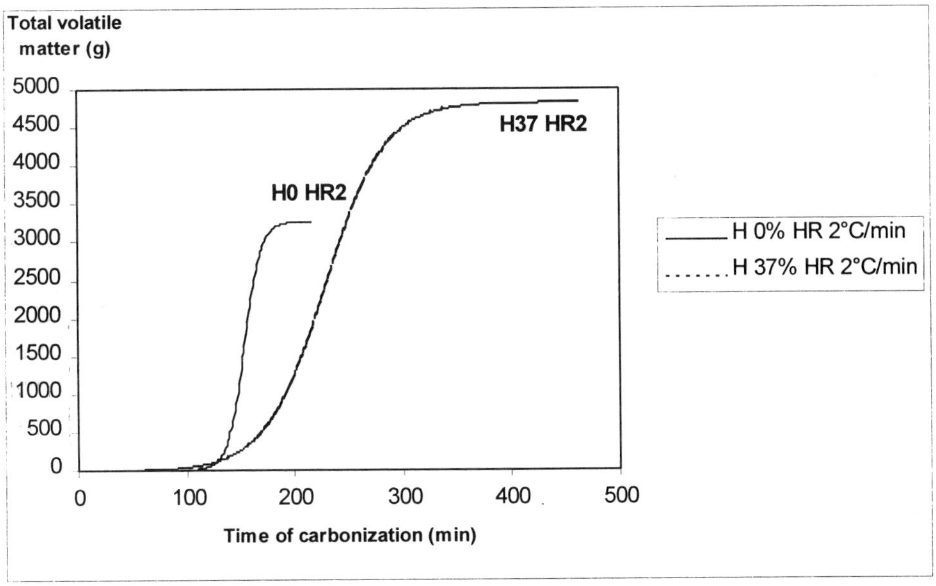

Figure 2 Evolution of the total volatile matter mass flows according to the moisture content of beech wood samples (anhydrous – H0, wet - H37) (heating rate : 2°C/min).

As expected, the factor M (maximal mass of matter volatilized) is significantly influenced by the moisture content of wood. The difference between anhydrous and wet wood samples is naturally explained by the quantity of water which has to be evaporated, contributing to an increase of the value of M for the symmetric logistic functions for wet wood samples (H37).

The two other factors a and b are also influenced by the moisture content. The mass flows are delayed and slowed down by an increasing moisture content of the wood, as shown in figure 2.

The effect of moisture content is so overwhelming that it could hide a possible effect of the other parameters. Consequently, a further analysis of the results has been conducted separately on the results for anhydrous and wet wood samples.

Anhydrous wood samples

The mean value of the parameters of the symmetric logistic function are given in table 3.

Table 3 Mean values and standard error (italic) of the parameters of the symmetric logistic function adjusted to the total volatile matter production from anhydrous wood samples (H0) carbonization.

	M	a	b
Beech cubes 4			
heating rate 2 °C/min	3 171 *(126)*	157 *(1)*	9.3 *(0.6)*
20 °C/min	3 170 *(29)*	47 *(1)*	7.1 *(0.6)*
Beech blocks 16			
heating rate 2° C/min	3 267 *(127)*	154 *(2)*	7.7 *(0.7)*
20 °C/min	3 255 *(49)*	48 *(2)*	6.1 *(0.3)*
Poplar cubes 4			
heating rate 2 °C/min	3 188 *(100)*	149 *(2)*	9.1. *(0.4)*
20 °C/min	3 190 *(51)*	46 *(1)*	6.2. *(0.3)*
Poplar blocks 16			
heating rate 2°C/min	3 175 *(30)*	156 *(1)*	9.4 *(0.6)*
20 °C/min	3 186 *(45)*	55 *(4)*	10.2 *(2.1)*

The factor M (maximal mass of total volatile matter) is not significantly influenced by the shape, the species or the heating rate. However, the factors a and b are significantly influenced.

The heating rate has of course a considerable effect on the factor a (time corresponding to 50 % of volatile matter production). The time is delayed by a factor of 3 for the heating rate 2 °C/min, compared to the heating rate 20 °C/min. The factor b (indicator of the reaction speed at the point of 50 % production) is also influenced, but significantly less. The reaction is more rapid in the case of heating rate 20 °C/min (b = 7.4) than of a 2 °C/min rate (b = 8.9).

We observe a significant effect of the shape of the wood pieces on factor a. This time is slightly delayed for the wood blocks 16 (a = 103) in comparison with cubes 4 (a = 100). In the case of factor b, it is the species that has a significant effect. The carbonization reaction is more rapid for beech (b = 7.5) then for poplar (b = 8.7). For both factors a and b, we observe a significant interaction between the species and the shape. Blocks 16 of poplar are significantly different from the other wood samples : factors a and b are higher. The result is that the carbonization process of blocks 16 of poplar is delayed and slowed down.

Wet wood samples

The mean values of the parameters of the symmetric logistic function are given in table 4.

Table 4 Mean values and standard error (italic) of the parameters of the symmetric logistic function adjusted to the total volatile matter production from wet wood samples (H37).

	M	a	b
Beech cubes 4			
heating rate 2 °C/min	5 636 *(72)*	236 *(1)*	37.6 *(0.3)*
20 °C/min	5 121 *(186)*	75 *(2)*	16.3 *(0.8)*
Beech blocks 16			
heating rate 2° C/min	4 834 *(81)*	228 *(2)*	27.6 *(2.2)*
20 °C/min	4 876 *(100)*	89 *(3)*	22.7 *(1.2)*
Poplar cubes 4			
heating rate 2 °C/min	5 035 *(340)*	224 *(6)*	29.0 *(5.1)*
20 °C/min	5 071 *(120)*	78 *(2)*	19.7 *(0.5)*
Poplar blocks 16			
heating rate 2°C/min	5 969 *(915)*	253 *(9)*	44.4 *(10.8)*
20 °C/min	4 749 *(80)*	86 *(3)*	24.8 *(2.2)*

As for anhydrous wood samples, the three factors - shape, species, heating rate - have no significant effect on the factor M. Also, the heating rate has of course a significant effect on the factors a and b : the production of volatile matter is delayed and slowed down for the rate 2 °C/min.

The shape of wood pieces significantly influences also the factor a. This one is equal in average to 153 min for cubes 4 and to 164 min for blocks 16. A significant interaction between the shape and the species is also observed for the factor a. As for anhydrous wood samples, blocks 16 of poplar differ significantly from other wood samples by a delayed volatilization of the material.

ENERGY FLOWS

Moisture content

As for mass flows, the analysis of variance shows a very highly significant effect of the moisture content of wood samples on energy flows (table 5).

Table 5 Effect of the moisture content on the parameters of the symmetric logistic function adjusted to the total energy flows of carbonization.

	M	a	b
H0	24 502	104	7.3
H37	29 123	175	20.4
Standard error	*745*	*1.8*	*0.7*

The maximal energy production is higher for wet wood samples than for anhydrous wood. The difference lies in the energy production by the non condensable fraction of gaseous effluents, while the energy production by the pyrolytic liquids is similar for anhydrous and wet wood samples [52]. Furthermore, energy flows are significantly delayed and slowed down by the water present in the wood.

Again, as for mass flow, the effect of moisture content can hide other significant effects of the other factors of variation. Consequently, a further analysis of variance has been conducted separately on anhydrous and wet wood samples. However, this supplementary analysis of variance shows two common observations for anhydrous and wet wood samples.

1° The factor M (maximal quantity of energy) is not significantly influenced by the shape, the species or the heating rate. Thus, only the moisture content has a significant effect on the total production of energy from the gaseous effluents.

2° Whatever the moisture content, the heating rate has a significant effect on the factor a (time corresponding to 50 % production) but not on the factor b (indicator of the reaction speed at the point of 50 % production). There is thus a delaying effect on the energy production when the heating rate is decreasing.

Anhydrous wood samples

The mean values of the parameters of the symmetric logistic function are given in table 6. We have identified significant effects of the shape and the species on factor a and b of the total energy production, as well as a significant interaction between shape and species. Blocks 16 of poplar are characterized by a total energy flow delayed and slowed down compared to other wood samples.

Table 6 Mean values and standard error (italic) of the parameters of the symmetric logistic function adjusted to the total energy production from anhydrous wood samples (H0) carbonization.

	M	a	b
Beech cubes 4			
heating rate 2 °C/min	27 120 *(2402)*	159 *(3)*	8.4 *(0.6)*
20 °C/min	21 488*(1067)*	47 *(0)*	6.6 *(0.2)*
Beech blocks 16			
heating rate 2° C/min	24 147 *(1012)*	155 *(3)*	6.6 *(0.4)*
20 °C/min	25 490*(571)*	49 *(2)*	5.5 *(0.1)*
Poplar cubes 4			
heating rate 2 °C/min	23 264*(1429)*	153 *(1)*	6.8 *(0.1)*
20 °C/min	24 273*(722)*	48 *(1)*	6.3 *(0.5)*
Poplar blocks 16			
heating rate 2°C/min	24 382*(1274)*	160 *(2)*	7.5 *(1.1)*
20 °C/min	25 491 *(1881)*	58 *(3)*	10.7 *(1.2)*

Wet wood samples

The mean values of the parameters of the symmetric logistic function are given in table 7.

Table 7 Mean values and standard error (italic) of the parameters of the symmetric logistic function adjusted to the total energy production from wet wood samples (H37) carbonization.

	M	a	b
Beech cubes 4			
heating rate 2 °C/min	34 991 *(244)*	248 *(1)*	22.9 *(0.9)*
20 °C/min	30 855 *(2089)*	80 *(1)*	12.6 *(1.2)*
Beech blocks 16			
heating rate 2° C/min	27 190 *(1766)*	241 *(6)*	19.2 *(2.8)*
20 °C/min	30 361 *(1352)*	105 *(5)*	19.9 *(1.8)*
Poplar cubes 4			
heating rate 2 °C/min	27 486 *(2229)*	239 *(7)*	19.5 *(4.5)*
20 °C/min	27 945 *(3177)*	100 *(3)*	19.8*(1.7)*
Poplar blocks 16			
heating rate 2°C/min	27 971 *(4940)*	270 *(7)*	26.8 *(3.2)*
20 °C/min	29 033 *(3115)*	110 *(7)*	23.3 *(3.5)*

The factor a is significantly influenced by the species and the shape of wood samples. Total energy flows are delayed for poplar in comparison with beech and for blocks 16 in comparison with cubes 4. However, there is no significant effect on factor b nor significant interaction.

DISCUSSION

The major fact resulting from the analysis of mass flow from wood carbonization is the general and considerable effect of moisture content on the process. The effect on the factor M of the symmetric logistic function (maximal mass of volatile matter produced at an infinite time) is expected : the water present in wood is evaporated and constitutes a part of the volatile matter. It is thus natural to observe a factor M higher for wet wood samples (H37) than for anhydrous samples (H0).

The effect of moisture content on the two other factors of the symmetric logistic function (a - time corresponding to 50 % production, b – indicator of the speed of reaction) is also noteworthy. Confirming the results of orientation experiments (moisture content * dimension) [52] these results demonstrate the delaying and slowing down effects of the water presents in wood on the carbonization process.

We expected also a natural effect of the heating rate on the factors a and b of the symmetric logistic function. This effect is observed : the production of volatile matter is delayed and slowed down for a heating rate of 2 °C/min in comparison with a 20 °C/min rate.

Compared with the effect of the moisture content, the effect of the species and the shape of wood samples is considerably lower. In fact, the interaction shape*species shows that blocks 16 of poplar are characterized by total volatile mass flows delayed and slowed down whatever the moisture content of the wood. It is not an effect of the shape or of the species : cubes 4 of poplar and blocks 16 of beech have factors a and b of the symmetric logistic function almost equal to those of cubes 4 of beech. It is definitely the blocks 16 poplar which show different results.

The cause of this difference could be a possible combined effect of the thermal conductance and the gas permeability. The thermal conductance is the lowest for blocks 16 poplar (see table 1) and the permeability to gases is the lowest for poplar. These two factors combined could contribute to delay and slow down the production of volatile matter for blocks 16 of poplar. However, we must stress the fact that the difference between these blocks 16 poplar and the other samples is not that big, although significant ; in the case of factor a, the difference is 5 to 6 min for anhydrous wood samples, 20 to 30 min for wet wood samples, when moisture content generates differences between anhydrous and wet samples of the order of 60 min.

The adjustment of the symmetric function to the energy flows from wood carbonization is an original and dynamic (and no more static) approach. The analysis of the symmetric logistic function demonstrates again the dramatic effect of water. As for mass flows, energy flows are delayed and slowed down for wet wood samples (H37). Water intervenes through the large quantities of heat it requires to be evaporated and eliminated from the solid matrix.

Regarding the effect of the shape and the species, the analysis must be conducted separately for anhydrous (H0) and wet (H37) samples. For anhydrous wood pieces (H0), we observe the same effects as for mass flows ; a delay of the energy flows for blocks 16 of poplar. There is a direct link between energy and mass flows.

However, for wet wood samples (H37), we observe a significant effect of the shape and of the species but without significant interaction between both factors. Energy flows are delayed for blocks 16 compared with cubes 4 and for poplar compared with beech. The effect of the lengthening of the wood blocks in the direction of the fibres as well as the effect of the species, act on the physical properties of wood samples in carbonization (decrease of the thermal conductance and of the permeability to gases). These effects could be enhanced by the moisture content of wood.

Regarding the factor M of the symmetric logistic function (maximal quantity of energy produced), we observe a very high variability of the value taken by the factor M. This high variability of experimental results does not allow us to demonstrate a determining effect of one of the factors of variation.

CONCLUSIONS

Moisture content of wood appears to be the most important physical parameter to take into account in wood carbonization. An increasing moisture content decreases the production of pyrolytic liquids and increases the production of non condensable gases by enhancing secondary reactions of pyrolysis inside the solid matrix. This effect is increased by the shape of wood samples ; wet blocks of 4 *4 * 16 cm produce less pyrolytic liquids and more non condensable gases than other wood samples. However, moisture content does not influence the chemical composition of carbonization products. This confirms the fact that water present in wood acts physically and not chemically in the carbonization of wood at low temperature and heating rate.

Moisture content delays and slows down mass and energy flows of carbonization. But for energy flows, it is difficult to demonstrate a determining effect of water on the maximal quantity of energy produced ; the values we obtain are very variable and do not lead to a reliable quantitative relation between the moisture content and the factor M of the symmetric logistic function.

Compared with the moisture content effect, the other parameters tested have a considerably lower effect on the carbonization process. If the heating rate also determines the mass and energy flows, as expected, it does not influence the carbonization yields nor the chemical composition of the products. This confirms the results of the literature : heating rates lower than 100 °C/min do not influence mass balances and products composition from the carbonization of wood.

The shape and the species of wood samples (cubes of 4 cm side and blocks of 4 * 4 * 16 cm // fibres) do not exert a significant and general effect on carbonization cycles. We simply observe a different behaviour of blocks 16 of poplar that show delayed and slowed down mass and energy flows. The cause is perhaps a combined effect of a low thermal conductance (lengthening in the direction of fibres) and of a low permeability to gases (poplar species, lengthening in the direction of fibres). Besides, the shape and the species of wood samples slightly influence energy flows from carbonization. Energy production cycles from gaseous effluents of carbonization are delayed for blocks 16 in comparison with cubes 4, for poplar in comparison with beech. This new approach of mass and energy flows of carbonization based on the symmetric logistic function can be used to define a phenomenological modelling of wood carbonization.

REFERENCES

[1] Gore W.T. (1982). Charcoal Production and Properties. *CSIR Special Report Hout 262, Pretoria, South Africa*, 117 p.

[2] Carré J, Hébert J., Lacrosse L. (1984). Analyse critique de la valorisation des matières ligneuses par voie sèche. *Rapport CE, DG VIII*, 113 p.

[3] Shah N., Girard P., Capart R. (1989) - Carbonization of Wood : Product Analysis and Energy Assessment. *Applied Energy*, **34**, 223-241.

[4] Paddon A.R., Harker A.P. (1979). The Production of Charcoal in a Portable Metal Kiln. *Tropical Products Institute - Report G119*, 28 p.

[5] Girard P. (1989). Fiche technique de carbonisation. *Document CTFT.*

[6] Deglise X., Richard C., Rolin A., François H. (1980). Fast Pyrolysis/Gasification of Lignocellulosic Materials at Short Residence Time. *In Proc. « Energy from Biomass », 1st EC Conference, PALZ, CHARTIER and HALL ed., Applied Science Pub.*, 548-553.

[7] Beaumont O., Schwob Y. (1984). Influence of Physical and Chemical Parameters on Wood Pyrolysis. *Ind. Eng. Chem. Process. Des. Dev.*, **23**, 637-641.

[8] Antal M.J., Croiset E., Dai X., De Almeida C., Mok W.S.L., Norberg N., Richard J.R., Majthoub M. (1996). High-Yield Biomass Charcoal. *Energy Fuels*, **Vol. 10**, 3, 652-658.

[9] Humphreys F.R., Ironside G.E. (1980). Charcoal from New South Wales Species of Timber. *In : GORE (1982).*

[10] Stamm A.J. (1964). Wood and Cellulose Science. *The Ronald Press Company, New York, USA*, 549 p.

[11] Graboski M. BAIN R. (1981). Properties of Biomass Relevant to Gasification. *In Reed T.B, 1981*, 41-71.

[12] Ragland K.W., Aerts D.J., Baker A.J. (1991). Properties of Wood for Combustion Analysis. *Bioresource Technology*, **37** (1991), 161-168.

[13] Gronli M. (1996). A Theoretical and Experimental Study of the Thermal Degradation of Biomass. *Ph.D Thesis, Institutt of Termisk Energi og Vannkraft, Trondheim, Norway*, 258 p + 2 annexes.

[14] Maclean J.D. (1941). *Heating, Piping and Air Conditioning*, **13**, 380. *In : STAMM, 1964.*
[15] Briane D., Doat J. (1985). Guide Technique de la Carbonisation. *Edisud, Aix-en-Provence, France*, 178 p.
[16] Dunlap F. (1912). *U.S. Dept. Agr. Forest, Serv. Bill. n°110, in STAMM A.J., 1964.*
[17] Salazar C.M. (1987). The Influence of Particle Size and Shape on the Mechanisms of Decomposition of Wood During Pyrolysis. *PhD Thesis, University of Melbourne, Australia* , 285 p.
[18] Koch P. (1969). Specific Heat of Ovendry Spruce Pine Wood and Bark. *Wood Sci.*, **1**, 4, 203-214.
[19] Ten Wolde A., Mcnatt J.D., Krahn L. (1988). Thermal Properties of Wood and Wood Panel Products for Use in Buildings. *DOE/USDA - 21697, Oak Ridge National Laboratory, USA. In : Ragland et al., 1991.*
[20] Smith D.N., Lee E. (1958). *Forest Products Res. Spec. Report 13 . Her Majesty's Stationery Office, London.*
[21] Comstock G.L. (1970). Directional Permeability of Softwoods. *Wood and Fiber*, **1**, 283-289.
[22] Roberts A.F. (1970). A Review of Kinetics Data for the Pyrolysis of Wood and Related Substances. *Combustion and Flame*, **14**, 261-272.
[23] Connor M.A. Viljoen M.H., Ilic J. (1997). Relationships between Wood Density, Wood Permeability and Charcoal Yield. In Proc. *"Developments in Thermochemical Biomass Conversion", BRIDGWATER A.V.? BOOCOCK D.G.B. ed, Blackie Academic & Professional, Glasgow*, UK, 82-96.
[24] Connor M.A. (1983). Heat and Mass Transfer Considerations in Fuel Production from Wood Wastes by Pyrolysis. *Reg. J. Energy Heat Mass Transfer*, **5** (3), 179-194.
[25] Martin G. (1984). Pyrolyse - gazéification du bois - Aspects physico-chimiques. *Thèse de doctorat d'ingénieur, Institut National Polytechnique de Lorraine, France, 2 vol.*, 240 p. et 148 p.
[26] Chrysostome G., Lemasle J.M. (1986) - Fluidised Bed Oxygen Gasification of Wood. *In Proc. : « Advanced Gasification », E.C. Series E, vol 8, Energy from Biomass, A.A.C.M. Beenackers et W. Van Swaay ed, D. Reidel Publishing G*, 28-71.
[27] Connor M.A., Salazar C.M. (1988). Factors Influencing the Decomposition Processes in Wood Particles During Low Temperature Pyrolysis. *In Proc. « Research in Thermochemical Biomass Conversion », BRIDGWATER A.V., KUESTER J.L. ed., Elsevier Applied Science pub., Londres, UK*, 164-178.
[28] Bilbao R., Millera A., Murillo M.B. (1994a). Heat Transfer and Weight Loss in the Thermal Decomposition of Large Wood Particles. *In Proc. « Advances in Thermochemical Biomass Conversion », BRIDGWATER A. ed, Blackie Academic & Professional, Glasgow, UK*, 833-858.
[29] Zaror C.A. (1982). Studies of the Pyrolysis of Wood at Low Temperatures. *PhD Thesis, Imperial College, Londres, UK*, 356 p.
[30] Chan W.R., Kelbon M., Krieger-Brockett B. (1988). Single - Particle Biomass Pyrolysis : Correlation of Reaction Products with Process Conditions. *Ind. Eng. Chem. Res.*, **27**, 2261-2275.
[31] Simmons G.M., Gentry M. (1986). Particle Size Limitations Due to Heat Transfer in Determining Pyrolysis Kinetics of Biomass. *Journal of Analytical and Applied Pyrolysis*, **10**, 117-127.
[32] Lucchesi A., Maschio G., Rizzo C., Stoppato G. (1988). A Pilot Plant for the Study of the Production of Hydrogen - Rich Syngas by Gasification of Biomass.

In Proc. « Research in Thermochemical Biomass Conversion », BRIDGWATER A.V., KUESTER J.L. ed., Elsevier Applied Science pub., London, UK, 642-654.

[33] Kumar M., Gupta R.C. (1997). Influence of Carbonization Conditions on the Pyrolytic Carbon Deposition in Acacia and Eucalyptus Wood Chars. *Energy sources*, **19**, 295-300.

[34] Bamford C.H., Crank J., Malan D.H. (1946). The Combustion of Wood. Part1. *Proc. of the Cambridge Philosophical Society*. **42**, 116-182.

[35] Roberts A.F., Clough G. (1963). Thermal Decomposition of Wood in an Inert Atmosphere. *9th Intl. Symposium on Combustion. The Combustion Institute*, 158-166.

[36] Lee C.K., Chaiken R.F., Singer J.M. (1976). Charring Pyrolysis of Wood in Fires by Laser Simulation. *16th Intern. Symp. on Combustion*, 1459-1470.

[37] Pyle D.L., Zaror C.A. (1984 b). Models for the Low Temperature Pyrolysis of Wood Particles. *In « Thermochemical Processing of Biomass », BRIDGWATER A.V. ed., Butterworths & Co pub., Londres, UK*, 201-216.

[38] Chan W.C.R. (1983). Analysis of Chemical and Physical Process during the Pyrolysis of Large Biomass Pellets. *PhD Thesis, University of Washington*, 197 p.

[39] Desrosiers R.E., Lin R.J. (1984). A Moving Boundary Model of Biomass Pyrolysis. *Solar Energy*, **33**, 2, 187-196.

[40] Saastamoinen J., Aho M. (1984). The Simultaneous Drying and Pyrolysis of Single Wood Particles and Pellets Made of Peat. *Intl. Symposium on Alternative Fuels and Hazardous Wastes, American Flame Research Committee*, 1-29.

[41] Capart R., Fagbemi L., Gelus M. (1985). Wood Pyrolysis : A Model Including Thermal Effect of the Reaction. *In Proc. 3rd EC Conference on « Energy from Biomass », Elsevier pub., Londres UK*, 842-846.

[42] Huff E.R. (1985). Effects of Several Parameters on Burning Times of Wood Pieces. *In Proc. « Fundamentals of Thermochemical Biomass Conversion », OVEREND R.P., MILNE T.A., MUDGE L.K. ed., Elsevier Applied Science pub., Londres, UK*, 761-775.

[43] Simmons W.W., Ragland K.W. (1985). Single Particle Combustion Analysis of Wood. *In Proc. « Fundamentals of Thermochemical Biomass Conversion », OVEREND R.P. MILNE T.A., MUDGE L.K. ed., Elsevier Applied Science pub., Londres, UK*, 777-792.

[44] Krieger-Brockett B., Glaister D.S. (1988). Wood Devolatization - Sensitivity to Feed Properties and Process Variables. *In Proc. « Research in Thermochemical Biomass Conversion », BRIDGWATER A.V., KUESTER J. ed., Elsevier Applied Sciences pub., London, UK*, 127-142.

[45] Bilbao R., Murillo M.B., Millera A., Mastral J.F. (1991). Thermal Decomposition of Lignocellulosic Materials : Comparison of the Results obtained in Different Experimental Systems. *Thermochimica Acta*, **190**, 163-173.

[46] Koufopanos C.A., Papayannakos N., MASCHIOG G., Lucchesi A. (1991). Modelling of the Pyrolysis of Biomass Particles. Studies on Kinetics, Thermal and Heat Transfer Effects. *Canadian Journal of Chemical Engineering*, **69**, 907-915.

[47] Connor M.A., Daria V., Ward J. (1994). Changes in Wood Structure During The Course of Carbonization. *In Proc. « Advances in Thermochemical Biomass Conversion », BRIDGWATER A.V. ed., Blackie Academic & Professional, Glasgow, UK*, 846-858.

[48] Briane D., Haberman A. (1984). Essais comparatifs de six systèmes de carbonisation artisanale. *Association Bois de Feu, Paris, France*, 188 p.

[49] Shah N. (1990). Carbonisation discontinue du bois en fours à combustion partielle. *Thèse de doctorat, Université de Compiègne, France*, 177 p.
[50] Fossum M., Hustad J.E. (1994). Biomass Gasification for Industrial Production of Tar and Charcoal. *In Proc. « Advances in Thermochemical Biomass Conversion », BRIDGWATER A.V. ed., Blackie Academic & Professional, Glasgow, UK,* 1242-1256.
[51] Buy Tien (1996) - Gasification of Wood : A Multi-Stage Approach. *PhD Thesis, Asian Institute of Technology, Bangkok, Thailand*, 113 p.
[52] Y. Schenkel (1999). *Modelization of mass and energy flows from wood carbonization in retort kilns* (PhD thesis in French). Gembloux (Belgium), Faculté Universitaire des Sciences Agronomiques, 328 p.

Influence of Temperature, Residence Time and Heating Rate on Pyrolytic Carbon Deposition in Beech Wood Chars

Y. Schenkel
CRA, Agricultural Engineering Department
Chaussée de Namur, 146 - B 5030 Gembloux Belgium

ABSTRACT : The concept of Pyrolytic Carbon Deposition is used to explain the difference measured between theoretical and observed mass yields of coal coke and wood char production. This concept has been applied on beech wood carbonization through a factorial experiment conducted on beech blocks of 2 cm side for 12 final temperatures (from 200 to 800° C), two residence times at final temperature (15 and 45 min) and three heating rates (2 - 5 - 10° C/min). Final temperature and residence time show significant effects. The Pyrolytic Carbon Deposit (PCD) increases with the final temperature of carbonization up to a peak corresponding to a temperature of 350° C, then decreases with increasing temperature. The PCD increase is related to an intensive volatilization phase of the wood material hence to a high level of recombination of the volatilized matter with the carbon structure (secondary reactions). The PCD decrease corresponds to a further thermal decomposition of the deposit itself. Residence time has also a significant effect but at low temperatures, below 400° C. A higher deposit is observed for the residence time at final temperature of 45 min, in accordance with a more developed volatilisation of the solid.

INTRODUCTION

For carbonization, the maximum theoretical mass yield corresponds to the entire conversion of the carbon content of lignocellulosic materials into char. In other words, no single atom of carbon should be volatilized, the total amount of carbon atoms is converted into a structure of pure carbon. The maximum theoretical mass yield of charcoal would be expected to range between 45 and 50 %, depending on the carbon content of the lignocellulosic material. This theoretical approach is unrealistic : large numbers of carbon atoms are volatilized during the carbonization process. Consequently, determining the maximum mass yield requires very accurate determination of the carbon content of both gaseous effluents and charcoal, according to the carbonization temperature.

In order to determine this maximum mass yield of carbonization, we have conducted an experiment on beech wood carbonization. The results of this experiment were then used in two different approaches to evaluate the maximum mass yield, the carbonization maximisation function and the pyrolytic carbon deposition.

MATERIALS AND METHODS

The carbonization experiments were conducted on anhydrous beech cubes of 2 cm side in a laboratory electrical kiln. The wood blocks were placed in crucibles closed by a cap. Samples were introduced into the kiln at room temperature.

The factors are :

(1) The final temperature of carbonization : 200, 250, 275, 300, 325, 350, 375, 400, 450, 500, 600 and 800 °C.
(2) The residence time at final temperature : 15 and 45 min.
(3) The heating rate of the kiln : 2 - 5 - 10 °C/min.

Four replications were conducted for each set of experimental conditions. Several characteristics are determined on the charcoal :

(1) Mass yield.
(2) Proximate and ultimate composition.
(3) Calorific value.

RESULTS

Table 1 summarises the average data observed for two major carbonization parameters, the mass yield and the fixed carbon content of charcoal. A highly significant effect of temperature was observed, as well as a significant effect of the residence time and a significant interaction between these two factors. The effect of heating rate was not significant.

MAXIMISATION FUNCTION OF CARBONIZATION

The percentage of carbon remaining in the charcoal and the percentage of hydrogen and oxygen eliminated from the wood, were determined as a function of the final carbonization temperature. This is the actual objective of carbonization : to maximise the carbon content of the residual solid char while maximising the elimination of hydrogen and oxygen from this solid. The result is two inverse functions - captured carbon and volatilized hydrogen and oxygen - that are shown in figure 1. The product function of these two functions is also determined. This function (C * (H,O)) can be defined as the maximisation function of the carbonization process (see figure 1).

Table 1 Mean values and standard error of the mass yield and fixed carbon content of charcoal, as a function of the carbonization temperature and the residence time at this temperature.

Temperature (°C)	Mass yield (% d.b.) Residence time (min) 15		45	
	Mean	*Standard error*	Mean	*Standard error*
200	99.91	*0.01*	99.59	*0.04*
250	98.73	*0.23*	94.37	*0.82*
375	40.70	*1.17*	36.76	*0.33*
400	35.32	*0.38*	34.01	*0.34*
475	31.15	*0.31*	29.98	*0.28*
800	23.26	*0.16*	23.08	*0.17*
Temperature (°C)	**Fixed carbon (% d.b.) Residence time (min) 15**		**45**	
	Mean	*Standard error*	Mean	*Standard error*
200	15.17	*0.23*	15.81	*0.13*
250	15.51	*0.18*	17.04	*0.30*
375	56.03	*2.00*	64.61	*0.43*
400	65.48	*0.76*	70.28	*0.75*
475	74.43	*0.59*	77.19	*0.25*
800	93.98	*0.22*	95.25	*0.24*

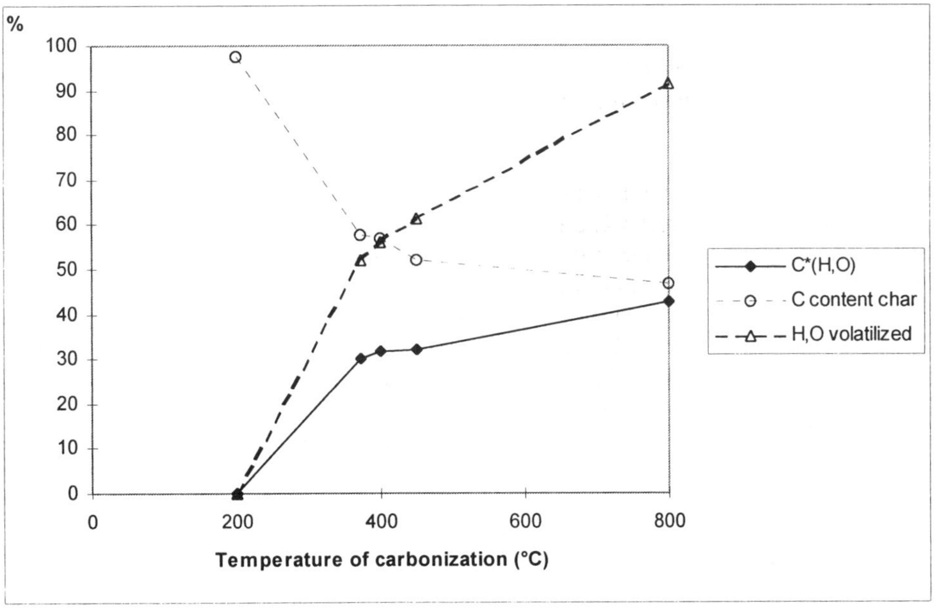

Figure 1 Evolution of the carbon content of charcoal, of the hydrogen and oxygen content of the volatile matter, and of the product of these two functions (C * (H, O)), as a function of the final carbonization temperature, for a residence time at final temperature of 45 min and a heating rate of 2 °C/min.

Figure 1 shows the rapid increase of the maximisation function of carbonization (C * (H, O)) up to a temperature of 400 °C, followed by a slight further increase up to 800 °C. Above 400 °C, hydrogen and oxygen continue to be eliminated while proportionally less carbon is lost. However, the marginal gain in carbon is low and is obtained by means of a high energy consumption.

Figures 2 and 3 show the maximisation function (C * (H, O)), for the three rates and the two residence times at final temperature, 45 and 15 minutes.

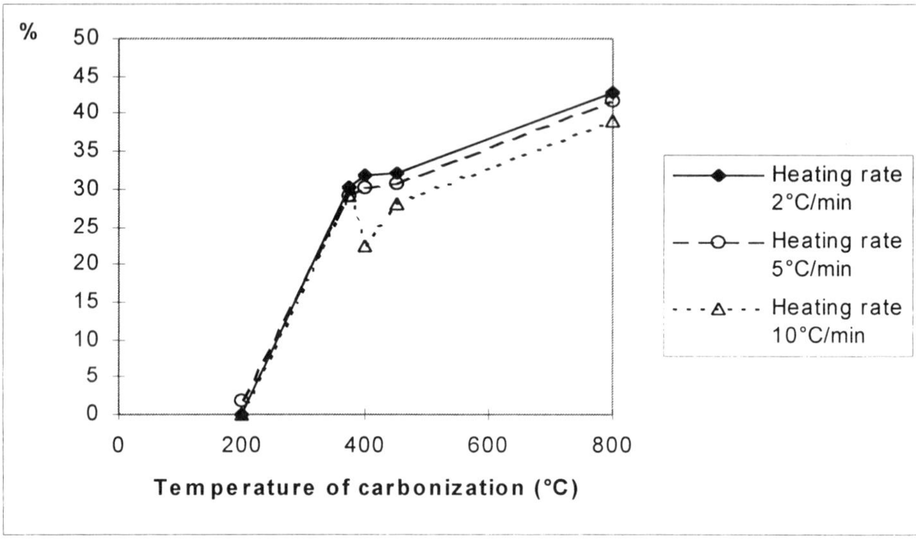

Figure 2 Maximisation function of carbonization as a function of final carbonization temperature, heating rate (2 - 5 - 10 °C/min) and for a residence time at final temperature of 45 minutes.

Again, a rapid increase of the maximisation function of carbonization can be observed up to 400 °C, followed by a slight increase at higher temperatures. It is also remarkable to note the tendency (although not significant) observed for the heating rates : the higher the heating rate, the lower the maximisation function, especially at temperatures higher than 400 °C. Analysing the data, we notice that this difference is mainly due to the carbon content of charcoal. This carbon content is higher for the 2 °C/min heating rate than for the 5 °C/min rate, which itself is higher than the 10 °C/min rate, at all temperature levels. The hydrogen and oxygen content of the volatile matter at the same temperature levels does not differ from one heating rate to another. These results confirm some results of the literature [2] - [6] : a low heating rate tends to increase the production of char to the detriment of pyrolysis oil and gases.

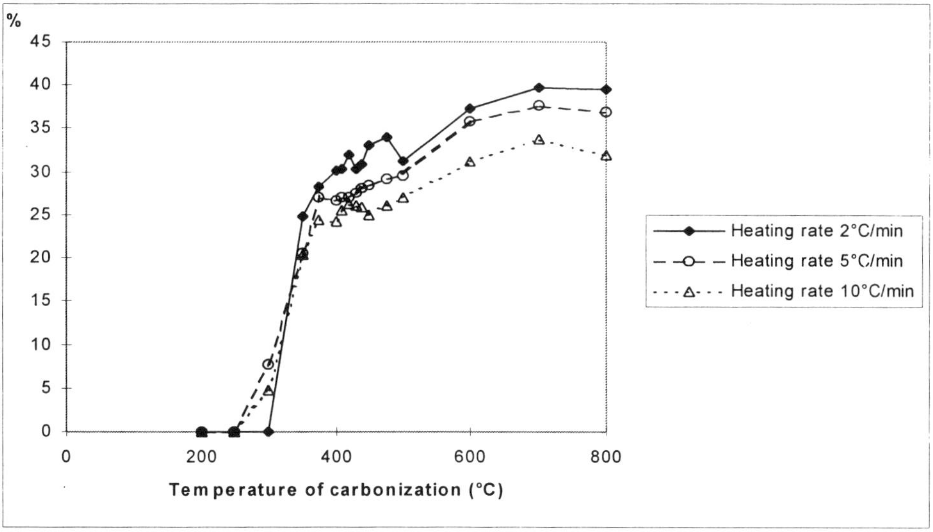

Figure 3 Maximisation function of carbonization according to final carbonization temperature, heating rate (2 - 5 - 10 °C/min) and for a residence time at final temperature of 15 min.

PYROLYTIC CARBON DEPOSITION

Another approach has been developed in the field of mineral coal research in coke production. This approach is based on the volatile contents of the non-thermally treated material and of the char. As a result of the process of carbonization, the whole volatile matter of wood must be eliminated, except the residual volatile matter contained in the char. Consequently, the theoretical mass yield of carbonization is the ratio between the non volatile matter of the wood on one hand and the non volatile matter of the char on the other hand. The formula is [7] :

$$MY_{cth} = \left[\frac{100 - VM_{aw}}{100 - VM_{ac}}\right] * 100$$

where

MY_{cth} = theoretical mass yield of carbonisation, %.

VM_{aw} = volatile content of anhydrous wood, % db.

VM_{ac} = volatile content of anhydrous charcoal, % db.

Classically the mass yield of charcoal is calculated as the ratio between the mass of charcoal produced and the mass of initial wood :

$$MY_{mc} = \frac{M_{ac}}{M_{aw}} * 100$$

where

MY_{ac} = mass yield of charcoal on anhydrous basis, %.

M_{ac} = mass of anhydrous charcoal, kg.

M_{aw} = mass of anhydrous wood, kg.

In practice, mass yields of carbonization (low heating rate) are always higher than the theoretical mass yields. The theory of pyrolytic carbon deposition (PCD) explains this phenomenon [6]-[12]. The level of PCD can be estimated by the means of a formula which calculates the PCD as the difference between the mass yield of carbonization (MY_{ac}) and the theoretical mass yield (MY_{cth}).

Table 2 summarises the value of PCD at 12 carbonization temperatures and two residence times. The volatile content of beech wood samples is 84.38 % d.b., determined on 9 samples. The evolution of the pyrolytic carbon deposit is shown in figure 4.

Table 2 Determination of the pyrolytic carbon deposition as a function of the final carbonization temperature and of two residence times at final temperature (beech cubes 2 cm side - mean values for the three heating rates 2 - 5 - 10 °C/min).

Temperature	VM_{ac}		MY_{cth}		MY_{ac}		PCD	
(°C)	RT15	RT45	RT15	RT45	RT15	RT45	RT15	RT45
200	84.50	83.87	99.91	96.84	99.91	99.59	0.00	2.75
250	84.18	82.63	98.73	89.93	98.73	94.37	0.01	4.44
275	82.64	80.33	89.99	79.40	93.47	87.40	3.49	8.01
300	79.17	75.56	74.97	63.93	84.79	75.46	9.82	11.54
325	69.99	65.76	52.04	45.63	65.53	58.95	13.49	13.33
350	56.08	52.05	35.56	32.58	51.98	46.81	16.41	14.23
375	43.04	34.44	27.42	23.83	40.70	36.80	13.28	12.97
400	33.64	28.78	23.54	21.93	35.32	34.01	11.78	12.08
450	24.44	26.80	20.67	19.97	31.15	29.99	10.48	10.02
500	19.03		19.29		28.55		9.26	
600	10.62		17.48		25.39		7.92	
800	4.60	3.57	16.37	16.20	23.26	23.08	6.89	7.88

VM_{ac} = volatile content of the charcoal, % d.b.

MY_{cth} = theoretical mass yield of carbonization, %.

MY_{ac} = mass yield of carbonization, %.

PCD = pyrolytic carbon deposition, %.

$RT15$ = residence time at final temperature of 15 min.

$RT45$ = residence time at final temperature of 45 min.

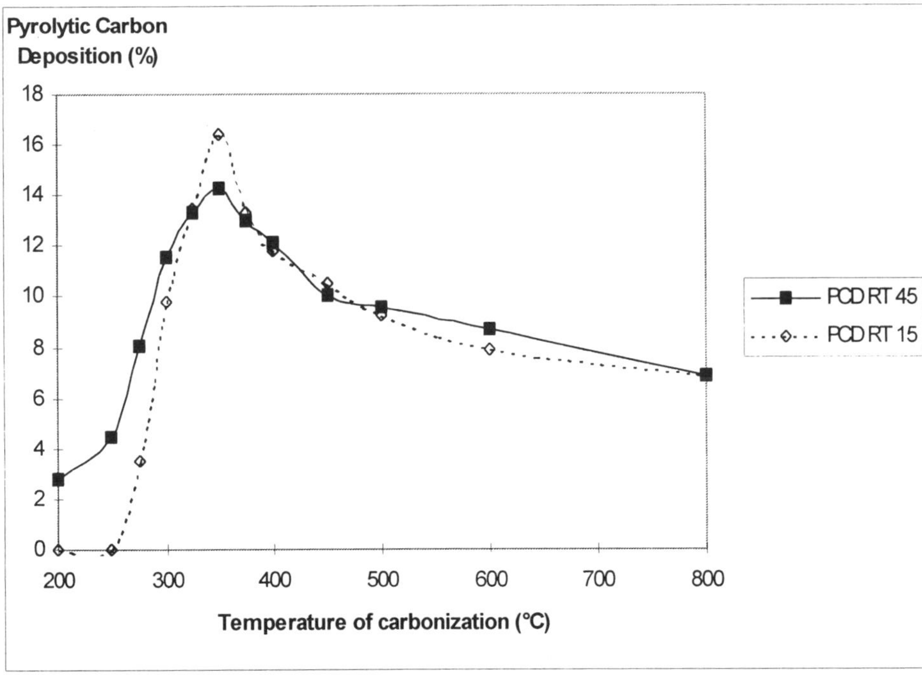

Figure 4 Evolution of the pyrolytic carbon deposition (PCD) as a function of final carbonization temperature and two residence times at final temperature (RT 15 and 45 min).

The pyrolytic carbon deposition is low at low temperatures (200 and 250 °C), then increases to reach a maximum around 350 °C and decreases at higher temperatures. The low values noticed at low temperatures correspond to a phase of pyrolysis during which the volatilization of the solid matrix is just beginning. The recombination of atoms of carbon with the carbonized structure is thus strongly reduced.

The maximum level of PCD at a temperature of 350 °C corresponds to an intensive phase of volatilization and hence to a high potential of recombination of the atoms of carbon with the solid matrix in pyrolysis. The decrease of the PCD with an increasing temperature above 350 °C is related to a thermal decomposition of the deposit, a secondary volatilization.

The data we observe are in contradiction with the data published by [7]. They observe an increase of the pyrolytic carbon deposition up to 800 °C followed by a decrease up to 1000 °C. But the authors [7] publish result only for three temperatures (600, 800 and 1000 °C) and give only one value for each test without any information on the experimental plan (number of replications for example).

We also observe a difference between PCD levels for the two residence time at final temperature. These differences appear at low temperatures, below 350 °C. The deposit is higher for the residence time 45 min, although the maximum peak is higher for the residence time 15 min. These differences are due to a more developed volatilization of the solid matrix for the residence time 45 min [1]. The results observed for the residence time 15 min show also a higher variability.

CONCLUSION

Two approaches have been developed to assess the theoretical or maximum mass yield of carbonization. The maximisation function of carbonization underscores the critical phase of low temperature (below 400 °C) in the carbonization process. This function also confirms the importance of a low heating rate to maximise the production of charcoal.

The evolution of the pyrolytic carbon deposition is mainly influenced by the temperature of carbonization. This deposition reaches a maximum peak at 350 °C, around 14 to 16 %. Then, the pyrolytic carbon deposition decreases regularly down to 7 °C at 800°C.

The residence time of the solid matter at final temperature is also a determining factor. A high residence time (45 min) leads to a higher pyrolytic carbon deposition, although the peak at 350 °C is high for the short residence time (15 min). The effect of the residence time appears mainly at low temperatures, below 450 °C.

REFERENCES

1. Schenkel Y. (1999) *Modelization of mass and energy flows from wood carbonization in retort kilns*. (PhD Thesis in French). Gembloux, Belgium, Faculté Universitaire des Sciences agronomiques, 328 p.

2. Doat J., Deglise X. (1982). Gazéification par pyrolyse éclair de quelques bois tropicaux. *Bois et Forêts des Tropiques*, **198**, 59-74.

3. Mackay D.M., Roberts P.V. (1982). The Influence of Pyrolysis Conditions on Yield and Microporosity of Lignocellulosic Chars. *Carbon*, **20**, 95-104

4. Beaumont O., Schwob Y. (1984). Influence of Physical and Chemical Parameters on Wood Pyrolysis. *Ind. Eng. Chem. Process. Des. Dev.*, **23**, 637-641.

5. Kumar M., Gupta R.C. (1993). Influence of Carbonization Conditions on Physical Properties of Acacia and Eucalyptus Wood Chars. *Transactions of the Indian Institute of Metals*, **46** (6), 345-352.

6. Kumar M., Gupta R.C. (1997). Influence of Carbonization Conditions on the Pyrolytic Carbon Deposition in Acacia and Eucalyptus Wood Chars. *Energy sources*, **19**, 295-300.

7. Chiu Y.F., Hong M.T. (1983) Influence of Volatile Matter and Deposited Carbon on Coke Yield from Coals. *Fuel*, **62**, 1150-1152.

8. Anthony D.B., Howard J.B., Hottel H.C., Meissner H.P. (1976). Rapid Devolatilisation and Hydrogasification of Bituminous Coal. *Fuel*, **55**, 121-128.

9. Koufopanos C.A., Papayannakos N., Maschiog G., Lucchesi A. (1991). Modelling of the Pyrolysis of Biomass Particles. Studies on Kinetics, Thermal and Heat Transfer Effects. *Canadian Journal of Chemical Engineering*, **69**, 907-915.

10. Martin G. (1984). Pyrolyse - gazéification du bois - Aspects physico-chimiques. *Thèse de doctorat d'ingénieur, Institut National Polytechnique de Lorraine, France, 2 vol.*, 240 p. et 148 p.

11. Di Blasi C., Russo G. (1994). Modeling of Transport Phenomena and Kinetics of Biomass Pyrolysis. In *Advances in Thermochemical Biomass Conversion* (Ed. BRIDGWATER A.V), Blackie Academic and Professional., Glasgow, UK, 906-921.

12. Antal M.J., Croiset E., Dai X., De Almeida C., Mok W.S.L., Norberg N., Richard J.R., Majthoub M. (1996). High-Yield Biomass Charcoal. *Energy Fuels*, **Vol. 10**, 3, 652-658.

Catalysed carbonisation of fine woodworking industry residues

J.Zandersons, A. Zhurinsh, A.Tardenaka, B.Spince
Latvian State Institute of Wood Chemistry, 27 Dzerbenes str., LV 1006, Riga, Latvia

ABSTRACT: Sawdust, shavings, veneer shorts and other fine residues of the woodworking industry are promising raw materials for production of charcoal. Studies performed at the Latvian State Institute of Wood Chemistry have demonstrated that engine oil refinery acid tar (EORA tar) is a good catalyst of wood carbonisation. Charcoal yield can be increased by 20 to 54 % in comparison with that in the case of non-catalysed pyrolysis charcoal.

The destruction and dehydration of hemicelluloses and cellulose starts already at 120 to 130°C. At 200°C, 70 % of acids and more than 80% of furfural are evolved. The formation of wood tar is decreased by an acidic catalyst promoting the process of cellulose dehydration. The dehydration of cellulose in the temperature range up to 280°C causes the formation of precursors of charcoal structures and provides a high charcoal yield.

The maximum of tar evaporation is observed at higher temperatures than usually. In the temperature range 350 to 450°C the organic matter of the EORA tar or the decomposition products of the sulphonic acids are distilled off. The lower heating value of this mixture of tars is 1.4 times higher than that of an ordinary wood tar. Therefore, the combined heating value of volatile products is maintained at a high level and will ensure the sufficiency of heat energy for the carbonisation process even if a damp wood should be pyrolysed.

The process utilises harmful industrial residues to produce a charcoal, which can be processed into activated carbon, carbonaceous construction materials and fuel briquettes.

INTRODUCTION

Fine residues of the woodworking industry such as sawdust, shavings and veneer shorts are in low demand and are used as a fuel at best. The amount of these residues is quite impressive. For example, veneer shorts comprise 15 to 20% of the amount of veneer blocks while the amount of sawdust is up to 12% of the volume of sawlogs. Nowadays the carbonisation of these wood residues is an industrial practice. However, the charcoal yield is only 15 to 20% on the o.d. wood basis if modern industrial carbonisation units such as a Herreshoff furnace (roaster) is used (2). One of the main reasons for the low charcoal yield of fine wood particles in industrial apparatus is the fast pyrolysis nature of the process, namely a high rate of heating at particle temperatures ranging from 200 to 280°C. If the heating rate exceeds 20-25°C min^{-1}, the depolymerisation reactions of cellulose prevail and a lot of soluble tar is formed whereas the yield of charcoal is low (1). The solution to this problem is in the promotion of the dehydration reactions of cellulose using catalysts during the preexothermic period of the heated wood particles.

Catalysts of wood pyrolysis are used to increase the yield of charcoal and to obtain a charcoal with modified properties or to produce definite valuable volatile products. The substances promoting condensation reactions are employed if high yields of charcoal are desired (4, 5, 7). Some catalysts increase the apparent density of charcoal, the pore dimensions and their adsorbency nature. Of all silvichemicals charcoal is currently the most in demand. Commercial interest is focused on deciduous wood charcoal or charcoal briquettes.

The Lewis and Brønsted acids lower the starting temperature of wood decomposition and charcoal formation and along with an increased char formation, diminish the yield of some volatile products of pyrolysis. The latter fact is an undesirable phenomenon if an energetic self-sufficiency in the carbonisation of wood is striven for (10). Therefore, the objective of the present investigation was to elucidate not only the effect of different available catalysts upon the yield of charcoal, but also that on the yield of volatile products.

EXPERIMENTAL SECTION

The grey alder (*Alnus incana*) wood was chosen as a test species in the present study for laboratory experiments, because it was also intended to elucidate the effect of catalysts on the mechanical properties of charcoal.

Grey alder wood specimens were prepared in the shape of blocks measuring 2×2×2 cm. The chosen amount of catalyst solution was soaked into wood and the specimens dried at ambient temperature to the moisture content 7 to 8%. An electrically heated thermoreactor was used. The heating rate was 3 to 4°C/min, the maximum temperature 500 to 520°C. To pyrolyse the sawdust and chips under isothermal conditions at the thermoreactor wall temperature 550°C, a pilot scale thermoreactor equipped with a two paddle rotating stirrer was used (11). Analyses of the volatile condensable wood thermodestruction products were done by using spectrophotometric methods developed at our laboratory (8).

The settled tar was extracted with chloroform from condensate (pyroligneous acid). An aliquot amount of extract was dried and the content of settled tar was determined gravimetrically. The dissolved tar was determined gravimetrically after settled tar extraction, by drying an aliquot amount of water phase solution.

For analysis of acids, esters, alcohols, ketones, aldehydes and furfural the test solution was prepared by dilution of a weighed sample of clear fraction of the condensate ~ 10 g to 100 ml with distillate water. The content of acids and esters was determined by potentiometric titration, using potassium hydroxide and hydrochloric acid standard volumetric solutions, c(KOH) = 0.1 mol/l; c(HCl) = 0.1 mol/l. An automatic potentiometric titration assembly RTS – 822 “Radiometer” (Denmark) was used.

Alcohols were determined using ammonium cerium (IV) nitrate reagent, which reacts with alcohols to form a stable colour complex. Colour intensity was measured spectrophotometrically at 490 nm. Content of alcohols was calculated against the standard solutions of methanol. Ketones were determined spectrophotometrically by reaction with salycilaldehyde in basic medium. Colour intensity of the formed complex was measured at 490 nm and content of ketones calculated against the standard solutions of acetone. Aldehydes forms yellow-orange complex with antron in 75 % H_2SO_4, but furfural – blue one. Intensity of colours was measured spectrophotometrically at two wawelenghts – 490 nm (λ_{max} of aldehydes complex) and 590 nm (λ_{max} of furfural complex). Content of aldehydes was calculated against the formaldehyde standard solutions, but amount of furfural – against furfural standard solutions.

Sulphuric acid, diammonium phosphate, zinc chloride and EORA tar from the Inchukalns deposit site were tested as carbonisation catalysts. The highly harmful waste acid tar is characterised by a good solubility in water: therefore, in an open air depository pond, it is found as a 50 % solution in water, containing 4.3 % sulphuric acid and 70 % sulphonic acids ($R\text{-}SO_3H$) both on the dry matter basis with an average sulphonic acid molecular mass of 610 (3,6).

RESULTS AND DISCUSSION

Under the effect of carbonisation catalysts, the thermal decomposition of wood components sets in at a temperature approximately by 100°C lower than in the case of non-catalysed carbonisation of wood. Table 1 lists the yield of the wood carbonisation solid residue recalculated on fixed carbon to exclude the effect of the fluctuations of volatiles and the ash content upon the charcoal yield. The fixed carbon content of charcoal in the experiments represented in Table 1 was 78% to 81%. The increased ash content of charcoal was caused by the catalysts, $(NH_4)_2HPO_4$ and $ZnCl_2$. The yield and overall heating value of the volatile products are given in Table 1.

This shows that the charcoal yield under the influence of the above-mentioned catalysts increases from 5 to 39%, but the decrease of tar yield is approximately proportional to the increase of the charcoal yield. The most effective catalysts (H_2SO_4 and $(NH_4)_2HPO_4$) cause a dramatic decrease in the dissolved tar yield (a decrease of 72.6 and 66.1%, respectively, on the basis of non-catalysed pyrolysis yield). The decrease of the settled tar yield is also significant, but less remarkable in the case of H_2SO_4 (37.8 and 71.1% respectively).

The decrease of the tar yield was the principal reason for the serious diminishing of the heating value of volatile products if sulphuric acid or $(NH_4)_2HPO_4$ were used as catalysts. The same interrelationship can be observed between the decrease of the gas yield and its heat energy.

It has been demonstrated that engine oil refinery acid tar is an excellent catalyst of wood carbonisation (9). The increase in charcoal yields ranges from 20 to 54% on the basis of non-catalysed pyrolysis charcoal yield. The structure of grey alder charcoal shown on the electron scanning microscope images (see Fig. 1) differs a lot from that of the charcoal obtained using the EORA tar as catalyst (see Fig. 2a). Fig. 2a shows that some vessels are filled with a glassy material, possibly a coke from the EORA tar. Fig. 2b demonstrates that the glassy material fills the vessel. However such coke inclusions are comparatively few, and we suppose that these by no means are the main cause of the remarkable increase in the charcoal yield.

Table 1. Yield (wt% on the o.d. wood basis) of grey alder (*Alnus incana*) wood catalytic pyrolysis products and the heating value of their volatile components.

Products of wood pyrolysis	Without catalyst	H_2SO_4 3.0%	$(NH_4)_2HPO_4$ 5.0%	$ZnCl_2$ 1.4%	Engine oil refinery acid tar 5%	Engine oil refinery acid tar 9%	Engine oil refinery acid tar 17%
Charcoal (as fixed carbon)	24.6	30.0	34.1	25.8	29.5	31.3	36.3
Dissolved tar	6.2	1,7	2.1	5.1	8,2	8,3	8,2
Settled tar	4.5	2.8	1.3	2.4			
Acids	7.2	7.6	6.0	7.3	6.3	6.5	7.6
Alcohols	1.9	1.5	1.5	3.3	3.2	3.0	1.2
Esters	2.3	0.7	0.8	0.9	1.3	1.2	1.1
Aldehydes	0.9	1.0	0.4	1.0	1.6	1.4	0.6
Ketones	1.8	1.0	1.3	1.1	1.4	0.8	0.4
Furfural	1.4	3.8	2.4	3.7	3.3	2.9	4.0
Noncondensable gases	17.1	13.7	11.9	17.8	13.5	13.0	14.1
Including combustible: CO	5.04	4.1	2.9	4.8	3.39	3.65	4.33
H_2	0.02	0.02	0.02	0.01	0.01	0.03	0.06
CH_4	1.04	0.67	0.60	0.90	0.43	0.73	1.23
Combined heating value;							
MJ/kg	7.19	4.91	3.80	6.06	7.16	7.00	7.22
%	100.0	68.0	52.8	84.3	99.6	97.4	100.4

*concentration of catalysts- wt % on the o.d. wood basis

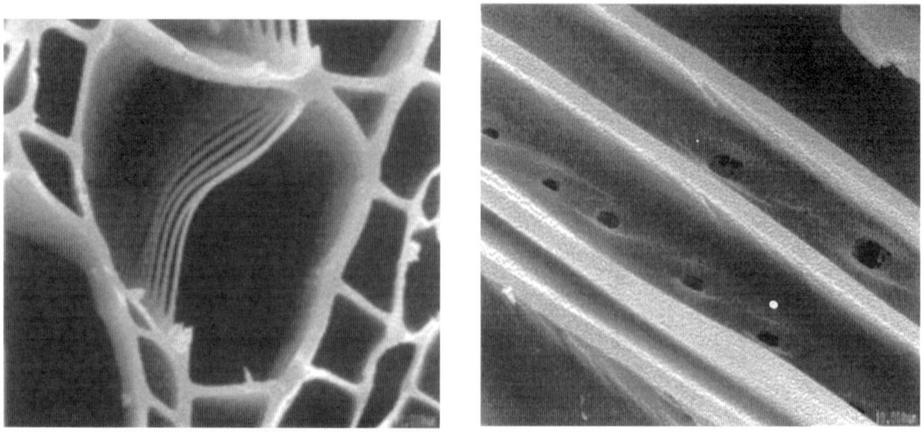

a. b

Fig. 1. Cross-section (a) and longitudinal section (b) of grey alder charcoal

It has been earlier demonstrated that sulphonic acids increase the charcoal yield. The amount of sulphuric acid in wood soaked with the EORA tar catalyst is no more than 0.7 to 0.8% on the o.d. wood basis. However the combined action of these cellulose dehydration promoters causes an increase in the charcoal yield.

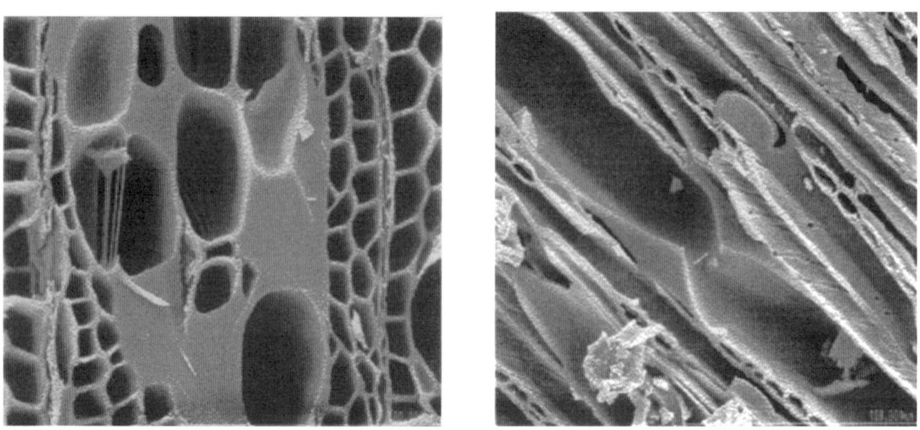

a b.

Fig. 2. Cross-section (a) and longitudinal section (b) of grey alder charcoal prepared using the EORA tar catalyst.

In this connection, the information about the formation of the main volatile condensable products during the pyrolysis process is of interest. This information is presented in Figures 3, 4 and 5.

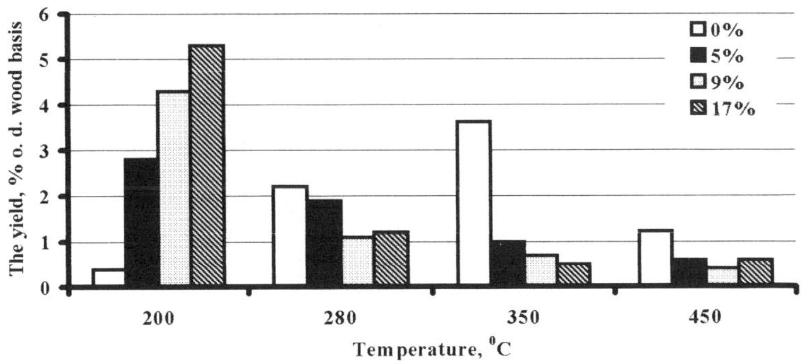

Fig. 3. Formation of acids during catalytic* pyrolysis

* in Fig. 3, 4 and 5 the engine oil refinery acid tar catalyst concentration in wood is expressed in wt % on the o. d. wood basis.

It has been shown by our experiments, that the destruction of hemicellulose and cellulose starts at approximately 120 to 130°C, and up to 200°C the overwhelming amount of acids and furfural evolved (if the EORA tar concentration in wood is 17% on the o.d. wood basis at the temperature 200°C up to 70% of acids and 82.5% of furfural are evolved, see Fig. 3, 4, 5).

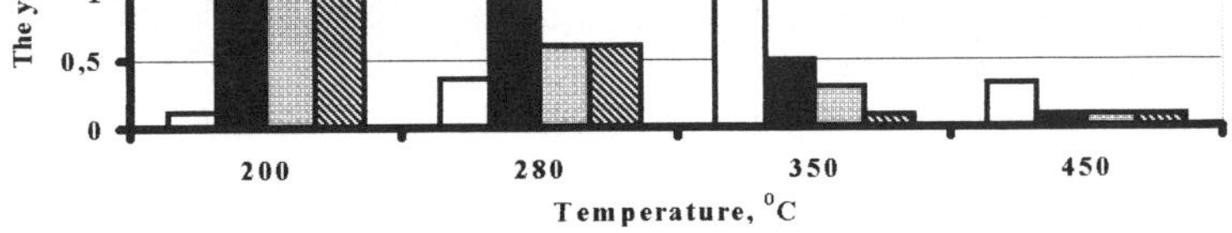

Fig. 4. Formation of furfural during catalytic pyrolysis

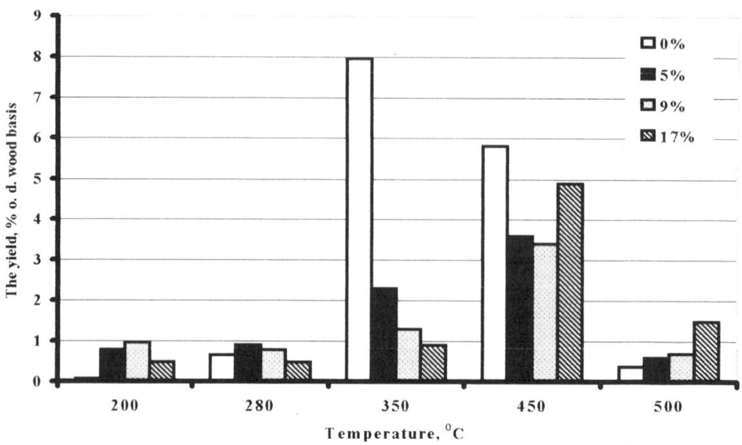

Fig. 5. Formation of tar during catalytic pyrolysis

At the same time at the temperature 200°C, the formation and evaporation of tar is observed. Although the amount of tar is only 0.8% on the o.d. wood basis (see Fig. 5), it is 5 times as high as that without the use of the catalyst but approximately the same as with sulphuric acid.

The maximum of tar formation is at higher temperatures than if non-catalysed pyrolysis is employed. It means that at temperatures 350°C to 450°C the organic matter soaked into wood as the EORA tar catalyst or its decomposition products is evaporated. The lower heat value of this tar is 1.4 times higher than that of the ordinary wood pyrolysis tar (41.5 MJ/kg and 29.6 MJ/kg, respectively). Owing to the organic matter introduced into wood with the sulphonic acids, the combined heating value of the volatile products is kept at a non-catalysed pyrolysis level (see Table 1). The charcoal obtained has apparent density as well as longitudinal crushing strength and modulus of elasticity to be 244 g/dm^3, 153.9 kg/cm^2 and 21.1; 277 g/dm^3, 158.6 kg/cm^2 and 23.0; 326g/dm^3, 182.7 kg/cm^2 and 23.2; 341 g/dm^3, 201 g/cm^2 and 31.7 for non-treated alder wood charcoal and that soaked with 5%, 9% and 17% EORA tar specimens, respectively. These properties of charcoal are important not only for technical-grade lump charcoal, but also for production of high-quality charcoal briquettes and carbon materials from fine charcoal particles.

Sawdust and veneer shorts are especially suitable raw materials for catalytic pyrolysis. In the present study the sawdust was treated with an appropriate amount of an EORA tar aqueous solution in a paddle mixer, the mixture was kept in a closed vessel for 48 hours, and the catalyst-soaked wet sawdust (the moisture content 45 to 50% on the wet basis) was pyrolysed in a pilot-scale thermoreactor equiped with a two paddle rotating stirer at a constant reactor wall temperature of 550 to 600°C. The charcoal yield calculated as fixed carbon was increased from 18 to 32.5% when the EORA tar concentration in the catalysed sawdust was 17% on the o.d. wood basis. The bulk density of the charcoal was also increased from 188 to 220 g/l. The duration of the process was decreased moderately, since the most time-consuming operations were drying of wood and heating of charcoal to achieve the desired fixed carbon content.

EORA tar is a promising catalyst to prepare charcoal for activated carbon production. The charcoal obtained in runs using the EORA tar catalyst was activated in half the time necessary to activate the charcoal prepared without the use of the catalyst. The adsorption efficiency of activated carbon was 1.3 to 1.6 times higher if the charcoal was produced in a catalysed process, and the pore size of the adsorbent was also uniform. The charcoal prepared in an EORA tar catalysed process contained 0.3 to 0.5% of sulphur. The activated carbon did not contain sulphur.

It stands to reason that the implementation of this catalyst in practice requires solution of the sulphur dioxide emission problem. The high yield of good-quality charcoal from dispersed wood residues and the utilisation of harmful wastes are arguments for the use of this technology. The level of sulphur dioxide emissions from a unit for catalytic pyrolysis of sawdust is the same as for a powerhouse using coal with low sulphur content. These emission levels should be tolerated to clean up the dangerously contaminated sites causing groundwater pollution.

CONCLUSIONS

Sawdust and other fine residues of the woodworking industry are in low demand and thus create disposal problems. To organize an economically effective charcoal production their yield should be improved. It has been disclosed that engine oil refinery acid tar can be used as a wood carbonisation catalyst. The charcoal yield of fine dispersed wood is increased up to 1.7 times, making the production practicable. In addition to that, the process self-suffiency in a heat carrier is ensured by the combined combustion of pyroligneous vapour and the EORA tar carbonisation volatiles. The superfluous residue of woodworking industry and the environmentally dangerous pollutant are used to advantage.

The quality of charcoal is improved because the apparent density, crushing strength and modulus of elasticity are increased. Though the charcoal contains 0.3 to 0.5 % sulphur, it is eliminated during the activation process. The activated carbon obtained demonstrates a uniform pore size and its activation time is less by half in comparison with non-catalyzed charcoal activation durability. The insignificant level of sulphur dioxide emission from a unit for catalytic pyrolysis should be tolerated to clean up the dangerously contaminated sites causing groundwater pollution and to manufacture a value-added product from a low in demand wood waste.

REFERENCES

1. Antal M.J., Jr. & Warhegyi G. (1995) Cellulose pyrolysis kinetics: The current state of knowledge. *Ind. Eng. Chem. Res.*, **34**, 703-717.
2. FAO Forestry Paper 63. (1985) Industrial Charcoal making. FAO UNO, Rome, 134 p.
3. Frolov A.F., Titova T.S., Karpova I.V., Denisova T.L. (1985) On a composition of the engine oil refinery acid tar. *Khimya I technologya topliva I masel (Chemistry and Technology of Fuel and Lubricants)* No 6, 37-38. (inRussian)
4. Kislitsyn A.N. (1990) Wood Pyrolysis. Chemistry, Kinetics, Products, Novel Processes. Publ. House "Lesnaja promyshlennost", Moscow, 312 p. (in Russian)

5. Kislitsyn A.N., Rodionova Z.M., Lebedeva Z.I., Kalugin E.N. (1984) Study of structure formation and modification of charcoal characteristics in the presence of chemical reagents. *Khimiya drevesiny (Wood Chemistry)*, No 1, p. 83-89. (in Russian)
6. Kukurs O. & Valdmanis J. (1999) Inčukalns acid tar repository. I Stratification and physical characteristics of repository. *Latvian Chemical Journal*, **4**, 100-3. (in Latvian)
7. Sekiguchi Y. & Shafizadeh F. (1984) The effect of inorganic additives on the formation, composition, and combustion of cellulosic char. *J. Appl. Polym. Sci.* **29**, 1267-1286.
8. Spince B., Zhurinsh A. & Zandersons J. (1998) Chemical analysis of wood pyrolysis liquid products. *Latvian Chemical Journal*, **3**, 92-5. (in Latvian)
9. Zandersons J., Zhurinsh A., Spince B. & Tardenaka A. (1996) Method for producing charcoal using a carbonisation catalyst. *Latvian pat.* No. 11995.
10. Zandersons J., Zhurinsh A. & Gravitis J. (1999) Carbonisation of wood residues by utilizing volatile thermolysis products combustion heat. *Cell. Chem. Techn.*, **33**, 147-60.
11. Zandersons J., Gravitis J., Kokorevics A., Zhurinsh A., Tardenaka A. & Spince B. (1999) Studies of the Brazilian sugarcane bagasse carbonisation process and products properties. *Biomass and Bioenergy*, **17**, 209-19.

Thermal Desorption Technology: Low Temperature Carbonisation of the Biomass for Manufacturing of Activated Carbon

G. E. Soméus
Thermal Desorption Technology Group L.L.C. of North America
European Branch: H-1222 Budapest, Szechenyi 59., Hungary

ABSTRACT: The Thermal Desorption Technology Recycle-Reduce-Reuse TDT-3R™ apparatus is an indirectly fired, horizontally arranged, patented (US 5,707,592), continuously operating rotary kiln design, whereas <600 °Celsius - material core - carbonisation temperature, vacuum (0-50 Pascal) in true reductive environment and separate process of steam activation is used. The efficient heat transfer is managed through the mantel. No internal moving parts are used for material throughput. Alternative continental basic materials: sawdust (hard and soft wood), straw, refuse grain and hardcoal. Pre-treatment of the basic material: grinding homogenisation to <5 mm, drying to <18 % moisture content. Main process: phase I. low temperature carbonisation, phase II. steam activation. Post treatment: screening and packaging. Gas-vapour phase stemming from the main process is separately treated, condensed for offsite utilisation of liquid and non-condensable gas is burned out, or all gas-vapours are directly burned off for onsite energy utilisation. Non-destructive Air Pollution Control Device "APCD" multi venturi off-gas scrubber is used. Process water is cleansed and recirculated. The TDT-3R™ rotary reactor and the comprehensive design does not containing any exotic technical constructions and/or exotic materials.

INTRODUCTION

The method of the low temperature carbonisation of the carboniferous materials is known since long time, whereas different apparatus technical solutions have been used. Many of these traditional carbonisation apparatus producing variating quality end products with too high burn off and does not meet the new environmental standards. For the innovative TDT-3R™ apparatus design the following main technical and engineering considerations have been taken for value added conversation of the biomass:

(1) As biomass is efficient heat insulator, the efficient indirect heat transfer to the material – through to core – is difficult and takes time.

(2) High initial moisture content have significant negative influence on the carbonisation process.

(3) Upon condensation of the gas-vapours for utilisation as bio-oil, the initial moisture content will be enriched in the liquids, which complicates its utilisation, especially over 25 % of moisture content.

(4) Flexible sealing between rotary kiln moving and stationary parts requires special solutions.

(5) The thermal engineering of the mechanical construction design requires special solutions.

(6) Pre and post treatment of material require comprehensive solutions.

(7) Biomass provides lower yield of activated carbon production versus hard coal, therefore special end product applications need to be developed in order to compensate specific production cost efficiency loss at similar investment.

Table 1. Fixed carbon content in weight % of raw materials (dry substance) used for the production of activated carbon

Material	Approximate carbon content (%)
Soft wood	35
Hard wood	40
Coconut shell	35
Grain and agro products	40
Hard coal	60-75
Anthracite	90

Large corporations have advanced industrialised activated carbon manufacturing technologies, however, these large units using mostly hard coal to maintain continuously supplied high yield production. Hard coal - as natural substance - may vary in chemical composition, and often including natural add-on inorganics and heavy metals, which might be disturbing elements at applications. Biomass - as natural substance – may also vary some in composition, but heavy metal contamination is less of care, only if man made contamination is the case. The efficient carbonisation of the biomass requires high efficient heat transfer equipment from the heat source to the material, which is one the main key elements of the TDT-3R™ technology equipment design.

THE KEY ELEMENTS OF THE TDT-3R™ TECHNOLOGY

The TDT-3R™ technology is an high efficient indirect heat up and indirect cool down process and apparatus of the carboniferous basic material to achieve value added refined end products.

The strategic components of the TDT-3R™ technology:

Indirectly heated, horizontally arranged, rotary kiln for Thermal Desorption and low temperature carbonisation of the 0 - 5 mm input basic material, containing the following technical characteristics:

(1) True reductive environment.

(2) Indirect heat transfer through the mantel to the ground material with extended surface area resulting efficient heat transfer to the feedstock, often with bad thermal conduction character.

(3) Advanced technical solution for the rotary kiln sealing between the moving and stationary parts for closed_input and output system.

(4) Advanced technical solution for the continuous throughput of the feedstock where no moving parts are used.

(5) Permanent operation and material throughput.

(6) Process under vacuum.

(7) Automatic process control of the operation.

(8) Simple construction, containing no exotic technical solutions and materials,

(9) Two main full scale reactor sizes are available for continuous feed stock, 0.8 m^3/h (6,000 m^3/year) and 2.5 m^3/h (18,000 m^3/year) with multi reactor installation opportunity.

Multi venturi off-gas scrubber for high efficient cleansing of contaminated flue gases stemming from thermal conversion of refuse and waste materials, containing the following advanced and innovative technical solutions:

(1) Multi venturi technical solution.

(2) Combined off-gas cleansing effect, high efficient counter stream water spray washing, drop out and removal effect.

(3) Combined water recirculation system.

(4) For variating flue gas volumes the system is externally infinitely variable.

(5) External cooling opportunity to avoid boiling up.

(6) Modular tower construction.

Solid material cooler:

(1) Indirect cool down of carbon products from 600 °C to 40 °C.

(2) Avoids oxidation - thus ash creation - during cooling process.

TDT-3R™ GENERAL CONCEPT DESCRIPTION

PRE-TREATMENT

Grinding and homogenisation to less than 5 mm of basic material. The ideal moisture content of the basic material > 5 % would be, however in practice 8-18 % is used. If necessary drying of the basic material is executed to decrease moisture content from >35 % to <10 %. Pre screening of the basic material is used, as the sizing of the input material will be reflected into the sizing of the end products. The closed input to the TDT-3R rotary kiln's internal area is managed by a mechanical device, which compress the feedstock to remove air from the inter space between particles and forcing material into the thermal processing area.

THE MAIN PROCESS

The main component of the TDT-3R™ is a specially designed, indirectly fired, rotary reactor in which the material in a reductive environment is carbonised - partially vaporised and/or gas-out - in low vacuum (0-50 Pascal) between the temperature ranges of 550°C - 600°C.

The reactor rotates around its symmetric axle, is horizontally arranged and cylindrically formed with no refractory line installed inside. The high quality steel mantel is heated from outside through the lined combustion chamber. Inside the reactor body, blades promote the transportation of the material. The reactor is a permanently working vessel; the basic material enters in at the input-end while the carbonised material and the pyrolysis gas-vapour phase are discharged at the output-end, separately from each other. Generally, carbon char and inorganics will remain in the solid phase while volatile organic compounds and water will be concentrated into the pyrolysis gas-vapour phase.

PYROLYSIS GAS-VAPOUR PHASE from the reductive decomposing process is either condensed for achievement of bio-oil and non-condensable gases for energy utilisation or all directly combusted on site at a minimum 850 °C 2 sec. residence time. Alternative post combustor has been developed for operating temperature of 1,250 °C 2 sec. residence time as well, with fast cooling and heat recovery opportunity. The remaining gas is cleansed in a high capture efficient, indirectly cooled, wet gas multi venturi scrubber prior to discharge. Scrubber process water is cleansed prior to discharge and the neutralised precipitate with from the water treatment is utilised.

CARBON PHASE is discharged for medium temperature steam activation. The end product is indirectly cooled, dried, and selected by our adjustable air selector for final sizing and utilisation.

TDT-3R™ PROCESS REACTOR AND IT'S OPERATION

The required amount of energy input is basically supplied from hot flue gases. These are produced in the combustion chamber for direct burn-off of the thermolysis gas vapours and heats the reactor body from outside the mantel. Heat transfer happens indirectly through the mantel to the material.

THE HEAT TRANSFER CONTAINS THREE PHASES:

Primarily: Energy transfer from heat under the material in the rotary kiln. The material, which is ground down to have extended surface area, is permanently mixed, which mixing process is promoted by the blades inside the reactor body. **The heated material is to be permanently replaced by cooler materials again and again. Therefore the thermal conductivity of the biomass material is of less importance and can be within a wider range**. This is a very important and vital technical factor, particularly in relation to the biomass refuses, which are usually bad thermal conductors and in practice are of mixed physical and chemical character.

Secondarily: Radiation heat transfer from the inner top surface of the reactor body.

Tertiary: Over the temperature of 275 °C (530 °F) an exothermic reaction starts during the decomposition of the material.

The exothermic process is a slow process, therefore the extended thermolysis gas-vapour production will not result in an explosive production of gas vapours. The thermolysis gas-vapour production is promptly removed from the reaction space by the master fan.

The thermal engineering design of the reactor is related to the throughput capacity of the reactor and the extremely qualitative variations of the input material. No matter if the basic material is of organic, inorganic and/or mixed character, the chemical components will be separated at a certain treatment temperature if the boiling point of the primary target contaminant component(s) are under 600 °C. The pyrolysis reactions are not only a sequenced series of reactions, but parallel series of reactions as well, with different levels of energy (1,2). The surface oxygen structures are quite stable at low temperatures below 200 °C, irrespectively of the temperature at which they are formed. However, upon heating up over 300 °C the surface oxygen structures produce CO_2 and over 500 °C CO and H_2 which is completed about 600 °C in vacuum.

To provide a successful Thermal Desorption process, three different factors are needed:

(1) Material, which is thermally decomposable.

(2) High enough temperature that the waste material pass the complicated decomposition phases.

(3) Long enough reaction time.

CARBONIZATION PHASE: There are four well distinguished phases concerning the TDT-3R™ thermolysis process inside the reactor:

(1) **Warm up phase**: up to 150 °C - 160 °C. Characterised by the evacuation of the free and most of the bounded water from the material.

(2) **Thermal decomposition phase**: up to 270 °C - 280 °C. Characterised by heavy discoloration of the material, and the evacuation of the remaining chemically bounded water, with simultaneous development of gasification.

(3) **Thermal desorption phase**: up to 380 °C - 400 °C. Characterised by self carbonisation with exothermic chemical reactions, heavy gasification process and escape of volatile compounds from the material. Thermolysis gas-vapour is continuously removed.

(4) **Stabilisation phase**: up to 600 °C. Remaining volatile compounds are completely removed from solid phase.

During carbonisation phase the basic material is gas out, e.g. devolatized. This process opens the pores where the distribution of the micro pores, meso pores and the macro pores are approximately reflections of the pore characteristics of the original basic material. This is most important that the carbonisation phase perfectly devolatize the basic material, through to core structure, as imperfect devolatization may result internal micro condensation of the tar, which may significantly decrease the expanded pore developments.

Figure 1.: TDT-3R™ rotary kiln

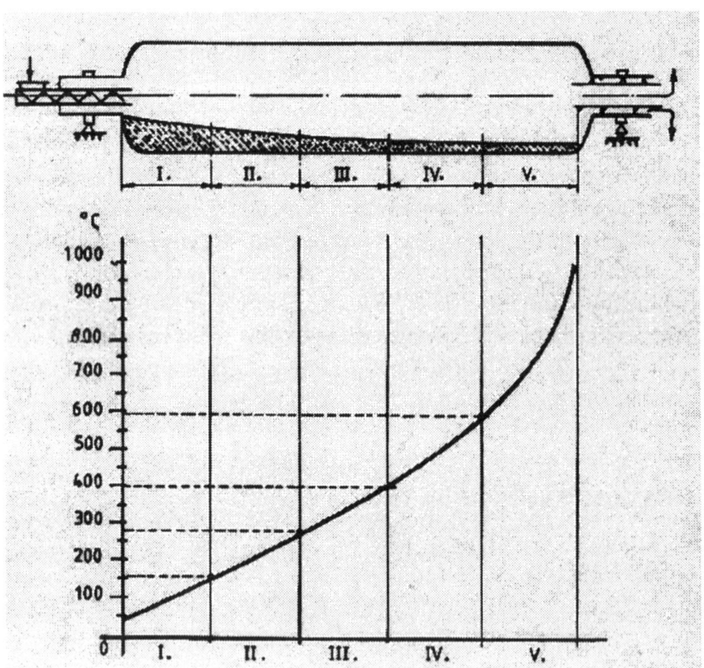

ACTIVATION PHASE: by steam – water vapour - activation for production of active sites and wider pores. In our testing program for biomass carbonisation we have experienced that medium temperature water vapour activation rather promotes meso and macro pore development, than micropores. The production of meso and macro pore structured activated carbon is our prime goal, as many of the ground water decontamination projects are related to hydrocarbon contamination's, where the contamination is consisting of large organic molecules to be removed. -

Chemical activated Field Demonstration Plant "FDP" tests has also been executed for the production of activated carbon from sawdust, involving mixing an inorganic chemical compound with the carbonaceous raw material such as activating agents phosphoric acid and zinc chloride at lower temperature. However, use of zinc chloride poses the danger of zinc traces in the end-product, therefore - although technically is available - steam activation is used.

The true reductive environment avoids ash creation, which ash may decrease the end product quality. There are some chemically bounded oxygen in the basic material and between the pressed feedstock as well, which will be directly burned off during the carbonisation process. In our experience the total ash content at the end product will remain at a range of approx. 3-6 %, depending also on the basic material natural inorganic composition characteristics as well.

Table 2 Biomass Raw Material Based Activated Carbon End Product Minimum Quality Characteristics (taken from industrial production)

Iodine Adsorption	mg/g	800
BET N2 Surface Area	m^2/g	750
Volatile Content	%	>1
Methylene blue adsorption	g/100g	10
Moisture	%	2
Ash content	%	3-6

POST TREATMENT, PROCESS RESIDUAL MATERIAL MANAGEMENT

Carbon end product is selected by an TDT-3R™ designed adjustable air selector in any size ranges less than 5 mm, whereas the lowest limit starts at 0-63 micron sizing range.

The characteristics of the TDT-3R™ emissions, whereas the process avoids:

(1) Creation of flux of soot and particles into gas-vapour phase, and

(2) Creation of high off-gas volumes, which are costly to treat, and

(3) Non-perfect burn out of organic components in the post combustion phase by reaching at least 850 °C for 2 sec. true residence time, and

(4) Flux of soot and particles into gas-vapour phase, and

(5) Creation of NOx, SOx, CO and CO2.

The TDT-3R™ technology meets the U.S. RCRA Miscellaneous Units 40 CFR 264 Subpart X for THERMAL DESORBERS and compatible E.U. Norms for THERMOLYSIS with the following main characteristics:

(a) Thermal Desorption Chamber

Indirect-fired heat source used for primary desorption chamber.

Relatively low operating temperature.

(b) <u>Air Pollution Control Devices (APCDs)</u>

Non-destructive APCDs used.

(c) <u>Waste Residual Management</u>

Treatment of residuals is separate from the thermal desorber.

The TDT-3R™ (a)primary desorption chamber, (b)condensation or burning of pyrolysis gas vapours and (c)non destructive APCD off gas scrubber are separate devices, whereas (1)treated solids, (2)condensate residuals, (3)APCD residuals, (4)organic air emission, (5)metal air emission, (6)acid gas emission treatment are according to all the relevant comprehensive U.S. and E.U. regulatory requirements for Operational Control, Residuals and Air Emission Parameters.

Figure 2.: TDT-3R™ Field Demonstration Plant

Table 3 Off-gas Emissions:

	Units	IPR 5/3 U.K. 1996 EU Dir. 89/369/EEC	(273 K, 101,3 kPa, 11 % O2) 17. BlmSchV□Germany	**TDT-3R™ Max. Values**
Dust	mg/Nm^3	30	10	**2**
THC (VOC)	mg/Nm^3	20	10	**1**
HCl	mg/Nm^3	30	10	**1**
HF	mg/Nm^3	2	1	**0,1**
SOx as SO2	mg/Nm^3	300	50	**5**
NOx as NO2	mg/Nm^3	350	100	**100**
CO	mg/Nm^3	100	50	**50**
Hg, Cd	mg/Nm^3	0.1	0,05	**0.01**
As, Cr, Cu, Pb, Ni,	mg/Nm^3	1	0,05	**0.001**
PCDD/PCDF	ng/Nm^3	1	0,1	**0.001**
Target		0.1	non detectable	**non detectable**

THE STATUS OF THE TDT-3R™ TECHNOLOGY

The status of the TDT-3R™ technology today is at the end of innovative phase and early commercialisation phase. Full scale applications are prepared on detailed execution manufacturing level.

FDP has been successfully tested with 0,5 m^3/h continuous throughput capacity. Based on the FDP experiences, up until the end of the 90's, 0,8 m^3/h and 2,5 m^3/h throughput capacity plant has been completely and detailed engineered and designed, for which project applications are under progress.

ECONOMICAL ASPECTS:

(1) At least 30 % decreased production cost even at smaller capacity installations.

(2) Provides high energy recovery profile.

(3) Low investment and O & M cost provides short amortisation time even at smaller capacity installations (approx. 3 years).

CONCLUSIONS

(1) The TDT-3R™ is utilizing refuse biomass for value added production of activated carbon.

(2) Indirectly heated horizontally arranged rotary kiln equipment is used for low temperature carbonization.

(3) Steam activation is used for activation.

(4) Gas-vapour phase either condensed, where bio-oil is utilized and non-condensable gases burned out, or directly burned for onsite energy recovery.

(5) The design is prepared to meet the new U.S. and E.U. technical requirements, environmental norms and standards.

(6) Complete solution, no need for offgas, process water, solid residue post processing.

(7) The process is characterized by flexible operation and simple maintenance.

(8) The process is characterized by modular design.

(9) The TDT-3R™ does not containing exotic technical solutions and construction materials.

(10) The TDT-3R™ is patented, original solution.

REFERENCES

1. Bansal R. C., Donnet J., Stoeckli F., (1988) *Active Carbon*
2. McEnancy B., Dovaston N., (1975) *Carbon*

Workshop Reports

This paper is a collation of the reports from the eight workshops held during the conference:

1. Analysis and characterisation of pyrolysis liquids
2. CHP systems
3. Combustion
4. Economics of bio-energy systems
5. Fundamentals of pyrolysis
6. Gas clean-up
7. Gasification tar management
8. Research needs

ANALYSIS AND CHARACTERISATION OF PYROLYSIS LIQUIDS

A. Oasmaa[a] and D. Meier[b]

[a] *VTT Energy, P.O.Box 1601, FIN-02044 VTT, Finland*
[b] *IWC, Leuschnerstrasse 91, D-21031 Hamburg, Germany*

INTRODUCTION

The first IEA (International Energy Agency) thermochemical Round Robin was organised in 1988 as part of the IEA Voluntary Standards Activity led by BC Research (1). The main conclusions were: the precision for carbon was excellent, while hydrogen, oxygen by difference and water were more variable, and oxygen by direct determination was poor. It was recommended to use a wider variety of samples in the future studies.

Since then considerable progress has been made both in the field of oil production and oil analysis. Therefore, two separate Round Robins were initiated in 1997: one within EU PyNE (Pyrolysis Network) and the other within IEA PYRA (Pyrolysis Activity). The objective of the EU PyNe Round Robin was to compare existing analytical methods without any restrictions. Two pine pyrolysis oils were analysed by eight laboratories for viscosity, water, heating value, elemental analysis, pH, solids, and density. The accuracy for hydrogen, water by Karl-Fischer, and density were good. The xylene-distillation method was stated to yield erroneous results. High variations were obtained for nitrogen, viscosity, pH, and solids. Ethanol was concluded to be more suitable for solids determination than acetone (2).

The main objective of the IEA PYRA Round Robin was to determine the interlaboratory precision and methods applied for elemental composition, water, pyrolytic lignin and main compounds. Two poplar oils were analysed by the IEA PYRA participants. It was concluded that the precision of carbon and hydrogen was very good, liquid sample handling plays a very important role in the C, H analysis, water by Karl-Fischer titration was acceptable, but should be checked carefully, and the method for the determination of pyrolytic lignin should be improved (3).

The latest Round Robin was carried out by 12 laboratories during January - March 2000. Four pyrolysis oils were provided: BTG softwood mixture (spruce and fir), Dynamotive pine (85% pine, 15% spruce), Ensyn hardwood mixture, and Pyrovac softwood bark (1/3 fir and 2/3 spruce with traces of hardwood bark). Based on the feedback from previous Round Robins it was decided to make a suggestion for the instructions on handling and analyses.

Summary

One oil was inhomogeneous due to a high water content causing phase-separation. This may lead to erroneous results. The oil producer pointed out some problems during production, which now have been at least partly solved. In general, the accuracy of all physical analyses was good. Some results were systematically on high side, which most probably is due to poor calibration of the equipment.

Water

Water addition method (4) was suggested for calibrating the Karl-Fischer titration method for pyrolysis oils.

Solids.

It was pointed out that the solids content does not indicate the absolute amount of solids in the oil, because the submicron particles of char present after filtration are difficult to measure. However, it was concluded that this analysis is accurate enough for its present purpose. The microscopic analysis of the oils showed a high amount of small particles below 1 μm. The pore size of filter paper may be reduced down to 1 μm. It was also stated that compounds originated from bark or needles do not dissolve well in alcohols. A more powerful solvent, like methanol-methylene chloride (1:1) can be used.

Carbon, hydrogen, nitrogen.

It was suggested to compare results with another standard, for example, for fossil fuel, in order to find out the normal deviations of the method. The standard deviations for carbon and hydrogen fall well into these limits (4), and it was concluded that the accuracy of these elements was good. Variation in the nitrogen content of white wood oils was obvious due to the similarity of nitrogen content and detection limit for nitrogen (4).

Viscosity.

Pyrolysis oils are Newtonian liquids (5) and hence, kinematic viscosity is applicable. The viscosity results at 20 and 40 °C were very consistent. The smaller standard deviation at 40 °C is logical because of the high temperature dependency of pyrolysis oils. Consequently, it was also suggested for the convenience of the end-users to determine the viscosity at two temperatures, i.e. 20 and 40 °C.

It was pointed out that inhomogeneity of the oil may lead to phase-separation in the capillary tube and hence, to erroneous results. Measuring the viscosity of inhomogeneous oil in a rotaviscotester is suggested.

Stability index.

The stability results varied a lot. Because the stability is measured as a change in viscosity the error in viscosity results yields an error in the stability index. Due to this it was discussed to exclude the stability results of one inhomogeneous oil.
A comment was presented that an increase in the water content of the oil improves its stability. Another comment was made on the correlation of viscosity and the stability index. More testing and further development of the stability test are needed. The stability index should not be the only measure of the storage stability of oil. Another simple test method may be necessary. The authors wish to bring out that, in some cases, the instructions had not been delivered to technicians, which has lead to erroneous results. Hence, the stability results will be re-checked.

It was commented that soot stabilises emulsions. On the other hand, soot in the pyrolysis oil most probably yields high particulates emissions.

Chemical characterisation.

The results of chemical characterisation were not very consistent. It was discussed to prepare standard solutions with known amounts of compounds for quantitative analyses. A question was also raised if the functional groups in pyrolysis oil should be analysed analogously with petroleum residues instead of individual compounds, as by quantitative ^{13}C-NMR. The amount of PAH (polyaromatic hydrocarbons) was extremely high for one pyrolysis oil, and it was discussed that more attention should be paid to the analysis of toxic compounds in the oils. The oil producer commented later that the high PAH may be due to contamination of other fuel and this will be checked.

Pyrolysis oil specification.

Fuel oil specifications (6) were only discussed very briefly because oil end-users were not present. pH was suggested to be added in order to help in material choices. It was mentioned that the volatility index or boiling point range distribution would be valuable. However, it has been proved that the boiling point cannot be determined for pyrolysis oil because of the thermal instability of the oil (7, 4). More feedback from end-users is needed for assessing the maximum allowable variation for each property.

CONCLUSIONS

In general, the accuracy of all physical analyses was good. Additionally, it was concluded that inhomogeneous oils may give erroneous results, at least considering kinematic viscosity and stability index. Good laboratory practice, like proper calibration of equipment, prevents systematic errors. Further, a high standard deviation may be due to the fact that the value is too close to the detection limit of the equipment. This is typical of the case with nitrogen.

The main conclusions of the Round Robin were: Karl-Fischer titration is recommended for analysing water in pyrolysis oils. Solids content as ethanol insolubles is accurate for white wood oils but a more powerful solvent, like a mixture of methanol and methylene chloride (1:1) is needed for extractive-rich oils. For the elemental analysis at least triplicates are recommended. Kinematic viscosity is an accurate method for pyrolysis oils. Stability index needs more clarification and testing. Results of chemical characterisation were not very consistent. It may be necessary to prepare standard solutions with known amounts of compounds for quantitative analyses. The final results of the Round Robin will be published in PyNe final report in April 2001.

The main conclusions were as follows:

- Karl-Fischer titration can be recommended for analysing water in pyrolysis oils. The method of water addition is suggested to be used for method calibration.
- The solids content using ethanol as a solvent is accurate for white wood oils. However, if the feedstock contains extractives, for example, from bark and/or needles, a mixture (1:1) of different solvents, like methanol and methylene chloride is recommended.
- For elemental analysis, at least triplicates are recommended. Nitrogen is not very accurate for white wood oils due to its low amount and the equipment limitations.
- Kinematic viscosity is applicable to pyrolysis oils because of its accuracy and Newtonian behaviour of pyrolysis oils. However, method calibration should be taken care of.
- Stability index needs more testing and its correlation with the water content and original viscosity should be studied. Another simple test method for stability may be needed. The results will be re-calculated because the determination was in some cases erroneous.
- In case of inhomogeneous oils, some analyses like kinematic viscosity and stability index cannot be applied due to possible phase-separation of oil during viscosity determination.
- The results of chemical characterisation were not very consistent. It may be necessary to prepare standard solutions with known amounts of compounds for quantitative analyses.

REFERENCES

1 Mckinley, J. W., Overend, R. P. & Elliott, D. C. 1994. "The ultimate analysis of biomass liquefaction products: The results of the IEA Round Robin #1". In: Proc. Biomass pyrolysis oil properties and combustion meeting, 26 - 28 September 1994, Estes Park, CO. Golden, CO: NREL. Pp. 34 - 53.(NREL-CP-430-7215.)

2 Meier, D. Technical PyNE Group Report. Characterisation and analysis. In: Minutes of 5th PyNe Meeting 28th February to 3rd March 1998, Salzburg, Austria.

3 Bridgwater, A., Czernik, S., Diebold, J., Meier, D., Oasmaa, A., Peacocke, G., Piskorz, J. & Radlein, D. "Fast pyrolysis of biomass: A handbook". Newbury: CPL Press., 1999. 188 p.

4 Oasmaa, A., Leppämäki, E., Koponen, P., Levander, J. & Tapola, E. "Physical characterisation of biomass-based pyrolysis liquids. Application of standard fuel oil analyses". Espoo: VTT Energy, 1997. 46 p. + app. 30 p. (VTT Publications 306.)

5 Leroy, J., Choplin, L. & Kallaguine, S. "Rheolological characterization of pyrolytic wood derived oils: Existence of a compensation effect". Chem. Eng. Comm., 1988, vol. 71, pp. 157 - 176.

6 Diebold, J. P., Milne, T. A., Czernik; S., Oasmaa, A., Bridgwater, A. V., Cuevas, A., Gust, S., Huffman, D. & Piskorz, J. "Proposed specifications for various grades of pyrolysis oils". In: Bridgwater, A. V. & Boocock, D. G. B. (eds.). Developments in Thermochemical Biomass Conversion, Banff, 20 - 24 May 1996. Vol. 1. London: Blackie Academic & Professional, 1997. Pp. 433 - 447.
7 Meier, D., Oasmaa, A. & Peacocke, G. V. C. "Properties of fast pyrolysis liquids: status of test methods. Characterisation of fast pyrolysis liquids". In: Bridgwater, A. V. & Boocock, D. G. B. (eds.). Developments in Thermochemical Biomass Conversion, Banff, 20 - 24 May 1996. Vol. 1. London: Blackie Academic & Professional, 1997. Pp. 391 - 408.

CHP SYSTEMS

B Jenkins [a] and J Brammer [b]

[a] *Dept. of Biological and Agricultural Engineering, One Shields Avenue, Davis, California, CA 95616-5294, USA*
[b] *Aston University, Aston Triangle, Birmingham B4 7ET, UK*

DISCUSSION:

This workshop was held to discuss problems and opportunities in the development and application of biomass fuelled combined heat and power systems (CHP), and to make recommendations for research, demonstration and implementation of CHP systems. Although the development of cohesive and coordinated policies relating to CHP was recognized as being important, the discussion focused on technical issues and recommendations.

Main findings of the discussion:

- The need for comparative studies: Studies have been done of the general feasibility of biomass CHP. However, there was a general consensus that more comprehensive comparisons were needed of the technical and economic performance of biomass fuelled CHP in competition with other energy systems, such as natural gas distribution for residential heating.
- The need for optimising CHP: Optimisation was seen as being important to CHP implementation to address issues of scale, heat and power partitioning, technology selection, fuel selection, fuel flexibility, quality of heat and power, valuation of heat and power, and the potential use of biomass in existing CHP systems using other fuels. The optimisation should be performed on the basis of economic performance for the various technology and fuel combinations, including the range of scales from the very small (e.g., farmer operated or other systems running Stirling engines, microturbines, organic Rankine cycles, or reciprocating engines fuelled perhaps with pyrolysis oil), to larger systems including co-feeding existing systems. Optimisation studies are needed to assist in decision-making relating to the specific configuration of any CHP system.

- The need for reliable and economic small scale systems: Although included under the general category of optimisation, there was a general consensus of the need for more effort on small-scale or micro-scale CHP systems. In part this stems from the perception that many small-capacity heat loads exist, and hence there are opportunities available if appropriate technologies can be identified. Current technologies were considered to be too expensive in general. Greater research and development is needed for the small-scale systems (< 500 kW_e, < 2 MW_t) to reduce cost, even if that means accepting a lower efficiency.
- The need for fuel standards: Standards defining the quality of biomass fuels and fuel blends are lacking, especially in relation to application in CHP systems. Technical and economic performance is improved when biomass fuels can meet specifications of the design, and designers need good information on fuel properties. Standards should address biomass quality, as well as processing and upgrading of biomass (e.g., pyrolysis oil, pellets, sewage sludge dewatering/conversion).
- The absence of environmental concerns specific to CHP: CHP systems were not perceived to have environmental concerns different from other power and heat systems.

RECOMMENDATIONS OF THE WORKSHOP:

1. Criteria should be established to aid decision making on the selection of a CHP system as opposed to separate generation of heat and power.
2. Studies should be made of the economics of building biomass-fuelled district heating networks for new residential/commercial communities, and in particular of the comparative costs of these and natural gas distribution networks for the supply of fuel for heating.
3. Protocols or methods should be developed for optimising biomass CHP systems, with particular reference to:
    - costs for electricity and heat
    - whether or not to upgrade fuels
    - potential economies of scale in capital and operating costs
    - the use of fuel blends
4. Studies should be continued into the properties of biomass fuels and blends, with reference to mitigating the impacts of fouling, corrosion, and other detrimental effects on plant efficiency, operation, and maintenance (particularly in co-fired systems).
5. Increased research should be encouraged on small-scale biomass CHP systems (< 500 kW_e), especially low-maintenance, simple, fuel-flexible systems.
6. Biomass CHP systems should be demonstrated at the commercial scale, including the potential production of chemicals as co-products of the heat and power system.
7. Partnerships should be encouraged between developed and developing countries for the transfer of biomass CHP technology.
8. A state-of-the-art review of biomass CHP systems should be prepared to aid in information dissemination and new program development.
9. The summary of recommendations from all the conference workshops should be made available as a stand-alone document targeted at policy makers and government program managers.

COMBUSTION

L Baxter [a] and B Jenkins [b]

[a] *Sandia National Laboratories, 7011 East Avenue, MS 9052, Livermore, CA 94550, USA*

[b] *Dept. of Biological and Agricultural Engineering, One Shields Avenue, Davis, California, CA 95616-5294, USA*

SUMMARY OF DISCUSSION:

A brief discussion was held of issues relating to biomass combustion. Main topics included:

- Principal emissions of concern are particulate matter and NOx, and work should continue on developing means to reduce these pollutants from combustion systems.
- Co-combustion of biomass with natural gas or gas from biomass gasification offers potential efficiency improvements relative to separate combustion of biomass, as well as potential emission reductions through reburning.
- Combustion of waste wood presents special problems from a regulatory standpoint, and research should continue on reducing pollutant emissions. There is some potential for blending with clean wood without substantially altering emission control systems in existing units. Better fuel characterization is needed for waste materials.
- Feedstock costs remain high in many cases, and logistics of fuel delivery are seen as a critical area for improvement.
- Improvements are needed in combustion technologies to improve combustion stability and to reduce the impact of transients, especially on emissions. Greater effort is needed to take advantage of new developments such as flameless oxidation and other advances.
- Fuel pretreatment is viewed as a way to improve combustion behavior and reduce a number of operating problems. Greater effort is needed on separation of undesirable constituents and components from biomass fuels.
- Ash disposal remains a concern and better uses for ash or means to mitigate special or hazardous characteristics of ash are needed.
- Methods to optimize blending of different feedstocks are needed, including predictions of inorganic behavior in fuel blends during combustion.
- Good models are needed for the design and operation of combustion systems. Expert systems, might, for example, contribute to diagnosis of operating problems or aid in the design of more complex systems.

RECOMMENDATIONS

1. Fuel properties and environmental impacts:
    a. Conduct research into the environmental impacts of biomass and waste fuels including the impacts of hazardous constituents and components.
    b. Continue to conduct research into the general properties of biomass with special emphasis on predicting the properties and behavior of fuel blends in combustion systems.
2. Fuel logistics:

a. Improve logistical support systems for using biomass and waste materials, including handling, transportation, blending, and feeding systems.
b. Develop test methods, standards, or other protocols for providing quality assurance in biomass fuels delivered to commercial users.

3. Modelling and design:
a. Generate comprehensive databases of information from commercial biomass plant operations appropriate for validating combustion models.
b. Develop more detailed comprehensive models for better understanding of biomass combustion phenomena.
c. Develop engineering models for improved design and operation of biomass combustion systems.

ECONOMICS OF BIO-ENERGY SYSTEMS

Yrjö Solantausta [a], David Beckman [b], Ian Burdon [c]

[a] *VTT Energy, Biologinkuja 5, PO Box 1601, Espoo FIN-02044 VTT, Finland*
[b] *ZETON Inc., 5325 Harvester Road, Burlington, Ontario L7L 5K4, Canada*
[c] *PB Power Ltd (Merz and McLellan), Amber Court, William Armstrong Drive, Newcastle upon Tyne NE4 7YQ, UK*

SCOPE OF THE WORKSHOP

Technoeconomic assessments are an integral part of any process development effort. Assessments vary in scope and size at different stages of process development work. However, it is recognised that uncertainties in input data may often be considerable, which may make use of results difficult.

TOPICS DISCUSSED

Comparisons of technologies on consistent basis

When new bioenergy technologies are compared to each other, it is of prime importance to carry out all stages of work on a consistent basis. In practise this means of carrying out rigorous performance analysis, sizing, and costing for all alternatives in detail.

It was also pointed out that perhaps cost analysis should not be used as a primary criterion, when processes at a very early development stage are assessed. In this case a rigorous performance analysis (and the respective efficiency) may be a better criterion in assessing a future potential of a given technology.

Objectives of studies and communication to politicians

Participants of the workshop expressed that it would be important to convey the messages from bioenergy feasibility studies to politicians and other decision-makers. However, it was not entirely clear in what format and from which platform such messages should be addressed. The IEA Bioenergy Executive Committee was mentioned as one possibility.

Uncertainties in the estimates

When analysing several new processes, typically not all processes are at the same level of development. This makes comparison difficult. In principle different contingencies for capital investment should be used. The problem is that reliable contingency factors are not usually available. One suggestion made at the workshop was to require considerable reduction in cost (for example, half the cost compared to existing alternatives) without any contingencies for new systems.

Economic estimates of new bioenergy processes tend to be positive. Reasons for this are that the economic feasibility studies are based on process data that is too uncertain, and results from too small scale equipment. Studies assessing the cost of new technologies have shown that the cost of the first demonstration plant and products are on average four times the cost of the original estimate 1.

The IEA Bioenergy techno-economic assessments

One platform that the IEA countries have used to help development of bioenergy applications has been the techno-economic assessments carried out during the past 20 years. An expert group has been assembled, which has developed standard procedures for the assessments. Lately the work has been carried out with industry, which typically has supplied cost data for the assessments.

CONCLUSIONS AND RECOMMENDATIONS

A consensus of the workshop was that an engineering analysis should be included early on in the development process. It was recognised that carrying out process evaluations from the start of the work may identify unknown aspects in a new process. In fact evaluations should be carried out during the whole time of process development.

1. Studies where the data is uncertain should first concentrate on process performance analysis first, such as efficiencies, and not on economic analysis.
2. Economic feasibility studies should not be done on processes where the basis of the data is too uncertain.

General process comparisons can be misleading. It is better to analyze a technology on a specific site and set of local conditions.

REFERENCES

1 Merrow, E., Chapel, S., Worthing, C., "A Review of Cost Estimation in New Technologies: Implications for Energy Process Plants". Prepared by RAND Corporation for the U.S. Department of Energy. Santa Monica, CA. RAND-R-2481-DOE. July 1979.

FUNDAMENTALS OF PYROLYSIS

B. Krieger-Brockett [a] and J. Piskorz [b]

[a] *University of Washington, Benson Hall BF-10, Seattle, WA 98195-1750, USA*
[b] *Resource Transforms International Ltd., 110 Baffin Pl., #5, Waterloo, ON N2V 1Z7, Canada*

INTRODUCTION

This is a summary of the PITBC workshop in which we discussed and reviewed selected aspects of fundamental research in the area of thermochemical biomass conversion. The workshop, attended by over 15 researchers from Europe, Americas and Asia, concentrated on needs, trends and outlook for biomass pyrolysis fundamental research. The remarks fell into these categories:

- the importance and generalities of true fundamental science, as well as the need for fundamental research in the field of biomass thermolysis/pyrolysis;
- the recent advances in light of a long tradition of biomass utilization studies conducted during the past century;
- the importance of communication, in a meaningful and systematic way, of significant accomplishments in the field to decision makers in the forest products and renewable energy industry and to technology developers.

"The formulation of a problem is often more essential than a solution, which may be merely a matter of mathematical or experimental skills. To raise new questions, new possibilities, to regard old questions from a new angle, requires creative imagination and marks real advances". Albert Einstein

DEFINITION

Mankind's fascination with pyrolysis started with the discovery of fire. Pyrolysis is usually understood to be thermal decomposition of organic matter occurring in oxygen-depleted or oxygen-free atmosphere, although there are other less general terms that are used to describe this same process (i.e., devolatilization, destructive distillation, carbonization, liquifaction, gasification). The pyrolysis step is of critical importance for all solid fuels and feedstocks since this step precedes any subsequent steps in any gas-phase combustion and biomass conversion schemes. The pyrolysis step, particularly in so-called "fast pyrolysis" happens in a few seconds or less depending on biomass particle size and heating rate.

On such short time scales, chemical reaction kinetics, mass transfer processes, phase transitions and heat transfer phenomena play important roles and can influence the ultimate conversion outcomes. Also, due to the same short time intervals, fundamental aspects of those transformations are not easily elucidated. An important objective is to obtain a comprehensive understanding of significant factors determining biopolymer thermolysis, and therefore to predispose or quantitatively predict certain conversion rates and products. There is no doubt that benefits to society from such work will come from technological developments and innovations leading to improved renewable fuel utilization, new high value chemicals, a sustainable high-quality environment and improved well being of all.

DISCUSSION

A thread throughout the discussion was the extreme importance of emphasizing the general applicability of *fundamental* experiments and models based on first principles. While the meaning of words such as "basic" and "fundamental" to diverse audiences was debated, there was general agreement that the results we label *fundamental* are *those that have general applicability*, that is, they are *NOT system-specific*. In this endeavour we agreed that vigorous attention must be paid to reporting *chemical* as well as *physical* details of our experiments and models in order for other researchers to be able to evaluate and use our results. The details that must be measured are numerous. Indeed, they are more numerous for biomass than other solid fuels owing to the high reactivity and heterogeneity of biomass (growth, storage, and reaction conditions make a difference in the conversion product slate). In addition to reaction conditions, we agreed that at a minimum the following reactant chemical specifics and physical parameters of the biomass will be useful for fundamental evaluations:

- chemical specifics: species names, compositions, and preparation/collection method
- physical parameters: particle sizes, particle heat transfer properties (thermal diffusivity), heats of reactions, particle mass transfer properties (effective diffusivity)

It was remarked that our research area has a long tradition based in the forest products industry and pulping chemistry. In those fields, somewhat slower and less thermally severe reactions were employed than we now study in biomass thermal conversion to gases, liquids, and energy. In many contemporary thermal conversion processes, the biomass particles and the fluids in reactors are NOT necessarily at the same conditions of temperature and composition and frequently have temperature and composition *gradients in both the particles and fluids*. Thus, in fundamental studies, we agreed that it will be useful to quantitatively report or estimate all fluid (gas or liquid) and solid compositions, gradients, and temperature-time histories, as well as flow characteristics that determine energy and material transport between the biomass particles and the fluids in the reactor.

Applauding the utility of *fundamental* pyrolysis data and predictions, Drs. Lede and Suuberg both commented that we should periodically critically synthesize our results. In particular, the results should be reported and evaluated in light of systematic approaches recognized by industry. Of necessity, these approaches incorporate both chemical and physical data and experiments *at increasing length scales*. It was suggested that reactor theory or the reaction engineering approach is a valuable tool in this regard. This approach systematically synthesizes the knowledge, often using dimensionless groups to categorize behaviour, from the molecular length scales (chemical composition), to the microscale, (the small homogeneous porous particles in which no gradients of temperature or composition exist), to the mesoscale (thermally thick, but practical feedstock particles in which temperature and composition gradients dominate), all the way to the industrial reactor length scale that must operate economically. It was further emphasized that economical biomass reactor technology must process a *heterogeneous collection* of mesoscale particles that are influenced by local fluid mechanics, temperature, and fluid composition.

The importance of collaboration with industry was underscored in order that we may incorporate realistic constraints of the technology into the range of experimental conditions and simulation predictions we study. For example, we discussed the lack of fundamental pyrolysis predictions suitable for use in practical large fixed or moving bed reactors, large fluidized beds, etc. employing *real* feedstocks. It was remarked by the industrial representatives that particular attention must be paid in pyrolysis studies to the often very different residence times and states of mixing for the fluids in contrast to the solids. The reaction engineering approach was suggested because it has previously been highly successful in finding novel ways of incorporating validated fundamental information into industrial processes, notably in the field of catalytic reactions, catalyst synthesis, and reactor design.

There was considerable discussion on the wealth of historical findings reported in the literature regarding biomass pyrolysis and the devolatilization preceeding biomass combustion. J. Piskorz listed relevant qualitative observations dating from 1914 regarding pyrolysis products from biomass. With such a wealth of literature to draw from, a discussion revealed that we now must emphasize that we *can be more quantitative* than in the past. In being more quantitative, we must also be more critical and complete in our research in order to put our new findings, and models derived from them, in the proper and meaningful perspective.

New instruments allow us to be increasingly quantitative and accurate, but it is our obligation to communicate how this often overwhelming amount of data can be meaningfully used. For example, it was once reported that there were of the order of 20 substances in wood tar when the most advanced (at that time) detectors and columns, perhaps FID and packed columns, were used. Now, with GC/MS or HPLC/MS systems available, it has been shown that hundreds of compounds are routinely found in tar. However, many of these compounds appear in concert with others or are "surrogates" for other compounds. Thus it is useful to "lump" or simplify the compounds into groups, which appear at the same reaction time, or under the same pyrolysis conditions, or from the same starting sample. These co-varying compounds also must be distinguished from those that are unique to particular experimental conditions or samples under study.

It was suggested that computer-based data analysis techniques (often involving multivariate statistical methods) can aid in this classification or simplification, as has been so profitable in other thermochemical conversion endeavours, for example, as applied to coal and petroleum. Again, it was emphasized that there is a need for a critical synthesis of the wealth of experimental data into regimes of behaviour, and simpler predictive equations or simulations, that are useful to the technologists in industry who are designing industrial scale reactors.

Regarding the fundamental, general applicability of what we measure, the workshop attendees agreed that it was important to communicate the importance of our findings, and that perhaps we need better marketing skills. Biomass is in a unique position. It is the only renewable source of carbon-based fuels and biopolymers, but it is in direct competition with fossil fuels and other renewable energy (solar energy, wind power, etc). Because the variety of options for renewable resource utilization is so large, the case for bio-based energy or materials has to be made more forcefully and concretely.

CONCLUSIONS AND RECOMMENDATIONS

General

- Renewable carbon-based biopolymers (biomass) are produced at a large annual rate in the biosphere.
- Conversion of this biomass by rapid thermal means, although presently in an incubation stage, appears to have a tremendous potential.
- Emerging renewable energy sources like wind, photovoltaic devices, and hydropower cannot displace the global utilization of biomass.
- Efforts should be directed to facilitate good interaction between scientific policy formulators, research organizations, and organizations wanting to commercialize pyrolysis conversion processes.
- Only concerted and cooperative efforts can accelerate the progress.

Specific

There is a very considerable literature, knowledge base, and prior art concerning thermal biomass conversion. Some historical examples include:

- R.C. Palmer (1914) reported increased (up to 42.5 wt%) charcoal yields when carbonizing wood under pressure of 150 psig.
- Klason (1914) accomplished a vacuum distillation of wood (cathode-light vacuum) and a short time. His results – tar yield 43.5wt%, char 19.4 wt%.
- Pictet, Sarasin discovered levoglucosan in 1918
- Bobrow mentioned hydroxyacetaldehyde in 1934.

Yet the substantial quantitative research progress of recent years could be easily illustrated by:

- Identification in pyrolysis oils of literally hundreds of chemicals (from thousands possible);
- Creation of national and international biomass utilization and conversion databases;
- Construction of a variety of pilot plants employing biomass pyrolysis;
- Increasing accuracy and predictive power of pyrolysis mathematical modeling.

Despite this substantial progress, biomass conversion activities suffer from lack of support by a mature industrial sector (there is none with the exception of the forest products industry, which often has conflicting goals). The oft-mentioned advances in "coal science" were likely due in part to the support of a mature coal-utilization industry. While biomass researchers can profit from models, tools and achievements of clean coal technologies, nevertheless, the government and industrial support is crucial to implementing the research advances in industrial practice. The European Union and Scandinavian countries are fulfilling this obligation to a significant degree already.

PITBC and similar conferences have an invaluable role to play in encouraging active discussion, in reporting developments of basic science, and in transferring this know-how to potential industry developers of the future.

GAS CLEAN-UP

JPA Neeft [a] and N Abatzoglou [b]

[a] *ECN - Netherlands Energy Research Foundation, PO Box 1, Petten, 1755 ZG, Netherlands*

[b] *Kemestrie inc., 4245 rue Garlock, Sherbrooke, Quebec, J1L 2C8, Canada*

INTRODUCTION[1]

Contaminants in the product or flue gases from thermochemical conversion of biomass are particles, tars, acid gases (HCl, COS, N_xO_y, SO_x, HCN), basic gases (NH_3), and metals (both alkali metals and heavy metals).

Several cleaning techniques exist for the removal of these contaminants (1, 2). These techniques can be subdivided into the three classes hot gas cleaning, dry scrubbing/removal techniques and wet scrubbing/removal techniques. Table 1 was used as a structure for the discussion in this workshop. On the basis of this table, the following discussion topics were chosen:

1. Catalytic converters for tar
2. Cost of gas cleaning
3. Catalytic removal of NH_3
4. Scrubbers / tar removal
5. Particle removal (all techniques)

Two questions were discussed for each of these five topics:

a. What are the results with this gas clean-up technology (removal efficiencies, experiences);
b. What are the needs for fundamental and applied research;

Table 1 Suitability of different gas cleaning methods for different contaminants

Gas cleaning technology	Contaminants							
	Particles	Tar + VOCs	HCl + HF	H_2S + COS	SOx	NH_3	NOx	Alkali metals
Cyclone	*							
Thermal cracker		*						
Catalytic converter		*				*	*	
Ceramic filter	*							*
Dry hot absorption			*	*	*	*		*
Fabric/baghouse filter	*							*
Rotating particle separator	*	*						*
Scrubber	*	*	*	*	*	*	*	*
Electrostatic precipitator	*	*						*
Granular filter Activated carbon Sand/sawdust	*	*	?	?		?		*
New ideas								

FINDINGS

Catalytic convertors

The discussion focused on Nickel catalysts. Commercial availability is very good as in Europe alone at least 4 manufacturers provide these catalysts. The pellet catalysts are less expensive than monolith supported catalysts. One of the large uncertainties of these catalysts is their deactivation rate under practical conditions. Deactivation occurs due to reactions with (heavy) metal-, sulfur- and chlorine compounds. Due to the lack of extensive data the costs for applying Ni-catalysed for tar removal in commercial biomass CHP systems are still unknown.

The intended application of these catalysts is in the medium to large scale. For small scales, heat recovery has to be applied such as in a reversed flow reactor, otherwise, catalysts can probably not be applied because of high costs.

Costs

The gas-cleanup depends on the biomass fuel, therefore, the cost of the cleanup section is case specific. However, these costs are related to the fuel flexibility: a simple and cheap gas-clean-up system will limit the fuel flexibility and might cause higher overall costs than a more advanced gas clean-up system in which also more contaminated so less expensive fuels can be used.

Catalytic removal of NH_3

This topic is covered by a large 5FP EU project (NNE5 99 00923). This topic did not raise a discussion but is seen as an important field of R&D in particular for the development of hot gas cleaning in biomass-IGCC.

Scrubbers / tar removal

Scrubbers for tar removal have a large disadvantage, which is the tar condensate and scrubber water. Several institutes have experiences with scrubbers (mostly wash towers and Venturi scrubbers), however, hardly any data can be found in open literature. Wash towers only remove part of the tars (typically 40%) whereas Venturi scrubbers can have much higher efficiencies of 80% or higher.

Techniques to clean the scrubber water have been researched. Proposed techniques for water cleaning are destruction by wet oxidation and absorption by activated carbon. Reference was made to a 1998 Verenum report (3). Scrubber solvents other than water have occasionally been used to remove tars from producer gases, the Technical University of Vienna has recent experiences (see also the presentation by Hermann Hofbauer in these proceedings).

The requirements for further development are the treatment or disposal of (tar containing) waste water, and the efficiency of scrubbers when scrubber solvents other than water are used.

Particle removal (all techniques)

Many techniques are used to remove particles from biomass fuel gases (see table), but only two techniques can possibly remove submicron particles: electrostatic precipitators (ESP) and rotating particle separators (RPS). There is hardly any experience with RPS for particle removal from biomass fuel gases.

Particle removal at high temperatures is possible with ceramic or sintered metal filters as has been demonstrated in the Värnamo biomass gasification plant. Filter blinding (reaction of tars in the filter pores) has occasionally been found to be a problem when testing high temperature filters. This phenomenon is, however, not reproducible nor understood;

Tars can also be removed as particulates (aerosols) which has only been shown for pyrolysis or updraft gasifier tars/oils using ESP's. Experience with the use of ESP for removal of tar aerosols from downdraft or CFB gasifiers is non-existing or is not reported.

NEEDS FOR FUNDAMENTAL AND APPLIED RESEARCH

- Assess the deactivation rate of nickel catalysts under practical conditions;
- Compare costs of gas cleaning in biomass gasification systems and assess the influence on fuel flexibility on these costs;
- Develop the catalytic removal of NH_3 which is important particularly in hot gas cleaning for IGCC application of fuel gases (activities are being performed in 5FP EU project);
- Further develop treatment or disposal techniques of tar containing waste water;
- Evaluate the efficiency of scrubbers for tar removal using water and other scrubber liquids;
- Assess the efficiency of techniques for removal of submicron particles from biomass fuel and flue gases, and of removal of submicron tar aerosols from biomass fuel gases;
- Learn to understand and predict blinding of high temperature ceramic/sintered metal filters;

REFERENCES

1 T. Milne, N. Abatzoglou and R.J. Evans: Biomass gasifier "tars": their nature, formation and conversion, Golden, CO (USA), NREL, NREL/TP-570-25357, 68 p. (1998).

2 J.P.A. Neeft, H.E.M. Knoef and P. Ojani: Behaviour of tar in biomass gasification systems. Tar related problems and their solutions, BTG/ECN, Novem-EWAB no.9919, 74 p. (1999).

3 P.Hasler, P.Morf, R.Bühler and T.Nussbaumer: Gas cleaning and waste water treatment for small scale biomass gasifiers, Zürich (Switzerland), Verenum, 102 p. (1998).

GASIFICATION TAR MANAGEMENT

Nicolas Abatzoglou [a] and Krister Sjostrom [b]

[a] *Kemestrie inc., 4245 rue Garlock, Sherbrooke, Quebec, J1L 2C8, Canada*
[b] *Royal Institute of Technology, Dept Chemical Technology, Teknikringen 42, Stockholm S-10044, Sweden*

INTRODUCTION

The Workshop focused on the utilization of catalysts at high temperature for tar transformation to combustible gases. The following points have been retained:

DOLOMITES

Dolomites: This naturally occurring catalysts are relatively efficient in converting tar to non-condensable combustible gases (CO, H_2) and CH_4, which is not reformed significantly. The efficiency varies from 50-90% depending on both tar and dolomites composition.

Dolomites age quite rapidly and replacement is required if efficiency is to be kept at the initial level. A fluidized bed catalytic reformer based on the use of dolomites has been commercialized by TPS.

The main problems associated with the use of the dolomites are:

- Aging due to coke deposition
- Fast erosion leading to size decrease and, in the case of the fluidized bed, entrainment out of the bed
- Decomposition of carbonates due to high CO_2 partial pressure
- Poisoning due to the deposition of Cl and S, two hetero-atoms responsible for the loss of the catalytic activity of the Ca and Mg oxides
- High Cl retention in the dolomite (i.e. when gasifying high Cl content straw) leads to the formation of $MgCl_2$ and CaCl2; these salts are melting at relatively low temperatures and can also form eutectic mixtures with other inorganic material (i.e. silica sands). Thus sintering can occur at relatively low temperatures.

METALLIC CATALYSTS

A wide variety of metallic and bi-metallic catalysts are available in the market. The most prominent are nickel catalysts supported in various matrices, namely acidic Al_2O_3. These catalysts are very efficient in reforming even heavy tar to CO and H_2. Simultaneous partial reforming of the synthetic gas cannot be avoided but it is not usually detrimental to the gas quality.

These catalysts are used in pellets, spheres or even deposited in monoliths (i.e. honeycomb matrices). In all cases the catalytic reactor configuration is a fixed bed running as a perfect Plug Flow Reactor (PFR). Fluidized bed configurations, as those in FCC systems utilized by the petrochemical industry for the production of reformulated gasoline, are not yet known for this application; the reason being that such a reactor involves an entrained bed configuration for continuously regenerating the catalyst and catalyst erosion becomes a considerable issue.

The fixed-bed are used in twin-beds configurations, where one of the vessels is operated as reformer while the other is in catalyst regeneration mode. Regeneration is achieved through mild conditions combustion of the deposited coke and S, or through steam reforming of the deposited coke.

A new avenue under study is the use of honeycomb monoliths and granular catalytic media both as reformers and particulate filters. Experimental work is underway in Sweden and Canada respectively.

Tar reforming has not yet reached the commercial application level; mainly because of the high cost and some peculiar technical problems associated with the use of the existing catalysts. These technical problems are reported below:

- The high temperature required for an efficient reforming (typically higher than 800°C) is an important aging factor for the catalyst; the supported metallic catalysts exposed at high temperature fields for long periods of time in presence of organic compounds are subjects to structural changes. The latter are responsible for the crystal lattices collapsing and the loss of the catalyst particles cohesion. Thus the catalyst tablets, spheres or the monolith are transformed to granules or powders, are subsequently entrained by the gas and lost.
- Coke formation, even though minimized by appropriate catalyst formulations during the last decade, cannot be avoided completely. Typically after 2-3 days on stream the catalysts require coke removal to regain their initial catalytic activity. Such regeneration cycles are introduced using twin-beds configurations as already explained. After each regeneration cycle the catalyst shows a slightly lower efficiency. The number of the possible regeneration cycles before the catalyst being considered useless determine the lifecycle of the catalyst and consequently the cost of the reforming process
- Heteroatomic contaminants present in the gas and the tar, like Cl and S, react with the metals at the surface of the catalyst. The process is known as catalyst poisoning because the salts formed are not showing catalytic activity towards tar reforming. Some of this poisoning, basically the one associated with the S can be reversible because high regeneration temperatures can transform the sulfides to sulfates and subsequently decompose the sulfates to SO_2/SO_3 and the corresponding metal oxides. This is not the case with Cl.
- Volatile metals, mainly alkalis, can also deposit on the surface of the catalyst and cause sintering or simply perturb the necessary for the catalytic action electric charges of the surface of the catalyst. This leads to a second poisoning due to chemical poisoning.

It is obvious that the presence of the Cl and S, as well as of volatile metals, is detrimental to any tar reforming catalytic system and that the success of this process pass through a successful removal of these contaminants prior of entering any catalytic reactor. Cl, S and metals getters have been already developed, tested and applied by the coal gasification industry. Nevertheless the biomass gasification field need to address this issue the faster possible in order to allow hot gas conditioning options combined with the use of gas turbines (i.e. in IGCC) to have success in the future. We must remember that the hot gas conditioning option is very promising for the following two reasons:

- Gas turbines efficiency is higher with high temperature gases

- Hot gas conditioning avoids the use of water scrubbing thus it does not create any contaminated liquid effluent which would need additional treatment prior to its final disposal.

CATALYSTS FOR METHANOL SYNTHESIS

Commercial catalyst are available for the production of methanol and other liquid fuels from synthetic gases. The main problem is the catalyst deactivation due to chemical poisoning from chlorine and sulphur.

The following conclusions were reached:

1. Need for R&D in the field of Cl, S and metals removal from hot synthetic gas stream
2. Need to test various existing catalysts at, at least, demonstration scale, in order to establish their lifecycles and consequently successfully address the technical barriers and economics of their utilization
3. Need for R&D in terms of developing catalysts that are robust enough to resist thermal restructuring (i.e. exceptional spinels)
4. Need for R&D in terms of catalysts regeneration (steam or controlled combustion). Appropriate profiles must be established as function of the catalyst composition and contamination level.

Table 1 gives a summary of the existing catalytic systems for tar removal as presented in 1998 by Abatzoglou in a meeting of the IEA Gasification Task Team.

Table 1 Summary of existing catalytic systems for tar removal (Abatzoglou)

PAH

- High temperature complete oxidation
- Thermal cracking (temperature well above 1100 °C); soot production; unsatisfactory conversion
- Catalytic cracking (enhanced by high temperature)
  - Zeolites (not efficient above 850 °C)
  - Dolomites
  - Nickel catalysts
  - Lewis acids
- Catalytic steam reforming
  - Basic matrices (MgO-CaO)/Dolomites
  - Alumina supported Ni and NiO catalysts

PCDD/F

- High temperature complete oxidation
- Low temperature (300 - 400 °C) catalytic oxidation
- Catalytic cracking (enhanced by high temperature)
  - Zeolites
  - Lewis acids
- Catalytic steam reforming
  - Basic matrices (MgO-CaO)/Dolomites
  - Alumina supported Ni and NiO catalysts

RESEARCH NEEDS IN THE AREA OF BIOMASS

A Segerborg-Fick [a] and S-Y Yokoyama [b]

[a] *European Commission, DG Research, Rue de la Loi 200, Brussels B-1049, Belgium*

[b] *National Institute for Resources and Environment, Tsukuba, Ibaraki 305, Japan*

RESEARCH STRATEGIES AND POLICIES

This workshop was attended by eight people, which shows that the main interest of this Conference audience was not about research strategies and policies. This was an interesting observation as most of the delegates were researchers themselves. Half of the group was from outside Europe, mainly from Asia (Thailand and Japan).

The discussion began by considering the European Research Area (ERA) that is a view on how to organise research on a European level in which creating centres of excellence is the main topic. The concerns on a European level is that there are now many different research programmes instead of emphasising collaboration with the resources and knowledge base available. It was interesting to discuss the similarities with, for example, MITI in Japan and US research programme. USA has had experience with centres of excellence for a long time in biomass research and the outcome is not always positive. One of the problems is that a huge amount of money is centralised at one place. This means a large increase in administration. There is also a risk that the flexibility to change research directions and content gets more difficult in large research groups. Generic research and unknown research groups could lose out in this type of research organisation. Some European countries were worried that within this ERA discussion, they would lose their research position in the biomass area because they would need to share money and knowledge instead of concentrating on being the best in their field.

RESEARCH AREAS IN THE BIOMASS FIELD

There was also discussion on what research areas should be emphasised. Feedstock security is a main issue in the biomass area. Therefore pre-normative research is very important.

Gasification for electricity production on both small-medium and large scale is interesting when co-generation is used. Gas cleaning is an important area to use research resources. Generic research is very important for understanding the chemical reactions in the gasification processes. The knowledge can be used for improving efficiency and the overall emissions. Emissions reduction for all kind of biomass incineration is important to satisfy current and future legislation.

Ash cleaning and recycling to replace the nutrients removed from the soil is an important area for research. In particular, the ash has to be clean and without heavy metals and other contaminants.

CONCLUSIONS

One of the conclusions from this workshop were that not all research should be done on a centralised level like centres of excellence. There is an impending risk that innovative research will be lost. One idea would be to give a small amount of money to a lot of researchers and let them prove their skills. The ones that succeeded should then continue to get more money and build up their promising research.

There was also criticism in joining well-developed research with less developed. Countries in the biomass frontline were worried that their research would lose momentum and not develop in a positive way.

Research money should be put on the following subjects:

- Gasification including co-generation,
- Gas cleaning with generic research including chemical reaction,
- Feed stock security,
- Pre-normative research,
- Emission reduction for all types of thermo-chemical conversion of biomass,
- Ash recycling and cleaning.

RESULTS

The main result of the workshop was that a European company found a partner in Thailand to help the to solve their bagasse problem. This shows that Europe already has a lead in RTD and that technology transfer can be achieved.

Author Index

Subject Index

The index is based on keywords for each paper. The page numbers shown after each index entry refer to the first page of the paper that considers that topic